Series in Analysis Vol. 6

Asymptotic Methods for Integrals

SERIES IN ANALYSIS **ISSN: 1793-1126**

Series Editor: Professor Roderick Wong
City University of Hong Kong, Hong Kong, China

Published

Vol. 1 Wavelet Analysis
edited by Ding-Xuan Zhou

Vol. 2 Differential Equations and Asymptotic Theory in Mathematical Physics
edited by Hua Chen and Roderick S.C. Wong

Vol. 3 Symmetrization and Applications
by S Kesavan

Vol. 5 Lecture Notes on Applied Analysis
by Roderick Wong

Vol. 6 Asymptotic Methods for Integrals
by Nico M. Temme

Forthcoming

Vol. 4 Mathematical Foundation and Applications of the P and H-P Finite Element Methods
by Benqi Guo

Series in Analysis Vol. 6

Asymptotic Methods for Integrals

Nico M. Temme

Emeritus Researcher
Retired from Centrum Wiskunde &
Informatica, The Netherlands

World Scientific

NEW JERSEY • LONDON • SINGAPORE • BEIJING • SHANGHAI • HONG KONG • TAIPEI • CHENNAI

Published by

World Scientific Publishing Co. Pte. Ltd.
5 Toh Tuck Link, Singapore 596224
USA office: 27 Warren Street, Suite 401-402, Hackensack, NJ 07601
UK office: 57 Shelton Street, Covent Garden, London WC2H 9HE

Library of Congress Cataloging-in-Publication Data
Temme, Nico M., 1940– author.
Asymptotic methods for integrals / Nico M. Temme, The Netherlands.
pages cm. -- (Series in analysis ; volume 6)
Includes bibliographical references and index.
ISBN 978-9814612159 (hardcover : alk. paper)
1. Integral equations--Asymptotic theory. 2. Differential equations--Asymptotic theory.
3. Functions, Special. I. Title.
QA308.T37 2015
515'.45--dc23

2014031039

British Library Cataloguing-in-Publication Data
A catalogue record for this book is available from the British Library.

To my grandson Nimrod

Preface

Large parameter problems occur in all branches of pure and applied mathematics, in physics and engineering, in statistics and probability theory. In many occasions the problems are presented in the form of integrals or differential equations, or both, but also finite sums, infinite series, difference equations, and implicit algebraic equations are encountered.

In this book the classical methods that are available for one-dimensional integrals are described: integrating by parts, the method of stationary phase, and the saddle point method and the related method of steepest descent. For two- and higher-dimensional integrals such methods are also available, and incidentally some of their elements are mentioned, but a more extensive treatment falls outside the scope of this book.

Integrals with large parameters occur in many problems from physics and statistics, and in particular they show up in the area of the classical functions of mathematical physics and probability theory, from which class many examples are taken to explain the classical methods.

Often given integral representations need to be modified, firstly by using the methods of complex analysis, and secondly by using transformations to obtain standard forms (Laplace-type integrals, for example), from which expansions can be obtained in a straightforward way.

A beginner in asymptotic analysis may need some time to get insight in the asymptotic phenomena of large-parameter problems and how to select certain representations, transformations or other preparations to obtain a standard from. The many examples in this book will certainly help to gain this insight.

In addition to the large parameter, other parameters may be present in the integrals to be studied. In that case questions may arise about the validity of a certain approximation or expansion when one or more additional parameters assume certain critical values. In the case that the validity is destroyed, different methods or representations are needed in order to find expansions in which these critical values do no harm.

This is the area of uniform expansions in which powerful expansions valid for several parameters can be obtained, but with more complicated methods because transformations for obtaining standard forms are usually not given in an explicit form and are difficult to analyze. Typically, error functions, Airy functions, and Bessel functions arise in uniform expansions. For uniform methods for integrals complete proofs on remainder estimates and precise descriptions of the domains of validity of the real or complex variables involved are still challenging problems. We concentrate on showing the methods for deriving uniform expansions; in a number of cases we give details on the mappings to locate the singularities, and we verify the results by numerical computations.

The study of asymptotic methods for integrals knows a long history with many prominent contributors. Riemann used in an unpublished note (1863) the method of steepest descent and Debye used this method in full extent to obtain approximations for Bessel functions. In other unpublished notes Riemann also gave the first steps for approximating the zeta function. Siegel used this method in 1932 to derive the Riemann–Siegel formula for the zeta function. The method of stationary phase was essential in Kelvin's work to describe the wave pattern behind a moving ship.

Excellent books are available now with details on these methods. Olver's book *Asymptotics and Special Functions* treats mainly second-order linear differential equations, but has sections on integrals as well. Books focusing on integrals are Wong: *Asymptotic Approximations of Integrals*, Paris & Kaminski: *Asymptotics and Mellin-Barnes Integrals*, and Paris: *Hadamard Expansions and Hyperasymptotic Evaluation.*

In the present work many methods and examples for integrals are described, usually in relation to the special functions from mathematical physics and probability theory. Many of these expansions are given in the *NIST Handbook of Mathematical Functions*, but many extra results are derived.

This book is dedicated to my grandson Nimrod, who was not yet born when his two elder brothers Ambrus and Fabian were mentioned in an earlier work; may the future generation take over! My deep gratitude goes to my wife, Gré, for her patience and understanding during the lengthy duration of this project, and for postponing part of the social enjoyment of retirement.

Nico M Temme
Abcoude, September 2014

Acknowledgments

The author acknowledges scientific support from his former institute CWI in Amsterdam and funding from the Spanish Government (*Ministerio de Economía y Competitividad*, projects MTM2006-09050, MTM2009-11686, MTM2012-34787) and from the Government of Navarra (projects MTM2007-63772 and MTM2010-21037). The collaboration and many contacts with Spanish colleagues Amparo Gil & Javier Segura (Santander), José Luis López & Ester Pérez Sinusía (Pamplona, Zaragoza) and Alfredo Deaño (Leuven, Belgium), have been very inspiring, pleasant and fruitful. Many results and examples in this book arose in the joint research with these colleagues.

I am indebted to Richard Paris (Dundee), and Adri Olde Daalhuis and Karen Ogilvie (Edinburgh); their remarks on the text have considerably improved the presentation. In spite of their and my best efforts, however, it is certain that some errors and misprints are bound to have crept into the text, and I ask for the reader's forgiveness for those that prove to be annoying.

My thanks are also due to Ms Lai Fun Kwong, Ms Ying-Oi Chiew and Mr Rajesh Babu of World Scientific Publishing Co., who have provided excellent support during the completion of this project.

Contents

PART 6 Uniform Examples for Special Functions 403

PART 1

Basic Methods for Integrals

Chapter 1

Introduction

1.1 Symbols used in asymptotic estimates

In asymptotic estimates we frequently use the big O-symbol, denoted by $\mathcal{O}$, and the little o-symbol. For estimating a function f with respect to g, both functions defined in domain $\mathcal{D} \in \mathbb{C}$, we assume that $g(z) \neq 0, z \in \mathcal{D}$ and z_0 is a limit point of $\mathcal{D}$. Possibly $g(z) \to 0$ as $z \to z_0$. Here, and in the following, we assume that the limit point z_0 is approached inside $\mathcal{D}$.

We use the $\mathcal{O}$-symbol in the form

$$f(z) = \mathcal{O}\left(g(z)\right), \quad z \in \mathcal{D}, \tag{1.1.1}$$

which means that there is a constant M such that $|f(z)| \leq M|g(z)|$ for all $z \in \mathcal{D}$.

Usually, for our asymptotic problems, z_0 is the point of infinity, $\mathcal{D}$ is an unbounded part of a sector, for example

$$\mathcal{D} = \{z : |z| \geq r,\ \alpha \leq \operatorname{ph} z \leq \beta\}, \tag{1.1.2}$$

where $\operatorname{ph} z$ denotes the phase of the complex number z, r is a nonnegative number. The estimate $f(z) = \mathcal{O}\left(1\right), z \in \mathcal{D}$, implies that $|f(z)|$ is bounded for all $z \in \mathcal{D}$.

For the little o-symbol we write

$$f(z) = o\left(g(z)\right), \quad z \to z_0, \quad z \in \mathcal{D}, \tag{1.1.3}$$

which means that

$$\lim_{z \to z_0} f(z)/g(z) = 0. \tag{1.1.4}$$

When we write $f(z) = o\left(1\right), z \to z_0, z \in \mathcal{D}$, we mean that $f(z)$ tends to zero when $z \to z_0$, $z \in \mathcal{D}$.

We write

$$f(z) \sim g(z), \quad z \to z_0, \quad z \in \mathcal{D}, \tag{1.1.5}$$

when $\lim\limits_{z \to z_0} f(z)/g(z) = 1$. In that case we say that the functions f and g are asymptotically equal at z_0.

In many cases we need compound approximations for f, in the sense that $f(z) = f_1(z) + f_2(z)$, where f_1 and f_2 may have different asymptotic behavior at z_0, which may be both relevant. In particular this occurs when f has an oscillatory behavior on the real line and z_0 is the point at infinity, as in the example $f(z) = \sin z$. The components $f_1(z) = e^{iz}/(2i)$ and $f_2(z) = -e^{-iz}/(2i)$ have completely different behavior in the complex plane.

The following theorem is useful when we know the behavior of a function f at infinity and we need estimates of the derivatives of f.

Theorem 1.1. *Let f be holomorphic*[1] *in a region containing a closed annular sector $\boldsymbol{S}$, and*

$$f(z) = \mathcal{O}\left(z^p\right), \quad (\text{or } f(z) = o\left(z^p\right)), \tag{1.1.6}$$

as $z \to \infty$ in $\boldsymbol{S}$, where p is an arbitrary fixed number. Then

$$f^{(m)}(z) = \mathcal{O}\left(z^{p-m}\right), \quad (\text{or } f^{(m)}(z) = o\left(z^{p-m}\right)), \tag{1.1.7}$$

as $z \to \infty$ in any closed annular sector $\boldsymbol{T}$ properly interior to $\boldsymbol{S}$ and having the same vertex.

Proof. For a proof we refer to Olver (1997, p. 9). □

1.2 Asymptotic expansions

The classical and simplest definition is for asymptotic power series.

Definition 1.1. Let F be a function of a real or complex variable z; let $\sum_{n=0}^{\infty} a_n z^{-n}$ denote a (convergent or divergent) formal power series, for which the sum of the first n terms is denoted by $S_n(z)$; let

$$R_n(z) = F(z) - S_n(z), \quad n = 0, 1, 2, \ldots. \tag{1.2.8}$$

That is,

$$F(z) = a_0 + \frac{a_1}{z} + \frac{a_2}{z^2} + \cdots + \frac{a_{n-1}}{z^{n-1}} + R_n(z), \quad n = 0, 1, 2, \ldots, \tag{1.2.9}$$

where we assume that when $n = 0$ we have $F(z) = R_0(z)$. Next, assume that for each $n = 0, 1, 2, \ldots$ the following relation holds

$$R_n(z) = \mathcal{O}\left(z^{-n}\right), \quad z \to \infty, \tag{1.2.10}$$

in some unbounded domain $\mathcal{D}$. Then $\sum_{n=0}^{\infty} a_n z^{-n}$ is called an asymptotic expansion of the function $F(z)$ and we denote this by

$$F(z) \sim \sum_{n=0}^{\infty} a_n z^{-n}, \quad z \to \infty, \quad z \in \mathcal{D}. \tag{1.2.11}$$

[1]That is, analytic and free from singularities.

This definition is due to Poincaré (1886). Analogous definitions can be given for $z \to 0$, or to other finite limit points.

Observe that we do not assume that the infinite series $\sum_{n=0}^{\infty} a_n z^{-n}$ converges for certain z-values. This is not relevant in asymptotics; in the definition only a property of $R_n(z)$ defined in (1.2.9) is requested, with n fixed.

1.3 A first example: Exponential integral

The classical example is the so-called exponential integral $E_1(z)$ written in the form

$$F(z) = z\,e^z E_1(z) = z \int_z^{\infty} t^{-1} e^{z-t}\, dt, \tag{1.3.12}$$

(for the special functions used in this section we refer to Temme (2010b)) where z is real and positive. Repeatedly using integration by parts, we obtain

$$F(z) = 1 - \frac{1}{z} + \frac{2!}{z^2} - \cdots + \frac{(-1)^{n-1}(n-1)!}{z^{n-1}} + (-1)^n n!\, z \int_z^{\infty} \frac{e^{z-t}}{t^{n+1}}\, dt. \tag{1.3.13}$$

In this case we have, since $t \ge z$,

$$(-1)^n R_n(z) = n!\, z \int_z^{\infty} \frac{e^{z-t}}{t^{n+1}}\, dt \le \frac{n!}{z^n} \int_z^{\infty} e^{z-t}\, dt = \frac{n!}{z^n}. \tag{1.3.14}$$

Indeed, $R_n(z) = \mathcal{O}(z^{-n})$ as $z \to \infty$. Hence

$$z \int_z^{\infty} t^{-1} e^{z-t}\, dt \sim \sum_{n=0}^{\infty} (-1)^n \frac{n!}{z^n}, \quad z \to \infty. \tag{1.3.15}$$

This series is divergent for any finite value of z. However, when z is sufficiently large and n is fixed, the finite part of the series $S_n(z)$ given by

$$S_n(z) = F(z) - R_n(z), \tag{1.3.16}$$

approximates the function $F(z)$ to any desired accuracy.

In this example we can derive the asymptotic expansion in a different way. We write (using the transformation $t = z(1+u)$) $F(z)$ as a Laplace integral

$$F(z) = z \int_0^{\infty} e^{-zu} f(u)\, du, \quad f(u) = 1/(1+u). \tag{1.3.17}$$

We now write

$$f(u) = 1 - u + u^2 - \cdots + (-1)^{n-1} u^{n-1} + (-1)^n u^n/(1+u), \tag{1.3.18}$$

and we obtain exactly the same expansion and we can obtain the same expression and upper bound for $|R_n(z)|$.

Later (see §2.1) we will explain how to extend the result for complex values of z. It will follow that the asymptotic expansion of the exponential integral

$$E_1(z) = \int_z^{\infty} e^{-t} \frac{dt}{t} \sim \frac{e^{-z}}{z} \sum_{n=0}^{\infty} (-1)^n \frac{n!}{z^n}, \quad z \to \infty, \tag{1.3.19}$$

is valid inside the sector $-\frac{3}{2}\pi < \operatorname{ph} z < \frac{3}{2}\pi$. This sector opening is wider than 2π, and this makes sense because $E_1(z)$ has a logarithmic singularity at the origin, and the expansion in (1.3.19) is valid for part of the analytic continuation of $E_1(z)$ outside the usual sector $-\pi < \operatorname{ph} z < \pi$.

This makes it possible to obtain expansions of the sine and cosine integrals defined by

$$\operatorname{si}(z) = -\int_z^\infty \frac{\sin t}{t}\,dt, \quad \operatorname{Ci}(z) = -\int_z^\infty \frac{\cos t}{t}\,dt, \tag{1.3.20}$$

with the relations

$$\begin{aligned} \operatorname{si}(z) &= \tfrac{1}{2}i\left(E_1(iz) - E_1(-iz)\right), \\ \operatorname{Ci}(z) &= -\tfrac{1}{2}\left(E_1(iz) + E_1(-iz)\right), \end{aligned} \tag{1.3.21}$$

which are valid for $|\operatorname{ph} z| < \frac{1}{2}\pi$.

By integrating by parts in (1.3.20), or by using the results for the exponential integral in (1.3.21), the following compound representation will follow:

$$\begin{aligned} \operatorname{si}(z) &= -\cos z\,\mathrm{f}(z) - \sin z\,\mathrm{g}(z), \\ \operatorname{Ci}(z) &= \sin z\,\mathrm{f}(z) - \cos z\,\mathrm{g}(z), \end{aligned} \tag{1.3.22}$$

where $\mathrm{f}(z)$ and $\mathrm{g}(z)$ have the asymptotic expansions

$$\begin{aligned} \mathrm{f}(z) &\sim \frac{1}{z}\left(1 - \frac{2!}{z^2} + \frac{4!}{z^4} - \frac{6!}{z^6} + \cdots\right), \\ \mathrm{g}(z) &\sim \frac{1}{z^2}\left(1 - \frac{3!}{z^3} + \frac{5!}{z^5} - \frac{7!}{z^7} + \cdots\right), \end{aligned} \tag{1.3.23}$$

as $z \to \infty$.

The expansions in (1.3.23) can also be derived from the Laplace integral representations

$$\mathrm{f}(z) = \int_0^\infty \frac{e^{-zt}}{t^2+1}\,dt, \quad \mathrm{g}(z) = \int_0^\infty \frac{te^{-zt}}{t^2+1}\,dt, \tag{1.3.24}$$

with $\Re z > 0$, and in §2.1 we will see that the expansions in (1.3.23) hold for the analytic continuation of these functions in the sector $|\operatorname{ph} z| < \pi$.

By differentiating the relations in (1.3.22) it follows that $\mathrm{f}'(z) = -\mathrm{g}(z)$ and $\mathrm{g}'(z) = \mathrm{f}(z) - 1/z$. These relations also follow from (1.3.24), and the derivatives of the asymptotic expansions in (1.3.23) satisfy the same relations.

Observe that when $\Im z$ is large, only one term of the representations is needed. When $y = \Im z \to +\infty$ we can write

$$\begin{aligned} \operatorname{si}(z) &= -\tfrac{1}{2}iE_1(-iz)\left(1 + \varepsilon(z)\right), \\ \operatorname{Ci}(z) &= -\tfrac{1}{2}E_1(-iz)\left(1 + \varepsilon(z)\right), \end{aligned} \tag{1.3.25}$$

where $\varepsilon(z) = E_1(iz)/E_1(-iz) = \mathcal{O}\left(e^{-2y}\right)$. This term is exponentially small with respect to unity, and we can use in this case

$$\operatorname{si}(z) \sim -\tfrac{1}{2}iE_1(-iz), \quad \operatorname{Ci}(z) \sim -\tfrac{1}{2}E_1(-iz), \tag{1.3.26}$$

as $\Im z \to +\infty$, with expansions following from (1.3.19). Similarly for the case $\Im z \to -\infty$.

For the other sine integral

$$\mathrm{Si}(z) = \int_0^z \frac{\sin t}{t}\,dt, \tag{1.3.27}$$

we can use the relation $\mathrm{Si}(z) = \frac{1}{2}\pi + \mathrm{si}(z)$ to obtain the behavior for large values of z.

From these examples we learn:

- expansions of certain special functions may follow from one basic result, in this case the expansion of the exponential integral;
- for oscillatory functions on the real line, such as $\mathrm{si}(z)$ and $\mathrm{Ci}(z)$, compound asymptotic representations are needed, and off the real line in the complex plane simpler approximations can be obtained.

These compound asymptotic representations occur frequently in asymptotic expansions of special functions, with as an important example the Bessel function $J_\nu(z)$.

1.4 Generalized asymptotic expansions

Asymptotic power series are most common in asymptotic analysis. In some cases we encounter expansions of a more general kind. A simple generalization happens when we have an expansion in, for example, powers of $1/\ln z$. More generally we introduce the concept of asymptotic scale.

Definition 1.2. Let $\{\psi_n\}_{n=0}^{\infty}$ be a sequence of continuous functions defined in some common domain $\mathcal{D}$ and z_0 a limit point of $\mathcal{D}$. Then the sequence constitutes an asymptotic scale as $z \to z_0$ in $\mathcal{D}$, if for $n = 0, 1, 2, \ldots$

$$\psi_{n+1}(z) = o\bigl(\psi_n(z)\bigr), \quad z \to z_0, \quad z \in \mathcal{D}. \tag{1.4.28}$$

Then the generalized asymptotic expansion is defined as follows.

Definition 1.3. Let $\{\phi_n\}_{n=0}^{\infty}$ be a sequence of continuous functions defined in $\mathcal{D}$. Then the formal series $\sum\limits_{n=0}^{\infty} \phi_n(z)$ is called a generalized asymptotic expansion of a function f with respect to the asymptotic scale $\{\psi_n\}$, as $z \to z_0$, if

$$f(z) = \sum_{n=0}^{N} \phi_n(z) + o\bigl(\psi_N(z)\bigr) \tag{1.4.29}$$

for every fixed $N \ge 0$. In this case we write

$$f(z) \sim \sum_{n=0}^{\infty} \phi_n(z); \quad \{\psi_n\}, \quad z \to z_0. \tag{1.4.30}$$

In many applications we use this with $\phi_n(z) = a_n\psi_n(z)$, a_n not depending on z, which gives a generalized Poincaré-type expansion.

We use these expansions mainly for notational purposes, and the concept is useful for describing certain expansions of special functions. However, the general character has some drawbacks. For more details, and a discussion about conspicuous features of this type of expansion, we refer to, for example, Wong (2001, §I.3).

1.5 Properties of asymptotic power series

First we observe that the coefficients a_n in Definition 1.1 follow from the limits

$$a_0 = \lim_{z\to\infty} f(z), \quad a_n = \lim_{z\to\infty} z^n \left(f(z) - \sum_{m=0}^{n-1} a_m z^{-m}\right). \tag{1.5.31}$$

From these limits we conclude that if f possesses the following asymptotic expansions

$$f(z) \sim \sum_{n=0}^{\infty} a_n z^{-n}, \quad f(z) \sim \sum_{n=0}^{\infty} b_n z^{-n}, \quad z \to \infty, \tag{1.5.32}$$

inside the same domain $\mathcal{D}$, then $a_n = b_n$, $n = 0, 1, 2, \ldots$.

On the other hand, two different functions may have the same asymptotic expansion. For example, let $\mathcal{D}$ be the sector $|\mathrm{ph}\, z| \le \frac{1}{2}\pi - \delta$, for some $\delta \in (0, \frac{1}{2}\pi)$. Then, if $f(z) \sim \sum_{n=0}^{\infty} a_n z^{-n}$ and $g(z) \sim \sum_{n=0}^{\infty} b_n z^{-n}$ as $z \to \infty$ in $\mathcal{D}$, with $b_n = 0, \forall n$, the functions f and $h = f + g$ have the same asymptotic expansion in $\mathcal{D}$.

In this case, when not all coefficients a_n vanish, we call the function g exponentially small (compared to f). A function like $g(z) = e^{-z}$ is such a function with a Poincaré-type asymptotic expansion for $z \to \infty$ in this $\mathcal{D}$ with all coefficients equal to 0. Such a function is called asymptotically equal to 0 to all orders (compared to a function with a non-vanishing Poincaré-type expansion).

Next we mention a few manipulations with asymptotic expansions. The proofs of the following two Lemmas follow immediately from Definition 1.1.

Lemma 1.1. *When the functions f and g possess the asymptotic expansions*

$$f(z) \sim \sum_{n=0}^{\infty} a_n z^{-n}, \quad g(z) \sim \sum_{n=0}^{\infty} b_n z^{-n}, \quad z \to \infty, \quad z \in \mathcal{D}, \tag{1.5.33}$$

then the linear combination $\alpha f(z) + \beta g(z)$, where α, β are constants, the product $f(z)g(z)$, and, if $a_0 \ne 0$, the reciprocal $1/f(z)$ possesses also an asymptotic expan-

sion in the same domain $\mathcal{D}$, and we have

$$\alpha f(z)+\beta g(z)\sim\sum_{n=0}^{\infty}(\alpha a_n+\beta b_n)\,z^{-n},$$
$$f(z)g(z)\sim\sum_{n=0}^{\infty}c_nz^{-n},\quad c_n=\sum_{m=0}^{n}a_mb_{n-m},\tag{1.5.34}$$
$$\frac{1}{f(z)}\sim\sum_{n=0}^{\infty}d_nz^{-n},$$

where

$$d_0=1/a_0,\quad a_0d_n=-\sum_{m=0}^{n-1}a_{n-m}d_m,\quad n=1,2,3,\ldots.\tag{1.5.35}$$

For integration of an asymptotic expansion, we have the following property.

Lemma 1.2. *If the function f is continuous in the domain $\mathcal{D}$ defined in* (1.1.2) *and $f(z)\sim\sum_{n=0}^{\infty}a_nz^{-n}$ as $z\to\infty$, $z\in\mathcal{D}$, then*

$$\int_z^{\infty}(f(t)-a_0-a_1/t)\,dt\sim\sum_{n=1}^{\infty}\frac{a_{n+1}}{n}z^{-n},\quad z\to\infty,\quad z\in\mathcal{D},\tag{1.5.36}$$

where the path of integration is the straight line in $\mathcal{D}$ joining z to ∞ with a fixed phase.

For taking derivatives of an asymptotic expansion we need more information about the function f. In general, differentiating an asymptotic expansion of a function f does not give the asymptotic expansion of f'. For example, consider $f(z)=e^{-z}\cos(e^z)$. This function has a vanishing asymptotic expansion: $f(z)\sim\sum_{n=0}^{\infty}a_nz^{-n}$, with all $a_n=0$ as $z\to\infty$ in the sector $|\mathrm{ph}\,z|\le\frac{1}{2}\pi-\delta$, $\delta\in(0,\frac{1}{2}\pi)$. But the termwise differentiated series is not the asymptotic expansion of

$$f'(z)=-\sin(e^z)-e^{-z}\cos(e^z),\tag{1.5.37}$$

which does not possess an asymptotic expansion as $z\to\infty$.

We have the following lemma.

Lemma 1.3. *Let $\mathcal{D}$ be the domain defined in* (1.1.2) *and assume that f possesses the expansion $f(z)\sim\sum_{n=0}^{\infty}a_nz^{-n}$ as $z\to\infty$, $z\in\mathcal{D}$. If f has a continuous derivative f', and if f' possesses an asymptotic expansion as $z\to\infty$ in $\mathcal{D}$, then*

$$f'(z)\sim-\sum_{n=1}^{\infty}na_nz^{-n-1},\quad z\to\infty,\quad z\in\mathcal{D}.\tag{1.5.38}$$

Proof. A proof can be found in Wong (2001, p. 7). □

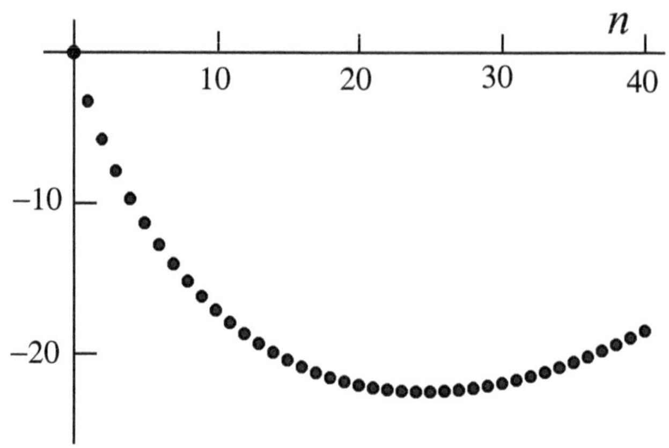

Fig. 1.1 Behavior of $\ln(n!/z^n)$, the logarithm of the absolute value of the terms in the asymptotic expansion (1.3.13) for $z = 25$ and $0 \le n \le 40$. The minimum value occurs for $n \sim z$; see (1.6.40).

When we know that f is analytic in the same $\mathcal{D}$, the same result holds, without assuming that f' has an asymptotic expansion. In more detail, we have the following lemma.

Lemma 1.4. *Let $\mathcal{D}$ be the domain defined in* (1.1.2) *and assume that f is analytic in $\mathcal{D}$. If $f(z) \sim \sum_{n=0}^{\infty} a_n z^{-n}$, uniformly in* $\operatorname{ph} z$ *as $z \to \infty$ in any closed sector inside $\mathcal{D}$, then*

$$f'(z) \sim -\sum_{n=1}^{\infty} n a_n z^{-n-1}, \quad z \to \infty, \tag{1.5.39}$$

also uniformly with respect to $\operatorname{ph} z$ *as $z \to \infty$ in any closed sector inside $\mathcal{D}$.*

Proof. For a proof we refer to Wong (2001, p. 8). □

1.6 Optimal truncation of asymptotic expansions

As a rule of thumb, when z is large, the absolute values of the terms $a_n z^{-n}$ in expansion (1.2.11) initially decrease in value, and ultimately they increase in value. Compare Figure 1.1 and the expansion in (1.3.15) for $z > 0$; we show $\ln(n!/z^n)$. The minimum occurs when n is roughly equal to z. We have, with $z = n$, by using Stirling's approximation,

$$\frac{n!}{z^n} \sim \sqrt{2\pi z}\, e^{-z}, \quad z \to \infty. \tag{1.6.40}$$

Because in (1.3.14) the absolute value of the remainder $R_n(z)$ is less than $n!/z^n$, we see that at the optimal truncation value $n \sim z$ the error is exponentially small.

It follows that we can determine the interval of z-values for which the asymptotic expansion in (1.3.15) can be used to obtain a certain precision. For example, if $z = 25$, we have $\sqrt{2\pi z}\, e^{-z} \doteq 1.74\text{e}{-10}$.

Hence, for $z \geq 25$ we can compute the exponential integral with a precision of about 10^{-10}, with a maximum of 26 terms in the asymptotic expansion. For z-values much larger than 25, fewer terms in the expansion are needed, and we can terminate the summation as soon as $n!/z^n$ is smaller than the required precision.

In our example of the exponential integral we obtained a relation between the remainder $R_n(z)$ and the first neglected term: the remainder is smaller and has the same sign. That is, for $n = 0, 1, \ldots$, we have

$$R_n(z) = \theta_n \, (-1)^n \frac{n!}{z^n}, \quad 0 < \theta_n < 1. \tag{1.6.41}$$

For many asymptotic expansions of special functions this type of strict upper bound for the remainder is known. When the large parameter z is complex, information is still available for many cases. In Olver (1997) upper bounds for remainders are derived for several kinds of Bessel functions and confluent hypergeometric functions (Whittaker functions) by using the differential equation satisfied by these functions. These bounds are also valid for complex values of parameters and argument. By using these bounds, reliable and efficient algorithms can be designed for the computation of a large class of special functions.

For asymptotic expansions derived from integrals, usually less detailed information on upper bounds of the remainder is available, in particular when the parameters are complex. In §6.3 we will derive upper bounds of the remainder for the expansion of $\ln\Gamma(z)$, and we also consider complex z.

In §2.6 we give more details about the optimal truncation of expansions and about re-expansions of the remainders, which may provide exponentially improved expansions.

Chapter 2

Expansions of Laplace-type integrals: Watson's lemma

We consider the large-z asymptotic expansions of Laplace-type integrals of the form

$$F_\lambda(z) = \int_0^\infty t^{\lambda-1} f(t) e^{-zt}\,dt, \quad \Re z > 0, \quad \Re\lambda > 0, \tag{2.0.1}$$

and other related forms, such as loop integrals. We also consider integrals showing the same asymptotic features, and that can be transformed into the standard form (2.0.1). We show several methods on how to calculate the coefficients in the expansions of these more general integrals.

In addition, we give a few examples in which the kernel e^{-zt} will be replaced by, for example, a Bessel function $K_\nu(z)$. We consider an example for deriving an expansion of a converging factor, which gives the first ideas of exponentially improved expansions. Finally we will show how we can deal with singularities of the function f in (2.0.1) .

2.1 Watson's lemma

We mention a very useful result from the theory of asymptotics for Laplace integrals, known as Watson's lemma (Watson, 1918b, p. 133) (see also Watson (1944, p. 236)). It is a rather simple result, but it has significant application within the theory on the asymptotic behavior of integrals. The second approach in §1.3 gives the set-up for the following result.

Theorem 2.1 (Watson's lemma). *Assume that:*

(i) *The function $f\colon \mathbb{R}^+ \to \mathbb{C}$ has a finite number of discontinuities.*

(ii) *As $t \to 0+$*

$$f(t) \sim \sum_{n=0}^{\infty} a_n t^n. \tag{2.1.2}$$

(iii) *The integral*

$$F_\lambda(z) = \int_0^\infty t^{\lambda-1} f(t) e^{-zt}\,dt, \quad \Re\lambda > 0, \tag{2.1.3}$$

is convergent for sufficiently large positive values of $\Re z$.

Then,

$$F_\lambda(z) \sim \sum_{n=0}^{\infty} a_n \frac{\Gamma(n+\lambda)}{z^{n+\lambda}}, \quad z \to \infty, \tag{2.1.4}$$

in the sector $|\mathrm{ph}\, z| \le \frac{1}{2}\pi - \delta < \frac{1}{2}\pi$, *where* $z^{n+\lambda}$ *has its principal value.*

Proof. For a proof we refer to Olver (1997, p. 113), where a more general condition (*ii*) is assumed. □

Remark 2.1. We make a few remarks on the conditions.

- When applying Watson's lemma in the theory of special functions, condition (i) often holds, since the function f is usually analytic in a domain containing $[0,\infty)$ in its interior. Compare the result in §1.3 for the exponential integral with $f(t) = 1/(1+t)$. In that case $f(t)$ is analytic in the sector $|\mathrm{ph}\, t| < \pi$.
- A larger z-sector than given below (2.1.4) can be obtained when we know that f is analytic in a certain domain of the complex plane. This extension will be considered in §2.1.1.
- When f is analytic in a certain domain, the initial condition $\Re\lambda > 0$ in (2.1.3) can usually be removed in the expansion in (2.1.4) when we use a normalizing gamma function in front of the integral. See Remark 2.2.

△

2.1.1 *Watson's lemma for extended sectors*

We next formulate a second theorem in which a much larger domain than in the previous theorem for the phase of the large parameter z is possible.

Theorem 2.2 (Watson's lemma, extended sectors). *Assume that:*

(*i*) *The function f is analytic inside a sector $\mathcal{D}$: $\alpha < \mathrm{ph}\, t < \beta$, where $\alpha < 0$ and $\beta > 0$.*

(*ii*) *For each $\delta \in (0, \frac{1}{2}\beta - \frac{1}{2}\alpha)$ (2.1.2) holds as $t \to 0$ in the sector*

$$\mathcal{D}_\delta : \alpha + \delta < \mathrm{ph}\, t < \beta - \delta; \tag{2.1.5}$$

for λ we again assume that $\Re\lambda > 0$.

(*iii*) *There is a real number σ such that $f(t) = \mathcal{O}(e^{\sigma|t|})$ as $t \to \infty$ in $\mathcal{D}_\delta$.*

Then the integral (2.1.3), or its analytic continuation, has the asymptotic expansion (2.1.4) in the sector

$$-\beta - \tfrac{1}{2}\pi + \delta \le \mathrm{ph}\, z \le -\alpha + \tfrac{1}{2}\pi - \delta. \tag{2.1.6}$$

The many-valued functions $t^{\lambda-1}$ in (2.1.3) and $z^{n+\lambda}$ in (2.1.4) have their principal values on the positive real axis and are defined by continuity elsewhere.

Proof. For the proof we refer to Olver (1997, p. 114). □

In the proof the path of integration in (2.1.3) is rotated around the origin as far as the singularities of f and its behavior at infinity allow in order to keep convergence when z assumes complex values.

To explain how the bounds in (2.1.6) arise, we allow the path of integration in (2.1.3) to rotate through an angle τ, and write $\operatorname{ph} t = \tau$ and $\operatorname{ph} z = \theta$, where $\alpha < \tau < \beta$. The condition for convergence in (2.1.3) is $\cos(\tau + \theta) > 0$, that is, $-\frac{1}{2}\pi < \tau + \theta < \frac{1}{2}\pi$. Combining this with the bounds for τ we obtain the bounds for θ in (2.1.6).

For the function f in (1.3.17) we can take $\alpha = -\pi$, $\beta = \pi$. Hence, the asymptotic expansion of the exponential integral given in §1.3 holds in the sector $|\operatorname{ph} z| \le \frac{3}{2}\pi - \delta$. As remarked below equation (1.3.19), this range is much larger than the usual domain of definition for the exponential integral, which reads: $|\operatorname{ph} z| < \pi$. The phrase *or its analytic continuation* is indeed relevant in this theorem.

The expansion in (2.1.2) can be written in the form

$$f(t) = \sum_{n=0}^{N-1} a_n t^n + t^N f_N(t), \quad N = 0, 1, 2, \ldots, \tag{2.1.7}$$

where $f_N(t) = \mathcal{O}(1)$ as $t \to 0$. Assuming that we can find $\sigma \ge 0$ and positive numbers M_N such that $|f_N(t)| \le M_N e^{\sigma t}$, $\forall N \ge 0$, it follows that the expansion in (2.1.4) can be written with remainders:

$$F_\lambda(z) = \sum_{n=0}^{N-1} a_n \frac{\Gamma(n+\lambda)}{z^{n+\lambda}} + R_N(z,\lambda), \quad N = 0, 1, 2, \ldots, \tag{2.1.8}$$

where, for sufficiently large $\Re z$,

$$|R_N(z,\lambda)| \le M_N \left| \frac{\Gamma(N+\lambda)}{(z-\sigma)^{N+\lambda}} \right|. \tag{2.1.9}$$

Remark 2.2. In Watson's lemma (Theorems 2.1 and 2.2) we assume that $\Re\lambda > 0$. However, by considering the normalized form

$$F_\lambda(z) = \frac{1}{\Gamma(\lambda)} \int_0^\infty t^{\lambda-1} e^{-zt} f(t)\, dt, \tag{2.1.10}$$

the expansion in (2.1.4) becomes

$$F_\lambda(z) \sim \sum_{n=0}^{\infty} a_n \frac{(\lambda)_n}{z^{n+\lambda}}, \quad z \to \infty, \tag{2.1.11}$$

where $(\lambda)_n$ denotes *Pochhammer's symbol*, or *shifted factorial*, defined by $(\lambda)_0 = 1$ and

$$(\lambda)_n = \frac{\Gamma(\lambda+n)}{\Gamma(\lambda)} = \lambda(\lambda+1)\cdots(\lambda+n-1), \quad n \ge 1. \tag{2.1.12}$$

The terms in the expansion in (2.1.11) are well defined for all complex λ, and the expansion may be valid left of the domain $\Re\lambda > 0$. The normalization as in

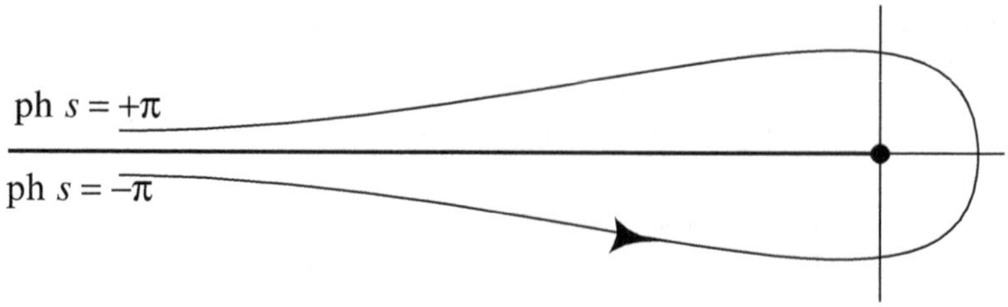

Fig. 2.1 Hankel contour for the integral in (2.2.14).

(2.1.10) with the gamma function happens frequently in representations of special functions. By using integration by parts (writing $t^{\lambda-1}\,dt = (1/\lambda)\,d\left(t^\lambda\right)$), it can be shown that analytic continuation of $F_\lambda(z)$ of (2.1.10) is possible into the domain $\Re\lambda \le 0$.

On the other hand, the integral in (2.1.10) can be written in the form of a loop integral as considered in Watson's lemma for loop integrals, see §2.2 (assuming that f is analytic in some domain around the positive axis), and then Theorem 2.3 can be applied in the form as given in Remark 2.3. △

2.1.2 *More general forms of Watson's lemma*

Several generalizations of Watson's lemma have been discussed in the literature, for example for the case that f of (2.1.3) has an expansion at the origin of the form

$$f(t) \sim \sum_{n=0}^{\infty} a_n t^{\lambda_n - 1}, \tag{2.1.13}$$

where $\Re\lambda_0 > 0$ and $\Re\lambda_{n+1} > \Re\lambda_n$, $n = 0, 1, 2, \ldots$. For this case we refer to Wong (2001, p. 22). For the case that the expansion of f at the origin contains powers of logarithms we refer to Wong and Wyman (1972).

2.2 Watson's lemma for loop integrals

A useful variant of Watson's lemma can be used for loop integrals of the form

$$G_\lambda(z) = \frac{1}{2\pi i}\int_{-\infty}^{(0+)} s^{\lambda-1} q(s) e^{zs}\,ds, \tag{2.2.14}$$

where λ is an unrestricted real or complex constant and the path runs around the branch cut of the multivalued function $s^{\lambda-1}$ as depicted in Figure 2.1. Along the lower side of the negative real axis we have $\operatorname{ph} s = -\pi$, along the upper side $\operatorname{ph} s = \pi$.

We have the following theorem.

Theorem 2.3 (Watson's lemma for loop integrals). *Assume that the integral in* (2.2.14) *converges for sufficiently large positive values of* $\Re z$, *that* q *is analytic* *inside a disk* $|s| \le r$ *for some* $r > 0$*, and that possible singularities of the function* q *are outside the contour. Let* a_n *denote the coefficients of the Maclaurin expansion*

$$q(s) = \sum_{n=0}^{\infty} a_n s^n, \quad |s| \le r. \tag{2.2.15}$$

Then

$$G(z) \sim \sum_{n=0}^{\infty} \frac{1}{\Gamma(1-\lambda-n)} \frac{a_n}{z^{n+\lambda}}, \quad z \to \infty, \tag{2.2.16}$$

in the sector $|\mathrm{ph}\, z| \le \frac{1}{2}\pi - \delta (< \frac{1}{2}\pi)$*. In this result all fractional powers have their principal values.*

Proof. For a proof of this theorem (with more general conditions on the function q) we refer to Olver (1997, p. 120). □

For the expansion in (2.2.16) a basic tool is the representation of the reciprocal gamma function along a so-called Hankel contour shown in Figure 2.1:

$$\frac{1}{\Gamma(z)} = \frac{1}{2\pi i} \int_{-\infty}^{(0+)} s^{-z} e^s \, ds, \quad z \in \mathbb{C}. \tag{2.2.17}$$

When $z = 0, -1, -2, \ldots$ this integral vanishes. When $z = n+1$, $n = 0, 1, 2, \ldots$, we can integrate along a circle $\mathcal{C}$ around the origin, and we obtain

$$\frac{1}{n!} = \frac{1}{2\pi i} \int_{\mathcal{C}} e^s \frac{ds}{s^{n+1}}, \quad n = 0, 1, 2, \ldots, \tag{2.2.18}$$

which is a well-known result, saying that the coefficient of s^n in the Maclaurin expansion of e^s equals $1/n!$.

Example 2.1 (Inversion of a Laplace transform). Consider the integral

$$\int_0^{\infty} e^{-st} J_0(t) \, dt = \frac{1}{\sqrt{s^2+1}}, \tag{2.2.19}$$

where $J_0(z)$ is the Bessel function. The inversion of the Laplace transform gives

$$J_0(t) = \frac{1}{2\pi i} \int_{c-i\infty}^{c+i\infty} e^{st} \frac{ds}{\sqrt{s^2+1}}, \quad c > 0. \tag{2.2.20}$$

We shift the contour to the left, around the two branch cuts, which run from $\pm i$ to $-\infty$. For the upper contour we shift the variable of integration and obtain

$$\frac{e^{it}}{2\pi i} \int_{-\infty}^{(0+)} e^{st} \frac{ds}{\sqrt{s(s+2i)}} = \frac{e^{it-\frac{1}{4}\pi i}}{\sqrt{2}} \frac{1}{2\pi i} \int_{-\infty}^{(0+)} s^{-\frac{1}{2}} q(s) e^{zs} \, ds, \tag{2.2.21}$$

where

$$q(s) = \frac{1}{\sqrt{s/(2i)+1}} = \sum_{n=0}^{\infty} a_n s^n, \quad |s| < 2. \tag{2.2.22}$$

The coefficients are given by

$$a_n = \binom{-\frac{1}{2}}{n}(2i)^{-n} = \left(\tfrac{1}{2}i\right)^n \frac{(\frac{1}{2})_n}{n!}, \quad n = 0, 1, 2, \ldots. \tag{2.2.23}$$

This gives the contributions from the upper branch, and we can take twice the real part to get the complete expansion:

$$J_0(t) \sim 2\Re\left(\frac{e^{it-\frac{1}{4}\pi i}}{\sqrt{2t}}\sum_{n=0}^{\infty}\frac{1}{\Gamma(\frac{1}{2}-n)}\frac{a_n}{t^n}\right), \quad t\to\infty. \tag{2.2.24}$$

The reciprocal gamma function can be removed by writing

$$\frac{1}{\Gamma(\frac{1}{2}-n)} = \frac{(-1)^n}{\pi}\Gamma\left(\tfrac{1}{2}+n\right) = \frac{(-1)^n}{\sqrt{\pi}}\left(\tfrac{1}{2}\right)_n, \quad n = 0, 1, 2, \ldots, \tag{2.2.25}$$

and the result in (2.2.24) can be written in the asymptotic form that we know for this Bessel function; see Chapter 9. ◇

Remark 2.3. The loop integral may be given in a different form, for example,

$$G_\lambda(z) = \frac{1}{2\pi i}\int_{+\infty}^{(0+)} s^{\lambda-1}q(s)e^{-zs}\,ds, \tag{2.2.26}$$

where the path of integration starts at $+\infty$, with $\operatorname{ph} s = 0$ along the upper side of the positive s-axis, turns anti-clockwise (we also say, in the positive sense) around the origin, and terminates at $+\infty$ along the lower side with $\operatorname{ph} s = 2\pi$.

In that case the same method can be used, now with the representation of the reciprocal gamma function in the form

$$\frac{e^{\pi i\lambda}}{z^\lambda\Gamma(1-\lambda)} = \frac{1}{2\pi i}\int_{+\infty}^{(0+)} s^{\lambda-1}e^{-zs}\,ds, \quad |\operatorname{ph} z| < \tfrac{1}{2}\pi. \tag{2.2.27}$$

A proof of this result can be obtained by assuming initially $\Re\lambda > 0$ and evaluating the integral by integrating along the upper and lower sides of the branch cut. This gives for the right-hand side

$$\frac{1}{2\pi i}\left(e^{2\pi i\lambda}-1\right)\int_0^\infty s^{\lambda-1}e^{-zs}\,ds = \frac{\sin(\pi\lambda)e^{\pi i\lambda}\Gamma(\lambda)}{\pi z^\lambda}. \tag{2.2.28}$$

By using

$$\frac{\sin(\pi\lambda)}{\pi\lambda} = \frac{1}{\Gamma(1-\lambda)\Gamma(1+\lambda)} \tag{2.2.29}$$

the form in (2.2.27) follows. Analytic continuation gives the result for general complex values of λ. △

2.3 More general forms of Laplace-type integrals

The class of integrals of the form in (2.1.3) is in fact the class of Laplace transforms which occur frequently in applications. Other forms of integrals that are not of Laplace transform type may look quite different, but after simple transformations they may be put into this standard form.

For example, consider

$$F_\lambda(z) = \int_0^1 t^{\lambda-1} e^{-z\,p(t)}\,dt, \quad p(t) = \frac{t}{1-t}, \quad \Re\lambda > 0. \tag{2.3.30}$$

By using the transformation $u = p(t)$, that is, $t = u/(1+u)$, we obtain

$$F_\lambda(z) = \int_0^\infty u^{\lambda-1} e^{-zu} f(u)\,du, \quad f(u) = (1+u)^{-1-\lambda}. \tag{2.3.31}$$

Application of Watson's lemma is straightforward now.

2.3.1 *Transformation to the standard form*

For a more general function p we need more steps than in the example in (2.3.30). We consider integrals of the form

$$F_\lambda(z) = \frac{1}{\Gamma(\lambda)} \int_0^\infty t^{\lambda-1} e^{-z\,p(t)} q(t)\,dt, \tag{2.3.32}$$

where $\Re\lambda > 0$. The integral should converge for sufficiently large values of $\Re z$. The functions p and q are assumed to be analytic inside a disk at the origin with radius $r > 0$ and inside a sector $\mathcal{D}$: $\alpha < \operatorname{ph} t < \beta$, where $\alpha < 0$ and $\beta > 0$. Also, $p'(t) > 0$ $(t \ge 0)$ and $q(0) \ne 0$.

We assume that we have available the coefficients p_n, q_n of the expansions

$$p(t) = \sum_{n=0}^\infty p_n t^n, \quad q(t) = \sum_{n=0}^\infty q_n t^n, \tag{2.3.33}$$

where $q_0 \ne 0$ and $p_1 > 0$.

We use the substitution

$$p(t) - p_0 = p_1 u, \tag{2.3.34}$$

which gives

$$F_\lambda(z) = \frac{e^{-\lambda p(0)}}{\Gamma(\lambda)} \int_0^\infty u^{\lambda-1} e^{-p_1 z u} f(u)\,du, \tag{2.3.35}$$

where

$$f(u) = \left(\frac{t}{u}\right)^{\lambda-1} q(t)\frac{dt}{du}. \tag{2.3.36}$$

By expanding $f(u) = \sum_{n=0}^{\infty} c_n u^n$ we can apply Watson's lemma, and obtain an expansion as in (2.1.4):

$$F_\lambda(z) \sim \frac{e^{-\lambda p_0}}{p_1^\lambda z^\lambda} \sum_{n=0}^{\infty} c_n \frac{(\lambda)_n}{p_1^n z^n}, \quad z \to \infty. \tag{2.3.37}$$

This expansion also holds for complex values of z. For determining the appropriate sectors for complex z we need more details on the singularities of the functions p and q. The transformation in (2.3.34) introduces new singularities for the function f of (2.3.36) in the u-plane.

2.4 How to compute the coefficients

We describe several methods how to handle the integral in (2.3.32). We demonstrate in each method the first steps in calculating the coefficients for the case of an incomplete gamma function. Later we derive the large-z expansion of $\Gamma(a, z)$, see (7.2.9), now we assume that $a < 0$ and that a depends on z.

Actually, we consider the form (see (7.1.5))

$$z^{\alpha z} e^z \Gamma(-\alpha z, z) = \int_0^\infty e^{-zp(t)} q(t)\, dt, \quad \alpha \geq 0, \tag{2.4.38}$$

where

$$p(t) = t + \alpha \ln(1+t), \quad q(t) = \frac{1}{1+t}. \tag{2.4.39}$$

We take $z \to \infty$ and will see later if complex values can be specified.

2.4.1 *Inversion method for computing the coefficients*

This method is in fact based on the transformation in (2.3.34), and we like to compute the coefficients c_n occurring in (2.3.37). For determining these c_n we need the inversion of the relation in (2.3.34), that is, we need the coefficients t_k in the expansion $t = \sum_{k=1}^{\infty} t_k u^k$. Because we have included p_1 on the right-hand side of (2.3.34), we know that $t_1 = 1$.

First we describe an analytical method that can be used because p is analytic near the origin. Then we can write

$$k\, t_k = \frac{1}{2\pi i} \int_{\mathcal{C}_u} \frac{dt/du}{u^k}\, du = \frac{1}{2\pi i} \int_{\mathcal{C}_t} \frac{dt}{u^k}, \quad k \geq 1, \tag{2.4.40}$$

where $\mathcal{C}_u$ and $\mathcal{C}_t$ are small circles around the origin. By using (2.3.34) in the second integral, we can obtain t_k by using residue calculus. To verify this, we write

$$k\, t_k = \frac{1}{2\pi i} \int_{\mathcal{C}_t} \left(\frac{p_1\, t}{p(t) - p_0}\right)^k \frac{dt}{t^k}. \tag{2.4.41}$$

Hence,

$$k\,t_k = \frac{1}{2\pi i}\int_{\mathcal{C}_t}\left(1+(p_2/p_1)t+(p_3/p_1)t^2+\ldots\right)^{-k}\frac{dt}{t^k}, \tag{2.4.42}$$

which shows that $k\,t_k$ equals the coefficient of t^{k-1} of the expansion of the binomial form in (2.4.42).

The first few coefficients are

$$t_1 = 1, \quad t_2 = -p_2/p_1, \quad t_3 = \left(2p_2^2 - p_1p_3\right)/p_1^2. \tag{2.4.43}$$

When we have t_k further expansions can be used to obtain the coefficients c_n of f defined in (2.3.36). This gives the first coefficients of the expansion in (2.3.37):

$$c_0 = q_0, \quad c_1 = \Bigl(p_1q_1 - p_2q_0(1+\lambda)\Bigr)/p_1. \tag{2.4.44}$$

Example 2.2 (Incomplete gamma function). For the functions p and q in (2.4.39) the substitution in (2.3.34) gives

$$\frac{dt}{du} = \frac{(1+\alpha)(1+t)}{1+\alpha+t}, \quad f(u) = \frac{1+\alpha}{1+\alpha+t}. \tag{2.4.45}$$

The coefficients p_n, q_n are given by $p_0 = 0, p_1 = 1+\alpha$, and

$$p_n = (-1)^{n+1}\frac{\alpha}{n}, \quad (n\ge 2), \quad q_n = (-1)^n, \quad (n \ge 0). \tag{2.4.46}$$

The first coefficients c_n are

$$c_0 = 1, \quad c_1 = -\frac{1}{1+\alpha}, \quad c_2 = \frac{2-\alpha}{2(1+\alpha)^2}, \quad c_3 = -\frac{\alpha^2-8\alpha+6}{6(1+\alpha)^3}. \tag{2.4.47}$$

This gives the expansion

$$z^{\alpha z}e^z\Gamma(-\alpha z, z) \sim \sum_{n=0}^{\infty} c_n \frac{n!}{\left((1+\alpha)\,z\right)^{n+1}}, \quad z\to\infty. \tag{2.4.48}$$

The function f in (2.4.45) is singular at the point $t_0 = -1-\alpha$ (observe that at this point the derivative of $p(t)$ vanishes). The corresponding points $u_\pm$ in the u-plane are given by

$$u_\pm = \frac{-1-\alpha+\alpha\ln\alpha \pm \alpha\pi i}{1+\alpha}. \tag{2.4.49}$$

In Figure 2.2 we show for $0 \le \alpha \le 100$ the location in the complex u-plane of the singular point u_+. The location of the singular point u_- follows from conjugation.

From Theorem 2.2 we conclude that the expansion in (2.4.48) holds for z in the sector $-\beta - \frac12\pi + \delta \le \operatorname{ph} z \le \beta + \frac12\pi - \delta$, where $\beta = \operatorname{ph} u_+$ and δ is a small positive number. ◊

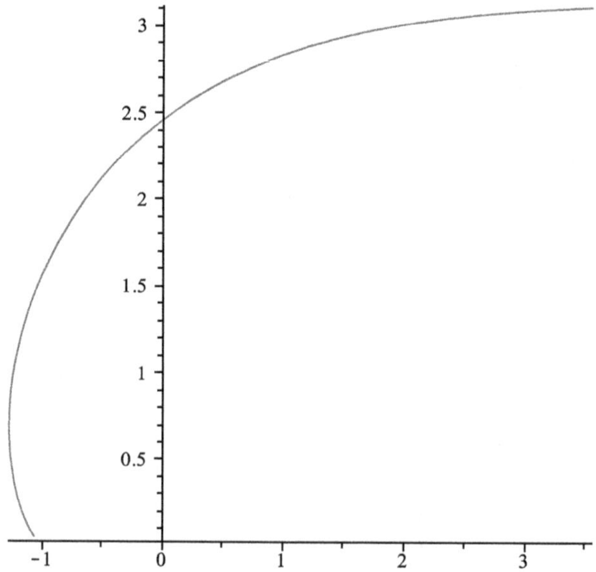

Fig. 2.2 Location in the complex plane of the singular point u_+ given in (2.4.49) of the function $f(u)$ defined in (2.4.45); $0 \le \alpha \le 100$.

2.4.2 *Integrating by parts*

When $\lambda = 1, 2, 3, \ldots$ we can use methods based on integration by parts. In these cases we do not need a transformation into the standard form. For general values of λ integration by parts seems not to be possible, unless $p(t) = p_0 + p_1 t$.

When $\lambda = 1$, consider

$$F_1(z) = \int_0^\infty e^{-z\,p(t)} q(t)\,dt = -\frac{1}{z}\int_0^\infty \frac{q(t)}{p'(t)}\,de^{-z\,p(t)}, \tag{2.4.50}$$

with expansions of $p(t)$ and $q(t)$ given in (2.3.33). This gives

$$F_1(z) = \frac{e^{-z\,p_0}}{z}\frac{q_0}{p_1} + \frac{1}{z}\int_0^\infty e^{-z\,p(t)} q_1(t)\,dt, \tag{2.4.51}$$

where

$$q_1(t) = \frac{d}{dt}\frac{q(t)}{p'(t)}. \tag{2.4.52}$$

This procedure can be repeated, and eventually gives the expansion

$$F_1(z) \sim \frac{e^{-z\,p_0}}{z\,p_1}\left(\sum_{n=0}^{N-1}\frac{q_n(0)}{z^n} + \frac{1}{z^{N-1}}\int_0^\infty e^{-z\,p(t)} q_N(t)\,dt\right), \tag{2.4.53}$$

$N = 0, 1, 2, \ldots$, where $q_0(t) = q(t)$ and

$$q_{n+1}(t) = \frac{d}{dt}\frac{q_n(t)}{p'(t)}, \quad n = 0, 1, 2, \ldots. \tag{2.4.54}$$

For $\lambda = 2, 3, 4, \ldots$ a similar method can be used. In that case the first $\lambda - 1$ terms of the series vanish.

Example 2.3 (Incomplete gamma function). For the functions p and q in (2.4.39) the functions q_n are easily generated by use of computer algebra. The first few are given by

$$q_0(t) = \frac{1}{1+t}, \quad q_1(t) = -\frac{1}{(1+t+\alpha)^2}, \quad q_2(t) = \frac{2+2t-\alpha}{(1+t+\alpha)^4}, \tag{2.4.55}$$

from which the coefficients $q_n(0)$ of the expansion

$$z^{\alpha z} e^z \Gamma(-\alpha z, z) \sim \frac{1}{z(1+\alpha)} \sum_{n=0}^{\infty} \frac{q_n(0)}{z^n} \tag{2.4.56}$$

follow, and the same expansion as in (2.4.48) is obtained.

The functions $q_n(t)$ are bounded functions on $\mathbb{R}^+$, and from mathematical induction it follows that $q_n(t) = \mathcal{O}\left(t^{-n-1}\right)$ as $t \to \infty$. Also, $q_n(0) = \mathcal{O}\left(\alpha^{-n-1}\right)$ as $\alpha \to \infty$, and in fact, large values of α improve the asymptotic nature of the expansion. ♢

2.4.3 *Manipulating power series*

For the functions p and q of (2.3.32) we take the expansions as in (2.3.33), and we introduce a function $G(u)$ by writing

$$G(u) = q(t) e^{-z(p(t)-p_0-p_1 t)}, \quad t = u/z. \tag{2.4.57}$$

By expanding the right-hand side for small values of t, using the variable u, and rearranging terms, we can obtain an expansion of the form

$$G(u) \sim \sum_{k=0}^{\infty} P_k(u) z^{-k}, \tag{2.4.58}$$

where the quantities P_k can be expressed in terms of the coefficients p_k, q_k.

Substitution into (2.3.32) gives

$$F_\lambda(z) \sim e^{-zp_0} \sum_{k=0}^{\infty} \Phi_k(\lambda) z^{-k-\lambda}, \tag{2.4.59}$$

where

$$\Phi_k(\lambda) = \frac{1}{\Gamma(\lambda)} \int_0^\infty u^{\lambda-1} e^{-p_1 u} P_k(u)\, du. \tag{2.4.60}$$

The P_k are polynomials of degree $2k$ in u. We have

$$\begin{aligned}
P_0(u) &= q_0, \\
P_1(u) &= u\left(q_1 - p_2 q_0 u\right), \\
P_2(u) &= u^2\left(q_2 - (p_2 q_1 + q_0 p_3) u + \tfrac{1}{2} q_0 p_2^2 u^2\right), \\
P_3(u) &= u^3\Big(q_3 - (p_2 q_2 + q_1 p_3 + q_0 p_4)\, u + \\
&\qquad \left(\tfrac{1}{2} q_1 p_2^2 + q_0 p_2 p_3\right) u^2 - \tfrac{1}{6} q_0 p_2^3 u^3\Big).
\end{aligned} \tag{2.4.61}$$

The polynomials $P_k(u)$ can be obtained as follows. By differentiating the relation in (2.4.57) with respect to u we obtain

$$zq(t)G'(u) = \Bigl(q'(t) - z\bigl(p'(t) - p_1\bigr)q(t)\Bigr)G(u). \tag{2.4.62}$$

Substituting power series

$$p'(t) = \sum_{n=1}^{\infty} np_n t^{n-1}, \quad q'(t) = \sum_{n=1}^{\infty} nq_n t^{n-1}, \quad t = u/z, \tag{2.4.63}$$

and those of $q(t)$ and $G(u)$ into (2.4.63), we obtain relations for the P_k when we compare equal powers of z. For example,

$$q_0 P_1'(u) = (q_1 - 2p_2q_0u)P_0(u), \tag{2.4.64}$$

giving (observe that from (2.4.57) it follows that $P_k(0) = 0, k \ge 1$)

$$P_1(u) = \int_0^u (q_1 - 2p_2q_0v)\,dv = u\left(q_1 - p_2q_0u\right). \tag{2.4.65}$$

For simple functions p and q this method may be a convenient alternative to the method based on inversion or integration by parts. The basic ideas of this method are taken from Lauwerier (1974, Chapter 10).

Example 2.4 (Incomplete gamma function). For the functions p and q in (2.4.39) the relation in (2.4.62) reads

$$\frac{z}{1+t}G'(u) = \left(\frac{-1}{(1+t)^2} - \frac{z}{1+t}\left(\frac{\alpha+1+t}{1+t} - (\alpha+1)\right)\right)G(u), \tag{2.4.66}$$

or

$$(z+u)G'(u) = (\alpha u - 1)G(u). \tag{2.4.67}$$

By substituting (2.4.58) it follows that $P_0(u) = 1$ and

$$P_k(u) = \alpha\int_0^u vP_{k-1}(v)\,dv - uP_{k-1}(u), \quad k = 1, 2, 3, \ldots. \tag{2.4.68}$$

This gives

$$\begin{aligned}
P_0(u) &= 1,\\
P_1(u) &= u\left(-1 + \tfrac{1}{2}\alpha u\right),\\
P_2(u) &= u^2\left(1 - \tfrac{5}{6}\alpha u + \tfrac{1}{8}\alpha^2u^2\right),\\
P_3(u) &= u^3\left(-1 + \tfrac{13}{12}\alpha u - \tfrac{7}{24}\alpha^2u^2 + \tfrac{1}{48}\alpha^3u^3\right).
\end{aligned} \tag{2.4.69}$$

In this example $\lambda = 1$ and the first few $\Phi_k(\lambda)$ of (2.4.60) are given by

$$\begin{aligned}
\Phi_0(\lambda) &= \frac{1}{1+\alpha}, & \Phi_1(\lambda) &= -\frac{1}{(1+\alpha)^3},\\
\Phi_2(\lambda) &= \frac{2-\alpha}{(1+\alpha)^5}, & \Phi_3(\lambda) &= -\frac{6-8\alpha+\alpha^2}{(1+\alpha)^7}.
\end{aligned} \tag{2.4.70}$$

Again, we find the same expansion as in (2.4.48).

◇

2.4.4 *Explicit forms of the coefficients in the expansion*

The methods described above are useful, but they do not give explicit forms of the coefficients in the asymptotic expansions. Because the transformation to the standard form may be not trivial, computations with the help of computer algebra are needed in practical problems.

Recently it was shown that the coefficients can be computed via a scheme that can be simply put into an algorithm for symbolic computations, and which avoids the inversion approach. We describe the method for Laplace-type integrals that can be expanded (after transformations) by using Watson's lemma. The method can also be used in other methods, as we will explain in Chapter 3 for Laplace's method.

Actually, in Wojdylo (2006a,b) the method has been derived for Laplace's method (for this case, see §3.3), and in this section we describe it for integrals of the type given in (2.3.32) with expansions of p and q as given in (2.3.33). In order to use Wojdylo's notation, we write these expansions in the form

$$p(t) = p(0) + \sum_{k=0}^{\infty} a_k t^{k+1}, \quad q(t) = \sum_{k=0}^{\infty} b_k t^k, \quad a_0 > 0, \quad b_0 \neq 0. \tag{2.4.71}$$

First we normalize the coefficients by writing

$$A_k = \frac{a_k}{a_0}, \quad B_k = \frac{b_k}{b_0}, \quad k \geq 0. \tag{2.4.72}$$

Then we define coefficients C_{jk} by writing

$$C_{j,0} = 0, \quad C_{0,k} = 0, \quad j, k \geq 1, \quad C_{0,0} = 1, \quad C_{1,1} = A_1, \tag{2.4.73}$$

and

$$C_{jk} = \sum_{m=k-1}^{j-1} A_{j-m} C_{m,k-1}, \quad j \geq k. \tag{2.4.74}$$

We take $C_{jk} = 0$ if $j < k$. We give a few examples:

$$C_{2,1} = A_2, \ C_{2,2} = A_1^2, \ C_{3,1} = A_3, \ C_{3,2} = 2A_1A_2, \ C_{3,3} = A_1^3. \tag{2.4.75}$$

These C_{jk} are combinatorial objects and are called partial ordinary Bell polynomials.

When we use the coefficients of p and q as given in (2.4.71), then the integral

$$F_\lambda(z) = \frac{1}{\Gamma(\lambda)} \int_0^\infty t^{\lambda-1} e^{-z\,p(t)} q(t)\, dt, \tag{2.4.76}$$

has the asymptotic expansion

$$F_\lambda(z) \sim b_0 \frac{e^{-\lambda p(0)}}{a_0^\lambda\, z^\lambda} \sum_{n=0}^{\infty} d_n \frac{(\lambda)_n}{a_0^n z^n}, \quad z \to \infty, \tag{2.4.77}$$

where

$$d_n = \sum_{j=0}^{n} B_{n-j} \sum_{k=0}^{j} \binom{-n-\lambda}{k} C_{jk}, \quad n = 0, 1, 2, \ldots. \tag{2.4.78}$$

The first coefficients are

$$d_0 = 1, \quad d_1 = \frac{a_0 b_1 - a_1(\lambda+1)}{a_0\, b_0}, \tag{2.4.79}$$

and these values correspond to those obtained with the inversion method in (2.4.44).

2.5 Other kernels

The method used in Watson's lemma can be applied in many other cases. For example, we can use Bessel functions. Consider the following integral:

$$F_{\lambda,\nu}(z) = \int_0^\infty t^{\lambda-1} K_\nu(zt) f(t)\, dt, \tag{2.5.80}$$

where f satisfies the conditions of Theorem 2.1. The Bessel function is singular at the origin, but when its argument is large, the behavior is comparable with that of the Laplace kernel e^{-zt}. We have (see Chapter 9)

$$K_\nu(z) \sim \sqrt{\frac{2}{\pi z}} e^{-z}, \quad -\tfrac{3}{2}\pi + \delta \le \operatorname{ph} z \le \tfrac{3}{2}\pi - \delta, \tag{2.5.81}$$

where δ is a small positive number. We can repeat the method described in Watson's lemma.

By expanding the function f as in (2.1.2) and using (see Olver and Maximon (2010, Eq. 10.43.19))

$$\int_0^\infty t^{\mu-1} K_\nu(t)\, dt = 2^{\mu-2}\, \Gamma\left(\tfrac{1}{2}\mu - \tfrac{1}{2}\nu\right) \Gamma\left(\tfrac{1}{2}\mu + \tfrac{1}{2}\nu\right), \quad |\Re\nu| < \Re\mu, \tag{2.5.82}$$

we obtain

$$F_{\lambda,\nu}(z) \sim \frac{2^{\lambda-2}}{z^\lambda} \sum_{n=0}^{\infty} a_n \frac{\Gamma\left(\frac{1}{2}\lambda + \frac{1}{2}n - \frac{1}{2}\nu\right) \Gamma\left(\frac{1}{2}\lambda + \frac{1}{2}n + \frac{1}{2}\nu\right)}{z^n}. \tag{2.5.83}$$

Other kernels also occur, for example $\sin(zt)$, $\cos(zt)$, $J_\nu(zt)$. In the case of the Bessel function, we may consider

$$F_{\lambda,\nu}(z) = \int_0^\infty t^{\lambda-1} J_\nu(zt) f(t)\, dt, \tag{2.5.84}$$

with conditions on f for convergence, in particular at infinity. In this example (and when we take the trigonometric functions), the moments

$$\int_0^\infty t^{\lambda+n-1} J_\nu(zt)\, dt, \tag{2.5.85}$$

do not exist for all n, and we need a different approach. When f is analytic in a sector of the complex plane that contains the positive real axis, we can use the representation of the Bessel function in terms of the Hankel functions:

$$J_\nu(z) = \tfrac{1}{2}\left(H_\nu^{(1)}(z) + H_\nu^{(2)}(z)\right), \tag{2.5.86}$$

and split up the integral in (2.5.84) accordingly. The Hankel functions have the following behavior in the complex plane (see (9.2.6))

$$\begin{aligned} H_\nu^{(1)}(z) &\sim \sqrt{\frac{2}{\pi z}} e^{i\omega}, & -\pi + \delta \le \operatorname{ph} z \le 2\pi - \delta, \\ H_\nu^{(2)}(z) &\sim \sqrt{\frac{2}{\pi z}} e^{-i\omega}, & -2\pi + \delta \le \operatorname{ph} z \le \pi - \delta, \end{aligned} \tag{2.5.87}$$

where

$$\omega = z - \tfrac{1}{2}\nu\pi - \tfrac{1}{4}\pi. \tag{2.5.88}$$

The moments follow from (2.5.82) and by using (see (9.2.5))

$$K_\nu(z) = \begin{cases} \tfrac{1}{2}\pi i e^{\nu\pi i/2}\, H_\nu^{(1)}\left(ze^{\pi i/2}\right), & -\pi \le \operatorname{ph} z \le \tfrac{1}{2}\pi, \\ -\tfrac{1}{2}\pi i e^{-\nu\pi i/2}\, H_\nu^{(2)}\left(ze^{-\pi i/2}\right), & -\tfrac{1}{2}\pi \le \operatorname{ph} z \le \pi, \end{cases} \tag{2.5.89}$$

in suitable sectors.

It will be clear that we can use a similar method for the kernels $\sin(zt)$ and $\cos(zt)$. A more systematic approach is possible, however, by using techniques based on the Mellin transform. For this we refer to Chapter 16.

2.6 Exponentially improved asymptotic expansions

When information on the remainder in an asymptotic expansion is available, it is possible to improve the accuracy of the expansion by re-expanding the remainder. In that case, both the large parameter, say z, and the truncation number that gives the smallest remainder, say n, should be considered as asymptotic variables.[1]

Consider, for example, the expansion of the incomplete gamma function in (7.2.11) with remainder given in

$$C_n(a,z) = z^{n+1-a}e^z \int_z^\infty t^{a-n-1}e^{-t}\,dt = z\int_0^\infty (1+u)^{a-1-n}e^{-zu}\,du. \tag{2.6.90}$$

We assume that z is positive and large. As in §1.6, the smallest term in the expansion roughly occurs when $n \sim z$. We write $n = z + \nu$, with $|\nu| \le \frac{1}{2}$, and obtain

$$C_n(a,z) = z\int_0^\infty (1+u)^{a-\nu-1}e^{-z(u+\ln(1+u))}\,du. \tag{2.6.91}$$

We take $v = u + \ln(1+u)$ as the new variable of integration, which gives

$$C_n(a,z) = z\int_0^\infty f(v)e^{-zv}\,dv, \quad f(v) = (1+u)^{\alpha-1}\frac{1+u}{2+u}, \tag{2.6.92}$$

where $\alpha = a - \nu$.

We can obtain an asymptotic expansion of this integral by applying Watson's lemma. First we need to invert the relation between u and v to obtain the expansion $u = \sum_{k=1}^\infty c_k v^k$. By using the relation $1 + u = (2+u)\dfrac{du}{dv}$ we easily find

$$c_1 = \tfrac{1}{2}, \quad 2kc_k = c_{k-1} - \sum_{j=1}^{k-1} jc_jc_{k-j}, \quad k = 2,3,4,\ldots. \tag{2.6.93}$$

[1]This section is based on Gil *et al.* (2007, §2.4.3).

This gives the first coefficients:

$$
\begin{aligned}
&c_1 = \tfrac{1}{2}, && c_2 = \tfrac{1}{16}, && c_3 = -\tfrac{1}{192}, \\
&c_4 = -\tfrac{1}{3072}, && c_5 = \tfrac{13}{61440}, && c_6 = -\tfrac{47}{1474560}.
\end{aligned}
\tag{2.6.94}
$$

For the function $f(v)$ defined in (2.6.92) we use the expansion $f(v) = \sum_{k=0}^{\infty} f_k v^k$ with the following first coefficients:

$$
\begin{aligned}
f_0 &= \tfrac{1}{2}, \\
f_1 &= \tfrac{1}{8}\left(-1 + 2\alpha\right), \\
f_2 &= \tfrac{1}{64}\left(1 - 6\alpha + 4\alpha^2\right), \\
f_3 &= \tfrac{1}{768}\left(1 + 14\alpha - 24\alpha^2 + 8\alpha^3\right), \\
f_4 &= \tfrac{1}{12288}\left(-13 - 10\alpha + 100\alpha^2 - 80\alpha^3 + 16\alpha^4\right), \\
f_5 &= \tfrac{1}{245760}\left(47 - 166\alpha - 240\alpha^2 + 520\alpha^3 - 240\alpha^4 + 32\alpha^5\right), \\
f_6 &= \tfrac{1}{5898240}\left(73 + 1274\alpha - 812\alpha^2 - 2240\alpha^3 + 2240\alpha^4 - 672\alpha^5 + 64\alpha^6\right).
\end{aligned}
\tag{2.6.95}
$$

Substituting the expansion of f in (2.6.92) we obtain

$$
C_n(a,z) \sim \sum_{k=0}^{\infty} f_k \frac{k!}{z^k}, \tag{2.6.96}
$$

which holds for large values of z and bounded values of $\alpha = a + z - n$.

To see the benefit of expanding $C_n(a,z)$, we take $a = 0.5$, $z = 6.25$, and $n = 6$. Taking $C_n(a,z) = 1$ in (7.2.11) gives $\Gamma(a,z) = 0.7223\text{e}{-3}$, with relative error $0.14\text{e}{-2}$. When we use the expansion given in (2.6.96), truncating after $k = 6$, we obtain $\Gamma(a,z) = 0.721303670366\text{e}{-3}$ with relative error $0.12\text{e}{-9}$.

In the asymptotic literature the quantity $C_n(a,z)$ (with the optimal choice of n when z is given) is called a *converging factor.* For many special functions converging factors have been developed, usually in a formal way. A rigorous treatment for the exponential integral $E_1(z)$ and the Kummer U-function for the case of complex z is given in Olver (1997, pp. 522–536).

The expansion of the converging factor for the expansion (7.2.11) can be followed by a new expansion of the converging factor for the expansion (2.6.96), when information on the remainders of this new expansion is available. In this way a further exponential improvement can be obtained, and this leads to what in the asymptotic literature has been called *hyperasymptotics*; see Berry and Howls (1991) for the first ideas and Olde Daalhuis (1992) for a detailed analysis of representations of successive remainders in this method, with application to the Kummer U-function.

For an application to the Euler gamma function, and for other details on hyperasymptotic methods, we refer to Paris and Kaminski (2001, §6.4), where new insight on the *Stokes phenomenon* is discussed as well (see also Chapter 5). This topic is important when asymptotic expansions are considered in the complex plane. The development of the theory of hyperasymptotic expansions has also been carried out for the solutions of a class of second-order ordinary differential equations; see Olde Daalhuis (1995) and Olde Daalhuis and Olver (1995).

2.7 Singularities of the integrand

In Theorem 2.1 (Watson's lemma) an expansion near the origin of the function f is assumed. When f has singularities that may approach the origin under the influence of an extra parameter, Watson's lemma cannot be applied, and for certain types of singularities we can modify Watson's lemma. We consider a few examples.

2.7.1 *A pole near the endpoint*

In this section we replace $f(t)$ in Theorem 2.1 by $f(t)/(t+\alpha)^\mu$ and assume that the function f is analytic in a domain $\mathcal{D}$ that contains the positive real axis. In particular, let f be analytic near the origin inside a disk with radius R, where R is a fixed positive number.

First we consider the simple case with $\lambda = 1$, $\mu = 1$:

$$F(\alpha, z) = \int_0^\infty e^{-z\,t} f(t) \frac{dt}{t+\alpha}, \tag{2.7.97}$$

where α is a complex number with $|\text{ph}\,\alpha| < \pi$. When α is small, or when α tends to zero, the standard method for Watson's lemma, that is, expanding $f(t)/(t+\alpha)$ in a power series, cannot be used. Because f is analytic in $\mathcal{D}$ (we also assume that $-\alpha \in \mathcal{D}$), we can write

$$\frac{f(t)}{t+\alpha} = \frac{f(-\alpha)}{t+\alpha} + g(t), \quad g(t) = \frac{f(t) - f(-\alpha)}{t+\alpha}, \tag{2.7.98}$$

where g is analytic in the same domain $\mathcal{D}$.

When using this splitting in (2.7.97) we obtain

$$F(\alpha, z) = f(-\alpha) e^{\alpha z} E_1(\alpha z) + \int_0^\infty e^{-z\,t} g(t)\, dt, \tag{2.7.99}$$

where $E_1(z)$ is the exponential integral

$$E_1(z) = \int_z^\infty \frac{e^{-t}}{t}\, dt. \tag{2.7.100}$$

We expand $g(t) = \sum\limits_{n=0}^{\infty} g_n t^n$ and obtain the asymptotic expansion

$$F(\alpha, z) \sim f(-\alpha) e^{\alpha z} E_1(\alpha z) + \sum_{n=0}^{\infty} g_n\, n!\, z^{-n-1}, \quad z \to \infty. \tag{2.7.101}$$

The sector of validity follows from the location of the singularities of g and other conditions; see Watson's lemma in Theorem 2.2.

The exponential integral $E_1(z)$ cannot be expressed in terms of elementary functions, and we consider the term in (2.7.101) with the exponential integral as the main approximant in the asymptotic behavior. This term describes the singular behavior of $F(\alpha, z)$ as $\alpha \to 0$.

We have the representation

$$E_1(z) = -\gamma - \ln z - \sum_{k=1}^{\infty} \frac{(-1)^k z^k}{k\, k!}, \quad |\mathrm{ph}(z)| < \pi, \tag{2.7.102}$$

where the series represents an analytic function, and γ is Euler's constant.

When in (2.7.99) $\alpha z \to \infty$ we can use the asymptotic expansion (see §1.3)

$$E_1(z) \sim \frac{e^{-z}}{z} \sum_{n=0}^{\infty} \frac{(-1)^n\, n!}{z^n}, \quad z \to \infty, \tag{2.7.103}$$

which holds for $|\mathrm{ph}(z)| < \frac{3}{2}\pi$.

2.7.1.1 *An application related to harmonic numbers*

As an example we consider the asymptotic behavior of the sum

$$S_n(z) = \sum_{k=1}^{n} \frac{z^k}{k}, \tag{2.7.104}$$

as $n \to \infty$, in particular for values of z close to 1. Observe that $H_n = S_n(1)$, where H_n are called the *harmonic numbers*, and that we have the well-known limit

$$\lim_{n\to\infty} (H_n - \ln n) = \gamma. \tag{2.7.105}$$

On the other hand, when $|z| < 1$ we have $S_n(z) \sim -\ln(1-z)$ as $n \to \infty$, and we notice a non-uniform behavior when $z \sim 1$.

We want information on $S_n(z)$ for $z \neq 1$ and introduce the remainder $R_n(z)$ in the expansion

$$-\ln(1-z) = S_n(z) + R_n(z), \quad n = 0, 1, \ldots, \tag{2.7.106}$$

where $R_n(z)$ has the representation

$$R_n(z) = \int_0^z \frac{t^n}{1-t}\, dt. \tag{2.7.107}$$

First we consider values of z, $z \neq 1$, in the principal sheet of the logarithmic function that has a branch cut from 1 to $+\infty$. Otherwise $R_n(z)$ is not well defined.

We investigate the asymptotic behavior of $R_n(z)$ as $n \to \infty$, for values of z close to 1. Writing $t = e^{-u}$ we transform (2.7.107) into

$$R_n(z) = \int_\alpha^\infty \frac{e^{-nu}}{e^u - 1}\, du = e^{-n\alpha} \int_0^\infty e^{-nv} f(v)\, dv, \quad \alpha = -\ln z, \tag{2.7.108}$$

excluding $z = 0$, where

$$f(v) = \frac{1}{e^{v+\alpha} - 1}. \tag{2.7.109}$$

Applying Watson's lemma we first expand $f(v) = \sum_{k=0}^{\infty} f_k v^k$, and obtain

$$R_n(z) \sim e^{-n\alpha} \sum_{k=0}^{\infty} \frac{k!\, f_k}{n^{k+1}}, \quad n \to \infty, \tag{2.7.110}$$

but this is not valid when $z \to 1$ ($\alpha \to 0$). For example, the first coefficients are

$$f_0 = \frac{1}{e^\alpha - 1}, \quad f_1 = -\frac{e^\alpha}{(e^\alpha - 1)^2}, \quad f_2 = \frac{e^\alpha(e^\alpha + 1)}{2(e^\alpha - 1)^3}, \tag{2.7.111}$$

and these coefficients are singular at $\alpha = 0$.

To obtain a uniform approximation, we split off the pole at $v = -\alpha$ and introduce the exponential integral:

$$R_n(z) = E_1(\alpha n) + z^n \int_0^\infty e^{-nv}\, g(v)\, dv, \tag{2.7.112}$$

where

$$g(v) = \frac{1}{e^{v+\alpha} - 1} - \frac{1}{v + \alpha}, \tag{2.7.113}$$

and we can expand $g(v) = \sum_{k=0}^{\infty} g_k v^k$ to obtain the expansion

$$\int_0^\infty e^{-nv}\, g(v)\, dv \sim \sum_{k=0}^{\infty} g_k \frac{k!}{n^{k+1}}, \quad n \to \infty. \tag{2.7.114}$$

This expansion holds uniformly for all bounded z in the principal sheet, even for $z = 1$. The first few coefficients are

$$g_0 = \frac{1}{e^\alpha - 1} - \frac{1}{\alpha}, \; g_1 = -\frac{e^\alpha}{(e^\alpha - 1)^2} + \frac{1}{\alpha^2}, \; g_2 = \frac{e^\alpha(e^\alpha + 1)}{2(e^\alpha - 1)^3} - \frac{2}{\alpha^3}. \tag{2.7.115}$$

All coefficients are regular at $\alpha = 0$.

For $S_n(z) = -\ln(1 - z) - R_n(z)$ we find

$$S_n(z) \sim -\ln(1 - z) - E_1(\alpha n) - z^n \sum_{k=0}^{\infty} g_k \frac{k!}{n^{k+1}}. \tag{2.7.116}$$

Considering (2.7.102) we find

$$-\ln(1 - z) - E_1(\alpha n) = -\ln(1 - z) + \ln(\alpha n) + \gamma + \sum_{k=1}^{\infty} \frac{(-1)^k (\alpha n)^k}{k\, k!}. \tag{2.7.117}$$

When we combine the logarithms we see that

$$\begin{aligned}\ln(\alpha n) - \ln(1-z) &= \ln n - \ln(1-z) + \ln(-\ln z) \\ &= \ln n + \ln\frac{\ln(1+(z-1))}{z-1} \\ &= \ln n + \ln\left(1 - \tfrac{1}{2}(z-1) + \mathcal{O}\left((z-1)^2\right)\right) \\ &= \ln n - \tfrac{1}{2}(z-1) + \mathcal{O}\left((z-1)^2\right),\end{aligned} \tag{2.7.118}$$

as $z \to 1$. In addition, the infinite series in (2.7.117) vanishes when $\alpha = 0$ ($z = 1$).

We conclude that when we combine in (2.7.116) the exponential integral with the logarithmic term, (2.7.116) gives a uniform expansion for $\alpha n \to 0$. For $z = 1$ the function g defined in (2.7.113) has the expansion (see (6.3.39))

$$g(v) = \frac{1}{e^v - 1} - \frac{1}{v} = -\frac{1}{2} + \sum_{k=1}^{\infty} \frac{B_{2k}}{(2k)!} v^{2k-1}, \tag{2.7.119}$$

where B_k are the Bernoulli numbers. This gives the complete expansion of the harmonic numbers

$$H_n \sim \ln n + \gamma + \frac{1}{2n} - \sum_{k=1}^{\infty} \frac{B_{2k}}{(2k)} \frac{1}{n^{2k}}, \quad n \to \infty, \tag{2.7.120}$$

or

$$H_n \sim \ln n + \gamma + \frac{1}{2n} - \frac{1}{12n^2} + \frac{1}{120n^4} - \frac{1}{504n^6} + \ldots. \tag{2.7.121}$$

This confirms the limit in (2.7.105).

2.7.2 *More general cases*

First we consider the case

$$F_\lambda(\alpha, z) = \frac{1}{\Gamma(\lambda)} \int_0^\infty t^{\lambda-1} e^{-zt} f(t) \frac{dt}{t+\alpha}, \quad \Re\lambda > 0, \quad |\mathrm{ph}\,\alpha| < \pi. \tag{2.7.122}$$

We can repeat the same method that we used for (2.7.97). We use a function $g(t)$ as in (2.7.98), we expand $g(t) = \sum_{n=0}^{\infty} g_n t^n$, and obtain the modification of (2.7.101):

$$F_\lambda(\alpha, z) \sim f(-\alpha) e^{\alpha z}\, \Gamma(1-\lambda, \alpha z) + \sum_{n=0}^{\infty} g_n\, n!\, z^{-n-1}, \tag{2.7.123}$$

where $\Gamma(a, z)$ is the incomplete gamma function considered in Chapter 7, and for which we have used the integral representation given in (7.1.6).

For the more general case

$$F_{\lambda,\mu}(\alpha, z) = \frac{1}{\Gamma(\lambda)} \int_0^\infty t^{\lambda-1} e^{-zt} f(t) \frac{dt}{(t+\alpha)^\mu}, \quad \Re\lambda > 0, \quad |\mathrm{ph}\,\alpha| < \pi, \tag{2.7.124}$$

we refer to §28.1.

Chapter 3

The method of Laplace

The standard form for Laplace's method is the integral

$$F(z) = \int_{-\infty}^{\infty} e^{-zt^2} f(t)\, dt, \quad \Re z > 0, \tag{3.0.1}$$

where the function f is assumed to be analytic inside a domain $\mathcal{D}$ of the complex plane that contains the real axis in its interior.

By splitting up the integral in two parts, for positive and negative t, and substituting $t = \pm\sqrt{u}$, two integrals arise that can be expanded for large values of z by applying Watson's lemma considered in §2.1.

On the other hand we can substitute a Maclaurin expansion to obtain the expansion:

$$f(t) = \sum_{k=0}^{\infty} f_k t^k \quad \Longrightarrow \quad F(z) \sim \sqrt{\frac{\pi}{z}} \sum_{k=0}^{\infty} f_{2k} \frac{\left(\frac{1}{2}\right)_k}{z^k}, \quad z \to \infty, \tag{3.0.2}$$

where the z-domain of validity depends on the location of the singularities of the function f in the complex plane.

Because the integrals considered in the saddle point method (and in several other methods) often can be transformed into the standard form (3.0.1), it is convenient to consider Laplace's method in addition to methods based on Watson's lemma. The starting point of Laplace's method is usually a contour integral.

The saddle point method will be discussed in more detail in Chapter 4. For a detailed analysis of Laplace's method we refer to Olver (1997, pp. 121–127), and first we give a theorem from that reference.

3.1 A theorem for the general case

We consider integrals of the type

$$I(z) = \int_a^b e^{-zp(t)}\, q(t)\, dt \tag{3.1.3}$$

for the case of a simple saddle point: the first derivative of $p(t)$ at the saddle point is zero and the second derivative is different from zero.

Theorem 3.1 (Laplace's method). *Assumptions:*

(*i*) $p(t)$ *and* $q(t)$ *are independent of* z*, and single valued and holomorphic in a domain* $\mathcal{D}$*.*

(*ii*) *The integration path* $\mathcal{P}$ *is independent of* z*. The endpoints* a *and* b *of* $\mathcal{P}$ *are finite or infinite, and* $\mathcal{P}$ *lies within* $\mathcal{D}$*.*

(*iii*) $p'(t)$ *has a simple zero at an interior point* t_0 *of* $\mathcal{P}$*.*

(*iv*) z *ranges along a ray or over an annular sector given by* $\theta_1 \le \theta \le \theta_2$ *and* $|z| \ge Z$*, where* $\theta = \operatorname{ph} z, \theta_2 - \theta_1 < \pi$*, and* $Z > 0$*. At* a *and* b*,* $I(z)$ *converges absolutely and uniformly with respect to* z*.*

(*v*) $\Re\left(e^{i\theta}p(t) - e^{i\theta}p(t_0)\right)$ *is positive on* $\mathcal{P}$*, except at* t_0*, and is bounded away from zero uniformly with respect to* θ *as* $t \to a$ *or* b *along* $\mathcal{P}$*.*

With the foregoing assumptions,

$$\int_a^b e^{-zp(t)}\, q(t)\, dt \sim \sqrt{\frac{\pi}{zp_2}} e^{-zp(t_0)} \sum_{n=0}^{\infty} a_{2n} \frac{(\frac12)_n}{z^n}, \tag{3.1.4}$$

as $z \to \infty$ *in the sector* $\theta_1 \le \operatorname{ph} z \le \theta_2$*.*

Proof. For a proof of this theorem (with more general conditions on the function q) we refer to Olver (1997, p. 121–125). ☐

Formulas for the first two coefficients are

$$a_0 = q_0, \quad a_2 = \frac{1}{p_2}\left(q_2 - \frac{3p_3q_1}{2p_2} + \left(\frac{15p_3^2}{8p_2^2} - \frac{3p_4}{2p_2}\right) q_0\right), \tag{3.1.5}$$

where p_n, q_n are the coefficients of the Taylor expansions of p, q at $t = t_0$. That is,

$$p(t) = \sum_{n=0}^{\infty} p_n\,(t-t_0)^n, \quad q(t) = \sum_{n=0}^{\infty} q_n\,(t-t_0)^n. \tag{3.1.6}$$

In forming $\sqrt{p_2}$ of (3.1.4) the branch of $\omega_0 = \operatorname{ph} p_2$ must satisfy $|\omega_0 + \theta + 2\omega| \le \frac12\pi$, where ω is the limiting value of $\operatorname{ph}(t-t_0)$ as $t \to t_0$ along $\mathcal{P}$.

In §3.2 we shall describe methods for calculating the coefficients in the expansion, one based on inversion methods and the other method gives explicit forms of the coefficients. In §3.4 we use Laplace's method to obtain an expansion of the complementary error function, and we verify the assumptions of the theorem.

Observe that the theorem considers a saddle point without discussing the properties of the path in connection with steepest descent, as considered in the saddle point method in Chapter 4. Condition (v) implies that the path runs into the valleys on both sides of the saddle point.

3.2 Constructing the expansion

We consider integrals of the type

$$F(z) = \int_{-\infty}^{\infty} e^{-zp(t)} q(t)\, dt. \tag{3.2.7}$$

Usually these integrals arise after some transformations, in particular in the saddle point method, and Laplace's method will give the asymptotic expansion of these integrals.

Initially we assume that z is a positive large parameter; for complex values we refer to Lemma 3.1 below. We assume that p is concave, and that its minimum is located at $t = 0$, that is, $p'(0) = 0$ and $p''(0) > 0$. In addition, we assume that the functions p and q are analytic in a neighborhood of the real axis.

We have the expansion

$$p(t) = p(0) + \tfrac{1}{2}p''(0)t^2 + \mathcal{O}\left(t^3\right), \quad t \to 0, \tag{3.2.8}$$

and we use the transformation

$$p(t) - p(0) = \tfrac{1}{2}p''(0)w^2, \tag{3.2.9}$$

with the assumption that $\operatorname{sign}(t) = \operatorname{sign}(w)$. For a proper interpretation of taking the square root, see (3.2.16).

The transformation gives

$$F(z) = e^{-zp(0)} \int_{-\infty}^{\infty} e^{-\frac{1}{2}zp''(0)w^2} f(w)\, dw, \quad f(w) = q(t)\frac{dt}{dw}. \tag{3.2.10}$$

We expand $f(w) = \sum_{k=0}^{\infty} c_k w^k$, and obtain

$$F(z) \sim e^{-zp(0)} \sum_{k=0}^{\infty} c_k \int_{-\infty}^{\infty} e^{-\frac{1}{2}zp''(0)w^2} w^k\, dw. \tag{3.2.11}$$

Evaluating the integrals, and observing that the integrals with odd k do not contribute, we obtain

$$F(z) \sim \sqrt{\frac{2\pi}{zp''(0)}} e^{-zp(0)} \sum_{k=0}^{\infty} \frac{2^k \left(\frac{1}{2}\right)_k c_{2k}}{\left(zp''(0)\right)^k}, \quad z \to \infty. \tag{3.2.12}$$

From Theorem 3.1 we conclude that the expansion in (3.2.12) holds for $z \to \infty$ in a certain sector $\theta_1 \le \operatorname{ph} z \le \theta_2$. We can specify this sector when we know the singularities of the function $f(w)$ in (3.2.10).

As in Watson's lemma for analytic functions, see Theorem 2.2, we assume that the function $f(w)$ defined in (3.2.10) is analytic inside a disk $|w| \le r$, $r > 0$ and in a sector $\alpha < \operatorname{ph} w < \beta$, where $\alpha < 0$ and $\beta > 0$. By assuming that f is an even function (which is no restriction) it follows that a similar condition holds around the negative axis. Inside this sector we also need a growth condition on f at infinity: we assume that there is a real number σ and $M > 0$ such that $|f(w)| \le M e^{\sigma |w^2|}$.

Under these conditions we can rotate the path of integration through an angle, taking into account the convergence at infinity and the singularities of the function f.

Write $z = re^{i\theta}$ and $w = \sigma e^{i\tau}$. Then, when rotating the path, the condition for convergence at infinity is $\cos(\theta+2\tau) > 0$, that is, $-\frac{1}{2}\pi < \theta+2\tau < \frac{1}{2}\pi$, which should be combined with $\alpha < \tau < \beta$ to remain inside the sector. Our conclusion is given in the following lemma.

Lemma 3.1. *The expansion in* (3.2.12) *holds uniformly inside the sector*

$$-2\beta - \tfrac{1}{2}\pi + \delta \le \operatorname{ph} z \le \tfrac{1}{2}\pi - 2\alpha - \delta, \tag{3.2.13}$$

for an arbitrary small positive number δ.

Remark 3.1. When in (3.2.10) the integral is over a finite interval, as may be the case in Theorem 3.1 for the integral in (3.1.4), then $\alpha = \beta = 0$, which means $\Re z \to +\infty$. △

3.2.1 *Inversion method for computing the coefficients*

We can compute the coefficients c_{2k} shown in Theorem 3.1 when we know the coefficients of the Maclaurin expansions of the functions p and q.

Let these expansions be given by (3.1.6), with $t_0 = 0$. Then, in order to obtain an expansion of t in powers of w, we substitute the expansion $t = \sum_{k=1}^{\infty} t_k w^k$ into (3.2.9), and obtain the coefficients t_k by comparing equal powers of w. Observe that from (3.2.9) and the assumption $\operatorname{sign}(t) = \operatorname{sign}(w)$, we know that $t_1 = 1$. The next few are

$$t_2 = -\frac{p_3}{2p_2}, \quad t_3 = \frac{5p_3^2 - 4p_2p_4}{8p_2^2}, \quad t_4 = \frac{3p_2p_3p_4 - 2p_3^3 - p_2^2p_5}{2p_2^3}. \tag{3.2.14}$$

When we have the coefficients t_k we can compute the coefficients c_k of the expansion of $f(w)$ defined in (3.2.10), and the first coefficients c_0 and c_2 of the expansion in (3.2.12) are the same (up to a factor) as given in (3.1.5).

Because we assume that the function $p(t)$ is analytic near the origin, we can write

$$k\, t_k = \frac{1}{2\pi i}\int_{\mathcal{C}_w} \frac{dt}{dw}\frac{dw}{w^k} = \frac{1}{2\pi i}\int_{\mathcal{C}_t} \frac{dt}{w^k}, \quad k = 1, 2, 3, \ldots, \tag{3.2.15}$$

where $\mathcal{C}_w$ and $\mathcal{C}_t$ are small circles around the origin in the w and t planes. We can replace w by using (3.2.9), taking the proper sign of the square root. This can be done by writing

$$w = t\sqrt{2\frac{p(t)-p(0)}{p''(0)\,t^2}} = t\sqrt{1 + t\frac{p_3}{p_2} + \mathcal{O}(t^2)}, \tag{3.2.16}$$

in which the square root always has the same (positive) sign. This gives

$$k\,t_k = \frac{1}{2\pi i}\int_{\mathcal{C}_t}\left(2\frac{p(t)-p(0)}{p''(0)\,t^2}\right)^{-k/2}\frac{dt}{t^k},\quad k=1,2,3,\ldots, \tag{3.2.17}$$

which says that $k\,t_k$ equals the coefficient of t^{k-1} in the expansion of the binomial form in (3.2.17).

3.3 Explicit forms of the coefficients in the expansion

We use Wojdylo's results (Wojdylo, 2006a,b) as we did in §2.4.4 for Laplace-type integrals. In order to use his notation, we write the expansions of p and q in the form

$$p(t) = p(0) + \sum_{k=0}^{\infty} a_k t^{k+2},\quad q(t) = \sum_{k=0}^{\infty} b_k t^k,\quad a_0 > 0,\quad b_0 \neq 0, \tag{3.3.18}$$

and normalize the coefficients by writing

$$A_k = \frac{a_k}{a_0},\quad B_k = \frac{b_k}{b_0},\quad k \geq 0. \tag{3.3.19}$$

Then, the integral

$$F(z) = \int_a^b e^{-zp(t)}\,q(t)\,dt,\quad a<0,\quad b>0,\quad p'(0)=0, \tag{3.3.20}$$

with similar conditions as in Theorem 3.1, has the asymptotic expansion

$$F(z) \sim b_0\sqrt{\frac{\pi}{a_0\,z}}e^{-zp(0)}\sum_{n=0}^{\infty}\left(\tfrac12\right)_n\frac{d_{2n}}{(a_0\,z)^n},\quad z\to\infty, \tag{3.3.21}$$

where

$$d_n = \sum_{j=0}^{n} B_{n-j}\sum_{k=0}^{j}\binom{-\frac12 n-\frac12}{k} C_{jk},\quad n=0,1,2,\ldots. \tag{3.3.22}$$

The coefficients C_{jk} are defined as in (2.4.73) and (2.4.74).

When in the integral in (3.3.20) $a=0$ we need all coefficients d_n in the expansion, and not only the ones with even index. We give the result for a more general integral of the form

$$F_\lambda(z) = \int_0^b t^{\lambda-1}e^{-zp(t)}\,q(t)\,dt,\quad b>0, \tag{3.3.23}$$

where p and q have expansions of the form

$$p(t) = p(0) + \sum_{k=0}^{\infty} a_k t^{k+\mu},\quad q(t) = \sum_{k=0}^{\infty} b_k t^k,\quad a_0 > 0,\quad b_0 \neq 0. \tag{3.3.24}$$

Hence, if μ is a positive integer, we assume that $p(t)$ has $\mu-1$ vanishing derivatives at $t=0$. However, μ may be an arbitrary positive real number.

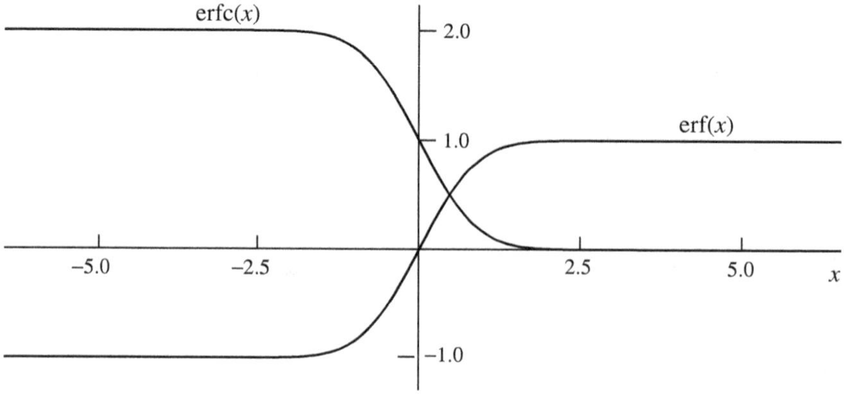

Fig. 3.1 The functions $\operatorname{erf} x$ and $\operatorname{erfc} x$.

Then, again according to the results of Wojdylo, $F_\lambda(z)$ has the asymptotic expansion

$$F_\lambda(z) \sim \frac{b_0\, e^{-zp(0)}}{\mu(a_0\, z)^{\lambda/\mu}} \sum_{n=0}^{\infty} \Gamma\left(\frac{\lambda+n}{\mu}\right) \frac{d_n}{(a_0\, z)^{n/\mu}}, \quad z \to \infty, \tag{3.3.25}$$

where

$$d_n = \sum_{j=0}^{n} B_{n-j} \sum_{k=0}^{j} \binom{-(n+\lambda)/\mu}{k} C_{jk}, \quad n = 0, 1, 2, \ldots. \tag{3.3.26}$$

This general form gives the results for Laplace-type integrals considered in §2.4.4 when we take $\mu = 1$. For more recent work on explicit formulas for the coefficients in the saddle point method and Laplace's method we refer to López and Pagola (2011a) and Nemes (2013).

3.4 The complementary error function

The error function and the complementary error function are defined by

$$\operatorname{erf} z = \frac{2}{\sqrt{\pi}} \int_0^z e^{-t^2}\, dt, \quad \operatorname{erfc} z = \frac{2}{\sqrt{\pi}} \int_z^\infty e^{-t^2}\, dt, \tag{3.4.27}$$

respectively, with the properties

$$\operatorname{erf} z + \operatorname{erfc} z = 1, \quad \operatorname{erf}(-z) = -\operatorname{erf} z, \quad \operatorname{erfc}(-z) = 2 - \operatorname{erfc} z. \tag{3.4.28}$$

See Figure 3.1. These functions are analytic functions and are essential in statistical distribution functions, where they are used as the normal or Gaussian distribution function (up to some scaling). We will also meet these functions in asymptotic problems when a saddle point coalesces with a pole (Chapter 21), or when a saddle point coalesces with an endpoint of the interval of integration (§22.1).

The large-z asymptotic expansion of $\operatorname{erfc} z$ can be obtained by integrating by parts in the integral representation, but here we derive the expansion by applying Laplace's method to the integral representation (see Temme (2010c, Eq. 7.7.1))

$$\operatorname{erfc} z = \frac{e^{-z^2}}{\pi} \int_{-\infty}^{\infty} \frac{e^{-z^2t^2}}{1+t^2}\, dt, \quad |\operatorname{ph} z| \le \tfrac{1}{4}\pi. \tag{3.4.29}$$

The expansion $1/\left(1+t^2\right) = 1 - t^2 + t^4 - \cdots$ gives

$$\operatorname{erfc} z \sim \frac{e^{-z^2}}{z\sqrt{\pi}} \sum_{n=0}^{\infty} (-1)^n \frac{\left(\frac{1}{2}\right)_n}{z^{2n}}, \quad z \to \infty, \tag{3.4.30}$$

with the first coefficients shown in

$$\operatorname{erfc} z \sim \frac{e^{-z^2}}{z\sqrt{\pi}} \left(1 - \frac{1}{2z^2} + \frac{3}{4z^4} - \frac{15}{8z^6} + \cdots\right), \quad z \to \infty. \tag{3.4.31}$$

Theorem 3.1 gives the z-domain $|\operatorname{ph} z| \le \frac{1}{4}\pi - \delta$. We can extend this domain by rotating the path of integration in (3.4.29), taking into account convergence at infinity and the singularities of the integrand at $\pm i$.

By using the notation of Lemma 3.1, we see that $-\alpha = \beta = \frac{1}{2}\pi$, from the lemma it follows that (observe that we have z^2 as the large parameter) the asymptotic expansion in (3.4.30) holds for $z \to \infty$, uniformly inside the sector $-\frac{3}{4}\pi + \delta \le \operatorname{ph} z \le \frac{3}{4}\pi - \delta$ for any small positive number δ.

An expansion for the sector $-\frac{3}{4}\pi + \delta \le \operatorname{ph}(-z) \le \frac{3}{4}\pi - \delta$ follows from the third relation in (3.4.28).

An expansion with remainder follows from writing

$$\frac{1}{1+t^2} = \sum_{n=0}^{N-1} (-1)^n t^{2n} + (-1)^N \frac{t^{2N}}{1+t^2}, \quad N = 0, 1, 2, \ldots. \tag{3.4.32}$$

This gives

$$\operatorname{erfc} z = \frac{e^{-z^2}}{z\sqrt{\pi}} \left(\sum_{n=0}^{N-1} (-1)^n \frac{\left(\frac{1}{2}\right)_n}{z^{2n}} + (-1)^N R_N(z) \right), \tag{3.4.33}$$

where

$$R_N(z) = \frac{z}{\sqrt{\pi}} \int_{-\infty}^{\infty} e^{-z^2t^2} \frac{t^{2N}}{1+t^2}\, dt, \quad N = 0, 1, 2, \ldots. \tag{3.4.34}$$

By using the integral representation of the incomplete gamma function in (7.1.6) it follows that

$$R_N(z) = z(-1)^N \left(\tfrac{1}{2}\right)_N e^{z^2} \Gamma\left(\tfrac{1}{2} - N, z^2\right). \tag{3.4.35}$$

To obtain a bound for $R_N(z)$ we first write (3.4.34) in the form where

$$R_N(z) = \frac{2}{\sqrt{\pi}} \frac{1}{z^{2N}} \int_0^{\infty} e^{-t^2} \frac{t^{2N}}{1+t^2/z^2}\, dt. \tag{3.4.36}$$

Let $z \in \mathbb{C}$ such that

$$|1 + t^2/z^2| \geq 1, \quad \forall t \geq 0. \tag{3.4.37}$$

Then

$$R_N(z) = \frac{\left(\frac{1}{2}\right)_N}{z^{2N}} \theta_N(z), \tag{3.4.38}$$

and $|\theta_N(z)| \leq 1$ for the z satisfying (3.4.37). This holds when $|\mathrm{ph}\, z^2| \leq \frac{1}{2}\pi, z \neq 0$. To see this, observe that the equality $|1 + \zeta| = 1$ in the ζ-plane is satisfied on the circle $(\xi + 1)^2 + \eta^2 = 1$, where $\zeta = \xi + i\eta$, and that for all ζ with $\xi \geq 0$ we have $|1 + \zeta| \geq 1$.

It follows that when $|\mathrm{ph}\, z| \leq \frac{1}{4}\pi$, the remainder $R_N(z)$ is bounded in magnitude by the first neglected term in (3.4.33) and has the same sign as this term when $z > 0$.

Remark 3.2. The remainder $R_N(z)$ defined in (3.4.34) can be re-expanded by using the method described in §2.6. △

Chapter 4

The saddle point method and paths of steepest descent

We have mentioned in Chapter 3 the connection between Laplace's method and the saddle point method. In this chapter we give several examples and details of the method. The integrals are presented as contour integrals in the complex plane, for which we select modifications of the contours before applying Laplace's method. Usually, these modified contours run through a saddle point of the integrand.

The starting point in the asymptotic analysis in this chapter is written in the form

$$F(\lambda) = \int_{\mathcal{C}} e^{-\lambda\phi(z)}\psi(z)\,dz, \tag{4.0.1}$$

where λ is a large real or complex parameter and ϕ, ψ are functions of the complex variable z and are analytic in a domain $\mathcal{D}$ of the complex plane. The integral is taken along a path $\mathcal{C}$ in $\mathcal{D}$ and avoids the singularities of the integrand. Integrals of this type arise naturally in the context of linear wave propagation and in other physical problems, and many special functions can be represented by such integrals.

Riemann sketched the saddle point method in 1863, and Debye used it in 1909 for Bessel functions of large orders. For a paper on the origin of the method of saddle points and steepest descent we refer to Petrova and Solov′ev (1997), where it is concluded that already in 1829 Cauchy used elements of this method.

Consider this problem from the viewpoint of numerical quadrature of the above integral. Assume that λ is real. Separating ϕ into its real and imaginary parts, writing $z = x + iy$, $\phi(z) = R(x,y) + iI(x,y)$, we know that, when λ is large, the evaluation of the integral is difficult because of the strong oscillatory behavior of the integrand, caused by the expression $e^{-i\lambda I(x,y)}$. This may give inaccurate numerical results because of cancellations in the computations.

Usually we have much freedom in choosing the path $\mathcal{C}$ in the complex plane (by invoking Cauchy's theorem). When the contour $\mathcal{C}$ can be chosen such that $I(x,y) = I_0$ (a constant) for $z = x + iy \in \mathcal{C}$, we can write

$$F(\lambda) = e^{-i\lambda I_0}\int_{\mathcal{C}} e^{-\lambda\Re\phi(z)}\psi(z)\,dz, \tag{4.0.2}$$

where the dominant part $e^{-\lambda R(x,y)}$ of the integral is non-oscillating (in some cases the new path $\mathcal{C}$ is split up into more paths, each path being defined by a different I_0,

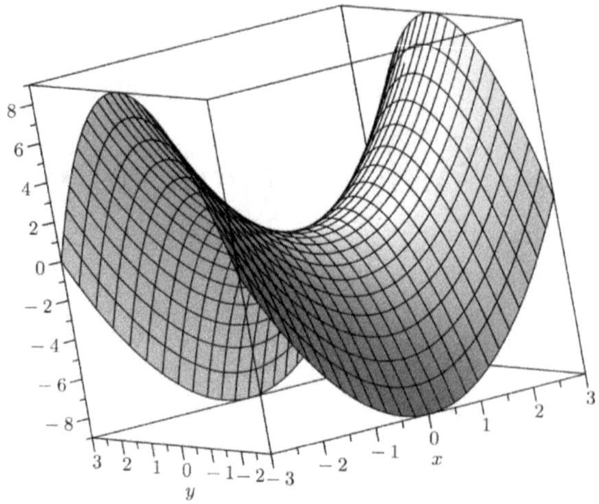

Fig. 4.1 The function $R(x, y) = \Re z^2 = x^2 - y^2$ for $|x| \le 3, |x| \le 3$.

resulting in a sum of integrals of the above type). From a numerical point of view the new representation of $F(\lambda)$ is very attractive. The question that remains is: which constant I_0 should be used? Luckily, *there is not much choice.*

Considering the real part of the phase function $R(x, y)$, and the landscape of mountains and valleys defined by $e^{-\lambda R(x,y)}$, we may assume that, if the original contour $\mathcal{C}$ extends to infinity in two directions, $\mathcal{C}$ will certainly run from one valley into another one.

These valleys will meet at a "mountain pass" where $R(x, y)$ has a saddle point. This is a point in the (x, y)-domain that is a stationary point of $R(x, y)$ where

$$\frac{\partial R(x,y)}{\partial x} = 0 \quad \text{and} \quad \frac{\partial R(x,y)}{\partial y} = 0, \tag{4.0.3}$$

but not a local extremum.

In Figure 4.1 we show the saddle point and the valleys of the function $R(x, y) = \Re z^2 = x^2 - y^2$ for $|x| \le 3, |x| \le 3$.

Summarizing, one tries to deform the contour $\mathcal{C}$ through one or more points where the dominant part of the integrand locally behaves like a bell-shaped Gaussian curve. These points are found at the saddle points of the integrand. When the contour is chosen through $z_0 = x_0 + iy_0$, one of these saddle points, the constant I_0 in (4.0.2) should be $I(x_0, y_0)$.

The method is best learned from clear examples, our focus in the remaining part of this chapter. Many general aspects and details can be found in De Bruijn

(1958, Chapters 5–6). In Gil *et al.* (2007, §5.5) also aspects of numerical quadrature related to the saddle point method are discussed.

4.1 The axis of the valley at the saddle point

Once we have decided which saddle point can be used for starting the asymptotic analysis, it is important to know the direction of the valley axis, that is, the direction into the valley along a steepest descent path from z_0. For the integral in (4.0.1) this can be decided by considering the phase of the second derivative of $\phi(z)$ at the saddle point. A local expansion at a saddle point z_0 gives (we take $\lambda > 0$, and for convenience $\lambda = 1$)

$$e^{-\phi(z)} = e^{-\phi(z_0)-\frac{1}{2}\phi''(z_0)(z-z_0)^2+\mathcal{O}(z-z_0)^3}. \tag{4.1.4}$$

Writing $\phi''(z_0) = re^{i\theta}$, $z - z_0 = Re^{i\sigma}$, we have

$$\phi''(z_0)(z-z_0)^2 = rR^2e^{i(\theta+2\sigma)}, \tag{4.1.5}$$

and for the axis of the valley at the saddle point z_0 this should be real and non-negative. Hence, we should have $\cos(\theta + 2\sigma) = 1$, and when we take $\sigma = -\frac{1}{2}\theta + k\pi$ $(k = 0, \pm 1)$ we run from the point z_0 into the valley along the path of steepest descent.

In other words, when $\theta = \operatorname{ph}(\phi''(z_0)) = 0$, then the axis of the valley runs parallel to the real axis, and otherwise it meets the real axis at an angle $-\frac{1}{2}\theta$ or $-\frac{1}{2}\theta \pm \pi$.

Compare this with the basic example e^{-z^2}. The axis of the saddle point and the steepest descent path coincide with the real axis. The imaginary axis is the path of steepest ascent.

4.2 Examples with simple exponentials

We demonstrate a few aspects of the saddle point method by considering integrals with simple exponentials. In later chapters many other examples will be given. For instance, for the reciprocal gamma function we refer to §6.2.

4.2.1 *A first example*

Let us consider the integral

$$F(\lambda) = \int_{-\infty}^{\infty} e^{-\lambda\phi(z)}\,dz, \tag{4.2.6}$$

with $\lambda > 0$ and $\phi(z) = \frac{1}{2}z^2 - iz$. There is a saddle point at $z = i$.

We write $z = x + iy$, which gives

$$\phi(z) = \tfrac{1}{2}(x^2 - y^2) + y + ix(y-1) = R(x,y) + iI(x,y). \tag{4.2.7}$$

The imaginary part $I(x, y)$ is constant, that is, $I(x, y) = c$, when $x(y-1) = c$. The original path $\mathcal{C} = \mathbb{R}$ can be shifted upwards to the path $y = 1$, corresponding to $c = 0$. This path runs through the saddle point at $z = i$. Observe that at the saddle point $I(x, y) = 0$.

Other choices on which $I(x, y)$ is constant, such as $x = 0$ or hyperbolas when $c \neq 0$, would give divergent integrals, and these paths do not run through the saddle point.

We shift the contour in (4.2.6) upwards through $z = i$, and obtain, substituting $z = x + i$,

$$F(\lambda) = \int_{-\infty+i}^{+\infty+i} e^{-\lambda(\frac{1}{2}z^2 - iz)}\, dz = e^{-\frac{1}{2}\lambda} \int_{-\infty}^{\infty} e^{-\frac{1}{2}\lambda x^2}\, dx = \sqrt{2\pi}\, e^{-\frac{1}{2}\lambda}. \qquad (4.2.8)$$

We see in this simple example that the result shows an exponentially small factor $e^{-\frac{1}{2}\lambda}$ in front of the x-integral, that is, after bringing the original contour into the path of steepest descent. No other choice of contour not running through the saddle point would have given this factor. In particular, a numerical quadrature scheme used for computing the integral in (4.2.6) as it is given, or every contour not passing through the saddle point, would suffer from serious numerical instabilities and cancellation of leading digits for large positive values of λ.

4.2.2 *A cosine transform*

Another example is the cosine transform of the Gaussian, that is,

$$I(\lambda) = \int_{-\infty}^{\infty} e^{-t^2} \cos(2\lambda t)\, dt = \sqrt{\pi} e^{-\lambda^2}. \qquad (4.2.9)$$

Taking $\lambda = 10$ we get $I(10) \doteq 0.6593662990\text{e}{-43}$.

When we write

$$I(\lambda) = \Re \int_{-\infty}^{\infty} e^{-t^2 + 2i\lambda t}\, dt, \qquad (4.2.10)$$

we see that there is a saddle point at $t = i\lambda$, and again we can shift the original path into the path steepest descent path $\Im t = \lambda$.

This example might be considered first by using the method of stationary phase, see Chapter 14, in particular §14.9, where we explain that when we use analytic integrands the saddle point method may be more successful than the method of stationary phase.

4.3 Steepest descent paths not through a saddle point

In the examples just considered we can change the path of integration into a path of steepest descent without passing singularities of the integrand. When the function $\psi(z)$ of the integral

$$F(\lambda) = \int_{\mathcal{L}} e^{-\lambda\phi(z)} \psi(z)\, dz \qquad (4.3.11)$$

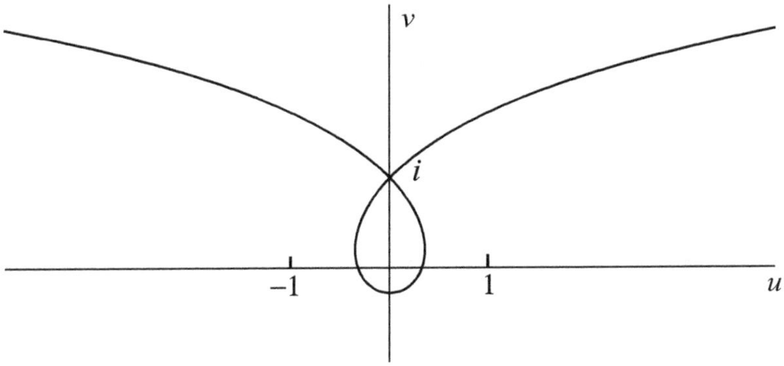

Fig. 4.2 Steepest descent path for the function $e^{-y\phi(w)}$ of the integral in (4.3.13).

has poles, which are passed during the modification of the contour, residues should be taken into account. Other types of singularities may yield extra integrals around branch cuts. In addition, more saddle points may occur, and it is necessary to investigate which one should be used for the saddle point analysis, and perhaps the new contour may pass through more than one saddle point.

It may happen that we cannot deform the contour into a path of steepest descent through one of the saddle points. This may occur when we have a contour integral with one or two fixed finite endpoints.

4.3.1 *A gamma function example*

First we consider the following example

$$\Gamma(1+iy) = \int_0^\infty e^{-t} t^{iy}\, dt \tag{4.3.12}$$

for large positive y. The phase of t is zero on the path of integration. There are many oscillations near the origin, and for asymptotic analysis we will deform the path.

We write the integral in the form

$$\Gamma(1+iy) = y^{1+iy} \int_0^\infty e^{-y\phi(w)}\, dw, \quad \phi(w) = w - i \ln w. \tag{4.3.13}$$

There is a saddle point at $w = i$, and $\phi(i) = i - i \ln i = i + \frac{1}{2}\pi$. We write $w = re^{i\theta}$. Then, the equation for a saddle point contour through $w = i$ reads

$$r \sin\theta - \ln r - 1 = 0. \tag{4.3.14}$$

In cartesian coordinates $w = u + iv$ we have the equation

$$v - 1 - \tfrac{1}{2} \ln\left(u^2 + v^2\right) = 0. \tag{4.3.15}$$

In Figure 4.2 we show the path defined by these equations, and observe that it does not pass through the origin. The path comes from $+\infty$, passes the saddle

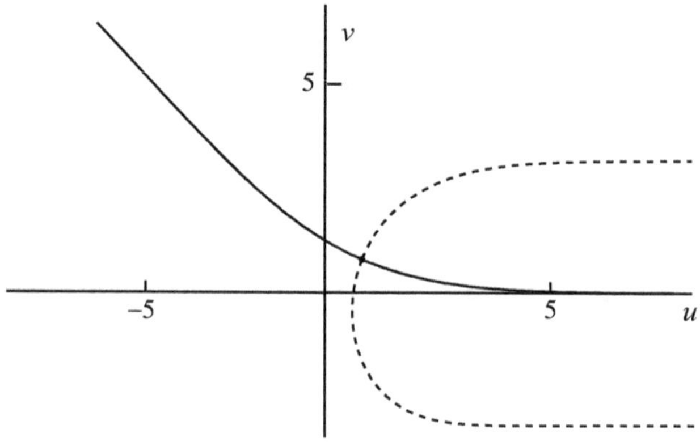

Fig. 4.3 Steepest descent path for the function $e^{-y\psi(w)}$ of the integral in (4.3.16).

point, and the path returns to the saddle point, but on a different Riemann sheet after crossing the negative axis. After crossing the saddle point for the second time, the path becomes a path of steepest ascent to $-\infty$.

When we visit $w = i$ for the second time, and we make a left turn at $w = i$, then we continue on the descent path. Each time we visit the saddle point again, we make a left turn, and after each turn the function $e^{-y\phi(w)}$ becomes smaller with a factor $e^{-2\pi y}$ because of the many-valued logarithm in $\phi(w)$.

When we use Laplace's method at the saddle point $w = i$, we can obtain a complete asymptotic expansion of $\Gamma(1+iy)$, although the path of steepest descent does not run through the origin.

A simple transformation $w \to e^w$ gives the integral

$$\Gamma(1+iy) = y^{1+iy} \int_{-\infty}^{\infty} e^{-y\psi(w)}\, dw, \quad \psi(w) = e^w - iw - w/y, \tag{4.3.16}$$

and now the original contour can be shifted into the steepest descent path through the saddle point at $\ln(1/y + i)$. In Figure 4.3 we show the new contour for $y = 1$. The dashed contour is the path of steepest ascent.

In Chapter 6 we consider the gamma function in more detail.

4.3.2 *An integral related to the error function*

In this example we consider

$$F_\alpha(\lambda) = \int_\alpha^{\infty+ic} e^{-\lambda(z^2-2iz)}\, dz, \quad \lambda > 0, \tag{4.3.17}$$

where $z = x + iy$, $\alpha = \beta + i\gamma$ may be any complex number and the upper endpoint means that the path tends to $+\infty + ic$ for some $c \in \mathbb{R}$; therefore, it runs down into the valley that contains the real positive axis. This integral can be written in terms of the error functions defined in (3.4.27).

As we have seen in §4.2, there is a saddle point at $z = i$ and on the saddle point contour $y = 1$ we have $I(x, y) = 0$ (see (4.2.7)). For general α, this contour $y = 1$ does not pass through the point α, the initial point of the contour. In order to integrate along a curve of constant phase, we can now consider the steepest descent path passing through $\alpha = \beta + i\gamma$; in that case, we would integrate along a path such that $I(x, y) = I(\beta, \gamma)$, that is, such that

$$x(y-1) = \beta(\gamma - 1). \tag{4.3.18}$$

We can consider several different possibilities for the path along which this relation is satisfied.

(1) When $\gamma = 1$ we integrate along $y = 1$, $x \geq \beta$.
(2) When $\beta = 0$ we integrate first from $z = i\gamma$ to $z = i$, then from this point to $z = +\infty + i$.
(3) When $\beta > 0$ we integrate from $x = \beta$ to $+\infty$ along the hyperbola branch defined in (4.3.18). The integral becomes

$$F_\alpha(\lambda) = e^{-i\lambda\beta(\gamma-1)} \int_\beta^\infty e^{-\lambda\psi(x)} \left(1 + i\frac{dy}{dx}\right) dx, \tag{4.3.19}$$

where

$$y = 1 + \frac{\beta(\gamma-1)}{x}, \quad \psi(x) = x^2 - y^2 + 2y = x^2 + 1 - \frac{\beta^2(\gamma-1)^2}{x^2}. \tag{4.3.20}$$

If we wish to obtain an asymptotic expansion for large values of λ by using this integral, we can transform it into a Laplace-type integral by substituting $u = \psi(x) - \psi(\beta)$ and use Watson's lemma.
(4) When $\beta < 0$ the steepest descent path defined by (4.3.18) runs into the other valley $(-\infty)$. In this case, we can write

$$F_\alpha(\lambda) = \int_{-\infty}^{\infty} e^{-\lambda(z^2-2iz)}\, dz - \int_{-\infty}^{\alpha} e^{-\lambda(z^2-2iz)}\, dz. \tag{4.3.21}$$

For the first integral see (4.2.8), while for the second one we can proceed as in the case $\beta > 0$.

Another example where one of the endpoints is finite, is provided by the *Scorer function* $\mathrm{Hi}(z)$, which has the integral representation

$$\mathrm{Hi}(z) = \frac{1}{\pi} \int_0^\infty e^{zw - \frac{1}{3}w^3}\, dw. \tag{4.3.22}$$

In Gil *et al.* (2001) it is shown that the original contour of integration can be continuously deformed into the steepest descent path when z is in the sector $\frac{2}{3}\pi \leq \mathrm{ph}\, z \leq \pi$, but that two pieces have to be considered when $0 \leq \mathrm{ph}\, z \leq \frac{2}{3}\pi$ (similar to what happened in the previous example (4.3.17) for $F_\alpha(\lambda)$).

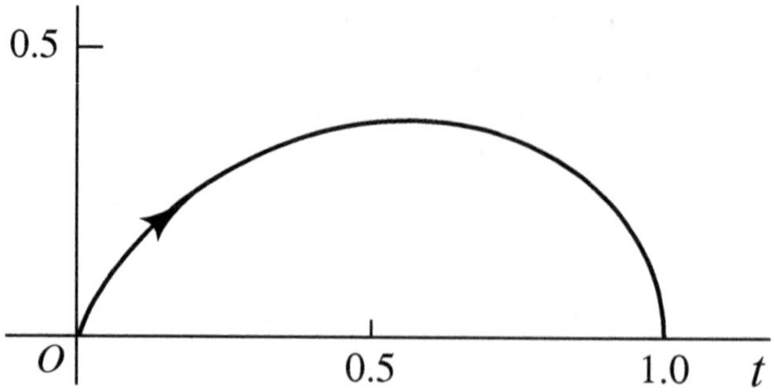

Fig. 4.4 Contour for the integral in (4.4.24).

4.4 An example with strong oscillations: A 100-digit challenge

This is another example in which an integral with strong oscillations can be transformed into one such that computations become easy. Consider the integral

$$I = \int_0^1 \cos\left(t^{-1}\ln t\right)\,\frac{dt}{t}. \tag{4.4.23}$$

Because of the behavior of the integrand at $t = 0$ this integral is difficult to evaluate in this form. We can deform the path of integration $[0,1]$ into a path in the complex plane on which it is safer for calculations.

We write

$$I = \Re\int_0^1 e^{\phi(t)}\,\frac{dt}{t}, \quad \phi(t) = i\frac{\ln t}{t}, \tag{4.4.24}$$

and introduce polar coordinates by writing $t = re^{i\theta}$. Then,

$$\phi(t) = \frac{1}{r}\left(\sin\theta\ln r - \theta\cos\theta\right) + \frac{i}{r}\left(\cos\theta\ln r + \theta\sin\theta\right). \tag{4.4.25}$$

Consider the relation

$$r(\theta) = e^{-\theta\tan\theta}, \quad 0 \le \theta < \tfrac{1}{2}\pi, \tag{4.4.26}$$

with $r(0) = 1$ and $r\left(\frac{1}{2}\pi\right) = 0$. Then this equation describes a path in the complex t-plane from 0 to 1, see Figure 4.4. On this path the imaginary part of $\phi(t)$ vanishes. For the real part of $\phi(t)$ we have using (4.4.26)

$$\Re\phi(t) = \frac{1}{r(\theta)}\left(-\theta\sin\theta\tan\theta - \theta\cos\theta\right) = -\frac{\theta}{r(\theta)\cos\theta}. \tag{4.4.27}$$

We integrate with respect to θ by using

$$\frac{1}{t}\frac{dt}{d\theta} = \frac{d}{d\theta}\ln t = \frac{d}{d\theta}\left(i\theta - \theta\tan\theta\right) = i - \frac{\sin\theta\cos\theta + \theta}{\cos^2\theta}, \tag{4.4.28}$$

and obtain

$$I = \int_0^{\pi/2} \frac{\sin\theta\cos\theta + \theta}{\cos^2\theta}\exp\left(-\frac{\theta}{r(\theta)\cos\theta}\right)d\theta, \tag{4.4.29}$$

with $r(\theta)$ given in (4.4.26). This is the desired integral without oscillations. It can be evaluated efficiently by a suitable quadrature method. Observe the very fast convergence of the integrand to zero at the point $\theta = \frac{1}{2}\pi$, which corresponds to the origin $t = 0$ in (4.4.23), where bad convergence causes the problem when staying on the real t-interval $[0, 1]$. The answer is $I = 0.3233674316777787614 0 \ldots$.

Remark 4.1. This example is given as one of the problems reviewed in Bornemann *et al.* (2004), where 10 "challenges" are considered presented in the January/February 2002 issue of SIAM News by Nick Trefethen, who warned that the problems were hard. The contest is known under the name "The SIAM 100-digit challenge". △

Remark 4.2. In §35.3 we consider another example of an oscillatory integral in which the integrand has an essential singularity at one endpoint, and we transform the integral into two integrals along contours in the complex plane. △

4.5 A Laplace inversion formula for erfc z

The complementary error function can be written as

$$\operatorname{erfc} z = \frac{1}{2\pi i}\int_{c-i\infty}^{c+i\infty} e^{s-2z\sqrt{s}}\,\frac{ds}{s}, \quad c > 0. \tag{4.5.30}$$

This contour integral follows from the inversion formula for the Laplace transform (see Temme (2010c, Eq. 7.14.4))

$$\int_0^\infty e^{-st}\operatorname{erfc}(z/\sqrt{t})\,dt = \frac{1}{s}e^{-2z\sqrt{s}}. \tag{4.5.31}$$

Again, we assume $z > 0$. With the transformation $s = z^2 w$, (4.5.30) becomes

$$\operatorname{erfc} z = \frac{1}{2\pi i}\int_{c-i\infty}^{c+i\infty} e^{z^2(w-2\sqrt{w})}\,\frac{dw}{w}, \quad c > 0, \tag{4.5.32}$$

with saddle point at $w = 1$. When $c = 1$ the vertical path intersects the real axis at the saddle point. The steepest descent path through $w = 1$ is given by $\Im(w-2\sqrt{w}) = 0$. In polar coordinates $w = re^{i\theta}$ the path is given by $r = 1/\cos^2\left(\frac{1}{2}\theta\right)$. In this way, we obtain

$$\operatorname{erfc} z = \frac{e^{-z^2}}{2\pi}\int_{-\pi}^{\pi} e^{-z^2\tan^2\left(\frac{1}{2}\theta\right)}\,d\theta. \tag{4.5.33}$$

The transformation $t = \tan\left(\frac{1}{2}\theta\right)$ gives the representation in (3.4.29).

4.6 A non-oscillatory integral for erfc z, $z \in \mathbb{C}$

We consider the defining integral of erfc z in (3.4.27) for complex values of $z = x+iy$. There is a saddle point at $t = 0$, but this point is not relevant for all z. In this example the fixed finite endpoint $t = z$ of the contour of integration needs attention, in a similar manner to that in §4.3.2.

When $y = 0$, the integrand is positive, and we do not need further steps to transform.

Assume next $x > 0, y > 0$, and write $t = u + iv$. Then, $t^2 = u^2 - v^2 + 2iuv$. We try to find a contour such that $\Im t^2$ is constant, this means, it should be equal to $\Im z^2$ (as it is at the finite endpoint of the contour). This gives for the wanted contour the equation $uv = xy$.

There are two analogous options for using this relation: integrating with respect to u (from x to ∞), or integrating with respect to v (from y to 0). We choose the latter option and obtain

$$\operatorname{erfc} z = \frac{2}{\sqrt{\pi}} e^{-2ixy} \int_y^0 e^{-u^2+v^2} \left(\frac{du}{dv} + i\right) dv, \quad u = \frac{xy}{v}. \tag{4.6.34}$$

This can be written in the form

$$\operatorname{erfc} z = \frac{2}{\sqrt{\pi}} e^{-z^2} \int_0^1 e^{-x^2(1/t^2-1)-y^2(1-t^2)} \left(\frac{x}{t^2} - iy\right) dt. \tag{4.6.35}$$

It is not difficult to verify that this representation also holds for $y < 0$ and can be transformed into the definition of erfc z (3.4.27) (with $z = x$) in the limit $y \to 0$.

When $x = 0$ we cannot use (4.6.35). In that case we split up the path of integration $\int_{iy}^{\infty}$ into iy to 0 and from 0 to $+\infty$. When we take just the limit $x = 0$ in (4.6.35), we miss the latter part. When $x < 0$ we can take the symmetry relation in (3.4.28).

After taking $t = \tanh s$, (4.6.35) can be written in the rather strange-looking form

$$\operatorname{erfc} z = \frac{2e^{-z^2}}{\sqrt{\pi}} \int_0^{\infty} e^{-\frac{x^2}{\sinh^2(s)} - \frac{y^2}{\cosh^2(s)}} \left(\frac{x}{\sinh^2(s)} - i\frac{y}{\cosh^2(s)}\right) ds, \tag{4.6.36}$$

which is valid for $x > 0$ and $y \in \mathbb{R}$.

Both (4.6.35) and (4.6.36) give representations of erfc z with non-oscillatory integrals. In addition, they give a simple splitting of e^{z^2} erfc z into real and imaginary parts.

4.7 The complex Airy function

In Chapter 8 we will give details on obtaining the asymptotic expansions of the Airy functions Ai(z) and Bi(z) starting from the integral representation[1]

$$\operatorname{Ai}(z) = \frac{1}{2\pi i} \int_{\mathcal{C}} e^{\frac{1}{3}w^3 - zw}\, dw, \tag{4.7.37}$$

[1]This example and the one in the next section are taken from Gil *et al.* (2007, §5.5).

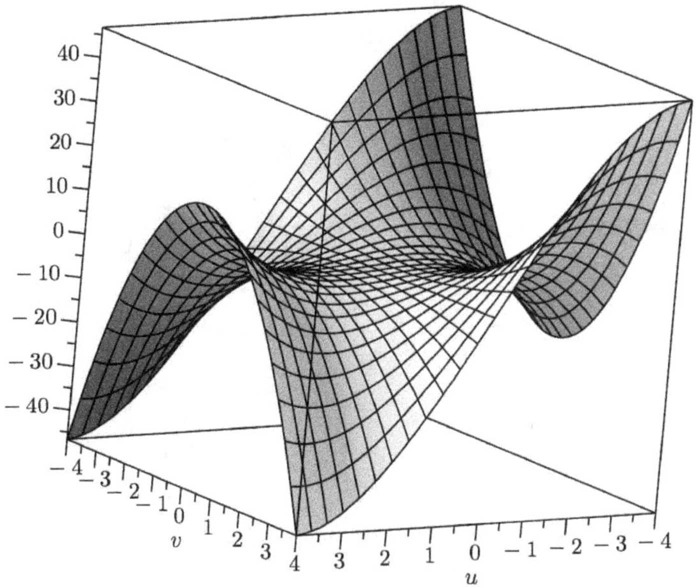

Fig. 4.5 The function $R(u,v) = \Re\left(\frac{1}{3}w^3 - zw\right) = \frac{1}{3}u^3 - uv^2 - zu$ for $z = 1$ and $|u| \le 4, |v| \le 4$. When $z = 0$ the surface becomes a "monkey saddle".

which is valid for all complex z. The contour $\mathcal{C}$ starts at $\infty e^{-\frac{1}{3}\pi i}$ and terminates at $\infty e^{+\frac{1}{3}\pi i}$ (the half-lines $\rho e^{\pm\frac{1}{3}\pi i}$, $\rho \ge 0$, are the two central lines of two of the three valleys of the amplitude of the integrand, the third one being the negative axis).

In Figure 4.5 we show the valleys of the function $R(u,v) = \Re\left(\frac{1}{3}w^3 - zw\right) = \frac{1}{3}u^3 - uv^2 - zu$ for $z = 1$ and $|u| \le 4, |v| \le 4$. There are saddle points at $w = \pm 1$. When $z = 0$ the surface becomes a "monkey saddle".

We consider (4.7.37) and describe the saddle point contours and paths of steepest descent for complex values of z. Let

$$\phi(w) = \tfrac{1}{3}w^3 - zw. \tag{4.7.38}$$

The saddle points are $w_0 = \sqrt{z}$ and $-w_0$. When $\operatorname{ph} z \in [0, \frac{2}{3}\pi]$, w_0 will be in the half-plane $\Re w > 0$ and $-w_0$ in $\Re w < 0$. We are interested in deforming the contour in (4.7.37) into the path of steepest descent through w_0.

This path is defined by

$$\Im\phi(w) = \Im\phi(w_0). \tag{4.7.39}$$

We write

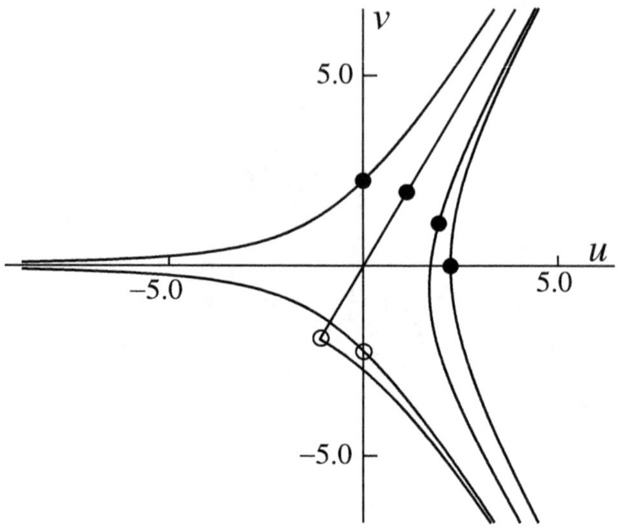

Fig. 4.6 Saddle point contours for ph $z = 0, \frac{1}{3}\pi, \frac{2}{3}\pi, \pi$ and $r = 5$.

$$z = x + iy = re^{i\theta}, \quad w = u + iv, \quad w_0 = u_0 + iv_0. \tag{4.7.40}$$

Then $w_0 = \sqrt{z}$ gives

$$u_0 = \sqrt{r}\cos\tfrac{1}{2}\theta, \quad v_0 = \sqrt{r}\sin\tfrac{1}{2}\theta, \quad x = u_0^2 - v_0^2, \quad y = 2u_0v_0. \tag{4.7.41}$$

The path of steepest descent through w_0 is given by the equation

$$u = u_0 + \frac{(v - v_0)(v + 2v_0)}{3\left(u_0 + \sqrt{\frac{1}{3}(v^2 + 2v_0v + 3u_0^2)}\right)}, \quad -\infty < v < \infty. \tag{4.7.42}$$

Integrating with respect to $\tau = v - v_0$ (and writing $\sigma = u - u_0$) we obtain

$$\operatorname{Ai}(z) = \frac{e^{-\zeta}}{2\pi i}\int_{-\infty}^{\infty} e^{\psi_r(\sigma,\tau)}\left(\frac{d\sigma}{d\tau} + i\right) d\tau, \quad \zeta = \tfrac{2}{3}z^{\frac{3}{2}}, \tag{4.7.43}$$

where

$$\sigma = \frac{\tau(\tau + 3v_0)}{3\left(u_0 + \sqrt{\frac{1}{3}(\tau^2 + 4v_0\tau + 3r)}\right)}, \quad -\infty < \tau < \infty, \tag{4.7.44}$$

and

$$\psi_r(\sigma,\tau) = \Re[\phi(w) - \phi(w_0)] = u_0(\sigma^2 - \tau^2) - 2v_0\sigma\tau + \tfrac{1}{3}\sigma^3 - \sigma\tau^2. \tag{4.7.45}$$

Details of the saddle point contours for $r = 5$ and a few θ-values are shown in Figure 4.6. The saddle points w_0 are located on the circle with radius $\sqrt{r}$ and four w_0 are indicated by small black dots (two saddles $-w_0$ are indicated by open dots). The saddle point on the positive real axis corresponds to the case ph $z = 0$. It is interesting to notice that when $\operatorname{ph} z = \frac{2}{3}\pi$ the saddle point contour passes through the two saddle points $\pm w_0$, and it makes a right turn at the saddle $-w_0$.

For z in the sector $\frac{2}{3}\pi < \operatorname{ph} z \le \pi$ the saddle point contour splits into two contours, each one passing through one saddle. The two saddles on the imaginary axis are for the case $\operatorname{ph} z = \pi$ and the corresponding two contours are also shown in Figure 4.6. Contributions from the two saddle points (when $\operatorname{ph} z = \pi$) give the oscillatory behavior of $\operatorname{Ai}(z)$ when $z < 0$; see §8.2.

The nonsmooth behavior of the path of integration as w_0 crosses the line $\operatorname{ph} z = \frac{2}{3}\pi$ is related to the Stokes phenomenon (see Chapter 5). The line defined by $\operatorname{ph} z = \frac{2}{3}\pi$ is a Stokes line. When w_0 crosses this line, we have $\Im\phi(w_0) = \Im\phi(-w_0)$, and indeed the path of integration runs through both w_0 and $-w_0$. Contributions from $-w_0$ become more and more relevant as $\operatorname{ph} z \to \pi$.

Remark 4.3. The integral representation obtained in (4.7.43) follows from the contour through the saddle point at $w_0 = +\sqrt{z}$. At this point, the phase function ϕ given in (4.7.38) has the value $\phi(w_0) = -\zeta$, with ζ defined in (4.7.43). At the other saddle point the phase function becomes $+\zeta$, and in particular for $z > 0$ this saddle point would give a larger contribution compared to the first one. The saddle point $-w_0$ may be viewed as a dominant saddle point, but for the Airy integral in (4.7.37) it is not the relevant saddle point. When $z > 0$, we cannot take the contour in (4.7.37) through $-w_0$ along a path on which $\Im\phi(w) = 0$ without disturbing convergence when modifying the initial contour. Of course, we can break our journey along the steepest descent loop and integrate from $\sqrt{z}$ along the steepest descent path to $-\infty$, picking up the large contribution at $-\sqrt{z}$, but we have to go back, along the same path, losing all we picked up on the journey out. △

4.8 A parabolic cylinder function

We consider the parabolic cylinder function $U(a,x)$ for positive x and a. For more details on this function we refer to Chapter 11 (for expansions for large argument) and to Chapter 30 (for expansions for large parameter).

The starting point is the integral (see Temme (2010a, Eq. 12.5.6))

$$U(a,x) = \frac{e^{\frac{1}{4}x^2}}{i\sqrt{2\pi}} \int_{\mathcal{C}} e^{-xs+\frac{1}{2}s^2} s^{-a} \frac{ds}{\sqrt{s}}, \tag{4.8.46}$$

where $\mathcal{C}$ is a vertical line on which $\Re s > 0$. On $\mathcal{C}$ we have $-\frac{1}{2}\pi < \operatorname{ph} s < \frac{1}{2}\pi$, and the many-valued function $s^{-a-1/2}$ assumes its principal value. The transformations

$$x = 2t\sqrt{a}, \quad s = \sqrt{a}\, w \tag{4.8.47}$$

give

$$U(a,x) = \frac{e^{\frac{1}{4}x^2} a^{\frac{1}{4}-\frac{1}{2}a}}{i\sqrt{2\pi}} \int_{\mathcal{C}} e^{a\phi(w)} \frac{dw}{\sqrt{w}}, \tag{4.8.48}$$

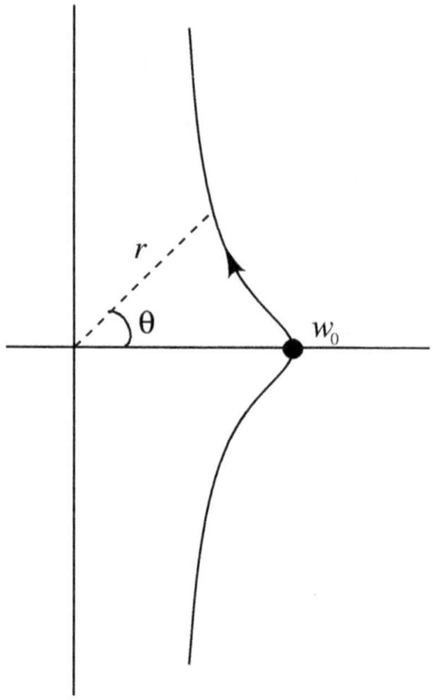

Fig. 4.7 Steepest descent contour for the integral in (4.8.48).

where

$$\phi(w) = \tfrac{1}{2}w^2 - 2tw - \ln w. \tag{4.8.49}$$

When a and x are possibly large the oscillatory behavior of the integrand follows from the exponential function. In this respect, the $\sqrt{w}$ factor is of no concern.

The saddle points follow from solving

$$\phi'(w) = \frac{w^2 - 2tw - 1}{w} = 0, \tag{4.8.50}$$

giving saddle points at $w = t \pm \sqrt{t^2+1}$. The steepest descent path follows from solving $\Im\phi(w) = \Im\phi(w_0)$ with $w_0 = t + \sqrt{t^2+1}$. The negative saddle point is not relevant when x and a are positive.

In the present case $\Im\phi(w_0) = 0$, and we obtain for the saddle point contour the equation

$$\tfrac{1}{2}r^2 \sin 2\theta - 2tr\sin\theta - \theta = 0, \quad \text{where} \quad w = re^{i\theta}, \tag{4.8.51}$$

which can be solved for $r = r(\theta)$,

$$r = \frac{t + \sqrt{t^2 + \theta\cot\theta}}{\cos\theta}, \quad -\tfrac{1}{2}\pi < \theta < \tfrac{1}{2}\pi, \tag{4.8.52}$$

giving the contour as shown in Figure 4.7. Then (4.8.48) can be written as

$$U(a,x) = \frac{e^{\frac{1}{4}x^2 + a\phi(w_0)}a^{\frac{1}{4}-\frac{1}{2}a}}{\sqrt{2\pi}} \int_{-\frac{1}{2}\pi}^{\frac{1}{2}\pi} e^{a\psi(\theta)}g(\theta)\,d\theta, \tag{4.8.53}$$

where

$$\psi(\theta) = \Re(\phi(w) - \phi(w_0)) = \tfrac{1}{2}r^2 \cos 2\theta - 2tr\cos\theta - \ln r - \phi(w_0), \tag{4.8.54}$$

and

$$\begin{aligned} g(\theta) = \Im\left(\frac{1}{\sqrt{w}}\frac{dw}{d\theta}\right) &= \Im\left(e^{\frac{1}{2}i\theta}\frac{1}{\sqrt{r}}\left(\frac{dr}{d\theta} + ir\right)\right) \\ &= \frac{(2\cos\theta + 1)r^2 - 2tr + 1}{4\sqrt{r}\cos\frac{1}{2}\theta\sqrt{t^2 + \theta\cot\theta}}. \end{aligned} \tag{4.8.55}$$

If we consider the quantity $\widetilde{\xi}$ defined by

$$\widetilde{\xi} = \tfrac{1}{2}\left(t\sqrt{t^2+1} + \ln(t + \sqrt{t^2+1})\right), \tag{4.8.56}$$

we have

$$\begin{aligned} \tfrac{1}{4}x^2 + a\phi(w_0) &= a\left(\tfrac{1}{2} - t\sqrt{t^2+1} - \ln(t + \sqrt{t^2+1})\right) \\ &= a\left(\tfrac{1}{2} - 2\widetilde{\xi}\right). \end{aligned} \tag{4.8.57}$$

This gives

$$U(a, x) = \frac{a^{\frac{1}{4}}\, e^{-2a\widetilde{\xi}}}{\sqrt{2\pi}\gamma(a)} \int_{-\frac{1}{2}\pi}^{\frac{1}{2}\pi} e^{a\psi(\theta)} g(\theta)\, d\theta, \tag{4.8.58}$$

where

$$\gamma(a) = e^{-\frac{1}{2}a} a^{\frac{1}{2}a}. \tag{4.8.59}$$

In Gil *et al.* (2006) the vertical line passing through the saddle point has been chosen as the path of integration for numerical quadrature. This path has the same direction as the steepest descent path when crossing the saddle point and runs into the same valleys. For numerical quadrature taking a path with the proper direction through the saddle point is most important for avoiding numerical cancellations, and choosing the correct path is of less concern.

Chapter 5

The Stokes phenomenon

An analytic function $f(z)$ may have different asymptotic approximations when $|z| \to \infty$ in different domains of the complex plane. Usually these approximations are related and an approximation in one domain may also play a role in a different domain, although it may be exponentially small inside that domain. In the following section we explain this for the Airy function, together with the role of several approximations, and how they become important when going from one domain to another domain. All this is related to the *Stokes phenomenon.*

Actually, this phenomenon concerns the abrupt change across certain rays in the complex plane, known as *Stokes lines*, exhibited by the coefficients multiplying exponentially subdominant terms in compound asymptotic expansions.

In the last two decades there has been much interest in the Stokes phenomenon, and it fits in the study of asymptotic expansions of integrals because it has to do with sudden changes in approximations when a certain parameter (in this case the phase of the large parameter) passes critical values. The recent interest and explanations of the Stokes phenomenon started with a paper by Michael Berry (Berry, 1989).

5.1 The Airy function

First we explain this phenomenon by using a simple example from differential equations (Meyer, 1989). Consider Airy's equation

$$\frac{d^2w(z)}{dz^2} = z\,w(z). \tag{5.1.1}$$

The solutions of this equation are analytic functions. When $|z|$ is large the solutions of (5.1.1) are approximated by linear combinations of (see §8.1)

$$u_\pm = z^{-\frac14}\, e^{\pm\zeta}, \quad \zeta = \tfrac{2}{3}z^{3/2}. \tag{5.1.2}$$

Obviously, $u_\pm$ are multivalued functions of the complex variable z with a branch point at $z = 0$. Therefore, as we go once around the origin, the solutions of (5.1.1) will return to their original values, but $u_\pm$ will not. It follows that the constants $c_\pm$ in the linear combination

$$w(z) \sim c_-u_-(z) + c_+u_+(z), \quad z \to \infty, \tag{5.1.3}$$

are domain-dependent. The constants change when we cross certain lines, the boundaries of certain sectors in the z-plane.

In the above example one of the terms $e^{\zeta}, e^{-\zeta}$ maximally dominates the other one at the rays $\operatorname{ph} z = 0, \operatorname{ph} z = \pm\frac{2}{3}\pi$. In this example these three rays are the Stokes lines. At the rays $\operatorname{ph} z = \pm\frac{1}{3}\pi$ and the negative z-axis the quantity ζ is purely imaginary, and, hence, the terms $e^{\zeta}, e^{-\zeta}$ are equal in magnitude. These three rays are called the *anti-Stokes lines*.[1]

For the Airy function $\operatorname{Ai}(z)$ we have the full asymptotic expansion (see Chapter 8)

$$\operatorname{Ai}(z) \sim c_- z^{-\frac{1}{4}} e^{-\zeta} \sum_{n=0}^{\infty} (-1)^n u_n \zeta^{-n} \quad c_- = \tfrac{1}{2}\pi^{-\frac{1}{2}}, \quad |\operatorname{ph} z| < \pi, \tag{5.1.4}$$

with coefficients given in (8.1.17).

On the other hand, in an other sector of the z-plane, we have

$$\operatorname{Ai}(z) \sim c_-\, z^{-\frac{1}{4}} \left(e^{-\zeta} \sum_{n=0}^{\infty} (-1)^n u_n \zeta^{-n} + i e^{\zeta} \sum_{n=0}^{\infty} u_n \zeta^{-n} \right), \tag{5.1.5}$$

in which exactly the same term (with the same constant c_-) is involved as in (5.1.4), and there is another term corresponding to u_+. We can rewrite this in a more familiar expansion

$$\operatorname{Ai}(-z) \sim \pi^{-\frac{1}{2}} z^{-\frac{1}{4}} \left(\sin\chi \sum_{n=0}^{\infty} (-1)^n \frac{c_{2n}}{\zeta^{2n}} - \cos\chi \sum_{n=0}^{\infty} (-1)^n \frac{c_{2n+1}}{\zeta^{2n+1}} \right), \tag{5.1.6}$$

where $\chi = \zeta + \frac{1}{4}\pi$ (see §8.2); ζ is defined in (5.1.2). in the sector $|\operatorname{ph} z| < \frac{2}{3}\pi$. In the overlapping domain of expansions (5.1.4) and (5.1.6), that is, when $\frac{1}{3}\pi < |\operatorname{ph} z| < \pi$, the term with u_+ is asymptotically small compared with u_-, and it suddenly appears in the asymptotic approximation when we cross with increasing values of $|\operatorname{ph} z|$ the Stokes lines at $\operatorname{ph} z = \pm\frac{2}{3}\pi$.

It seems that, when going from (5.1.4) to (5.1.5), the constant multiplying u_+ changes discontinuously from a zero value (when $|\operatorname{ph} z| < \frac{2}{3}\pi$) to a non-zero value when we cross the Stokes line. This sudden appearance of the term u_+ does not have much influence on the asymptotic behavior near the Stokes lines at $|\operatorname{ph} z| = \frac{2}{3}\pi$, because u_+ is dominated maximally by u_- on these rays. However, see §5.3 below.

5.2 The recent interest in the Stokes phenomenon

This phenomenon of the *discontinuity of the constants* was discovered by Stokes and was discussed by him in a series of papers (on Airy functions in 1857, on Bessel functions in 1868). It is a phenomenon which is not confined to Airy or Bessel

[1]This terminology is not the same in all branches of applied mathematics and mathematical physics: sometimes one sees a complete interchange of the names 'Stokes line' and 'anti-Stokes line'.

functions. The discovery by Stokes was, as Watson says, apparently one of those which are made at three o'clock in the morning. Stokes wrote in a 1902 retrospective paper: "The inferior term enters as it were into a mist, is hidden for a little from view, and comes out with its coefficients changed".

In 1989 the mathematical physicist Michael Berry provided a deeper explanation (Berry, 1989). He suggested that the coefficients of the subdominant expansion should be regarded not as a discontinuous constant but, for fixed $|z|$, as a continuous function of $\operatorname{ph} z$. Berry's innovative and insightful approach was followed by a series of papers by himself and other writers. In particular, Olver put the formal approach by Berry on a rigorous footing in papers with applications to confluent hypergeometric functions (including Airy functions, Bessel functions, and Weber parabolic functions).

At the same time interest arose in earlier work by Stieltjes, Airy, Dingle,..., to expand remainders of asymptotic expansions at optimal values of the summation variable. This resulted in *exponentially improved* asymptotic expansions, a method of improving asymptotic approximations by including small terms in the expansion that are in fact negligible compared with other terms in the expansion.

The Stokes phenomenon and the topic of exponential asymptotics are connected with uniform expansions of integrals, in particular, with approximations which are uniformly valid with respect to variations in the phase of the large parameter. We mention the contributions on a better understanding of the asymptotics of the gamma function in Berry (1991), Paris and Wood (1992) and Boyd (1994). More general papers are Howls (1992), Berry and Howls (1991, 1994). For applications to the Kummer U-function we mention Olde Daalhuis (1992, 1993). In Boyd (1990) new results for the modified K-Bessel function have been given. In Jones (1990) a method has been devised for estimating the optimal remainder in an asymptotic approximation which is uniform with respect to variations in the phase of the large parameter.

For an overview of the Stokes phenomenon and related matters we refer to Paris and Wood (1995) (an introductory paper) and Olver and Wong (2010, §2.11). See also Paris and Kaminski (2001, Chapter 6) for a treatment of the Stokes phenomenon using Mellin–Barnes integrals. In a recent paper by Farid Khwaja and Olde Daalhuis (2014) many aspects of the Stokes phenomenon are explained in the large-parameter asymptotic analysis of the hypergeometric functions by using integral representations.

5.3 Exponentially small terms in the Airy expansions

We explain the relation between the Stokes phenomenon and the exponentially small terms in the asymptotic expansion of the Airy function. Consider the terms in the expansions in (5.1.4)–(5.1.6). They have the asymptotic form

$$u_n\zeta^{-n} = \mathcal{O}\left(\Gamma(n)\,(2\zeta)^{-n}\right), \quad n \to \infty. \tag{5.3.7}$$

When z is large the terms decrease at first and then increase. The least term of the first series of (5.1.5) is near $n = n^* = \lfloor |2\zeta| \rfloor$ and its size is of order $e^{-2|\zeta|}$. At the Stokes lines at $|\mathrm{ph}\, z| = \frac{2}{3}\pi$ the quantity ζ is negative and the exponential term in front of the first series in (5.1.5) equals $e^{|\zeta|}$. Hence, the order of magnitude of $e^{-\zeta}c_{n^*}\zeta^{-n^*}$ is roughly of the same size as the second part in (5.1.5), that is, of the size of e^{ζ} that is present in front of the second series. It follows that near the Stokes lines (and of course when z turns to the negative axis) the second series in (5.1.5) is not at all negligible when we truncate the first series at the least term with index n^*.

At present we know, after Berry's observations, that near the Stokes lines one of the constants $c_\pm$ in the asymptotic representation in (5.1.3) in fact is a rapidly changing function of z. In the case of (5.1.5) we can write

$$\mathrm{Ai}(z) \sim c_-\, z^{-\frac{1}{4}} \left(e^{-\zeta}\sum_{n=0}^{\infty}(-1)^n u_n\zeta^{-n} + iS(z)\, e^{\zeta}\sum_{n=0}^{\infty} u_n\zeta^{-n}\right), \tag{5.3.8}$$

where $S(z)$ switches rapidly but smoothly from 0 to 1 across the Stokes line at $\mathrm{ph}\, z = \frac{2}{3}\pi$. A good approximation to $S(z)$ involves the error function (see §3.4), which can describe the fast transition in this asymptotic problem. We will see the error function in the problems considered in Chapters 21, 36, and later chapters, where it is used to describe similar fast transitions.

Many writers have contributed recently in this field, both for the Stokes phenomenon of integrals and that of differential equations.

5.4 Expansions in connection with the Stokes phenomenon

As mentioned earlier, in Berry (1989) the Stokes phenomenon has been given a new interpretation. In Olver (1991a,b) Berry's approach is rigorously treated for integrals representing the Kummer U-function (see also Olver (1994)). Olver showed that the exponential integral

$$E_p(z) = z^{p-1}\int_z^{\infty}\frac{e^{-t}}{t^p}\,dt = z^{p-1}\Gamma(1-p, z), \tag{5.4.9}$$

where $\Gamma(a, z)$ is the incomplete gamma function, plays an important role in Berry's smooth interpretation of the Stokes phenomenon for certain integrals and special functions. Olver (1991b) investigates $E_p(z)$ in particular at the Stokes lines $\mathrm{ph} z = \pm\pi$ and the results are used in Olver (1991a) for the Kummer U-function. We give a few details of Olver's results.

Let

$$F_p(z) = \frac{\Gamma(p)}{2\pi}\,\frac{E_p(z)}{z^{p-1}} \tag{5.4.10}$$

and $z = \rho e^{i\theta}, \alpha = n - \rho + p$ with ρ a large parameter, p fixed. Then[2]

$$F_{n+p}(z) \sim (-1)^n i e^{-p\pi i}\left(\tfrac{1}{2}\operatorname{erfc}\left(c(\theta)\sqrt{\rho/2}\right) - i\frac{e^{-\frac{1}{2}\rho(c(\theta))^2}}{\sqrt{2\pi\rho}}\sum_{s=0}^{\infty} g_{2s}(\theta,\alpha)\left(\tfrac{1}{2}\right)_s 2^s\rho^{-s}\right), \tag{5.4.11}$$

uniformly with respect to $\theta \in [-\pi+\delta, 3\pi-\delta]$ and bounded values of $|\alpha|$; δ denotes an arbitrarily small positive constant. Furthermore,

$$c(\theta) = \sqrt{2\left(e^{i\theta} + i(\theta-\pi) + 1\right)}, \tag{5.4.12}$$

with the choice of branch of the square root that implies $c(\theta) \sim (\pi - \theta)$ as $\theta \to \pi$; the coefficients $g_{2s}(\theta,\alpha)$ are continuous functions of θ and α. A similar expansion of $F_{n+p}(z)$ is given when $\theta \in [-3\pi+\delta, \pi-\delta]$. As Olver remarks, this expansion quantifies the Stokes phenomenon, that is, the rapid but smooth change in form of other expansions as θ passes through the common interval of validity of the other expansions.

5.4.1 *Applications to a Kummer function*

By using these results Olver (1991a) gives a detailed treatment of the Kummer function $U(a,b,z)$.

Let $R_n(a,b,z)$ be the remainder in the expansion

$$U(a, a-b+1; z) = z^{-a}\sum_{s=0}^{n-1}(-1)^s\frac{(a)_s(b)_s}{s!\,z^s} + R_n(a,b,z), \tag{5.4.13}$$

where

$$n = |z| - a - b + 1 + \alpha, \tag{5.4.14}$$

$|z|$ being large, a and b being fixed real or complex parameters, and $|\alpha|$ being bounded. Then

$$R_n(a,b,z) = (-1)^n 2\pi\frac{z^{b-1}e^z}{\Gamma(a)\Gamma(b)}\left(\sum_{s=0}^{m-1}(-1)^s\frac{(1-a)_s(1-b)_s}{s!}\frac{F_{n-s+a+b-1}(z)}{z^s} + (1-a)_m(1-b)_m R_{m,n}(a,b,z)\right), \tag{5.4.15}$$

where $F_p(z)$ is the incomplete gamma function (or exponential integral, see (5.4.9) and (5.4.10)), m is an arbitrary fixed integer, and

$$R_{m,n}(a,b,z) = \begin{cases} \mathcal{O}\left(e^{-z-|z|}z^{-m}\right), & \text{if} \quad |\mathrm{ph}\, z| \le \pi, \\ \mathcal{O}\left(z^{-m}\right), & \text{if} \quad \pi \le |\mathrm{ph}\, z| \le \frac{5}{2}\pi - \delta. \end{cases} \tag{5.4.16}$$

---

[2]This expansion is related to the expansions derived in Chapter 37 for the incomplete gamma functions.

Furthermore, these sectors of validity are maximal. Observe that the expansion in (5.4.13) starts with the Poincaré-type expansion as given in (10.2.18). For other details on the expansion we refer to Olver's paper.

In later papers by Olver, Olde Daalhuis, etc., many results for the Kummer U-function and other special functions are obtained by methods based on differential equations.

PART 2

Basic Methods: Examples for Special Functions

Chapter 6

The gamma function

The Euler gamma function is usually defined by the integral representation

$$\Gamma(z) = \int_0^\infty t^{z-1} e^{-t}\, dt, \quad \Re z > 0, \tag{6.0.1}$$

and is an analytic function in this domain. By using the functional relation

$$\Gamma(z+1) = z\Gamma(z) \tag{6.0.2}$$

we can extend the domain of analyticity to all finite complex numbers, except for the nonpositive integers, where $\Gamma(z)$ has first-order poles.

The integral representation for the reciprocal gamma function has been given in (2.2.17):

$$\frac{1}{\Gamma(z)} = \frac{1}{2\pi i}\int_{-\infty}^{(0+)} s^{-z} e^{s}\, ds, \quad z \in \mathbb{C}, \tag{6.0.3}$$

where we integrate along a Hankel contour shown in Figure 2.1. This function is analytic.

From the reflection formula

$$\Gamma(z)\,\Gamma(1-z) = \frac{\pi}{\sin \pi z}, \quad z \notin \mathbb{Z}, \tag{6.0.4}$$

it easily follows that

$$\lim_{z\to -n} (z+n)\Gamma(z) = \frac{(-1)^n}{n!}, \quad n = 0, 1, 2, \ldots. \tag{6.0.5}$$

It will appear that the gamma function can be computed very efficiently by using asymptotic expansions. The recurrence relation (6.0.2) is useful when z is not large enough to apply the asymptotic expansion. The reflection formula (6.0.4) can be used when $\Re z < 0$.

In calculus *Stirling's formula* is usually given for the factorial function

$$n! \sim \sqrt{2\pi n}\, n^n e^{-n}, \quad n \to \infty; \tag{6.0.6}$$

see Stirling (1730, p. 135) and Tweddle (2003). In many applications Stirling's formula proves to be extremely useful.

For the gamma function the first approximation reads

$$\Gamma(z) \sim \sqrt{2\pi} z^{z-\frac{1}{2}} e^{-z}, \quad z \to \infty. \tag{6.0.7}$$

We will derive several expansions of the gamma function.

- In §6.1: the expansion of $\Gamma(z)$ by Laplace's method.
- In §6.2: the expansion of $1/\Gamma(z)$ by the saddle point method.
- In §6.3: an expansion of $\ln\Gamma(z)$ by Watson's lemma.
- In §6.4: expansions of $\Gamma(z+a)$ and $1/\Gamma(z+a)$.
- In §6.5: two expansions of the ratio $\dfrac{\Gamma(z+a)}{\Gamma(z+b)}$ by Watson's lemma.
- In §6.6: an expansion of $\dbinom{n}{m}$ by using the saddle point method.
- In §6.7: an expansion of the product of two gamma functions with complex parameters by using uniform methods.
- In §6.8: expansions of ratios of three gamma functions.
- In §25.6.1: an expansion of the ratio of two gamma functions by using uniform methods.

6.1 $\Gamma(z)$ by Laplace's method

We derive the well-known expansion of the gamma function. We start with the integral

$$\Gamma(z+1) = \int_0^\infty u^z\, e^{-u}\, du, \quad \Re z > -1. \tag{6.1.8}$$

To write it in the form (3.2.7) we need a few transformations. Initially we assume that z is positive. We write $u = zt$, which gives

$$\Gamma(z) = z^z e^{-z} \int_0^\infty e^{-zp(t)}\, dt, \quad p(t) = t - \ln t - 1. \tag{6.1.9}$$

Because

$$p'(t) = \frac{t-1}{t}, \quad p''(t) = \frac{1}{t^2}, \tag{6.1.10}$$

the function p is concave and has a minimum at $t = 1$, with $p(1) = 0$ and $p''(1) = 1$.

We verify the assumptions of Theorem 3.1. The domain $\mathcal{D}$ is the complex plane without the nonpositive reals. We take a branch cut for the logarithm in $p(t)$ from 0 to $-\infty$, and take $|\mathrm{ph}\, t| < \pi$. In assumption (iv) we take $\theta_1 = -\frac{1}{2}\pi + \delta$, $\theta_2 = \frac{1}{2}\pi - \delta$, with δ a small positive number. In this way the integral in (6.1.9) converges absolutely and uniformly with respect to z at both endpoints 0 and ∞. The other assumptions are clearly satisfied.

We substitute (see (3.2.9))

$$\tfrac{1}{2}w^2 = p(t), \quad \mathrm{sign}(w) = \mathrm{sign}(t-1). \tag{6.1.11}$$

This transformation can be rewritten as

$$w = (t-1)\sqrt{\frac{2(t-1-\ln t)}{(t-1)^2}}, \tag{6.1.12}$$

where the square root is positive for positive values of t. For complex values of t the mapping is defined by analytic continuation. The positive t-axis is mapped in the w-plane onto the real axis. More details on the transformation are given in §6.1.2.

The substitution in (6.1.11) gives

$$\Gamma(z) = z^z e^{-z} \int_{-\infty}^{\infty} e^{-\frac{1}{2}zw^2} f(w)\, dw, \quad f(w) = \frac{dt}{dw} = \frac{wt}{t-1}. \tag{6.1.13}$$

By expanding $f(w) = \sum_{k=0}^{\infty} a_k w^k$ as in §3.2 we obtain the expansion

$$\Gamma(z) \sim \sqrt{2\pi}\, z^{z-\frac{1}{2}} e^{-z} \sum_{k=0}^{\infty} \frac{2^k (\frac{1}{2})_k\, a_{2k}}{z^k}, \quad z \to \infty, \tag{6.1.14}$$

and after evaluating the coefficients (see §6.1.1) we have the familiar expansion, the generalization of Stirling's formula,

$$\Gamma(z) \sim \sqrt{2\pi}\, z^{z-\frac{1}{2}} e^{-z} \left(1 + \tfrac{1}{12} z^{-1} + \tfrac{1}{288} z^{-2} - \tfrac{139}{51840} z^{-3} + \cdots \right), \tag{6.1.15}$$

as $z \to \infty$. From details on the transformation given in §6.1.2, we conclude that f defined in (6.1.13) is analytic inside the sectors $|\mathrm{ph}(\pm z)| < \frac{1}{4}\pi$. Hence, $-\alpha = \beta = \frac{1}{4}\pi$, and it follows from Lemma 3.1 that the expansion in (6.1.14) is valid inside the sector $|\mathrm{ph}\, z| \le \pi - \delta$.

On the other hand, by using the reflection formula (6.0.4) in the form

$$\Gamma(z) = -\frac{\pi}{z \sin \pi z} \frac{1}{\Gamma(-z)}, \tag{6.1.16}$$

we can also extend the domain of validity of the asymptotic expansion in (6.1.15). For example, when $\Im z \to +\infty$ and $\Re z < 0$, use (6.1.15) and (6.1.16) with $-z = e^{-\pi i} z$, and the dominant part of $\sin \pi z$. Similarly when $\Im z \to -\infty$ with $-z = e^{+\pi i} z$.

This method can also be used to derive the expansion

$$\frac{1}{\Gamma(z)} \sim \frac{1}{\sqrt{2\pi}} z^{-z+\frac{1}{2}} e^{z} \left(1 - \tfrac{1}{12} z^{-1} + \tfrac{1}{288} z^{-2} + \tfrac{139}{51840} z^{-3} + \cdots \right), \quad z \to \infty, \tag{6.1.17}$$

again valid in the sector $|\mathrm{ph}\, z| \le \pi - \delta < \pi$. Note that the coefficients of the even powers of the series are the same as in (6.1.15), those of the odd powers have opposite signs.

Error bounds for the asymptotic expansion of $\Gamma(z)$ are given in Olver (1968, 1970b) and Boyd (1994).

6.1.1 *Calculating the coefficients*

Because of the special form of the functions p and q (with $q(t) = 1$), we can use several methods for calculating the coefficients.

To compute the coefficients of the inverse expansion $t = \sum_{k=0}^{\infty} t_k w^k$ we use (see (6.1.13))

$$(t-1)\frac{dt}{dw} = wt. \tag{6.1.18}$$

Substitution of the expansion of t gives

$$\sum_{k=0}^{\infty} t_{k+1} w^k \sum_{k=0}^{\infty} (k+1) t_{k+1} w^k = \sum_{k=0}^{\infty} t_k w^k, \tag{6.1.19}$$

which leads to the recurrence relation

$$\sum_{j=0}^{k} (j+1) t_{j+1} t_{k-j+1} = t_k, \quad k \ge 0. \tag{6.1.20}$$

It easily follows from (6.1.18) that $t_0 = 1$ and $t_1 = 1$. This gives

$$(k+1)t_k = t_{k-1} - \sum_{j=2}^{k-1} j t_j t_{k-j+1}, \quad k \ge 2. \tag{6.1.21}$$

We find for later coefficients

$$t_2 = \tfrac{1}{3},\ t_3 = \tfrac{1}{36},\ t_4 = -\tfrac{1}{270},\ t_5 = \tfrac{1}{4320},\ t_6 = \tfrac{1}{17010}. \tag{6.1.22}$$

Because $f(w) = \dfrac{dt}{dw}$, the coefficients a_k of the expansion $f(w) = \sum_{k=0}^{\infty} a_k w^k$ are given by $a_k = (k+1)t_{k+1}$, and we obtain the expansion

$$\Gamma(z) \sim \sqrt{2\pi}\, z^{z-\frac{1}{2}} e^{-z} \sum_{k=0}^{\infty} \frac{2^k (\frac{1}{2})_k (2k+1) t_{2k+1}}{z^k}, \quad z \to \infty, \tag{6.1.23}$$

or, by evaluating the coefficients, the expansion given in (6.1.15).

In §3.3 we have explained how to obtain explicit forms of the coefficients.

6.1.2 *Details on the transformation*

We discuss some details about the transformation in (6.1.11) because it gives insight into the singularities of the function $f(w)$ given in (6.1.13). Then we can extend the domain for z of the expansion in (6.1.15) that follows from Theorem 3.1.

The transformation is also used in other asymptotic problems related to the gamma function, for example for uniform expansions of the incomplete gamma functions $\gamma(a,z)$ and $\Gamma(a,z)$ when a and z are both large. For this we refer to Chapter 37.

We obtain more insight in the mapping $t \mapsto w(t)$ and its inverse when we draw images of the half-lines ℓ_θ defined by

$$\ell_\theta = \{t \mid t = re^{i\theta}, r > 0\}, \tag{6.1.24}$$

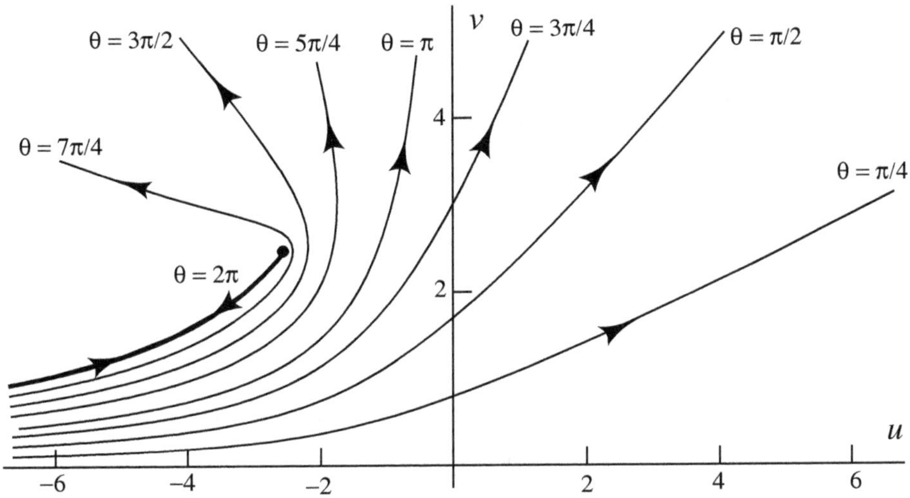

Fig. 6.1 Images of the half-lines defined in (6.1.24) in the w-plane under the mapping in (6.1.11) for several values of θ.

for fixed real θ, $\theta \in [-2\pi, 2\pi]$. This is outside the usual domain for specifying domains in the complex domain, but because the function $p(t)$ in (6.1.11) contains a logarithmic term it will make sense.

Writing $w = u + iv$, we see that the image of ℓ_θ in the w-plane is governed by the equations

$$\begin{aligned} \tfrac{1}{2}(u^2 - v^2) &= r\cos\theta - 1 - \ln r, \\ uv &= r\sin\theta - \theta. \end{aligned} \tag{6.1.25}$$

In Figure 6.1 we show the images of the half-lines ℓ_θ for several values of θ. The half-line ℓ_π is mapped onto the hyperbola $uv = -\pi$; $v \ge 0$, $u \le 0$. The image of the half-line $\ell_{2\pi}$ is folded around parts of the hyperbola $uv = -2\pi$. The dot indicates the singular point $2\sqrt{\pi}e^{3\pi i/4}$ that corresponds in the t-plane to the point $e^{2\pi i}$. In this way we have discovered the singularities $w_\pm = 2\sqrt{\pi}e^{\pm 3\pi i/4}$ of the function $f(w)$ in (6.1.13). The function $f(w)$ has a branch point at $w_\pm$ and the shown images of the lines $\ell_{\pm 2\pi}$ can be taken as a branch cut for this function.

The negative t-axis is mapped to two hyperbolas corresponding to $\theta = \pm\pi$, with $u < 0$. The point $t = 0$ is mapped to $-\infty$, the point $t = -\infty$ to $\pm i\infty$. It follows that the domain $\mathcal{D}$ where $p(t)$ is assumed to be analytic is mapped in the w-plane to a domain on the right of these two hyperbolas.

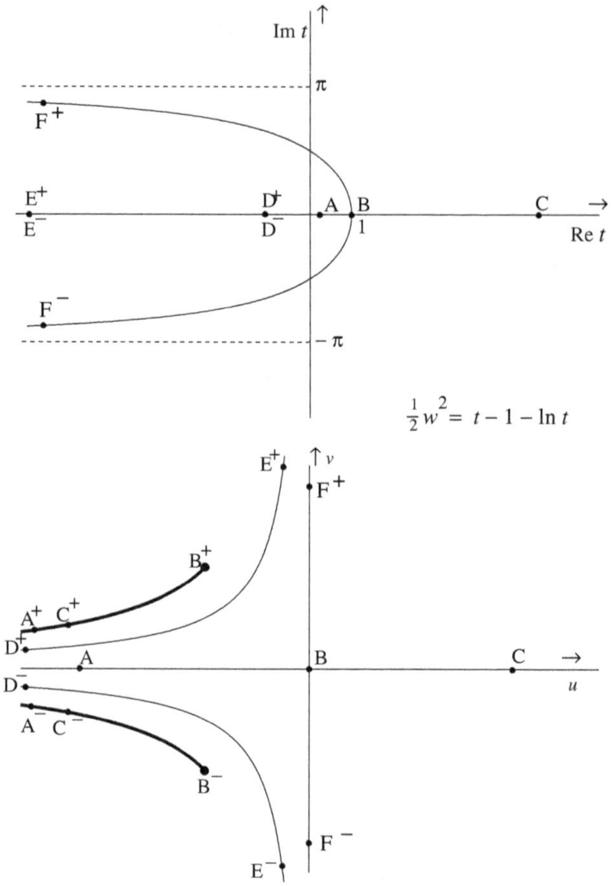

Fig. 6.2 Corresponding points in the t- and w-planes for the transformation $\frac{1}{2}w^2 = t - \ln t - 1$ in (6.1.11). The points $B^{\pm}$ are singular points for the function $f(w)$ defined in (6.1.13). In the t-plane they correspond to $\exp(\pm 2\pi i)$. The w-points $D^{\pm}, E^{\pm}$ correspond to t-points $D^{\pm}, E^{\pm}$ having phases $\pm\pi i$; in the w-plane they are on the images of the curves $\ell_{\pm\pi}$ (hyperbolas defined by $uv = \mp\pi$). The thick parts of the hyperbolas are defined by $uv = \pm 2\pi$, and are branch cuts of $f(w)$. The path of steepest descent F^-BF^+ in the t-plane is used in §6.2 and is defined in (6.2.27); this path corresponds to the imaginary axis in the w-plane.

In Figure 6.2 we give a more global description of the transformation with corresponding points in the t- and w-planes. The images of the half-lines $\ell_{\pm 2\pi}$ correspond to parts of the hyperbola $uv = \pm 2\pi$. The endpoints of these curves, indicated by $B^{\pm}$, are the points $w_{\pm}$, the singular points of the mapping.

The picture also shows the curve F^-BF^+, which in the t-plane is a path of steepest ascent for the integral in (6.1.9). In the w-plane its image is the imaginary axis. These paths are important in §6.2, where we give the saddle point analysis for an integral representation of $1/\Gamma(z)$.

6.2 $1/\Gamma(z)$ by the saddle point method

Hankel's contour integral (see Hankel (1863)) is one of the beautiful representations of the gamma function. In fact it is an integral for the reciprocal gamma function, and we have used it in §2.2 in Watson's lemma for contour integrals. Hankel's contour is shown in Figure 2.1, and the integral of $1/\Gamma(z)$ is given in (6.0.3).

We use Hankel's contour integral for applying the saddle point method. We consider positive values of z. A first transformation $s = zt$ in (6.0.3) gives

$$\frac{1}{\Gamma(z)} = \frac{e^z z^{1-z}}{2\pi i} \int_{-\infty}^{(0+)} e^{zp(t)}\, dt, \quad p(t) = t - 1 - \ln t, \tag{6.2.26}$$

and the contour can be taken the same as in Figure 2.1.

Observe that we have the same function $p(t)$ as in (6.1.9). The saddle point is at $t = 1$. Writing $t = re^{i\theta}$, we find that the equation $\Im p(t) = \Im p(1) = 0$ is satisfied when the polar coordinates of t satisfy

$$r = \frac{\theta}{\sin\theta}, \quad -\pi < \theta < \pi. \tag{6.2.27}$$

This defines the path of steepest descent; in Figure 6.2 it is the curve F^-BF^+ in the top figure.

The transformation $\frac{1}{2}w^2 = p(t)$ is the same as in §6.1, see (6.1.11). The saddle point contour described in (6.2.27) is mapped onto the whole imaginary axis. So we can write

$$\frac{1}{\Gamma(z)} = \frac{e^z z^{1-z}}{2\pi i} \int_{-i\infty}^{+i\infty} e^{\frac{1}{2}zw^2} f(w)\, dw, \tag{6.2.28}$$

where, see also (6.1.13),

$$f(w) = \frac{dt}{dw} = \frac{tw}{t-1}. \tag{6.2.29}$$

It will be clear that we obtain an asymptotic expansion of the reciprocal gamma function in which the series is as in (6.1.14) with a_{2k} replaced by $(-1)^k a_{2k}$. That is,

$$\frac{1}{\Gamma(z)} \sim \frac{e^z z^{\frac{1}{2}-z}}{\sqrt{2\pi}} \sum_{k=0}^{\infty} (-1)^k \frac{2^k (\frac{1}{2})_k\, a_{2k}}{z^k}, \quad z \to \infty. \tag{6.2.30}$$

It follows again (see also §6.1) that in the asymptotic expansions of $\Gamma(z)$ and $1/\Gamma(z)$ the same coefficients γ_k occur. We have for both functions

$$\begin{aligned} \frac{1}{\Gamma(z)} &\sim \frac{e^z z^{\frac{1}{2}-z}}{\sqrt{2\pi}} \sum_{n=0}^{\infty} \gamma_n z^{-n}, \\ \Gamma(z) &\sim \sqrt{2\pi}\, e^{-z} z^{z-\frac{1}{2}} \sum_{n=0}^{\infty} (-1)^n \gamma_n z^{-n}, \end{aligned} \tag{6.2.31}$$

where the first coefficients γ_n are

$$\gamma_0 = 1, \quad \gamma_1 = -\tfrac{1}{12}, \quad \gamma_2 = \tfrac{1}{288}, \quad \gamma_3 = \tfrac{139}{51840}, \quad \gamma_4 = -\tfrac{571}{2488320}. \tag{6.2.32}$$

These expansions are uniformly valid inside the sector $|\mathrm{ph}\, z| \le \pi - \delta < \pi$.

6.2.1 *Another integral representation of* $1/\Gamma(z)$

The saddle point analysis of the reciprocal gamma function gives a further interesting integral. When we integrate in the t-plane over the saddle point contour with respect to θ, using (6.2.27), we have

$$dt = \frac{d\left(re^{i\theta}\right)}{d\theta}\,d\theta = e^{i\theta}\left(\frac{dr}{d\theta}+ir\right)\,d\theta = (i+h(\theta))\,d\theta, \tag{6.2.33}$$

where $h(\theta)$ is an odd function of θ. It follows that

$$\frac{1}{\Gamma(z)} = \frac{e^z z^{1-z}}{2\pi}\int_{-\pi}^{\pi} e^{-z\Phi(\theta)}\,d\theta, \tag{6.2.34}$$

where

$$\Phi(\theta) = -\Re p(t) = 1-\theta\cot\theta+\ln\frac{\theta}{\sin\theta}. \tag{6.2.35}$$

To evaluate $\Phi(\theta)$ for small values of θ we have

$$\Phi(\theta) = \tfrac{1}{2}\theta^2+\tfrac{1}{36}\theta^4+\tfrac{1}{405}\theta^6+\tfrac{1}{4200}\theta^8+\tfrac{1}{42525}\theta^{10}\ldots, \tag{6.2.36}$$

where all coefficients are positive. In general we have

$$\Phi(\theta) = -\sum_{n=1}^{\infty}(-1)^n\frac{2n+1}{2n}\frac{2^{2n}B_{2n}}{(2n)!}\theta^{2n}, \tag{6.2.37}$$

which follows from well-known expansions of the trigonometric functions.

Representation (6.2.34) is very useful when one wants to evaluate the gamma function by means of a simple quadrature rule. As explained in Gil *et al.* (2007, Chapter 5), the trapezoidal rule gives extremely high accuracy in this case.

6.3 The logarithm of the gamma function

We start with the representation (Whittaker and Watson, 1996, p. 249)

$$\ln\Gamma(z) = \ln\left(\sqrt{2\pi}z^{z-\frac{1}{2}}e^{-z}\right)+\int_0^{\infty}\beta(t)e^{-zt}\,dt, \tag{6.3.38}$$

where

$$\beta(t) = \frac{1}{t}\left(\frac{1}{e^t-1}-\frac{1}{t}+\frac{1}{2}\right) = \frac{1}{t^2}\left(\tfrac{1}{2}t\coth\tfrac{1}{2}t-1\right). \tag{6.3.39}$$

The expansion of $\beta(t)$ reads

$$\beta(t) = \tfrac{1}{12}-\tfrac{1}{720}t^2+\tfrac{1}{30240}t^4+\ldots = \sum_{n=1}^{\infty}\frac{B_{2n}}{(2n)!}t^{2n-2}, \quad |t|<2\pi, \tag{6.3.40}$$

where B_n are the Bernoulli numbers. This expansion can be viewed as a special case of the generating function for the generalized Bernoulli polynomials that we will use in (6.5.70).

Table 6.1 Approximating $\Gamma(z)$ via the expansions in (6.3.42) and (6.1.15).

z	(6.3.42)	(6.1.15)
1	1.0002878	0.9997110
2	1.0000036	0.9999927
3	2.0000005	1.9999995
4	6.0000002	6.0000009
5	24.0000002	24.0000028

Using Watson's lemma (see Theorem 2.2) we obtain an asymptotic expansion of the logarithm of the gamma function (Stirling's series):

$$\ln\Gamma(z) \sim \ln\left(\sqrt{2\pi}z^{z-\frac{1}{2}}e^{-z}\right) + \sum_{n=1}^{\infty} \frac{B_{2n}}{2n\,(2n-1)}\frac{1}{z^{2n-1}}, \tag{6.3.41}$$

as $z \to \infty$. Since the singularities (poles) of $\beta(t)$ are located on the imaginary axis, this expansion holds uniformly for $|\mathrm{ph}\, z| \le \pi - \delta < \pi$. The first terms are shown in the expansion

$$\begin{aligned}\ln\Gamma(z) \sim \ln\left(\sqrt{2\pi}z^{z-\frac{1}{2}}e^{-z}\right) + \tfrac{1}{12}z^{-1} - \tfrac{1}{360}z^{-3} + \tfrac{1}{1260}z^{-5} - \tfrac{1}{1680}z^{-7} + \\ \tfrac{1}{1188}z^{-9} - \tfrac{691}{360360}z^{-11} + \tfrac{1}{156}z^{-13} - \tfrac{3617}{122400}z^{-15} + \cdots\,,\end{aligned} \tag{6.3.42}$$

as $z \to \infty$, $|\mathrm{ph}\, z| < \pi$. Taking the exponential of this result, we get the generalization of Stirling's formula in (6.1.15).

A remarkable feature is that in $\ln\Gamma(z)$ only odd powers of $1/z$ occur, whereas in the expansion of $\Gamma(z)$ both even and odd powers can be seen. This explains that (6.3.42) is much more efficient for numerical calculations than (6.1.15).

In Table 6.1 we show the results of applying (6.3.42) (with terms up to and including the term $1/1260z^{-5}$) and (6.1.15) (with terms up to and including $-139/51840z^{-3}$), for $z = 1, 2, 3, 4, 5$. It follows that the accuracy is already quite interesting for these small values of the large asymptotic parameter.

6.3.1 *Estimations of the remainder*

Because of the importance of (6.3.42) for numerical applications we now investigate the remainder, and we construct upper bounds. First we introduce a different representation of the function $\beta(t)$ that is used in (6.3.38). In Temme (1996a, §3.6.1) we have shown that this function can be written in the form

$$\beta(t) = P_N(t) + (-1)^N t^{2N} R_N(t), \tag{6.3.43}$$

where

$$P_N(t) = \sum_{n=1}^{N} \frac{B_{2n}}{(2n)!}t^{2n-2}, \quad R_N(t) = \sum_{k=1}^{\infty} \frac{2}{(t^2 + 4k^2\pi^2)(2\pi k)^{2N}}. \tag{6.3.44}$$

We assume that $t \neq \pm 2\pi i, \pm 4\pi i, \ldots$.

When $N = 0$ we have the well-known result

$$\beta(t) = \sum_{k=1}^{\infty} \frac{2}{t^2 + 4k^2\pi^2}, \quad t \neq \pm 2\pi i, \pm 4\pi i, \ldots. \tag{6.3.45}$$

This is one of the many examples of partial fraction decomposition for a class of trigonometric functions, see the representation (6.3.39).

We write (6.3.38) in the form

$$\ln\Gamma(z) = \ln\left(\sqrt{2\pi}z^{z-\frac{1}{2}}e^{-z}\right) + \Phi(z), \tag{6.3.46}$$

with

$$\Phi(z) = \int_0^\infty \beta(t)e^{-zt}\,dt, \quad \Re z > 0. \tag{6.3.47}$$

Using (6.3.43) we obtain

$$\Phi(z) = \sum_{n=1}^{N} \frac{B_{2n}}{2n\,(2n-1)} \frac{1}{z^{2n-1}} + S_N, \tag{6.3.48}$$

with

$$S_N = (-1)^N \int_0^\infty e^{-zt} t^{2N} R_N(t)\,dt = \frac{2(-1)^N}{z^{2N+1}} \int_0^\infty e^{-u} u^{2N} T_N(z,u)\,du, \tag{6.3.49}$$

and

$$T_N(z,u) = \sum_{k=1}^{\infty} \frac{4\pi^2k^2z^2}{u^2 + 4\pi^2k^2z^2} \frac{1}{(2\pi k)^{2N+2}}. \tag{6.3.50}$$

If $z > 0$ we can write $T_N(z,u)$ in the form

$$T_N(z,u) = \theta_N \sum_{k=1}^{\infty} \frac{1}{(2\pi k)^{2N+2}} = \theta_N \frac{(-1)^{N+2} B_{2N+2}}{2\,(2N+2)!}, \tag{6.3.51}$$

with $0 < \theta_N < 1$. Hence, in this case we have

$$S_N = \theta_N \frac{B_{2N+2}}{2(N+1)(2N+1)} \frac{1}{z^{2N+1}}. \tag{6.3.52}$$

In other words, S_N has the sign of the first neglected term in (6.3.48) and its absolute value is smaller than that term. Moreover, because the series is alternating, for each n, the value of $\Phi(z)$ lies always between the value of the sum of n terms and that of the sum of $n+1$ terms of the series in (6.3.48).

This is an ideal situation in asymptotics. In these circumstances one verifies for a real z-value which term in (6.3.42) is smaller than the required precision, and one knows that that term, and all subsequent terms, can be neglected. For example, when $z \geq 10$, all terms in (6.3.42) after that of z^{-11} can be neglected to obtain an accuracy of 1.92e−14, or less.

To obtain insight for complex values of z we introduce the quantity

$$K_z = \max_{s\geq 0}\left|\frac{z^2}{s^2+z^2}\right|. \tag{6.3.53}$$

Observe that K_z does not change when in $z^2/(s^2+z^2)$ the variables z and/or s are multiplied by arbitrary real numbers ($\neq 0$). Now we obtain

$$|S_N| < \frac{K_z|B_{2N+2}|}{2(N+1)(2N+1)|z|^{2N+1}}. \tag{6.3.54}$$

To determine K_z we use

$$K_z^{-2} = \min_{u\geq 0}\frac{\left(u^2+(x^2-y^2)\right)^2+4x^2y^2}{(x^2+y^2)^2}, \quad z = x+iy, \tag{6.3.55}$$

and consequently

$$K_z^{-2} = \begin{cases} 1, & \text{if} \quad x^2 \geq y^2; \\ 4x^2y^2/(x^2+y^2)^2, & \text{if} \quad x^2 < y^2. \end{cases} \tag{6.3.56}$$

Hence, if $|\mathrm{ph}\, z| < \frac{1}{4}\pi$ then (as in the case of real $z = x$) $K_z = 1$. From this it follows that when $|\mathrm{ph}\, z| < \frac{1}{4}\pi$, the absolute value of the remainder R_n of (6.3.48) is again smaller than the absolute value of the first neglected term in the series. When $\frac{1}{4}\pi < \mathrm{ph}\, z < \frac{1}{2}\pi$ the above method gives an increasingly unfavorable estimate of R_N as z approaches the imaginary axis. In Spira (1971) the following simple result is derived:

$$|R_N| \leq \begin{cases} 2|B_{2N}/(2N-1)|\,|\Im z|^{1-2N}, & \text{if } \Re z < 0,\ \Im z \neq 0; \\ |B_{2N}/(2N-1)|\,|z|^{1-2N}, & \text{if } \Re z \geq 0. \end{cases} \tag{6.3.57}$$

Of course, for computations with $\Re z < 0$, we can use the reflection formula (6.0.4).

6.4 Expansions of $\Gamma(z+a)$ and $1/\Gamma(z+a)$

By repeating certain steps in §6.1 and §6.2, and generalizing the integrands of the integrals representing the gamma functions, it is an easy exercise to obtain the following expansions

$$\Gamma(z+a) \sim \sqrt{2\pi}e^{-z}z^{z+a-\frac{1}{2}}\sum_{k=0}^{\infty}\frac{a_k(a)}{z^k}, \tag{6.4.58}$$

and

$$\frac{1}{\Gamma(z+b)} \sim \frac{1}{\sqrt{2\pi}}e^{z}z^{-z-b+\frac{1}{2}}\sum_{k=0}^{\infty}\frac{b_k(b)}{z^k}, \tag{6.4.59}$$

The first coefficients are

$$
\begin{aligned}
&a_0(a) = b_0(b) = 1,\\
&a_1(a) = \tfrac{1}{12}\left(1 - 6a + 6a^2\right), \quad b_1(b) = -\tfrac{1}{12}\left(1 - 6b + 6b^2\right),\\
&a_2(a) = \tfrac{1}{288}\left(1 - 36a + 120a^2 - 120a^3 + 36a^4\right),\\
&b_2(b) = \tfrac{1}{288}\left(1 + 12b - 24b^2 - 24b^3 + 36b^4\right),\\
&a_3(a) = \tfrac{1}{51840}\left(-139 - 450a + 8190a^2 - 20160a^3 + \right.\\
&\qquad\qquad \left. 18900a^4 - 7560a^5 + 1080a^6\right),\\
&b_3(a) = \tfrac{1}{51840}\left(139 - 270b - 1710b^2 + 1440b^3 + \right.\\
&\qquad\qquad \left. 2700b^4 - 1080b^5 - 1080b^6\right).
\end{aligned}
\tag{6.4.60}
$$

For $a = \frac{1}{2}$ we have $b_k\left(\frac{1}{2}\right) = (-1)^k a_k\left(\frac{1}{2}\right)$, that is,

$$
\begin{aligned}
\Gamma\left(z + \tfrac{1}{2}\right) &\sim \sqrt{2\pi}\, e^{-z} z^z \sum_{k=0}^{\infty} \frac{\gamma_k}{z^k},\\
\frac{1}{\Gamma\left(z + \frac{1}{2}\right)} &\sim \frac{e^z z^{-z}}{\sqrt{2\pi}} \sum_{k=0}^{\infty} (-1)^k \frac{\gamma_k}{z^s},
\end{aligned}
\tag{6.4.61}
$$

and the first few coefficients are

$$
\gamma_0 = 1, \quad \gamma_1 = -\tfrac{1}{24}, \quad \gamma_2 = \tfrac{1}{1152}, \quad \gamma_3 = \tfrac{1003}{414720}. \tag{6.4.62}
$$

6.5 The ratio of two gamma functions

In applications one frequently meets expressions with the ratio of two gamma functions. When the arguments of both functions are large it is not always possible to use numerical approximations of both functions, since they may become too large for the computer's number system. Moreover, loss of accuracy may occur when we divide two large expressions that follow from (6.0.7). This is due to the inaccuracy with which this dominant term in the expansion of the gamma function will be computed when z is large.

It is very useful when an algorithm for computing the scaled gamma function

$$
\Gamma^*(z) = \frac{\Gamma(z)}{\sqrt{2\pi}\, z^{z-\frac{1}{2}} e^{-z}}, \quad \Re z > 0, \tag{6.5.63}
$$

is available. From (6.0.7) it follows that $\Gamma^*(z) = 1 + \mathcal{O}(1/z)$, as $z \to \infty$.

Assume that we need to compute $\Gamma(z + a)/\Gamma(z + b)$ for large values of z. Then we can find an expansion of this ratio by multiplying the expansions in (6.4.58) and (6.4.59). This gives an expansion of the form

$$
\frac{\Gamma(z + a)}{\Gamma(z + b)} \sim z^{a-b} \sum_{k=0}^{\infty} \frac{C_k(a, b)}{z^k}. \tag{6.5.64}
$$

As we will see, the coefficients can be expressed in terms of generalized Bernoulli polynomials, and the expansion follows rather easily from Watson's lemma, without complicated transformations. We also give a more efficient expansion.

Remark 6.1. In the numerical computation of the ratio of gamma functions in (6.5.64) for large values of z, it may be attractive to use the expansion of $\Gamma^*(z)$ defined in (6.5.63), and to avoid the expansions of the $\Gamma(z+a)$, $1/\Gamma(z+b)$, or the expansion of the ratios in (6.5.64). Indeed, from (6.5.63) it follows that

$$\frac{\Gamma(z+a)}{\Gamma(z+b)} = z^{a-b}\frac{\Gamma^*(z+a)}{\Gamma^*(z+b)}Q(z,a,b), \tag{6.5.65}$$

where

$$\begin{aligned} Q(z,a,b) = \left(1+\frac{a}{z}\right)^{a-\frac12}\left(1+\frac{b}{z}\right)^{-b+\frac12} &\times \\ \exp\left(z\left(\ln\left(1+\frac{a}{z}\right)-\frac{a}{z}-\ln\left(1+\frac{b}{z}\right)+\frac{b}{z}\right)\right)&. \end{aligned} \tag{6.5.66}$$

It is not difficult to verify that $Q(z,a,b) = 1+\mathcal{O}(1/z)$, as $z\to\infty$; (6.5.65) shows quite well which contributions play a role in the ratio of the gamma functions.

Although $Q(z,a,b)$ is composed of elementary functions, one should be careful when evaluating the above expression when z is large. The point is that, for small values of z, the function $\ln(1+z)-z$ cannot be accurately computed directly from the log-function (a loss in *relative* accuracy occurs). However, it is rather easy to write a code for the function $\ln(1+z)-z$ for small values of $|z|$. △

6.5.1 *A simple expansion*

From the beta integral

$$B(p,q) = \frac{\Gamma(p)\Gamma(q)}{\Gamma(p+q)} = \int_0^1 t^{p-1}(1-t)^{q-1}\,dt, \quad \Re p>0, \quad \Re q>0, \tag{6.5.67}$$

we obtain

$$\begin{aligned} \frac{\Gamma(z+a)}{\Gamma(z+b)} &= \frac{1}{\Gamma(b-a)}\int_0^1 t^{z+a-1}(1-t)^{b-a-1}\,dt, \\ &= \frac{1}{\Gamma(b-a)}\int_0^\infty u^{b-a-1}e^{-zu}f(u)\,du, \end{aligned} \tag{6.5.68}$$

where $\Re(b-a)>0$ and

$$f(u) = e^{-au}\left(\frac{1-e^{-u}}{u}\right)^{b-a-1}. \tag{6.5.69}$$

From the generating series for the generalized Bernoulli polynomials (see (15.6.38))

$$e^{xz}\left(\frac{z}{e^z-1}\right)^{\sigma} = \sum_{n=0}^{\infty}\frac{B_n^{(\sigma)}(x)}{n!}z^n, \quad |z|<2\pi, \tag{6.5.70}$$

it follows that

$$f(u) = \sum_{n=0}^{\infty} c_n(a,b)u^n, \quad c_n(a,b) = (-1)^n \frac{B_n^{(a-b+1)}(a)}{n!}. \tag{6.5.71}$$

Watson's lemma (in the form of Theorem 2.2) gives the result (under the condition $\Re(b-a) > 0$, but see Remark 6.2)

$$\frac{\Gamma(z+a)}{\Gamma(z+b)} \sim z^{a-b} \sum_{n=0}^{\infty} c_n(a,b)\,(b-a)_n \,\frac{1}{z^n}, \qquad z \to \infty. \tag{6.5.72}$$

The first coefficients c_n are

$$\begin{aligned} c_0(a,b) &= 1, \quad c_1(a,b) = \tfrac{1}{2}(1-a-b), \\ c_2(a,b) &= \tfrac{1}{24}(3a^2+6ab+3b^2-7a-5b+2), \\ c_3(a,b) &= \tfrac{1}{48}(1-a-b)(a^2+2ab+b^2-3a-b). \end{aligned} \tag{6.5.73}$$

The singular points of the function $f(u)$ occur on the imaginary axis, at the points $2k\pi i$, $k \in \mathbb{Z}$, $k \neq 0$. We conclude that this expansion holds uniformly inside the sector $|\mathrm{ph}\, z| \leq \pi - \delta < \pi$.

6.5.2 *A more efficient expansion*

By representing $f(u)$ in a different way a more efficient expansion is possible. To obtain that expansion we write

$$f(u) = e^{-u(b+a-1)/2} \left(\frac{\sinh(u/2)}{u/2} \right)^{b-a-1}. \tag{6.5.74}$$

Again, using (6.5.70), we can write

$$f(u) = e^{-u(b+a-1)/2} \sum_{n=0}^{\infty} (-1)^n C_n(\rho) u^{2n}, \quad C_n(\rho) = (-1)^n \frac{B_{2n}^{(2\rho)}(\rho)}{(2n)!}, \tag{6.5.75}$$

where $\rho = (a-b+1)/2$. Application of Watson's lemma now gives

$$\frac{\Gamma(z+a)}{\Gamma(z+b)} \sim w^{a-b} \sum_{n=0}^{\infty} (-1)^n C_n(\rho)\,(b-a)_{2n}\,\frac{1}{w^{2n}}, \qquad z \to \infty, \tag{6.5.76}$$

uniformly in the sector $|\mathrm{ph}\, z| \leq \pi - \delta < \pi$, with $w = z + (a+b-1)/2$ and $\Re(b-a) > 0$.

Comparing the new expansion with the one in (6.5.72), we note three favorable features.

- Only even powers of $1/w$ occur in the new series. Hence, the new expansion is more efficient than the one in (6.5.72).
- For real a, b and z, with $0 < a-b+1 < 1$ the remainder can be estimated. Let $N = 0, 1, 2, \ldots$ and let R_N be defined by writing

$$\frac{\Gamma(z+a)}{\Gamma(z+b)} = w^{a-b} \left(\sum_{n=0}^{N-1} (-1)^n C_n(\rho)\,(b-a)_{2n}\,\frac{1}{w^{2n}} + R_N \right). \tag{6.5.77}$$

Then, when $z+\min(a,(a+b-1)/2)>0$ and $0<a-b+1<1$, we have (see Frenzen (1987))

$$R_N=\theta_N(-1)^N C_N(\rho)\,(b-a)_{2N}\,\frac{1}{w^{2N}},\quad 0<\theta_N<1. \tag{6.5.78}$$

- The coefficients $C_n(\rho)$ are polynomials of one variable. Hence, they are simpler than $c_n(a,b)$, which are polynomials of two variables.

We give the first coefficients $C_n(\rho)$ of (6.5.76).

$$\begin{aligned}
&C_0(\rho)=1,\quad C_1(\rho)=\frac{\rho}{12},\quad C_2(\rho)=\frac{\rho(5\rho+1)}{1440},\\
&C_3(\rho)=\frac{\rho(35\rho^2+21\rho+4)}{362880},\\
&C_4(\rho)=\frac{\rho(5\rho+2)(35\rho^2+28\rho+9)}{87091200},\\
&C_5(\rho)=\frac{\rho(385\rho^4+770\rho^3+671\rho^2+286\rho+48)}{11496038400}.
\end{aligned} \tag{6.5.79}$$

Remark 6.2. The expansions in (6.5.72) and (6.5.76) are derived in Tricomi and Erdélyi (1951) and Fields (1966), respectively. The results are given without our initial condition $\Re(b-a)>0$. We have used this condition for using (6.5.68), and to derive the expansions in (6.5.72) and (6.5.76). In the cited references the starting point is a loop integral, and then the condition can be dropped for both results. See also Remark 2.2, where we have explained this phenomenon for general cases.

The expansions remain valid when $a-b=-k$, $k=1,2,\ldots$ (the coefficients in both expansions remain well defined in this case). When this happens we have $\Gamma(z+a)/\Gamma(z+a+k)=1/(z+a)_k$. On the other hand, when $a-b=k$, $k=0,1,2,\ldots$, then the expansions are finite because

$$(-k)_n=\begin{cases}0, & \text{if}\quad n>k,\\ (-1)^n k!/(k-n)! & \text{if}\quad n\le k,\end{cases} \tag{6.5.80}$$

where k,n are nonnegative integers. △

Remark 6.3. To obtain the loop integral in the present case, we write the second integral in (6.5.68) along a path around the positive real axis. This gives (see also Remark 2.3)

$$\frac{\Gamma(z+a)}{\Gamma(z+b)}=\frac{e^{(a-b)\pi i}\Gamma(1+a-b)}{2\pi i}\int_{+\infty}^{(0+)} u^{b-a-1}e^{-zu}f(u)\,du. \tag{6.5.81}$$

where $f(u)$ is defined in (6.5.69), the initial and final values of $\operatorname{ph} u$ being 0 and 2π, respectively. The contour cuts the imaginary axis between $-2\pi i$ and $+2\pi i$. This representation holds for all a and b with the trivial exception $a-b\ne -1,-2,\ldots$. By expanding $f(u)$ in powers of u and using (6.0.3) we again obtain the expansion in (6.5.72), without the condition $\Re(b-a)>0$. If we wish we can write (6.5.81) in the form used in Watson's lemma for loop integrals, see §2.2), by putting $u = se^{\pi i}$. This gives (the initial and final values of ph s being $-\pi$ and π, respectively)

$$\frac{\Gamma(z+a)}{\Gamma(z+b)} = \frac{\Gamma(1+a-b)}{2\pi i} \int_{-\infty}^{(0+)} s^{b-a-1} e^{zs} f(-s)\, ds. \tag{6.5.82}$$

When z is complex the path may be rotated and then ph s will have values outside the interval $[-\pi, \pi]$. △

6.6 A binomial coefficient

For the binomial coefficient we can obtain asymptotic representations by using the familiar relation to the gamma function:

$$\binom{n}{m} = \frac{n!}{(n-m)!\, m!} = \frac{\Gamma(n+1)}{\Gamma(n-m+1)\,\Gamma(m+1)}. \tag{6.6.83}$$

Without referring to this relation we can use the representation (Cauchy integral)

$$\binom{n}{m} = \frac{1}{2\pi i} \int \frac{(1+z)^n}{z^{m+1}}\, dz, \tag{6.6.84}$$

where the integral is taken along a circle around the origin. This follows from the binomial expansion

$$(1+z)^n = \sum_{m=0}^{n} \binom{n}{m} z^m. \tag{6.6.85}$$

Clearly, we use integral values of n and m, with $0 \le m \le n$. We write the integral in the form

$$\binom{n}{m} = \frac{1}{2\pi i} \int e^{n\phi(z)}\, \frac{dz}{z}, \quad \phi(z) = \ln(1+z) - \mu \ln z, \quad \mu = \frac{m}{n}. \tag{6.6.86}$$

The saddle point z_0 follows from $\phi'(z) = 0$ giving $z_0 = \mu/(1-\mu)$, and we assume that $\mu \in [\delta, 1-\delta]$, where δ is a fixed small positive number.

At the saddle point z_0 the imaginary part of ϕ is zero, and we try to determine a saddle point contour through z_0 by solving the equation $\Im\phi(z) = 0$ in the complex plane. Writing $z = x + iy$, it follows that

$$\Im\phi(z) = \arctan \frac{y}{1+x} - \mu \arctan \frac{y}{x}. \tag{6.6.87}$$

Hence, $\Im\phi(z) = 0$ can be replaced by the equation

$$\frac{y}{1+x} = \tan\left(\mu \arctan \frac{y}{x}\right). \tag{6.6.88}$$

The trivial solution $y = 0$ gives the positive x-axis, which cannot be used for integration (it is a path of steepest ascent). To solve the equation it is convenient

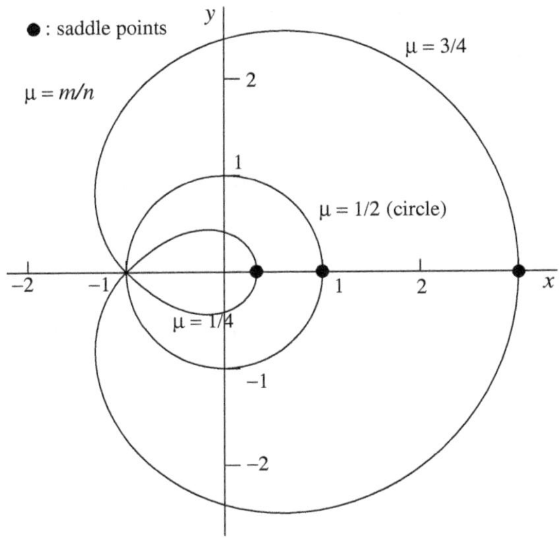

Fig. 6.3 Saddle point contours defined by $r(\theta)$ of (6.6.89) for $\mu = \frac{1}{4}, \frac{1}{2}, \frac{3}{4}$.

to introduce polar coordinates, writing $z = re^{i\theta}$. After some manipulations we find that equation (6.6.88) can be solved in the form

$$r(\theta) = \frac{\sin \mu\theta}{\sin(1-\mu)\theta}, \quad -\pi \le \theta \le \pi. \tag{6.6.89}$$

This defines a curve that goes through the saddle point z_0 ($\theta = 0$) and through the point $z = -1$ ($\theta = \pm\pi$). In Figure 6.3 we show the curves for $\mu = \frac{1}{4}, \frac{1}{2}, \frac{3}{4}$.

The real part of ϕ along the saddle point contour is given by

$$\begin{aligned} \Re\phi(z) &= \tfrac{1}{2} \ln(1 + 2r\cos\theta + r^2) - \mu \ln r \\ &= \ln \frac{\sin\theta}{\sin(1-\mu)\theta} - \mu \ln \frac{\sin \mu\theta}{\sin(1-\mu)\theta}, \end{aligned} \tag{6.6.90}$$

where the second form is obtained by using $r = r(\theta)$ given in (6.6.89). We can integrate with respect to θ along the contour, using

$$\frac{dz}{z} = \frac{1}{re^{i\theta}} \frac{d\, re^{i\theta}}{d\theta}\, d\theta = \left(\frac{r'}{r} + i\right) d\theta. \tag{6.6.91}$$

Because $r' = dr/d\theta$ is an odd function of θ the term r'/r does not contribute and we obtain

$$\binom{n}{m} = \frac{1}{\pi} \int_0^\pi \left(\frac{\sin\theta}{\sin(1-\mu)\theta}\right)^n \left(\frac{\sin(1-\mu)\theta}{\sin\mu\theta}\right)^m d\theta, \quad \mu = \frac{m}{n}. \tag{6.6.92}$$

As a result of the transformations, the integrand has a Gaussian (bell-shaped) form, and is positive. The integral can be used as a starting point for a further asymptotic analysis. It can be used to obtain an asymptotic expansion that is valid for large values of n and m.

An asymptotic expansion can also be obtained from (6.6.86) by using the transformation $z \to w(z)$ with $w(z_0) = 0$:

$$\phi(z) = \phi(z_0) + \tfrac{1}{2}w^2. \tag{6.6.93}$$

For positive z and real w we assume the condition $\operatorname{sign}(z - z_0) = \operatorname{sign}(w)$.

The steepest descent contour through the saddle point z_0 is mapped to the imaginary axis in the w-plane, and we obtain

$$\binom{n}{m} = \frac{(1-\mu)^{m-n}}{\mu^m} \frac{1}{2\pi i} \int_{-i\infty}^{i\infty} e^{\frac{1}{2}nw^2} g(w)\, dw, \tag{6.6.94}$$

where

$$g(w) = \frac{1}{z}\frac{dz}{dw} = \frac{1}{1-\mu}\frac{(1+z)w}{z - z_0}. \tag{6.6.95}$$

Expanding g in powers of w gives an asymptotic expansion. First we need the coefficients c_k in the expansion $z = \sum_{k=0}^{\infty} z_k w^k$. We have

$$z_0 = \frac{\mu}{1-\mu}, \quad z_1 = \sqrt{\frac{\mu}{(1-\mu)^3}}, \quad z_2 = \frac{1+\mu}{3(1-\mu)^2}. \tag{6.6.96}$$

Once the coefficients z_k are available, we can find the coefficients g_k in the expansion $g(w) = \dfrac{1}{\sqrt{\mu(1-\mu)}} \sum_{k=0}^{\infty} g_k w^k$. The first coefficients are

$$g_0 = 1, \quad g_1 = \frac{2\mu - 1}{3\sqrt{\mu(1-\mu)}}, \quad g_2 = \frac{\mu^2 - \mu + 1}{12\mu(1-\mu)}. \tag{6.6.97}$$

The asymptotic expansion reads

$$\binom{n}{m} \sim \frac{1}{\sqrt{2\pi n}} \frac{(1-\mu)^{m-n}}{\mu^m \sqrt{\mu(1-\mu)}} \sum_{k=0}^{\infty} (-1)^k 2^k g_{2k} \left(\tfrac{1}{2}\right)_k n^{-k}, \tag{6.6.98}$$

as $n \to \infty$. Clearly, the expansion is not valid when μ or $1 - \mu$ tend to zero.

The same expansion can be found by using the asymptotic expansions of the gamma functions in the representation

$$\binom{n}{m} = \frac{\Gamma(n+1)}{\Gamma(\mu n)\,\Gamma((1-\mu)n)}, \tag{6.6.99}$$

with n large. Also from this starting point, we conclude that (6.6.98) holds if n is large, and m and $n - m$ need to be large as well. In §6.6.1 we give an expansion in which we accept small values of m.

6.6.1 *A uniform expansion of the binomial coefficient*

It is possible to construct a different expansion in which $\mu \to 0$ is allowed. Observe that the graph of the left-hand side of (6.6.93) for $0 < \mu < 1$ has a quadratic character for z near z_0. When $\mu \to 0$ the saddle point z_0 tends to zero, and in that case the left-hand side of (6.6.93) loses its quadratic character near z_0, although on the right-hand side always a quadratic function is used. This explains why the asymptotic expansion given in (6.6.98) is not valid when $\mu \to 0$.

We try another transformation, this time by writing

$$\ln(1+z) - \mu \ln z = w - \mu \ln w + A(\mu). \tag{6.6.100}$$

The left-hand side has the same saddle point as before: $z_0 = \mu/(1-\mu)$, and the right-hand side at $w_0 = \mu$. Both sides have a quadratic character near the saddle points, but the functions behave in the same manner when $\mu \to 0$.

The quantity $A(\mu)$ follows from prescribing that the saddle points in both variable z and w should correspond. Substituting $z = z_0$ and $w = \mu$ in (6.6.100) gives

$$A(\mu) = -\mu - (1-\mu)\ln(1-\mu). \tag{6.6.101}$$

Another assumption is that for positive z and w we have $\operatorname{sign}(z-z_0) = \operatorname{sign}(w-w_0)$.

The transformation in (6.6.100) gives for the representation in (6.6.86)

$$\binom{n}{m} = e^{-m}(1-\mu)^{m-n}\frac{1}{2\pi i}\int \frac{e^{nw}}{w^{m+1}}\, f(w)\, dw, \tag{6.6.102}$$

where the contour of integration is a small circle around the origin and

$$f(w) = \frac{w}{z}\frac{dz}{dw}, \quad \frac{dz}{dw} = \frac{z(1+z)(w-\mu)}{(1-\mu)(z-z_0)w}. \tag{6.6.103}$$

A first approximation follows from replacing $f(w)$ by $f(\mu)$. Applying l'Hôpital's rule to dz/dw we find

$$\left.\frac{dz}{dw}\right|_{w=\mu} = \frac{1}{(1-\mu)^{3/2}} \quad \Longrightarrow \quad f(\mu) = \frac{1}{\sqrt{1-\mu}}. \tag{6.6.104}$$

This gives

$$\binom{n}{m} \sim \frac{e^{-m}n^m}{m!}(1-\mu)^{m-n} f(\mu), \quad n \to \infty. \tag{6.6.105}$$

Observe that this approximation gives an exact result 1 for $m = 0$ and that it is not valid when $\mu \to 1$.

To compute higher-order approximations we use integration by parts, writing

$$\begin{aligned}\binom{n}{m} &= \frac{e^{-m}n^m}{m!}(1-\mu)^{m-n} f(\mu) + \\ &\quad e^{-m}(1-\mu)^{m-n}\frac{1}{2\pi i\, n}\int \frac{f(w)-f(\mu)}{w-\mu}\, de^{n(w-\mu\ln w)}, \\ &= \frac{e^{-m}n^m}{m!}(1-\mu)^{m-n} f(\mu) + \\ &\quad e^{-m}(1-\mu)^{m-n}\frac{1}{2\pi i\, n}\int \frac{e^{nw}}{w^{m+1}} f_1(w)\, dw,\end{aligned} \tag{6.6.106}$$

where

$$f_1(w) = -w\frac{d}{dw}\frac{f(w)-f(\mu)}{w-\mu}. \tag{6.6.107}$$

More terms in the approximation follow in the same way. We have

$$\begin{aligned}\binom{n}{m} = \frac{e^{-m}n^m}{m!}(1-\mu)^{m-n}\sum_{k=0}^{K-1}\frac{f_k(\mu)}{n^k} + \\ e^{-m}(1-\mu)^{m-n}\frac{1}{2\pi i\, n^K}\int \frac{e^{nw}}{w^{m+1}} f_K(w)\, dw,\end{aligned} \tag{6.6.108}$$

where $f_0 = f$ defined in (6.6.103) and

$$f_k(w) = -w\frac{d}{dw}\frac{f_{k-1}(w)-f_{k-1}(\mu)}{w-\mu}, \quad k = 1,2,3,\ldots. \tag{6.6.109}$$

The first coefficients are

$$\begin{aligned} f_0(\mu) &= \frac{1}{\sqrt{1-\mu}}, \quad f_1(\mu) = -\frac{\mu}{12(1-\mu)^{3/2}}, \quad f_2(\mu) = \frac{\mu^2}{288(1-\mu)^{5/2}} \\ f_3(\mu) &= \frac{\mu(432-432\mu+139\mu^2)}{51840(1-\mu)^{7/2}}, \quad f_4(\mu) = -\frac{\mu^2(1728-1728\mu+571\mu^2)}{2488320(1-\mu)^{9/2}}. \end{aligned} \tag{6.6.110}$$

The expansion in (6.6.108) is valid if n is large, with $\mu \in [0, 1-\delta]$, where δ is a small positive fixed number. In Chapter 34 we use a similar method to obtain a uniform expansion of Stirling numbers.

The result of this section can also be obtained by writing

$$\binom{n}{m} = \frac{\Gamma(n+1)}{m!\,\Gamma\left((1-\mu)n\right)}, \tag{6.6.111}$$

and by dividing the two expansions of the gamma functions.

However, the representation in (6.6.102) has an extra element. When we expand $f(w) = \sum_{k=0}^{\infty} c_k(\mu)w^k$ (with f defined in (6.6.103)), and substitute this expansion into (6.6.102), then we have the finite exact representation

$$\binom{n}{m} = e^{-m}(1-\mu)^{m-n}n^m\sum_{k=0}^{m}\frac{c_k(\mu)}{(m-k)!}n^{-k}. \tag{6.6.112}$$

This expansion is valid of all $\mu \in [0,1]$, and has an asymptotic property for large n and all $m = 0,1,2,\ldots,n$.

To determine the coefficients $c_k(\mu)$, we need the coefficients $b_k(\mu)$ in the expansion $z = \sum_{k=1}^{\infty} b_k(\mu)w^k$. The coefficient $b_1(\mu)$ easily follows from (6.6.100), by writing this relation in the form

$$\frac{z}{w} = \exp\left(\frac{\ln(1+z) - w - A(\mu)}{\mu}\right). \tag{6.6.113}$$

This gives

$$b_1(\mu) = \exp\left(-\frac{A(\mu)}{\mu}\right) = e(1-\mu)^{(1-\mu)/\mu}, \tag{6.6.114}$$

with $b_1(0) = 1$. From the equation for dz/dw in (6.6.103) we obtain

$$\begin{aligned} b_2(\mu) &= \frac{b_1(b_1-1)}{\mu}, \quad b_3(\mu) = -\frac{b_1(-1+4b_1+\mu b_1^2-3b_1^2)}{2\mu^2}, \\ b_4(\mu) &= \frac{b_1(-1+12b_1+(9\mu-27)b_1^2+(2\mu^2-12\mu+16)b_1^3}{6\mu^3}, \end{aligned} \tag{6.6.115}$$

which have finite limits when $\mu \to 0$. From these $b_k(\mu)$ we can obtain the first coefficients $c_k(\mu)$:

$$\begin{aligned} c_0(\mu) &= 1, \quad c_1(\mu) = \frac{b_1-1}{\mu}, \quad c_2(\mu) = -\frac{b_1(2+(\mu-2)b_1)}{\mu^2}, \\ c_3(\mu) &= \frac{b_1(3+6(\mu-2)b_1+(2\mu-3)(\mu-3)b_1^2}{2\mu^3}. \end{aligned} \tag{6.6.116}$$

For $\mu = 0$ we have $c_1(0) = \frac{1}{2}$, $c_2(0) = \frac{1}{12}$, $c_3(0) = 0$.

We observe that the finite exact representation in (6.6.112) is constituted by terms containing powers of e, while the result becomes a positive integer. Indeed, all powers of e disappear after summation and multiplying by e^{-m}.

6.7 Asymptotic expansion of a product of gamma functions

The product of gamma functions $\Gamma\left(\mu+\frac{1}{2}-i\tau\right)\Gamma\left(\mu+\frac{1}{2}+i\tau\right)$ occurs in representations of certain Legendre functions (in particular, conical functions, see §29.4). When μ is an integer ($\mu = m \ge 0$) this product can be written as

$$\Gamma\left(m+\tfrac{1}{2}-i\tau\right)\Gamma\left(m+\tfrac{1}{2}+i\tau\right) = \frac{\pi}{\cosh(\pi\tau)}\prod_{n=1}^{m}\left((m-n+\tfrac{1}{2})^2+\tau^2\right). \tag{6.7.117}$$

In this section we derive an asymptotic expansion that holds for large values of μ, uniformly with respect to $\tau \in \mathbb{R}$.

Starting point is the beta integral. We recall

$$B(p,q) = \frac{\Gamma(p)\Gamma(q)}{\Gamma(p+q)} = \int_0^1 t^{p-1}(1-t)^{q-1}\,dt, \quad \Re p > 0, \quad \Re q > 0, \tag{6.7.118}$$

which gives the representation

$$\frac{\Gamma\left(\mu+\frac{1}{2}-i\tau\right)\Gamma\left(\mu+\frac{1}{2}+i\tau\right)}{\Gamma(2\mu+1)} = \int_0^1 e^{-\mu\phi(t)}\frac{dt}{\sqrt{t(1-t)}}, \tag{6.7.119}$$

where

$$\phi(t) = -(1-i\beta)\ln t - (1+i\beta)\ln(1-t), \quad \beta = \frac{\tau}{\mu}. \tag{6.7.120}$$

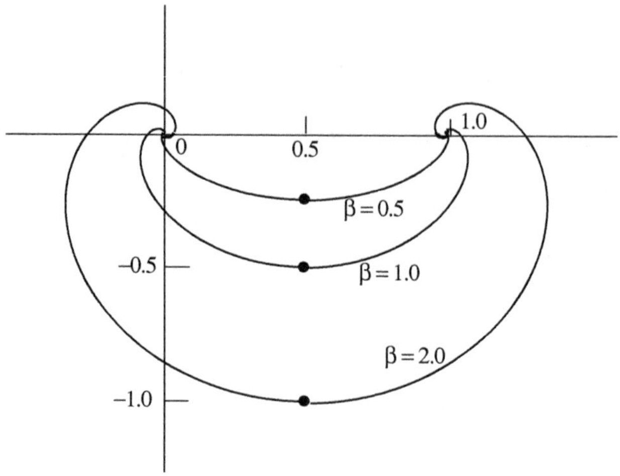

Fig. 6.4 Saddle point contours through the saddle points of the integral in (6.7.119). The black dots are the saddle points for $\beta = 0.5, 1.0, 2.0$.

The saddle point, the solution of $\phi'(t) = 0$, occurs at $t = t_0 = \frac{1}{2} - \frac{1}{2}i\beta$. We have

$$\phi(t_0) = -\ln\frac{1+\beta^2}{4} + 2\beta\arctan\beta, \quad \phi''(t_0) = \frac{8}{1+\beta^2}. \tag{6.7.121}$$

In Figure 6.4 we show the saddle point contours through the saddle points (indicated by black dots) for $\beta = 0.5, 1.0, 2.0$.

We use the saddle point method to obtain an asymptotic expansion. We substitute

$$\phi(t) = \phi(t_0) + \tfrac{1}{2}\phi''(t_0)\, s^2, \tag{6.7.122}$$

and obtain

$$\frac{\Gamma\left(\mu+\frac{1}{2}-i\tau\right)\Gamma\left(\mu+\frac{1}{2}+i\tau\right)}{\Gamma(2\mu+1)} = e^{-\mu\phi(t_0)}\int_{-\infty}^{\infty} e^{-\frac{1}{2}\mu\phi''(t_0)s^2} f(s)\, ds, \tag{6.7.123}$$

where

$$f(s) = \frac{dt}{ds}\frac{1}{\sqrt{t(1-t)}} = \frac{\phi''(t_0)}{\phi'(t)}\frac{s}{\sqrt{t(1-t)}} = \frac{s\phi''(t_0)\sqrt{t(1-t)}}{2t-1+i\beta}. \tag{6.7.124}$$

This function is bounded on $\mathbb{R}$. It is analytic in a domain that contains the real s-axis. This follows from locating the singularities of the mapping defined in (6.7.122). These come from points $t_k = t_0 e^{2\pi i k}$, $k = \pm1, \pm2, \ldots$, and points $\widetilde{t_k} = 1 - t_k$. These groups give corresponding points s_k and $\widetilde{s_k}$ in the s-plane, which follow from (6.7.122). At these points the function ϕ' has zeros, and, hence, the function f defined in (6.7.124) is singular.

For example, we have

$$\phi(t_k) = \phi(t_0) - (1-i\beta)2\pi i k = \phi(t_0) + \frac{4}{1+\beta^2}s_k^2, \tag{6.7.125}$$

and we see that the points s_k are bounded away from the saddle point at $s=0$ and from the real axis for all $\beta\in\mathbb{R}$.

For $\beta=0$ we have the explicit relation

$$t=\frac{1\pm\sqrt{1-\exp(-4s^2)}}{2},\quad \operatorname{sign}\left(t-\tfrac12\right)=\operatorname{sign}(s), \tag{6.7.126}$$

and t as a function of s is singular at the points satisfying $-4s^2=2\pi ik$, $k=\pm1,\pm2,\ldots$, which confirms the general case $\beta\neq0$.

We conclude that the expansion holds as $\mu\to\infty$, uniformly with respect to $\beta\in\mathbb{R}$.

At the saddle point, the relation between t and s is

$$t=t_0+s+\frac{4i\beta}{3\left(\beta^2+1\right)}s^2-\frac{13\beta^2+9}{9\left(\beta^2+1\right)^2}s^3+\ldots. \tag{6.7.127}$$

We expand $f(s)=\sum_{k=0}^{\infty}f_ks^k$, and obtain the required expansion

$$\frac{\Gamma\left(\mu+\frac12-i\tau\right)\Gamma\left(\mu+\frac12+i\tau\right)}{\Gamma(2\mu+1)}\sim\sqrt{\frac{\pi}{\mu}}\left(\frac{\beta^2+1}{4}\right)^{\mu}e^{-2\tau\arctan\beta}\sum_{k=0}^{\infty}\frac{c_k(\beta)}{\mu^k}. \tag{6.7.128}$$

The first few coefficients are

$$c_0(\beta)=1,\quad c_1(\beta)=-\frac{\beta^2+3}{24\left(\beta^2+1\right)},\quad c_2(\beta)=\frac{\left(\beta^2+3\right)^2}{1152\left(\beta^2+1\right)^2},$$
$$c_3(\beta)=\frac{139\beta^6+387\beta^4-5751\beta^2+2035}{414720\left(\beta^2+1\right)^3}. \tag{6.7.129}$$

When $\tau=0$ the left-hand side of (6.7.128) becomes

$$\frac{\Gamma\left(\mu+\frac12\right)\Gamma\left(\mu+\frac12\right)}{\Gamma(2\mu+1)}=\sqrt{\pi}\,2^{-2\mu}\frac{\Gamma\left(\mu+\frac12\right)}{\Gamma(\mu+1)}, \tag{6.7.130}$$

and the expansion reduces to (see also §6.5, in particular (6.5.72) with $a=\frac12$ and $b=1$)

$$\frac{\Gamma\left(\mu+\frac12\right)}{\Gamma(\mu+1)}\sim\frac{1}{\sqrt{\mu}}\sum_{k=0}^{\infty}\frac{c_k(0)}{\mu^k}. \tag{6.7.131}$$

As a final step we can use in (6.7.128) the standard asymptotic expansion of $\Gamma(2\mu+1)=2\mu\Gamma(2\mu)$ (see §6.1). This gives

$$\Gamma\left(\mu+\tfrac12-i\tau\right)\Gamma\left(\mu+\tfrac12+i\tau\right)\sim2\pi(\mu^2+\tau^2)^{\mu}e^{-2\mu-2\tau\arctan\beta}\sum_{k=0}^{\infty}\frac{d_k(\beta)}{\mu^k}. \tag{6.7.132}$$

The first coefficients are

$$d_0(\beta)=1,\quad d_1(\beta)=-\frac{1}{12\left(\beta^2+1\right)},\quad d_2(\beta)=\frac{1}{288\left(\beta^2+1\right)^2},$$
$$d_3(\beta)=\frac{247-756\beta^2}{51840\left(\beta^2+1\right)^3},\quad d_4(\beta)=\frac{3024\beta^2-1003}{2488320\left(\beta^2+1\right)^4}. \tag{6.7.133}$$

We observe that the shown coefficients $d_k(\beta)$ $(k \ge 1)$ vanish as $|\beta| \to \infty$.

Finally, we can keep μ fixed, and let τ act as the large parameter. Consider for example $\mu = \frac{1}{2}$. For large τ, the arctan in (6.7.132) can be expanded as

$$\arctan(2\tau) = \tfrac{1}{2}\pi - \tfrac{1}{2}\tau^{-1} + \mathcal{O}\left(\tau^{-3}\right). \tag{6.7.134}$$

When we replace the series in (6.7.132) by unity, the right-hand side of (6.7.132) becomes $2\pi\,\tau e^{-\pi\tau}$, which is a perfect match (when τ is large) with the left-hand side, which equals $\pi\tau/\sinh(\pi\tau)$.

We have used the expansions of this section in an algorithm for computing the P-conical function, see Gil *et al.* (2009), where we needed the computation of $\Gamma\left(\mu+\frac{1}{2}-i\tau\right)\Gamma\left(\mu+\frac{1}{2}+i\tau\right)$ for large μ. It was convenient to have a real expansion of this real product.

6.8 Expansions of ratios of three gamma functions

First we mention the expansion

$$\frac{\Gamma(\nu+c)}{\Gamma(\nu+a)\Gamma(\nu+b)} = \sum_{k=0}^{\infty} \frac{(a-c)_k(b-c)_k}{k!\,\Gamma(\nu+a+b-c+k)}, \tag{6.8.135}$$

which converges if $\Re(\nu+c) > 0$, and is an easy consequence of the known value of the ${}_2F_1$-function at $z=1$; see (12.0.2). For large values of ν the series (whether or not convergent) can be used as an asymptotic expansion in the sector $|\mathrm{ph}\,\nu| \le \pi-\delta$. We give this simple expansion, because, in some sense, it is related to the following one, which has attracted attention in the literature.

We have

$$\frac{\Gamma(\nu+a)\Gamma(\nu+b)}{\Gamma(\nu+c)} \sim \sum_{k=0}^{\infty} (-1)^k \frac{(c-a)_k(c-b)_k}{k!}\Gamma(\nu+a+b-c-k), \tag{6.8.136}$$

as $\nu \to \infty$, uniformly in the sector $|\mathrm{ph}\,\nu| \le \pi-\delta$. Several proofs are available and for a proof based on Barnes' lemma (see Whittaker and Watson (1996, p. 289)) and contour integration, we refer to Olver (1995). That proof provides an error bound of the remainder (which is of the order of the first neglected term), and is based on the form with $c=1$. However, a simple shift in the parameter ν gives the more general form.

The expansion in (6.8.136) is important in the theory of the Stokes phenomenon and in the related exponentially improved asymptotic solutions of linear differential equations (particularly of Kummer's equation (10.1.1)).

In this section we use a different approach which is based on the ideas of Chapter 15, where we have obtained asymptotic expansions of coefficients of a power series by using an approach related to Darboux's method.

First we evaluate the integral

$$F(\nu) = \frac{1}{2\pi i}\int_{-\infty}^{(0+)} {}_2F_1\left(\begin{matrix} a,b \\ c \end{matrix};s\right)\frac{ds}{s^{\nu+1}}, \tag{6.8.137}$$

along a Hankel contour as shown in Figure 2.1, with the indicated phases of s. The contour cuts the real axis inside the interval $(0,1)$. For convergence at infinity we need the conditions $\Re(a+\nu)>0$ and $\Re(b+\nu)>0$, which follows from using the connection formula

$$\begin{aligned} {}_2F_1\left({a,b \atop c};z\right) = A(-z)^{-a}{}_2F_1\left({a,1-c+a \atop 1-b+a};\frac{1}{z}\right) + \\ B(-z)^{-b}{}_2F_1\left({b,1-c+b \atop 1-a+b};\frac{1}{z}\right), \end{aligned} \tag{6.8.138}$$

where $|\mathrm{ph}(1-z)|<\pi$ and

$$A=\frac{\Gamma(c)\Gamma(b-a)}{\Gamma(b)\Gamma(c-a)},\quad B=\frac{\Gamma(c)\Gamma(a-b)}{\Gamma(a)\Gamma(c-b)}. \tag{6.8.139}$$

When, in addition, $\Re\nu<0$ and $\Re a$ and $\Re b$ are large enough, we can integrate along the negative axis, taking into account the phase of s. The result is

$$F(\nu)=-\frac{\sin(\nu\pi)}{\pi}\int_0^\infty {}_2F_1\left({a,b \atop c};-t\right)\frac{dt}{t^{\nu+1}}, \tag{6.8.140}$$

and this is a known integral (see Olde Daalhuis (2010b, Eq. 15.14.1)). We obtain

$$F(\nu)=\frac{\Gamma(c)\Gamma(\nu+a)\Gamma(\nu+b)}{\Gamma(a)\Gamma(b)\Gamma(\nu+c)\Gamma(\nu+1)}, \tag{6.8.141}$$

and using the principle of analytic continuation, we can drop the conditions $\Re(a+\nu)>0$, $\Re(b+\nu)>0$, $\Re\nu<0$, and assume large values of $\Re\nu$. We can use this result in the sector $|\mathrm{ph}\,\nu|\le\pi-\delta$.

To obtain an asymptotic expansion of $F(\nu)$ from (6.8.137), we concentrate on the behavior of the hypergeometric function at the point $s=1$. This is the only finite singular point of this function, and we will explain in Chapter 15 how to do this for coefficients of a power series. Observe that when $\nu=n=0,1,2,\ldots$,

$$F(n)=\frac{(a)_n(b)_n}{n!\,(c)_n}, \tag{6.8.142}$$

which is the coefficient of the power series of the ${}_2F_1$-function in (6.8.137), and $F(n)$ has a representation as in (6.8.137) where the contour is a circle around the origin with radius less than 1.

The ${}_2F_1$-function in (6.8.137) is analytic in the complex plane except for a singularity at $s=1$, with a branch cut from 1 to $+\infty$. We note that the function has only algebraic growth at infinity of an order determined by the fixed numbers a and b (see also (6.8.138)) . Hence, when $\Re\nu$ is sufficiently large, we can deform the contour around the branch cut, and obtain

$$F(\nu)=\frac{1}{2\pi i}\int_{+\infty}^{(1+)} {}_2F_1\left({a,b \atop c};s\right)\frac{ds}{s^{\nu+1}}, \tag{6.8.143}$$

where the contour cuts the real axis inside the interval $(0,1)$. Along the lower part of the cut we have $\mathrm{ph}(1-s)=\pi$, and along the upper side it is $-\pi$.

After this, we replace the hypergeometric function in (6.8.137) by using the connection formula

$$
\begin{aligned}
{}_2F_1\left({a,b \atop c};s\right) &= C\,{}_2F_1\left({a,b \atop a+b-c+1};1-s\right) \\
&\quad + D(1-s)^{c-a-b}\,{}_2F_1\left({c-a,c-b \atop c-a-b+1};1-s\right),
\end{aligned}
\tag{6.8.144}
$$

where $|\mathrm{ph}(1-s)| < \pi$ and

$$
C = \frac{\Gamma(c)\Gamma(c-a-b)}{\Gamma(c-a)\Gamma(c-b)}, \qquad D = \frac{\Gamma(c)\Gamma(a+b-c)}{\Gamma(a)\Gamma(b)}. \tag{6.8.145}
$$

We observe that the first term on the right-hand side is analytic in the half-plane $\Re s > 0$; hence, it does not give a contribution to $F(\nu)$. It follows that

$$
F(\nu) = D\frac{1}{2\pi i}\int_{+\infty}^{(1+)} (1-s)^{c-a-b}\,{}_2F_1\left({c-a,c-b \atop c-a-b+1};1-s\right)\frac{ds}{s^{\nu+1}}. \tag{6.8.146}
$$

The substitution $s = e^t$ gives a new loop integral, and we can apply Watson's lemma for this type of integral (see §2.2). In this way we will obtain an asymptotic expansion in negative powers of ν. This is the approach that we will use in Chapter 15 to obtain expansions of coefficients of a power series.

In the present case we expand the hypergeometric function in powers of $(1-s)$ and use the integral (see Temme (1996a, p. 75))

$$
\frac{1}{2\pi i}\int_{c-i\infty}^{c+i\infty} s^{-p}(1-s)^{-q-1}\,ds = \frac{\Gamma(p+q)}{\Gamma(p)\Gamma(q+1)}, \tag{6.8.147}
$$

where $\Re(p+q) > 0$ and $0 < c < 1$. After a few manipulations with the gamma functions we obtain the asymptotic expansion given in (6.8.136).

Chapter 7

Incomplete gamma functions

In Tricomi (1950) we find the remark "Seit einiger Zeit pflege ich die unvollständige Gammafunktion $\gamma(\alpha, x)$ das *Aschenbrödel der Funktionen* zu nennen".[1] In this and other chapters we take care of Tricomi's complaint.

Here we first consider the large-z expansion of $\Gamma(a, z)$ with a fixed. In later sections we assume that a may be large, but the ratio $\lambda = z/a$ should be bounded away from unity. In that case the expansions will be in terms of elementary functions. These results are in terms of elementary functions and based on Temme (1994a). A complete uniform expansion (also valid near $\lambda = 1$) will be considered in Chapter 37, together with a similar approach for $\gamma(-a, -z)$ and $\Gamma(-a, -z)$. In the uniform expansions the complementary error function is used as main approximant.

7.1 Integral representations

The incomplete gamma functions are defined by (see Paris (2010, §8.2(i)))

$$\gamma(a, z) = \int_0^z e^{-t} t^{a-1}\, dt, \quad \Gamma(a, z) = \int_z^\infty e^{-t} t^{a-1}\, dt, \tag{7.1.1}$$

where in the first integral $\Re a > 0$ and in both integrals $|\mathrm{ph}\, z| < \pi$. Together they form the gamma function:

$$\gamma(a, z) + \Gamma(a, z) = \Gamma(a). \tag{7.1.2}$$

In probability theory the ratios

$$P(a, z) = \frac{\gamma(a, z)}{\Gamma(a)}, \quad Q(a, z) = \frac{\Gamma(a, z)}{\Gamma(a)} \tag{7.1.3}$$

are more common. We also use the function

$$\gamma^*(a, z) = \frac{z^{-a}}{\Gamma(a)} \gamma(a, z) = \frac{1}{\Gamma(a)} \int_0^1 e^{-zu} u^{a-1}\, du, \tag{7.1.4}$$

which is an analytic function of z and a.

[1] For some time, I used to call the incomplete gamma function $\gamma(\alpha, x)$ the *Cinderella of special functions.*

We can write the integral for $\Gamma(a,z)$ in the form

$$\Gamma(a,z) = z^a e^{-z} \int_0^\infty (1+t)^{a-1} e^{-zt}\,dt, \quad \Re z > 0. \tag{7.1.5}$$

Another useful integral is

$$\Gamma(a,z) = \frac{e^{-z}}{\Gamma(1-a)} \int_0^\infty \frac{e^{-zt}\,t^{-a}}{1+t}\,dt, \quad \Re a < 1, \quad \Re z > 0, \tag{7.1.6}$$

which can be verified by differentiating the right-hand side with respect to z.

In §37.2 we give representations of $P(a,z)$ and $Q(a,z)$ in the form of loop integrals, which can be obtained from (7.1.6), see also Remark 37.1. In this way we can remove conditions on a.

7.2 $\Gamma(a,z)$: Asymptotic expansion for $x \gg a$

We use the integral in (7.1.5) and expand

$$f(t) = (1+t)^{a-1} = \sum_{k=0}^\infty \binom{a-1}{k} t^k, \quad |t| < 1. \tag{7.2.7}$$

For this function $\alpha = -\pi$, $\beta = \pi$ (see Theorem 2.2 and (2.1.5)), and we obtain

$$\Gamma(a,z) \sim z^{a-1} e^{-z} \sum_{k=0}^\infty \binom{a-1}{k} \frac{k!}{z^k}, \quad z \to \infty, \quad |\mathrm{ph}\, z| \le \tfrac{3}{2}\pi - \delta. \tag{7.2.8}$$

The expansion in (7.2.8) is usually written in the form

$$\Gamma(a,z) \sim z^{a-1} e^{-z} \sum_{k=0}^\infty \frac{(-1)^k (1-a)_k}{z^k}, \quad |\mathrm{ph}\, z| \le \tfrac{3}{2}\pi - \delta. \tag{7.2.9}$$

Observe that the exponential integral considered in §1.3 is a special case. We have $E_1(z) = \Gamma(0,z)$, and more general (see (5.4.9)) $E_p(z) = z^{p-1}\Gamma(1-p,z)$.

The expansion in (7.2.9) can be obtained in several other ways, for example, by repeatedly using the recurrence relation

$$\begin{aligned}\Gamma(a,z) &= z^{a-1}e^{-z} + (a-1)\Gamma(a-1,z) \\ &= z^{a-1}e^{-z}\left(1 + \frac{a-1}{z}\right) + (a-1)(a-2)\Gamma(a-2,z).\end{aligned} \tag{7.2.10}$$

This gives the expansion with explicit remainder

$$\Gamma(a,z) = z^{a-1}e^{-z}\left(\sum_{n=0}^{N-1} \frac{(-1)^n(1-a)_n}{z^n} + \frac{(-1)^N(1-a)_N}{z^N} C_N(a,z)\right), \tag{7.2.11}$$

where $N = 0, 1, 2, \ldots$ and

$$C_N(a,z) = z^{N+1-a} e^z \Gamma(a-N,z) = z \int_0^\infty (1+t)^{a-N-1} e^{-zt}\,dt. \tag{7.2.12}$$

When a and z are positive and $N \ge a-1$ we have the upper bound $C_N(a,z) \le 1$.

7.3 $\gamma(a,x)$: Asymptotic expansion for $a > x$

The expansion

$$\gamma(a,z) = \frac{z^a e^{-z}}{a} \sum_{n=0}^{\infty} \frac{z^n}{(a+1)_n}, \tag{7.3.13}$$

can be viewed as an asymptotic expansion, and it is very useful for large a with fixed z. In this section we derive an expansion for positive a and $z = x$ that will be valid for large a and x such that $\lambda = x/a$ satisfies $0 \le \lambda \le \lambda_0 < 1$, where λ_0 is a fixed number in $(0,1)$.

We use the integral representation

$$\gamma(a,x) = x^a e^{-x} \int_0^1 e^{-a\phi(t)} \frac{dt}{1-t}, \tag{7.3.14}$$

which follows from (7.1.1), with

$$\phi(t) = -\lambda t - \ln(1-t), \quad \lambda = \frac{x}{a}. \tag{7.3.15}$$

Because $\phi(t)$ is monotonic in $(0,1)$ we can transform $u = \phi(t)$, giving

$$\gamma(a,x) = x^a e^{-x} \int_0^{\infty} e^{-au} f(u)\, du, \tag{7.3.16}$$

where

$$f(u) = \frac{1}{(1-t)\phi'(t)} = \frac{1}{\lambda t + 1 - \lambda}. \tag{7.3.17}$$

By expanding $f(u) = \sum_{k=0}^{\infty} a_k(\lambda) u^k / k!$ we obtain

$$\gamma(a,x) \sim \frac{x^a e^{-x}}{a} \sum_{k=0}^{\infty} \frac{a_k(\lambda)}{a^k}, \tag{7.3.18}$$

where

$$a_0(\lambda) = \frac{1}{1-\lambda}, \quad a_k(\lambda) = -\frac{\lambda}{1-\lambda} \frac{d}{d\lambda} a_{k-1}(\lambda), \quad k \ge 1. \tag{7.3.19}$$

This relation easily follows from differentiating the expansion in (7.3.18) with respect to x. We have

$$\begin{aligned}
a_0(\lambda) &= \frac{1}{1-\lambda}, \\
a_1(\lambda) &= -\frac{\lambda}{(1-\lambda)^3}, \\
a_2(\lambda) &= \frac{\lambda(1+2\lambda)}{(1-\lambda)^5}, \\
a_3(\lambda) &= -\frac{\lambda(1+8\lambda+6\lambda^2)}{(1-\lambda)^7}, \\
a_4(\lambda) &= \frac{\lambda(1+22\lambda+58\lambda^2+24\lambda^3)}{(1-\lambda)^9}, \\
a_5(\lambda) &= -\frac{\lambda(1+52\lambda+328\lambda^2+444\lambda^3+120\lambda^4)}{(1-\lambda)^{11}}.
\end{aligned} \tag{7.3.20}$$

7.3.1 *Singularity of the integrand*

Applying Watson's lemma to obtain expansion (7.3.18), we need to know more about $f(u)$. Clearly, when $\lambda \uparrow 1$ a singularity of f approaches the origin in the u-plane. In §2.7.2 we have mentioned how to deal with special forms of singularities. In the present case it is not clear at first sight which type of singularity arises, and we give a few details.

The function f is singular at the value u_0 that corresponds to $t = t_0$, where $t_0 = 1 - 1/\lambda$, which satisfies $\phi'(t_0) = 0$. If $\lambda \in (0,1)$ the singularity t_0 is negative. The point u_0 is given by $u_0 = \phi(t_0) = -\lambda + 1 + \ln\lambda$, which is negative as well if $\lambda \in (0,1)$, and it approaches the origin when $\lambda \to 1$.

We can give more details about the singularity of $f(u)$ at $u = u_0$ by first expanding t at this point. When we expand $\phi(t)$ in powers of $(t - t_0)$, we have

$$u = u_0 + \tfrac{1}{2}\phi''(t_0)(t-t_0)^2 + \mathcal{O}\left((t-t_0)^3\right), \quad \phi''(t_0) = \lambda^2, \tag{7.3.21}$$

because $\phi'(t_0) = 0$.

We invert the relation $u = \phi(t)$ by substituting a complete expansion

$$t = t_0 + \sum_{k=1}^{\infty} c_k (u-u_0)^{k/2}, \quad u_0 = -\lambda + 1 + \ln\lambda. \tag{7.3.22}$$

Comparison of equal powers of $(u - u_0)$ gives

$$c_1 = \frac{\sqrt{2}}{\lambda}, \quad c_2 = -\frac{2}{3\lambda}, \quad c_3 = \frac{\sqrt{2}}{18\lambda}, \quad c_4 = \frac{2}{135\lambda}. \tag{7.3.23}$$

By using the expansion of t in $f(u)$ of (7.3.17) it follows that

$$f(u) = \frac{1}{\sqrt{2(u-u_0)}} \sum_{k=0}^{\infty} d_k (u-u_0)^{k/2}, \tag{7.3.24}$$

where

$$d_0 = 1, \quad d_1 = \tfrac{1}{3}\sqrt{2}, \quad d_2 = \tfrac{1}{6}, \quad d_3 = \tfrac{4}{135}\sqrt{2}. \tag{7.3.25}$$

This describes the singularity of the integrand in (7.3.16). The square root singularity is quite common in transformations like $u = \phi(t)$ at a point where $\phi'(t)$ vanishes.

It follows that we can write

$$f(u) = f_1(u) + \frac{1}{\sqrt{u-u_0}} f_2(u), \tag{7.3.26}$$

where f_1 and f_2 are analytic at u_0 and in a fixed neighborhood of the origin for all $\lambda \in (0,1]$. Using Maclaurin expansions $f_j(u) = \sum_{k=0}^{\infty} a_k^{(j)}(\lambda) u^k$ we obtain from (7.3.16)

$$\gamma(a,x) \sim \frac{x^a e^{-x}}{a} \sum_{k=0}^{\infty} \frac{a_k^{(1)}(\lambda)}{a^k} + x^a e^{-x} \sum_{k=0}^{\infty} a_k^{(2)}(\lambda)\Phi_k, \tag{7.3.27}$$

where

$$\Phi_k = \int_0^\infty e^{-au} u^k \frac{du}{\sqrt{u-u_0}}. \tag{7.3.28}$$

These functions can be expressed in terms of a Kummer function (see (10.1.5))

$$\Phi_k = (-u_0)^{k+\frac{1}{2}} k!\, U\left(k+1, k+\tfrac{3}{2}, -au_0\right), \tag{7.3.29}$$

with first element (see Temme (2010c, Eq. 7.7.4))

$$\Phi_0 = \sqrt{\frac{\pi}{a}} e^{-au_0} \operatorname{erfc}\left(\sqrt{-au_0}\right). \tag{7.3.30}$$

Other Φ_k can be obtained by differentiating the relation in (7.3.28) with respect to a, or by using recurrence relations of the Kummer functions.

It will not be an easy job to obtain the coefficients $a_k^{(j)}(\lambda)$ in the series in (7.3.27), and the present discussion is only for showing that singularities of functions (like that of f in (7.3.17)) obtained after transformations are not always presented on a plate.

Luckily, there are more elegant ways to obtain expansions of the incomplete gamma functions that are valid when $\lambda \sim 1$, and for this we refer to Chapter 37.

Problem 7.1. *Verify that the coefficients a_k of the expansion in* (7.3.18) *can also be obtained by using integration by parts, starting with* (7.3.14). *Let*

$$g_0(t) = \frac{1}{1-t}, \quad g_k(t) = \frac{d}{dt}\frac{g_{k-1}(t)}{\phi'(t)}, \quad k = 1, 2, 3, \ldots, \tag{7.3.31}$$

where $\phi(t)$ is given in (7.3.15). *Show that*

$$a_k(\lambda) = \frac{g_k(0)}{\phi'(0)}, \quad k = 0, 1, 2, \ldots. \tag{7.3.32}$$

In this way obtain an expansion with remainder:

$$\gamma(a,x) = \frac{x^a e^{-x}}{a}\left(\sum_{k=0}^{K-1} \frac{a_k(\lambda)}{a^k} + \frac{R_K(a,x)}{a^K}\right), \tag{7.3.33}$$

where

$$R_K(a,x) = \int_0^1 e^{-a\phi(t)} g_K(t)\, dt, \quad K = 0, 1, 2, \ldots. \tag{7.3.34}$$

Observe that the recurrence relation in (7.3.19) *does not follow from the recursion for the $g_k(t)$. However, see also Remark 7.1.* ♡

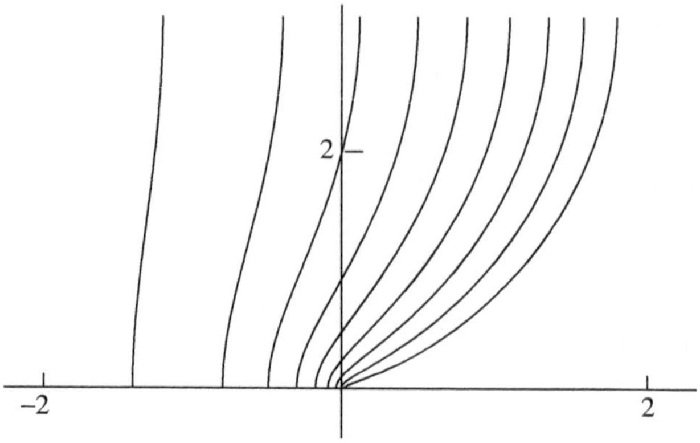

Fig. 7.1 Graphs of the curves of u_0 for $\lambda = re^{i\theta}$ for several values of r and θ.

7.3.2 *More details on the transformation* $u = \phi(t)$

The expansion in (7.3.18) is valid for complex values of a and x. To describe some details it is convenient to assume that we consider $\gamma(a, \lambda a)$, where λ does not depend on a. In that case we can view the singular point $u_0 = -\lambda + 1 + \ln \lambda$ as independent of a as well.

In Figure 7.1 the curves show u_0 for $\lambda = re^{i\theta}$. On each curve, r is fixed and $\theta \in [0, \pi]$. We have taken $r = n/10$, $n = 1, 2, 3, \ldots, 9$, and the curves from left to right are with increasing values of n. For each r we have $\Im u_0 = \pi$ if $\theta = \pi$. For $\theta \in [-\pi, 0]$ the curves follow from complex conjugation.

From the location of the complex singular point u_0 it is possible to define the sector for complex a where the expansion in (7.3.18) is valid for complex a. From Theorem 2.2 we conclude that this sector follows from (see (2.1.6))

$$-\beta - \tfrac{1}{2}\pi + \delta \le \operatorname{ph} a \le \beta + \tfrac{1}{2}\pi - \delta, \tag{7.3.35}$$

where $\beta = |\operatorname{ph} u_0|$.

Next we verify if indeed no other singular points can be found for $-2\pi \le \operatorname{ph} t \le 2\pi$. The value u_0 corresponds to the zero $t_0 = 1 - 1/\lambda$ of the denominator of $f(u)$ in (7.3.17). It seems quite obvious to conclude that the corresponding u_0 is the only singularity of $f(u)$. However, when we use a transformation like $u = \phi(t)$, in which a logarithm is involved, one may expect other singular points. In fact, all points $u_n = u_0 + 2n\pi i$, $n \in \mathbb{Z}$ are singular points, and these correspond to values t_n in the t-plane satisfying

$$1 - t_k = \frac{1}{\lambda} e^{2\pi i k}, \quad k \in \mathbb{Z}. \tag{7.3.36}$$

For each k then we have $\phi'(t_k) = 0$, and the logarithm in $\phi(t)$ gives different singular points $u_k = 1 - \lambda + \ln \lambda - 2\pi i k$.

In Figure 7.2 we give details of the transformation $u = \phi(t)$ that is used to obtain (7.3.16). We show the images of the half-lines that run from $t = 1$ to infinity. We

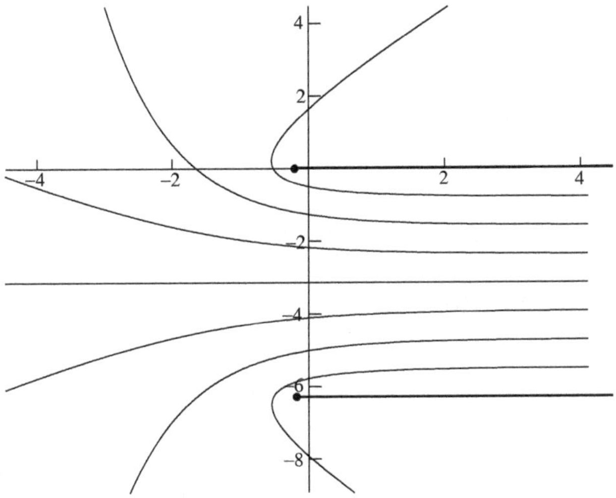

Fig. 7.2 Graphs in the u-plane of the images of the half-lines $t = 1 - re^{i\theta}$ under the transformation $u = \phi(t)$, for $r > 0$, $\theta = \frac{1}{4}\pi n$, $n = 0, 1, 2, \ldots, 8$, and $\lambda = \frac{1}{2}$.

take $\lambda = \frac{1}{2}$ and we parametrize the lines by writing $1 - t = re^{i\theta}$, and each line has a fixed value of θ, with $\theta = \frac{1}{4}\pi n\pi$, $n = 0, 1, 2, \ldots, 8$.

For $n = 0$ the line in the t-plane runs from 1 to $t_0 = -1$, and the image starts at $+\infty$ to $u_0 = \frac{1}{2} - \ln 2$ (indicated by a black dot on the negative axis); then t goes from t_0 to $-\infty$, and the image u_0 of this part runs from u_0 back to $+\infty$. Because t_0 is a singular point, the part $[u_0, +\infty)$ is run twice. See also the square root singularity in the local expansion in (7.3.22).

The same happens for $n = 8$, with $\theta = 2\pi$. The image is a horizontal half-line that runs from the singular point $u_{-1} = u_0 - 2\pi i$ to $+\infty - 2\pi i$. Between these two special cases the images are shown for $n = 1, 2, 3, \ldots, 7$. We see that between the values $\theta = 0$ and $\theta = 2\pi$ no singular points are observed; similarly for $\theta \in [-2\pi, 0]$.

7.4 $\Gamma(a, x)$: Asymptotic expansion for $x > a$

The well-known expansion in (7.2.9) cannot be used when a also becomes large, say of order $\mathcal{O}(x)$. We derive an expansion similar to the one given for $\gamma(a.x)$ in §7.3. In this case we take x as the large parameter, and a may be large as well, but the ratio $\alpha = a/x$ should be bounded away from unity: $\alpha \le \alpha_0 < 1$. We even allow negative values of α.

We use (7.1.5) and write

$$\Gamma(a, x) = x^a e^{-x} \int_0^\infty e^{-x\psi(t)} \frac{dt}{1+t}, \tag{7.4.37}$$

where

$$\psi(t) = t - \alpha \ln(1+t), \quad \alpha = \frac{a}{x}. \tag{7.4.38}$$

Because $\psi(t)$ is monotonic in $(0,\infty)$ (recall that $\alpha<1$), we can transform $v=\psi(t)$, giving

$$\Gamma(a,x)=x^a e^{-x}\int_0^\infty e^{-xv}\,g(v)\,dv, \tag{7.4.39}$$

where

$$g(v)=\frac{1}{1+t}\frac{dt}{dv}=\frac{1}{(1+t)\psi'(t)}=\frac{1}{1+t-\alpha}. \tag{7.4.40}$$

Expanding $g(v)=\sum_{k=0}^\infty b_k(\alpha)v^k/k!$, we obtain

$$\Gamma(a,x)\sim x^{a-1}e^{-x}\sum_{k=0}^\infty \frac{b_k(\alpha)}{x^k}. \tag{7.4.41}$$

By comparing the transformations to obtain (7.3.16) and (7.4.39) it is clear that the coefficients $a_k(\lambda)$ and $b_k(\alpha)$ are closely related. We have

$$b_k(\alpha)=-\lambda^{k+1}a_k(\lambda),\quad k\ge 0,\quad \lambda=1/\alpha. \tag{7.4.42}$$

By using the relation in (7.3.19), it follows that the coefficients satisfy the recurrence relation

$$b_k(\alpha)=\frac{1}{\alpha-1}\left(kb_{k-1}(\alpha)+\alpha b'_{k-1}(\alpha)\right),\quad b_0=\frac{1}{1-\alpha}. \tag{7.4.43}$$

The expansion in (7.4.41) is valid for complex values of a and x. Similar as in the case of $\gamma(a,x)$ it is convenient to assume that we consider $\Gamma(\alpha z,z)$ where α does not depend on z. The singular point v_0 of the transformation used to obtain (7.4.39) corresponds to $t_0=\alpha-1$ and is given by $v_0=\alpha-1-\alpha\ln\alpha$. Hence, the expansion is valid in the sector given by

$$-\beta-\tfrac12\pi+\delta\le \operatorname{ph} z\le \beta+\tfrac12\pi-\delta, \tag{7.4.44}$$

where $\beta=|\operatorname{ph} v_0|$.

Remark 7.1. The recursion relation in (7.4.43) follows from one for $a_k(\lambda)$ in (7.3.19), which is easily obtained by differentiating the expansion in (7.3.18) and the simple differential equation of $\gamma(a,x)$; see also a different approach in Problem 7.1. The availability of a differential equation is one of the benefits when deriving recurrence relations for the coefficients of asymptotic expansions. In the present case we can find the recurrence relation for $b_k(\alpha)$ without referring to this advantage. Let $B_k(\alpha)=b_k(\alpha)/k!$. Then, by (7.4.40),

$$B_k(\alpha)=\frac{1}{2\pi i}\int_{\mathcal{C}_v} g(v)\frac{dv}{v^{k+1}}=\frac{1}{2\pi i}\int_{\mathcal{C}_t}\frac{dt}{(1+t)\bigl(t-\alpha\ln(1+t)\bigr)^{k+1}}, \tag{7.4.45}$$

where $\mathcal{C}_v$ and $\mathcal{C}_t$ are small circles around the origin in the v-plane and t-plane, respectively. Next,

$$B'_{k-1}(\alpha)=\frac{k}{2\pi i}\int_{\mathcal{C}_t}\frac{\ln(1+t)}{(1+t)\bigl(t-\alpha\ln(1+t)\bigr)^{k+1}}\,dt. \tag{7.4.46}$$

Hence,

$$\begin{aligned} B_{k-1}(\alpha) &= \frac{1}{2\pi i}\int_{\mathcal{C}_t} \frac{t-\alpha\ln(1+t)}{(1+t)\bigl(t-\alpha\ln(1+t)\bigr)^{k+1}}\,dt \\ &= -\frac{\alpha}{k}B'_{k-1}(\alpha) + (\alpha-1)B_k(\alpha) + \\ &\quad \frac{1}{2\pi i}\int_{\mathcal{C}_t} \frac{t+1-\alpha}{(1+t)\bigl(t-\alpha\ln(1+t)\bigr)^{k+1}}\,dt. \end{aligned} \tag{7.4.47}$$

The integral can be written in the form

$$\frac{-1}{2\pi i k}\int_{\mathcal{C}_t} d\bigl(t-\alpha\ln(1+t)\bigr)^{-k}. \tag{7.4.48}$$

We conclude that the integral vanishes, and (7.4.43) follows. △

Remark 7.2. In Chapter 37 we will derive uniform expansions of the ratios $P(a,x)$ and $Q(a,x)$ defined in (7.1.3). There it will become clear why the expansions of $\gamma(a,x)$ and $\Gamma(a,x)$ in this and the previous section are so similar, with almost the same coefficients. △

Chapter 8

The Airy functions

Airy functions are special cases of Bessel functions of order $\pm\frac{1}{3}$ and are named after G. B. Airy (1838), a British astronomer, who used them when studying rainbow phenomena. They occur in many other problems from physics, for example as solutions to boundary value problems in quantum mechanics and electromagnetic theory.

We mention a few properties of the Airy functions, for more details we refer to Olver (2010). Airy functions are solutions of Airy's differential equation

$$\frac{d^2w}{dz^2} = zw. \tag{8.0.1}$$

Two independent solutions are denoted by $\mathrm{Ai}(z)$ and $\mathrm{Bi}(z)$, which are analytic functions and real when z is real. They are oscillatory for $z < 0$ and decrease ($\mathrm{Ai}(z)$) or increase ($\mathrm{Bi}(z)$) exponentially fast for $z > 0$, see Figure 8.1. The Wronskian relation is

$$\mathrm{Ai}(z)\mathrm{Bi}'(z) - \mathrm{Ai}'(z)\mathrm{Bi}(z) = \frac{1}{\pi}. \tag{8.0.2}$$

For real $z = x$ we have the integral representations

$$\begin{aligned}
\mathrm{Ai}(x) &= \frac{1}{\pi}\int_0^\infty \cos\left(\tfrac{1}{3}t^3 + xt\right)\, dt,\\
\mathrm{Bi}(x) &= \frac{1}{\pi}\int_0^\infty \sin\left(\tfrac{1}{3}t^3 + xt\right)\, dt + \frac{1}{\pi}\int_0^\infty e^{-\frac{1}{3}t^3 + xt}\, dt.
\end{aligned} \tag{8.0.3}$$

These integrals are not suitable to obtain asymptotic expansions, and we use contour integrals for this purpose.

The change in behavior of the solutions of the equation in (8.0.1) as z crosses the origin (from oscillatory to exponential behavior) can be interpreted by observing that $z = 0$ is a turning point for this equation. Airy's equation is the simplest second-order linear differential equation showing such a turning point (at $z = 0$). We will see Airy functions acting as main approximants in uniform asymptotic expansions, cf. Chapter 23, when integrals have two coalescing saddle points. This feature corresponds to the turning point problems in differential equations.

In §4.7 we have become acquainted with some asymptotic details of the Airy function $\mathrm{Ai}(z)$, and in this chapter we give more information, also for $\mathrm{Bi}(z)$. In

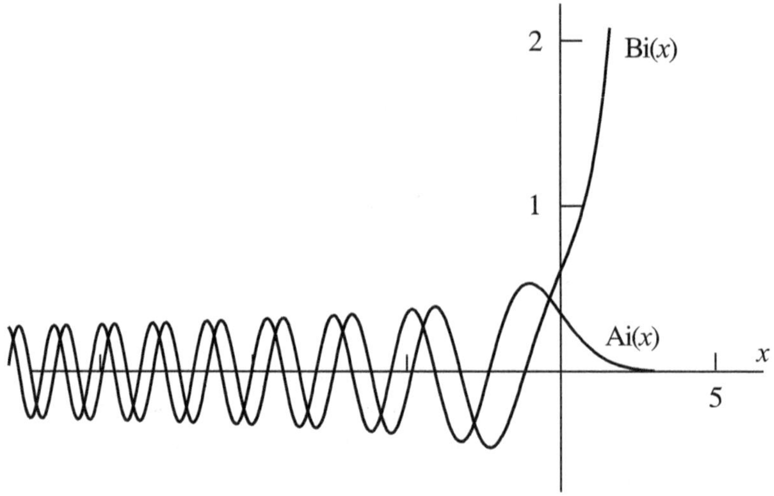

Fig. 8.1 Graphs of the Airy functions $\mathrm{Ai}(x)$ and $\mathrm{Bi}(x)$ on the real line.

addition, we derive expansions of functions related to integrals of Airy functions. For information on error bounds for the remainders in the asymptotic expansions and on exponentially-improved expansions we refer to Olver (2010, §9.7).

8.1 Expansions of $\mathrm{Ai}(z), \mathrm{Bi}(z)$

A standard integral representation for $\mathrm{Ai}(z)$ is

$$\mathrm{Ai}(z) = \frac{1}{2\pi i}\int_{\mathcal{C}_1} e^{-zt+\frac{1}{3}t^3}\,dt. \tag{8.1.4}$$

where the contour $\mathcal{C}_1$ starts at the point at infinity with $\mathrm{ph}\,t = -\frac{1}{3}\pi$ and ends at the point at infinity with $\mathrm{ph}\,t = \frac{1}{3}\pi$. It is the contour on the right in Figure 8.2.

This is clearly an example where a contour runs from one valley into another valley, and to obtain the asymptotic behavior we will find the saddle point where this should happen.

8.1.1 *Transforming the saddle point contour*

Initially we consider large positive values of z. To get a saddle point independent of z we transform in the integral $t = w\sqrt{z}$, and obtain

$$\mathrm{Ai}(z) = \frac{\sqrt{z}}{2\pi i}\int_{\mathcal{C}} e^{-\zeta\phi(w)}\,dw, \tag{8.1.5}$$

where the contour $\mathcal{C}$ is similar to that in (8.1.4) and

$$\zeta = \tfrac{2}{3}z\sqrt{z}, \quad \phi(w) = \tfrac{3}{2}\left(w - \tfrac{1}{3}w^3\right). \tag{8.1.6}$$

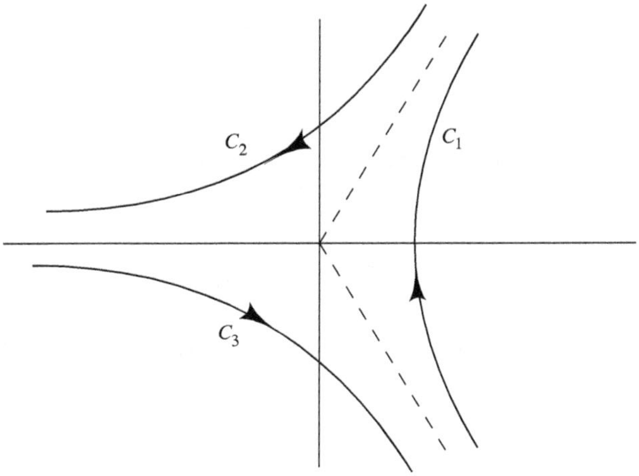

Fig. 8.2 Three contours of integration for the Airy integrals; C_1 is used in (8.1.4).

The saddle points follow from $\phi'(w) = \frac{3}{2}\left(1 - w^2\right) = 0$, giving $w = \pm 1$ with

$$\phi(-1) = -1, \quad \phi(1) = 1, \quad \Im\phi(\pm 1) = 0. \tag{8.1.7}$$

Furthermore, writing $w = u + iv$, we have

$$\Im\phi(w) = \tfrac{3}{2}v\left(1 - u^2 + \tfrac{1}{3}v^2\right). \tag{8.1.8}$$

The equation $\Im\phi(w) = 0$ is solved by

$$v = 0, \quad \text{or} \quad 1 - u^2 + \tfrac{1}{3}v^2 = 0. \tag{8.1.9}$$

The first solution gives for the saddle point at $w = 1$ a path of steepest ascent, the second one defines a hyperbola that runs into the valleys of the integrand and that passes through the saddle point at $w = 1$. It follows that we can deform the contour $\mathcal{C}$ into the hyperbola. On the hyperbola we have $\Re\phi(w) = -3u + 4u^3$.

In Remark 4.3 we have explained why we do not use the saddle point at $w = -1$, although the value of the integrand of the integral in (8.1.5) at $w = -1$ is much larger than the value at $w = 1$.

Writing

$$\mathrm{Ai}(z) = \frac{\sqrt{z}e^{-\zeta}}{2\pi i}\int_{\mathcal{C}} e^{-\zeta(\phi(w)-\phi(1))}\,dw, \tag{8.1.10}$$

integrating on the hyperbola with respect to v, observing that $dw = (du/dv + i)dv$, and that du/dv is an odd function of v, we finally obtain the real representation

$$\mathrm{Ai}(z) = \frac{\sqrt{z}e^{-\zeta}}{2\pi}\int_{-\infty}^{\infty} e^{-\zeta p(v)}\,dv, \tag{8.1.11}$$

where

$$p(v) = (u-1)(2u+1)^2, \quad u = \sqrt{1 + \tfrac{1}{3}v^2}. \tag{8.1.12}$$

We now can apply Laplace's method, and make the substitution

$$s^2 = p(v), \quad \operatorname{sign}(s) = \operatorname{sign}(v), \tag{8.1.13}$$

and obtain

$$\operatorname{Ai}(z) = \frac{\sqrt{z}e^{-\zeta}}{2\pi} \int_{-\infty}^{\infty} e^{-\zeta s^2} f(s)\, ds, \tag{8.1.14}$$

where

$$f(s) = \frac{dv}{ds} = \frac{6us}{v(9+4v^2)}. \tag{8.1.15}$$

We expand $f(s) = \sum_{n=0}^{\infty} c_n s^n$, and eventually obtain the well-known asymptotic expansion:

$$\operatorname{Ai}(z) \sim \frac{e^{-\zeta}}{2\sqrt{\pi}\, z^{\frac14}} \sum_{n=0}^{\infty} (-1)^n \frac{u_n}{\zeta^n}, \quad u_n = (-1)^n \left(\tfrac12\right)_n \frac{c_{2n}}{c_0}, \tag{8.1.16}$$

where ζ is defined in (8.1.6), $u_0 = 1$, and for $n \ge 1$

$$u_n = \frac{2^n}{3^{3n}\,(2n)!} \left(\tfrac12\right)_{3n} = \frac{(2n+1)(2n+3)(2n+5)\cdots(6n-1)}{(216)^n\, n!}. \tag{8.1.17}$$

In §8.3 we explain how to derive this representation of u_n.

The function $f(s)$ is a bounded function for real s and its singularities occur on the imaginary axis. To verify this, observe that ds/dv vanishes if $v^2 = -\frac94$, that is, if $u = \frac12$, giving $s^2 = -2$. As in Theorem 2.2 or Lemma 3.1 we conclude that the expansion holds for $|\operatorname{ph}\zeta| < \frac32\pi$. Therefore, the expansion holds uniformly inside the sector $|\operatorname{ph} z| \le \pi - \delta < \pi$.

The result for the Airy function $\operatorname{Bi}(z)$ follows by using the connection formula

$$\operatorname{Bi}(z) = e^{-\pi i/6} \operatorname{Ai}\left(z e^{-2\pi i/3}\right) + e^{\pi i/6} \operatorname{Ai}\left(z e^{2\pi i/3}\right). \tag{8.1.18}$$

In this way the following expansion can be obtained:

$$\operatorname{Bi}(z) \sim \pi^{-\frac12} z^{-\frac14} e^{\zeta} \sum_{n=0}^{\infty} \frac{u_n}{\zeta^n}, \tag{8.1.19}$$

and this time the expansion is valid only in the sector $|\operatorname{ph} z| \le \frac13\pi - \delta < \frac13\pi$.

Problem 8.1. *By differentiating the expansions in* (8.1.16) *and* (8.1.19) *verify that the following expansions are valid for the derivatives of the Airy functions:*

$$\begin{aligned} \operatorname{Ai}'(z) &\sim -\frac{z^{\frac14} e^{-\zeta}}{2\sqrt{\pi}} \sum_{n=0}^{\infty} (-1)^n \frac{v_n}{\zeta^n}, \quad && |\operatorname{ph} z| \le \pi - \delta, \\ \operatorname{Bi}'(z) &\sim \frac{z^{\frac14} e^{-\zeta}}{\sqrt{\pi}} \sum_{n=0}^{\infty} \frac{v_n}{\zeta^n}, && |\operatorname{ph} z| \le \tfrac13\pi - \delta, \end{aligned} \tag{8.1.20}$$

where

$$v_0 = 1, \quad v_n = -\frac{6n+1}{6n-1} u_n, \quad n = 1, 2, 3, \ldots. \tag{8.1.21}$$

♡

8.2 Expansions of $\mathrm{Ai}(-z), \mathrm{Bi}(-z)$

From the obvious result

$$\frac{1}{2\pi i}\int_{\mathcal{C}} e^{-zt+\frac{1}{3}t^3}\,dt = 0, \tag{8.2.22}$$

where the contour $\mathcal{C}$ is formed by the three contours in Figure 8.2, that is, $\mathcal{C} = \mathcal{C}_1 \cup \mathcal{C}_2 \cup \mathcal{C}_3$, taking into account the directions of integration on the contours, we obtain

$$e^{-2\pi i/3}\mathrm{Ai}\left(ze^{-2\pi i/3}\right) + e^{2\pi i/3}\mathrm{Ai}\left(ze^{2\pi i/3}\right) + \mathrm{Ai}(z) = 0. \tag{8.2.23}$$

From this relation and (8.1.18), and observing that the asymptotic expansion (8.1.16) holds for $|\mathrm{ph}\, z| < \pi$, we can obtain expansions of $\mathrm{Ai}(z)$ and $\mathrm{Bi}(z)$ in a domain that contains the negative axis. That is, we can write

$$\begin{aligned}\mathrm{Ai}(-z) &= \frac{1}{\sqrt{\pi}\, z^{\frac14}}\left(\sin(\zeta+\tfrac14\pi)P(z) - \cos(\zeta+\tfrac14\pi)Q(z)\right),\\ \mathrm{Bi}(-z) &= \frac{1}{\sqrt{\pi}\, z^{\frac14}}\left(\cos(\zeta+\tfrac14\pi)P(z) + \sin(\zeta+\tfrac14\pi)Q(z)\right),\end{aligned} \tag{8.2.24}$$

where $P(z)$ and $Q(z)$ have the expansions

$$P(z) \sim \sum_{n=0}^{\infty}(-1)^n\frac{u_{2n}}{\zeta^{2n}}, \quad Q(z) \sim \sum_{n=0}^{\infty}(-1)^n\frac{u_{2n+1}}{\zeta^{2n+1}}, \tag{8.2.25}$$

as $z \to \infty$ in the sector $|\mathrm{ph}\, z| \le \frac23\pi - \delta < \frac23\pi$, with $\zeta = \frac23 z\sqrt{z}$.

Problem 8.2. *Verify the following expansions for the derivatives:*

$$\begin{aligned}\mathrm{Ai}'(-z) &= -\frac{z^{\frac14}}{\sqrt{\pi}}\left(\cos(\zeta+\tfrac14\pi)R(z) + \sin(\zeta+\tfrac14\pi)S(z)\right),\\ \mathrm{Bi}'(-z) &= \frac{z^{\frac14}}{\sqrt{\pi}}\left(\sin(\zeta+\tfrac14\pi)R(z) - \cos(\zeta+\tfrac14\pi)S(z)\right),\end{aligned} \tag{8.2.26}$$

where

$$R(z) \sim \sum_{n=0}^{\infty}(-1)^n\frac{v_{2n}}{\zeta^{2n}}, \quad S(z) \sim \sum_{n=0}^{\infty}(-1)^n\frac{v_{2n+1}}{\zeta^{2n+1}}, \quad |\mathrm{ph}\, z| \le \tfrac23\pi - \delta, \tag{8.2.27}$$

with v_n defined in (8.1.21). *By using the Wronskian in* (8.0.2) *verify the exact relation*

$$P(z)R(z) + Q(z)S(z) = 1. \tag{8.2.28}$$

♡

Problem 8.3. *Give the saddle point analysis of the Airy function integral defined in* (8.1.4) *for the case that* $z < 0$. *Take as starting point*

$$\mathrm{Ai}(-z) = \frac{1}{2\pi i} \int_{\mathcal{C}_1} e^{zt+\frac{1}{3}t^3}\, dt \tag{8.2.29}$$

and replace $\mathcal{C}_1$ *by* $\mathcal{C}_2 \cup \mathcal{C}_3$ *(see Figure 8.2), traversed in the other direction. Substitute* $t = w\sqrt{z}$, *and obtain*

$$\mathrm{Ai}(-z) = \frac{\sqrt{z}}{2\pi i} \int_{-\mathcal{C}_2 \cup \mathcal{C}_3} e^{\lambda\phi(w)}\, dw, \tag{8.2.30}$$

cf. (8.1.5), *with* $\phi(w) = \frac{1}{3}w^3 + w$. *Show that the path of steepest descent through* $w = i$ *is given by*

$$u = (v-1)\sqrt{\frac{v+2}{3v}}, \qquad w = u + iv, \qquad v > 0, \tag{8.2.31}$$

and that there is a path through $w = -i$ *which is the reflection of this path with respect to the* u*-axis.* ♡

8.3 Two simple ways to obtain the coefficients

We have not explained how to find the explicit form of the coefficients as given in (8.1.17). Actually, the transformation (8.1.13) used to obtain (8.1.14), the straightforward inversion of the relation between s and v, and calculating the coefficients in the expansion of $f(s)$, does not give the explicit form in an easy way.

Knowing the structure of the expansion, we can use the Airy differential equation to obtain this form. In §8.4 we use a slightly different transformation, and we explain how to find the explicit form of the coefficients when we consider a more general Airy-type integral; see (8.4.39) and take $\alpha = 0$. But first we show two other methods to obtain explicit forms of the coefficients in an easy way by using integrals.

First, in (8.1.11) we put $v = \sqrt{3}\sinh(r/3)$. Then $u = \cosh(r/3)$ and

$$p(v) = 4u^3 - 3u - 1 = \cosh r - 1. \tag{8.3.32}$$

This gives

$$\mathrm{Ai}(z) = \frac{\sqrt{z}}{\pi\sqrt{3}} \int_0^\infty e^{-\zeta\cosh r} \cosh\left(\tfrac{1}{3}r\right)\, dr = \frac{\sqrt{z}}{\pi\sqrt{3}} K_{\frac{1}{3}}(\zeta), \tag{8.3.33}$$

which is the well-known representation in terms of the modified Bessel function. The asymptotic expansion of this function (see (9.1.3)) easily follows from one of the many integral representations of this function, for example, the one considered in (9.1.1).

A more direct way to obtain the coefficients in (8.1.16) follows from the representation

$$\mathrm{Ai}(z) = \frac{e^{-\zeta}}{\pi} \int_0^\infty e^{-\sqrt{z}v^2} \cos\left(\tfrac{1}{3}v^3\right)\, dv. \tag{8.3.34}$$

This can be obtained by integrating in (8.1.4) along the vertical through the saddle point by substituting $t = \sqrt{z} + iv$. The expansion in (8.1.16) can be obtained by expanding the cosine in powers of v. We cannot rotate the path because then the exponential functions in the cosine become dominant. The first version of Watson's lemma (Theorem 2.1) is applicable after substituting $v^2 = t$, and we find the domain $\operatorname{ph}\sqrt{z} \in (-\frac{1}{2}\pi, \frac{1}{2}\pi)$, so again, $|\operatorname{ph} z| \le \pi - \delta < \pi$.

This approach shows that a saddle point analysis with determination of steepest descent paths is not really needed. Often, a path through the saddle point that runs into the valleys is all that is needed.

8.4 A generalized form of the Airy function

We consider the Airy-type integrals

$$A_\alpha(z) = \frac{1}{2\pi i}\int_{\mathcal{C}} e^{\frac{1}{3}w^3 - zw} w^\alpha\, dw, \quad \alpha \in \mathbb{C}, \tag{8.4.35}$$

for large values of z. For general α we take a branch cut along the negative axis, and assume that $w^\alpha = 1$ if $w = 1$. The contour $\mathcal{C}$ is as in (8.1.4), and it should cross the real axis at a positive point.

For nonnegative integer values of α the functions $A_\alpha(z)$ are derivatives of the Airy function:

$$A_n(z) = (-1)^n \frac{d^n}{dz^n}\operatorname{Ai}(z), \quad n = 0, 1, 2, \ldots. \tag{8.4.36}$$

For negative integer values of α they can be viewed as antiderivatives. We have, for example,

$$A_{-1}(z) = \int_z^\infty \operatorname{Ai}(t)\, dt. \tag{8.4.37}$$

We will prove the following lemma.

Lemma 8.1. *The function $A_\alpha(z)$ defined in (8.4.35) has the following asymptotic expansion:*

$$A_\alpha(z) \sim \frac{z^{\alpha/2-1/4}e^{-\zeta}}{2\sqrt{\pi}} \sum_{n=0}^\infty (-1)^n \frac{u_n^{(\alpha)}}{\zeta^n}, \quad z \to \infty, \quad |\operatorname{ph} z| < \pi, \tag{8.4.38}$$

where $\zeta = \frac{2}{3}z^{\frac{3}{2}}$ and, in terms of the Gauss hypergeometric function,

$$u_n^{(\alpha)} = \frac{2^n}{3^{3n}\,(2n)!}\left(\tfrac{1}{2}\right)_{3n}\, {}_2F_1\left(\begin{matrix} -\alpha, -2n \\ \frac{1}{2} - 3n \end{matrix}; 3\right). \tag{8.4.39}$$

Compare this form with (8.1.17): $u_n^{(0)} = u_n$, and the hypergeometric function gives all the extras when $\alpha \ne 0$.

Proof. Assume initially that $z > 0$ and substitute $w = s\sqrt{z}$ in (8.4.35). Write the integral in the form

$$A_\alpha(z) = \frac{z^{\frac{1}{2}(\alpha+1)}e^{-\zeta}}{2\pi i}\int_{\infty e^{-\frac{1}{3}\pi i}}^{\infty e^{\frac{1}{3}\pi i}} e^{\frac{3}{2}\zeta\psi(s)}s^\alpha\, ds, \tag{8.4.40}$$

where

$$\psi(s) = \tfrac{1}{3}s^3 - s + \tfrac{2}{3} = \tfrac{1}{3}(s+2)(s-1)^2. \tag{8.4.41}$$

The substitution $t = (s-1)\sqrt{(s+2)/3}$ gives

$$A_\alpha(z) = \frac{z^{\frac{1}{2}(\alpha+1)}e^{-\zeta}}{2\pi i}\int_{-i\infty}^{i\infty} e^{\frac{3}{2}\zeta t^2}g(t)\, dt, \quad g(t) = s^\alpha\frac{ds}{dt}. \tag{8.4.42}$$

By expanding $g(t) = \sum_{n=0}^{\infty} g_n t^n$,

$$A_\alpha(z) \sim \frac{z^{\frac{1}{2}\alpha-\frac{1}{4}}e^{-\zeta}}{2\pi}\sum_{n=0}^{\infty}(-1)^n\left(\tfrac{2}{3}\right)^n g_{2n}\frac{\Gamma\left(n+\frac{1}{2}\right)}{\zeta^n}. \tag{8.4.43}$$

The coefficients g_n can be represented as a Cauchy integral:

$$g_n = \frac{1}{2\pi i}\int_C g(t)\frac{dt}{t^{n+1}}, \tag{8.4.44}$$

where C is a small circle around the origin. By integrating in the s-plane, it follows that

$$g_n = \frac{1}{2\pi i}\int_C s^\alpha\frac{ds}{((s+2)/3)^{(n+1)/2}(s-1)^{n+1}}, \tag{8.4.45}$$

where C is a circle around $s = 1$ with radius less than 1. By using Olde Daalhuis (2010b, Eq. 15.6.3) it follows that

$$g_n = \frac{(-1)^n\,\Gamma(\frac{3}{2}n+\frac{1}{2}-\alpha)}{3^n n!\,\Gamma(\frac{1}{2}n+\frac{1}{2}-\alpha)}\,{}_2F_1\left(\begin{matrix}-\alpha, -n\\ \frac{1}{2}n+\frac{1}{2}-\alpha\end{matrix}; -2\right). \tag{8.4.46}$$

This form of the Gauss hypergeometric function has to be modified when $\frac{1}{2}(n+1) = \alpha$, but using the connection formula (6.8.144), and taking into account that n is an integer, we can write

$${}_2F_1\left(\begin{matrix}-\alpha, -n\\ \frac{1}{2}n+\frac{1}{2}-\alpha\end{matrix}; -2\right) = \frac{\Gamma(\frac{3}{2}n+\frac{1}{2})\,\Gamma(\frac{1}{2}n+\frac{1}{2}-\alpha)}{\Gamma(\frac{1}{2}n+\frac{1}{2})\,\Gamma(\frac{3}{2}n+\frac{1}{2}-\alpha)}\,{}_2F_1\left(\begin{matrix}-\alpha, -n\\ \frac{1}{2}-\frac{3}{2}n\end{matrix}; 3\right). \tag{8.4.47}$$

This gives

$$g_{2n} = \frac{1}{3^{2n}(2n)!}\frac{\Gamma(3n+\frac{1}{2})}{\Gamma(n+\frac{1}{2})}\,{}_2F_1\left(\begin{matrix}-\alpha, -2n\\ \frac{1}{2}-3n\end{matrix}; 3\right), \tag{8.4.48}$$

and using this form in (8.4.43) we have proved the lemma. □

Chapter 9

Bessel functions: Large argument

For Bessel functions with large argument detailed information is available about estimates for the remainders in the large-z expansions; see Olver (1997, pp. 266–270). These bounds are obtained by using the differential equation of the Bessel functions.

In this chapter we give a few details about obtaining large-z expansions from integrals. We will see that, once we have the expansion of the modified Bessel function $K_\nu(z)$, we can use several connection formulas for the ordinary and modified Bessel functions to obtain their asymptotic expansion for large z. This is the approach that we have used in Temme (1996a, §9.7). In this chapter we will give a few details.

We also give integral representations of some Bessel functions by using saddle point methods. From these we may also derive the standard large-z expansions, but a more interesting feature is their use in numerical computations.

The large-ν expansions in terms of elementary functions of the Bessel functions are in the form of Debye's expansions. We derive these from integrals for the modified Bessel functions. In §23.5 we derive an Airy-type expansion of the Bessel function $J_\nu(z)$ by using an integral representation and in §9.6 we give the steepest descent paths of the modified Bessel function with purely imaginary order. We conclude with an expansion of an integral containing the J-Bessel function.

9.1 The modified Bessel function $K_\nu(z)$

We start with the integral representation

$$K_\nu(z) = \frac{\sqrt{\pi}(2z)^\nu e^{-z}}{\Gamma\left(\nu+\frac{1}{2}\right)} \int_0^\infty e^{-2zt} t^{\nu-\frac{1}{2}} (1+t)^{\nu-\frac{1}{2}}\, dt, \tag{9.1.1}$$

where $\Re z > 0$ and $\Re\nu > -\frac{1}{2}$.

By expanding

$$(1+t)^{\nu-\frac{1}{2}} = \sum_{k=0}^{\infty} \binom{\nu-\frac{1}{2}}{k} t^k, \tag{9.1.2}$$

the standard asymptotic expansion described by Watson's lemma follows. That is,

$$K_\nu(z) \sim \sqrt{\frac{\pi}{2z}}\, e^{-z} \sum_{k=0}^{\infty} \frac{a_k(\nu)}{z^k}, \quad z \to \infty, \tag{9.1.3}$$

where $a_k(\nu)$ are defined in terms of the Pochhammer symbols (see (2.1.12)) as

$$a_k(\nu) = (-1)^k \frac{(\frac{1}{2}-\nu)_k(\frac{1}{2}+\nu)_k}{2^k\, k!}, \quad k = 0, 1, 2, \ldots. \tag{9.1.4}$$

This expansion holds for bounded ν uniformly inside the sector $|\mathrm{ph}\, z| \le \frac{3}{2}\pi - \delta < \frac{3}{2}\pi$; see Theorem 2.2. It terminates and is exact when $\nu = n + \frac{1}{2}$, $n \in \mathbb{Z}$.

9.2 The ordinary Bessel functions

The Hankel functions can be expressed in terms of this modified Bessel function. We have (see Olver and Maximon (2010, Eq. 10.27.7))

$$\begin{aligned} H_\nu^{(1)}(z) &= -\frac{2i}{\pi} e^{-\frac{1}{2}\nu\pi i} K_\nu\left(z e^{-\frac{1}{2}\pi i}\right), & -\tfrac{1}{2}\pi < \mathrm{ph}\, z \le \pi, \\ H_\nu^{(2)}(z) &= \frac{2i}{\pi} e^{\frac{1}{2}\nu\pi i} K_\nu\left(z e^{\frac{1}{2}\pi i}\right), & -\pi < \mathrm{ph}\, z \le \tfrac{1}{2}\pi. \end{aligned} \tag{9.2.5}$$

The expansion of $K_\nu(z)$ in (9.1.3) can be used to obtain expansions of the Hankel functions. We have

$$\begin{aligned} H_\nu^{(1)}(z) &\sim \sqrt{\frac{2}{\pi z}} e^{i\omega} \sum_{k=0}^{\infty} i^k \frac{a_k(\nu)}{z^k}, & -\pi + \delta \le \mathrm{ph}\, z \le 2\pi - \delta, \\ H_\nu^{(2)}(z) &\sim \sqrt{\frac{2}{\pi z}} e^{-i\omega} \sum_{k=0}^{\infty} (-i)^k \frac{a_k(\nu)}{z^k}, & -2\pi + \delta \le \mathrm{ph}\, z \le \pi - \delta, \end{aligned} \tag{9.2.6}$$

where

$$\omega = z - \left(\tfrac{1}{2}\nu + \tfrac{1}{4}\right)\pi. \tag{9.2.7}$$

From the expansions of the Hankel functions we can obtain expansions of the other ordinary Bessel functions. For $J_\nu(z)$, $Y_\nu(z)$ we can use

$$\begin{aligned} J_\nu(z) &= \tfrac{1}{2}\left(H_\nu^{(1)}(z) + H_\nu^{(2)}(z)\right), \\ Y_\nu(z) &= -\frac{1}{2} i \left(H_\nu^{(1)}(z) - H_\nu^{(2)}(z)\right), \end{aligned} \tag{9.2.8}$$

and the asymptotic representations are usually written in the form

$$\begin{aligned} J_\nu(z) &= \sqrt{\frac{2}{\pi z}}\left(\cos\omega\, P(z,\nu) - \sin\omega\, Q(z,\nu)\right), \\ Y_\nu(z) &= \sqrt{\frac{2}{\pi z}}\left(\sin\omega\, P(z,\nu) + \cos\omega\, Q(z,\nu)\right), \end{aligned} \tag{9.2.9}$$

with expansions

$$P(z,\nu) \sim \sum_{k=0}^{\infty} (-1)^k \frac{a_{2k}(\nu)}{z^{2k}}, \quad Q(z,\nu) \sim \sum_{k=0}^{\infty} (-1)^k \frac{a_{2k+1}(\nu)}{z^{2k+1}}. \tag{9.2.10}$$

The representations and expansions for $J_\nu(z)$ and $Y_\nu(z)$ are valid in the sector $-\pi + \delta \le \operatorname{ph} z \le \pi - \delta$.

Error bounds for the remainders in the expansions of $H_\nu^{(1)}(z)$ and $H_\nu^{(2)}(z)$ are summarized in Olver and Maximon (2010, §10.17).

9.3 The modified Bessel function $I_\nu(z)$

Consider the integral representation

$$I_\nu(z) = \frac{(2z)^\nu e^z}{\sqrt{\pi}\,\Gamma\left(\nu+\frac12\right)} \int_0^1 e^{-2zt} t^{\nu-\frac12} (1-t)^{\nu-\frac12}\,dt, \quad \Re\nu > -\tfrac12. \tag{9.3.11}$$

To obtain an asymptotic expansion for large z we cannot apply Theorem 2.2 because $(1-t)^{\nu-\frac12}$ is not analytic in a sector that contains the positive reals, but we can use Theorem 2.1 if $\Re\nu \ge \frac12$.

We substitute the expansion

$$(1-t)^{\nu-\frac12} = \sum_{k=0}^{\infty} \binom{\nu-\frac12}{k} (-t)^k, \tag{9.3.12}$$

and interchange the order of summation and integration. When we evaluate the resulting integrals not over $[0,1]$ but over $[0,\infty)$, we obtain (cf. (2.1.4))

$$I_\nu(z) \sim \frac{e^z}{\sqrt{2\pi z}\,\Gamma\left(\nu+\frac12\right)} \sum_{k=0}^{\infty} (-1)^k \binom{\nu-\frac12}{k} \frac{\Gamma(k+\nu+\frac12)}{z^k}, \quad z \to \infty. \tag{9.3.13}$$

This expansion is usually written in the form

$$I_\nu(z) \sim \frac{e^z}{\sqrt{2\pi z}} \sum_{k=0}^{\infty} (-1)^k \frac{a_k(\nu)}{z^k}, \tag{9.3.14}$$

where $a_k(\nu)$ is defined in (9.1.4). This expansion holds for bounded ν and is uniformly valid in the sector $|\operatorname{ph} z| \le \frac12\pi - \delta < \frac12\pi$; see Theorem 2.1. It terminates and is not exact when $\nu = n + \frac12$, $n \in \mathbb{Z}$.

9.3.1 *A compound expansion of $I_\nu(z)$*

The sector for z in (9.3.14) is rather limited, and we will show now how to extend this domain by using a compound expansion.

We can use connection formulas with the $K_\nu(z)$ or the Hankel functions, but it is instructive to describe a different method, because this method can also be used for other integrals.

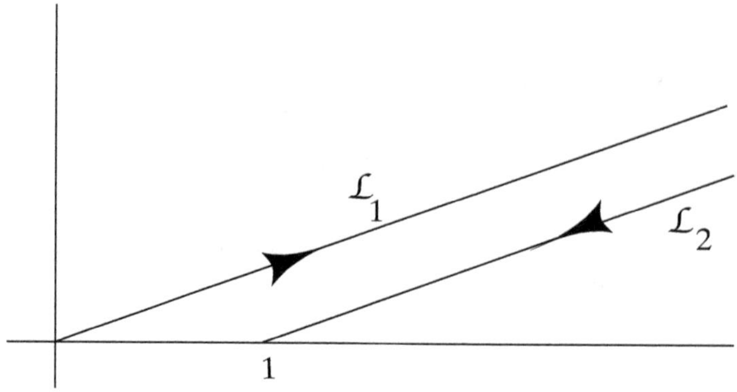

Fig. 9.1 The path $\mathcal{L}_1 \cup \mathcal{L}_2$ for obtaining the compound expansion of $I_\nu(z)$ in (9.3.18).

We change the interval $[0,1]$ in (9.3.11) into the path $\mathcal{L}_1 \cup \mathcal{L}_2$, where $\mathcal{L}_1$ starts in $t=0$ and ends at infinity, and $\mathcal{L}_2$ is the path back to $t=1$; see Figure 9.1. The contributions for $I_\nu(z)$ over these paths are noted by $I_\nu^{(j)}(z)$, $j=1,2$, and we have $I_\nu(z) = I_\nu^{(1)}(z) + I_\nu^{(2)}(z)$. For convergence at $t=0$ and $t=1$ we need $\Re\nu > -\frac{1}{2}$.

The path $\mathcal{L}_1$ can be taken as a straight line:

$$\mathcal{L}_1 = \{t : t = \tau e^{i\sigma},\ \tau \ge 0\}, \tag{9.3.15}$$

for some fixed $\sigma > 0$. In this way the singularity of the integrand at $t=1$ is avoided; we also need $\sigma < 2\pi$, but it depends on the value of $\theta = \operatorname{ph} z$ which value of σ gives convergence at infinity. For this we need the condition $\cos(\sigma+\theta) > 0$, that is, $-\frac{1}{2}\pi - \theta < \sigma < \frac{1}{2}\pi - \theta$. Initially, we take $\theta \in (-\frac{1}{2}\pi, \frac{1}{2}\pi)$ and σ a small positive number such that $\cos(\sigma+\theta) > 0$.

The function $I_\nu^{(1)}(z)$ has the asymptotic expansion as given in (9.3.14), but now, by invoking Theorem 2.2, valid in the sector $-2\pi - \frac{1}{2}\pi < \theta < \frac{1}{2}\pi$.

In the integral for $I_\nu^{(2)}(z)$ we take the path $\mathcal{L}_2$ above the interval $(1,+\infty)$, a branch cut for the integrand. We assume that $\operatorname{ph}(1-t) = 0$ when $t \in (0,1)$. Hence, above the branch cut we have $(1-t)^{\nu-\frac{1}{2}} = (t-1)^{\nu-\frac{1}{2}} e^{-\pi i(\nu-\frac{1}{2})}$. This gives

$$I_\nu^{(2)}(z) = \frac{(2z)^\nu\, e^{z-\pi i(\nu-\frac{1}{2})}}{\sqrt{\pi}\,\Gamma\left(\nu+\frac{1}{2}\right)} \int_\infty^1 e^{-2zt} t^{\nu-\frac{1}{2}} (t-1)^{\nu-\frac{1}{2}}\, dt. \tag{9.3.16}$$

Looking at (9.1.1) we conclude that

$$I_\nu^{(2)}(z) = \frac{e^{-\pi i(\nu+\frac{1}{2})}}{\pi} K_\nu(z), \tag{9.3.17}$$

but we can forget about this relation and apply Watson's lemma again, now to (9.3.16).

Anyhow, we see that the asymptotic expansion of $I_\nu^{(2)}(z)$ follows from (9.1.3), and is valid in the sector $|\operatorname{ph} z| < \frac{3}{2}\pi$. Combining both expansions we obtain for $I_\nu(z)$ the compound expansion

$$I_\nu(z) \sim \frac{e^z}{\sqrt{2\pi z}} \sum_{k=0}^\infty (-1)^k \frac{a_k(\nu)}{z^k} + \frac{e^{-z-\pi i(\nu+\frac{1}{2})}}{\sqrt{2\pi z}} \sum_{k=0}^\infty \frac{a_k(\nu)}{z^k}, \tag{9.3.18}$$

and this expansion is valid uniformly in the sector $-\frac{3}{2}\pi+\delta \le \operatorname{ph} z \le \frac{1}{2}\pi-\delta$.

Repeating this approach, splitting the interval $[0,1]$ of (9.3.11) into $\mathcal{L}_1 \cup \mathcal{L}_2$, now with $\mathcal{L}_1$ (see (9.3.15)) with $\sigma < 0$, we find

$$I_\nu(z) \sim \frac{e^z}{\sqrt{2\pi z}} \sum_{k=0}^{\infty} (-1)^k \frac{a_k(\nu)}{z^k} + \frac{e^{-z+\pi i(\nu+\frac{1}{2})}}{\sqrt{2\pi z}} \sum_{k=0}^{\infty} \frac{a_k(\nu)}{z^k}, \tag{9.3.19}$$

and this expansion is valid uniformly in the sector $-\frac{1}{2}\pi+\delta \le \operatorname{ph} z \le \frac{3}{2}\pi-\delta$. Both series terminate when $\nu = n+\frac{1}{2}$, $n \in \mathbb{Z}$; in that case (9.3.19) becomes an exact expansion.

9.4 Saddle point method for $K_\nu(z)$, $z \in \mathbb{C}$

The modified Bessel function $K_\nu(z)$ has the integral representation (see (9.6.69))

$$K_\nu(z) = \frac{1}{2} \int_{-\infty}^{\infty} e^{-z\cosh w - \nu w}\, dw, \tag{9.4.20}$$

where we assume $\Re z \ge 0$ and $z \ne 0$.

We derive the path of steepest descent through the saddle point at the origin for complex z, and then we can take $\operatorname{ph} z \in (-\pi, \pi)$. The order ν is fixed and does not play a role in the asymptotic analysis.

We write

$$z = x+iy = re^{i\theta}, \quad w = u+iv. \tag{9.4.21}$$

Then the function $\phi(w) = z\cosh w$ can be written as

$$\phi(w) = x\cosh u\cos v - y\sinh u\sin v + i\left(y\cosh u\cos v + x\sinh u\sin v\right). \tag{9.4.22}$$

The path of steepest descent through the saddle point $w=0$ is defined by $\Im\phi(w) = \Im\phi(0)$, which gives the equation

$$y\cosh u\cos v + x\sinh u\sin v = y. \tag{9.4.23}$$

We can solve for v in the form

$$\sin v = -\frac{\sin\theta\sinh u}{\cosh u+\cos\theta}, \quad v \in (-\theta,\theta), \quad u \in \mathbb{R}. \tag{9.4.24}$$

Solving (9.4.24) for v when $|\theta| \le \frac{1}{2}\pi$, we take the principal branch of arcsin that satisfies $\arcsin(0)=0$. When $\frac{1}{2}\pi < |\theta|$ we need other branches. To avoid the arcsin function outside its principal domain, we can also solve for u in the form

$$\sinh u = -\frac{\sin\theta\sin v}{\cos v-\cos\theta}, \quad v \in (-\theta,\theta), \quad u \in \mathbb{R}. \tag{9.4.25}$$

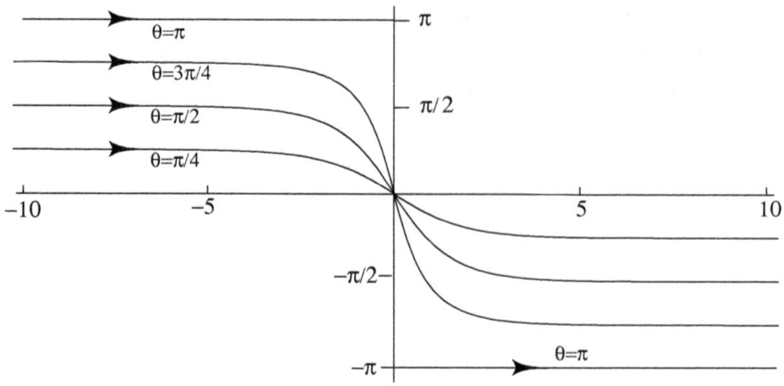

Fig. 9.2 Saddle point contours defined in (9.4.24) for the integral in (9.4.20) for several values of $\theta = \operatorname{ph} z$.

In Figure 9.2 we show the saddle point contours for $\theta = k/\pi$, $k = 1, 2, 3, 4$. When $\theta = \pi$ the saddle point contour consists of three line segments:

- from $-\infty + i\pi$ to $i\pi$;
- from $+i\pi$ to $-i\pi$;
- from $-i\pi$ to $+\infty - i\pi$.

Similarly for $\theta = -\pi$.

We find for the integral in (9.4.20), when we integrate with respect to u,

$$K_\nu(z) = \tfrac{1}{2}e^{-z}\int_{-\infty}^{\infty} e^{-r\psi(u)} f(u)\,du, \tag{9.4.26}$$

where

$$\psi(u) = \frac{\sinh^2 u}{\cosh u + \cos\theta}, \tag{9.4.27}$$

and

$$f(u) = \left(1 + i\frac{dv}{du}\right)e^{-\nu(u+iv)}, \quad \frac{dv}{du} = -\frac{\sin\theta}{\cosh u + \cos\theta}. \tag{9.4.28}$$

For large values of u we have

$$v \sim -\theta\,\operatorname{sign}(u), \quad \psi(u) \sim \cosh u, \tag{9.4.29}$$

which shows the rate of convergence of the integral in (9.4.26).

If $|\theta| > 0$ we can choose to integrate the integral in (9.4.20) with respect to v, and we have

$$K_\nu(z) = \tfrac{1}{2}e^{-z}\int_{-\theta}^{\theta} e^{-r\phi(v)} g(v)\,dv, \tag{9.4.30}$$

where

$$\phi(v) = \frac{\sin^2 v}{\cos v - \cos\theta}, \tag{9.4.31}$$

and

$$g(v) = -\left(\frac{du}{dv} + i\right) e^{-\nu(u+iv)}, \quad \frac{du}{dv} = -\frac{\sin\theta}{\cos v - \cos\theta}. \tag{9.4.32}$$

Remark 9.1. The integrals in (9.4.26) and (9.4.30) can be used for obtaining the asymptotic expansion of $K_\nu(z)$ by using Laplace's method. In this way the same expansion as in (9.1.3) will be obtained. The construction of the coefficients will be more complicated, however, whereas the coefficients in (9.1.3) are obtained only from a simple binomial expansion in (9.1.2). △

Remark 9.2. The integrals in this section are more relevant for numerical evaluations of these functions for a wide range of the phase of $\nu \in \mathbb{C}$ and z with $|\mathrm{ph}\, z| < \pi$, $z \neq 0$. When $|z| \geq \delta > 0$ and $|\nu|$ of moderate size, numerical quadrature may replace the use of the asymptotic expansion. Observe that the dominant factor e^{-z} of $K_\nu(z)$ in (9.4.26) and (9.4.30) is in front of the integrals. The integrands are simple and converge quickly. The trapezoidal rule for this type of integral is a very efficient tool, as explained in Gil *et al.* (2007, Chapter 5). The same holds for the integrals given below for the Bessel functions with possible large x and moderate $\nu \in \mathbb{C}$. △

9.4.1 *Integral representations from saddle point analysis*

Viewed in the light of Remark 9.2 we give a few extra integral representations that may be useful for numerical evaluations of the ordinary Bessel functions.

We use the connection formulas in (9.2.5) for $z = x > 0$ and obtain from (9.4.26) the integral representation

$$H_\nu^{(1)}(x) = \frac{\sqrt{2}\, e^{i\chi}}{\pi} \int_{-\infty}^{\infty} e^{-x\frac{\sinh^2 u}{\cosh u} - \nu u} h_1(u)\, du, \tag{9.4.33}$$

where

$$h_1(u) = e^{-i\nu \arcsin \tanh u} \frac{\cosh u + i}{(1+i)\cosh u}, \quad \chi = x - \left(\tfrac{1}{2}\nu + \tfrac{1}{4}\right)\pi. \tag{9.4.34}$$

The result for the other Hankel function follows from changing the sign of i throughout, that is

$$H_\nu^{(2)}(x) = \frac{\sqrt{2}\, e^{-i\chi}}{\pi} \int_{-\infty}^{\infty} e^{-x\frac{\sinh^2 u}{\cosh u} - \nu u} h_2(u)\, du, \tag{9.4.35}$$

where

$$h_2(u) = e^{i\nu \arcsin \tanh u} \frac{\cosh u - i}{(1-i)\cosh u}. \tag{9.4.36}$$

The relations in (9.2.8) give the representations

$$\begin{aligned} J_\nu(x) &= \frac{\sqrt{2}}{\pi} \int_{-\infty}^{\infty} e^{-x\frac{\sinh^2 u}{\cosh u} - \nu u} f_\nu(u)\, du, \\ Y_\nu(x) &= \frac{\sqrt{2}}{\pi} \int_{-\infty}^{\infty} e^{-x\frac{\sinh^2 u}{\cosh u} - \nu u} g_\nu(u)\, du, \end{aligned} \tag{9.4.37}$$

where

$$
\begin{aligned}
f_\nu(u) &= \frac{(\cosh u+1)\cos\chi' + (\cosh u-1)\sin\chi'}{2\cosh u},\\
g_\nu(u) &= \frac{(\cosh u+1)\sin\chi' - (\cosh u-1)\cos\chi'}{2\cosh u},
\end{aligned}
\tag{9.4.38}
$$

with $\chi' = \chi - \nu \arcsin\tanh u$.

9.4.2 *Saddle point method for $J_\nu(x)$, $x \le \nu$*

Although the representation in (9.4.37) for $J_\nu(x)$ is valid for all complex values of ν, when ν becomes large a different approach is needed. When the argument and order of the Bessel functions are both large, usually uniform expansions are needed, in particular when the argument and order are nearly equal; see §23.5 for an expansion in terms of the Airy function.

We use the integral (see Olver and Maximon (2010, Eq. 10.9.17))

$$
J_\nu(z) = \frac{1}{2\pi i}\int_{\infty-\pi i}^{\infty+\pi i} e^{z\sinh w - \nu w}\,dw, \quad |\mathrm{ph}\, z| < \tfrac12\pi. \tag{9.4.39}
$$

As explained in Temme (1996a, §9.2), by changing the endpoints at infinity, this contour integral can be used for other phases of z as well.

We describe a saddle point analysis for $J_\nu(x)$ with $x \le \nu$, and write $x = \nu/\cosh\alpha$, $\alpha \ge 0$. Then

$$
J_\nu(x) = \frac{1}{2\pi i}\int_{\infty-\pi i}^{\infty+\pi i} e^{\nu\phi(w)}\,dw, \quad \phi(w) = \frac{\sinh w}{\cosh\alpha} - w. \tag{9.4.40}
$$

A saddle point is located at $w_0 = \alpha$; there are other saddle points, for example at $-\alpha$, but only through w_0 we can choose a path on which $\Im\phi(w)$ is constant and on which we preserve convergence. We have $\Im\phi(w_0) = 0$ and the saddle point contour is defined by (we write $w = u + iv$)

$$
\cosh u = \cosh\alpha\,\frac{v}{\sin v}, \quad u \ge \alpha, \quad -\pi < v < \pi. \tag{9.4.41}
$$

We integrate the integral in (9.4.40) on this contour with respect to v and obtain

$$
J_\nu(x) = \frac{1}{2\pi i}\int_{-\pi}^{\pi} e^{\nu\psi(v)}\left(\frac{du}{dv} + i\right)dv, \quad \psi(v) = \frac{\sinh u\cos v}{\cosh\alpha} - u. \tag{9.4.42}
$$

We can delete the term du/dv because this is an odd function of v, and $\psi(v)$ is even. We can also put the dominant factor $e^{\nu\psi(0)}$ with $\psi(0) = \tanh\alpha - \alpha$ in front of the integral, and obtain

$$
J_\nu(x) = \frac{e^{\nu(\tanh\alpha-\alpha)}}{2\pi}\int_{-\pi}^{\pi} e^{\nu(\psi(v)-\psi(0))}\,dv, \quad x = \frac{\nu}{\cosh\alpha}. \tag{9.4.43}
$$

By using Laplace's method we can obtain Debye's expansion (see Olver and Maximon (2010, Eq. 10.19.3)) of $J_\nu(x)$, with $\nu\to\infty$ and $\alpha > 0$. The integral representation holds for $\alpha \ge 0$. When $\alpha = 0$ we have

$$
J_\nu(\nu) = \frac{1}{2\pi}\int_{-\pi}^{\pi} e^{\nu(\sinh u\cos v - u)}\,dv, \quad u = \operatorname{arccosh}\frac{v}{\sin v}. \tag{9.4.44}
$$

For small positive values of v we expand the function in the exponent (observe that it is an even function of v):

$$\sinh u \cos v - u = -\tfrac{4\sqrt{3}}{27}\left|v^3\right|\left(1+\tfrac{2}{525}v^4+\tfrac{4}{14175}v^6+\ldots\right), \quad v \to 0. \tag{9.4.45}$$

This cubic behavior is typical of the case when the two saddle points at $\pm\alpha$ coalesce when $\alpha = 0$. We shall say more about this in Chapter 23.

9.5 Debye-type expansions of the modified Bessel functions

For an overview of Debye-type expansions of the ordinary and modified Bessel we refer to Olver and Maximon (2010, §10.19(ii)) and for Debye's paper to Debye (1909). For a recent detailed study on Debye-type expansions obtained from integrals, including error bounds, we refer to Nemes (2014).

In this section we derive the Debye-type expansion of the modified Bessel functions by using integrals. For $K_\nu(z)$ we consider positive z and ν and take the integral in (9.4.20). We write it in the form

$$K_\nu(\nu z) = \int_{-\infty}^{\infty} e^{-\nu\phi(s)}\,ds, \quad \phi(s) = z\cosh s - s. \tag{9.5.46}$$

There is a saddle point at $s_0 = \operatorname{arcsinh}(1/z)$, and we apply Laplace's method by writing

$$\phi(s) - \phi(s_0) = \tfrac{1}{2}w^2, \quad \operatorname{sign}(s - s_0) = \operatorname{sign} w. \tag{9.5.47}$$

This gives

$$K_\nu(\nu z) = \tfrac{1}{2}e^{-\nu\eta(z)}\int_{-\infty}^{\infty} e^{-\frac{1}{2}\nu w^2} f(w)\,dw, \tag{9.5.48}$$

where

$$f(w) = \frac{ds}{dw} = \frac{w}{z\sinh s - 1}, \tag{9.5.49}$$

and $\eta(z) = \phi(s_0)$, that is,

$$\eta(z) = \sqrt{1+z^2} + \ln\frac{z}{1+\sqrt{1+z^2}}. \tag{9.5.50}$$

Application of l'Hôpital's rule yields

$$f(0) = \left.\frac{ds}{dw}\right|_{w=0} = \frac{1}{(z\cosh s_0)\left.\frac{ds}{dw}\right|_{w=0}}. \tag{9.5.51}$$

This gives, by taking into account the condition on the transformation given in (9.5.47),

$$f(0) = \frac{1}{(z^2+1)^{\frac{1}{4}}}. \tag{9.5.52}$$

By expanding $f(w)$ in powers of w we obtain an expansion, which we write in the form

$$K_\nu(\nu z) = \sqrt{\frac{\pi}{2\nu}} \frac{e^{-\nu\eta(z)}}{(1+z^2)^{1/4}} P(\nu, z), \quad P(\nu, z) \sim \sum_{k=0}^{\infty} (-1)^k \frac{u_k(t)}{\nu^k}, \tag{9.5.53}$$

as $\nu \to \infty$, uniformly with respect to $z \geq 0$. The coefficients will be given below.

First we consider the function $I_\nu(z)$ and use the integral representation (see (9.6.69))

$$I_\nu(z) = \frac{1}{2\pi i} \int_{\infty - \pi i}^{\infty + \pi i} e^{z \cosh s - \nu s} \, ds, \tag{9.5.54}$$

which we write in the form

$$I_\nu(\nu z) = \frac{1}{2\pi i} \int_{\infty - \pi i}^{\infty + \pi i} e^{\nu \phi(s)} \, ds, \tag{9.5.55}$$

where $\phi(s)$ is given in (9.5.46).

Write $s = u + iv$. Then the saddle point contour through s_0 follows from $\Im \phi(s) = 0$ (when z and ν are positive), and is given by

$$u = \operatorname{arcsinh}\left(\frac{v}{z \sin v}\right), \quad -\pi < v < \pi. \tag{9.5.56}$$

This path is the steepest ascent path through s_0 for the integral in (9.5.46), and on this path $\phi(s)$ is real. We can use the same transformation as in (9.5.47), and this time we obtain the expansion

$$I_\nu(\nu z) = \frac{1}{\sqrt{2\pi\nu}} \frac{e^{\nu\eta(z)}}{(1+z^2)^{1/4}} Q(\nu, z), \quad Q(\nu, z) \sim \sum_{k=0}^{\infty} \frac{u_k(t)}{\nu^k}, \quad \nu \to \infty. \tag{9.5.57}$$

The coefficients $u_k(t)$ are polynomials in $t = 1/\sqrt{z^2+1}$, and the first few are

$$u_0(t) = 1, \quad u_1(t) = \frac{3t - 5t^3}{24}, \quad u_2(t) = \frac{81t^2 - 462t^4 + 385t^6}{1152}, \tag{9.5.58}$$

with recurrence relation

$$u_{k+1}(t) = \tfrac{1}{2} t^2 \left(1 - t^2\right) u_k'(t) + \tfrac{1}{8} \int_0^t \left(1 - 5s^2\right) u_k(s) \, ds, \quad k = 0, 1, 2, \ldots. \tag{9.5.59}$$

This recursion formula follows from the method based on differential equations, see Olver (1997, Chapter 10, §7). Once we have the form of the expansions it is not difficult to verify the recurrence relation by using the differential equation for the modified Bessel functions:

$$z^2 \frac{d^2 w}{dz^2} + z \frac{dw}{dz} - \left(z^2 + \nu^2\right) w = 0. \tag{9.5.60}$$

By differentiating the expansions we obtain

$$\begin{aligned} K_\nu'(\nu z) &= -\sqrt{\frac{\pi}{2\nu}} \frac{(1+z^2)^{1/4}}{z} e^{-\nu\eta(z)} R(\nu, z), \\ I_\nu'(\nu z) &= \frac{1}{\sqrt{2\pi\nu}} \frac{(1+z^2)^{1/4}}{z} e^{\nu\eta(z)} S(\nu, z), \end{aligned} \tag{9.5.61}$$

where

$$R(\nu,z)\sim\sum_{k=0}^{\infty}(-1)^k\frac{v_k(t)}{\nu^k},\quad S(\nu,z)\sim\sum_{k=0}^{\infty}\frac{v_k(t)}{\nu^k}. \tag{9.5.62}$$

For the coefficients we have $v_0(t)=1$ and

$$v_k(t)=u_k(t)+t\left(t^2-1\right)\left(\tfrac12 u_{k-1}(t)+tu'_{k-1}(t)\right),\quad k=1,2,3,\ldots. \tag{9.5.63}$$

The first coefficients are

$$v_0(t)=1,\quad v_1(t)=\frac{-9t+7t^3}{24},\quad v_2(t)=\frac{-135t^2+594t^4-455t^6}{1152}. \tag{9.5.64}$$

From the Wronskian relation

$$K_\nu(z)I'_\nu(z)-K'_\nu(z)I_\nu(z)=\frac{1}{z}, \tag{9.5.65}$$

we obtain

$$P(\nu,z)S(\nu,z)+Q(\nu,z)R(\nu,z)=2, \tag{9.5.66}$$

which gives, after substituting the expansions, extra relations between the coefficients:

$$\sum_{j=0}^{2k}(-1)^j u_j(t)v_{2k-j}(t)=0,\quad k=1,2,3,\ldots. \tag{9.5.67}$$

Remark 9.3. The Debye-type expansions of the modified Bessel functions given here have the so-called double asymptotic property. In this case, this means that they are derived and valid for large ν and hold uniformly with respect to $z\ge 0$. But they are valid for large values of z and $\nu\ge 0$ as well. In fact, we could have started with the integral representation

$$K_{\nu z}(z)=\tfrac12\int_{-\infty}^{\infty}e^{-z\psi(s)}\,ds,\quad \psi(s)=\cosh s-\nu s, \tag{9.5.68}$$

which can be expanded as before, giving an asymptotic expansion that can be rearranged in the form as given in (9.5.53). △

9.6 Modified Bessel functions of purely imaginary order

For these function we use the integral representations (see Olver and Maximon (2010, §10.32))[1]

$$\begin{aligned}K_{i\nu}(x)&=\tfrac12\int_{-\infty}^{\infty}e^{-x\cosh t+i\nu t}\,dt,\\ I_{i\nu}(x)&=\frac{1}{2\pi i}\int_{\infty-\pi i}^{\infty+\pi i}e^{x\cosh t-i\nu t}\,dt.\end{aligned} \tag{9.6.69}$$

[1]This section is based on Temme (1994b).

We assume that $x > 0,\ \nu \ge 0$. $K_{i\nu}(x)$ is real and an even function with respect to ν; $I_{i\nu}(x)$ is complex when $\nu > 0$. We define

$$L_{i\nu}(x) = \frac{I_{i\nu}(x) + I_{-i\nu}(x)}{2}, \tag{9.6.70}$$

which is real and even with respect to ν. In fact, $L_{i\nu}(x)$ is the real part of $I_{i\nu}(x)$ and $K_{i\nu}(x)$ is the imaginary part of $I_{i\nu}(x)$ (up to a factor):

$$I_{i\nu}(x) = L_{i\nu}(x) - i\,\frac{\sinh \pi\nu}{\pi}\,K_{i\nu}(x). \tag{9.6.71}$$

The function $K_{i\nu}(x)$ plays an important role in potential problems for a wedge. It is the kernel of the Kontorovich–Lebedev transform. We have the pair of transforms:

$$\begin{aligned} g(y) &= \int_0^\infty f(x) K_{ix}(y)\,dx, \\ f(x) &= \frac{2x}{\pi^2}\sinh(\pi x)\int_0^\infty y^{-1} K_{ix}(y) g(y)\,dy. \end{aligned} \tag{9.6.72}$$

See Lebedev (1972, §6.5) or Yakubovich (1996, Chapters 2–5).

In describing saddle point contours for the integrals in (9.6.69) we distinguish between the two cases $x \ge \nu$ and $x \le \nu$. In the second case the saddle point contour for $K_{i\nu}(x)$ is rather exotic, and when the parameters are large the functions $K_{i\nu}(x)$ and $L_{i\nu}(x)$ are rapidly oscillating. For $\nu \sim x$ (both large) Airy-type expansions are needed for the asymptotic representations. For uniform asymptotic expansions for large ν we refer to Dunster (1990), where the role of these functions as approximants in uniform asymptotic expansions of solutions of certain second-order linear differential equations is considered as well. In Shi and Wong (2010) error bounds for the remainders in the Airy-type expansions are derived. Numerical aspects are considered in Gil *et al.* (2002, 2004).

9.6.1 *The monotonic case: $x \ge \nu \ge 0$*

In this case we write

$$\nu = x\sin\theta, \quad 0 \le \theta \le \tfrac{1}{2}\pi. \tag{9.6.73}$$

9.6.1.1 *The contour for $K_{i\nu}(x)$*

When $\nu = 0$ the real axis is the path of steepest descent of the first integral in (9.6.69), with a saddle point at $t = 0$. When $\nu > 0$ the path and saddle point shift upwards in the complex plane. In that case we write

$$K_{i\nu}(x) = \tfrac{1}{2}\int_{-\infty}^{\infty} e^{-\phi(t)}\,dt, \quad \phi(t) = x(\cosh t - it\sin\theta). \tag{9.6.74}$$

The saddle point t_0 follows from solving the equation $\phi'(t) = 0$, giving $t_0 = i\theta$. It suffices to consider only this saddle point, but there are more. Since $\Im\phi(t_0) = 0$ the

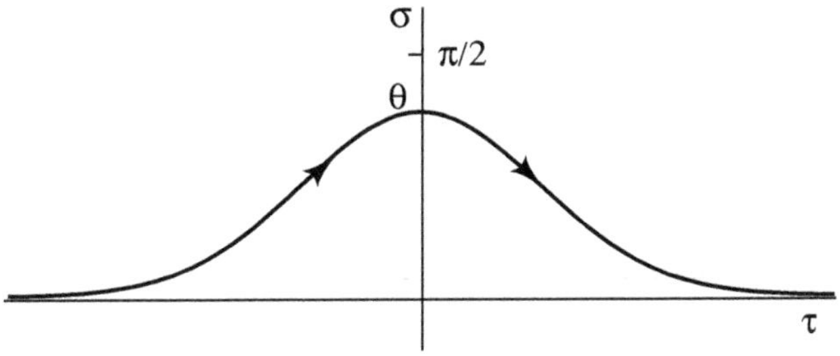

Fig. 9.3 Steepest descent path (9.6.76), $\theta = \frac{3}{8}\pi$.

equation $\Im\phi(t) = \Im\phi(t_0)$ gives the following relation between the real and imaginary parts of t (we write $t = \tau + i\sigma,\ t_0 = \tau_0 + i\sigma_0$, with $\tau_0 = 0, \sigma_0 = \theta$)

$$\sin\sigma = \sin\theta\,\frac{\tau}{\sinh\tau}, \quad -\infty < \tau < \infty, \quad 0 < \sigma \le \sigma_0 \le \tfrac{1}{2}\pi. \tag{9.6.75}$$

Hence, the path of steepest descent through the point $t_0 = i\theta$ is given by

$$\sigma(\tau) = \arcsin\left(\sin\theta\,\frac{\tau}{\sinh\tau}\right), \quad -\infty < \tau < \infty. \tag{9.6.76}$$

This path is shown in Figure 9.3.

Other solutions of the equation in (9.6.75) are contours of steepest descent through the saddle points $i(\pm\pi - \theta)$, and are given by

$$\sigma_-(\tau) = -\pi - \sigma(\tau), \quad \sigma_+(\tau) = \pi - \sigma(\tau). \tag{9.6.77}$$

These solutions will be used in the next subsection.

Integration with respect to τ on the path described by (9.6.76) gives the representation

$$K_{i\nu}(x) = \frac{1}{2}\int_{-\infty}^{\infty} e^{-\psi(\tau)}\,\frac{dt}{d\tau}\,d\tau, \quad \psi(\tau) = x\cosh\tau\cos\sigma + \nu\sigma, \tag{9.6.78}$$

where σ as function of τ is given in (9.6.76). The function $\psi(\tau)$ is an even function of τ. Observe that $\dfrac{dt}{d\tau} = 1 + i\dfrac{d\sigma}{d\tau}$ and that $\dfrac{d\sigma}{d\tau}$, being an odd function of τ, does not give a contribution in (9.6.78). Hence, we can write

$$K_{i\nu}(x) = \int_0^{\infty} e^{-\psi(\tau)}\,d\tau, \quad \psi(\tau) = x\cosh\tau\cos\sigma + \nu\sigma. \tag{9.6.79}$$

From (9.6.69) it follows that $K_{i\nu}(x)$ is positive when $x \ge 0$. It may be convenient to extract the dominant factor in the representation in (9.6.79), by writing

$$K_{i\nu}(x) = e^{-\psi(0)}\int_0^{\infty} e^{-(\psi(\tau)-\psi(0))}\,d\tau, \tag{9.6.80}$$

where

$$\psi(0) = \sqrt{x^2-\nu^2} + \nu\arcsin\frac{\nu}{x} = x(\cos\theta + \theta\sin\theta). \tag{9.6.81}$$

When $x = \nu$ the function $\psi(\tau)$ is not analytic at $\tau = 0$. We have when $x = \nu$, as $\tau \to 0$ (through real values):

$$\psi(\tau) = \tfrac{1}{2}\pi\nu + \sqrt{3}\nu\left|\tau^3\right|\left(\tfrac{4}{27} + \tfrac{8}{14175}\tau^4 - \tfrac{16}{382725}\tau^6 + \mathcal{O}\left(\tau^8\right)\right). \tag{9.6.82}$$

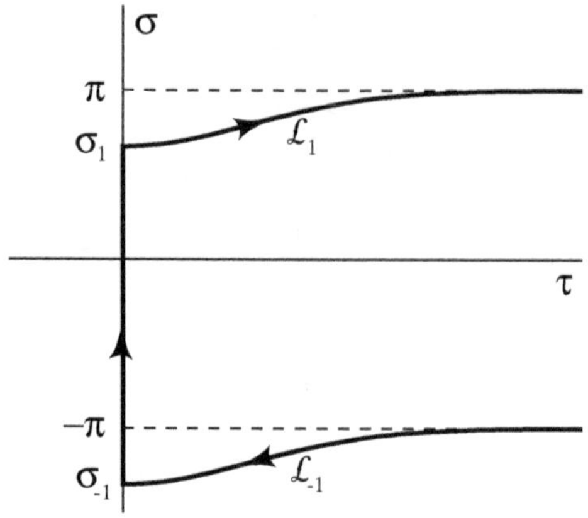

Fig. 9.4 Steepest descent path used in (9.6.84) for $\theta = \frac{1}{4}\pi$.

9.6.1.2 *The contours for $I_{i\nu}(x)$ and $L_{i\nu}(x)$*

In this case we need three saddle points. The derivative of the function $\phi(t) = x\cosh t - i\nu t$ occurring in the second integral of (9.6.69) vanishes at the points

$$t_k = i\sigma_k = i\left((-1)^k\theta + k\pi\right), \quad k \in \mathbb{Z}. \tag{9.6.83}$$

The saddle point contour now passes through the saddle points t_{-1}, t_0, t_1, and the path of integration of the second integral in (9.6.69) is split up in three parts: $\mathcal{L}_{-1} \cup \mathcal{L}_0 \cup \mathcal{L}_1$, where

- $\mathcal{L}_{-1}$ runs from $+\infty - i\pi$ to t_{-1},
- $\mathcal{L}_0$ runs from t_{-1} to t_1 (a segment of the imaginary axis),
- $\mathcal{L}_1$ runs from t_1 to $+\infty + i\pi$.

These paths are shown in Figure 9.4. On each path we have $\Im\phi(t) = 0$. On $\mathcal{L}_{\pm 1}$ the relation between the real and imaginary parts of $t = \tau + i\sigma$ is given by (9.6.77) with $\tau \ge 0$. Thus we obtain

$$\begin{aligned} I_{i\nu}(x) = & \frac{1}{2\pi i}\int_{\mathcal{L}_{-1}} e^{x\cosh\tau\cos\sigma_- + \nu\sigma_-}\left(1 + i\frac{d\sigma_-}{d\tau}\right)d\tau \,+ \\ & \frac{1}{2\pi}\int_{\sigma_{-1}}^{\sigma_1} e^{x\cos\sigma + \nu\sigma}\,d\sigma \,+ \\ & \frac{1}{2\pi i}\int_{\mathcal{L}_1} e^{x\cosh\tau\cos\sigma_+ + \nu\sigma_+}\left(1 + i\frac{d\sigma_+}{d\tau}\right)d\tau. \end{aligned} \tag{9.6.84}$$

Hence, by using (9.6.71) and separating the real part,

$$\begin{aligned} L_{i\nu}(x) &= \frac{1}{2\pi}\int_{\sigma_{-1}}^{\sigma_1} e^{x\cos\sigma + \nu\sigma}\,d\sigma - \frac{\sinh\pi\nu}{\pi}\int_0^\infty e^{-\psi(\tau)}\frac{d\sigma}{d\tau}\,d\tau \\ &= \frac{1}{2\pi}\int_{\sigma_{-1}}^{\sigma_1} e^{x\cos\sigma + \nu\sigma}\,d\sigma - \frac{\sinh\pi\nu}{\pi}\int_0^{\sigma_0} e^{-\psi(\tau)}\,d\sigma, \end{aligned} \tag{9.6.85}$$

where $x \geq \nu \geq 0$, $\psi(\tau)$ is given in (9.6.79), the relation between τ and σ in (9.6.75), and

$$\frac{d\sigma}{d\tau} = \tan\sigma\left(\frac{1}{\tau} - \coth\tau\right). \tag{9.6.86}$$

This gives the desired representation of $L_{i\nu}(x)$ along the saddle point contours for these functions. Observe that the integrals are non-oscillating. The integral on the interval $[\sigma_{-1}, \sigma_1]$ gives the main contribution, especially when the parameters are large. The integrand peaks at the point $\sigma = \sigma_0 = \theta$. At this point the integrand assumes the value $e^{\psi(0)}$, compare (9.6.80).

9.6.2 *The oscillatory case: $\nu \geq x > 0$*

Under the present condition we write

$$\nu = x\cosh\mu, \quad \mu \geq 0. \tag{9.6.87}$$

9.6.2.1 *The contour for $K_{i\nu}(x)$*

For the contour of $K_{i\nu}(x)$ we need an infinite number of saddle points for obtaining the steepest descent path. However, a simple summation procedure reduces the number of saddle points to just a few, as in the case of $I_{i\nu}(x)$ in the previous section.

We write $\phi(t)$ of (9.6.74) in the form $\phi(t) = x(\cosh t - it\cosh\mu)$. It follows that the saddle points of the integrands in (9.6.69) are now given by

$$t_k^{\pm} = \pm\tau_0 + \tfrac{1}{2}\pi i + 2k\pi i, \quad \tau_0 = \mu, \quad k \in \mathbb{Z}. \tag{9.6.88}$$

The saddle point contour through the saddle $t_k^{\pm}$ is defined by the equation $\Im\phi(t) = \Im\phi(t_k^{\pm}) = \pm x(\sinh\mu - \mu\cosh\mu)$, that is by:

$$\sin\sigma = \cosh\mu\,\frac{\tau}{\sinh\tau} \pm \frac{\sinh\mu - \mu\cosh\mu}{\sinh\tau}, \tag{9.6.89}$$

which is independent of k. The contours cannot have common points with the imaginary axis (where $\tau = 0$). In fact this axis separates two groups of contours: those through t_k^+ from those through t_k^-, the first group corresponding to positive values of τ, the second one to negative values.

The parabola-shaped curves satisfying (9.6.89) are shown in Figure 9.5. On each 'parabola' two saddle points are located. Hence, a complete parabola cannot serve as a path of descent, since always one branch extending to infinity is a path of ascent. Only parts of the parabola shaped contours are used for the steepest descent paths.

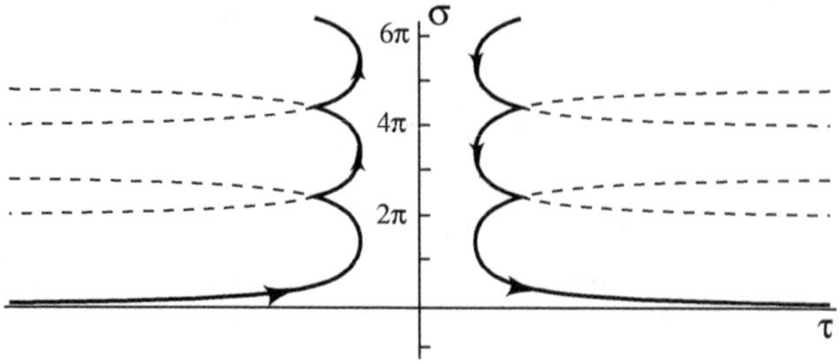

Fig. 9.5 Steepest descent path used in (9.6.90).

First the original contour in (9.6.69) for $K_{i\nu}(x)$ is split up in two parts: $\mathcal{L}^- \cup \mathcal{L}^+$, where

- $\mathcal{L}^-$ runs from $-\infty$ to 0, and from 0 to $+i\infty$,
- $\mathcal{L}^+$ runs from $+i\infty$ to 0, and from 0 to $+\infty$.

Since $\nu > 0$ the integrals along $\mathcal{L}^\pm$ are convergent at $+i\infty$. Next we deform $\mathcal{L}^\pm$ along the thickened parts of the saddle point contours shown in Figure 9.5. Owing to the symmetry we can consider $\mathcal{L}^+$ only: we take twice the real part of this integral. On the lower branch of $\mathcal{L}^+$ (running from t_0^+ to $+\infty$) we integrate with respect to τ, on the upper branch (running from $+i\infty$ to t_0^+) we integrate with respect to σ. The result is:

$$K_{i\nu}(x) = \Re\left(e^{-i\chi}\left(\int_{\tau_0}^{\infty} e^{-\psi(\tau)}\left(1+i\frac{d\sigma}{d\tau}\right)d\tau - \int_{\frac{1}{2}\pi}^{\infty} e^{-\psi(\tau)}\left(\frac{d\tau}{d\sigma}+i\right)d\sigma\right)\right), \tag{9.6.90}$$

where

$$\chi = \Im\phi(t_0^+) = x\sinh\tau_0 - \nu\tau_0, \quad \psi(\tau) = x\cosh\tau\cos\sigma + \nu\sigma, \tag{9.6.91}$$

and the relation between τ and σ is given in (9.6.89) (with the + sign).

The second integral in (9.6.90) can be reduced to an integral over a finite interval. The fact is that the function

$$P(\sigma) = e^{-x\cosh\tau\cos\sigma}\left(\frac{d\tau}{d\sigma}+i\right) \tag{9.6.92}$$

is periodic with respect to σ with period 2π: $P(\sigma) = P(\sigma+2\pi)$. It follows that

$$\int_{\frac{1}{2}\pi}^{\infty} P(\sigma)e^{-\nu\sigma}\,d\sigma = \frac{1}{1-e^{-2\pi\nu}}\int_{\frac{1}{2}\pi}^{\frac{5}{2}\pi} P(\sigma)e^{-\nu\sigma}\,d\sigma. \tag{9.6.93}$$

Thus we obtain the representation

$$K_{i\nu}(x) = \cos\chi\left(\int_{\tau_0}^{\infty} e^{-\psi(\tau)}\,d\tau - \frac{1}{1-e^{-2\pi\nu}}\int_{\frac{1}{2}\pi}^{\frac{5}{2}\pi} e^{-\psi(\tau)}\frac{d\tau}{d\sigma}\,d\sigma\right) + \sin\chi\left(\int_{\tau_0}^{\infty} e^{-\psi(\tau)}\frac{d\sigma}{d\tau}\,d\tau - \frac{1}{1-e^{-2\pi\nu}}\int_{\frac{1}{2}\pi}^{\frac{5}{2}\pi} e^{-\psi(\tau)}\,d\sigma\right). \tag{9.6.94}$$

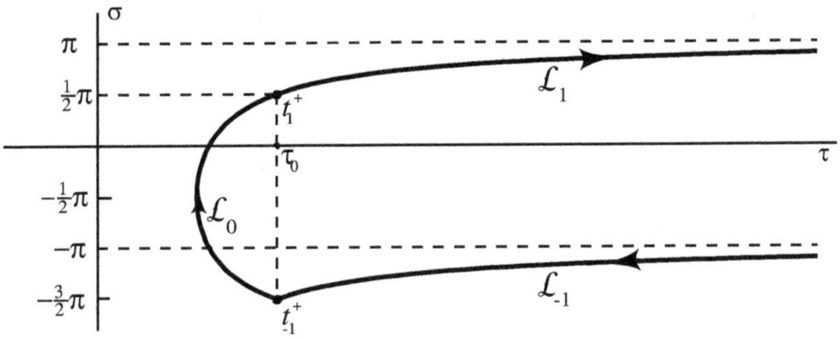

Fig. 9.6 Steepest descent path used in (9.6.95).

When the parameters x, ν are large the main contribution comes from the point $t_0^+ = \tau_0 + \frac{1}{2}\pi i$. At this point we have $e^{-\psi(\tau_0)} = e^{-\frac{1}{2}\pi\nu}$. This quantity gives a proper estimate of the maximal absolute value of the oscillating function $K_{i\nu}(x)$.

9.6.2.2 *The contours for $I_{i\nu}(x)$ and $L_{i\nu}(x)$*

The second integral in (9.6.69) can be split up into three parts:

- $\mathcal{L}_{-1}$ runs from $+\infty - \pi i$ to t_{-1}^+,
- $\mathcal{L}_0$ runs from t_{-1}^+ to t_0^+,
- $\mathcal{L}_1$ runs from t_0^+ to $+\infty + \pi i$.

These contours are shown in Figure 9.6.

The integrals along $\mathcal{L}_{-1}$ and $\mathcal{L}_1$ are transformed to integrals along the path from t_0^+ to ∞, which has been used for $K_{i\nu}(x)$. This is done by changing σ into $-\pi - \sigma, \pi - \sigma$, respectively. The integral along $\mathcal{L}_0$ is shifted upwards by changing σ into $-2\pi + \sigma$; afterwards we reverse the direction of integration from $\frac{1}{2}\pi$ to $\frac{5}{2}\pi$ into $\frac{5}{2}\pi$ to $\frac{1}{2}\pi$; the two operations for $\mathcal{L}_0$ can be established by the single transformation $\sigma \to \pi - \sigma$. In this way we obtain integrals as used for $K_{i\nu}(x)$. Corresponding τ values are not changed during these transformations, as follows from the relation (9.6.89). Corresponding values of $d\sigma/d\tau$ do change sign, however.

Thus we obtain

$$
\begin{aligned}
I_{i\nu}(x) = \frac{e^{i\chi}}{2\pi i}\Biggl(& 2\sinh \pi\nu \int_{\tau_0}^{\infty} e^{-\psi(\tau)} \left(1 - i\frac{d\sigma}{d\tau}\right) d\tau \; + \\
& e^{\pi\nu} \int_{\frac{1}{2}\pi}^{\frac{5}{2}\pi} e^{-\psi(\tau)} \left(-\frac{d\tau}{d\sigma} + i\right) d\sigma\Biggr).
\end{aligned}
\tag{9.6.95}
$$

The quantities χ and $\psi(\tau)$ are given in (9.6.91). The imaginary part of $I_{i\nu}(x)$ corresponds to (9.6.94) (see also (9.6.71)). Separation of the real parts gives

$$\begin{aligned} L_{i\nu}(x) = &\frac{\cos\chi}{2\pi}\left(-2\sinh\pi\nu\int_{\tau_0}^{\infty} e^{-\psi(\tau)}\frac{d\sigma}{d\tau}\,d\tau + e^{\pi\nu}\int_{\frac12\pi}^{\frac52\pi} e^{-\psi(\tau)}\,d\sigma\right) + \\ &\frac{\sin\chi}{2\pi}\left(2\sinh\pi\nu\int_{\tau_0}^{\infty} e^{-\psi(\tau)}\,d\tau - e^{\pi\nu}\int_{\frac12\pi}^{\frac52\pi} e^{-\psi(\tau)}\frac{d\tau}{d\sigma}\,d\sigma\right). \end{aligned} \tag{9.6.96}$$

When the parameters x, ν are large the main contribution again comes from the point $t_0^+ = \tau_0 + \frac12\pi i$. At this point we have $\psi(\tau_0) = \frac12\pi\nu$. Hence, the quantity $e^{\frac12\pi\nu}$ gives a proper estimate of the maximal absolute value of the oscillating functions $I_{i\nu}(x)$ and $L_{i\nu}(x)$.

9.7 A J-Bessel integral

As an application of the expansions of the Bessel functions and the relations with the Hankel functions, we investigate the asymptotic behavior of the integral

$$J_{\mu,\nu}(z) = \int_0^z t^\mu J_\nu(t)\,dt, \quad \Re(\mu+\nu) > -1, \tag{9.7.97}$$

for large values of z. By using the expansion

$$J_\nu(z) = \frac{(z/2)^\nu}{\Gamma(\nu+1)}\sum_{k=0}^{\infty}\frac{(-\frac14 z^2)^k}{(\nu+1)_k\,k!}, \tag{9.7.98}$$

it follows that

$$J_{\mu,\nu}(z) = \frac{z^{\mu+\nu+1}}{2^\nu\,\Gamma(\nu+1)}\sum_{k=0}^{\infty}\frac{(-\frac14 z^2)^k}{(\mu+\nu+2k+1)(\nu+1)_k\,k!}, \tag{9.7.99}$$

or, in terms of a generalized hypergeometric function,

$$J_{\mu,\nu}(z) = \frac{z^{\mu+\nu+1}}{(\mu+\nu+1)2^\nu\,\Gamma(\nu+1)}\,{}_1F_2\left({\frac12(\mu+\nu+1) \atop \frac12(\mu+\nu+3),\ \nu+1}\,;\,-\tfrac14 z^2\right). \tag{9.7.100}$$

Asymptotic expansions of generalized hypergeometric functions can be found in Askey and Olde Daalhuis (2010, §16.11), but for the present case we give a direct approach.

We use (see Olver and Maximon (2010, Eq. 10.22.43))

$$\int_0^\infty t^\mu J_\nu(t)\,dt = 2^\mu\frac{\Gamma\left(\frac12\nu+\frac12\mu+\frac12\right)}{\Gamma\left(\frac12\nu-\frac12\mu+\frac12\right)}, \quad \Re(\mu+\nu) > -1, \quad \Re\mu < \tfrac12, \tag{9.7.101}$$

and write

$$J_{\mu,\nu}(z) = 2^\mu\frac{\Gamma\left(\frac12\nu+\frac12\mu+\frac12\right)}{\Gamma\left(\frac12\nu-\frac12\mu+\frac12\right)} - \int_z^\infty t^\mu J_\nu(t)\,dt. \tag{9.7.102}$$

In this integral we can substitute the standard asymptotic expansion of the J-Bessel function, but it is easier to use the Hankel function. That is, we use (9.2.8) and write

$$J_{\mu,\nu}(z) = 2^{\mu}\frac{\Gamma\left(\frac12\nu+\frac12\mu+\frac12\right)}{\Gamma\left(\frac12\nu-\frac12\mu+\frac12\right)} - \Re\, H_{\mu,\nu}(z), \tag{9.7.103}$$

where

$$H_{\mu,\nu}(z) = \int_z^{\infty} t^{\mu} H_{\nu}^{(1)}(t)\, dt. \tag{9.7.104}$$

To dispose of the condition $\Re\mu < \frac12$ we take the path of integration along a half-line in the complex plane with $\Im t \to +\infty$. Because of the asymptotic behavior of the Hankel function this gives a convergent integral in (9.7.104) for all μ and ν. We derive the large-z asymptotic expansion of $H_{\mu,\nu}(z)$.

We recall the large-z expansion (see (9.2.6))

$$H_{\nu}^{(1)}(z) \sim \sqrt{\frac{2}{\pi z}}\, e^{i\omega} \sum_{k=0}^{\infty} \frac{c_k(\nu)}{z^k}, \tag{9.7.105}$$

where $\omega = z - \frac12\nu\pi - \frac14\pi$ and $c_k(\nu) = i^k a_k(\nu)$; $a_k(\nu)$ is defined in (9.1.4).

Integrating the asymptotic expansion, we find

$$H_{\mu,\nu}(z) \sim \sqrt{\frac{2}{\pi}}\, e^{i\omega} \sum_{k=0}^{\infty} c_k(\nu)\Phi_k, \tag{9.7.106}$$

where

$$\Phi_k = e^{-iz}\int_z^{\infty} t^{\mu-\frac12-k} e^{it}\, dt = e^{-iz} i^{\mu-k+\frac12}\Gamma\left(\mu-k+\tfrac12, -iz\right). \tag{9.7.107}$$

Here, $\Gamma(a,z)$ is the incomplete gamma function with large-z asymptotic expansion given in (7.2.9). For the Φ_k the expansion reads

$$\Phi_k \sim i z^{\mu-k-\frac12} \sum_{m=0}^{\infty} d_m(\mu,k) z^{-m}, \quad d_m(\mu,k) = \left(\tfrac12 + k - \mu\right)_m (-i)^m. \tag{9.7.108}$$

Using this in (9.7.106) we obtain

$$H_{\mu,\nu}(z) \sim i\, z^{\mu-\frac12}\sqrt{\frac{2}{\pi}}\, e^{i\omega} \sum_{k=0}^{\infty} h_k(\mu,\nu) z^{-k}, \tag{9.7.109}$$

where

$$h_k(\mu,\nu) = \sum_{j=0}^{k} c_j(\nu) d_{k-j}(\mu,j). \tag{9.7.110}$$

The first coefficients are

$$\begin{aligned} h_0(\mu,\nu) &= 1, \\ h_1(\mu,\nu) &= \tfrac18 i\left(-5+8\mu+4\nu^2\right), \\ h_2(\mu,\nu) &= \tfrac{1}{128}\left(-129+272\mu-128\mu^2+136\nu^2-64\nu^2\mu-16\nu^4\right). \end{aligned} \tag{9.7.111}$$

All coefficients $h_{2k}(\mu,\nu)$ and $h_{2k+1}(\mu,\nu)/i$ are real, and we write

$$H_{\mu,\nu}(z) = i\, z^{\mu-\frac{1}{2}}\sqrt{\frac{2}{\pi}}e^{i\omega}\left(P(\mu,\nu;z) + iQ(\mu,\nu;z)\right), \qquad (9.7.112)$$

giving

$$\Re H_{\mu,\nu}(z) = -z^{\mu-\frac{1}{2}}\sqrt{\frac{2}{\pi}}\left(P(\mu,\nu;z)\sin\omega + Q(\mu,\nu;z)\cos\omega\right), \qquad (9.7.113)$$

with expansions

$$P(\mu,\nu;z) \sim \sum_{k=0}^{\infty}\frac{h_{2k}(\mu,\nu)}{z^{2k}}, \quad Q(\mu,\nu;z) \sim \sum_{k=0}^{\infty}\frac{h_{2k+1}(\mu,\nu)}{i\, z^{2k+1}}. \qquad (9.7.114)$$

Having the form of the expansion of $H_{\mu,\nu}(z)$ in (9.7.109) we can obtain a recurrence relation for the coefficients h_k by differentiating (9.7.104) and (9.7.109). So we obtain

$$h_k(\mu,\nu) = c_k(\nu) + i\left(\mu+\tfrac{1}{2}-k\right)h_{k-1}(\mu,\nu), \quad k = 1,2,3,\ldots. \qquad (9.7.115)$$

Using the recursion

$$c_{k+1}(\nu) = i\lambda_k c_k(\nu), \quad \lambda_k = \frac{\nu^2-(k+\frac{1}{2})^2}{2(k+1)}, \qquad (9.7.116)$$

we can eliminate $c_k(\nu)$, and by writing

$$h_k(\mu,\nu) = i^k H_k(\mu,\nu), \qquad (9.7.117)$$

we obtain a real recursion for the $H_k(\mu,\nu)$:

$$H_{k+1}(\mu,\nu) = \left(\lambda_k+\mu-\tfrac{1}{2}-k\right)H_k(\mu,\nu) - \lambda_k\left(\mu+\tfrac{1}{2}-k\right)H_{k-1}(\mu,\nu). \quad (9.7.118)$$

With $H_{-1}(\mu,\nu) = 0, H_0(\mu,\nu) = 1$ we can use this for $k \ge 0$.

By using the $H_k(\mu,\nu)$, the series in (9.7.114) can be written in the form

$$\begin{aligned} P(\mu,\nu;z) &\sim \sum_{k=0}^{\infty}(-1)^k\frac{H_{2k}(\mu,\nu)}{z^{2k}},\\ Q(\mu,\nu;z) &\sim \sum_{k=0}^{\infty}(-1)^k\frac{H_{2k+1}(\mu,\nu)}{z^{2k+1}}. \end{aligned} \qquad (9.7.119)$$

Summarizing, the final result for $J_{\mu,\nu}(z)$ of (9.7.97) reads

$$\begin{aligned} J_{\mu,\nu}(z) = {} & 2^{\mu}\frac{\Gamma\left(\frac{1}{2}\nu+\frac{1}{2}\mu+\frac{1}{2}\right)}{\Gamma\left(\frac{1}{2}\nu-\frac{1}{2}\mu+\frac{1}{2}\right)} + \\ & z^{\mu-\frac{1}{2}}\sqrt{\frac{2}{\pi}}\Bigl(P(\mu,\nu;z)\sin\omega + Q(\mu,\nu;z)\cos\omega\Bigr), \end{aligned} \qquad (9.7.120)$$

where $P(\mu,\nu;z)$ and $Q(\mu,\nu;z)$ have the expansions given in (9.7.119). These expansions can be used for large values of z inside the sector $|\mathrm{ph}\, z| \le \pi-\delta$, and with fixed ν and μ which should satisfy $\Re(\mu+\nu) > -1$.

Chapter 10

Kummer functions

We consider asymptotic forms of this important class of functions, which are also known as confluent hypergeometric functions. The expansions in this chapter are for large argument z or for large parameters with other variables fixed, or in bounded domains. For other combinations of the parameters and argument we need uniform expansions. These will be given in other chapters as applications of general standard forms and as examples for certain methods; see Chapter 22 (in particular, §22.5) and Chapter 27.

For easy reference we give a few properties, for which we refer to Olde Daalhuis (2010a).

10.1 General properties

The confluent hypergeometric functions are solutions of the differential equation

$$zw'' + (c - z)w' - aw = 0. \tag{10.1.1}$$

The standard solutions are denoted by

$$ {}_1F_1\left({a \atop c} ; z \right) = \sum_{k=0}^{\infty} \frac{(a)_k}{(c)_k} \frac{z^k}{k!} \quad \text{and} \quad U(a, c, z), \tag{10.1.2}$$

with the usual exception $c \neq 0, -1, -2, \ldots$. The ratio ${}_1F_1(a; c; z)/\Gamma(c)$ is well defined for these values. We have

$$\lim_{c \to -m} \frac{1}{\Gamma(c)} {}_1F_1\left({a \atop c} ; z \right) = \frac{(a)_{m+1}\, z^{m+1}}{(m+1)!} {}_1F_1\left({a + m + 1 \atop m + 2} ; z \right). \tag{10.1.3}$$

The Kummer functions have integral representations

$$ {}_1F_1\left({a \atop c} ; z \right) = \frac{\Gamma(c)}{\Gamma(a)\,\Gamma(c - a)} \int_0^1 e^{zt} t^{a-1} (1 - t)^{c-a-1}\, dt, \tag{10.1.4}$$

valid for $\Re a > 0$ and $\Re(c - a) > 0$, and

$$U(a, c, z) = \frac{1}{\Gamma(a)} \int_0^\infty e^{-zt} t^{a-1} (1 + t)^{c-a-1}\, dt, \tag{10.1.5}$$

valid for $\Re a > 0$ and $\Re z > 0$.

The following contour integrals are available as well, and they are important for deriving asymptotic expansions for large parameters. We have (see Olde Daalhuis (2010a, §13.4(ii)))

$$ {}_1F_1\left(\begin{matrix} a \\ c \end{matrix}; z\right) = \frac{\Gamma(c)\Gamma(1+a-c)}{2\pi i\,\Gamma(a)} \int_0^{(1+)} e^{zt} t^{a-1}(t-1)^{c-a-1}\,dt, \tag{10.1.6} $$

where $\Re a > 0$ and $c - a \neq 1, 2, 3, \ldots$;

$$ {}_1F_1\left(\begin{matrix} a \\ c \end{matrix}; z\right) = \frac{z^{1-c}\Gamma(c)}{2\pi i} \int_{-\infty}^{(1+)} e^{zt} t^{-c}\left(1 - \frac{1}{t}\right)^{-a} dt, \tag{10.1.7} $$

where $|\mathrm{ph}\, z| < \frac{1}{2}\pi$;

$$ U(a, c, z) = \frac{\Gamma(1-a)}{2\pi i} \int_{-\infty}^{(0+)} e^{zt} t^{a-1}(1-t)^{c-a-1} dt, \tag{10.1.8} $$

where $a \neq 1, 2, 3, \ldots$ and $|\mathrm{ph}\, z| < \frac{1}{2}\pi$. The contour cuts the real axis between 0 and 1. At this point the fractional powers are determined by $\mathrm{ph}(1-t) = 0$ and $\mathrm{ph}\, t = 0$.

Other contour integrals follow from the relations

$$ {}_1F_1\left(\begin{matrix} a \\ c \end{matrix}; z\right) = e^z\, {}_1F_1\left(\begin{matrix} c-a \\ c \end{matrix}; -z\right), \tag{10.1.9} $$

which is called Kummer's transformation, and

$$ U(a, c, z) = z^{1-c} U(1+a-c, 2-c, z). \tag{10.1.10} $$

Connection formulas between the Kummer functions are

$$ \frac{1}{\Gamma(c)}\, {}_1F_1\left(\begin{matrix} a \\ c \end{matrix}; z\right) = \frac{e^{\mp\pi i a}}{\Gamma(c-a)} U(a, c, z) + \frac{e^{\pm\pi i (c-a)}}{\Gamma(a)} e^z U\left(c-a, c, z e^{\pm\pi i}\right), \tag{10.1.11} $$

and, when c is not an integer,

$$ U(a, c, z) = \frac{\Gamma(1-c)}{\Gamma(a-c+1)}\, {}_1F_1\left(\begin{matrix} a \\ c \end{matrix}; z\right) + \frac{z^{1-c}\Gamma(c-1)}{\Gamma(a)}\, {}_1F_1\left(\begin{matrix} a-c+1 \\ 2-c \end{matrix}; z\right). \tag{10.1.12} $$

Bessel functions, incomplete gamma functions, and parabolic cylinder functions are special cases of the Kummer functions. For the modified Bessel functions we have

$$ \begin{aligned} I_\nu(z) &= \frac{\left(\frac{1}{2}z\right)^\nu e^{\pm z}}{\Gamma(\nu+1)}\, {}_1F_1\left(\begin{matrix} \nu + \frac{1}{2} \\ 2\nu + 1 \end{matrix}; \mp 2z\right), \\ K_\nu(z) &= \sqrt{\pi}(2z)^\nu e^{-z}\, U\left(\nu + \tfrac{1}{2}, 2\nu+1, 2z\right). \end{aligned} \tag{10.1.13} $$

The Whittaker functions are an alternative pair of the Kummer functions, with relations

$$ \begin{aligned} M_{\kappa,\mu}(z) &= e^{-\frac{1}{2}z} z^{\frac{1}{2}+\mu}\, {}_1F_1\left(\begin{matrix} \frac{1}{2} + \mu - \kappa \\ 1 + 2\mu \end{matrix}; z\right), \\ W_{\kappa,\mu}(z) &= e^{-\frac{1}{2}z} z^{\frac{1}{2}+\mu}\, U\left(\tfrac{1}{2} + \mu - \kappa, 1 + 2\mu, z\right). \end{aligned} \tag{10.1.14} $$

These functions are solutions of the differential equation

$$w'' + \left(-\frac{1}{4} + \frac{\kappa}{z} + \frac{\frac{1}{4} - \mu^2}{z^2}\right) w = 0. \tag{10.1.15}$$

For $a = -n$ (n a nonnegative integer) it follows from (10.1.12) that

$$\begin{aligned} U(-n, c, z) &= \frac{\Gamma(1-c)}{\Gamma(-n-c+1)} {}_1F_1\left({-n \atop c}; z\right) \\ &= (-1)^n (c)_n {}_1F_1\left({-n \atop c}; z\right), \end{aligned} \tag{10.1.16}$$

which is the case of the Laguerre polynomials, because

$$L_n^{(\alpha)}(z) = \frac{(-1)^n}{n!} U(-n, \alpha+1, z) = \binom{n+\alpha}{n} {}_1F_1\left({-n \atop \alpha+1}; z\right). \tag{10.1.17}$$

10.2 Asymptotic expansions for large z

The asymptotic expansion for large z of $U(a, c, z)$ follows as in §9.1 for the expansion of $K_\nu(z)$. We have

$$U(a, c, z) \sim z^{-a} \sum_{k=0}^{\infty} \frac{(a)_k (a-c+1)_k}{k!\,(-z)^k}, \quad |\operatorname{ph} z| \le \tfrac{3}{2}\pi - \delta. \tag{10.2.18}$$

For ${}_1F_1(a; c; z)$ we need first a simple transformation $t \to 1-t$ in (10.1.4), which gives

$${}_1F_1\left({a \atop c}; z\right) = \frac{\Gamma(c)\, e^z}{\Gamma(a)\, \Gamma(c-a)} \int_0^1 e^{-zt} t^{c-a-1} (1-t)^{a-1}\, dt, \tag{10.2.19}$$

which corresponds to the relation in (10.1.9).

As in §9.3 for the special case $I_\nu(z)$, we obtain by expanding $(1-t)^{a-1}$ in powers of t the result

$${}_1F_1\left({a \atop c}; z\right) \sim \frac{e^z z^{a-c} \Gamma(c)}{\Gamma(a)} \sum_{k=0}^{\infty} \frac{(c-a)_k (1-a)_k}{k!\, z^k}, \quad |\operatorname{ph} z| \le \tfrac{1}{2}\pi - \delta. \tag{10.2.20}$$

We have explained for $I_\nu(z)$ how the sector of z can be extended by making a compound expansion. We can repeat the analysis for the function ${}_1F_1(a; c; z)$, but we can also use the connection formulas in (10.1.11) and use the expansion of $U(a, c, z)$ in (10.2.18). In this way,

$$\begin{aligned} {}_1F_1\left({a \atop c}; z\right) \sim\ & \frac{e^z z^{a-c} \Gamma(c)}{\Gamma(a)} \sum_{k=0}^{\infty} \frac{(c-a)_k (1-a)_k}{k!\, z^k} + \\ & \frac{e^{\pm\pi i a} z^{-a} \Gamma(c)}{\Gamma(c-a)} \sum_{k=0}^{\infty} \frac{(a)_k (a-c+1)_k}{k!\, (-z)^k}, \end{aligned} \tag{10.2.21}$$

and these expansions are valid in the sectors $-\frac{1}{2}\pi+\delta \le \pm\mathrm{ph}\, z \le \frac{3}{2}\pi-\delta$, where the upper and lower signs correspond to those in (10.2.21).

For the asymptotic expansion of $U(a,c,z)$ converging factors are developed; see §2.6 for details on this topic and references. For converging factors of the expansion of ${}_1F_1(a;c;z)$, see Slater (1960, §4.2).

In §17.2 we will give alternatives for the asymptotic expansion of $U(a,c,z)$ in the form of a convergent expansion in terms of other Kummer U-functions.

10.3 Expansions for large a

For large a and bounded values of z and c we derive expansions for $a \to \pm\infty$ in terms of Bessel functions, and these expansions are valid in bounded z-domains that contain the origin. For the U-functions we exclude $z \le 0$ because this function is singular at the origin and has a branch cut for these values.

Earlier, asymptotic expansions for large a are given in Slater (1960, §4.4.1, §4.6.1), where the results are for ${}_1F_1\left(\frac{1}{4}u^2+\frac{1}{2}c;c;z^2\right)$ and $U\left(\frac{1}{4}u^2+\frac{1}{2}c,c;z^2\right)$ for large u. This choice of notation for the a-parameter has some advantages (the coefficients in the expansions are simpler). Slater derived expansions by using the Kummer differential equation (10.1.1), and the coefficients follow from recurrence relations. In §10.3.6 we will point out that Slater's result for the U-function is not correct.

10.3.1 *Tricomi's function $E_\nu(z)$*

We will obtain expansions of the Kummer F-function in terms of the Bessel functions $I_\nu(z)$ and $J_\nu(z)$, and these functions have an algebraic singularity at $z=0$. The F-function is analytic in z and we can modify the asymptotic representations that we will derive by using the function $E_\nu(z)$ used by Tricomi; see Tricomi (1947, p. 34). We have

$$E_\nu(z)=z^{-\frac{1}{2}\nu}J_\nu\left(2\sqrt{z}\right)=\sum_{k=0}^{\infty}(-1)^k\frac{z^k}{k!\,\Gamma(\nu+k+1)}. \tag{10.3.22}$$

It is an analytic function of z. In terms of the modified Bessel function we can write

$$E_\nu(-z)=z^{-\frac{1}{2}\nu}I_\nu\left(2\sqrt{z}\right)=\sum_{k=0}^{\infty}\frac{z^k}{k!\,\Gamma(\nu+k+1)}. \tag{10.3.23}$$

We will use the integral representations (see Olver and Maximon (2010, Eq. 10.9.19))

$$J_\nu(z)=\frac{\left(\frac{1}{2}z\right)^\nu}{2\pi i}\int_{-\infty}^{(0+)} e^{t-z^2/(4t)}\,\frac{dt}{t^{\nu+1}}, \tag{10.3.24}$$

and

$$I_\nu(z)=\frac{\left(\frac{1}{2}z\right)^\nu}{2\pi i}\int_{-\infty}^{(0+)} e^{t+z^2/(4t)}\,\frac{dt}{t^{\nu+1}}, \tag{10.3.25}$$

which follows from the previous one when z is replaced by $e^{\frac{1}{2}\pi i}z$.

For $E_\nu(z)$ we have the integral representation

$$E_\nu(z) = \frac{1}{2\pi i}\int_{-\infty}^{(0+)} e^{t-z/t}\frac{dt}{t^{\nu+1}}, \tag{10.3.26}$$

which is valid for all finite complex values of ν and z.

10.3.2 *Expansion of* $U(a,c,z)$*,* $a \to +\infty$

For the U-function we start with (10.1.5). By writing $t/(1+t) = e^{-s}$ the integral becomes after a few steps

$$U(a,c,z) = \frac{e^{\frac{1}{2}z}}{\Gamma(a)}\int_0^\infty e^{-as-z/s}s^{-c}f(z,s)\,ds, \tag{10.3.27}$$

where

$$f(z,s) = e^{zg(s)}\left(\frac{s}{1-e^{-s}}\right)^c, \quad g(s) = \frac{1}{s} - \frac{1}{e^s-1} - \frac{1}{2}. \tag{10.3.28}$$

The function f is analytic in the strip $|\Im s| < 2\pi$ and it can be expanded for $|s| < 2\pi$ into

$$f(z,s) = \sum_{k=0}^{K-1} c_k(z)s^k + s^K f_K(z,s), \quad K = 0,1,2,\ldots, \tag{10.3.29}$$

where the coefficients $c_k(z)$ are combinations of Bernoulli numbers and Bernoulli polynomials; see §15.6. We have

$$\left(\frac{s}{1-e^{-s}}\right)^c = \sum_{k=0}^\infty \frac{B_k^c(0)}{k!}(-s)^k, \quad g(s) = -\sum_{k=1}^\infty \frac{B_{2k}}{(2k)!}s^{2k-1}. \tag{10.3.30}$$

The first few are

$$\begin{aligned} &c_0(z) = 1, \quad c_1(z) = \tfrac{1}{12}\left(6c - z\right),\\ &c_2(z) = \tfrac{1}{288}\left(-12c + 36c^2 - 12zc + z^2\right),\\ &c_3(z) = \tfrac{1}{51840}\left(-5z^3 + 90z^2c + (-540c^2 + 180c + 72)z + 1080c^2(c-1)\right). \end{aligned} \tag{10.3.31}$$

We substitute the expansion (10.3.29) into (10.3.27) and obtain

$$U(a,c,z) = \frac{e^{\frac{1}{2}z}}{\Gamma(a)}\left(\sum_{k=0}^{K-1} c_k(z)\Psi_k(z) + R_K(a,c,z)\right), \tag{10.3.32}$$

where

$$R_K(a,c,z) = \int_0^\infty e^{-as-z/s}s^{K-c}f_K(z,s)\,ds, \tag{10.3.33}$$

and, in terms of the modified Bessel function $K_\nu(z)$,

$$\Psi_k(z) = \int_0^\infty e^{-as-z/s}s^{k-c}\,ds = 2\left(\frac{z}{a}\right)^{\frac{1}{2}(k+1-c)} K_{c-k-1}\left(2\sqrt{az}\right). \tag{10.3.34}$$

This representation follows from (see Olver and Maximon (2010, Eq. 10.32.10))

$$K_\nu(z) = \tfrac{1}{2}\left(\tfrac{1}{2}z\right)^\nu \int_0^\infty e^{-t-z^2/(4t)} \frac{dt}{t^{\nu+1}}, \quad |\mathrm{ph}\, z| < \tfrac{1}{4}\pi. \tag{10.3.35}$$

This function is even with respect to the order: $K_\nu(z) = K_{-\nu}(z)$.

In Temme (1981) we have constructed a bound for the remainder R_K and we have shown that the sequence $\{\Psi_k\}$ constitutes an asymptotic sequence (see §1.4) for $a \to \infty$ in the sense that

$$\frac{\Psi_k(z)}{\Psi_{k-1}(z)} = \mathcal{O}\left(\frac{1+\sqrt{az}}{a}\right), \quad a \to \infty, \tag{10.3.36}$$

uniformly in bounded c-intervals and bounded z-intervals ($z > 0$), but these intervals can be extended to complex domains. This shows the asymptotic nature of the expansion in (10.3.32).

The complete expansion in terms of the K-function reads

$$U(a,c,z) \sim 2\left(\frac{z}{a}\right)^{\frac{1}{2}(1-c)} \frac{e^{\frac{1}{2}z}}{\Gamma(a)} \sum_{k=0}^\infty c_k(z) \left(\frac{z}{a}\right)^{\frac{1}{2}k} K_{c-k-1}\left(2\sqrt{az}\right), \tag{10.3.37}$$

as $a \to +\infty$, uniformly in bounded c- and z-domains. We can give c every finite complex value, and for z we exclude real negative values and $z = 0$. For the condition on a we refer to Remark 10.2. In §10.3.5 we give an expansion for $a \to -\infty$.

By using the recurrence relation

$$K_{\nu+1}(z) = K_{\nu-1}(z) + \frac{2\nu}{z} K_\nu(z), \tag{10.3.38}$$

and rearranging the expansion in (10.3.37) we can obtain an expansion of the form

$$\begin{aligned} &U(a,c,z) \sim 2\left(\frac{z}{a}\right)^{\frac{1}{2}(1-c)} \frac{e^{\frac{1}{2}z}}{\Gamma(a)} \times \\ &\left(K_{c-1}\left(2\sqrt{az}\right) \sum_{k=0}^\infty \frac{a_k(z)}{a^k} + \sqrt{\frac{z}{a}} K_c\left(2\sqrt{az}\right) \sum_{k=0}^\infty \frac{b_k(z)}{a^k}\right), \end{aligned} \tag{10.3.39}$$

as $a \to \infty$ inside the sector $-\pi+\delta \le \mathrm{ph}\, a \le \pi - \delta$ (see Remark 10.2). The coefficients can be expressed in terms of $c_k(z)$, and the first relations are

$$\begin{aligned} a_0(z) &= c_0(z) = 1, \quad b_0(z) = c_1, \\ a_1(z) &= (1-c)c_1 + zc_2, \\ b_1(z) &= (2-c)c_2 + zc_3, \\ a_2(z) &= (c^2 - 3c + 2)c_2 + (4z - 2zc)c_3 + z^2 c_4, \\ b_2(z) &= (c^2 - 5c + 6)c_3 + (6z - 2zc)c_4 + z^2 c_5, \end{aligned} \tag{10.3.40}$$

where $c_k = c_k(z)$ are defined in (10.3.29) and the first few are given in (10.3.31).

Remark 10.1. Explicit representations of these coefficients are[1]

$$\begin{aligned} a_k(z) &= \sum_{m=0}^{k} \binom{k}{m} (1-c)_{k-m} z^m c_{k+m}(z), \\ b_k(z) &= \sum_{m=0}^{k} \binom{k}{m} (2-c)_{k-m} z^m c_{k+m+1}(z), \end{aligned} \tag{10.3.41}$$

$k = 0, 1, 2, \ldots$. △

Remark 10.2. Integrals as given in (10.3.27) can be defined for a large domain of the complex parameters by modifying the path of integration. For example, such that the path in (10.3.27) meets the origin at an angle equal to $\operatorname{ph} z$. In this way we can define this integral for all $\operatorname{ph} z \in [-\pi, \pi]$, $z \neq 0$. When we can do this, that is, when $z \neq 0$, we can take any complex value of c. We can allow complex values of a by modifying the contour at infinity, just as in Watson's lemma. The functions $f(z, s)$ and $f_K(z, s)$ (see (10.3.29)) have singularities on the imaginary axis, and for a we can allow complex values in the sector $-\pi + \delta \le \operatorname{ph} a \le \pi - \delta$. △

Remark 10.3. We can obtain similar expansions for $U(a + \alpha, c, z)$ with α a fixed number. In that case (10.3.27) becomes

$$U(a+\alpha, c, z) = \frac{e^{\frac{1}{2}z}}{\Gamma(a+\alpha)} \int_0^\infty e^{-as-z/s} s^{-c} \widetilde{f}(z, s)\, ds, \tag{10.3.42}$$

where $\widetilde{f}(z, s) = e^{-\alpha s} f(z, s)$. For certain applications this modification may be convenient. The construction of the expansions can be repeated, with different gamma function and coefficients $a_k(z), b_k(z), c_k(z)$ in (10.3.39). The special choice $\alpha = \frac{1}{2}c$ gives Slater's form of the a-parameter, mentioned in the beginning of §10.3, and the function $\widetilde{f}(z, s)$ is given by

$$\widetilde{f}(z, s) = e^{zg(s)} \left(\frac{\frac{1}{2}s}{\sinh\left(\frac{1}{2}s\right)} \right)^c. \tag{10.3.43}$$

The expansion based on this case is related to the one given in (10.3.75); they are not the same because of the factor $C(a, b)$ in that result. △

10.3.3 *Expansion of* ${}_1F_1(a; c; z)$, $a \to +\infty$

For an expansion of the F-function we start with the integral in (10.1.6) with as contour the circle $|t - 1| = 1$. The transformation $t = w/(w - 1)$ transforms this

[1]With thanks to Hanyou Chu, Sao Paolo, who obtained this result after considering the first values. A proof follows from induction.

circle into itself. To verify this we write $w = t/(t-1)$. With $t = 1 + e^{i\theta}$, $\theta \in [0, 2\pi)$, we obtain $w = 1 + e^{-i\theta}$. The result of the substitution is

$$ {}_1F_1\left(\begin{matrix} a \\ c \end{matrix}; z\right) = \frac{\Gamma(c)\Gamma(1+a-c)}{2\pi i\,\Gamma(a)} \int_{\mathcal{C}} e^{zw/(w-1)} w^{a-1} (w-1)^{-c}\, dw, \tag{10.3.44} $$

where $\mathcal{C}$ is the circle $|w-1| = 1$, again with the positive orientation.

Next we take $w = e^s$. With $w = 1 + e^{i\theta}$, $\theta \in [0, 2\pi)$, we see that the circle $\mathcal{C}$ becomes a path in the s-plane described by

$$ s = u + iv, \quad u = \ln(2\cos v), \quad -\tfrac{1}{2}\pi < v < \tfrac{1}{2}\pi, \quad u \le \ln 2. \tag{10.3.45} $$

After some manipulations we obtain

$$ {}_1F_1\left(\begin{matrix} a \\ c \end{matrix}; z\right) = \frac{\Gamma(c)\Gamma(1+a-c)\, e^{\frac{1}{2}z}}{\Gamma(a)\; 2\pi i} \int_{\mathcal{L}} e^{as+z/s} s^{-c} f(z, -s)\, ds, \tag{10.3.46} $$

where f is defined in (10.3.28) and $\mathcal{L}$ can be taken as the Hankel contour shown in Figure 2.1. Below and above the branch cut along the negative axis the phase of s is $-\pi$ and $+\pi$, respectively. This representation is valid for all complex z and c; initially it is valid for $\Re a > 0$, but we can rotate the contour to include values inside the sector $|\operatorname{ph} a| < \pi$.

Upon substituting the expansion in (10.3.29) we obtain

$$ \begin{aligned} &\frac{1}{\Gamma(c)}\, {}_1F_1\left(\begin{matrix} a \\ c \end{matrix}; z\right) = \\ &\qquad \frac{\Gamma(1+a-c)e^{\frac{1}{2}z}}{\Gamma(a)} \left(\sum_{k=0}^{K-1} (-1)^k c_k(z) \Phi_k(z) + S_K(a, c, z)\right), \end{aligned} \tag{10.3.47} $$

where

$$ S_K(a, c, z) = \frac{(-1)^K}{2\pi i} \int_{\mathcal{L}} e^{as+z/s} s^{K-c} f_K(z, -s)\, ds, \tag{10.3.48} $$

and, in terms of the modified Bessel function $I_\nu(z)$ (see (10.3.25)),

$$ \Phi_k(z) = \frac{1}{2\pi i} \int_{\mathcal{L}} e^{as+z/s} s^{k-c}\, ds = \left(\frac{z}{a}\right)^{\frac{1}{2}(k+1-c)} I_{c-k-1}\left(2\sqrt{az}\right). \tag{10.3.49} $$

When we use the function $E_\nu(z)$ we can write the function Φ_k in the form (see (10.3.23))

$$ \Phi_k(z) = a^{c-k-1} E_{c-k-1}(-az), \tag{10.3.50} $$

and the complete expansion has the form

$$ \frac{1}{\Gamma(c)}\, {}_1F_1\left(\begin{matrix} a \\ c \end{matrix}; z\right) \sim a^{c-1} \frac{\Gamma(1+a-c)e^{\frac{1}{2}z}}{\Gamma(a)} \sum_{k=0}^{\infty} (-1)^k \frac{c_k(z)}{a^k} E_{c-k-1}(-az), \tag{10.3.51} $$

as $a \to +\infty$, uniformly in bounded c- and z-domains.

In the above results we require $\Re a \ge 0$. For $c = 0, -1, -2, \ldots$, the left-hand side of (10.3.47) can be interpreted by using (10.1.3).

We can rearrange the expansions of the F-function as we did in (10.3.39). We use

$$I_{\nu-1}(z) = I_{\nu+1}(z) + \frac{2\nu}{z} I_\nu(z), \tag{10.3.52}$$

and obtain

$$\begin{aligned} \frac{1}{\Gamma(c)}\,{}_1F_1\left({a \atop c}; z\right) &\sim \left(\frac{z}{a}\right)^{\frac12(1-c)} \frac{\Gamma(1+a-c)e^{\frac12 z}}{\Gamma(a)} \times \\ &\left(I_{c-1}\left(2\sqrt{az}\right) \sum_{k=0}^{\infty} \frac{a_k(z)}{a^k} - \sqrt{\frac{z}{a}} I_c\left(2\sqrt{az}\right) \sum_{k=0}^{\infty} \frac{b_k(z)}{a^k}\right). \end{aligned} \tag{10.3.53}$$

These expansions of ${}_1F_1(a;c;z)$ are valid for $a \to \infty$ inside the sector $-\pi + \delta \le \operatorname{ph} a \le \pi - \delta$. The coefficients $a_k(z)$ and $b_k(z)$ are the same as in (10.3.39).

Remark 10.4. For an expansion of ${}_1F_1(a+\alpha;c;z)$ for $a \to \infty$ and α fixed, the same approach can be used as described in Remark 10.3. △

Remark 10.5. An expansion of ${}_1F_1(a;c;z)$ for $a \to \infty$ was first given by Perron (1921) in terms of elementary functions. This expansion follows from the one given in (10.3.53) by expanding the modified Bessel function for large argument (assuming that z is bounded away from the origin). △

10.3.4 *Expansion of* ${}_1F_1(a;c;z)$, $a \to -\infty$

For the F-function we can use (10.1.9), and this gives an expansion as in (10.3.47) with $c-a$ as the large parameter. However, we can write, using (10.3.46),

$$\begin{aligned} {}_1F_1\left({-a \atop c}; -z\right) &= e^{-z}\,{}_1F_1\left({c+a \atop c}; z\right) \\ &= \frac{\Gamma(c)\Gamma(1+a)e^{-\frac12 z}}{2\pi i\,\Gamma(a+c)} \int_{\mathcal{L}} e^{as+z/s} s^{-c} f(-z,s)\,ds, \end{aligned} \tag{10.3.54}$$

because $e^{cs} f(z,-s) = f(-z,s)$ with f given in (10.3.28). By expanding f as in (10.3.29) this representation gives

$$\frac{1}{\Gamma(c)}\,{}_1F_1\left({-a \atop c}; -z\right) = \frac{\Gamma(1+a)e^{-\frac12 z}}{\Gamma(a+c)} \left(\sum_{k=0}^{K-1} c_k(-z)\Phi_k(z) + T_K(a,c,z)\right), \tag{10.3.55}$$

where Φ_k is defined in (10.3.49) and

$$T_K(a,c,z) = \frac{1}{2\pi i} \int_{\mathcal{L}} e^{as+z/s} s^{K-c} f_K(-z,s)\,ds. \tag{10.3.56}$$

In terms of the function $E_\nu(z)$ we can write the complete expansion in the form (changing the sign of z)

$$\frac{1}{\Gamma(c)}\,{}_1F_1\left({-a \atop c}; z\right) \sim a^{c-1} \frac{\Gamma(1+a)e^{\frac12 z}}{\Gamma(a+c)} \sum_{k=0}^{\infty} \frac{c_k(z)}{a^k} E_{c-k-1}(az), \tag{10.3.57}$$

as $a \to \infty$. Again, in these results we require $\Re a \geq 0$ and bounded $|z|$.

We can also write

$$\frac{1}{\Gamma(c)}\,{}_1F_1\left(\begin{matrix}-a\\ c\end{matrix};\, z\right) \sim \left(\frac{z}{a}\right)^{\frac{1}{2}(1-c)} \frac{\Gamma(1+a)e^{\frac{1}{2}z}}{\Gamma(a+c)} \times \left(J_{c-1}\left(2\sqrt{az}\right)\sum_{k=0}^{\infty}\frac{a_k(z)}{a^k} - \sqrt{\frac{z}{a}}\,J_c\left(2\sqrt{az}\right)\sum_{k=0}^{\infty}\frac{b_k(z)}{a^k}\right). \tag{10.3.58}$$

These expansions of ${}_1F_1(-a;c;z)$ are valid for $a \to \infty$ inside the sector $-\pi+\delta \leq \operatorname{ph} a \leq \pi - \delta$. The coefficients $a_k(z)$ and $b_k(z)$ are the same as in (10.3.39) and (10.3.53).

For $a = n$ (a nonnegative integer) we have for the Laguerre polynomials (see (10.1.17))

$$L_n^{(\alpha)}(z) \sim n^{\alpha} e^{\frac{1}{2}z}\left(E_\alpha(nz)\sum_{k=0}^{\infty}\frac{a_k(z)}{n^k} - zE_{\alpha+1}(nz)\sum_{k=0}^{\infty}\frac{b_k(z)}{n^k}\right), \tag{10.3.59}$$

where the coefficients are as in (10.3.58), with $c = \alpha+1$. This expansion is uniformly valid for z in a bounded domain containing the origin and for bounded $|\alpha|$.

Remark 10.6. For an expansion of ${}_1F_1(-a-\alpha;c;z)$ for $a \to \infty$ and α fixed, the same approach can be used as described in Remark 10.3. △

Remark 10.7. The expansions of this section are related to the convergent expansion given by Buchholz, see Buchholz (1969, §7.4). In §15.8.2 we will give details on the coefficients in this expansion. △

10.3.5 *Expansion of $U(a,c,z)$, $a \to -\infty$*

In this case we can again use connection formulas, for example, the relation (10.1.12) in the form (using (10.1.9))

$$U(-a,c,z) = \frac{\Gamma(1-c)}{\Gamma(-a-c+1)}\,{}_1F_1\left(\begin{matrix}-a\\ c\end{matrix};\, z\right) + \frac{\Gamma(c-1)}{\Gamma(-a)}z^{1-c}e^{z}\,{}_1F_1\left(\begin{matrix}1+a\\ 2-c\end{matrix};\, -z\right), \tag{10.3.60}$$

which form has a real representation. However, although the U-function is well defined for integer values of c, a nasty limiting procedure is needed in that case. In addition, the second F-function has $2-c$ at the c-place, which gives different coefficients compared with the ones for the first F-function.

The best approach is by using the two forms given in (10.1.11). We write

$$\frac{1}{\Gamma(a+c)}U(-a,c,z) = \frac{e^{\mp\pi i a}}{\Gamma(c)}\,{}_1F_1\left(\begin{matrix}-a\\ c\end{matrix};\, z\right) - \frac{e^{z\pm\pi i c}}{\Gamma(-a)}U\left(a+c,c,ze^{\pm\pi i}\right). \tag{10.3.61}$$

We will see in (10.3.67) that this representation also gives a real form of the asymptotic result.

For the U-function on the right-hand side we have the representation

$$U(a+c,c,z)=\frac{e^{\frac12 z}}{\Gamma(a+c)}\int_0^\infty s^{-c}e^{-as-z/s}f(z,-s)\,ds, \tag{10.3.62}$$

where $f(z,s)$ is defined in (10.3.28), and using (10.3.34) we obtain

$$\begin{aligned}&\frac{\Gamma(a+c)}{\Gamma(-a)}\,e^{z}U\left(a+c,c,ze^{\pm\pi i}\right)\sim\\ &\frac{e^{\frac12 z}}{\Gamma(-a)}\sum_{k=0}^{\infty}(-1)^k c_k(z)\left(\frac{ze^{\pm\pi i}}{a}\right)^{\frac12(k+1-c)}K_{c-k-1}\left(2e^{\pm\frac12\pi i}\sqrt{az}\right).\end{aligned} \tag{10.3.63}$$

We express the modified Bessel functions in this case in terms of ordinary Bessel functions by using

$$e^{\pm\frac12\nu\pi i}K_\nu\left(ze^{\pm\frac12\pi i}\right)=-\tfrac12\pi\left(Y_\nu(z)\pm iJ_\nu(z)\right),\quad |\mathrm{ph}\,z|\le\tfrac12\pi, \tag{10.3.64}$$

which follows from (9.2.5) and (9.2.8). This gives

$$\begin{aligned}&\frac{\Gamma(a+c)}{\Gamma(-a)}\,e^{z\pm\pi ic}U\left(a+c,c,ze^{\pm\pi i}\right)\sim\\ &\tfrac12\pi\frac{e^{\frac12 z}}{\Gamma(-a)}\sum_{k=0}^{\infty}c_k\left(\frac{z}{a}\right)^{\frac12(k+1-c)}\left(Y_{c-k-1}(\zeta)\pm iJ_{c-k-1}(\zeta)\right),\end{aligned} \tag{10.3.65}$$

where $\zeta=2\sqrt{az}$.

For the F-function in (10.3.57) we use the expansion given in (10.3.55), and we replace z by $-z$. Then the expansion can be written in terms of the J-Bessel function; see §10.3.1. That is, we can write (10.3.55) in the form

$$\frac{1}{\Gamma(c)}\,{}_1F_1\left(\begin{matrix}-a\\ c\end{matrix};z\right)\sim\left(\frac{z}{a}\right)^{\frac12(1-c)}\frac{\Gamma(1+a)e^{\frac12 z}}{\Gamma(a+c)}\sum_{k=0}^{\infty}c_k(z)\left(\frac{z}{a}\right)^{\frac12 k}J_{c-k-1}(\zeta). \tag{10.3.66}$$

Using the results in (10.3.65) and (10.3.66) in (10.3.61) we obtain

$$\begin{aligned}U(-a,c,z)\sim\left(\frac{z}{a}\right)^{\frac12(1-c)}&\Gamma(a+1)e^{\frac12 z}\sum_{k=0}^{\infty}c_k(z)\left(\frac{z}{a}\right)^{\frac12 k}\times\\ &\left(\cos(\pi a)J_{c-k-1}(\zeta)+\sin(\pi a)Y_{c-k-1}(\zeta)\right),\end{aligned} \tag{10.3.67}$$

as $a\to\infty$, where $\zeta=2\sqrt{az}$. We assume that $\Re a\ge0$ and that z and c are bounded complex numbers, $z\ne0$. Observe that the condition for the argument of the Bessel functions in (10.3.64) gives for z in the above result $|\mathrm{ph}\,z|\le\pi$.

Again we can rearrange the expansion as in previous cases, and we have

$$\begin{aligned}&U(-a,c,z)\sim\left(\frac{z}{a}\right)^{\frac12(1-c)}\Gamma(a+1)e^{\frac12 z}\times\\ &\left(C_{c-1}(\zeta)\sum_{k=0}^{\infty}\frac{a_k(z)}{a^k}-\sqrt{\frac{z}{a}}C_c(\zeta)\sum_{k=0}^{\infty}\frac{b_k(z)}{a^k}\right),\end{aligned} \tag{10.3.68}$$

where

$$C_\nu(\zeta) = \cos(\pi a) J_\nu(\zeta) + \sin(\pi a) Y_\nu(\zeta). \tag{10.3.69}$$

These expansions of $U(-a, c, z)$ are valid for $a \to \infty$ inside the sector $-\pi + \delta \le \operatorname{ph} a \le \pi - \delta$. The coefficients $a_k(z)$ and $b_k(z)$ are the same as in (10.3.39).

For $a = n$ (a nonnegative integer) we obtain from (10.3.68) the expansion given in (10.3.59) for the Laguerre polynomials (see (10.1.17)).

10.3.6 *Slater's results for large a*

Slater's expansions of the Kummer functions for large a are given in Slater (1960, §4.6.1), and are in terms of the large parameter a written in the form

$$a = \tfrac{1}{4}u^2 + \tfrac{1}{2}b, \tag{10.3.70}$$

where $u > 0$ if a and b are real with $a > \frac{1}{2}b$.

In the introduction of §10.3 we have mentioned that we can repair Slater's expansion of the function $U\left(\frac{1}{4}u^2 + \frac{1}{2}c, c; z^2\right)$ for large u. Because we have not used this choice of parameters for the a-parameter, we cannot compare Slater's results with our results for the U-function. For details we refer to Temme (2013a), where we have used Slater's notation.

Slater's results for large a are written in the form

$$\begin{aligned} e^{-\frac{1}{2}z^2} z^b {}_1F_1\left(\begin{matrix} a \\ b \end{matrix}; z^2\right) &\sim \Gamma(b) u^{1-b} 2^{b-1} \times \\ &\left(z I_{b-1}(uz) \sum_{k=0}^{\infty} \frac{A_k(z)}{u^{2k}} + \frac{z}{u} I_b(uz) \sum_{k=0}^{\infty} \frac{B_k(z)}{u^{2k}} \right), \end{aligned} \tag{10.3.71}$$

with the incorrect result (we repair this in (10.3.78))

$$\begin{aligned} e^{-\frac{1}{2}z^2} z^b U\left(a, b, z^2\right) &\sim \frac{2^{2-b} u^{b-1}}{\Gamma(a)} \times \\ &\left(z K_{b-1}(uz) \sum_{k=0}^{\infty} \frac{A_k(z)}{u^{2k}} - \frac{z}{u} K_b(uz) \sum_{k=0}^{\infty} \frac{B_k(z)}{u^{2k}} \right). \end{aligned} \tag{10.3.72}$$

The coefficients are given by $A_0 = 1$ and

$$\begin{aligned} B_k(z) &= -\tfrac{1}{2}A_k'(z) + \int_0^z \left(\tfrac{1}{2}t^2 A_k(t) - \frac{b-\frac{1}{2}}{t} A_k'(t) \right) dt, \\ A_{k+1}(z) &= \frac{b-\frac{1}{2}}{z} B_k - \tfrac{1}{2}B_k'(z) + \tfrac{1}{2} \int^z t^2 B_k(t)\, dt + K_k, \end{aligned} \tag{10.3.73}$$

and K_k is chosen so that $A_{k+1}(z) \to 0$ as $z \to 0$. In fact,[2]

$$
\begin{aligned}
A_0(z) &= 1, \quad B_0(z) = \tfrac{1}{6}z^3, \\
A_1(z) &= \tfrac{1}{6}(b-2)z^2 + \tfrac{1}{72}z^6, \\
B_1(z) &= -\tfrac{1}{3}b(b-2)z - \tfrac{1}{15}z^5 + \tfrac{1}{1296}z^9, \\
A_2(z) &= -\tfrac{1}{120}(5b-12)(b+2)z^4 + \tfrac{1}{6480}(5b-52)z^8 + \tfrac{1}{31104}z^{12}, \\
B_2(z) &= \tfrac{1}{90}(5b-12)(b+2)(b+1)z^3 - \\
&\qquad \tfrac{1}{45360}(175b^2 - 350b - 1896)z^7 - \tfrac{7}{12960}z^{11} + \tfrac{1}{933120}z^{15}.
\end{aligned}
\tag{10.3.74}
$$

Slater's expansions are based on Olver's method for differential equations; see Olver (1997, Chapter 12) (Slater has referred to earlier papers by Olver). This method is very powerful, it gives expansions valid in large domains of the parameters and recurrence relations for the coefficients. In addition, the method provides realistic error bounds for remainders in the expansions.

In Olver's method, the expansions are first given for two linear independent solutions of the differential equation. In the case of the Kummer functions this equation is a transformed version of Kummer's differential equation given in (10.1.1). Then the expansions of ${}_1F_1(a;b;z)$ and $U(a,b,z)$ follow from linear combinations of these solutions, and the coefficients in these combinations follow from certain known limiting forms of the Kummer functions (in the present case for $z \to 0$).

On the other hand, when the recurrence relations for the coefficients in the expansions are derived, these recursions usually include constants of integration. In the present case these are the quantities K_k used by Slater in (10.3.73). A certain choice of these constants generates a formal solution of the differential equation.

These two steps have to be taken into account when constructing the expansions of the functions ${}_1F_1(a;b;z)$ and $U(a,b,z)$, and it appears that Slater has not used the correct steps for the U-function.

When working with integrals these difficulties are not present: we always start with a representation of the function to be considered. Admittedly, we can usually not construct recurrence relations for the coefficients, and the construction of error bounds or estimates for remainders is more difficult, but there will never be a misunderstanding about the correct form of the representations.

We can repair Slater's expansion by including a factor containing a ratio of gamma functions in the representation in (10.3.72). That is, Slater's expansion can been written in the form

$$
\begin{aligned}
e^{-\frac{1}{2}z^2} z^b U\left(a,b,z^2\right) &\sim \frac{2^{2-b}u^{b-1}}{\Gamma(a)}\, C(a,b)\, \times \\
&\left(zK_{b-1}(uz) \sum_{k=0}^{\infty} \frac{A_k(z)}{u^{2k}} - \frac{z}{u} K_b(uz) \sum_{k=0}^{\infty} \frac{B_k(z)}{u^{2k}} \right),
\end{aligned}
\tag{10.3.75}
$$

[2]In Slater's formula (4.6.46) the x should be z.

with the relation (10.3.70) for a and coefficients $A_k(z)$ and $B_k(z)$ as in (10.3.73). The correction factor $C(a,b)$ is given by

$$C(a,b)=\left(\tfrac{1}{4}u^2\right)^{1-b}\frac{\Gamma(a)}{\Gamma(1+a-b)}=\left(\tfrac{1}{4}u^2\right)^{1-b}\frac{\Gamma\left(\frac{1}{4}u^2+\frac{1}{2}b\right)}{\Gamma(1+\frac{1}{4}u^2-\frac{1}{2}b)}, \tag{10.3.76}$$

and it has the asymptotic expansion (see (6.5.76))

$$C(a,b)\sim\sum_{n=0}^{\infty}C_n(\rho)(1-b)_{2n}\frac{(-1)^n}{w^{2n}},\quad \rho=\tfrac{1}{2}b,\quad w=\tfrac{1}{4}u^2. \tag{10.3.77}$$

The coefficients $C_n(\rho)$ can be expressed in terms of Bernoulli coefficients.

By replacing in (10.3.75) $C(a,b)$ by the first relation in (10.3.76), it follows that

$$\begin{aligned}e^{-\frac{1}{2}z^2}z^bU\left(a,b,z^2\right)\sim&\frac{2^bu^{1-b}}{\Gamma(1+a-b)}\times\\&\left(zK_{b-1}(uz)\sum_{k=0}^{\infty}\frac{A_k(z)}{u^{2k}}-\frac{z}{u}K_b(uz)\sum_{k=0}^{\infty}\frac{B_k(z)}{u^{2k}}\right).\end{aligned} \tag{10.3.78}$$

This expansion (or (10.3.75)) should have been the form of Slater's expansion.

10.4 Expansions for large c

Expansions for bounded $|a|$ will be considered. For the F-function $|z|$ should be bounded, for the U-function we give two expansions: one for bounded $|z|$ and one for $z > c$.

10.4.1 *Expansion of* ${}_1F_1(a;c;z)$, $c\to+\infty$

For the F-function an asymptotic representation follows from the Maclaurin series given in (10.1.2). If we wish an expansion in negative powers of c we can start with the integral representation given in (10.1.4) written in the form

$${}_1F_1\left({a\atop c};z\right)=\frac{\Gamma(c)}{\Gamma(a)\Gamma(c-a)}\int_0^{\infty}e^{-ct}f(t)\,dt, \tag{10.4.79}$$

where

$$f(t)=\left(1-e^{-t}\right)^{a-1}e^{at+z\left(1-e^{-t}\right)}. \tag{10.4.80}$$

By expanding $f(t)=t^{a-1}\sum_{k=0}^{\infty}c_k(a,z)t^k$ and using Watson's lemma, we obtain the expansion

$${}_1F_1\left({a\atop c};z\right)\sim\frac{\Gamma(c)}{\Gamma(c-a)}\sum_{k=0}^{\infty}c_k(a,z)\frac{(a)_k}{c^{a+k}},\quad c\to\infty. \tag{10.4.81}$$

By using an expansion of the ratio $\Gamma(c)/\Gamma(c-a)$ given in (6.5.72), we can obtain an expansion of the form

$$ {}_1F_1\left(\begin{matrix} a \\ c \end{matrix}; z\right) \sim \sum_{k=0}^{\infty} \frac{d_k(a,z)}{c^k}, \quad c \to \infty. \tag{10.4.82}$$

From Watson's lemma (see Chapter 2) we conclude that this expansion is valid for $|\mathrm{ph}\, c| \le \pi - \delta$. This follows from the location of the singularities of the function f on the imaginary axis and from the validity of the expansion of the ratio of the gamma functions, which is valid in the same sector; z and a can be every complex fixed number.

However, for the expansion in (10.4.82) we can assume $\Re c \ge 0$ and use the results of §10.4.5 for $\Re c \le 0$.

By expanding $1/(c)_k$ in the series in (10.1.2) in negative powers of c and rearranging, we obtain the same expansion. By using Kummer's transformation given in (10.1.9) we can obtain expansions valid for large c and bounded values of $c-a$ and z.

10.4.2 *Expansion of $U(a,c,z)$, $c \to +\infty$, $z < c$*

We write $z = c\zeta$ and assume that $0 < \zeta \le \zeta_0 < 1$ and start with the integral in (10.1.5). We write it in the form

$$U(a,c,c\zeta) = \frac{1}{\Gamma(a)} \int_0^{\infty} f(t) e^{-c\phi(t)}\, dt, \tag{10.4.83}$$

where

$$f(t) = t^{a-1}(1+t)^{-a-1}, \quad \phi(t) = \zeta t - \ln(1+t). \tag{10.4.84}$$

The derivative of the function ϕ vanishes at $t_0 = 1/\zeta - 1$, and we can use Laplace's method. When c and ζ are positive, with $\zeta \in (0,1)$ the saddle point t_0 is positive and we transform

$$\phi(t) - \phi(t_0) = \tfrac{1}{2}s^2, \quad \mathrm{sign}(s) = \mathrm{sign}(t - t_0). \tag{10.4.85}$$

This gives

$$U(a,c,c\zeta) = \left(\frac{c}{z}\right)^c \frac{e^{z-c}}{\Gamma(a)} \int_{s_0}^{\infty} g(s) e^{-\frac{1}{2}cs^2}\, ds, \tag{10.4.86}$$

where

$$s_0 = -\sqrt{2\,(\zeta - 1 - \ln\zeta)}, \quad g(s) = f(t)\frac{dt}{ds}. \tag{10.4.87}$$

The sign of the square root follows from the condition given in the transformation in (10.4.85).

When ζ approaches 1, the point s_0 approaches 0, and for this case we can derive an expansion in terms of parabolic cylinder functions; see Chapter 22. In the present analysis we assume that $0 < \zeta \le \zeta_0 < 1$, where ζ_0 is a fixed number. In that case we apply Laplace's method: we expand $g(s) = \sum_{k=0}^{\infty} c_k s^k$, replace $s_0$by $-\infty$, and obtain

$$U(a,c,c\zeta) \sim \left(\frac{c}{z}\right)^c \frac{e^{z-c}}{\Gamma(a)}\sqrt{\frac{2\pi}{c}}\sum_{k=0}^{\infty} c_{2k}\frac{2^k\left(\frac{1}{2}\right)_k}{c^k}, \quad c\to\infty. \tag{10.4.88}$$

For computing the coefficients we first need the expansion

$$t = \frac{1}{\zeta}\left(1-\zeta+s+\tfrac{1}{3}s^2+\tfrac{1}{36}s^3+\ldots\right). \tag{10.4.89}$$

In this way we obtain

$$z^{c-1}U(a,c,z) \sim \sqrt{2\pi}\, c^{c-\frac{3}{2}}(1-\zeta)^{a-1}\frac{e^{z-c}}{\Gamma(a)}\sum_{k=0}^{\infty}\frac{d_k}{c^k}, \quad d_k = \frac{c_{2k}}{c_0}2^k\left(\tfrac{1}{2}\right)_k, \tag{10.4.90}$$

where $\zeta = z/c$, uniformly for $\zeta \in [0, 1-\delta]$, where δ is a small fixed positive number. The first coefficients are

$$\begin{aligned}
d_0 &= 1,\\
d_1 &= \frac{1}{12(1-\zeta)^2}\left(13-2\zeta+\zeta^2+6a\zeta(\zeta-4)+6a^2\zeta^2\right),\\
d_2 &= \frac{1}{288(1-\zeta)^4}\left(524\zeta+30\zeta^2+313+\zeta^4-4\zeta^3+\right.\\
&\quad 12\zeta(3\zeta^3-14\zeta^2+9\zeta-148)a+60\zeta^2(2\zeta^2-10\zeta+29)a^2+\\
&\quad \left.120\zeta^3(\zeta-4)a^3+36\zeta^4a^4\right).
\end{aligned} \tag{10.4.91}$$

Remark 10.8. If $\Re c > -1$ and $|\mathrm{ph}\, z| < \pi$, then the left-hand side of (10.4.90) tends to $\Gamma(c-1)/\Gamma(a)$, as easily follows from the standard integral representation of the U-function. When $\zeta \to 0$, the coefficients d_k reduce to the coefficients $a_k(-1)$ in the expansion of $\Gamma(z-1)$ in (6.4.58). △

10.4.3 *Expansion of* $U(a,c,z)$, $c \to +\infty$, $z > c$

The expansion of this section is related to the one given of the incomplete gamma function in §7.4. We have the relation

$$\Gamma(a,z) = z^a e^{-z} U(1, a+1, z). \tag{10.4.92}$$

When $c < z$, that is, $\zeta > 1$, then the saddle point $t_0 = 1/\zeta - 1$ of $\phi(t)$ defined in (10.4.84) is negative, and we can use Watson's lemma. We substitute $w = \phi(t)$ and obtain

$$U(a,c,c\zeta) = \frac{1}{\Gamma(a)}\int_0^{\infty} w^{a-1}h(w)e^{-cw}\,dw, \tag{10.4.93}$$

where

$$h(w) = \left(\frac{t}{w}\right)^{a-1}(1+t)^{-a-1}\frac{dt}{dw}. \tag{10.4.94}$$

The expansion $h(w) = (\zeta - 1)^{-a} \sum_{k=0}^{\infty} h_k w^k$ gives

$$U(a, c, c\zeta) \sim (z - c)^{-a} \sum_{k=0}^{\infty} h_k \frac{(a)_k}{c^k}, \quad c \to \infty, \tag{10.4.95}$$

uniformly for $\zeta \geq 1 + \delta$. The first coefficients are

$$\begin{aligned} h_0 &= 1, \\ h_1 &= \frac{(a+1)(1-2\zeta)}{2(1-\zeta)^2}, \\ h_2 &= \frac{(a+2)(3a - 12a\zeta + 12a\zeta^2 - 4\zeta + 1 + 12\zeta^2)}{24(1-\zeta)^4}, \\ h_3 &= \frac{(a+3)(-4\zeta^2(4\zeta+1) + (1-2\zeta)(12\zeta^2 - 4\zeta + 1)a + (1-2\zeta)^3 a^2)}{48(1-\zeta)^6}. \end{aligned} \tag{10.4.96}$$

10.4.4 *Expansion of* $U(a, c, z)$, $c \to -\infty$

Again we start with the integral in (10.1.5), and write

$$U(a, -c, c\zeta) = \frac{1}{\Gamma(a)} \int_0^{\infty} t^{a-1} (1+t)^{-a-1} e^{-c\psi(t)}\, dt, \tag{10.4.97}$$

where

$$\psi(t) = \zeta t + \ln(1+t). \tag{10.4.98}$$

When $\zeta > 0$ the function $\psi(t)$ is monotonic on $[0, \infty)$, and we substitute $s = \psi(t)$, which gives

$$U(a, -c, c\zeta) = \frac{1}{\Gamma(a)} \int_0^{\infty} s^{a-1} e^{-cs} g(s)\, ds, \tag{10.4.99}$$

where

$$g(s) = \left(\frac{t}{s}\right)^{a-1} (1+t)^{-a-1} \frac{dt}{ds}. \tag{10.4.100}$$

Expanding $g(s) = (1+\zeta)^{-a} \sum_{k=0}^{\infty} g_k s^k$ and using Watson's lemma, we obtain

$$U(a, -c, z) \sim (c + z)^{-a} \sum_{k=0}^{\infty} g_k \frac{(a)_k}{c^k}, \quad c \to \infty, \tag{10.4.101}$$

where g_k follow from the first coefficients are

$$\begin{aligned} g_0 &= 1, \\ g_1 &= -\frac{(a+1)(2\zeta+1)}{2(1+\zeta)^2}, \\ g_2 &= \frac{(a+2)(3a + 12a\zeta + 12a\zeta^2 + 4\zeta + 1 + 12\zeta^2)}{24(1+\zeta)^4}, \\ g_3 &= -\frac{(a+3)(4\zeta^2(4\zeta-1) + (2\zeta+1)(12\zeta^2 + 4\zeta + 1)a + (2\zeta+1)^3 a^2)}{48(1+\zeta)^6}, \end{aligned} \tag{10.4.102}$$

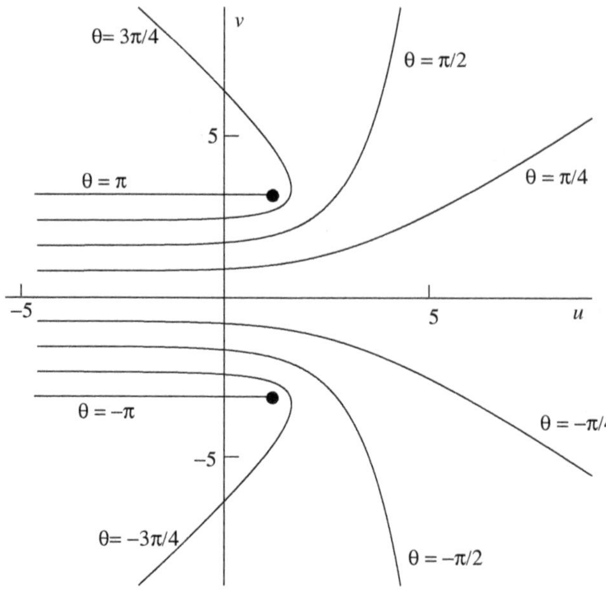

Fig. 10.1 Images in the s-plane of the half-lines $t = -1 + re^{i\theta}$, $r > 0$, where $\theta = \pm\frac{1}{4}\pi k$, $k = 1, 2, 3, 4$ for $\zeta = 0.1$.

with $\zeta = z/c$. The coefficients follow from h_k in the previous section by replacing ζ by $-\zeta$ and changing the sign of the odd coefficients. In other words, the expansion in (10.4.101) follows from (10.4.95) by changing $c \to -c$ and $\zeta \to -\zeta$.

The expansion in (10.4.101) is valid for all $\zeta \geq 0$. For complex c we need to know the location of the singularities of the function $g(s)$ defined in (10.4.100). The transformation of the t-plane to the s-plane is singular at the point t_0 where $\psi'(t)$ vanishes. That is, at $t_0 = -1 - 1/\zeta$. The corresponding point in the s-plane is $s_0 = -1 - \zeta - \ln(-\zeta)$. When $\zeta > 0$ this gives two singular points

$$s_\pm = -1 - \zeta - \ln(|\zeta|) \pm \pi i. \tag{10.4.103}$$

When $\zeta \to 0$ these points tend to $+\infty$, when $\zeta \to \infty$ they tend to $-\infty$.

From Watson's lemma (see Theorem 2.2) it follows that the expansion in (10.4.101) is valid inside the sector $-\beta - \frac{1}{2}\pi + \delta \leq \operatorname{ph} c \leq -\alpha + \frac{1}{2}\pi - \delta$, where $\alpha = \operatorname{ph} s_-$ and $\beta = \operatorname{ph} s_+$. A similar result follows for complex values of ζ.

In Figure 10.1 we give details of the transformation $s = \psi(t)$. We show the images in the s-plane of the half-lines $t = -1 + re^{i\theta}$, $r > 0$, where $\theta = \pm\frac{1}{4}\pi k$, $k = 1, 2, 3, 4$ for $\zeta = 0.1$. The black dots indicate the singular points at $s_\pm = 1.20 \pm \pi i$. The images of the half-lines are given by the relations (we write $s = u + iv$)

$$\begin{aligned} u &= \zeta r \cos\theta + \ln r, \\ v &= \zeta r \sin\theta + \theta. \end{aligned} \tag{10.4.104}$$

The images fold around the singular points when θ approaches $\pm\pi$. For $\theta = \pm\pi$ the images are half-lines $u \le \Re s_\pm$, $v = \pm\pi$ (traversed twice, with r from 0 to $1/\zeta$, and from $1/\zeta$ to 0).

10.4.5 *Expansion of* ${}_1F_1(a;c;z)$, $c \to -\infty$

For this case we write the connection formula given in (10.1.12) in the form

$$\begin{aligned}\frac{1}{\Gamma(-c)}\,{}_1F_1\left(\begin{matrix}a\\-c\end{matrix};z\right) &= \Gamma(a+c+1)\times\\ &\left(-\frac{\sin(\pi c)}{\pi}U(a,-c,z)+\frac{z^{1+c}e^z}{\Gamma(a)\Gamma(c+2)}\,{}_1F_1\left(\begin{matrix}1-a\\c+2\end{matrix};-z\right)\right).\end{aligned} \tag{10.4.105}$$

In this formula integer values of c are allowed if we use the limit shown in (10.1.3). In fact, together with Kummer's relation (10.1.9), the formula reduces to the limit when $c \to m$. When c is not a positive integer, the asymptotic expansion follows from the result for $U(a,-c,z)$ from §10.4.4 because the results coming from the F-function can be neglected due to $1/\Gamma(c+2)$.

10.5 Uniform expansions of the Kummer functions

The expansions given in this chapter are for large values of a or c, and usually we need restrictions on the z-domain.

Expansions for large values of z and unrestricted positive λ-values of $U(\lambda,\lambda+\sigma+1,z)$ and ${}_1F_1(\lambda+1;\lambda+1-\sigma;z)$ are considered in §25.4. The expansions are in terms of elementary functions.

Uniform expansions for large c in terms of parabolic cylinder functions are given in Chapter 22, with fixed a of $U(a,c,c\lambda)$ (see §22.5.1) and of ${}_1F_1(a;c;c\lambda)$ (see §22.5.2). These expansions are valid uniformly with respect to $\lambda \ge 0$.

In Chapter 27 again we consider expansions of the Kummer functions in terms of Bessel functions and the results are related to those of the present chapter, with less restrictions on the z-domain. We obtain expansions for large a and unbounded z. In addition we extend the method for $U(a,-c,az)$ (see §27.5.1) and for ${}_1F_1(a+c;c;az)$, ${}_1F_1(-a;-c;-az)$ (see §27.5.2), by allowing c to become large as well.

Chapter 11

Parabolic cylinder functions: Large argument

Parabolic cylinder functions are associated with the differential equation

$$\frac{d^2w}{dz^2} + \left(pz^2 + qz + r\right) w = 0, \quad p \neq 0. \tag{11.0.1}$$

In the field of special functions we are usually concerned with the equation

$$\frac{d^2w}{dz^2} - \left(\tfrac{1}{4}z^2 + a\right) w = 0, \tag{11.0.2}$$

which follows from the more general equation by a simple change of variable.

Solutions of these equations are used in mathematical physics in boundary value problems in or outside a parabolic cylinder; see for an historical introduction Weber (1869). Many properties in connection with physical applications are given in Buchholz (1969).

Parabolic cylinder functions are special cases of the Kummer functions (see Temme (2010a, §12.7(iv))) introduced in Chapter 10, however, it is of interest to consider these functions separately.

In this chapter we consider large variable expansions of Poincaré-type. For large values of the parameter we refer to Chapter 30. General information on these functions can be found in Temme (2010a).

11.1 A few properties of the parabolic cylinder functions

As in Miller (1955), Olver (1959) and Temme (2010a), we denote two standard solutions of (11.0.2) by $U(a,z), V(a,z)$. Another notation found in the literature is $D_\nu(z) = U\left(-\nu - \frac{1}{2}, z\right)$, which is useful in describing the relation with the Hermite polynomials:

$$D_n(z) = U\left(-n - \tfrac{1}{2}, z\right) = 2^{-n/2} e^{-\frac{1}{4}z^2} H_n\left(z/\sqrt{2}\right), \quad n = 0, 1, 2, \ldots. \tag{11.1.3}$$

It will be clear that $U(a,-z)$, $V(a,-z)$, $U(-a,\pm iz)$, $V(-a,\pm iz)$ are solutions of equation (11.0.2) as well, and there should be connection formulas between the solutions. The following ones are used in the next sections and in Chapter 30.

We have

$$U(a,z) = \frac{\Gamma(\frac{1}{2}-a)}{\cos(\pi a)}\bigl(V(a,-z) - \sin(\pi a)\, V(a,z)\bigr), \tag{11.1.4}$$

$$V(a,z) = \frac{\Gamma(\frac{1}{2}+a)}{\pi}\bigl(\sin(\pi a)\, U(a,z) + U(a,-z)\bigr), \tag{11.1.5}$$

$$U(a,-z) = ie^{i\pi a}U(a,z) + \frac{\sqrt{2\pi}}{\Gamma\left(\frac{1}{2}+a\right)}e^{i\pi(\frac{1}{2}a-\frac{1}{4})}U(-a,-iz), \tag{11.1.6}$$

and a similar one when i is replaced by $-i$.

The Wronskian relations between the solution pairs $\{U(a,z), V(a,z)\}$ and $\{U(a,z), U(a,-z)\}$ of (11.0.2) read

$$\begin{aligned} &U(a,z)V'(a,z) - U'(a,z)V(a,z) = \sqrt{2/\pi}, \\ &U(a,z)U'(a,-z) + U'(a,z)U(a,-z) = -\frac{\sqrt{2\pi}}{\Gamma\left(\frac{1}{2}+a\right)}. \end{aligned} \tag{11.1.7}$$

The first relation shows that $U(a,z)$ and $V(a,z)$ are two linearly independent solutions of (11.0.2) for all values of a.

11.2 The functions $U(a,z)$

We start with the integral representation (see Temme (2010a, §12.5(i)))

$$U(a,z) = \frac{e^{-\frac{1}{4}z^2}}{\Gamma\left(\frac{1}{2}+a\right)}\int_0^\infty w^{a-\frac{1}{2}}e^{-\frac{1}{2}w^2-zw}\,dw, \quad \Re a > -\tfrac{1}{2}, \tag{11.2.8}$$

and expanding the exponential function $e^{-\frac{1}{2}w^2}$ we obtain

$$U(a,z) \sim \frac{e^{-\frac{1}{4}z^2}}{z^{a+\frac{1}{2}}}\sum_{k=0}^{\infty}(-1)^k\frac{\left(a+\frac{1}{2}\right)_{2k}}{k!\,(2z^2)^k}, \quad z\to\infty, \quad |\mathrm{ph}\, z| \le \tfrac{3}{4}\pi - \delta. \tag{11.2.9}$$

In this case the sector for z is determined by condition (iii) of Theorem 2.2, in which case we have $\alpha = -\frac{1}{4}\pi$, $\beta = \frac{1}{4}\pi$. As explained in Remark 2.2, we can drop the condition $\Re a > -\frac{1}{2}$ mentioned in (11.2.8).

The dominant exponential function $e^{-\frac{1}{2}w^2}$ in (11.2.8) may be treated differently by substituting $w = zt$, assuming that $z > 0$ for the moment. This gives

$$U(a,z) = \frac{z^{a+\frac{1}{2}}e^{-\frac{1}{4}z^2}}{\Gamma\left(\frac{1}{2}+a\right)}\int_0^\infty t^{a-\frac{1}{2}}e^{-z^2\left(\frac{1}{2}t^2+t\right)}\,dt. \tag{11.2.10}$$

We transform the integral into the standard form for applying Watson's lemma by substituting

$$s = \tfrac{1}{2}t^2 + t \quad \Longrightarrow \quad t = \sqrt{1+2s} - 1. \tag{11.2.11}$$

This gives

$$U(a,z) = \frac{z^{a+\frac{1}{2}}e^{-\frac{1}{4}z^2}}{\Gamma\left(\frac{1}{2}+a\right)} \int_0^\infty s^{a-\frac{1}{2}} e^{-z^2 s} f(s)\, ds, \tag{11.2.12}$$

where

$$f(s) = \left(\frac{t}{s}\right)^{a-\frac{1}{2}} \frac{dt}{ds} = \left(\frac{t}{s}\right)^{a-\frac{1}{2}} \frac{1}{\sqrt{1+2s}}. \tag{11.2.13}$$

In this case we have to expand the function $f(s)$, and luckily we have a special case of the Gauss hypergeometric function (see Olde Daalhuis (2010b, Eq. 15.4.18))

$${}_2F_1\left(\begin{matrix}\alpha, \frac{1}{2}+\alpha \\ 2\alpha\end{matrix}; z\right) = \frac{1}{\sqrt{1-z}}\left(\tfrac{1}{2}+\tfrac{1}{2}\sqrt{1-z}\right)^{1-2\alpha}, \tag{11.2.14}$$

which gives, with $z = -2s$ and $\alpha = \frac{1}{4}+\frac{1}{2}a$,

$$f(s) = \sum_{k=0}^{\infty} \frac{\left(\frac{1}{4}+\frac{1}{2}a\right)_k \left(\frac{3}{4}+\frac{1}{2}a\right)_k}{k!\left(\frac{1}{2}+a\right)_k}(-2s)^k. \tag{11.2.15}$$

By using the duplication formula of the gamma function:

$$\sqrt{\pi}\,\Gamma(2z) = 2^{2z-1}\Gamma(z)\Gamma\left(z+\tfrac{1}{2}\right), \tag{11.2.16}$$

we obtain

$$f(s) = \sum_{k=0}^{\infty} \frac{2^{-2k}\left(\frac{1}{2}+a\right)_{2k}}{k!\left(\frac{1}{2}+a\right)_k}(-2s)^k, \tag{11.2.17}$$

and substituting this expansion into (11.2.12) we recover the expansion in (11.2.9).

This time the sector for z follows from the singularity of $f(s)$ on the negative axis, giving $|\mathrm{ph}\, z^2| \le \frac{3}{2}\pi$.

For a compound expansion including other sectors for z, we refer to (11.3.26). For these expansions it is convenient to introduce functions $P(a,z)$ and $Q(a,z)$ by writing

$$\begin{aligned} U(a,z) &= z^{-a-\frac{1}{2}} e^{-\frac{1}{4}z^2} P(a,z), \\ U(-a,-iz) &= z^{a-\frac{1}{2}} e^{\frac{1}{4}z^2} e^{-\frac{1}{2}\pi i(a-\frac{1}{2})} Q(a,z), \end{aligned} \tag{11.2.18}$$

with expansions

$$P(a,z) \sim \sum_{k=0}^{\infty} (-1)^k \frac{\left(a+\frac{1}{2}\right)_{2k}}{k!\,(2z^2)^k}, \quad Q(a,z) \sim \sum_{k=0}^{\infty} \frac{\left(\frac{1}{2}-a\right)_{2k}}{k!\,(2z^2)^k}, \tag{11.2.19}$$

as $z \to \infty$. When we use the expansion for $U(a,z)$ the sector is as in (11.2.9).

11.3 The function $U(a,-z)$

Starting from (11.2.8), we write in this case

$$U(a,-z)=\frac{e^{-\frac14 z^2}z^{a+\frac12}}{\Gamma\left(\frac12+a\right)}\int_0^\infty t^{a-\frac12}e^{-z^2\left(\frac12 t^2-t\right)}\,dt,\quad \Re a>-\tfrac12, \tag{11.3.20}$$

with an interior saddle point at $t=1$. A simple transformations gives

$$U(a,-z)=\frac{e^{\frac14 z^2}z^{a+\frac12}}{\Gamma\left(\frac12+a\right)}\int_{-1}^\infty (1+s)^{a-\frac12}e^{-\frac12 z^2 s^2}\,ds. \tag{11.3.21}$$

Expansion of the binomial gives

$$U(a,-z)\sim\frac{\sqrt{2\pi}\,e^{\frac14 z^2}z^{a-\frac12}}{\Gamma\left(\frac12+a\right)}\sum_{k=0}^\infty\frac{\left(\frac12-a\right)_{2k}}{k!\,(2z^2)^k},\quad |\mathrm{ph}\,z|\le\tfrac14\pi-\delta. \tag{11.3.22}$$

This can also be written in the form

$$U(a,-z)\sim\frac{\sqrt{2\pi}\,e^{\frac14 z^2}z^{a-\frac12}}{\Gamma\left(\frac12+a\right)}Q(a,z),\quad |\mathrm{ph}\,z|\le\tfrac14\pi-\delta, \tag{11.3.23}$$

where $Q(a,z)$ is introduced in (11.2.18) (we cannot use an equality sign in this relation).

The sector for z is determined by the singularity at $s=-1$ (see Lemma 3.1), and we observe that the two expansions in (11.2.9) and (11.3.22) are not sufficient for describing the asymptotic behavior of $U(a,z)$ for all z. The diagonals $\mathrm{ph}\,z=\pm\frac34\pi$ are not covered. In addition, the expansion in (11.2.9) is valid for every a, although we started with the condition $\Re a>-\frac12$ (see also Remark 2.2). However, for the expansion in (11.3.22) we cannot drop the condition $\Re a>-\frac12$.

For a better description of the behavior of $U(a,-z)$ we can use the method described in §9.3.1 for the modified Bessel function $I_\nu(z)$. The limited sector in (11.3.22) is due to the finite part $[-1,0]$ of the interval of integration in (11.3.21). We can replace this part by two lines, one from -1 to $-\infty$, and one from $-\infty$ back to the origin. In this way we can obtain larger sectors for z, which depend on our choice of taking these lines above or under the branch cut along $(-\infty,-1]$. In so doing, we discover the connection formula given in (11.1.6).

We return to (11.1.6) and take the expansions of the two U-functions on the right-hand side from (11.2.9). This gives the compound representation

$$U(a,-z)=\frac{\sqrt{2\pi}\,e^{\frac14 z^2}z^{a-\frac12}}{\Gamma\left(\frac12+a\right)}Q(a,z)+\frac{e^{i\pi(a+\frac12)-\frac14 z^2}}{z^{a+\frac12}}P(a,z), \tag{11.3.24}$$

and we can use the expansions given in (11.2.19). By taking the intersection of the sectors for z of the series for $Q(a,z)$ (coming from $U(-a,-iz)$) and for the series for $P(a,z)$ (coming from $U(a,z)$), it follows that the expansion in (11.3.24) is valid in the sector $-\frac14\pi+\delta\le\mathrm{ph}\,z\le\frac34\pi-\delta$.

We can change the sign of i in (11.1.6), and obtain a similar result:

$$U(a,-z) = \frac{\sqrt{2\pi}\, e^{\frac14 z^2} z^{a-\frac12}}{\Gamma\left(\frac12+a\right)} Q(a,z) + \frac{e^{-i\pi(a+\frac12)-\frac14 z^2}}{z^{a+\frac12}} P(a,z), \qquad (11.3.25)$$

and when we take the expansions given in (11.2.19), we conclude that the result holds when $-\frac34\pi+\delta \le \operatorname{ph} z \le \frac14\pi-\delta$.

We observe that these compound expansions give overlapping sectors with the expansion in (11.2.9), and moreover, there is no restriction on a.

When $|\operatorname{ph} z| \le \frac14\pi-\delta$ the function $U(a,-z)$ is exponentially large, unless a has one of the values $-\frac12, -\frac32, -\frac52, \ldots$. In that case the first term in (11.3.24) vanishes, and the second term becomes real and is exponentially small. This corresponds to the relation with the Hermite polynomials given in (11.1.3).

In the literature the compound expansions in (11.3.24) and (11.3.25) are usually given in a different form. When we replace in (11.3.24) z by $ze^{\pi i}$ and in (11.3.25) z by $ze^{-\pi i}$, we obtain

$$U(a,z) \sim \frac{e^{-\frac14 z^2}}{z^{a+\frac12}} P(a,z) \pm i \frac{\sqrt{2\pi}}{\Gamma\left(\frac12+a\right)} e^{\mp i\pi a} e^{\frac14 z^2} z^{a-\frac12} Q(a,z). \qquad (11.3.26)$$

With the expansions given in (11.2.19) this holds for $\frac14\pi+\delta \le \pm\operatorname{ph} z \le \frac54\pi-\delta$, where the upper and lower signs correspond to those in (11.3.26). This result is not valid for $z>0$, but together with (11.2.9), they give the expansions for all directions in the z-plane.

11.4 The function $V(a, z)$

For the function $V(a,z)$ we combine the connection formulas given in (11.1.5) and (11.1.6) into the relation

$$V(a,z) = \sqrt{\frac{2}{\pi}} e^{\frac12\pi i(a-\frac12)} U(-a,-iz) + \frac{i}{\Gamma\left(\frac12-a\right)} U(a,z), \qquad (11.4.27)$$

and a similar form by changing the sign of i throughout. We obtain from (11.2.18)

$$V(a,z) = \sqrt{\frac{2}{\pi}} e^{\frac14 z^2} z^{a-\frac12} Q(a,z) \pm \frac{i\, e^{-\frac14 z^2}}{z^{a+\frac12}\,\Gamma\left(\frac12-a\right)} P(a,z). \qquad (11.4.28)$$

When we use the expansions given in (11.2.19), we conclude that the asymptotic result holds for $-\frac14\pi+\delta \le \pm\operatorname{ph} z \le \frac34\pi-\delta$, where the upper and lower signs correspond to those in (11.4.28).

Using the connection formula given in (11.1.4), we obtain

$$V(a,-z) = \frac{e^{\pm\pi i a-\frac14 z^2}}{z^{a+\frac12}\,\Gamma\left(\frac12-a\right)} P(a,z) + \sqrt{\frac{2}{\pi}} \sin(\pi a) z^{a-\frac12} e^{\frac14 z^2} Q(a,z), \qquad (11.4.29)$$

and with the expansions given in (11.2.19) it follows that the asymptotic result is valid for $-\frac14\pi+\delta \le \pm\operatorname{ph} z \le \frac34\pi-\delta$.

11.5 Expansions of the derivatives

We introduce the functions $R(a,z)$ and $S(a,z)$ by writing (cf. (11.2.18))

$$\begin{aligned} U'(a,z) &= -\tfrac{1}{2}z^{\frac{1}{2}-a}e^{-\frac{1}{4}z^2}R(a,z), \\ U'(-a,-iz) &= -\tfrac{1}{2}z^{a+\frac{1}{2}}e^{\frac{1}{4}z^2}e^{-\frac{1}{2}\pi i\left(a+\frac{1}{2}\right)}S(a,z). \end{aligned} \tag{11.5.30}$$

By straightforward differentiating the asymptotic expansion for $U(a,z)$ we obtain (cf. (11.2.19))

$$\begin{aligned} R(a,z) &\sim 1-\sum_{k=1}^{\infty}(-1)^k\frac{2k-a+\frac{1}{2}}{2k+a-\frac{1}{2}}\frac{\left(a+\frac{1}{2}\right)_{2k}}{k!\,(2z^2)^k}, \\ S(a,z) &\sim 1-\sum_{k=1}^{\infty}\frac{2k+a+\frac{1}{2}}{2k-a-\frac{1}{2}}\frac{\left(\frac{1}{2}-a\right)_{2k}}{k!\,(2z^2)^k}, \end{aligned} \tag{11.5.31}$$

as $z\to\infty$, in sectors that follow from the expansions of $U(a,z)$ and $U(-a,-iz)$.

When we differentiate (11.1.6) it follows that

$$U'(a,-z) = -\tfrac{1}{2}\frac{\sqrt{2\pi}}{\Gamma\left(a+\frac{1}{2}\right)}z^{\frac{1}{2}+a}e^{\frac{1}{4}z^2}S(a,z)+\tfrac{1}{2}ie^{\pi ia}z^{\frac{1}{2}-a}e^{-\frac{1}{4}z^2}R(a,z), \tag{11.5.32}$$

and using the expansions given in (11.5.31) we obtain an asymptotic result that is valid in the sector $-\frac{1}{4}\pi+\delta\le \operatorname{ph} z\le \frac{3}{4}\pi-\delta$. Changing the sign of i gives an expansion in the sector $-\frac{3}{4}\pi+\delta\le \operatorname{ph} z\le \frac{1}{4}\pi-\delta$.

Similarly, by differentiating (11.1.5) and using the results for $U'(a,\pm z)$,

$$V'(a,z) = \tfrac{1}{2}\sqrt{\frac{2}{\pi}}z^{\frac{1}{2}+a}e^{\frac{1}{4}z^2}S(a,z)\mp\tfrac{1}{2}i\frac{z^{\frac{1}{2}-a}e^{-\frac{1}{4}z^2}}{\Gamma\left(\frac{1}{2}-a\right)}R(a,z). \tag{11.5.33}$$

With the expansions in (11.5.31) the asymptotic result holds for $-\frac{1}{4}\pi+\delta\le \pm\operatorname{ph} z\le \frac{3}{4}\pi-\delta$, where the upper and lower signs match those in (11.5.33).

Finally,

$$V'(a,-z) = -\tfrac{1}{2}\sqrt{\frac{2}{\pi}}\sin(\pi a)z^{a+\frac{1}{2}}e^{\frac{1}{4}z^2}S(a,z)+\tfrac{1}{2}\frac{e^{\pm\pi ia-\frac{1}{4}z^2}}{\Gamma\left(\frac{1}{2}-a\right)}z^{\frac{1}{2}-a}R(a,z), \tag{11.5.34}$$

and with the expansions given in (11.5.31) it follows that the asymptotic result is valid for $-\frac{1}{4}\pi+\delta\le \pm\operatorname{ph} z\le \frac{3}{4}\pi-\delta$, with signs corresponding to those in (11.5.34).

When we use the representations in (11.2.18), (11.5.33), (11.5.30) and (11.4.28) in the first Wronskian relation in (11.1.7), it follows that

$$P(a,z)S(a,z)+Q(a,z)R(a,z)=2. \tag{11.5.35}$$

The same result follows when we use the second Wronskian relation in (11.1.7).

Chapter 12

The Gauss hypergeometric function

The Gauss hypergeometric function is defined by

$$ {}_2F_1\left({a,b \atop c};z\right) = \sum_{n=0}^{\infty} \frac{(a)_n(b)_n}{n!\,(c)_n} z^n, \quad |z| < 1, \tag{12.0.1} $$

where the Pochhammer symbol $(\lambda)_n$ is defined in (2.1.12). For c we assume the usual condition $c \neq 0, -1, -2, \ldots$. If $\Re(c-a-b) > 0$, then

$$ {}_2F_1\left({a,b \atop c};1\right) = \frac{\Gamma(c)\Gamma(c-a-b)}{\Gamma(c-a)\Gamma(c-b)}. \tag{12.0.2} $$

For details on this function we refer to Olde Daalhuis (2010b); several results in that reference are for the function $\mathbf{F}(a,b;c;z) = {}_2F_1(a,b;c;z)/\Gamma(c)$, which is an analytic function of c.

The standard integral representation is (see Olde Daalhuis (2010b, Eq. 15.6.1))

$$ {}_2F_1\left({a,b \atop c};z\right) = \frac{\Gamma(c)}{\Gamma(b)\,\Gamma(c-b)} \int_0^1 \frac{t^{b-1}(1-t)^{c-b-1}}{(1-zt)^a}\,dt, \tag{12.0.3} $$

where $|\mathrm{ph}(1-z)| < \pi$ and $\Re c > \Re b > 0$. Many other representations are available for wider ranges of the parameters. For example,

$$ {}_2F_1\left({a,b \atop c};z\right) = \frac{\Gamma(c)\Gamma(1+b-c)}{2\pi i\,\Gamma(b)} \int_0^{(1+)} \frac{t^{b-1}(t-1)^{c-b-1}}{(1-zt)^a}\,dt, \tag{12.0.4} $$

valid for $c - b \neq 1, 2, 3, \ldots$, with again $|\mathrm{ph}(1-z)| < \pi$ and $\Re b > 0$.

Asymptotic aspects for large values of z (with other parameters fixed) are not interesting, because several connection formulas are available for these functions, with argument

$$ 1/z, \quad 1-z, \quad 1/(1-z), \quad z/(z-1), \quad (z-1)/z, \tag{12.0.5} $$

which enable computation for a large part of the complex plane. As explained in Gil *et al.* (2007, §2.3), for the computation small environments of the points $e^{\pm\pi i/3}$ have to be excluded, but for these points other expansions can be used as well; see also López and Temme (2012). For a complete list of connection formulas, of which we will use several in the cases considered in this chapter, we refer to Olde Daalhuis (2010b, §15.10). Simple formulas are

$$\begin{aligned} {}_2F_1\left({a, b \atop c} ; z \right) &= (1-z)^{-a} {}_2F_1\left({a, c-b \atop c} ; \frac{z}{z-1} \right) \\ &= (1-z)^{-b} {}_2F_1\left({c-a, b \atop c} ; \frac{z}{z-1} \right) \\ &= (1-z)^{c-a-b} {}_2F_1\left({c-a, c-b \atop c} ; z \right), \end{aligned} \tag{12.0.6}$$

and two other examples are given in (6.8.138) and (6.8.144).

Far more interesting from an asymptotic point of view is the case of large parameters a, b and c, whether or not in combination with large or small values of z. However, as we will see, there are also trivial subcases in the large parameter cases.

In this chapter we consider a few cases with large parameters. We start with $c \to \pm\infty$. Note the symmetry with respect to a and b, so we concentrate on $b \to \pm\infty$. In all cases we consider z bounded or bounded away from the origin, which cases will be indicated with $|z| \le z_0$ and $|z| \ge z_0$, respectively, with z_0 a fixed positive number. In some cases we need to pay special attention to the critical point $z = 1$.

We consider only a few cases; in §12.3 we give references to the literature for more information on expansions of ${}_2F_1(a+\varepsilon_1\lambda, b+\varepsilon_2\lambda; c+\varepsilon_3\lambda; z)$ for $|\lambda| \to \infty$ with ε_j equal to $0, \pm 1$ and complex values of z.

12.1 Large values of c

For bounded values of z the asymptotic forms are rather simple, for z in unbounded domains we need Kummer functions as main approximants; see §28.2.

12.1.1 *Large positive c;* $|z| \le z_0$

For large positive values of c the series in (12.0.1) is an excellent asymptotic representation for fixed values of a, b, and z, even for bounded z outside the unit disk. In that case the series can be viewed as a non-convergent asymptotic expansion. For conditions on complex values of c and z we refer to Wagner (1988) or Olde Daalhuis (2010b, §15.12(ii)).

When we wish a Poincaré-type asymptotic expansion, we can start from (12.0.3) in the form

$${}_2F_1\left({a, b \atop c+\lambda} ; z \right) = \frac{\Gamma(c+\lambda)}{\Gamma(b)\,\Gamma(c+\lambda-b)} \int_0^1 t^{b-1}(1-t)^{c+\lambda-b-1}(1-zt)^{-a}\,dt, \tag{12.1.7}$$

and substitute $t \to 1 - e^{-t}$. This gives

$${}_2F_1\left({a, b \atop c+\lambda} ; z \right) = \frac{\Gamma(c+\lambda)}{\Gamma(b)\,\Gamma(c+\lambda-b)} \int_0^\infty t^{b-1} f(t) e^{-\lambda t}\,dt, \tag{12.1.8}$$

where

$$f(t) = \left(\frac{e^t - 1}{t}\right)^{b-1} e^{(1-c)t} \left(1 - z + ze^{-t}\right)^{-a}. \tag{12.1.9}$$

By expanding

$$f(t) = \sum_{k=0}^{\infty} q_k(z) t^k, \tag{12.1.10}$$

we obtain

$${}_2F_1\left({a, b \atop c + \lambda} ; z \right) \sim \frac{\Gamma(c + \lambda)}{\Gamma(c + \lambda - b)} \sum_{k=0}^{\infty} \frac{q_k(z)(b)_k}{\lambda^{k+b}}. \tag{12.1.11}$$

To find the sector for complex λ, we observe that f has singularities at $\pm 2\pi i k$, $k = 1, 2, 3, \ldots$. The other singularity is at $t_0 = \ln(z/(z-1))$. It is easily verified that if $\Re z \le \frac{1}{2}$ then $\Re t_0 \le 0$. Hence, in that case the expansion is uniformly valid in the sector $|\mathrm{ph}\,\lambda| \le \pi - \delta$. This follows from Theorem 2.2. For $\Re z > \frac{1}{2}$ the sector for c is determined by $\mathrm{ph}(t_0 \pm 2\pi i k)$, $k = 1, 2, 3, \ldots$. In addition, z should be bounded, because t_0 tends to zero if $z \to \infty$. That case will be considered in §28.2.

For further details we refer to Wagner (1988) or Olde Daalhuis (2010b, §15.12(ii)).

12.1.2 *Large negative c; $|z| \le z_0$*

As in the case with $c > 0$, we might use the series in (12.0.1), but the expansion is not very attractive. In addition, a complication arises when c is a large negative integer. However, this case is again simple, because of the connection formula

$$\begin{aligned} {}_2F_1\left({a, b \atop c} ; z \right) = {} & \frac{\Gamma(a - c + 1)\Gamma(b - c + 1)}{\Gamma(1 - c)\Gamma(a + b - c + 1)} w_3(z) - \\ & \frac{\Gamma(a - c + 1)\Gamma(b - c + 1)\Gamma(c - 1)}{\Gamma(1 - c)\Gamma(a)\Gamma(b)} w_2(z), \end{aligned} \tag{12.1.12}$$

where the functions $w_2(z)$ and $w_3(z)$ have the form

$$\begin{aligned} w_2(z) &= z^{1-c}(1 - z)^{c-a-b} {}_2F_1\left({1 - a, 1 - b \atop 2 - c} ; z \right), \\ w_3(z) &= {}_2F_1\left({a, b \atop a + b - c + 1} ; 1 - z \right). \end{aligned} \tag{12.1.13}$$

In both hypergeometric functions the argument is bounded and with $-c$ on the c-place the series in (12.0.1) or the expansions in §12.1.1 can be used.

12.1.3 *Large positive c; $|z| \ge z_0$*

We use the integral representation in (12.0.3) with z replaced by $-z$, assuming that $|\mathrm{ph}(1+z)| < \pi$, and use the change of variable $1 - t = e^{-u}$. This gives

$$ {}_2F_1\left(\begin{matrix} a, b \\ c+\lambda \end{matrix}; -z\right) = \frac{\Gamma(c+\lambda)}{\Gamma(b)\Gamma(c+\lambda-b)} \int_0^\infty \frac{\left(1-e^{-u}\right)^{b-1} e^{(b-c)u}}{\left(1+z\left(1-e^{-u}\right)\right)^a} e^{-\lambda u}\, du. \quad (12.1.14) $$

This can be written in the form

$$ {}_2F_1\left(\begin{matrix} a, b \\ c+\lambda \end{matrix}; -z\right) = \frac{z^{-a}\Gamma(c+\lambda)}{\Gamma(b)\Gamma(c+\lambda-b)} \int_0^\infty u^{b-1}(u+\alpha)^{-a} e^{-\lambda u} f(u)\, du, \quad (12.1.15) $$

where

$$ f(u) = e^{(a+b-c)u} \left(\frac{1-e^{-u}}{u}\right)^{b-1} \left(\frac{e^{u+\alpha}-1}{u+\alpha}\right)^{-a}, \quad \alpha = \ln\frac{1+z}{z}. \quad (12.1.16) $$

Observe that $z > 0$ implies $\alpha > 0$, and that $\alpha \sim 1/z$ as $z \to \infty$. The singular points of f are

$$ u_k = \pm 2\pi i k, \quad k = 1, 2, 3, \ldots, \quad \text{and} \quad u_k - \alpha. \quad (12.1.17) $$

Integrals of this type, with large λ and $\alpha \to 0$, will be considered in Chapter 28, see §28.1, in which case the Kummer U-functions are used as main approximants. For an application we refer to §29.4, where we consider the case of toroidal functions, which are special forms of Legendre functions (which for their part can be written in terms of Gauss hypergeometric functions).

12.1.4 *Large negative c; $|z| \ge z_0$*

We use the connection formula

$$ {}_2F_1\left(\begin{matrix} a, b \\ c \end{matrix}; z\right) = \frac{\Gamma(c)\Gamma(b-a)}{\Gamma(b)\Gamma(c-a)} w_5(z) + \frac{\Gamma(c)\Gamma(a-b)}{\Gamma(a)\Gamma(c-b)} w_6(z), \quad (12.1.18) $$

where the functions $w_5(z)$ and $w_6(z)$ have the form

$$ \begin{aligned} w_5(z) &= e^{(c-b)\pi i} z^{b-c} (1-z)^{c-a-b} {}_2F_1\left(\begin{matrix} 1-b, c-b \\ a-b+1 \end{matrix}; \frac{1}{z}\right), \\ w_6(z) &= e^{(c-a)\pi i} z^{a-c} (1-z)^{c-a-b} {}_2F_1\left(\begin{matrix} 1-a, c-a \\ b-a+1 \end{matrix}; \frac{1}{z}\right). \end{aligned} \quad (12.1.19) $$

In both hypergeometric functions the large negative parameter occurs at the b-place, and this case will be considered in §12.2.2 and §28.3. In the asymptotic expansion the ${}_1F_1$ functions are used.

12.2 Large values of b

For the large-b asymptotics we can concentrate on $b < 0$, because for $b > 0$ we can use the third relation in (12.0.6).

12.2.1 Large negative b; $|z| \ge z_0$

This case is again rather simple, because when $|z|$ is not small, we can use the connection formula (see Olde Daalhuis (2010b, Eqs. 15.10.30, 15.10.33))

$$\begin{aligned} {}_2F_1\left({a,b \atop c};z\right) = e^{(c-a)\pi i}\frac{\Gamma(c)\Gamma(1-b)}{\Gamma(a)\Gamma(c-a-b+1)}w_4(z) + \\ e^{-a\pi i}\frac{\Gamma(c)\Gamma(1-b)}{\Gamma(a-b+1)\Gamma(c-a)}w_5(z), \end{aligned} \tag{12.2.20}$$

where we use the functions $w_4(z)$ and $w_5(z)$ in the form

$$\begin{aligned} w_4(z) &= z^{a-c}(1-z)^{c-a-b}{}_2F_1\left({1-a, c-a \atop c-a-b+1};1-\frac{1}{z}\right), \\ w_5(z) &= e^{a\pi i}z^{-a}{}_2F_1\left({a, a-c+1 \atop a-b+1};\frac{1}{z}\right). \end{aligned} \tag{12.2.21}$$

In the representations of the Gauss function in (12.2.20) and (12.2.21) the parameter b occurs only at the c-place, and it follows that for negative b we can use the approach described in §12.1.1.

For details on complex values of b we refer to Wagner (1984), where the asymptotics is considered for large values of a.

12.2.2 Large b, $|z| \le z_0$

When $z = \mathcal{O}(1/b)$, the large-b asymptotic behavior cannot be described in terms of elementary functions. To understand the nature of the asymptotics for large positive b and $z \to 0$, we recall the well-known limit

$$\lim_{b\to+\infty} {}_2F_1\left({a,b \atop c};\frac{z}{b}\right) = {}_1F_1\left({a \atop c};z\right), \tag{12.2.22}$$

which may be used to define the Kummer function (see Chapter 10).

This limit formally follows from (12.0.1) by computing termwise the limit

$$\lim_{b\to+\infty} \frac{(b)_n}{b^n} = 1. \tag{12.2.23}$$

For a detailed study on expansions of the generalized hypergeometric functions related to this type of limit we refer to Nagel (2004). An example that follows from this paper is the convergent expansion[1]

$${}_2F_1\left({a,b \atop c};\frac{z}{b}\right) = \sum_{s=0}^{\infty} {}_2F_0\left({-s,b \atop -};\frac{1}{b}\right)\frac{(a)_s(-z)^s}{(c)_s\, s!}{}_1F_1\left({a+s \atop c+s};z\right), \quad |z|<|b|. \tag{12.2.24}$$

When $b \to \infty$ all terms of the series vanish, except the one with $s = 0$, which becomes ${}_1F_1(a;c;z)$. This corresponds to the limit in (12.2.22). In each term the

[1] With thanks to Adri Olde Daalhuis.

Table 12.1 The four cases.

	ε_1	ε_2	ε_3
A	0	0	+1
B	0	−1	+1
C	+1	−1	0
D	+1	+2	0

${}_2F_0$-function is a rational function of b of order $\mathcal{O}\left(b^{-s/2}\right)$ (or of lower order) as $b \to \infty$, and when we rearrange the series, we can obtain a Poincaré-type asymptotic expansion for large b, with a, c and z in bounded complex domains.

In §28.3 we describe a uniform method for obtaining an expansion of a ${}_2F_1$-function in terms of the ${}_1F_1$-function with z not necessarily small. However, the same or similar results can be obtained when we use the connection formula in (12.2.20), with the functions $w_4(z)$ and $w_5(z)$ in the same form as in (12.2.21). For both hypergeometric functions with large b and large argument (when z is small), the case described in §12.1.3 applies, with approximations in terms of the Kummer U-functions, and the connection formula should give the ${}_1F_1$-function.

12.3 Other large parameter cases

In this chapter we have considered the asymptotic expansion of ${}_2F_1(a + \varepsilon_1\lambda, b + \varepsilon_2\lambda; c + \varepsilon_3\lambda; z)$ for the case that only one ε_j is different from zero. It has been explained in Watson (1918a), Luke (1969, Chapter VII) and Temme (2003), that the many cases with $\varepsilon_j = -1, 0, 1$, can be reduced to only three cases, together with the case ${}_2F_1(a + \lambda, b + 2\lambda; c; z)$. In Table 12.1 we summarize the four cases.

We have considered the simple Case A (with large c) in §12.1. The remaining cases have been considered in Watson (1918a). In Paris (2013a,b) expansions are given of the ${}_2F_1$-functions with large parameters when the ε_j are not restricted to $0, \pm1$. These expansions are in terms of elementary functions, with attention to complex parameters.

In Olde Daalhuis (2010b, §15.12) Cases B, C and D are considered, with further references. These expansions are in fact uniform expansions and are given in terms of special functions. For Case B the parabolic cylinder function $U(a, z)$ is used, for Case C modified Bessel function $I_\nu(z)$, and for Case D the Airy function $\mathrm{Ai}(z)$. For a more recent overview and new results for expansions of ${}_2F_1(a+\varepsilon_1\lambda, b+\varepsilon_2\lambda; c+\varepsilon_3\lambda; z)$ with $\varepsilon_j = 0, -1, 1$ we refer to Farid Khwaja and Olde Daalhuis (2014).

In several sections of this chapter we have explained that, for some cases, we need uniform expansions when we allow that $z \to 0$ (with example ${}_2F_1(a, b - \lambda; c; z)$

in §12.2.2) or when we allow z in unbounded domains (with examples ${}_2F_1(a,b;c\pm\lambda;z)$ in §12.1.3 and §12.1.4). In these cases Kummer functions are used as main approximants; again, see Farid Khwaja and Olde Daalhuis (2014) as an addition to Olde Daalhuis (2010b, §15.12).

12.3.1 *Jacobi polynomials of large degree*

The Jacobi polynomials have the representation

$$P_n^{(\alpha,\beta)}(x) = \binom{n+\alpha}{n}\,{}_2F_1\left({-n,\alpha+\beta+n+1 \atop \alpha+1};\tfrac12-\tfrac12x\right), \tag{12.3.25}$$

and we see that the asymptotic problem for large values of n corresponds to Case C of Table 12.1. In Chapter 15 we will consider the Legendre and Gegenbauer polynomials (which are special cases of the Jacobi polynomials) by using a method related to Darboux's method, but for the Jacobi polynomials the generating function is rather complicated, and we will explain here how to handle the case for $x\in(-1,1)$.

We take the integral representation in (12.0.4) and write

$$P_n^{(\alpha,\beta)}(x) = \frac{\Gamma(\alpha+n+1)\Gamma(\beta+n+1)}{\Gamma(\alpha+\beta+n+1)\,n!}\frac{1}{2\pi i}\int_0^{(1+)}\frac{t^{\alpha+\beta}}{(t-1)^{\beta+1}}e^{-n\phi(t)}\,dt, \tag{12.3.26}$$

where

$$\phi(t) = \ln(t-1)-\ln t-\ln(1-zt), \quad z=\tfrac12-\tfrac12x. \tag{12.3.27}$$

We write $x=\cos\theta$ with $\theta\in[0,\pi]$. Then the saddle points are

$$t_+ = \frac{2}{1-e^{i\theta}} = \frac{e^{\frac12 i(\pi-\theta)}}{\sin\left(\frac12\theta\right)}, \quad t_- = \frac{2}{1-e^{-i\theta}} = \frac{e^{-\frac12 i(\pi-\theta)}}{\sin\left(\frac12\theta\right)}, \tag{12.3.28}$$

and

$$\phi\left(t_\pm\right) = \pm i\theta, \quad \phi''\left(t_\pm\right) = \frac{4\sin^4\left(\frac12\theta\right)}{\sin\theta}e^{\pm i(\theta-\frac12\pi)}. \tag{12.3.29}$$

The saddle point contour runs in the lower half-plane from the origin through the saddle point t_- to $t=1/z>1$, and returns to the origin in the upper half-plane through t_+. For the lower part we use the transformation

$$\phi(t)-\phi\left(t_-\right) = \tfrac12\phi''\left(t_-\right)w^2. \tag{12.3.30}$$

For the coefficients in the expansion

$$t = t_- + w + \sum_{k=2}^{\infty} c_k w^k \tag{12.3.31}$$

we find for the first few

$$c_2 = \tfrac12 i\tan\left(\tfrac12\theta\right), \quad c_3 = \tfrac1{16}\left(e^{-2i\theta}-3\right)\tan^2\left(\tfrac12\theta\right). \tag{12.3.32}$$

Further steps are straightforward and for the complete expansion we can take twice the real part of the contribution from the saddle point t_-. This gives

$$\begin{aligned} P_n^{(\alpha,\beta)}(x) = &\frac{1}{\sqrt{\pi n}} \sin^{-\alpha-\frac12}\left(\tfrac12\theta\right) \cos^{-\beta-\frac12}\left(\tfrac12\theta\right) \times \\ &\frac{\Gamma(\alpha+n+1)\Gamma(\beta+n+1)}{\Gamma(\alpha+\beta+n+1)\, n!} (\cos\chi P + \sin\chi Q), \end{aligned} \tag{12.3.33}$$

where

$$\chi = \tfrac12 (2n+1+\alpha+\beta)\,\theta - \tfrac14\pi - \tfrac12\pi\alpha, \tag{12.3.34}$$

and P and Q have the expansions

$$P \sim \sum_{k=0}^{\infty} \frac{p_k}{n^k}, \quad Q \sim \sum_{k=0}^{\infty} \frac{q_k}{n^k}, \quad n \to \infty. \tag{12.3.35}$$

The expansion is valid uniformly with respect to $x \in [-1+\delta, 1-\delta]$. If we wish we can expand the ratio of gamma functions in (12.3.33) by using the results of §6.5. We have

$$\frac{\Gamma(\alpha+n+1)\Gamma(\beta+n+1)}{\Gamma(\alpha+\beta+n+1)\, n!} \sim 1 - \frac{\alpha\beta}{n} + \frac{\alpha(\alpha+1)\beta(\beta+1)}{2n^2} + \ldots. \tag{12.3.36}$$

For the first coefficients in the series in (12.3.35) we find

$$\begin{aligned} p_0 &= 1, \quad q_0 = 0, \\ p_1 &= \tfrac14(2\alpha\beta - \alpha - \beta - 1), \\ q_1 &= -\frac{2\alpha^2 - 2\beta^2 + \left(2\alpha^2 + 2\beta^2 - 1\right) x}{8\sin\theta}, \\ p_2 &= \frac{p_2^{(0)} + p_2^{(1)} x + p_2^{(2)} x^2}{128 \sin^2\theta}, \\ q_2 &= \tfrac14(2\alpha\beta - 3\alpha - 3\beta - 3) q_1, \end{aligned} \tag{12.3.37}$$

where

$$\begin{aligned} p_2^{(0)} &= 20\left(\alpha^2+\beta^2\right) - 8\left(\alpha^4+\beta^4\right) - 9 - p_2^{(2)}, \\ p_2^{(1)} &= -4\left(2\alpha^2 - 5 + 2\beta^2\right)\left(\alpha^2 - \beta^2\right), \\ p_2^{(2)} &= -4((\alpha-\beta)^2+2)(\alpha+\beta)^2 + 24(2\alpha\beta-1)(\alpha+\beta) - \\ &\quad 32\alpha^2\beta^2 + 40\alpha\beta - 13. \end{aligned} \tag{12.3.38}$$

For a different form of the large-degree asymptotic expansion of the Jacobi polynomial we refer to Hahn (1980); the coefficients of the expansion are given in explicit form. See also Koornwinder *et al.* (2010, §18). In Meijer (1985) a different expansion is derived for $x \in (-1, 1)$ by using the saddle point method and a complete expansion is given, with coefficients explicitly defined in terms of finite sums. Special cases are considered for Gegenbauer and Legendre polynomials. For the latter the expansion reduces to the result of Stieltjes given in §15.2. Henk Meijer's starting point is an integral due to Tom Koornwinder; in an appendix a proof of the integral is given by Henk van Haeringen, who showed that Koornwinder's integral is the same as in (12.3.26), up to a transformation.

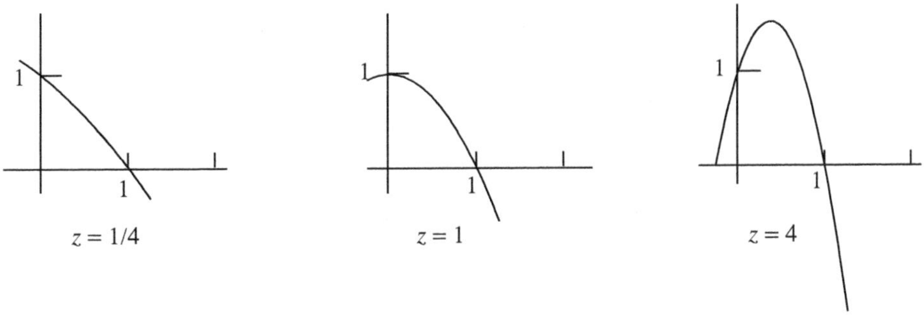

Fig. 12.1 The function $(1-t)(1+zt)$ for a few values of z.

12.3.2 *An example of the case* ${}_2F_1(a, b-\lambda; c+\lambda; z)$

We consider a simple form of Case B of Table 12.1.[2] First we show how we can use the complementary error function (see §3.4), which is a special case of the parabolic cylinder function.

Consider

$$ {}_2F_1\left(\begin{matrix} -n, 1 \\ n+2 \end{matrix}; -z\right) = (n+1)\int_0^1 \left((1-t)(1+zt)\right)^n \, dt, \tag{12.3.39}$$

for large positive n, with z near the point 1. This function reduces to a polynomial when $n = 0, 1, 2, \ldots$, but for the analysis this restriction is not needed.

We write

$$\left((1-t)(1+zt)\right)^n = e^{-n\phi(t)}, \quad \phi(t) = -\ln\left((1-t)(1+zt)\right), \tag{12.3.40}$$

and we have

$$\phi'(t) = \frac{2tz + 1 - z}{(1-t)(1+zt)}. \tag{12.3.41}$$

We see that the integrand has a peak value at $t_0 = (z-1)/(2z)$. So, if $z = 1$ the peak is at $t = 0$, if $z > 1$ then $t_0 \in (0,1)$, and if $z < 1$ then $t_0 < 0$. See Figure 12.1.

The same situation occurs for the integral

$$\int_0^\infty e^{-n\left(\frac{1}{2}w^2 - \alpha w\right)} \, dw, \tag{12.3.42}$$

the integrand having a peak value at $w_0 = \alpha$. The integral is a complementary error function and in a uniform expansion of (12.3.39) that holds for large n with z in a neighborhood of 1 we need this error function. When we take more general values of a, b and c in ${}_2F_1(a, b-\lambda; c+\lambda; z)$, parabolic cylinder functions should be used for $z \sim 1$. For details on this asymptotic feature we refer to Chapter 22.

We give a few details for the integral in (12.3.39). We transform by writing

$$(1-t)(1+zt) = e^{-\frac{1}{2}w^2 + \alpha w}, \tag{12.3.43}$$

with the conditions

[2]This section is based on Temme (2003).

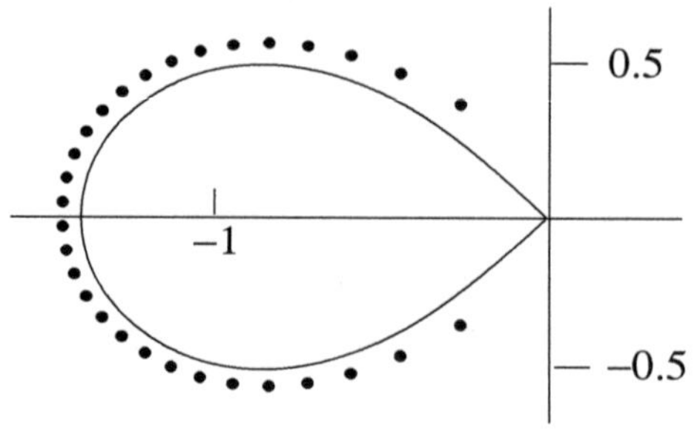

Fig. 12.2 Zeros of $P_n^{(\alpha,\beta)}(z),\ n=30, \alpha=-\beta=n+1$.

$$t=0 \Longleftrightarrow w=0, \quad t=1 \Longleftrightarrow w=\infty, \quad t=t_0 \Longleftrightarrow w=\alpha. \tag{12.3.44}$$

The quantity α follows from satisfying the matching of t_0 with α:

$$(1-t_0)(1+zt_0)=e^{-\frac{1}{2}w_0^2+\alpha w_0}. \tag{12.3.45}$$

This gives

$$\tfrac{1}{2}\alpha^2=-\ln\left(\frac{4z}{(z+1)^2}\right), \quad \operatorname{sign}(\alpha)=\operatorname{sign}(z-1). \tag{12.3.46}$$

We obtain

$${}_2F_1\left({-n,1 \atop n+2};-z\right)=(n+1)\int_0^\infty e^{-n\left(\frac{1}{2}w^2-\alpha w\right)}f(w)\,dw, \tag{12.3.47}$$

where

$$f(w)=\frac{dt}{dw}=\frac{w-\alpha}{t-t_0}\,\frac{(1-t)(1+zt)}{2z}. \tag{12.3.48}$$

A first approximation follows by replacing $f(w)$ by

$$f(\alpha)=\frac{1+z}{2\sqrt{2}\,z}, \tag{12.3.49}$$

which gives, as $n\to\infty$, uniformly with respect to z in a neighborhood of $z=1$,

$$\begin{aligned}{}_2F_1\left({-n,1 \atop n+2};-z\right) &\sim nf(\alpha)\int_0^\infty e^{-n\left(\frac{1}{2}w^2-\alpha w\right)}\,dw \\ &=\sqrt{\pi n}\,\frac{1+z}{4z}e^{\frac{1}{2}n\alpha^2}\operatorname{erfc}\left(-\alpha\sqrt{n/2}\right).\end{aligned} \tag{12.3.50}$$

This asymptotic approximation is in agreement with the general case ${}_2F_1(a,b-\lambda;c+\lambda;-z)$ considered in Olde Daalhuis (2003).

When in the Gauss hypergeometric function the parameter a or b equals a non-negative integer, then we can write this function in terms of a Jacobi polynomial. In the present case this relation is:

$${}_2F_1\left({-n,1 \atop n+2};-z\right)=\frac{(n+1)!}{2^{2n}\left(\frac{3}{2}\right)_n}(1+z)^nP_n^{(n+1,-n-1)}\left(\frac{1-z}{1+z}\right). \tag{12.3.51}$$

If $\alpha > -1, \beta > -1$ then the zeros of $P_n^{(\alpha,\beta)}(x)$ are located in the interval $(-1, 1)$ of orthogonality of these polynomials. In the present case the zeros are complex. See Figure 12.2 for the distribution of the zeros of

$$P_n^{(\alpha,\beta)}(z), \quad n = 30, \quad \alpha = n + 1, \quad \beta = -n - 1. \tag{12.3.52}$$

As $n \to \infty$, the zeros approach the curve $\left|1 - z^2\right| = 1$. This follows from the fact (see §42.2.3 and Table 42.2) that the zeros of the function $\operatorname{erfc}\left(-\alpha\sqrt{n/2}\right)$ are near the diagonals $\operatorname{ph}(-\alpha) = \pm\frac{3}{4}\pi$.

Chapter 13

Examples of ${}_3F_2$-polynomials

Large parameter cases of the ${}_3F_2$-functions arise frequently and there are no systematic methods to approach the problem. In Chapter 12 we have explained that we have suitable integral representations and connection formulas for the ${}_2F_1$-functions for obtaining all kinds of expansions for large parameters. For the ${}_3F_2$-functions these convenient starting points are missing, and even for the polynomial cases with argument ± 1 the approach has to be based on ad hoc methods.

For the ${}_3F_2$-functions many connection formulas and special values are available, and these may be helpful to write the functions in a suitable form; see Prudnikov *et al.* (1990, §7.7).

In this chapter we consider two examples that are treated by different methods. In one example it appears that a double integral gives a suitable representation for obtaining a rather simple asymptotic expansion, and we also describe a generating function approach. In the second case we first use a discrete form of Watson's lemma, and in a second approach again a generating function. A more complicated example can be found in Lin and Wong (2013), where uniform asymptotic expansions are derived for discrete Chebyshev polynomials, that can be written in terms of a ${}_3F_2$-polynomial. In that case a double integral is used by writing the ${}_3F_2$-polynomial as an integral containing a ${}_2F_1$-function, and by using an integral representation of this function.

13.1 A ${}_3F_2$ associated with the Catalan–Larcombe–French sequence

We consider the problem of finding the large-n asymptotics of[1]

$$f(n) = {}_3F_2\left({-n, \frac{1}{2}, \frac{1}{2} \atop \frac{1}{2} - n, \frac{1}{2} - n} ; -1 \right). \tag{13.1.1}$$

[1]This section is based on Temme (2006).

Peter Larcombe[2] communicated this problem and conjectured that $\lim_{n\to\infty} f(n) = 2$. Tom Koornwinder gave a proof, based on dominated convergence. For details of the proof, see Larcombe (2006), where a different representation of $f(n)$ is also considered in the form

$$f(n) = 2^n {}_3F_2\left({-n, -\frac{1}{2}n, \frac{1}{2} - \frac{1}{2}n \atop \frac{1}{2} - n, \frac{1}{2} - n} ; 1\right). \tag{13.1.2}$$

The equivalence of these two forms follows from a quadratic transformation of the ${}_3F_2$-functions as given in Prudnikov *et al.* (1990, p. 497), that is,

$$\begin{aligned} {}_3F_2\left({a, b, c \atop 1+a-b, 1+c-c} ; z\right) = \\ (1-z)^{-a} {}_3F_2\left({\frac{1}{2}a, \frac{1}{2} + \frac{1}{2}a, 1+a-b-c \atop 1+a-b, 1+a-c} ; \frac{-4z}{(1-z)^2}\right), \end{aligned} \tag{13.1.3}$$

with $a = -n$, $b = c = \frac{1}{2}$, and $z = -1$. Another form is given by (see Larcombe and French (2004))

$$f(n) = \frac{n!}{2^n \left(\frac{1}{2}\right)_n} {}_3F_2\left({-n, -n, \frac{1}{2} \atop 1, \frac{1}{2} - n} ; -1\right). \tag{13.1.4}$$

In Clark (2004) an asymptotic expansion of $\frac{1}{2}f(n)$ has been derived. The asymptotic analysis is based on the representation

$$P_n = \frac{1}{n!} \sum_{p+q=n} \binom{2n}{p} \binom{2q}{q} \frac{(2p)!\,(2q)!}{p!\,q!}. \tag{13.1.5}$$

By using the relation

$$(2n)! = 2^{2n} n! \left(\tfrac{1}{2}\right)_n, \quad n = 0, 1, 2, \ldots, \tag{13.1.6}$$

it is straightforward to verify that (13.1.5) can be written as

$$P_n = \frac{2^{4n}}{n!} \sum_{p=0}^{n} \frac{\left(\frac{1}{2}\right)_p \left(\frac{1}{2}\right)_p \left(\frac{1}{2}\right)_{n-p} \left(\frac{1}{2}\right)_{n-p}}{p!\,(n-p)!}. \tag{13.1.7}$$

By using

$$(a)_{n-k} = (-1)^k \frac{(a)_n}{(1-a-n)_k}, \tag{13.1.8}$$

it follows that

$$P_n = \frac{2^{4n} \left(\frac{1}{2}\right)_n \left(\frac{1}{2}\right)_n}{n!\,n!} \sum_{p=0}^{n} (-1)^p \frac{(-n)_p \left(\frac{1}{2}\right)_p \left(\frac{1}{2}\right)_p}{p!\,(\frac{1}{2}-n)_p (\frac{1}{2}-n)_p}, \tag{13.1.9}$$

that is,

$$P_n = \frac{2^{4n} \left(\frac{1}{2}\right)_n \left(\frac{1}{2}\right)_n}{n!\,n!} {}_3F_2\left({-n, \frac{1}{2}, \frac{1}{2} \atop \frac{1}{2} - n, \frac{1}{2} - n} ; -1\right), \tag{13.1.10}$$

---

[2]Private communications.

which gives the relation to $f(n)$ by using (13.1.1):

$$P_n = \frac{2^{4n}\left(\frac{1}{2}\right)_n\left(\frac{1}{2}\right)_n}{n!\,n!}\,f(n) = \binom{2n}{n}^2 f(n). \tag{13.1.11}$$

The numbers P_n are, for $n = 0, 1, 2, \ldots$, known as the elements of the sequence (A053175) $\{1, 8, 80, 896, 10816, \ldots\}$, called the *Catalan–Larcombe–French* sequence,[3] which was originally discussed in Catalan (1887).

We derive a complete asymptotic expansion of the numbers P_n by using integral representations of the corresponding ${}_3F_2$-functions. Our results are the same as those obtained in (Clark, 2004), who used the binomial sum in (13.1.5) without reference to the ${}_3F_2$-functions.

13.1.1 *Transformations*

We derive an integral representation of the ${}_3F_2$-function of (13.1.1) by using several transformations for special functions. We start with the beta integral

$$B(x,y) = \frac{\Gamma(x)\Gamma(y)}{\Gamma(x+y)} = \int_0^1 t^{x-1}(1-t)^{y-1}\,dt \tag{13.1.12}$$

and use it with $x = k + \frac{1}{2}$, $y = n - k + \frac{1}{2}$, together with

$$\Gamma\left(\tfrac{1}{2} - z\right)\Gamma\left(\tfrac{1}{2} + z\right) = \frac{\pi}{\cos \pi z}, \quad z \notin \mathbb{Z}. \tag{13.1.13}$$

It follows that

$$\frac{\left(\frac{1}{2}\right)_k}{\left(\frac{1}{2} - n\right)_k} = \frac{(-1)^k n!}{\sqrt{\pi}\,\Gamma\left(n + \frac{1}{2}\right)} \int_0^1 t^{k-\frac{1}{2}}(1-t)^{n-k-\frac{1}{2}}\,dt, \quad k = 0, 1, 2, \ldots. \tag{13.1.14}$$

We substitute this in the representation of the ${}_3F_2$-function in (13.1.1):

$${}_3F_2\left(\begin{matrix} -n, \frac{1}{2}, \frac{1}{2} \\ \frac{1}{2} - n, \frac{1}{2} - n \end{matrix}; -1\right) = \sum_{k=0}^{n} (-1)^k \frac{(-n)_k \left(\frac{1}{2}\right)_k \left(\frac{1}{2}\right)_k}{k!\,(\frac{1}{2} - n)_k (\frac{1}{2} - n)_k}. \tag{13.1.15}$$

This gives after performing the k-summation

$$f(n) = \frac{n!}{\pi\left(\frac{1}{2}\right)_n} \int_0^1 t^{-\frac{1}{2}}(1-t)^{n-\frac{1}{2}}\,{}_2F_1\left(\begin{matrix} -n, \frac{1}{2} \\ \frac{1}{2} - n \end{matrix}; \frac{t}{1-t}\right) dt. \tag{13.1.16}$$

We substitute $t = \sin^2(\theta/2)$ and obtain

$$f(n) = \frac{n!}{\pi\left(\frac{1}{2}\right)_n} \int_0^\pi \cos^{2n}(\theta/2)\,{}_2F_1\left(\begin{matrix} -n, \frac{1}{2} \\ \frac{1}{2} - n \end{matrix}; \tan^2(\theta/2)\right) d\theta. \tag{13.1.17}$$

We apply a quadratic transformation (see Olde Daalhuis (2010b, Eq. 15.8.15)) to obtain

$$f(n) = \frac{n!}{\pi\left(\frac{1}{2}\right)_n} \int_0^\pi {}_2F_1\left(\begin{matrix} -\frac{1}{2}n, \frac{1}{2} - \frac{1}{2}n \\ \frac{1}{2} - n \end{matrix}; \sin^2\theta\right) d\theta, \tag{13.1.18}$$

[3] See the *On-Line Encyclopedia of Integer Sequences*, https://oeis.org/A053175.

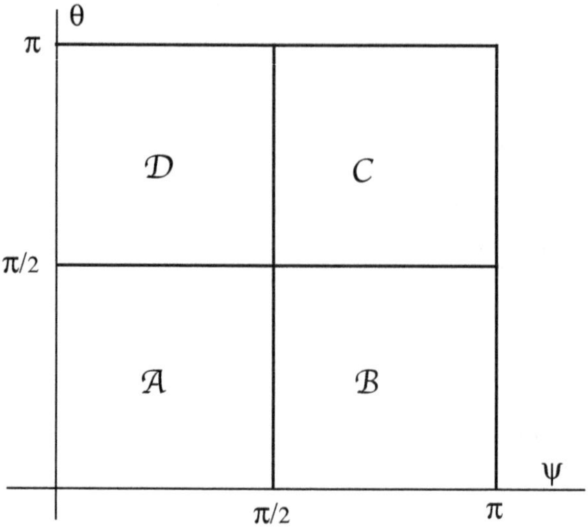

Fig. 13.1 The domain of integration of the integral in (13.1.22) and subdomains $\mathcal{A}$, $\mathcal{B}$, $\mathcal{C}$ and $\mathcal{D}$.

and use the representation of the Legendre polynomial

$$P_n(x) = \frac{(2n)!}{2^n n!\, n!} x^n {}_2F_1\left({-\tfrac{1}{2}n, \tfrac{1}{2} - \tfrac{1}{2}n \atop \tfrac{1}{2} - n} ; x^{-2}\right). \tag{13.1.19}$$

This gives

$$f(n) = \frac{2^{-n} n!\, n!}{\pi \left(\frac{1}{2}\right)_n \left(\frac{1}{2}\right)_n} \int_0^\pi \sin^n\theta \; P_n\left(\frac{1}{\sin\theta}\right) d\theta. \tag{13.1.20}$$

Next, consider

$$P_n(z) = \frac{1}{\pi} \int_0^\pi \left(z + \sqrt{z^2-1}\cos\psi\right)^n d\psi, \quad n = 0, 1, 2, \ldots, \tag{13.1.21}$$

which gives the double integral

$$f(n) = \frac{n!\, n!}{\pi^2 \left(\frac{1}{2}\right)_n \left(\frac{1}{2}\right)_n} \int_0^\pi \int_0^\pi \left(\frac{1+\cos\theta\cos\psi}{2}\right)^n d\theta\, d\psi. \tag{13.1.22}$$

For the relations of the Legendre polynomials we refer to Temme (1996a, Chapter 8).

13.1.2 *Asymptotic analysis*

The landscape of the integrand in (13.1.22) shows peaks at the boundary points $(0,0)$ and (π,π), where it assumes the value 1. Along the interior lines $\theta = \frac{1}{2}\pi$ and $\psi = \frac{1}{2}\pi$ the integrand has the value 2^{-n}. Inside the squares $\mathcal{A}$ and $\mathcal{C}$, see Figure 13.1, the value of the integrand is between 2^{-n} and 1, in the squares $\mathcal{B}$ and $\mathcal{D}$ it is between 0 and 2^{-n}. In addition, the contributions from $\mathcal{A}$ and $\mathcal{C}$ are the same, and those from $\mathcal{B}$ and $\mathcal{D}$ are the same as well.

From an asymptotic point of view it follows that the integral over the full square equals twice the integral over $\mathcal{A}$, with an error that is of order $\mathcal{O}(2^{-n})$, while the total integral is of order $\mathcal{O}(1)$, as n is large. Hence, we concentrate on the integral over $\mathcal{A}$, and write for large values of n

$$f(n) = 2\frac{2^{-n}n!\,n!}{\pi^2\left(\frac{1}{2}\right)_n\left(\frac{1}{2}\right)_n}\left(\int_0^{\frac{1}{2}\pi}\int_0^{\frac{1}{2}\pi}(1+\cos\theta\cos\psi)^n\,d\theta\,d\psi + E_n\right), \tag{13.1.23}$$

where $E_n = \mathcal{O}(2^{-n})$. Next, we neglect E_n, put $u = \sin(\theta/2)$, $v = \sin(\psi/2)$, and obtain

$$f(n) \sim \frac{8\,n!\,n!}{\pi^2\left(\frac{1}{2}\right)_n\left(\frac{1}{2}\right)_n}\int_0^{\frac{1}{2}\sqrt{2}}\int_0^{\frac{1}{2}\sqrt{2}}\frac{\left(1-u^2-v^2+2u^2v^2\right)^n}{\sqrt{1-u^2}\,\sqrt{1-v^2}}\,du\,dv. \tag{13.1.24}$$

For the integrals in (13.1.23) and (13.1.24) asymptotic expansions can be obtained by using Laplace's method for double integrals; see Wong (2001, § VIII.10)). In our case a simpler approach is based on neglecting a part of square $\mathcal{A}$ by introducing polar coordinates

$$u = r\cos t, \quad v = r\sin t, \quad 0 \le r \le \tfrac{1}{2}\sqrt{2}, \quad 0 \le t \le \tfrac{1}{2}\pi. \tag{13.1.25}$$

This gives (again we make an error in the integral that is of order $\mathcal{O}(2^{-n})$)

$$f(n) \sim \frac{8\,n!\,n!}{\pi^2\left(\frac{1}{2}\right)_n\left(\frac{1}{2}\right)_n}\int_0^{\pi/2}\int_0^{\frac{1}{2}\sqrt{2}}\frac{(1-r^2+2r^4\cos^2 t\sin^2 t)^n}{\sqrt{(1-r^2\cos^2 t)(1-r^2\sin^2 t)}}r\,dr\,dt. \tag{13.1.26}$$

We change r^2 into r, and obtain

$$f(n) \sim \frac{4\,n!\,n!}{\pi^2\left(\frac{1}{2}\right)_n\left(\frac{1}{2}\right)_n}\int_0^{\pi/2}\int_0^{\frac{1}{2}}\frac{(1-r+2r^2\cos^2 t\sin^2 t)^n}{\sqrt{(1-r\cos^2 t)(1-r\sin^2 t)}}\,dr\,dt. \tag{13.1.27}$$

First the standard method for obtaining asymptotic expansions of a Laplace-type integral can be used (for the r-integral). The second step is done by integrating the coefficients of this expansion with respect to t.

For the r-integral we transform the variable of integration by putting

$$w = -\ln\left(1-r+2r^2\cos^2 t\sin^2 t\right). \tag{13.1.28}$$

This mapping is one-to-one for $r \in [0, \frac{1}{2}]$, uniformly with respect to $t \in [0, \frac{1}{2}\pi]$, with corresponding w-interval $[0, w_0]$, where $w_0 = w\left(\frac{1}{2}\right)$.

We obtain

$$f(n) \sim \frac{4\,n!\,n!}{\pi^2\left(\frac{1}{2}\right)_n\left(\frac{1}{2}\right)_n}\int_0^{\pi/2}\int_0^{w_0} e^{-nw}F(w,t)\,dw\,dt, \tag{13.1.29}$$

where

$$F(w,t) = \frac{1}{\sqrt{(1-r\cos^2 t)(1-r\sin^2 t)}}\frac{dr}{dw}. \tag{13.1.30}$$

13.1.3 *Asymptotic expansion*

We obtain the asymptotic expansion of the w-integral in (13.1.29) by using Watson's lemma (see Chapter 2).

The function $F(w,t)$ is analytic in a neighborhood of the origin of the w-plane. We expand

$$F(w,t) = \sum_{k=0}^{\infty} c_k(t) w^k \tag{13.1.31}$$

and substitute this expansion in (13.1.29). Interchanging the order of summation and integration, and replacing the interval of the w-integrals by $[0,\infty)$, we obtain

$$f(n) \sim \frac{4\,n!\,n!}{n\pi^2\,(\frac12)_n\,(\frac12)_n} \sum_{k=0}^{\infty} C_k \frac{k!}{n^k}, \quad n \to \infty, \tag{13.1.32}$$

where

$$C_k = \int_0^{\pi/2} c_k(t)\,dt, \quad k = 0,1,2,\ldots. \tag{13.1.33}$$

The coefficients $c_k(t)$ can be obtained by the following method. First we need the inverse of the transformation defined in (13.1.28). That is, we need coefficients b_k in the expansion

$$r(w) = \sum_{k=1}^{\infty} b_k(t) w^k. \tag{13.1.34}$$

We can find $r(w)$ from (13.1.28) as a solution of a quadratic equation, with the condition $r(w) \sim w$ as $w \to 0$, that is, $b_1(t) = 1$. However, we can also differentiate (13.1.28) with respect to r and substitute the expansion (13.1.34), and solve for the coefficients $b_k(t)$. When we have these coefficients we can expand $F(w,t)$ of (13.1.30) and find $c_k(t)$.

The first few coefficients $c_k(t)$ are

$$\begin{aligned}
c_0(t) &= 1,\\
c_1(t) &= \tfrac12(-1+8s^2-8s^4),\\
c_2(t) &= \tfrac18(1-28s^2+220s^4-384s^6+192s^8),\\
c_3(t) &= \tfrac{1}{48}(-1+92s^2-1628s^4+10752s^6-24576s^8+\\
&\quad 23040s^{10}-7680s^{12}),\\
c_4(t) &= \tfrac{1}{384}(1-280s^2+10024s^4-130848s^6+773904s^8-\\
&\quad 2054400s^{10}+2691840s^{12}-1720320s^{14}+430080s^{16}),\\
c_5(t) &= \tfrac{1}{3840}(-1+848s^2-55328s^4+1259040s^6-13396560s^8+\\
&\quad 73983360s^{10}-215329920s^{12}+349224960s^{14}-\\
&\quad 319549440s^{16}+154828800s^{18}-30965760s^{20}),
\end{aligned} \tag{13.1.35}$$

where $s = \sin^2 t$. For the corresponding C_k we have

$$C_0 = \tfrac{1}{2}\pi, \quad C_1 = 0, \quad C_2 = \tfrac{1}{8}\pi, \quad C_3 = \tfrac{1}{8}\pi, \quad C_4 = \tfrac{55}{384}\pi, \quad C_5 = \tfrac{11}{64}\pi. \quad (13.1.36)$$

As a next step we can replace in (13.1.32) the ratios $n!/\left(\frac{1}{2}\right)_n$ by the asymptotic expansion (see §6.5.1)

$$\frac{n!}{\left(\frac{1}{2}\right)_n} = \sqrt{\pi}\frac{\Gamma(n+1)}{\Gamma(n+\frac{1}{2})} \sim \sqrt{\pi n}\sum_{k=0}^{\infty}\frac{\gamma_k}{n^k}, \quad (13.1.37)$$

where

$$\begin{aligned}&\gamma_0 = 1, \quad \gamma_1 = \tfrac{1}{8}, \quad \gamma_2 = \tfrac{1}{128}, \quad \gamma_3 = -\tfrac{5}{1024},\\ &\gamma_4 = -\tfrac{21}{32768}, \quad \gamma_5 = \tfrac{399}{262144}.\end{aligned} \quad (13.1.38)$$

This finally gives

$$f(n) \sim 2\left(1 + \frac{1}{4n} + \frac{17}{32n^2} + \frac{207}{128n^3} + \frac{14875}{2048n^4} + \frac{352375}{8192n^5} + \ldots\right). \quad (13.1.39)$$

13.1.4 *An alternative method*

The numbers P_n were proposed as *Catalan numbers* by an associate of Catalan. They appear as coefficients in the series expansion of an elliptic integral of the first kind

$$K(k) = \int_0^{\frac{1}{2}\pi} \frac{1}{\sqrt{1-k^2\sin^2 t}}\,dt, \quad (13.1.40)$$

which is transformed and written as a power series in k (through an intermediate variable); this gives a generating function for the sequence $\{P_n\}$. For details we refer to Jarvis *et al.* (2004).

In Temme (2003) a generating function for the numbers P_n is given in terms of the square of a modified Bessel function, and we use this approach to obtain an asymptotic expansion of $f(n)$. See also Larcombe and French (2004) for details on this generating function.

We consider numbers F_n defined as coefficients in the generating function

$$\left(e^{w/2}\, I_0(w/2)\right)^2 = \sum_{n=0}^{\infty} F_n w^n. \quad (13.1.41)$$

By considering the relation of the Bessel function to the confluent hypergeometric functions (see (10.1.13)), we can write (13.1.41) in the form

$$\left({}_1F_1\left(\begin{matrix}\frac{1}{2}\\ 1\end{matrix};\, w\right)\right)^2 = e^{2w}\left({}_1F_1\left(\begin{matrix}\frac{1}{2}\\ 1\end{matrix};\, -w\right)\right)^2 = \sum_{n=0}^{\infty} F_n w^n. \quad (13.1.42)$$

This gives the representation for F_n:

$$F_n = \sum_{k=0}^{n} \frac{\left(\frac{1}{2}\right)_k}{k!\,k!}\,\frac{\left(\frac{1}{2}\right)_{n-k}}{(n-k)!\,(n-k)!}. \quad (13.1.43)$$

By using (13.1.8) it follows that

$$F_n = \frac{\left(\frac12\right)_n}{n!\,n!}\sum_{k=0}^{n}(-1)^k\frac{(-n)_k(-n)_k\left(\frac12\right)_k}{\left(\frac12-n\right)_k k!\,k!}, \tag{13.1.44}$$

or

$$F_n = \frac{\left(\frac12\right)_n}{n!\,n!}\,{}_3F_2\left(\begin{matrix}-n,-n,\frac12\\ 1,\frac12-n\end{matrix};-1\right). \tag{13.1.45}$$

It follows from (13.1.4) that

$$f(n) = \frac{n!\,n!\,n!}{2^n\left(\frac12\right)_n\left(\frac12\right)_n}\,F_n. \tag{13.1.46}$$

From (13.1.41) we obtain

$$F_n = \frac{1}{2\pi i}\int_{\mathcal{C}}\frac{\left(e^{w/2}\,I_0(w/2)\right)^2}{w^{n+1}}\,dw = \frac{1}{2\pi i}\int_{\mathcal{C}}\frac{e^{2w}}{w^{n+1}}\,h(w)\,dw, \tag{13.1.47}$$

where

$$h(w) = \left(e^{-w/2}\,I_0(w/2)\right)^2, \tag{13.1.48}$$

and the contour $\mathcal{C}$ is a circle around the origin, or every contour that can be obtained from this circle by using Cauchy's theorem.

In Chapter 15 we will consider the asymptotic expansion of coefficients of power series, where we pick up the main contributions from singularities nearest to the origin. In the present case, the function $h(w)$ is analytic, and as mentioned in that chapter, we may consider the saddle point method.

The main contribution to the integral in (13.1.47) will come from the saddle point of e^{2w}/w^n, that is from $w = w_0 = n/2$. In the standard saddle point method (see Chapter 4) a quadratic transformation is used to bring the main part of the integrand in the form of a Gaussian. In the present case, however, it is not very clear how to proceed. We can obtain an expansion by just expanding the function $h(w)$ (which is slowly varying for $w > 0$) at the saddle point w_0. But the derivatives of $h(w)$ evaluated at w_0 will depend on the large parameter, and the effect of this will not be predictable (although it works when using several expansions of Kummer functions).

Because it is not so difficult to obtain the large-w expansion of $h(w)$, we propose the following approach. First we expand (see (9.3.14))

$$e^{-w/2}\,I_0(w/2) \sim \frac{1}{\sqrt{\pi w}}\sum_{k=0}^{\infty}\frac{\left(\frac12\right)_k\left(\frac12\right)_k}{k!\,w^k}, \tag{13.1.49}$$

and this gives

$$h(w) \sim \frac{1}{\pi w}\sum_{k=0}^{\infty}\frac{c_k}{w^k}, \tag{13.1.50}$$

where

$$c_0 = 1, \quad c_1 = \tfrac{1}{2}, \quad c_2 = \tfrac{5}{8}, \quad c_3 = \tfrac{21}{16}, \quad c_4 = \tfrac{507}{128}, \quad c_5 = \tfrac{4035}{256}. \tag{13.1.51}$$

When we substitute the expansion in (13.1.50) into (13.1.47) we obtain

$$F_n \sim \frac{2^{n+1}}{\pi(n+1)!}\sum_{k=0}^{\infty}\frac{2^k c_k}{(n+2)_k}, \quad n\to\infty. \tag{13.1.52}$$

By expanding this in negative powers of n, using (13.1.46) and the expansion in (13.1.37), we obtain for f_n the same expansion as in (13.1.39).

13.2 An integral of Laguerre polynomials

We are interested in the large-k behavior of the integral

$$I_k = \int_0^\infty x^{\frac{1}{2}} e^{-x}\left(L_k^{(0)}(x)\right)^2 dx, \tag{13.2.53}$$

where $L_k^{(0)}(x)$ is the Laguerre polynomial. Integrals of this type arose in a study on the marginal distribution of an eigenvalue of a Wishart matrix, and Edmundo J. Huertas Cejudo and Alfredo Deaño[4] presented the integral I_k in the form

$$I_k = \tfrac{1}{2}\sqrt{\pi}\frac{\left(-\frac{1}{2}\right)_k\left(-\frac{1}{2}\right)_k}{k!\,k!}\,{}_3F_2\left({-k,-k,\frac{3}{2} \atop \frac{3}{2}-k,\frac{3}{2}-k};1\right). \tag{13.2.54}$$

In Prudnikov *et al.* (1988, p. 478) the following more general form is given:

$$\begin{aligned}&\int_0^\infty x^{\alpha-1}e^{-cx}L_m^{(\gamma)}(cx)L_n^{(\lambda)}(cx)\,dx = \\ &\qquad \frac{(1+\gamma)_m(\lambda-\alpha+1)_n\Gamma(\alpha)}{m!\,n!\,c^\alpha}\,{}_3F_2\left({-m,\alpha,\alpha-\lambda \atop \gamma+1,\alpha-\lambda-n};1\right),\end{aligned} \tag{13.2.55}$$

valid for $\Re\alpha > 0$ and $\Re c > 0$ (this c is not needed). This gives

$$I_k = \tfrac{1}{2}\sqrt{\pi}\frac{\left(-\frac{1}{2}\right)_k}{k!}\,{}_3F_2\left({-k,\frac{3}{2},\frac{3}{2} \atop 1,\frac{3}{2}-k};1\right), \tag{13.2.56}$$

and the relation between the two representations may follow from one of the many connection formulas for the ${}_3F_2$ polynomials, of which we have seen another example in (13.1.3).

[4]Universidade de Coimbria, Portugal and KU Leuven, Belgium.

13.2.1 A first approach

The terms of the expansion of the ${}_3F_2$ in (13.2.54) are all positive and increasing. We write it in reversed order:

$${}_3F_2\left(\begin{matrix}-k,-k,\frac{3}{2}\\ \frac{3}{2}-k,\frac{3}{2}-k\end{matrix};1\right)=\sum_{n=0}^{k}\frac{(-k)_{k-n}(-k)_{k-n}\left(\frac{3}{2}\right)_{k-n}}{\left(\frac{3}{2}-k\right)_{k-n}\left(\frac{3}{2}-k\right)_{k-n}(k-n)!}. \qquad (13.2.57)$$

Using

$$\frac{(-k)_{k-n}}{\left(\frac{3}{2}-k\right)_{k-n}}=\frac{k!}{\left(-\frac{1}{2}\right)_k}\frac{\left(-\frac{1}{2}\right)_n}{n!}, \qquad (13.2.58)$$

we obtain

$$I_k=\tfrac{1}{2}\sqrt{\pi}\sum_{n=0}^{k}\frac{\left(-\frac{1}{2}\right)_n\left(-\frac{1}{2}\right)_n\left(\frac{3}{2}\right)_{k-n}}{n!\,n!\,(k-n)!}. \qquad (13.2.59)$$

We note parenthetically that this can also be written in the form

$$I_k=\tfrac{1}{2}\sqrt{\pi}\frac{\left(\frac{3}{2}\right)_k}{k!}\,{}_3F_2\left(\begin{matrix}-k,-\frac{1}{2},-\frac{1}{2}\\ 1,-\frac{1}{2}-k\end{matrix};1\right), \qquad (13.2.60)$$

but we shall make no use of this fact.

We expand the dominant term in the sum in (13.2.59) for large k. That is,

$$\frac{\left(\frac{3}{2}\right)_{k-n}}{(k-n)!}=\frac{1}{\Gamma\left(\frac{3}{2}\right)}\frac{\Gamma\left(\frac{3}{2}-n+k\right)}{\Gamma(1-n+k)}\sim 2\sqrt{\frac{k}{\pi}}\sum_{m=0}^{\infty}\frac{c_m(n)}{k^m}, \qquad (13.2.61)$$

for fixed n, where

$$c_0(n)=1,\quad c_1(n)=-\tfrac{1}{8}(4n-3),\quad c_2(n)=-\tfrac{1}{128}\left(16n^2-24n+7\right). \qquad (13.2.62)$$

For details of this expansion, see §6.5.

Using the expansion in (13.2.59) we obtain

$$I_k\sim\sqrt{k}\sum_{m=0}^{\infty}\frac{d_m}{k^m},\quad d_m=\sum_{n=0}^{k}\frac{\left(-\frac{1}{2}\right)_n\left(-\frac{1}{2}\right)_n}{n!\,n!}c_m(n). \qquad (13.2.63)$$

Because $c_0(n)=1$, the series for d_0 converges as $k\to\infty$, and the limit is $4/\pi$, we propose as a first approximation

$$I_k\sim\frac{4\sqrt{k}}{\pi}. \qquad (13.2.64)$$

The series for d_1 is also convergent as $k\to\infty$, with limit $1/\pi$, and we add this to the earlier estimate:

$$I_k\sim\frac{4\sqrt{k}}{\pi}\left(1+\frac{1}{4k}+\ldots\right),\quad k\to\infty. \qquad (13.2.65)$$

The series for d_m, $m\ge 2$, do not converge as $k\to\infty$, and we need to find the large-k behavior of d_m for fixed m. The next contribution to the asymptotic expansion will contain a term with $\ln k$. This follows from writing the coefficient $c_2(n)$ in the form

$$c_2(n) = -\tfrac{1}{128}\left(16n^2 - 24n + 7\right) = -\tfrac{7}{128} + \tfrac{1}{16}n - \tfrac{1}{8}n(n-1). \tag{13.2.66}$$

This gives three contributions in the term d_2 of which the first two correspond to convergent expansions. The divergence comes from the part

$$\sum_{n=2}^{k} \frac{\left(-\frac12\right)_n\left(-\frac12\right)_n}{n!\,n!}n(n-1) = \frac{1}{4\pi}\sum_{n=0}^{k-2}\frac{\Gamma\left(\frac32+n\right)\Gamma\left(\frac32+n\right)}{n!\,(n+2)!}. \tag{13.2.67}$$

The convergence can be controlled by writing

$$\sum_{n=0}^{k-2}\left(\frac{\Gamma\left(\frac32+n\right)\Gamma\left(\frac32+n\right)}{n!\,(n+2)!} - \frac{1}{n+1}\right) + H_{k-1}, \tag{13.2.68}$$

where H_n are the harmonic numbers introduced in §2.7.1.1, with asymptotic expansion given in (2.7.121).

The limit of the sum can be found by writing

$$\sum_{n=0}^{k-2}\left(\frac{\Gamma\left(\frac32+n\right)\Gamma\left(\frac32+n\right)}{n!\,(n+2)!}x^n - \frac{x^n}{n+1}\right), \tag{13.2.69}$$

and if $0 < x < 1$ this converges (as $k \to \infty$) to

$${}_2F_1\left(\begin{matrix}\frac32, \frac32\\ 3\end{matrix}; x\right) + \frac{1}{x}\ln(1-x). \tag{13.2.70}$$

By using connection formulas of the hypergeometric function we can expand it at the point $x = 1$. However, in the present case this relation has a special form because the a, b and c parameters satisfy a special relation: $c = a + b + m$ with $m = 0$. In that case the formula has a logarithmic term. We have, for $m = 0, 1, 2, \ldots$,

$$\begin{gathered}\frac{1}{\Gamma(a+b+m)}F\left(\begin{matrix}a, b\\ a+b+m\end{matrix}; z\right) = \frac{1}{\Gamma(a+m)\,\Gamma(b+m)} \times \\ \sum_{n=0}^{m-1} A_n(z-1)^n - \frac{(z-1)^m}{\Gamma(a)\,\Gamma(b)}\sum_{n=0}^{\infty} B_n(1-z)^n\left(\ln(1-z) + C_n\right),\end{gathered} \tag{13.2.71}$$

where

$$\begin{aligned} A_n &= \frac{(a)_n(b)_n(m-n-1)!}{n!}, \\ B_n &= \frac{(a+m)_n(b+m)_n}{n!(n+m)!}, \\ C_n &= -\psi(n+1) - \psi(n+m+1) + \psi(a+n+m) + \psi(b+n+m), \end{aligned} \tag{13.2.72}$$

and $\psi(z) = \Gamma'(z)/\Gamma(z)$.

Using the expansion of the ${}_2F_1$-function in (13.2.70) with $x \to 1$, we see that the term with the logarithm is cancelled and we can obtain the proper limit. For the coefficients d_m with $m \geq 3$ we have not investigated if a similar approach is possible.

It should be observed that the step from (13.2.59) to (13.2.63) is heuristic. It looks like a discrete application of Watson's lemma for Laplace integrals. In the next section we consider a complete approach.

13.2.2 *A generating function approach*

We recognize in (13.2.59) a convolution of coefficients of two power series governed by

$$\sum_{n=0}^{\infty} \frac{\left(-\frac{1}{2}\right)_n \left(-\frac{1}{2}\right)_n}{n!\, n!} = {}_2F_1\!\left({-\frac{1}{2}, -\frac{1}{2} \atop 1} ; z \right), \quad \sum_{n=0}^{\infty} \frac{\left(\frac{3}{2}\right)_n}{n!} = (1-z)^{-\frac{3}{2}}. \tag{13.2.73}$$

Hence,

$$I_k = \frac{\frac{1}{2}\sqrt{\pi}}{2\pi i} \int_{\mathcal{C}} G(z)\, \frac{dz}{z^{k+1}}, \quad G(z) = {}_2F_1\!\left({-\frac{1}{2}, -\frac{1}{2} \atop 1} ; z \right) (1-z)^{-\frac{3}{2}}, \tag{13.2.74}$$

where $\mathcal{C}$ is a circle around the origin with radius less than 1. The ${}_2F_1$-function can be expressed in terms of elliptic integrals, but this is not used here.

In Chapter 15 we will explain how to obtain asymptotic expansions of coefficients of power series by using this type of integral. We deform the contour around the branch cut of the singular point that is nearest to the origin, and use Watson's lemma for loop integrals after some substitutions.

In the present case, the function $G(z)$ has one singular point at $z = 1$, where we have an algebraic and logarithmic singularity, and we substitute the expansion that follows from (13.2.71) with $m = 2$. After substituting the expansion we use the Cauchy integral

$$f_k(\alpha) = \frac{1}{2\pi i} \int_{\mathcal{C}} (1-z)^{\alpha}\, \frac{dz}{z^{k+1}}, \tag{13.2.75}$$

$k = 0, 1, 2, \ldots$, and we can use one of the forms

$$f_k(\alpha) = (-1)^k \binom{\alpha}{k} = (-1)^k \frac{\Gamma(\alpha+1)}{k!\,\Gamma(\alpha+1-k)} = \frac{(-\alpha)_k}{k!} = \frac{\Gamma(k-\alpha)}{k!\,\Gamma(-\alpha)}. \tag{13.2.76}$$

For the logarithmic terms we use the latter from, and differentiate with respect to α. This gives

$$\begin{aligned} &\frac{1}{2\pi i} \int_{\mathcal{C}} \ln(1-z)(1-z)^{\alpha}\, \frac{dz}{z^{k+1}} = \phi_k(\alpha), \\ &\phi_k(\alpha) = f_k(\alpha)\left(\psi(-\alpha) - \psi(k-\alpha)\right). \end{aligned} \tag{13.2.77}$$

We split up the function $G(z)$ into two parts, $G(z) = G_1(z) + G_2(z)$, where

$$\begin{aligned} G_1(z) &= \frac{4}{\pi} \sum_{n=0}^{1} A_n (-1)^n (1-z)^{n-\frac{3}{2}}, \\ G_2(z) &= -\frac{1}{4\pi} \sum_{n=0}^{\infty} B_n (1-z)^{n+\frac{1}{2}} \left(\ln(1-z) + C_n\right). \end{aligned} \tag{13.2.78}$$

Using (13.2.77) we obtain the corresponding contributions $I_k^{(j)}$, $j = 1, 2$, in the form

$$\begin{aligned} I_k^{(1)} &= \frac{4}{\pi} \frac{\Gamma\left(k+\frac{3}{2}\right)}{k!} \left(1 - \frac{1}{8\left(k+\frac{1}{2}\right)}\right), \\ I_k^{(2)} &\sim -\frac{1}{8k!\,\sqrt{\pi}} \sum_{n=0}^{\infty} B_n \left(-n-\tfrac{1}{2}\right)_k \times \\ &\quad \left(C_n + \psi\left(-n-\tfrac{1}{2}\right) - \psi\left(k-n-\tfrac{1}{2}\right)\right), \end{aligned} \tag{13.2.79}$$

where the series has an asymptotic character for large k. The coefficients B_n and C_n are given by

$$\begin{aligned} B_n &= \frac{\left(\frac{3}{2}\right)_n \left(\frac{3}{2}\right)_n}{n!(n+2)!}, \\ C_n &= -\,\psi(n+1) - \psi(n+3) + 2\,\psi\left(n+\tfrac{3}{2}\right). \end{aligned} \tag{13.2.80}$$

By using the recursion formula $\psi(z+1) = \psi(z) + 1/z$ and the values $\psi(1) = -\gamma$, $\psi\left(\frac{1}{2}\right) = -\gamma - 2\ln 2$, we obtain

$$C_{n+1} = C_n - \frac{1}{n+1} - \frac{1}{n+3} + \frac{4}{2n+3}, \quad C_0 = \tfrac{5}{2} - 4\ln 2. \tag{13.2.81}$$

For large values of z we have the asymptotic expansion

$$\psi(z) \sim \ln z - \frac{1}{2z} - \sum_{n=1}^{\infty} \frac{B_{2n}}{2n\, z^{2n}}, \tag{13.2.82}$$

which follows from differentiating the expansion in (6.3.41). This gives the logarithmic terms $\ln k$ in the expansion of I_k announced in the previous section.

The term $I_k^{(1)}$ gives the main contribution to the asymptotic behavior of $I_k = I_k^{(1)} + I_k^{(2)}$. By using the expansion in (13.2.61) with $n = 0$ we see that the first terms in the asymptotic expansion of $I_k^{(1)}$ are as in (13.2.65). When we expand all terms in (13.2.79) for large k we find

$$\begin{aligned} I_k = \frac{4\sqrt{k}}{\pi} \Bigg(1 + \frac{1}{4k} - \frac{1 + 2\gamma + 12\ln 2 + 2\ln k}{256k^2} + \\ \frac{-17 + 6\gamma + 36\ln 2 + 6\ln k}{1024k^3} + \mathcal{O}\left(\frac{\ln k}{k^4}\right)\Bigg), \quad k \to \infty. \end{aligned} \tag{13.2.83}$$

PART 3

Other Methods for Integrals

Chapter 14

The method of stationary phase

In the method of stationary phase integrals of the type

$$F(\omega) = \int_a^b e^{i\omega\phi(t)}\psi(t)\,dt, \tag{14.0.1}$$

are considered, where ω is a real large parameter, a, b and ϕ are real; $a = -\infty$ or/and $b = +\infty$ are allowed.

Clearly, if ϕ and ψ are analytic functions in domains containing the interval $[a, b]$, then we can try the method of steepest descent considered in Chapter 4. However, the idea of the method of stationary phase, as originally developed by Stokes and Kelvin, antecedes the method of steepest descent, and there are reasons for considering it separately.

14.1 Critical points

The asymptotic character of the integral in (14.0.1) is completely determined if the behavior of the functions ϕ and ψ is known in the vicinity of the *critical points.* These are

- stationary points: t-values in $[a, b]$ such that $\phi'(t) = 0$;
- the finite endpoints a and b;
- t-values for which $\phi(t)$ and/or $\psi(t)$ are singular.

The following integral shows all these types of critical points:

$$F(\omega) = \int_{-1}^{1} e^{i\omega t^2}\sqrt{|t-c|}\,dt, \quad -1 < c < 1. \tag{14.1.2}$$

Several questions arise:

- Which of these critical points give the relevant contributions?
- What to do when no critical points exist?
- What to do when several relevant critical points exist?
- What to do when critical points coalesce?

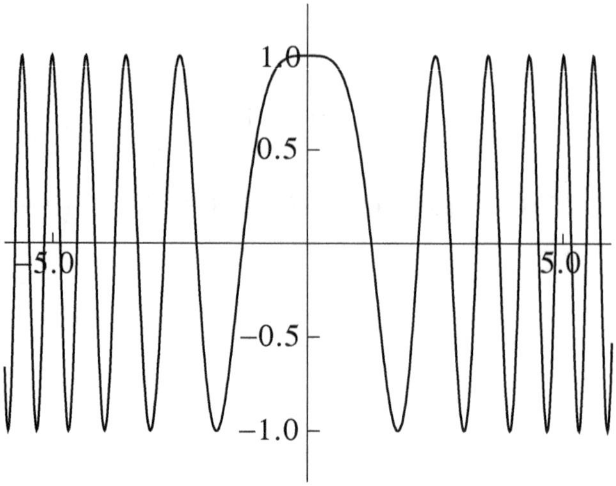

Fig. 14.1 Real part of $e^{i\omega x^2}$ for $\omega = 1$. At the stationary point $x = 0$ less oscillations occur.

The latter situation occurs when in (14.1.2) $c \to 0$ or $c \to \pm 1$.

In Figure 14.1 we see why a stationary point may give a contribution: less oscillations occur at the stationary point compared with other points in the interval, where the oscillations neutralize each other.

In this chapter we explain that for integrals of the form (14.0.1) the method of stationary phase can be a useful tool for obtaining the asymptotic behavior, and we give several examples.

On the other hand, as we will also demonstrate, in many cases that involve special functions, and where the functions ϕ and ψ have analytic extensions in the complex plane, the saddle point method may provide elegant methods with results that cannot be obtained by using the method of stationary phase.

14.2 Integrating by parts: No stationary points

For integrals on a finite interval $[a, b]$ we discuss a method based on integration by parts. If in (14.0.1) ϕ has no stationary point, that is, $\phi'(t) \neq 0$ in the interval $[a, b]$, then contributions from the endpoints follow from integrating by parts. We have

$$\begin{aligned} F(\omega) &= \int_a^b e^{i\omega\phi(t)}\psi(t)\,dt = \frac{1}{i\omega}\int_a^b \psi(t)\frac{de^{i\omega\phi(t)}}{\phi'(t)} \\ &= \frac{e^{i\omega\phi(b)}}{i\omega\phi'(b)}\psi(b) - \frac{e^{i\omega\phi(a)}}{i\omega\phi'(a)}\psi(a) + \frac{1}{i\omega}\int_a^b e^{i\omega\phi(t)}\psi_1(t)\,dt, \end{aligned} \tag{14.2.3}$$

where

$$\psi_1(t) = -\frac{d}{dt}\frac{\psi(t)}{\phi'(t)}. \tag{14.2.4}$$

The integral in this result has the same form as the one in (14.0.1), and when ϕ and ψ are sufficiently smooth, we can continue this procedure.

In this way we obtain for $N = 0, 1, 2, \ldots$ the compound expansion

$$\begin{aligned} F(\omega) = \frac{e^{i\omega\phi(b)}}{i\omega\phi'(b)} \sum_{n=0}^{N-1} \frac{\psi_n(b)}{(i\omega)^n} - \frac{e^{i\omega\phi(a)}}{i\omega\phi'(a)} \sum_{n=0}^{N-1} \frac{\psi_n(a)}{(i\omega)^n} + \\ \frac{1}{(i\omega)^N} \int_a^b e^{i\omega\phi(t)} \psi_N(t)\, dt, \end{aligned} \tag{14.2.5}$$

where for $N = 0$ the terms in the first line vanish. The integral can be viewed as a remainder of the expansion. We have $\psi_0 = \psi$ and

$$\psi_{n+1}(t) = -\frac{d}{dt} \frac{\psi_n(t)}{\phi'(t)}, \quad n = 0, 1, 2, \ldots. \tag{14.2.6}$$

This expansion can be obtained when $\phi, \psi \in C^N[a, b]$. When we assume that we can find positive numbers M_N such that $|\psi_N(t)| \le M_N$ for $t \in [a, b]$, we can find an upper bound for the remainder in (14.2.5), and this estimate will be of order $\mathcal{O}\left(\omega^{-N}\right)$.

14.3 Three critical points: A formal approach

We consider

$$F(\omega) = \int_a^b e^{i\omega t^2} f(t)\, dt, \quad \omega > 0, \quad -\infty < a < 0 < b < \infty, \tag{14.3.7}$$

and we assume that $f \in C^\infty([a, b])$. We proceed in a formal way, just to show how to obtain the complete expansion. In later sections we will give more details on rigorous methods.

We write

$$F(\omega) = F_s(\omega) + F_{a,b}(\omega) \tag{14.3.8}$$

where $F_s(\omega)$ denotes the contribution from the stationary point at the origin and $F_{a,b}(\omega)$ those from the endpoints a, b.

For $F_s(\omega)$ we expand $f(t) = \sum_{n=0}^{\infty} c_n t^n$, substitute this in (14.3.7), extend the domain to $\mathbb{R}$, and obtain

$$F_s(\omega) \sim \sum_{n=0}^{\infty} c_n \int_{-\infty}^{\infty} t^n e^{i\omega t^2}\, dt \sim e^{\frac{1}{4}\pi i} \sqrt{\frac{\pi}{\omega}} \sum_{n=0}^{\infty} i^n \left(\tfrac{1}{2}\right)_n \frac{c_{2n}}{\omega^n}. \tag{14.3.9}$$

The divergent integrals in (14.3.9) can be evaluated by rotating the path of integration through an angle $\frac{1}{4}\pi$ into the complex plane. In this formal approach we do not verify the validity of this step.

For $F_{a,b}(\omega)$ we use straightforward integration by parts:

$$F_{a,b}(\omega) = \int_a^b e^{i\omega t^2} f(t)\, dt = \frac{1}{2i\omega} \int_a^b \frac{f(t)}{t}\, d\left(e^{i\omega t^2}\right), \tag{14.3.10}$$

and proceed as if the singularity at $t = 0$ were not present. This gives

$$F_{a,b}(\omega) \sim \frac{1}{2i\omega}\frac{f(b)}{b}e^{i\omega b^2} - \frac{1}{2i\omega}\frac{f(a)}{a}e^{i\omega a^2} + \frac{1}{2i\omega}\int_a^b e^{i\omega t^2} f_1(t)\,dt, \tag{14.3.11}$$

where

$$f_1(t) = -\frac{d}{dt}\frac{f(t)}{t}. \tag{14.3.12}$$

Proceeding in this way we obtain

$$F_{a,b}(\omega) \sim \frac{e^{i\omega b^2}}{2ib\omega}\sum_{n=0}^{\infty}\frac{f_n(b)}{(2i\omega)^n} - \frac{e^{i\omega a^2}}{2ia\omega}\sum_{n=0}^{\infty}\frac{f_n(a)}{(2i\omega)^n}, \tag{14.3.13}$$

where $f_0 = f$ and

$$f_n(t) = -\frac{d}{dt}\frac{f_{n-1}(t)}{t}, \quad n = 1, 2, 3, \ldots. \tag{14.3.14}$$

Taking all contributions into account, we find for (14.3.7) the expansion

$$\begin{aligned} F(\omega) \sim\; & e^{\frac{1}{4}\pi i}\sqrt{\frac{\pi}{\omega}}\sum_{n=0}^{\infty} i^n \left(\tfrac{1}{2}\right)_n \frac{c_{2n}}{\omega^n} + \\ & \frac{e^{i\omega b^2}}{2ib\omega}\sum_{n=0}^{\infty}\frac{f_n(b)}{(2i\omega)^n} - \frac{e^{i\omega a^2}}{2ia\omega}\sum_{n=0}^{\infty}\frac{f_n(a)}{(2i\omega)^n}, \quad \omega \to +\infty. \end{aligned} \tag{14.3.15}$$

14.4 On the use of neutralizers

When, as in the integral (14.1.2), several critical points have to be considered, the finite or infinite interval can be split up into a number of subintervals containing less critical points. In this way, it is not always possible to obtain intervals on which methods based on straightforward integration by parts or local expansions can be used. For example, in the case of

$$F(\omega) = \int_{-1}^{1} e^{i\omega t^2} f(t)\,dt, \quad \omega > 0, \tag{14.4.16}$$

we can split up $[-1, 1]$ into $[-1, 0]$ and $[0, 1]$, but each subinterval has two critical points. To handle this, J. G. van der Corput (1948) introduced neutralizers in order to get intervals in which only one critical point exists.

Definition 14.1. A *neutralizer* N_a at a point a is a $C^\infty(\mathbb{R})$ function such that:

(1) $N_a(a) = 1$, and all its derivatives vanish at a.
(2) There is a positive number d such that $N_a(x) = 0$ outside $(a - d, a + d)$.

See Figure 14.2, where we have shown a neutralizer at $a = 0$. This one can be used for a stationary point at the origin, but in a similar way we can use neutralizers active at an endpoint.

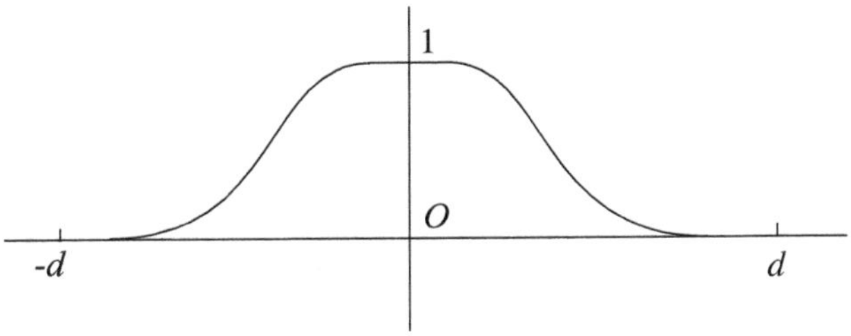

Fig. 14.2 A neutralizer N_0 at the origin.

We can give explicit forms of such a neutralizer. For instance, take $\alpha < \alpha' < \beta' < \beta$, then the function $N(x)$ that satisfies $N(x) = 1$ if $\alpha \le x \le \alpha'$, $N(x) = 0$ if $\beta' \le x \le \beta$, and

$$N(x) = \frac{\displaystyle\int_x^{\beta'} \exp\left(-\frac{1}{u-\alpha'} - \frac{1}{\beta'-u}\right) du}{\displaystyle\int_{\alpha'}^{\beta'} \exp\left(-\frac{1}{u-\alpha'} - \frac{1}{\beta'-u}\right) du}, \quad \alpha' < x < \beta', \tag{14.4.17}$$

is a neutralizer at α in the interval $[\alpha, \beta]$.

With a neutralizer N_0 at the origin as shown in Figure 14.2 with $d = 1$, we can write the integral in (14.4.16) in the form

$$F(\omega) = F_1(\omega) + F_2(\omega), \tag{14.4.18}$$

where

$$\begin{aligned} F_1(\omega) &= 2\int_0^1 e^{i\omega t^2} f(t)(1 - N_0(t))\, dt, \\ F_2(\omega) &= 2\int_0^T e^{i\omega t^2} f(t) N_0(t)\, dt. \end{aligned} \tag{14.4.19}$$

We assume that $f \in C^{2m}([0,1])$ and we may assume that f and the neutralizer are even. For the upper limit T we can take every number satisfying $T > 1$, because the neutralizer vanishes for $t \ge 1$.

In the integral for $F_1(\omega)$ the stationary point at $t = 0$ is harmless (it is neutralized), and we can integrate by parts to obtain the contributions from $t = 1$, as we have done in §14.2.

Because all derivatives of $1 - N_0(t)$ vanish at $t = 1$, the asymptotic terms for $F_1(\omega)$ do not depend on the neutralizer N_0, but the remainders in the successive steps do depend on the neutralizer.

In the integral for $F_2(\omega)$ we substitute

$$f(t) = \sum_{n=0}^{2m-2} c_{2n} t^{2n} + t^{2m} R_m(t). \tag{14.4.20}$$

This gives the expansion

$$F_2(\omega) = \sum_{n=0}^{2m-2} c_{2n}\Phi_{2n} + S_m(\omega), \tag{14.4.21}$$

where

$$\Phi_{2n} = 2\int_0^T e^{i\omega t^2} t^{2n} N_0(t)\, dt, \tag{14.4.22}$$

and

$$S_m(\omega) = 2\int_0^T e^{i\omega t^2} t^{2m} R_m(t) N_0(t)\, dt. \tag{14.4.23}$$

For $n > 0$ we integrate by parts in the integrals of Φ_{2n}. This gives after n steps

$$\Phi_{2n} = 2\int_0^T e^{i\omega t^2} t^{2n} N_0(t)\, dt = \frac{i^n}{\omega^n}\left(\tfrac{1}{2}\right)_n \Phi_0 + \mathcal{O}\left(\omega^{-\nu}\right), \tag{14.4.24}$$

for every $\nu > 0$. The $\mathcal{O}$-term comes from integrals with derivatives of $N_0(t)$, and these integrals are exponentially small compared to the Φ_{2n}.

In the integral for Φ_0 we take $T = +\infty$, write $N_0(t) = 1 + (N_0(t) - 1)$ and the first term gives

$$\Phi_{2n} = i^n e^{\frac{1}{4}\pi i}\sqrt{\frac{\pi}{\omega}}\left(\tfrac{1}{2}\right)_n \frac{1}{\omega^n} + \mathcal{O}\left(\omega^{-\nu}\right), \quad n = 0, 1, 2, \ldots, \quad \forall \nu > 0. \tag{14.4.25}$$

We can neglect the contributions from the integral with $N_0(t) - 1$, because integration by parts shows that contributions at the origin and at infinity vanish, and after many steps the remaining integral can be included in the $\mathcal{O}$-term in (14.4.24).

So far, all steps are safe. The remaining problem is the estimation of $S_m(\omega)$ in (14.4.23). For a proper interpretation of the expansion in (14.4.21), considering (14.4.25), it should satisfy $S_m(\omega) = \mathcal{O}\left(\omega^{-m-\frac{1}{2}}\right)$. In §14.5 we show how to bound remainders in a rather simple way; we give a complete asymptotic description of the integral in (14.4.16), and we show how neutralizers can be avoided in this case.

In the literature the rigorous approach is based on different methods. In Wong (2001, §II.3) a method of Erdélyi (Erdélyi, 1955) is used, and in Olver (1974) infinite integrals are handled with extra factors, such as $e^{-\varepsilon t^2}$, $\varepsilon > 0$, to ensure convergence. Olver gives error bounds for the expansion as well. In §14.8.1 we describe a method for more general forms of the integral given in (14.4.16) by using a different integration by parts method.

14.5 How to avoid neutralizers?

We can avoid the use of neutralizers by using an integration by parts method which is frequently used in the construction of uniform expansions. We will show this for the integral

$$F(\omega) = \int_a^b e^{i\omega t^2} f(t)\, dt, \quad \omega > 0, \quad -\infty < a < 0 < b < \infty, \tag{14.5.26}$$

which has a stationary point at the origin, and the finite endpoints will also give contributions. Because we need derivatives of f in the expansions, we assume that $f \in C^N[a,b]$ for some nonnegative integer N.

We write $f(t) = f(0) + (f(t) - f(0))$ and obtain

$$F(\omega) = f(0)\Phi_{a,b}(\omega) + \int_a^b e^{i\omega t^2} \left(f(t) - f(0)\right) dt, \tag{14.5.27}$$

where

$$\Phi_{a,b}(\omega) = \int_a^b e^{i\omega t^2}\, dt, \tag{14.5.28}$$

which can be expressed in terms of Fresnel integrals or complementary error functions. We give more details later.

Because the integrand vanishes at the origin we can integrate by parts in (14.5.27), and write

$$F(\omega) = f(0)\Phi_{a,b}(\omega) + \frac{1}{2i\omega}\int_a^b \frac{f(t)-f(0)}{t}\, de^{i\omega t^2}. \tag{14.5.29}$$

This gives

$$\begin{aligned} F(\omega) = {} & \frac{e^{i\omega b^2}}{2i\omega}\frac{f(b)-f(0)}{b} - \frac{e^{i\omega a^2}}{2i\omega}\frac{f(a)-f(0)}{a} + \\ & f(0)\Phi_{a,b}(\omega) + \frac{1}{2i\omega}\int_a^b f_1(t)e^{i\omega t^2}\, dt, \end{aligned} \tag{14.5.30}$$

where

$$f_1(t) = -\frac{d}{dt}\frac{f(t)-f(0)}{t}. \tag{14.5.31}$$

The integral in (14.5.30) can be expanded in the same way, and we obtain

$$\begin{aligned} F(\omega) = {} & \frac{e^{i\omega b^2}}{2i\omega}\sum_{n=0}^{N-1}\frac{C_n(b)}{(2i\omega)^n} - \frac{e^{i\omega a^2}}{2i\omega}\sum_{n=0}^{N-1}\frac{C_n(a)}{(2i\omega)^n} + \\ & \Phi_{a,b}(\omega)\sum_{n=0}^{N-1}\frac{f_n(0)}{(2i\omega)^n} + \frac{1}{(2i\omega)^N}\int_a^b f_N(t)e^{i\omega t^2}\, dt, \end{aligned} \tag{14.5.32}$$

where $N \ge 0$ and for $n = 0, 1, 2, \ldots, N$ we define

$$C_n(t) = \frac{f_n(t)-f_n(0)}{t}, \quad f_{n+1}(t) = -\frac{d}{dt}\frac{f_n(t)-f_n(0)}{t}. \tag{14.5.33}$$

For $N = 0$ the three sums in (14.5.32) are empty.

The integral in (14.5.32) can be viewed as a remainder of the expansion. When we assume that we can find positive numbers M_N such that $|f_N(t)| \le M_N$ for $t \in [a,b]$, we can find an upper bound for this remainder, which will be of order $\mathcal{O}\left(\omega^{-N}\right)$.

14.5.1 *A few details about the Fresnel integral*

For more information on the special functions in this section we refer to Temme (2010c). The function $\Phi_{a,b}(\omega)$ defined in (14.5.28) can be written as

$$\begin{aligned}\Phi_{a,b}(\omega) &= \int_{-\infty}^{\infty} e^{i\omega t^2}\,dt - \int_{-\infty}^{a} e^{i\omega t^2}\,dt - \int_{b}^{\infty} e^{i\omega t^2}\,dt \\ &= \sqrt{\frac{\pi}{\omega}}e^{\frac14\pi i} - \int_{-a}^{\infty} e^{i\omega t^2}\,dt - \int_{b}^{\infty} e^{i\omega t^2}\,dt.\end{aligned} \tag{14.5.34}$$

In terms of the Fresnel integral

$$\mathcal{F}(z) = \int_{z}^{\infty} e^{\frac12\pi i t^2}\,dt, \tag{14.5.35}$$

we have

$$\Phi_{a,b}(\omega) = \sqrt{\frac{\pi}{\omega}}e^{\frac14\pi i} - \sqrt{\frac{\pi}{2\omega}}\mathcal{F}\left(-a\sqrt{\frac{2\omega}{\pi}}\right) - \sqrt{\frac{\pi}{2\omega}}\mathcal{F}\left(b\sqrt{\frac{2\omega}{\pi}}\right). \tag{14.5.36}$$

The asymptotic expansion of $\mathcal{F}(z)$ can be obtained by straightforward integrating by parts in (14.5.35) and also follows from the representation

$$\mathcal{F}(z) = e^{\frac12\pi i z^2}(\mathrm{g}(z) + i\,\mathrm{f}(z)) \tag{14.5.37}$$

and from the asymptotic expansions of the auxiliary functions f, g

$$\begin{aligned}\mathrm{f}(z) &\sim \frac{1}{\pi z}\sum_{m=0}^{\infty}(-1)^m\frac{1\cdot3\cdot5\cdots(4m-1)}{(\pi z^2)^{2m}}, \\ \mathrm{g}(z) &\sim \frac{1}{\pi^2 z^3}\sum_{m=0}^{\infty}(-1)^m\frac{1\cdot3\cdot5\cdots(4m+1)}{(\pi z^2)^{2m}}.\end{aligned} \tag{14.5.38}$$

Because the Fresnel integrals in (14.5.36) have large positive arguments we can obtain a complete asymptotic expansion of $\Phi_{a,b}(\omega)$. But we can also accept the function $\Phi_{a,b}(\omega)$ as given in (14.5.28) or (14.5.36). This becomes of interest when a or b are small, or even tend to zero under the influence of a certain parameter. In that case we keep $\Phi_{a,b}(\omega)$ as it is given, and the expansion in (14.5.32) remains valid when $a \to 0$ or $b \to 0$. It is in fact a uniform expansion with respect to small values of a and b.

The Fresnel integral can also be written in terms of the complementary error function (see §3.4). We have

$$\mathcal{F}(z) = \frac{e^{\frac14\pi i}}{\sqrt{2}}\operatorname{erfc}\zeta, \quad \zeta = \sqrt{\tfrac12\pi}\,e^{-\frac14\pi i}\,z, \tag{14.5.39}$$

and the asymptotic expansion of $\mathcal{F}(z)$ also follows from (3.4.30).

Remark 14.1. The coefficients $f_n(0)$ in the expansion in (14.5.32) follow from the Maclaurin coefficients of $f_0 = f$. Let $f_n(s) = \sum_{k=0}^{\infty} c_k^{(n)} s^k$. Then, from (14.5.33),

$$c_k^{(n)} = -(k+1)c_{k+2}^{(n-1)} = (-1)^n(k+1)(k+3)\cdots(k+2n-1)c_{k+2n}^{(0)}, \tag{14.5.40}$$

that is,

$$c_k^{(n)} = (-2)^n \left(\tfrac{1}{2} + \tfrac{1}{2}k\right)_n c_{k+2n}^{(0)}. \tag{14.5.41}$$

Hence

$$f_n(0) = c_0^{(n)} = (-2)^n \left(\tfrac{1}{2}\right)_n c_{2n}^{(0)}. \tag{14.5.42}$$

This relation can be used by comparing the expansion in (14.5.32) with the one obtained by a formal approach in §14.3. △

14.6 Algebraic singularities at both endpoints: Erdélyi's example

Consider the following integral

$$F(\omega) = \int_\alpha^\beta e^{i\omega t}(t-\alpha)^{\lambda-1}(\beta-t)^{\mu-1} f(t)\,dt, \tag{14.6.43}$$

where $\Re\lambda > 0$, $\Re\mu > 0$. For this type of integral a straightforward approach using integration by parts is not possible. However, a less straightforward method will be considered in §14.6.2.

For this class of integrals we have the following theorem.

Theorem 14.1 (Erdélyi (1955)). *If f is N times continuously differentiable in the finite interval $[\alpha,\beta]$, then*

$$F(\omega) = A_N(\omega) + B_N(\omega) + \mathcal{O}\left(1/\omega^N\right), \quad \omega\to\infty, \tag{14.6.44}$$

where

$$A_N(\omega) = \sum_{n=0}^{N-1} \frac{\Gamma(n+\lambda)}{n!\,\omega^{n+\lambda}} e^{i(\frac{1}{2}\pi(n+\lambda)+\alpha\omega)} \frac{d^n}{dt^n}\left((\beta-t)^{\mu-1}f(t)\right)\bigg|_{t=\alpha}, \tag{14.6.45}$$

and

$$B_N(\omega) = \sum_{n=0}^{N-1} \frac{\Gamma(n+\mu)}{n!\,\omega^{n+\mu}} e^{i(\frac{1}{2}\pi(n-\mu)+\beta\omega)} \frac{d^n}{dt^n}\left((t-\alpha)^{\lambda-1}f(t)\right)\bigg|_{t=\beta}. \tag{14.6.46}$$

Erdélyi's proof is based on the use of neutralizers (see §14.4). All the same, it is quite easy to obtain the expansions of this theorem, just by using an expansion in powers of $(t-\alpha)$ of $(\beta-t)^{\mu-1}f(t)$ (and extending the interval of integration from $[\alpha,\beta]$ to $[\alpha,\infty]$), which gives the expansion of $A_N(\omega)$. The expansion of $B_N(\omega)$ follows from an expansion in powers of $(\beta-t)$ of $(t-\alpha)^{\lambda-1}f(t)$ (and extending the interval of integration from $[\alpha,\beta]$ to $[-\infty,\alpha]$).

A meaning of the integrals

$$\int_\alpha^\infty e^{i\omega t}(t-\alpha)^{\lambda+n-1}\,dt \quad\text{and}\quad \int_{-\infty}^\beta e^{i\omega t}(\beta-t)^{\mu+n-1}\,dt \tag{14.6.47}$$

for $n = 0, 1, 2, \ldots$ follows from rotating the path of integration into the complex plane in order to obtain convergence for all n.

Problem 14.1. *Take in (14.6.43) $\alpha = 0$, $\beta = 1$, and $f = 1$. Obtain the representation of the Kummer function (see Chapter 10)*

$$ {}_1F_1\left({\lambda \atop \lambda+\mu}; i\omega\right) = \frac{\Gamma(\lambda+\mu)}{\Gamma(\lambda)\Gamma(\mu)} \int_0^1 e^{i\omega t} t^{\lambda-1}(1-t)^{\mu-1}\,dt, \tag{14.6.48} $$

and by using Theorem 14.1 obtain the expansion (see (10.2.21))

$$ \begin{aligned} {}_1F_1\left({\lambda \atop \lambda+\mu}; i\omega\right) \sim\ & \frac{\Gamma(\lambda+\mu)}{\Gamma(\lambda)} e^{i\omega} \sum_{n=0}^{\infty} e^{-\frac12\pi i(n+\mu)} \frac{(\mu)_n(1-\lambda)_n}{n!\,\omega^{n+\mu}} + \\ & \frac{\Gamma(\lambda+\mu)}{\Gamma(\mu)} \sum_{n=0}^{\infty} e^{\frac12\pi i(n+\lambda)} \frac{(\lambda)_n(1-\mu)_n}{n!\,\omega^{n+\lambda}}, \quad \omega \to +\infty. \end{aligned} \tag{14.6.49} $$

♡

14.6.1 *Application: A conical function*

Conical functions are a class of Legendre function with a special notation and range of the parameters. We consider the integral representation

$$ \begin{aligned} & \Gamma\left(\tfrac12+\mu\right)\sqrt{\tfrac12\pi}\,\sinh^\mu\beta\,P^{-\mu}_{-\frac12+i\tau}(\cosh\beta) = \\ & \qquad \int_0^\beta (\cosh\beta-\cosh t)^{\mu-\frac12}\cos\tau t\,dt, \quad \Re\mu > -\tfrac12, \end{aligned} \tag{14.6.50} $$

which is given in Magnus *et al.* (1966, p. 184).

The large-τ asymptotics follows from applying Theorem 14.1. We have the following result

$$ \Gamma\left(\tfrac12+\mu\right)\sqrt{\tfrac12\pi}\,\sinh^\mu\beta\,P^{-\mu}_{-\frac12+i\tau}(\cosh\beta) \sim \sum_{n=0}^{\infty}\frac{A_n}{\tau^{n+1}} + \sum_{n=0}^{\infty}\frac{B_n}{\tau^{n+\mu+\frac12}}, \tag{14.6.51} $$

where

$$ A_n = -\sin\left(\tfrac12 n\pi\right)\frac{d^n}{dt^n}(\cosh\beta-\cosh t)^{\mu-\frac12}\Big|_{t=0}, \tag{14.6.52} $$

$$ B_n = \cos\chi_n\,\frac{\Gamma(n+\mu+\frac12)}{n!}\,\frac{d^n}{dt^n}\left(\frac{\cosh\beta-\cosh t}{\beta-t}\right)^{\mu-\frac12}\Bigg|_{t=\beta}, \tag{14.6.53} $$

and

$$ \chi_n = \tfrac12\left(n-\mu-\tfrac12\right)+\beta\tau. \tag{14.6.54} $$

The coefficients A_n vanish for even n, because of the sine function. They vanish too for odd n, because in that case the derivatives vanish at $t = 0$.

The expansion in (14.6.51) can be written in the form

$$P^{-\mu}_{-\frac{1}{2}+i\tau}(\cosh\beta) \sim \sqrt{\frac{2}{\pi\sinh\beta}} \sim \sum_{n=0}^{\infty} \cos\chi_n \left(\mu+\tfrac{1}{2}\right)_n \frac{b_n}{\tau^{n+\mu+\frac{1}{2}}}, \tag{14.6.55}$$

where the first few coefficients b_n are given by

$$\begin{aligned} b_0 &= 1,\\ b_1 &= \frac{(2\mu-1)x}{4\sinh\beta},\\ b_2 &= \frac{(2\mu-1)(-8+(6\mu-1)x^2)}{96\sinh^2\beta},\\ b_3 &= \frac{(2\mu-1)x((-1+4\mu^2)x^2+16-16\mu)}{384\sinh^3\beta}. \end{aligned} \tag{14.6.56}$$

In Magnus *et al.* (1966, p. 202) the following representation is given, which is an alternative to the asymptotic expansion,

$$\begin{aligned} &\sqrt{2\pi\sinh\beta}\, P^{-\mu}_{-\frac{1}{2}+i\tau}(\cosh\beta) = \\ &\quad e^{-i\beta\tau}\frac{\Gamma(-i\tau)}{\Gamma\left(\frac{1}{2}+\mu-i\tau\right)}\,{}_2F_1\left(\begin{matrix}\frac{1}{2}+\mu,\frac{1}{2}-\mu\\ 1+i\tau\end{matrix};\frac{1}{1-e^{2\beta}}\right)+\\ &\quad e^{i\beta\tau}\frac{\Gamma(i\tau)}{\Gamma\left(\frac{1}{2}+\mu+i\tau\right)}\,{}_2F_1\left(\begin{matrix}\frac{1}{2}+\mu,\frac{1}{2}-\mu\\ 1-i\tau\end{matrix};\frac{1}{1-e^{2\beta}}\right). \end{aligned} \tag{14.6.57}$$

The standard series of the hypergeometric functions are convergent if $\beta > \frac{1}{2}\ln 2$ and for $\beta \ge \delta > 0$ they can be used as asymptotic expansions as $\tau \to \infty$.

14.6.2 *Avoiding neutralizers in Erdélyi's example*

Theorem 14.1 gives two expansions, one from each critical point α and β. By a different integrating by parts procedure we can obtain expansions which take contributions from both critical points in each step. We write (see also the method used in §14.5)

$$f(t) = a_0 + b_0(t-\alpha) + (t-\alpha)(\beta-t)g_0(t), \tag{14.6.58}$$

where a_0, b_0 follow from substituting $t=\alpha$ and $t=\beta$. This gives

$$a_0 = f(\alpha), \quad b_0 = \frac{f(\beta)-f(\alpha)}{\beta-\alpha}, \tag{14.6.59}$$

and for (14.6.43) we obtain

$$F(\omega) = a_0\Phi + b_0\Psi + \int_\alpha^\beta e^{i\omega t}(t-\alpha)^\lambda(\beta-t)^\mu g_0(t)\,dt, \tag{14.6.60}$$

where (see also Problem 14.1)

$$\begin{aligned} \Phi &= (\beta-\alpha)^{\lambda+\mu-1}e^{i\omega\alpha}\frac{\Gamma(\lambda)\Gamma(\mu)}{\Gamma(\lambda+\mu)}\,{}_1F_1\left(\begin{matrix}\lambda\\ \lambda+\mu\end{matrix};i(\beta-\alpha)\omega\right),\\ \Psi &= (\beta-\alpha)^{\lambda+\mu}e^{i\omega\alpha}\frac{\Gamma(\lambda+1)\Gamma(\mu)}{\Gamma(\lambda+\mu+1)}\,{}_1F_1\left(\begin{matrix}\lambda+1\\ \lambda+\mu+1\end{matrix};i(\beta-\alpha)\omega\right). \end{aligned} \tag{14.6.61}$$

Now we integrate by parts and obtain, observing that the integrated terms will vanish,

$$F(\omega) = a_0\Phi + b_0\Psi + \frac{1}{i\omega}\int_\alpha^\beta e^{i\omega t}(t-\alpha)^{\lambda-1}(\beta-t)^{\mu-1}f_1(t)\,dt, \tag{14.6.62}$$

where

$$f_1(t) = -(t-\alpha)^{1-\lambda}(\beta-t)^{1-\mu}\frac{d}{dt}\left((t-\alpha)^\lambda(\beta-t)^\mu g_0(t)\right). \tag{14.6.63}$$

We can continue with this integral in the same manner, and obtain

$$F(\omega) = \Phi\sum_{n=0}^{N-1}\frac{a_n}{(i\omega)^n} + \Psi\sum_{n=0}^{N-1}\frac{b_n}{(i\omega)^n} + R_N, \quad N = 0,1,2,\ldots, \tag{14.6.64}$$

where

$$R_N = \frac{1}{(i\omega)^N}\int_\alpha^\beta e^{i\omega t}(t-\alpha)^{\lambda-1}(\beta-t)^{\mu-1}f_N(t)\,dt. \tag{14.6.65}$$

The coefficients are defined by

$$a_n = f_n(\alpha), \quad b_n = \frac{f_n(\beta)-f_n(\alpha)}{\beta-\alpha}, \tag{14.6.66}$$

and the functions f_n follow from the recursive scheme

$$\begin{aligned} f_n(t) &= a_n + b_n(t-\alpha) + (t-\alpha)(\beta-t)g_n(t),\\ f_{n+1}(t) &= -(t-\alpha)^{1-\lambda}(\beta-t)^{1-\mu}\frac{d}{dt}\left((t-\alpha)^\lambda(\beta-t)^\mu g_n(t)\right), \end{aligned} \tag{14.6.67}$$

with $f_0 = f$.

The expansion in (14.6.64) contains confluent hypergeometric functions, and Erdélyi's expansion in Theorem 14.1 is in terms of elementary functions. In Erdélyi's approach error estimates cannot be obtained because these depend on the choice of the neutralizers. The present approach avoids neutralizers, and Erdélyi's expansion follows by using the expansions of the confluent hypergeometric function given in (14.6.49). Estimates of R_N follow when we know more about the function f, in particular when we have bounds on the derivatives of f.

In addition, the expansion remains valid when $\beta-\alpha$ becomes small. Of course, when $\omega(\beta-\alpha)$ is not large, the ${}_1F_1$-functions should not be expanded.

14.7 Transformation to standard form

Consider again the general form

$$F(\omega) = \int_a^b e^{i\omega\phi(t)}\psi(t)\,dt. \tag{14.7.68}$$

Assume that the real function ϕ has a first-order stationary point at $t_0 \in [a,b]$, that is, $\phi'(t_0) = 0$ and $\phi''(t_0) \neq 0$. When t_0 is the only stationary point the usual step is to transform the integral into a standard form by writing

$$\phi(t) - \phi(t_0) = \tfrac{1}{2}\phi''(t_0)s^2, \quad \operatorname{sign}(s) = \operatorname{sign}(t - t_0). \tag{14.7.69}$$

This gives

$$F(\omega) = e^{i\omega\phi(t_0)} \int_c^d e^{\frac{1}{2}i\phi''(t_0)\omega s^2} f(s)\, ds, \quad f(s) = \psi(t)\frac{dt}{ds}, \tag{14.7.70}$$

where $c \le 0$ and $d \ge 0$ follow from the transformation.

When both c and d are finite and different from zero, there are three critical points. In that case the endpoints will give contributions of order $\mathcal{O}(1/\omega)$ and the origin of order $\mathcal{O}(1/\sqrt{\omega})$, and to obtain a complete expansion of the integral in (14.7.70) we can use neutralizers (see §14.4), or an integration by parts method (see §14.5).

When $c = -\infty$ and $d = \infty$, then a first-order approximation is given by

$$F(\omega) = \int_{-\infty}^{\infty} e^{\frac{1}{2}i\phi''(t_0)\omega s^2} f(s)\, ds \sim e^{i\sigma\pi/4}\sqrt{\frac{2\pi}{|\phi''(t_0)|\omega}}\, f(0), \quad \omega \to +\infty, \tag{14.7.71}$$

where $\sigma = \operatorname{sign}(\phi''(t_0))$. Assuming proper conditions on the function f, we can obtain higher-order approximations by expanding this function in powers of s.

Example 14.1 (Bessel function $J_n(z)$). Consider

$$J_n(x) = \frac{1}{\pi}\int_0^{\pi} \cos(nt - x\sin t)\, dt = \Re \frac{1}{\pi}\int_0^{\pi} e^{i(nt - x\sin t)}\, dt, \tag{14.7.72}$$

with $n = 0, 1, 2, \ldots$, and x positive and large. There is a stationary point at $t = \frac{1}{2}\pi$, and we transform this to the origin:

$$J_n(x) = \Re \frac{e^{\frac{1}{2}n\pi i}}{\pi}\int_{-\frac{1}{2}\pi}^{\frac{1}{2}\pi i} e^{ix\phi(t)}\psi(t)\, dt, \tag{14.7.73}$$

where

$$\phi(t) = -\cos t, \quad \phi''(0) = 1, \quad \psi(t) = e^{int}. \tag{14.7.74}$$

Now, $\cos t = 1 - 2\sin^2 \frac{1}{2}t$, and the substitution in (14.7.69) reads $s = 2\sin\frac{1}{2}t$. This gives

$$J_n(x) = \frac{1}{\pi}\Re e^{\frac{1}{2}n\pi i - ix}\int_c^d e^{\frac{1}{2}ixs^2} f(s)\, ds, \tag{14.7.75}$$

where

$$c = -\sqrt{2}, \quad d = \sqrt{2}, \quad f(s) = \frac{e^{2in\arcsin\left(\frac{1}{2}s\right)}}{\sqrt{1 - \frac{1}{4}s^2}}, \tag{14.7.76}$$

which is of the form (14.7.70). A first approximation can be obtained by replacing $f(s)$ by $f(0) = 1$, and extending the interval of integration to the real line. This gives, see (14.7.71),

$$J_n(x) \sim \frac{1}{\pi}\Re e^{\frac{1}{2}n\pi i - ix + \frac{1}{4}\pi i}\sqrt{\frac{2\pi}{x}} = \sqrt{\frac{2}{\pi x}}\cos\left(x - \tfrac{1}{2}n\pi - \tfrac{1}{4}\pi\right), \quad x \to \infty. \quad (14.7.77)$$

It should be noted that other methods and representations are more convenient for obtaining an asymptotic expansion of the Bessel function; see §9.2. First, the representation is only valid for integer values of n, and the method of stationary phase assumes real x, although this result holds for general order n and complex argument x, $|\mathrm{ph}\, x| < \pi$. In addition, it is not at all easy to obtain the general terms in the complete expansion (see the method in §14.3) of this Bessel function by using representation (14.7.75).

When using that approach it follows that the endpoints c and d will give contributions of order $\mathcal{O}(x^{-n-1})$, $n = 0, 1, 2, \ldots$. However, it should be noted that contributions from the endpoints cancel each other. This can be seen by comparing the complete expansion of $J_n(x)$ that follows from (9.2.9), in which only terms occur of order $\mathcal{O}(x^{-n-\frac{1}{2}})$, $n = 0, 1, 2, \ldots$. ◇

14.8 General order stationary points

More general integrals of the form

$$F(\omega) = \int_c^d e^{\frac{1}{\rho}i\omega s^\rho} f(s)\, ds, \quad \rho = 3, 4, 5, \ldots, \quad (14.8.78)$$

can be treated in the same way.

For example, the Bessel function $J_n(x)$ with $n = x$ has the integral representation (see Example 14.1)

$$J_n(n) = \Re\frac{1}{\pi}\int_0^\pi e^{in(u - \sin u)}\, du, \quad (14.8.79)$$

and the phase function $\phi(u) = u - \sin u$ has a stationary point at $u = 0$ with $\phi''(0) = 0$. In fact, $\phi(u) = \frac{1}{6}u^3 + \mathcal{O}(u^5)$ as $u \to 0$. In this case we can transform the integral into (14.8.78) with $\rho = 3$ and $c = 0$.

14.8.1 *Integrating by parts*

The integration by parts procedure of §14.5 works well when we replace $f(s)$ in (14.8.78) by $f(s) - \sum_{j=0}^{\rho-1} f^{(j)}(0)s^j/j!$ and we need asymptotic expansions of the functions $\int_c^d e^{\frac{1}{\rho}i\omega s^\rho} s^j\, ds$, but these can easily be obtained.

The integration by parts method should be modified when in (14.8.78) the number ρ is not an integer. For the case $\rho \ge 1$ and with an algebraic singularity at the origin in, for example,

$$F_{\lambda,\rho}(\omega) = \int_0^b t^{\lambda-1} e^{i\omega t^\rho} f(t)\,dt, \quad 0 < \lambda \le 1, \tag{14.8.80}$$

we find an approach in Wong (2001, §II.3), where it is assumed that $f \in C^\infty[0,b]$ and the integral is written as the difference of the integrals over $[0,\infty)$ and $[b,\infty)$. The function f is extended to a C^∞-function on $[0,\infty)$, with $f(t) = 0$ in a neighborhood of infinity.

To avoid extending the interval of integrating and defining f outside $[0,b]$, we write

$$f(t) = a_0 + a_1 t + \cdots + a_{n-1} t^{n-1} + t^n g(t), \tag{14.8.81}$$

where we assume that $\rho > 0$ and take $n \ge \max(0, \rho - \lambda)$. We integrate by parts:

$$F_{\lambda,\rho}(\omega) = \sum_{j=0}^{n-1} a_j \Phi_j + \frac{1}{i\omega\rho}\int_0^b t^{\lambda_1} g(t)\,d\,e^{i\omega t^\rho}, \quad \lambda_1 = \lambda + n - \rho, \tag{14.8.82}$$

where the Φ_j can be written in terms of incomplete gamma functions. We obtain

$$F_{\lambda,\rho}(\omega) = \sum_{j=0}^{n-1} a_j \Phi_j + \frac{g(b) b^{\lambda_1} e^{i\omega b^\rho} - \sigma_0 g(0)}{i\omega\rho} + \frac{1}{i\omega\rho}\int_0^b t^{\lambda_1 - 1} e^{i\omega t^\rho} f_1(t)\,dt, \tag{14.8.83}$$

where $\sigma_0 = 1$ if $\lambda_1 = 0$, $\sigma_0 = 0$ if $\lambda_1 > 0$, and

$$f_1(t) = -t^{1-\lambda_1}\frac{d}{dt}\left(t^{\lambda_1} g(t)\right). \tag{14.8.84}$$

The integral has the same form as the one in (14.8.80), and we can continue.

14.9 The method fails: Examples

In some cases it is not possible to obtain quantitative information. For instance, when ϕ does not have a stationary point in $[a,b]$, and ψ is a C^∞-function in (a,b), which vanishes with all its derivatives at a and b. The function $F(\omega)$ in (14.0.1) may be exponentially small in that case, and we have

$$F(\omega) = o\left(\omega^{-n}\right), \quad \omega \to \infty, \quad \forall n. \tag{14.9.85}$$

An example is

$$F(\omega) = \int_{-\infty}^{\infty} e^{i\omega t}\frac{dt}{1+t^2} = \pi e^{-\omega}. \tag{14.9.86}$$

The integral

$$F(\omega) = \int_{-\frac12\pi}^{\frac12\pi} e^{i\omega t}\, e^{-1/\cos t}\,dt \tag{14.9.87}$$

is exponentially small as well, and in this case we can use the saddle point method to obtain all asymptotic information.

Problem 14.2. *Explain why the integrals in* (14.9.86) *and* (14.9.87) *are not exponentially small when the intervals of integration are* $[0,\infty)$ *and* $[0,\frac12\pi)$*, respectively.* ♡

14.9.1 *The Airy function*

Another example is the Airy function (see Chapter 8 and (8.0.3)) when we use the real integral

$$\mathrm{Ai}(x) = \frac{1}{\pi}\int_0^\infty \cos\left(\tfrac{1}{3}t^3 + xt\right)\, dt = \Re\frac{1}{2\pi}\int_{-\infty}^\infty e^{i\left(\frac{1}{3}t^3+xt\right)}\, dt, \quad x \in \mathbb{R}. \tag{14.9.88}$$

We know that for $x \to +\infty$ this function is exponentially small; see the results of §8.1. The expansion given in (8.1.16) cannot be obtained by using the method of stationary phase for the representation

$$\mathrm{Ai}(x) = \Re\frac{\sqrt{x}}{2\pi}\int_{-\infty}^\infty e^{ix^{3/2}\left(\frac{1}{3}t^3+t\right)}\, dt, \quad x > 0. \tag{14.9.89}$$

There is no real stationary point. However, when we modify the contour of integration we can apply the saddle point method. Saddle points are $\pm i\sqrt{x}$ and we can deform the contour of integration so that it runs through the saddle point $+i\sqrt{x}$ and terminates in the valleys of the exponential function at $\infty e^{5\pi i/6}$ and $\infty e^{\pi i/6}$. This gives a representation similar to that in (8.1.5).

When x is negative we can write

$$\mathrm{Ai}(-x) = \Re\frac{\sqrt{x}}{\pi}\int_0^\infty e^{ix^{3/2}\left(\frac{1}{3}t^3-t\right)}\, dt, \quad x > 0, \tag{14.9.90}$$

and there is a stationary point at $t = 1$. To handle this case we may try to use the method of stationary phase, but again, we prefer here a modification of the interval of integration.

We split up the contour in (14.9.90) into two parts. The first part is the negative imaginary axis. This integral becomes purely imaginary and can be neglected. The second part starts at $-i\infty$, runs through the saddle point $+1$ into the valley at $\infty e^{1\pi i/6}$. By using the saddle point method the expansion as given in §8.2 can be obtained.

14.9.2 *A more complicated example*

A less obvious example is the following integral[1]

$$I(a,b) = \int_a^b \left(\frac{x+a}{x-a}\right)^{2ai}\left(\frac{b-x}{b+x}\right)^{2bi}\frac{dx}{x}; \tag{14.9.91}$$

a and b are large positive parameters with $b/a = c$, a constant greater than 1. We can write $I(a,b)$ in the form

$$I(a,b) = \int_1^c e^{2ai\psi(x)}\,\frac{dx}{x}, \quad \psi(x) = \ln\frac{x+1}{x-1} + c\ln\frac{c-x}{c+x}. \tag{14.9.92}$$

Can we apply the method of stationary phase? Is there a point $x_0 \in [1, c]$ where the exponential phase function $\psi(x)$ is stationary? The answer to both questions is

[1]This section is based on Temme (1989).

negative. Moreover, there is a complication: the endpoints of integration coincide with singularities of $\psi(x)$. As we will see, $I(a,b)$ is exponentially small and has the expansion

$$I(a,b) \sim -\frac{\pi i}{3\sinh 2\pi a}, \quad a \to \infty. \tag{14.9.93}$$

We give a few steps to explain how to handle this case. We use in (14.9.91) the transformation $t = (x-a)/(b-x)$. Then we obtain

$$I(a,b) = A(a,b)\int_0^\infty \frac{t^{-2ai}(t+t_1)^{2ai}\,dt}{(t+t_2)^{2bi}(1+t)(1+ct)}, \tag{14.9.94}$$

where $t_1 = 2/(c+1)$, $t_2 = (c+1)/(2c)$, and

$$A(a,b) = (2c)^{-2bi}(c+1)^{2ai}(c-1)^{2bi-2ai+1}. \tag{14.9.95}$$

Recall that $c > 1$. Hence we have the inequalities

$$0 < \frac{1}{c} < \frac{2}{c+1} < \frac{c+1}{2c} < 1. \tag{14.9.96}$$

This shows that the five singular points $0, -1/c, -t_1, -t_2, -1$ of the above integral satisfy

$$-1 < -t_2 < -t_1 < -\frac{1}{c} < 0. \tag{14.9.97}$$

Since c is assumed to be fixed, no confluence of singularities can happen. Otherwise the problem would be much more difficult.

We compute the saddle point(s) of the integrand, and write

$$\frac{t^{-2ai}(t+t_1)^{2ai}}{(t+t_2)^{2bi}} = e^{-2ai\phi(t)}, \quad \phi(t) = \ln t - \ln(t+t_1) + c\ln(t+t_2). \tag{14.9.98}$$

It is straightforward to verify that

$$\phi'(t) = \frac{c(t+1/c)^2}{t(t+t_1)(t+t_2)}. \tag{14.9.99}$$

Hence, ϕ has a (double) saddle point at $-1/c$. We expand

$$\phi(t) = \phi(-1/c) + \tfrac{1}{6}\phi'''(-1/c)(t+1/c)^3 + \mathcal{O}(t+1/c)^4. \tag{14.9.100}$$

A few computations give

$$\phi'''(-1/c) = -\frac{4c^4(c+1)}{(c-1)^2}. \tag{14.9.101}$$

We observe that (14.9.94) has a double saddle point outside the interval of integration, and that this point coincides with a single pole of the integrand. We need a slight modification of the saddle point method to handle this case.

14.9.2.1 *A loop integral*

The final preparatory step is to introduce a suitable loop integral of which the path of integration can be shifted to the saddle point at $-1/c$. We introduce

$$J(a,b) = \int_{-\delta-i\infty}^{-\delta+i\infty} \frac{(-t)^{-2ai}(t+t_1)^{2ai}\,dt}{(t+t_2)^{2bi}(1+t)(1+ct)}, \tag{14.9.102}$$

where δ is a positive number satisfying $0<\delta<1/c$. It is not difficult to verify that the integral is convergent at ∞. The phase of the complex parameter t is, initially, between $\frac{1}{2}\pi$ and $\frac{3}{2}\pi$. The minus-sign in $(-t)$ is interpreted as $e^{-\pi i}$ and we assume that for negative values of t the phase of $-t$ equals zero; the phases of $t+t_1$ and $t+t_2$ are zero for positive values of t.

To recover $I(a,b)$ from the complex integral, we bend the vertical path around the interval $[0,\infty)$. On the upper side of this interval we have $(-t)^{-2ai} = e^{-2\pi a}|t|^{-2ai}$, on the lower side we have $(-t)^{-2ai} = e^{2\pi a}|t|^{-2ai}$. So we arrive at the result

$$J(a,b) = -2\sinh(2\pi a)I(a,b)/A(a,b), \tag{14.9.103}$$

where $A(a,b)$ is given in (14.9.95).

We slightly change the phase function introduced in (14.9.98) by writing

$$\phi(t) = \ln(e^{-\pi i}t) - \ln(t+t_1) + c\ln(t+t_2). \tag{14.9.104}$$

The formulas (14.9.99), (14.9.100) and (14.9.101) also hold for this new ϕ. We have

$$J(a,b) = \int_{\mathcal{L}} e^{-2ia\phi(t)} \frac{dt}{(1+t)(1+ct)}, \tag{14.9.105}$$

where $\mathcal{L}$ is the above introduced vertical, now with $\delta = 1/c$ and with a small semicircle at the right of the pole at $-1/c$. We introduce the transformation of variables (see (14.9.100))

$$\tfrac{1}{6}\phi'''(-1/c)w^3 = \phi(t) - \phi(-1/c), \tag{14.9.106}$$

and we choose the branch that satisfies $w \sim t + 1/c$ in a neighborhood of the stationary point $-1/c$. Upon using (14.9.104), (14.9.106) in (14.9.102), we obtain

$$J(a,b) = B(a,b)\int_{\mathcal{C}} e^{\frac{1}{3}i\mu a w^3} f(w)\frac{dw}{w}, \tag{14.9.107}$$

where $\mu = -\phi'''(-1/c)$ and

$$B(a,b) = e^{-2ai\phi(-1/c)} = (c-1)/A(a,b), \quad f(w) = \frac{w}{(1+t)(1+ct)}\frac{dt}{dw}. \tag{14.9.108}$$

Since μ is positive (see (14.9.101)), the "best" path $\mathcal{C}$ in (14.9.107) is the steepest descent path defined by the rays

$$\operatorname{ph} w = -\tfrac{1}{2}\pi, \quad \operatorname{ph} w = \tfrac{1}{6}\pi. \tag{14.9.109}$$

Locally, the same holds for the t-plane near $t=-1/c$. In order to avoid the pole, the path in the w-plane has a small circular arc near the origin. The integration runs from $-i\infty$ to $\infty\exp(\frac{1}{6}\pi i)$, and the pole at the origin is on the left-hand side of the contour.

We substitute the Maclaurin series

$$f(w) = \sum_{k=0}^{\infty} c_k w^k \tag{14.9.110}$$

in (14.9.107), and the result is the asymptotic expansion

$$J(a,b) \sim B(a,b) \sum_{k=0}^{\infty} c_k F_k, \quad F_k = \int_{\mathcal{C}} e^{\frac{1}{3}i\mu a w^3} w^{k-1}\, dw, \tag{14.9.111}$$

as $a \to \infty$. To compute F_k we use the path described by (14.9.109). F_0 needs some special care. We write

$$F_0 = \left(\int_{-i\infty}^{-ir} + \int_{r\exp(\frac{1}{6}\pi i)}^{\infty\exp(\frac{1}{6}\pi i)} \right) e^{\frac{1}{3}i\mu a w^3} \frac{dw}{w} + i \int_{-\pi/2}^{\pi/6} e^{\frac{1}{3}i\mu a r^3 e^{3i\theta}}\, d\theta, \tag{14.9.112}$$

where r is a small positive number. The first two integrals cancel. The third one assumes in the limit $r \to 0$ the value $\frac{2}{3}\pi i$. Hence $F_0 = \frac{2}{3}\pi i$. The remaining integrals follow straightforwardly:

$$\begin{aligned} F_k &= \left(e^{\pi i k/6} - e^{-\pi i k/2}\right) \int_0^{\infty} e^{-\frac{1}{3}\mu a w^3} w^{k-1}\, dw \\ &= \tfrac{2}{3} e^{\pi i/2 - \pi i k/6} \sin\left(\tfrac{1}{3}k\pi\right) \Gamma\left(\tfrac{1}{3}k\right) \left(\tfrac{1}{3}\mu a\right)^{-k/3}, \quad k = 1,2,3,\ldots. \end{aligned} \tag{14.9.113}$$

Observe that this result can also be interpreted for $k = 0$, giving again $F_0 = \frac{2}{3}\pi i$.

Combining (14.9.103), (14.9.108) and (14.9.111) we obtain the final result

$$I(a,b) \sim -\frac{c-1}{2\sinh(2\pi a)} \sum_{k=0}^{\infty} c_k F_k, \quad a \to \infty. \tag{14.9.114}$$

To evaluate $c_0 = f(0)$ we use (14.9.108) and l'Hôpital's rule. This gives

$$c_0 = \left.\frac{dt}{dw}\right|_{w=0} \frac{c}{c-1} \lim_{w\to 0} \frac{w}{1+ct} = \frac{1}{c-1}, \tag{14.9.115}$$

and

$$I(a,b) \sim -\frac{\pi i}{3\sinh 2\pi a}, \quad a \to \infty. \tag{14.9.116}$$

We conclude by giving the first few coefficients c_k. These are obtained by using (14.9.106), (14.9.108), and (14.9.110), and the first few are

$$c_0 = \frac{1}{c-1}, \quad c_1 = 0, \quad c_2 = -\frac{2c^2(c^2+1)}{5(c-1)^3}, \quad c_3 = \frac{4c^3(c^2+1)}{5(c-1)^4}. \tag{14.9.117}$$

Remark 14.2. The integral in (14.9.91) appeared in Lekner (1987, Eq. 6.64), where it was used for describing the Rayleigh approximation for a reflection amplitude in the theory of electromagnetic and particle waves. Mahler (1986) showed that the integral tends more rapidly to 0 than every finite negative power of a and he gave an upper bound of the integral. He concluded that his results do not imply estimates of the form $I(a,b) = \mathcal{O}(e^{-a})$; his technique was based on real integrals, the method of stationary phase, and integration by parts. △

Chapter 15

Coefficients of a power series. Darboux's method

A generating function for a special function $F_n(z)$ usually has the form of the convergent series

$$G(z,w) = \sum_{n=0}^{\infty} F_n(z)w^n. \tag{15.0.1}$$

The radius of convergence may be finite or infinite, and may depend on the variable z. The classical orthogonal polynomials, and many other special functions, have such a generating function. For example, the Laguerre polynomials satisfy the relation

$$(1-w)^{-\alpha-1}e^{-wz/(1-w)} = \sum_{n=0}^{\infty} L_n^{(\alpha)}(z)w^n, \quad |w|<1; \tag{15.0.2}$$

α and z may assume every finite complex value.

From the generating function a representation of the function $F_n(z)$ in the form of a Cauchy integral is as follows:

$$F_n(z) = \frac{1}{2\pi i}\int_{\mathcal{C}} G(z,w)\,\frac{dw}{w^{n+1}}, \tag{15.0.3}$$

where $\mathcal{C}$ is a circle around the origin inside the domain where $G(z,w)$ is w-analytic.

The same holds, of course, for the coefficients in a Maclaurin series of a function that is analytic in a neighborhood of the origin, and even for an analytic function with a Laurent expansion

$$f(z) = \sum_{n=-\infty}^{\infty} a_n z^n \tag{15.0.4}$$

in an annulus $0<|z|<r$, with

$$a_n = \frac{1}{2\pi i}\int_{\mathcal{C}} f(z)\,\frac{dz}{z^{n+1}}, \tag{15.0.5}$$

in which $\mathcal{C}$ is a simple closed contour inside the annulus.

Cauchy integrals of the form (15.0.3) and (15.0.5) can be used for obtaining asymptotic expansion of $F_n(z)$ and a_n for large values of n.

In this chapter we give several examples and show how to deal with integrals of the type (15.0.3). The method is, briefly, to move the contour $\mathcal{C}$ away from the origin as far as possible. The method works well for functions $G(z,w)$ of algebraic growth at infinity, with a finite number of poles and branch points. Then the ideal contour consists of small circles around the poles and loop integrals along branch cuts radiating away from the origin of the w-plane.

When the generating function has an essential singularity as in (15.0.2) the method has to be modified. In §15.5 we give a few comments on this case regarding Laguerre polynomials, and in §15.8.2 we consider a more complicated case in detail. Expansions in terms of Bessel functions are derived for the cases with this type of essential singularity.

The asymptotic expansions coming from the algebraic singularities on the circle of convergence of the series in (15.0.1) will give the dominant contributions. The method works when there is a finite number of singularities on this circle. When, under the influence of parameters, two or more singularities coalesce with each other methods from uniform asymptotic analysis are needed. For this we refer to Wong and Zhao (2005); Liu *et al.* (2013).

The method is related to *Darboux's method*, in which again the asymptotic behavior is considered of the coefficients of a power series $f(z)=\sum a_n z^n$. A comparison function g, say, is needed with the same relevant singular point(s) as f. When g has an expansion $g(z)=\sum b_n z^n$, in which the coefficients b_n have known asymptotic behavior, then, under certain conditions on $f(z)-g(z)$ near the singularity, it is possible to find asymptotic forms for the coefficients a_n. For an introduction to Darboux's method and examples for orthogonal polynomials, we refer to Szegő (1975, §8.4).

When $G(z,w)$ is an analytic function of w alternative methods have to be considered, and usually the saddle point method can give the asymptotic expansion. An example is the generating function for the Hermite polynomials:

$$e^{2zw-w^2}=\sum_{n=0}^{\infty}\frac{H_n(z)}{n!}w^n,\quad w,z\in\mathbb{C}. \tag{15.0.6}$$

More complicated methods are needed when z in (15.0.3) or other parameters have influence on the large-n asymptotic behavior of the coefficients $F_n(z)$, in which case usually methods from uniform expansions are needed. For example, in §23.4 we give an example starting with the generating function in (15.0.6) to obtain a uniform expansion of the Hermite polynomials in terms of the Airy function.

15.1 A first example: A binomial coefficient

Consider as a simple example with one algebraic singularity the expansion

$$(1-w)^{\alpha}=\sum_{n=0}^{\infty}a_n w^n,\quad |w|<1, \tag{15.1.7}$$

where

$$a_n = (-1)^n \binom{\alpha}{n} = \frac{(-\alpha)_n}{n!} = \frac{\Gamma(n-\alpha)}{n!\,\Gamma(-\alpha)}. \tag{15.1.8}$$

The asymptotic expansion follows from §6.5, but we continue explaining the method of this chapter. The Cauchy integral in (15.0.5) becomes

$$a_n = \frac{1}{2\pi i}\int_{\mathcal{C}} (1-w)^{\alpha}\,\frac{dw}{w^{n+1}}, \tag{15.1.9}$$

and $\mathcal{C}$ can be deformed into a loop around the branch cut from 1 to $+\infty$. We take the phase of $1-w$ under the cut equal to π, and above the cut equal to $-\pi$ (in this way, when circling around the point $w=1$, $\mathrm{ph}(1-w)=0$ if $w\in(0,1)$). When $-1<\Re\alpha<n$ we can close the contour around the branch cut and obtain

$$a_n = -\frac{\sin\pi\alpha}{\pi}\int_1^{\infty}\frac{(x-1)^{\alpha}}{x^{n+1}}\,dx, \tag{15.1.10}$$

which, of course, can be expressed in terms of the binomial coefficient or Pochhammer symbol shown in (15.1.8), or in terms of the beta integral.

To obtain an asymptotic expansion of a_n by the method as we have in mind for this chapter, we substitute $x=e^t$, giving

$$a_n = -\frac{\sin\pi\alpha}{\pi}\int_0^{\infty} t^{\alpha}h(t)e^{-nt}\,dt,\quad h(t)=\left(\frac{e^t-1}{t}\right)^{\alpha}. \tag{15.1.11}$$

Using Watson's lemma (see §2.1) we can obtain the asymptotic expansion of a_n for large values of n by expanding $h(t)$ in powers of t. When the condition $\Re\alpha>-1$ for the integral in (15.1.10) does not hold, the integral in (15.1.11) can be changed into a loop integral around $(0,\infty)$, and in that case we can use Watson's lemma for loop integrals, see §2.2.

In §6.6 we have considered the binomial coefficient $\displaystyle\binom{n}{m}$ in which both n and m are large. In that case we used the saddle point method.

15.2 Legendre polynomials of large degree

The Legendre polynomials have the generating function

$$\frac{1}{\sqrt{1-2xw+w^2}} = \sum_{n=0}^{\infty} P_n(x)w^n,\quad -1\le x\le 1,\quad |w|<1. \tag{15.2.12}$$

We write $x=\cos\theta$, $0\le\theta\le\pi$, and obtain the Cauchy integral representation

$$P_n(\cos\theta) = \frac{1}{2\pi i}\int_{\mathcal{C}}\frac{1}{\sqrt{1-2\cos\theta\,w+w^2}}\,\frac{dw}{w^{n+1}}, \tag{15.2.13}$$

where $\mathcal{C}$ is a circle around the origin with radius less than 1. There are two singular points on the unit circle: $w_{\pm}=e^{\pm i\theta}$. When $\theta\in(0,\pi)$ we can deform the contour $\mathcal{C}$ into two loops $\mathcal{C}_{\pm}$ around the two branch cuts.

We assume that the square root in (15.2.13) is positive for real values of w and that branch cuts run from each $w_\pm$ parallel to the real axis, with $\Re w \to +\infty$. For $\mathcal{C}_+$ we substitute $w = w_+ e^s$, and obtain a similar contour $\mathcal{C}_+$ around the origin in the s-plane. We start the integration along $\mathcal{C}_+$ at $+\infty$, with $\operatorname{ph} s = 0$, turn around the origin clockwise, and return to $+\infty$ with $\operatorname{ph} s = -2\pi$. The contribution from $\mathcal{C}_+$ becomes

$$P_n^+(\cos\theta) = \frac{e^{-(n+\frac12)i\theta+\frac14\pi i}}{\pi\sqrt{2\sin\theta}} \int_0^\infty e^{-ns} f^+(s)\,\frac{ds}{\sqrt{s}}, \tag{15.2.14}$$

where

$$f^+(s) = \sqrt{\frac{s}{e^s-1}\,\frac{1-e^{-2i\theta}}{e^s-e^{-2i\theta}}}. \tag{15.2.15}$$

We expand $f^+(s) = \sum_{k=0}^{\infty} c_k s^k$ and use Watson's lemma to obtain the large-n expansion of $P_n^+(\cos\theta)$:

$$P_n^+(\cos\theta) \sim \frac{e^{-(n+\frac12)i\theta+\frac14\pi i}}{\sqrt{2\pi n\sin\theta}} \sum_{k=0}^{\infty} \left(\tfrac12\right)_k \frac{c_k}{n^k}, \quad n\to\infty. \tag{15.2.16}$$

The first coefficients are

$$\begin{aligned} c_0 &= 1, \quad c_1 = \frac{-2\sin\theta + i\cos\theta}{2\sin\theta},\\ c_2 &= \frac{4\sin^2\theta - 3\cos^2\theta - 6i\sin\theta\cos\theta}{12\sin^2\theta}. \end{aligned} \tag{15.2.17}$$

If $\theta\in[\theta_0, \pi-\theta_0]$, where θ_0 is a small positive number, then the conditions are satisfied to apply Watson's lemma. For small values of θ and $\pi-\theta$ we need uniform expansions in terms of the J-Bessel function; see §28.2.1.

The contribution from the singularity at w_- can be obtained in the same way. It is the complex conjugate of the contribution from w_+, and we have $P_n(\cos\theta) = 2\Re P_n^+(\cos\theta)$.

In this example we can obtain a simpler expansion valid for large values of n. When we substitute $e^{-s} = t$, the representation in (15.2.14) becomes

$$P_n^+(\cos\theta) = \frac{e^{-(n+1)i\theta+\frac12\pi i}}{\pi} \int_0^1 \frac{t^n}{\sqrt{1-t}}\,\frac{dt}{\sqrt{1-te^{-2\pi i}}}, \tag{15.2.18}$$

which can be written in terms of the Gauss hypergeometric function. Using (12.0.3) we obtain

$$P_n^+(\cos\theta) = \frac{e^{-(n+1)i\theta+\frac12\pi i}}{\pi} \frac{\Gamma(n+1)\Gamma\left(\frac12\right)}{\Gamma\left(n+\frac32\right)}\, {}_2F_1\!\left(\begin{matrix}\frac12, n+1\\ n+\frac32\end{matrix}; e^{-2i\theta}\right). \tag{15.2.19}$$

The first transformation formula in (12.0.6) gives for $P_n(\cos\theta)$:

$$P_n(\cos\theta) = \frac{4\,n!}{\pi\left(\frac32\right)_n\sqrt{2\sin\theta}} \Re\left(e^{-(n+\frac12)i\theta+\frac14\pi i}\, {}_2F_1\!\left(\begin{matrix}\frac12, \frac12\\ n+\frac32\end{matrix}; \frac{ie^{-i\theta}}{2\sin\theta}\right)\right). \tag{15.2.20}$$

We can expand the F-function if, on $[0,\pi]$, $|2\sin\theta|>1$. It follows that, if $\frac{1}{6}\pi<\theta<\frac{5}{6}\pi$, we have the convergent expansion

$$P_n(\cos\theta)=\frac{4\,n!}{\pi\left(\frac{3}{2}\right)_n\sqrt{2\sin\theta}}\times \sum_{k=0}^{\infty}\frac{\left(\frac{1}{2}\right)_k\left(\frac{1}{2}\right)_k}{k!\left(n+\frac{3}{2}\right)_k}\frac{\cos\left(\left(n+k+\frac{1}{2}\right)\theta-\frac{1}{2}\pi\left(k+\frac{1}{2}\right)\right)}{(2\sin\theta)^k}. \tag{15.2.21}$$

As explained in §12.1.1, this expansion can be viewed as an asymptotic expansion. In fact we have

$$P_n(\cos\theta)\sim\frac{4\,n!}{\pi\left(\frac{3}{2}\right)_n\sqrt{2\sin\theta}}\times \sum_{k=0}^{\infty}\frac{\left(\frac{1}{2}\right)_k\left(\frac{1}{2}\right)_k}{k!\left(n+\frac{3}{2}\right)_k}\frac{\cos\left(\left(n+k+\frac{1}{2}\right)\theta-\frac{1}{2}\pi\left(k+\frac{1}{2}\right)\right)}{(2\sin\theta)^k}, \tag{15.2.22}$$

for large n, uniformly for all $\theta\in[\theta_0,\pi-\theta_0]$, where θ_0 is a fixed small positive number.

This expansion has been given by Stieltjes (1890) with a simple upper bound for the remainder of the asymptotic expansion; see also Szegő (1975, §8.21).

15.2.1 *A paradox in asymptotics*

A related expansion reads (see Koornwinder *et al.* (2010, Eq. 18.15.12))

$$P_n(\cos\theta)\sim\sqrt{\frac{2}{\sin\theta}}\times \sum_{k=0}^{\infty}\binom{-\frac{1}{2}}{k}\binom{k-\frac{1}{2}}{n}\frac{\cos\left(\left(n-k+\frac{1}{2}\right)\theta+\pi\left(n-\frac{1}{2}k-\frac{1}{4}\right)\right)}{(2\sin\theta)^k}, \tag{15.2.23}$$

again with convergence when $\frac{1}{6}\pi<\theta<\frac{5}{6}\pi$.

There is a paradox in connection with this expansion, which does not happen for the expansion in (15.2.22). We verify what happens by observing that, in the case of convergence, we can write the expansion in (15.2.23) in the form

$$\sqrt{\frac{2}{\pi\sin\theta}}\frac{\Gamma\left(n+\frac{1}{2}\right)}{n!}\Re\left(e^{i\left(n+\frac{1}{2}\right)\theta-\frac{1}{4}\pi i}{}_2F_1\left(\begin{matrix}\frac{1}{2},\frac{1}{2}\\ \frac{1}{2}-n\end{matrix};\frac{ie^{-i\theta}}{2\sin\theta}\right)\right). \tag{15.2.24}$$

Observe that the large parameter n now occurs at the c-place with a negative sign; this is different from (15.2.20), which is an exact relation.

By using the connection formula given in (12.1.12), (15.2.24) can be written as

$$\frac{4\,n!}{\pi\left(\frac{3}{2}\right)_n\sqrt{2\sin\theta}}\left(e^{\left(n+\frac{1}{2}\right)i\theta-\frac{1}{4}\pi i}{}_2F_1\left(\begin{matrix}\frac{1}{2},\frac{1}{2}\\ \frac{3}{2}+n\end{matrix};\frac{-ie^{i\theta}}{2\sin\theta}\right)+ e^{-\left(n+\frac{1}{2}\right)i\theta+\frac{1}{4}\pi i}{}_2F_1\left(\begin{matrix}\frac{1}{2},\frac{1}{2}\\ \frac{3}{2}+n\end{matrix};\frac{ie^{-i\theta}}{2\sin\theta}\right)\right), \tag{15.2.25}$$

and, paradoxically, this is twice the right-hand side of (15.2.21). In Olver (1970a) this paradox is explained in a different way that, in the case of convergence, the expansion in (15.2.23) is not $P_n(\cos\theta)$ but $2P_n(\cos\theta)$. The expansion in (15.2.23) is also given in Szegő (1975, §8.4(3)), without the observation that, in the case of convergence, the sum is $2P_n(\cos\theta)$.

15.3 Gegenbauer polynomials of large degree

The Gegenbauer polynomials have the generating function

$$\frac{1}{(1-2xw+w^2)^\gamma} = \sum_{n=0}^{\infty} C_n^{(\gamma)}(x)w^n, \quad -1 \le x \le 1, \quad |w|<1. \tag{15.3.26}$$

We write $x = \cos\theta$, $0 \le \theta \le \pi$, and obtain the Cauchy integral representation

$$C_n^{(\gamma)}(\cos\theta) = \frac{1}{2\pi i}\int_{\mathcal{C}} \frac{1}{(1-2\cos\theta w + w^2)^\gamma}\frac{dw}{w^{n+1}}. \tag{15.3.27}$$

The contribution from the singularity $w_+ = e^{i\theta}$ is (compare with (15.2.14))

$$C_n^{(\gamma)^+}(\cos\theta) = \frac{\sin(\pi\gamma)\, e^{-(n+\gamma)i\theta+\frac12\gamma\pi i}}{\pi\,(2\sin\theta)^\gamma}\int_0^\infty e^{-ns} f^+(s)\,\frac{ds}{s^\gamma}, \tag{15.3.28}$$

where

$$f^+(s) = \left(\frac{s}{e^s-1}\,\frac{1-e^{-2i\theta}}{e^s-e^{-2i\theta}}\right)^\gamma. \tag{15.3.29}$$

This representation is valid only if $\Re\gamma < 1$, for other values we can use a loop integral. Watson's lemma can be used to obtain the asymptotic expansion.

The analogue of (15.2.19) becomes

$$C_n^{(\gamma)^+}(\cos\theta) = e^{-(n+2\gamma)i\theta+\pi i\gamma}\frac{\Gamma(n+\gamma+1)}{\Gamma(\gamma)\Gamma(n+2\gamma)}\,{}_2F_1\!\left(\begin{matrix}\gamma, n+2\gamma\\ n+\gamma+1\end{matrix}; e^{-2i\theta}\right), \tag{15.3.30}$$

and now we can take $\Re\gamma > 0$. The transformation formula in (12.0.6) gives for the Gegenbauer polynomial

$$\begin{aligned} C_n^{(\gamma)}(\cos\theta) = {} & \frac{2\Gamma(2\gamma)(2\gamma)_n}{(2\sin\theta)^\gamma\Gamma(\gamma)\Gamma(\gamma+1)(\gamma+1)_n}\times \\ & \Re\left(e^{-(n+\gamma)i\theta+\frac12\pi i\gamma}\,{}_2F_1\!\left(\begin{matrix}\gamma, 1-\gamma\\ n+\gamma+1\end{matrix}; \frac{ie^{-i\theta}}{2\sin\theta}\right)\right). \end{aligned} \tag{15.3.31}$$

The hypergeometric function can be expanded as in (15.2.21), again with a convergent series if $\frac16\pi < \theta < \frac56\pi$. When we take a finite expansion, a bound on the remainder in this expansion is available in Szegő (1975, p. 197).

By expanding the ${}_2F_1$-function in (15.3.31), a large-n asymptotic expansion follows that is valid uniformly in a compact subinterval of $(0,\pi)$. An expansion that is valid for $\theta \sim \pi$ is mentioned in §28.3.1.4.

15.4 Jacobi polynomials of large degree

We can continue with the Jacobi polynomials, but the generating function is quite complicated. It is simpler to use representations in terms of the hypergeometric function or to use the *Rodrigues formula*

$$P_n^{(\alpha,\beta)}(x) = \frac{(-1)^n}{2^n\,n!}\frac{1}{(1-x)^\alpha(1+x)^\beta}\left(\frac{d}{dx}\right)^n\left((1-x)^{n+\alpha}(1+x)^{n+\beta}\right), \quad (15.4.32)$$

for $n = 0,1,2,\ldots$. This gives the Cauchy-type integral

$$P_n^{(\alpha,\beta)}(x) = \frac{(-1)^n}{2\pi i}\frac{2^{-n}}{(1-x)^\alpha(1+x)^\beta}\int_{\mathcal{C}}(1-z)^{n+\alpha}(1+z)^{n+\beta}\,\frac{dz}{(z-x)^{n+1}}, \quad (15.4.33)$$

where $\mathcal{C}$ is a simple closed contour encircling the point $z = x$, but not the points $z = \pm 1$. Observe that the large parameter n shows up in more places, which is not assumed in the approach in this chapter.

In fact, (15.4.33) leads to (see Olde Daalhuis (2010b, Eq. 15.6.2))

$$\begin{aligned} P_n^{(\alpha,\beta)}(x) = (-1)^n\left(\frac{2}{1-x}\right)^\alpha\binom{n+\beta}{n}\ \times \\ {}_2F_1\left({-n-\alpha, n+\beta+1 \atop \beta+1}\,;\,\frac{1+x}{2}\right), \end{aligned} \quad (15.4.34)$$

The forms in (12.3.25) and (15.4.34) follow from each other by using $P_n^{(\beta,\alpha)}(-x) = (-1)^n P_n^{(\alpha,\beta)}(x)$ and the third relation in (12.0.6).

We used the representation in (12.3.25) to derive the large-degree asymptotics of the Jacobi polynomials for x strictly inside $(-1,1)$. In §28.4 we use (15.4.33) to obtain an expansion in terms of Laguerre polynomials, which can be used for $x \sim 1$.

15.5 Laguerre polynomials of large degree

The generating function is given in (15.0.2), which gives the Cauchy-type integral representation

$$L_n^{(\alpha)}(z) = \frac{1}{2\pi i}\int_{\mathcal{C}}(1-t)^{-\alpha-1}e^{-tz/(1-t)}\,\frac{dt}{t^{n+1}}, \quad (15.5.35)$$

where α and z may assume every finite complex value and $\mathcal{C}$ is a circle around the origin with radius less than unity.

An explicit representation is

$$L_n^{(\alpha)}(z) = \sum_{k=0}^{n}(-1)^k\binom{n+\alpha}{n-k}\frac{z^k}{k!}, \quad n = 0,1,2,\ldots. \quad (15.5.36)$$

When $\alpha > -1$, these polynomials are orthogonal on $[0,\infty)$ with weight function $z^\alpha e^{-z}$.

The large-n behavior of $L_n^{(\alpha)}(z)$ can be obtained from the integral representation in (15.5.35). However, the method that we used earlier in this chapter is not applicable because of the essential singularity of the integrand in the integral representation at $t = 1$.

When we modify the method of Darboux for the integral in (15.5.35) we obtain an expansion that we have found earlier in Chapter 10, where we have considered the confluent hypergeometric functions or Kummer functions. The relation to the Kummer functions is

$$L_n^{(\alpha)}(z) = \binom{n+\alpha}{n} {}_1F_1\left(\begin{matrix}-n\\ \alpha+1\end{matrix};z\right) = \frac{(-1)^n}{n!}U(-n,\alpha+1,z), \tag{15.5.37}$$

and to obtain the large-n asymptotic behavior of $L_n^{(\alpha)}(z)$ we can use the results of §10.3.4 and §10.3.5.

In (10.3.59) we have given a complete expansion in terms of the Bessel functions (for the function $E_\nu(z)$ we refer to §10.3.1). This expansion is valid as $n \to \infty$ with α and z in bounded domains of the complex plane. When z is bounded away from the origin we can expand the Bessel functions for large values of its argument.

In §32.1 we give more details on this expansion of the Laguerre polynomials. In §15.8.2 we consider another example in which the Cauchy integral shows an essential singularity, and again we need Bessel functions for the expansion.

15.6 Generalized Bernoulli polynomials $B_n^{(\mu)}(z)$

Bernoulli polynomials of degree n, complex order μ and complex argument z, denoted by $B_n^{(\mu)}(z)$, are usually defined by a generating function. We have[1]

$$\frac{w^\mu e^{wz}}{(e^w-1)^\mu} = \sum_{n=0}^{\infty} \frac{B_n^{(\mu)}(z)}{n!} w^n, \quad |w| < 2\pi. \tag{15.6.38}$$

We use the contour integral that follows from this generating function, and we consider n as the large parameter, with the other parameters fixed. We exclude $\mu = 0, -1, -2, \ldots$, because in that case the generating function in (15.6.38) is an analytic function of w, and the Bernoulli polynomials behave quite differently in that case.

Bernoulli numbers B_n and the standard Bernoulli polynomials are special cases:

$$B_n = B_n^{(1)}(0), \quad B_n(z) = B_n^{(1)}(z), \quad n = 0, 1, 2, \ldots. \tag{15.6.39}$$

The Bernoulli polynomials and their simpler forms show up in many places of mathematics. Historically, they were important in the calculus of finite differences. In fact, the coefficients in all the usual central-difference formulas for interpolation, numerical differentiation and integration, and differences in terms of derivatives can be expressed in terms of these polynomials (see Milne-Thomson (1951); Nörlund (1924)). For a bibliography on Bernoulli and Euler polynomials we refer to Dilcher *et al.* (1991); see also Dilcher (2010).

[1]This section and the next one on Euler polynomials are based on López and Temme (2010b).

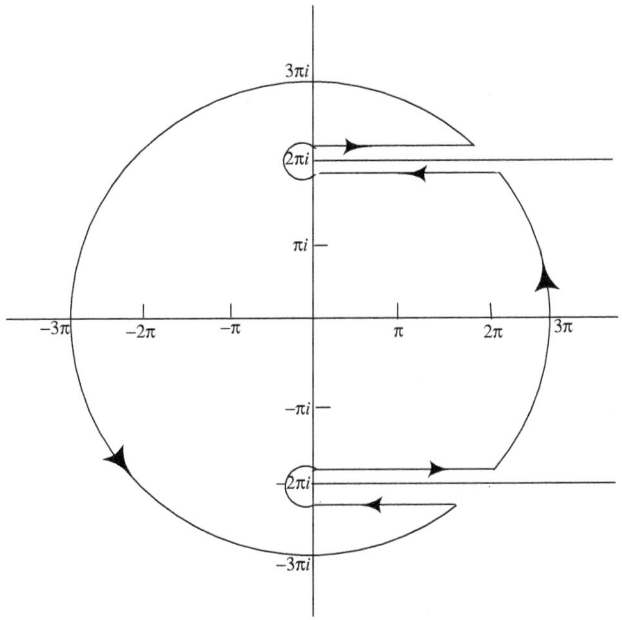

Fig. 15.1 Contour for (15.6.40) for general μ.

15.6.1 *Asymptotic expansions for large degree*

We use (15.6.38) and obtain the representation

$$B_n^{(\mu)}(z) = \frac{n!}{2\pi i}\int_{\mathcal{C}} \frac{w^\mu e^{wz}}{(e^w-1)^\mu}\frac{dw}{w^{n+1}}, \tag{15.6.40}$$

where $\mathcal{C}$ is a circle around the origin, with radius less than 2π. The singularities at $\pm 2\pi i$ are the sources for the main asymptotic contributions. We integrate around a circle with radius 3π, avoiding branch cuts running from $\pm 2\pi i$ to $+\infty$. See Figure 15.1. The contribution from the circular arc is $\mathcal{O}((3\pi)^{-n})$, which is exponentially small with respect to the main contributions.

We denote the loops by $\mathcal{C}_\pm$ and the contributions from the loops by $I_\pm$. For the upper loop we substitute $w = 2\pi i e^s$. This gives

$$I_+ = \frac{n!}{2\pi i}\frac{e^{2\pi i z}}{(2\pi i)^n}\int_{\mathcal{C}_+} g(s)s^{-\mu}e^{-ns}\,ds, \tag{15.6.41}$$

where

$$g(s) = \left(\frac{2\pi i s}{e^p-1}\right)^\mu e^{zp+\mu s}, \quad p = 2\pi i\,(e^s-1). \tag{15.6.42}$$

We start the integration along the contour $\mathcal{C}_+$ at $s = +\infty$, with $\operatorname{ph} s = 0$, turn around the origin clockwise, and return to $+\infty$ with $\operatorname{ph} s = -2\pi$.

To obtain an asymptotic expansion we apply Watson's lemma for loop integrals, see §2.2. We expand $g(s) = \sum_{k=0}^{\infty} g_k s^k$, substitute this into (15.6.41), and interchange summation and integration. This gives

$$I_+ \sim n!\,\frac{e^{2\pi i z}}{(2\pi i)^n}\sum_{k=0}^{\infty} g_k F_k, \quad F_k = \frac{1}{2\pi i}\int_{\mathcal{C}_+} s^{k-\mu} e^{-ns}\,ds. \tag{15.6.43}$$

To evaluate the integrals we turn the path by writing $s = e^{\pi i}t$, and use the representation of the reciprocal gamma function in terms of the Hankel contour; see (2.2.17). The result is

$$F_k = n^{\mu-k-1}e^{\pi i\mu}\frac{(-1)^k}{\Gamma(\mu-k)} = n^{\mu-k-1}e^{\pi i\mu}\frac{(1-\mu)_k}{\Gamma(\mu)}. \tag{15.6.44}$$

This gives the expansion

$$I_+ \sim \frac{n!\,n^{\mu-1}}{(2\pi)^n\,\Gamma(\mu)}\,e^{i\chi}\sum_{k=0}^{\infty}\frac{(1-\mu)_k g_k}{n^k}, \quad \chi = 2\zeta - \tfrac12 n\pi, \quad \zeta = \left(z + \tfrac12\mu\right)\pi. \tag{15.6.45}$$

The result $I_+ + I_-$ can be obtained by taking twice the real part of I_+ (for convenience we do not consider z and μ as complex numbers). We write $g_k = g_k^{(r)} + i g_k^{(i)}$ (with $g_k^{(r)}, g_k^{(i)}$ real when z and μ are real), and obtain

$$\begin{aligned} B_n^{(\mu)}(z) \sim{}& \frac{2\,n!\,n^{\mu-1}}{(2\pi)^n\,\Gamma(\mu)}\times \\ &\left(\cos\chi\sum_{k=0}^{\infty}\frac{(1-\mu)_k g_k^{(r)}}{n^k} - \sin\chi\sum_{k=0}^{\infty}\frac{(1-\mu)_k g_k^{(i)}}{n^k}\right), \end{aligned} \tag{15.6.46}$$

as $n\to\infty$, with z and μ fixed complex numbers, with $\mu \ne 0, -1, -2, \ldots$.

The first few coefficients $g_k^{(r)}, g_k^{(i)}$ are

$$\begin{aligned} g_0^{(r)} &= 1, \quad g_0^{(i)} = 0, \quad g_1^{(r)} = \tfrac12\mu, \quad g_1^{(i)} = 2\zeta, \\ g_2^{(r)} &= \tfrac{1}{24}(3\mu^2 + (4\pi^2-1)\mu - 48\zeta^2), \quad g_2^{(i)} = (1+\mu)\zeta, \\ g_3^{(r)} &= \tfrac{1}{48}(\mu^3 + (4\pi^2-1)\mu^2 + 8(\pi^2-6\zeta^2)\mu - 96\zeta^2), \\ g_3^{(i)} &= \tfrac{1}{12}\zeta(3\mu^2 + (4\pi^2+5)\mu - 16\zeta^2 + 4). \end{aligned} \tag{15.6.47}$$

The first-order approximation reads

$$B_n^{(\mu)}(z) = \frac{2\,n!\,n^{\mu-1}}{(2\pi)^n\,\Gamma(\mu)}\left(\cos\pi(2z+\mu-\tfrac12 n) + \mathcal{O}(1/n)\right), \tag{15.6.48}$$

as $n\to\infty$. Nörlund (1961, p. 39) described the same method of this section and only gives the first-order approximation.

Remark 15.1. If $\mu = 0, -1, -2, \ldots$ the expansion vanishes because of the factor $1/\Gamma(\mu)$ in front of the expansion in (15.6.46). For these values of μ, the generating function in (15.6.38) has no finite singular points, and the Bernoulli polynomials have a completely different behavior; for details see López and Temme (2010b, §2.1). △

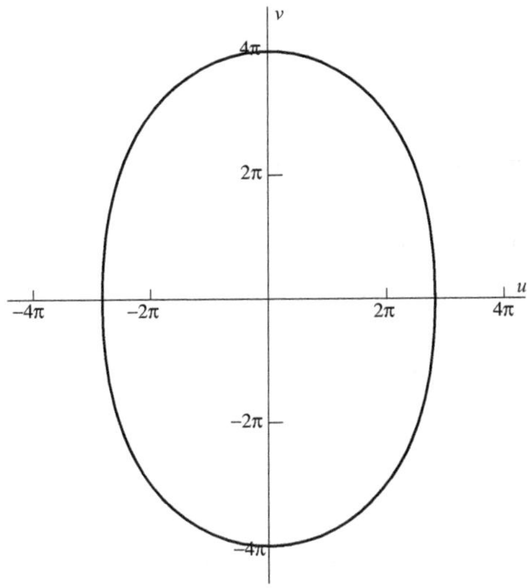

Fig. 15.2 Cassini oval defined by (15.6.52), $w = u + iv$.

15.6.2 *An alternative expansion*

As observed in the previous method, the main contributions to (15.6.40) come from the singular points of the integrand at $\pm 2\pi i$. In this section we expand part of the integrand of (15.6.40) in a two-point Taylor expansion. In this way a simpler asymptotic representation can be obtained. For more details on this topic we refer to Chapter 18.

We write

$$B_n^{(\mu)}(z) = \frac{n!}{2\pi i} 2^{3\mu} \pi^{2\mu} \int_{\mathcal{C}} \frac{f(w)}{(w^2 + 4\pi^2)^\mu} \frac{dw}{w^{n+1}}, \tag{15.6.49}$$

where

$$f(w) = 2^{-3\mu} \pi^{-2\mu} \left(w^2 + 4\pi^2\right)^\mu \left(\frac{w}{e^w - 1}\right)^\mu e^{wz}, \tag{15.6.50}$$

and expand

$$f(w) = \sum_{k=0}^{\infty} (\alpha_k + w\beta_k) \left(w^2 + 4\pi^2\right)^k. \tag{15.6.51}$$

The series converges inside the Cassini oval defined by

$$\left|w^2 + 4\pi^2\right| = 12\pi^2. \tag{15.6.52}$$

See Theorem 18.1 and Figure 15.2.

The coefficients α_0 and β_0 can be found by substituting $w = \pm 2\pi i$. This gives

$$\begin{aligned} \alpha_0 &= \frac{f(2\pi i) + f(-2\pi i)}{2} = \cos 2\zeta, \\ \beta_0 &= \frac{f(2\pi i) - f(-2\pi i)}{4\pi i} = \frac{1}{2\pi} \sin 2\zeta, \end{aligned} \tag{15.6.53}$$

where ζ is defined in (15.6.45). The next coefficients are

$$\begin{aligned}
\alpha_1 &= -\frac{1}{16\pi^2}(3\mu\cos 2\zeta + 2\pi\eta\sin 2\zeta),\\
\beta_1 &= \frac{1}{32\pi^3}(2\pi\eta\cos 2\zeta + (2-3\mu)\sin 2\zeta),\\
\alpha_2 &= \frac{1}{1536\pi^4}((-12\pi^2\eta^2 + 4\mu\pi^2 - 33\mu + 27\mu^2)\cos 2\zeta +\\
&\qquad 12\pi\eta(3\mu-1)\sin 2\zeta),\\
\beta_2 &= \frac{1}{3072\pi^5}(-36\pi\eta(\mu-1)\cos 2\zeta +\\
&\qquad (36 - 69\mu + 27\mu^2 + 4\mu\pi^2 - 12\pi^2\eta^2)\sin 2\zeta)\,,
\end{aligned} \tag{15.6.54}$$

where $\zeta = \left(z + \frac{1}{2}\mu\right)\pi$ and $\eta = \mu - 2z$.

Substituting the expansion in (15.6.51) into (15.6.49) we obtain

$$B_n^{(\mu)}(z) = n!\,2^{3\mu}\pi^{2\mu}\sum_{k=0}^{\infty}\left(\alpha_k\Psi_k^{(n)} + \beta_k\Psi_k^{(n-1)}\right), \tag{15.6.55}$$

where

$$\Phi_k^{(n)} = \frac{1}{2\pi i}\int_{\mathcal{C}}\left(w^2+4\pi^2\right)^{k-\mu}\frac{dw}{w^{n+1}}. \tag{15.6.56}$$

We have $\Phi_k^{(2n+1)} = 0$ and

$$\Phi_k^{(2n)} = (2\pi)^{2k-2\mu-2n}\binom{k-\mu}{n} = (-1)^n(2\pi)^{2k-2\mu-2n}\frac{(\mu-k)_n}{n!}. \tag{15.6.57}$$

Hence,

$$\begin{aligned}
B_{2n}^{(\mu)}(z) &= (2n)!\,2^{3\mu}\pi^{2\mu}\sum_{k=0}^{\infty}\alpha_k\Phi_k^{(2n)},\\
B_{2n+1}^{(\mu)}(z) &= (2n+1)!2^{3\mu}\pi^{2\mu}\sum_{k=0}^{\infty}\beta_k\Phi_k^{(2n)}.
\end{aligned} \tag{15.6.58}$$

These convergent expansions have an asymptotic character for large n. This follows from

$$\frac{\Phi_{k+1}^{(2n)}}{\Phi_k^{(2n)}} = 4\pi^2\frac{\mu-k}{\mu-k-1+n} = \mathcal{O}\left(n^{-1}\right), \quad n\to\infty, \tag{15.6.59}$$

where k is fixed.

We compare the first-term approximations given in (15.6.48) and those from (15.6.58). From (15.6.48) we obtain

$$B_{2n}^{(\mu)}(z) \sim (-1)^n\frac{(2n)!\,2^\mu n^{\mu-1}}{(2\pi)^{2n}\,\Gamma(\mu)}\cos\pi(2z+\mu) + \ldots, \tag{15.6.60}$$

and from (15.6.58)

$$B_{2n}^{(\mu)}(z) = (-1)^n \frac{(2n)!\,2^\mu}{(2\pi)^{2n}\,\Gamma(\mu)} \frac{\Gamma(n+\mu)}{n!} \cos\pi(2z+\mu) + \ldots. \tag{15.6.61}$$

Because $\Gamma(n+\mu)/n! \sim n^{\mu-1}$ as $n\to\infty$, we see that the first approximations give the same asymptotic estimates.

Remark 15.2. The expansions in (15.6.58) do not vanish when μ is a negative integer, as the expansion in (15.6.46) does; see also Remark 15.1. We have when $\mu = -m = 0, -1, -2, \ldots$,

$$\Phi_k^{(2n)} = \begin{cases} (2\pi)^{2k-2m-2n} \dbinom{k-m}{n}, & k \ge n+m, \\ 0, & k < n+m. \end{cases} \tag{15.6.62}$$

Hence, the summation in (15.6.58) starts with $k = n+m$, and when n is large we need coefficients α_k, β_k in (15.6.58) with large k. In this way, the expansion is no longer of any use, because the coefficients become quite complicated for large k. As mentioned earlier, for results for integer values of μ, we refer to López and Temme (2010b). △

15.7 Generalized Euler polynomials $E_n^{(\mu)}(z)$

The derivation of the asymptotic expansions of the generalized Euler polynomials follows the same steps as in the Bernoulli case. Generalized Euler polynomials of degree n, complex order μ and complex argument z, denoted by $E_n^{(\mu)}(z)$, can be defined by the generating function

$$\frac{2^\mu e^{wz}}{(e^w+1)^\mu} = \sum_{n=0}^{\infty} \frac{E_n^{(\mu)}(z)}{n!} w^n, \quad |w| < \pi. \tag{15.7.63}$$

Again we exclude $\mu = 0, -1, -2, \ldots$, because then the generating function is an analytic function of w, and the Euler polynomials show a completely different behavior in that case; see (López and Temme, 2010b) for more details.

15.7.1 *Asymptotic expansions for large degree*

Again we use the Cauchy integral

$$E_n^{(\mu)}(z) = \frac{n!}{2\pi i} \int_{\mathcal{C}} \frac{2^\mu e^{wz}}{(e^w+1)^\mu} \frac{dw}{w^{n+1}}, \tag{15.7.64}$$

where $\mathcal{C}$ is a circle around the origin, with radius less than π. This follows from (15.7.63).

The analysis is as in §15.6.1. We use a contour for the integral (15.7.64) as in Figure 15.1, now with loops around the branch points $\pm\pi i$, and with radius of the large circle smaller than 3π. We denote the integrals around the loops by $I_\pm$. After the substitution $w = \pi i \exp(s)$ we obtain for the upper loop

$$I_+ = \frac{2^\mu n!}{2\pi i}\frac{e^{\pi i z - \mu\pi i}}{(\pi i)^{n+\mu}} \int_{\mathcal{C}_+} h(s) s^{-\mu} e^{-ns}\, ds, \tag{15.7.65}$$

where

$$h(s) = e^{zp}\left(\frac{\pi i s}{e^p - 1}\right)^\mu, \quad p = \pi i\,(e^s - 1). \tag{15.7.66}$$

We expand $h(s) = \sum_{k=0}^\infty h_k s^k$ and interchange summation and integration in (15.7.65). By using (15.6.43) and (15.6.44) we obtain the result

$$\begin{aligned} E_n^{(\mu)}(z) \sim \frac{2^{\mu+1}\, n!\, n^{\mu-1}}{\pi^{n+\mu}\,\Gamma(\mu)} \times \\ \left(\cos\chi \sum_{k=0}^\infty \frac{(1-\mu)_k h_k^{(r)}}{n^k} - \sin\chi \sum_{k=0}^\infty \frac{(1-\mu)_k h_k^{(i)}}{n^k}\right), \end{aligned} \tag{15.7.67}$$

as $n \to \infty$, where z and μ are fixed complex numbers, $\mu \ne 0, -1, -2, \ldots$, and

$$\chi = \zeta - \tfrac12 n\pi, \quad \zeta = \left(z - \tfrac12\mu\right)\pi. \tag{15.7.68}$$

The first few coefficients $h_k^{(r)}, h_k^{(i)}$ are

$$\begin{aligned} &h_0^{(r)} = 1, \quad h_0^{(i)} = 0, \quad h_1^{(r)} = -\tfrac12\mu, \quad h_1^{(i)} = \zeta, \\ &h_2^{(r)} = \tfrac{1}{24}(3(1-2\pi^2)\mu^2 + (13\pi^2 - 12\zeta\pi - 1)\mu - 12\zeta^2), \\ &h_2^{(i)} = \tfrac12(1-\mu)\zeta, \\ &h_3^{(r)} = \tfrac{1}{48}z(-\mu^3 + (1-\pi^2)\mu^2 + 2(\pi^2 + 6\zeta^2)\mu - 24\zeta^2), \\ &h_3^{(i)} = \tfrac{1}{24}\zeta(3\mu^2 + (\pi^2 - 7)\mu - 4\zeta^2 + 4). \end{aligned} \tag{15.7.69}$$

The first-order approximation reads

$$E_n^{(\mu)}(z) = \frac{2^{\mu+1}\, n!\, n^{\mu-1}}{\pi^{n+\mu}\,\Gamma(\mu)}\left(\cos\pi(z - \tfrac12\mu - \tfrac12 n) + \mathcal{O}(1/n)\right), \tag{15.7.70}$$

as $n \to \infty$.

15.7.2 *An alternative expansion*

We repeat the steps of §15.6.2. We write

$$E_n^{(\mu)}(z) = \frac{n!}{2\pi i}(4\pi)^\mu \int_{\mathcal{C}} \frac{g(w)}{(w^2+\pi^2)^\mu}\frac{dw}{w^{n+1}}, \tag{15.7.71}$$

$$g(w) = \left(\frac{w^2+\pi^2}{2\pi}\right)^\mu \left(\frac{1}{e^w+1}\right)^\mu e^{wz}, \tag{15.7.72}$$

and expand

$$g(w) = \sum_{k=0}^{\infty} (\gamma_k + w\delta_k)\left(w^2+\pi^2\right)^k. \tag{15.7.73}$$

We have

$$\begin{aligned}
\gamma_0 &= \cos\zeta, \quad \delta_0 = \frac{1}{\pi}\sin\zeta,\\
\gamma_1 &= -\frac{1}{4\pi^2}(\mu\cos\zeta + \pi\eta\sin\zeta), \quad \delta_1 = \frac{1}{4\pi^3}(\pi\eta\cos\zeta + (2-\mu)\sin\zeta),\\
\gamma_2 &= \frac{1}{96\pi^4}((-9\mu - 3\pi^2\eta^2 + \pi^2\mu + 3\mu^2)\cos\zeta + 6\pi\eta(\mu-1)\sin\zeta),\\
\delta_2 &= \frac{1}{96\pi^5}(6\pi\eta(3-\mu)\cos\zeta + (36 - 21\mu + 3\mu^2 + \pi^2\mu - 3\pi^2\eta^2)\sin\zeta),
\end{aligned} \tag{15.7.74}$$

where $\eta = \mu - 2z$.

Substituting the expansion in (15.7.73) into (15.7.71) we obtain

$$E_n^{(\mu)}(z) = (4\pi)^\mu n! \sum_{k=0}^{\infty}\left(\gamma_k \Psi_k^{(n)} + \delta_k \Psi_k^{(n-1)}\right), \tag{15.7.75}$$

where

$$\Psi_k^{(n)} = \frac{1}{2\pi i}\int_{\mathcal{C}} \left(w^2+\pi^2\right)^{k-\mu} \frac{dw}{w^{n+1}}. \tag{15.7.76}$$

We have $\Psi_k^{(2n+1)} = 0$ and

$$\Psi_k^{(2n)} = \pi^{2k-2\mu-2n}\binom{k-\mu}{n} = (-1)^n \pi^{2k-2\mu-2n}\frac{(\mu-k)_n}{n!}. \tag{15.7.77}$$

Hence,

$$\begin{aligned}
E_{2n}^{(\mu)}(z) &= (2\pi)^\mu (2n)! \sum_{k=0}^{\infty} \gamma_k \Psi_k^{(2n)},\\
E_{2n+1}^{(\mu)}(z) &= (2\pi)^\mu (2n+1)! \sum_{k=0}^{\infty} \delta_k \Psi_k^{(2n)}.
\end{aligned} \tag{15.7.78}$$

These convergent expansions have an asymptotic character for large n. This follows from

$$\frac{\Psi_{k+1}^{(2n)}}{\Psi_k^{(2n)}} = \pi^2 \frac{\mu-k}{\mu-k-1+n} = \mathcal{O}\left(n^{-1}\right), \quad n\to\infty. \tag{15.7.79}$$

Comparing the first-term approximations given in (15.7.70) and those from (15.7.78) we obtain from (15.7.70)

$$E_{2n}^{(\mu)}(z) \sim (-1)^n \frac{(2n)!\,2^{2\mu} n^{\mu-1}}{\pi^{2n+\mu}\,\Gamma(\mu)} \cos\pi(z - \tfrac{1}{2}\mu) + \ldots, \tag{15.7.80}$$

and from (15.7.78)

$$E_{2n}^{(\mu)}(z) = (-1)^n \frac{(2n)!\,2^{2\mu}}{\pi^{2n+\mu}\,\Gamma(\mu)} \frac{\Gamma(n+\mu)}{n!} \cos\pi(z - \tfrac{1}{2}\mu) + \ldots, \tag{15.7.81}$$

and we see that the first approximations give the same asymptotic estimates.

Remark 15.3. The expansions in (15.7.78) do not vanish when μ is a negative integer, as the expansion in (15.7.67) does. We have when $\mu = -m = 0, -1, -2, \ldots$,

$$\Psi_k^{(2n)} = \begin{cases} \pi^{2k-2m-2n} \binom{k-m}{n}, & k \geq n+m, \\ 0, & k < n+m. \end{cases} \tag{15.7.82}$$

Hence, the summation in (15.7.78) starts with $k = n + m$. △

15.8 Coefficients of expansions of the ${}_1F_1$-function

We consider expansions of the Kummer ${}_1F_1$-function, see Chapter 10, in terms of the Bessel functions. Tricomi (1947, §1.8) has given two expansions in terms of Bessel functions. Buchholz (1969, §7.4) has given a third example. In §15.8.1 we consider the asymptotic behavior of the coefficients of one of Tricomi's expansions, and in §15.8.2 those of Buchholz's expansion.[2]

15.8.1 *Coefficients of Tricomi's expansion*

This expansion reads

$${}_1F_1\left(\begin{matrix} a \\ c \end{matrix}; z\right) = \Gamma(c) e^{\frac{1}{2}z} \sum_{n=0}^{\infty} 2^{-n} z^n A_n(a,c) E_{c-1+n}(\kappa z), \tag{15.8.83}$$

where $E_\nu(z)$ is introduced in §10.3.1, and can be expressed in terms of the J- or I-Bessel functions, and

$$\kappa = \tfrac{1}{2}c - a. \tag{15.8.84}$$

The first few coefficients A_n are given by

$$A_0(a,c) = 1, \ A_1(a,c) = 0, \ A_2(a,c) = \tfrac{1}{2}c, \ A_3(a,c) = -\tfrac{1}{3}(c - 2a). \tag{15.8.85}$$

They satisfy the recurrence relation

$$(n+1)A_{n+1} = (n+c-1)A_{n-1} - 2\kappa A_{n-2}, \quad n \geq 2, \tag{15.8.86}$$

and have the generating function

$$e^{2\kappa w}(1-w)^{\kappa - \frac{1}{2}c}(1+w)^{-\kappa - \frac{1}{2}c} = \sum_{n=0}^{\infty} A_n(a,c) w^n, \quad |w| < 1. \tag{15.8.87}$$

When $c = 2a$, that is, $\kappa = 0$, the right-hand side of (15.8.83) becomes the power series expansion of the modified Bessel function. This follows from $E_{2a-1+n}(0) = 1/\Gamma(n+2a)$ and

$$A_{2n}(a, 2a) = \frac{(a)_n}{n!}, \quad A_{2n+1}(a, 2a) = 0, \quad n = 0, 1, 2, \ldots. \tag{15.8.88}$$

[2]This section is based on López and Temme (2010a).

This gives

$$ {}_1F_1\left(\begin{matrix} a \\ 2a \end{matrix};z\right) = \Gamma\left(a+\tfrac12\right) e^{\frac12 z} \left(\tfrac14 z\right)^{\frac12 - a} I_{a-\frac12}\left(\tfrac12 z\right), \tag{15.8.89} $$

which is a known identity for the Kummer function, as we have seen in (10.1.13).

Remark 15.4. The expansions in (15.8.83) is in fact for the ratio ${}_1F_1(a;c;z)/\Gamma(c)$, which is well defined for $c=-m$, $m=0,1,2,\ldots$; see (10.1.3). △

Remark 15.5. The behavior of the coefficients $A_n(a,c)$ will become completely different when in (15.8.87) $\kappa - \frac12 c = N$ and $-\kappa - \frac12 c = M$, where M and N are nonnegative integers. That is, when $a=-N$ (Laguerre case, see (10.1.17)) and $c = M - N$. For details on this case we refer to López and Temme (2010a), where also an alternative expansion with a large-n asymptotic property is given. △

Remark 15.6. The expansion in (15.8.83) is mentioned in Abramowitz and Stegun (1964, p. 506) and Slater (1960, §3.8), and Slater also gives a proof. Analytical aspects of the expansion in (15.8.83) are discussed in Chiccoli *et al.* (1989). Tricomi's expansion (15.8.83) is used in Maino *et al.* (1981) for the computation of parabolic cylinder functions. △

To obtain an asymptotic expansion we use the Cauchy integral

$$ A_n(a,c) = \frac{1}{2\pi i}\int_{\mathcal{C}} e^{(p-q)w}(1-w)^p(1+w)^q \frac{dw}{w^{n+1}}, \tag{15.8.90} $$

where

$$ p=-a, \quad q=a-c, \tag{15.8.91} $$

and $\mathcal{C}$ is a circle around the origin with radius less than unity.

We proceed as for the Bernoulli and Euler polynomials. The loops around the branch cuts are denoted by $\mathcal{L}_\pm$ and the contributions from the loops by $I_\pm$.

For the loop around the singular point $w=1$ we substitute $w=e^s$, and obtain

$$ I_+ = \frac{2^q e^{p-q}}{2\pi i}\int_{\mathcal{C}_+} g(s)\left(se^{\pi i}\right)^p e^{-ns}\,ds, \tag{15.8.92} $$

where

$$ g(s) = \left(\frac{e^s-1}{s}\right)^p \left(\frac{e^s+1}{2}\right)^q e^{(p-q)(e^s-1)}, \tag{15.8.93} $$

and $\mathcal{C}_+$ is the image of $\mathcal{L}_+$. $\mathcal{C}_+$ is a contour that encircles the origin clockwise. For the asymptotic analysis we extend $\mathcal{C}_+$ to $+\infty$. That is, we start the integration along the contour $\mathcal{C}_+$ at $s=+\infty$, with $\operatorname{ph} s = 2\pi$, turn around the origin clockwise, and return to $+\infty$ with $\operatorname{ph} s = 0$.

We apply Watson's lemma for loop integrals, see §2.2. We expand

$$g(s) = \sum_{k=0}^{\infty} g_k(p,q)s^k, \tag{15.8.94}$$

and substitute this into (15.8.92). This gives

$$I_+ \sim \frac{2^q e^{p-q}}{2\pi i} \sum_{k=0}^{\infty} g_k(p,q) F_k, \quad F_k = \frac{1}{2\pi i}\int_{\mathcal{C}_+} \left(se^{\pi i}\right)^p s^k e^{-ns}\, ds. \tag{15.8.95}$$

To evaluate the integrals we rotate the path by writing $s = e^{-\pi i}t$, and use the representation of the reciprocal gamma function in terms of the Hankel contour given in (2.2.17). The result is

$$F_k = n^{-p-k-1}\frac{(-1)^k}{\Gamma(-p-k)} = n^{-p-k-1}\frac{(1+p)_k}{\Gamma(-p)}. \tag{15.8.96}$$

This gives the expansion

$$I_+ \sim \frac{2^q e^{p-q}}{n^{p+1}\Gamma(-p)} \sum_{k=0}^{\infty} \frac{(1+p)_k}{n^k}\, g_k(p,q), \quad n\to\infty. \tag{15.8.97}$$

The expansion of I_-, the contribution from the loop around the branch point $w=-1$, follows in a similar manner. It also follows by using the substitution $w\to -w$ in (15.8.90) and observing the symmetry with the previous case: change the sign of κ (which involves interchanging p and q), and include a factor $(-1)^n$. This gives

$$I_- \sim (-1)^n \frac{2^p e^{q-p}}{n^{q+1}\Gamma(-q)} \sum_{k=0}^{\infty} \frac{(1+q)_k}{n^k}\, g_k(q,p), \quad n\to\infty. \tag{15.8.98}$$

These two expansions give the compound expansion of A_n:

$$A_n(a,c) \sim I_+ + I_-, \quad n\to\infty, \tag{15.8.99}$$

and this holds for fixed values of p and q (or κ and c).

The first few coefficients $g_k(p,q)$ are

$$\begin{aligned} g_0(p,q) &= 1, \quad g_1(p,q) = \tfrac{1}{2}\left(3p-q\right), \\ g_2(p,q) &= \tfrac{1}{24}\left(13p-9q+3(3p-q)^2\right), \\ g_3(p,q) &= \tfrac{1}{48}\left(8(p-q)+(3p-q)(13p-9q)+(3p-q)^3\right), \\ g_4(p,q) &= \tfrac{1}{5760}\left(30(13p-9q)(3p-q)^2+15(3p-q)^4 + \right. \\ &\qquad \left. 238p-270q-3090pq+2285p^2+885q^2\right). \end{aligned} \tag{15.8.100}$$

15.8.2 *Coefficients of Buchholz's expansion*

We use the function $E_\nu(z)$ as introduced in §10.3.1.[3] Then, Buchholz's expansion reads (see Buchholz (1969, §7.4))

$$ {}_1F_1\left(\begin{matrix}a\\c\end{matrix};z\right)=\Gamma(c)e^{\frac12 z}\sum_{n=0}^{\infty}2^{-n}z^nP_n(c;z)E_{c-1+n}(\kappa z), \tag{15.8.101}$$

where $\kappa=\frac12 c-a$, the same as used in (15.8.83). The first few coefficients P_n are given by

$$\begin{aligned} &P_0(c;z)=1,\quad P_1(c;z)=\tfrac16 z,\quad P_2(c;z)=\tfrac{1}{72}\left(z^2+12c-24\right),\\ &P_3(c;z)=\tfrac{1}{6480}z\left(5z^2+180c-432\right).\end{aligned} \tag{15.8.102}$$

The $P_n(c;z)$ are polynomials of degree n in z and satisfy the recurrence relation (Abad and Sesma, 1999) (we write $P_n(z)=P_n(c;z)$)

$$\begin{aligned} P_n(z)&=z^{-\frac12 n}\int_0^z u^{\frac12 n-1}Q_n(u)\,du,\\ Q_n(u)&=\tfrac14 uP_{n-1}(u)+(c-2)P'_{n-1}(u)-uP''_{n-1}(u),\end{aligned} \tag{15.8.103}$$

where $n\ge 1$.

The polynomials $P_n(c;z)$ have the generating function

$$ e^{-\frac12 z(\coth w-1/w)}\left(\frac{\sinh w}{w}\right)^{c-2}=\sum_{n=0}^{\infty}(-1)^nP_n(c;z)w^n, \tag{15.8.104}$$

where $|w|<\pi$. Recall that in the Laguerre case, see §15.5, the generating function has an essential singularity, and in the present case there are essential singularities at $\pm i\pi$.

From the expansion in (15.8.104) it follows that

$$ P_n(c;-z)=(-1)^nP_n(c;z). \tag{15.8.105}$$

We observe that for $z=0$ we have

$$ P_{2n+1}(c;0)=0,\quad P_{2n}(c;0)=2^{2n}\frac{B_{2n}^{(-\mu)}(-\frac12\mu)}{(2n)!},\quad \mu=c-2, \tag{15.8.106}$$

where $B_n^{(\mu)}(z)$ is the generalized Bernoulli polynomial considered in §15.6.

Remark 15.7. In Abad and Sesma (1999, 1997) Buchholz' expansion is modified to give an asymptotic expansion of the U-Kummer function in terms of K-Bessel functions. In Abad and Sesma (1999) several properties of the polynomials $P_n(c;z)$ are discussed, including the useful recurrence relation in (15.8.103). △

---

[3]This section is based on López and Temme (2010a).

To obtain an asymptotic expansion we use the Cauchy integral

$$(-1)^n P_n(c;z) = \frac{1}{2\pi i}\int_{\mathcal{C}} e^{-\frac{1}{2}z(\coth w - 1/w)} \left(\frac{\sinh w}{w}\right)^{c-2} \frac{dw}{w^{n+1}}, \tag{15.8.107}$$

where $\mathcal{C}$ is a circle around the origin with radius smaller than π. The singularities of the integrand of (15.8.107) nearest to the origin occur at $w = \pm\pi i$, and because of the $\coth w$ in the exponential function we have essential singularities. As in the case of Laguerre polynomials we need Bessel functions for the asymptotic form.

We first transform $w \to iw$, because we like to have the singularities on the real line, as for $A_n(a,c)$ in §15.8.1. We obtain

$$P_n(c;z) = i^n P_n^{(\mu)}(iz), \tag{15.8.108}$$

where

$$P_n^{(\mu)}(z) = \frac{1}{2\pi i}\int_{\mathcal{C}} e^{z(\cot w - 1/w)/2} \left(\frac{\sin w}{w}\right)^{\mu} \frac{dw}{w^{n+1}}, \quad \mu = c-2. \tag{15.8.109}$$

We proceed as for the Bernoulli and Euler polynomials. The loops around the branch cuts from $\pm\pi$ to $\pm\infty$ are referred to as $\mathcal{L}_\pm$ and the corresponding contributions as $I_\pm$.

For the singularity at $+\pi$ we substitute $w = \pi e^s$. We have, as $s \to 0$,

$$\begin{aligned} \sin w &= \sin\pi\,(1-e^s) = -\pi s - \tfrac{1}{2}\pi s^2 + \mathcal{O}(s^3), \\ \cot w &= \cot\pi\,(e^s - 1) = \frac{1}{\pi s} - \frac{1}{2\pi} + \mathcal{O}(s). \end{aligned} \tag{15.8.110}$$

After these preparations we can write

$$I_+ = \frac{e^{-3z/(4\pi)}\pi^{-n}}{2\pi i}\int_{\mathcal{L}_+} e^{\phi(s)} f(s)(-s)^{\mu} e^{z/(2\pi s) - ns}\, ds, \tag{15.8.111}$$

where

$$\begin{aligned} \phi(s) &= \tfrac{1}{2}z\left(\cot\pi\,(e^s-1) - \frac{1}{\pi s} + \frac{1}{2\pi} - \frac{1}{\pi}\left(e^{-s}-1\right)\right), \\ f(s) &= \left(\frac{\sin\pi(1-e^s)}{-s\pi e^s}\right)^{\mu}. \end{aligned} \tag{15.8.112}$$

By writing $s = e^{\pi i}t$ we rotate the contour onto the negative real axis and obtain

$$I_+ = \frac{e^{-3z/(4\pi)}\pi^{-n}}{2\pi i}\int_{\mathcal{L}} g(t) t^{\mu} e^{-z/(2\pi t) + nt}\, dt, \tag{15.8.113}$$

where $\mathcal{L}$ is the Hankel contour and $g(t) = e^{\phi(-t)} f(-t)$. That is,

$$\begin{aligned} g(t) = \exp\left(-\tfrac{1}{2}z\left(\cot\pi\,(1-e^{-t}) - \frac{1}{\pi t} - \frac{1}{2\pi} + \frac{1}{\pi}\left(e^{t}-1\right)\right)\right) \times \\ \left(\frac{e^t \sin\pi\,(1-e^{-t})}{t\pi}\right)^{\mu}. \end{aligned} \tag{15.8.114}$$

We use integration by parts and the integral representation of the J-Bessel function given in (10.3.24), with notation (see also (10.3.22))

$$\Phi_\mu(z) = \left(\frac{z}{2\pi n}\right)^{(\mu+1)/2} J_{-\mu-1}\left(\sqrt{2nz/\pi}\right) = n^{-\mu-1} E_{-\mu-1}\left(\frac{nz}{2\pi}\right). \tag{15.8.115}$$

This shows that $\Phi_\mu(z)$ is an analytic function of its argument z.

We write

$$g(t) = g(0) + g'(0)t + t^2 h(t), \tag{15.8.116}$$

where the derivative is with respect to t. Substituting this in (15.8.113) we obtain, after integrating by parts,

$$\begin{aligned} I_+ = \frac{e^{-3z/(4\pi)}}{\pi^n} \Biggl(& g(0)\Phi_\mu(z) + g'(0)\Phi_{\mu+1}(z) + \\ & \frac{1}{2\pi i\, n} \int_{\mathcal{L}} g_1(t) t^\mu e^{-z/(2\pi t)+nt}\, dt \Biggr), \end{aligned} \tag{15.8.117}$$

where

$$g_1(t) = -\left(t^2 h'(t) + \frac{z}{2\pi} h(t) + (\mu+2) t h(t)\right). \tag{15.8.118}$$

Continuing this procedure, we obtain

$$I_+ \sim \frac{e^{-3z/(4\pi)}}{\pi^n} \left(\Phi_\mu(z) \sum_{k=0}^\infty \frac{A_k(\mu,z)}{n^k} + \Phi_{\mu+1}(z) \sum_{k=0}^\infty \frac{B_k(\mu,z)}{n^k} \right), \tag{15.8.119}$$

where $A_k(\mu,z) = g_k(0)$ and $B_k(\mu,z) = g_k'(0)$ and the functions $g_k(t)$ are defined by

$$g_{k+1}(t) = -\left(t^2 h_k'(t) + \frac{z}{2\pi} h_k(t) + (\mu+2) t h_k(t)\right), \tag{15.8.120}$$

for $k = 0, 1, 2, \ldots$, with $g_0(t) = g(t)$. The functions $h_k(t)$ follow from

$$g_k(t) = g_k(0) + g_k'(0)t + t^2 h_k(t), \quad k = 0, 1, 2, \ldots. \tag{15.8.121}$$

For the singularity at $-\pi$ we can use a symmetry argument by observing that we can transform in (15.8.109) $w \to -w$, which is equivalent to changing the sign of z and adding a front factor $(-1)^n$. This gives

$$\begin{aligned} & I_- \sim (-1)^n \frac{e^{3z/(4\pi)}}{\pi^n} \times \\ & \left(\Phi_\mu(-z) \sum_{k=0}^\infty \frac{A_k(\mu,-z)}{n^k} + \Phi_{\mu+1}(-z) \sum_{k=0}^\infty \frac{B_k(\mu,-z)}{n^k} \right). \end{aligned} \tag{15.8.122}$$

The first few coefficients in the expansion (15.8.119) are given by

$$A_0(\mu,z) = 1, \quad B_0(\mu,z) = \frac{1}{24\pi}\left(12\mu\pi - 13z + 4z\pi^2\right), \tag{15.8.123}$$

$$\begin{aligned} A_1(\mu,z) = & -\frac{z}{2304\pi^3} \left(-96z\pi^3 - 288z\pi - 104z^2\pi^2 - 312z\mu\pi \right. \\ & \left. +169z^2 + 16z^2\pi^4 + 96z\pi^3\mu + 144\mu^2\pi^2 - 192\mu\pi^4 + 48\mu\pi^2\right). \end{aligned} \tag{15.8.124}$$

Using (15.8.119) and (15.8.122), we obtain

$$P_n^{(\mu)}(z) = \frac{e^{-3z/(4\pi)}}{\pi^n} Q_n^{(\mu)}(z) + (-1)^n \frac{e^{3z/(4\pi)}}{\pi^n} Q_n^{(\mu)}(-z), \tag{15.8.125}$$

where

$$Q_n^{(\mu)}(z) \sim \Phi_\mu(z) \sum_{k=0}^{\infty} \frac{A_k(\mu, z)}{n^k} + \Phi_{\mu+1}(z) \sum_{k=0}^{\infty} \frac{B_k(\mu, z)}{n^k}, \tag{15.8.126}$$

and $\Phi_\mu(z)$ is defined in (15.8.115). This expansion holds for $n \to \infty$, uniformly with respect to bounded complex z and μ. In particular the expansion is valid for small z. When z is bounded away from 0 we can use asymptotic expansions of the Φ-functions, that is, of the Bessel functions (see (15.8.115)) for large argument.

For the expansion in (15.8.101) we need the relation in (15.8.108), and we have

$${}_1F_1\left(\begin{matrix} a \\ c \end{matrix}; z\right) = \Gamma(c) e^{\frac{1}{2}z} \sum_{n=0}^{\infty} 2^{-n} (iz)^n P_n^{(\mu)}(iz) E_{c-1+n}(\kappa z). \tag{15.8.127}$$

Chapter 16

Mellin–Barnes integrals and Mellin convolution integrals

In §2.5 we have given examples in which the idea of Watson's lemma can be used for integrals of the form

$$F_\lambda(x) = \int_0^\infty t^{\lambda-1} h(xt) f(t)\, dt. \qquad (16.0.1)$$

This integral reduces to the standard form of Watson's lemma when the kernel $h(t) = e^{-t}$. For different kernels we have explained how the method can be modified. We have observed that in some cases the moments

$$\int_0^\infty t^{\lambda+n-1} h(xt)\, dt, \qquad (16.0.2)$$

may not exist for all n. By using certain analytic properties of special kernels, such as $J_\nu(t)$, it is still possible to obtain an asymptotic expansion. However, a more systematic approach will be welcome.

In this chapter we explain how certain integrals of the type (16.0.1) can be expanded for large (or small) values of the parameter x. We need a few conditions on f, which will be given later, but first we give an idea how the method works by giving an example.

First we introduce the Mellin transform of a function.

Definition 16.1. Let f be locally integrable on $(0, \infty)$. We denote by $M[f; z]$ the Mellin transform of f:

$$M[f; z] = \int_0^\infty t^{z-1} f(t)\, dt, \qquad (16.0.3)$$

(when this integral exists) or its analytic continuation as a function of z. The inverse transform is

$$f(t) = \frac{1}{2\pi i} \int_{c-i\infty}^{c+i\infty} t^{-z} M[f; z]\, dz, \qquad (16.0.4)$$

where c is such that the vertical line of integration lies inside the z-domain where $M[f; z]$ is defined. For conditions on f we refer to the literature (see Roy *et al.* (2010, §1.14(iv))).

16.1 Mellin–Barnes integrals

To describe the use of Mellin–Barnes integrals we consider

$$F_\nu(x) = \int_0^\infty \frac{J_\nu(xt)}{1+t}\,dt, \quad \Re\nu > -1, \tag{16.1.5}$$

and we shall derive expansions for large and small values of x. We have (see Olver and Maximon (2010, Eq. 10.22.43))

$$\int_0^\infty t^{z-1} J_\nu(t)\,dt = \frac{2^{z-1}\Gamma\left(\frac12\nu+\frac12 z\right)}{\Gamma\left(1+\frac12\nu-\frac12 z\right)}, \quad \Re(z+\nu)>0, \quad \Re z < \tfrac32, \tag{16.1.6}$$

with inverse

$$J_\nu(t) = \frac{1}{2\pi i}\int_{c-i\infty}^{c+i\infty} t^{-z}\frac{2^{z-1}\Gamma\left(\frac12\nu+\frac12 z\right)}{\Gamma\left(1+\frac12\nu-\frac12 z\right)}\,dz, \quad c+\Re\nu>0. \tag{16.1.7}$$

Usually the Mellin transform is defined inside a strip in the complex plane, as indicated by the conditions in (16.1.6), where the first condition comes from the convergence at $t=0$ and the second one from convergence at ∞. But the right-hand side of (16.1.6) is analytic for $\Re(z+\nu)>0$, and we can choose c without regard for the condition $\Re z < \frac32$.

We substitute the representation of the Bessel function into (16.1.5) and obtain, after interchanging the order of integration,

$$F_\nu(x) = \frac{1}{2\pi i}\int_{c-i\infty}^{c+i\infty} \frac{2^{z-1}\Gamma\left(\frac12\nu+\frac12 z\right)}{\Gamma\left(1+\frac12\nu-\frac12 z\right)}\left(\int_0^\infty \frac{t^{-z}}{1+t}\,dt\right) x^{-z}\,dz. \tag{16.1.8}$$

The inner integral is again a Mellin transform, and in this case we can find an explicit form of it by using the beta integral (see (6.5.67)) in the form

$$B(p,q) = \int_0^\infty t^{p-1}(1+t)^{-p-q}\,dt, \tag{16.1.9}$$

where $\Re p>0$ and $\Re q>0$. This gives

$$\int_0^\infty \frac{t^{-z}}{1+t}\,dt = \Gamma(1-z)\Gamma(z) = \frac{\pi}{\sin\pi z}, \quad 0<\Re z<1, \tag{16.1.10}$$

and

$$F_\nu(x) = \frac{1}{2i}\int_{c-i\infty}^{c+i\infty} \frac{2^{z-1}\Gamma\left(\frac12\nu+\frac12 z\right)}{\Gamma\left(1+\frac12\nu-\frac12 z\right)}\frac{x^{-z}\,dz}{\sin\pi z}, \quad 0<c<1. \tag{16.1.11}$$

The asymptotic expansion for large positive x can be obtained by shifting the contour to the right and picking up the residues at the poles $z=1,2,3,\ldots$. This gives

$$F_\nu(x) \sim \sum_{n=1}^\infty (-1)^{n+1}2^{n-1}\Gamma\left(\tfrac12\nu+\tfrac12 n\right)\Gamma\left(1+\tfrac12\nu-\tfrac12 n\right)x^{-n}, \tag{16.1.12}$$

as $x\to+\infty$. By estimating the remainders with a finite number of terms it can be shown that this is a valid asymptotic expansion for $x\to+\infty$.

An expansion for small x follows from shifting the contour to the left. In that case we have two series of poles: at $z = 0, -1, -2, \ldots$ and at $z = -\nu - 2n$ ($n = 0, 1, 2, \ldots$). Some of these poles coincide when ν is an integer, and in that case we have to calculate the residues from a double pole. This gives

$$F_\nu(x) \sim \tfrac{1}{2}\sum_{n=0}^{\infty}\left(-\tfrac{1}{2}x\right)^n \frac{\Gamma\left(\frac{1}{2}\nu - \frac{1}{2}n\right)}{\Gamma\left(1 + \frac{1}{2}\nu + \frac{1}{2}n\right)} - \frac{\pi}{\sin \pi\nu} J_\nu(x), \tag{16.1.13}$$

where we have used the series representation of the Bessel function:

$$J_\nu(z) = \left(\tfrac{1}{2}z\right)^\nu \sum_{n=0}^{\infty} \frac{\left(-\frac{1}{4}z^2\right)^n}{n!\,\Gamma(1 + \nu + n)}. \tag{16.1.14}$$

The other series in (16.1.13) is convergent as well. When ν is a nonnegative integer, we need a limiting process in (16.1.13), and series with logarithmic terms $\ln x$ occur. When ν is a negative integer we can use in (16.1.5) the relation $J_{-n}(x) = (-1)^n J_n(x)$. For further details on this example, see Paris and Kaminski (2001, §5.1.3).

We conclude from this example

- When the Mellin transforms of the functions f and h of (16.0.1) are known, we may be able to obtain a different integral representation of the function $F_\lambda(x)$ as a contour integral in the complex plane in terms of gamma functions.
- Two types of asymptotic (perhaps convergent) expansions can be obtained.
- In certain cases the expansion contains logarithmic terms of the large or small variable.

The expansion for small x and for large x follow from one method because we can write (16.0.1) in the form

$$F_\lambda(x) = x^{-\lambda}\int_0^\infty s^{\lambda-1} h(t) f(t/x)\,dt. \tag{16.1.15}$$

We see that large and small values of x can be interchanged, if f and h allow.

Integrals of the type used in (16.1.7) and (16.1.11) are called *Mellin–Barnes integrals*, and they play an important role in defining certain hypergeometric functions, in particular, the Meijer G-function (see Askey and Olde Daalhuis (2010, §16.17)). Their use in asymptotic analysis is extensively described in Paris and Kaminski (2001). The integrals in (16.1.7) and (16.1.11) are examples of Meijer G-functions.

We will show in the next section how we can obtain these expansions without needing explicit forms of the Mellin transforms and how logarithmic terms follow from the behavior of the functions f and h at the origin or at infinity. We describe the method as given in López (2008). See also the pioneering approach in Bleistein and Handelsman (1975, Chapters 4, 6) and the approach in Wong (2001, Chapter VI), with a summary in Olver and Wong (2010, §§2.5,2.6).

For applications to Stieltjes and Hilbert transforms we again refer to Wong (2001, Chapter VI). For an application to Appell's function F_2, see Garcia and López (2010).

16.2 Mellin convolution integrals

Mellin convolution integrals are of the form[1]

$$I(x) = \int_0^\infty f(t)h(xt)\,dt, \tag{16.2.16}$$

where f and h belong to function classes described in the following definitions.

Definition 16.2. We denote by $\mathcal{F}$ the set of locally integrable functions on $(0,\infty)$ satisfying:

(*i*) f has an asymptotic expansion at infinity:

$$f(t) = \sum_{k=0}^{n-1} \frac{a_k}{t^{\alpha_k}} + f_n(t), \qquad n = 1,2,3,\ldots, \tag{16.2.17}$$

where $\{a_k\}$ and $\{\alpha_k\}$ are sequences of complex and real numbers, respectively, with α_k strictly increasing and $f_n(t) = \mathcal{O}(t^{-\alpha_n})$ as $t \to \infty$.

(*ii*) $f(t) = \mathcal{O}(t^{-a})$ as $t \to 0^+$ with $a \in \mathbb{R}$.

Definition 16.3. We denote by $\mathcal{H}$ the set of locally integrable functions on $(0,\infty)$ satisfying:

(*i*) h has an asymptotic expansion at $t = 0^+$:

$$h(t) = \sum_{k=0}^{n-1} b_k t^{\beta_k} + h_n(t), \quad n = 1,2,3,\ldots, \tag{16.2.18}$$

where $\{b_k\}$ and $\{\beta_k\}$ are sequences of complex and real numbers respectively, with β_k strictly increasing and $h_n(t) = \mathcal{O}(t^{\beta_n})$ as $t \to 0^+$.

(*ii*) $h(t) = \mathcal{O}\left(t^{-b}\right)$ when $t \to \infty$ with $b \in \mathbb{R}$.

Remark 16.1. In the foregoing discussion we require for the parameters a, b, α_0 and β_0 to satisfy the following relations:

Condition I : $a - \beta_0 < 1 < b + \alpha_0$.
Condition II : $-\beta_0 < b$ and $a < \alpha_0$.

The first condition ensures convergence of the integral in (16.2.16). △

We define $\alpha_{-1} = a$ and $\beta_{-1} = -b$ and observe that $\alpha_{-1} < \alpha_0$ and $\beta_{-1} < \alpha_0$.

Lemma 16.1. *For every nonnegative integer n there exists a nonnegative integer m such that $\alpha_{n-1} - \beta_m < 1 < \alpha_n - \beta_{m-1}$.*

[1] This section is based on López (2008), where proofs of the lemma and theorems of this section can be found. The author appreciates the contributions from José Luis López.

The main results of this section are stated in the following two theorems.

Theorem 16.1. *Let $f \in \mathcal{F}$ and $h \in \mathcal{H}$. Then, for every n, $m \in \mathbb{N}$ such that $\alpha_{n-1} - \beta_m < 1 < \alpha_n - \beta_{m-1}$,*

$$\begin{aligned}\int_0^\infty h(xt)f(t)\,dt = \sum_{k=0}^{n-1} a_k M[h; 1-\alpha_k]x^{\alpha_k-1} + \sum_{j=0}^{m-1} b_j M[f;\beta_j+1]x^{\beta_j} + \\ \int_0^\infty f_n(t)h_m(xt)\,dt.\end{aligned} \tag{16.2.19}$$

If $\alpha_k - \beta_j = 1$ for some couple (k, j) then, in this formula, the sum of terms

$$a_k M[h; 1-\alpha_k]x^{\alpha_k-1} + b_j M[f;\beta_j+1]x^{\beta_j} \tag{16.2.20}$$

must be replaced by

$$\begin{aligned}&\lim_{z\to 0}\left(x^{\beta_j}\left(a_k x^{-z} M[h;1+z-\alpha_k] + b_j M[f;z+\beta_j+1]\right)\right) = \\ &x^{\beta_j}\left(\lim_{z\to 0}\left(a_k M[h;1+z-\alpha_k] + b_j M[f;z+\beta_j+1]\right) - a_k b_j \ln x\right).\end{aligned} \tag{16.2.21}$$

Theorem 16.2. *With the hypothesis of Theorem 16.1, the expansion* (16.2.19) *is an asymptotic expansion for small x. If $\alpha_n \neq \beta_m + 1$, then*

$$\int_0^\infty f_n(t)h_m(xt)\,dt = \mathcal{O}(x^{\beta_m} + x^{\alpha_n-1}), \quad x \to 0; \tag{16.2.22}$$

if $\alpha_n = \beta_m + 1$, then

$$\int_0^\infty f_n(t)h_m(xt)\,dt = \mathcal{O}(x^{\beta_m}\ln x), \quad x \to 0. \tag{16.2.23}$$

Remark 16.2. The above theorem also applies to integrals of the form $\int_c^\infty h(xt)f(t)\,dt$ with $0 < c < \infty$ and with the same hypotheses for f and h except the one regarding the asymptotic behavior of f at $t = 0$. We write

$$\int_c^\infty h(xt)f(t)\,dt = \int_0^\infty h(xt)f_c(t)\,dt, \tag{16.2.24}$$

with $f_c(t) = f(t)\chi_{(c,\infty)}(t)$, $\chi_{(c,\infty)}(t)$ being the characteristic function of the interval $(0,\infty)$: $\chi_{(c,\infty)}(t) = 1$ if $t \in (c,\infty)$ and $\chi_{(c,\infty)}(t) = 0$ if $t \notin (c,\infty)$. In this case we have $f_c(t) = \mathcal{O}(t^{-a})$ as $t \to 0^+$ with $a < 0$ and $|a|$ as large as we wish. The theorem also applies to integrals of the form $\displaystyle\int_c^d h(xt)f(t)\,dt$ by writing:

$$\int_c^d h(xt)f(t)\,dt = \int_c^\infty h(xt)f(t)\,dt - \int_d^\infty h(xt)f(t)\,dt, \tag{16.2.25}$$

and using the earlier understanding. △

16.3 Error bounds

Let the remainder $f_n(t)$ in the expansion (16.2.17) satisfy the bound $|f_n(t)| \leq F_n t^{-\alpha_n}$ for all $t \in (0,\infty)$ and the remainder $h_m(t)$ in (16.2.18) satisfy the bound $|h_m(t)| \leq H_m t^{\beta_m}$ for all $t \in (0,\infty)$ for some positive constants F_n and H_m.

If $\beta_m + 1 \neq \alpha_n$, then

$$\left|\int_0^\infty f_n(t) h_m(xt)\, dt\right| \leq \begin{cases} C^1_{n,m} x^{\alpha_n - 1} \text{ if } & \beta_m > \alpha_n - 1, \\ C^2_{n,m} x^{\beta_m} \quad \text{ if } & \beta_m < \alpha_n - 1, \end{cases} \tag{16.3.26}$$

with

$$C^1_{n,m} = F_n \left(\frac{H_m}{1 + \beta_m - \alpha_n} + \frac{|b_{m-1}| + H_{m-1}}{\alpha_n - \beta_{m-1} - 1} \right) \tag{16.3.27}$$

and

$$C^2_{n,m} = H_m \left(\frac{F_n}{\alpha_n - \beta_m - 1} + \frac{|a_{n-1}| + F_{n-1}}{\beta_m + 1 - \alpha_{n-1}} \right). \tag{16.3.28}$$

If $\beta_m + 1 = \alpha_n$, then

$$\left|\int_0^\infty f_n(t) h_m(xt)\, dt\right| \leq \left(C^3_{n,m} + F_n H_m |\ln x|\right) x^{\beta_m}, \tag{16.3.29}$$

with

$$C^3_{n,m} = F_n \frac{|b_{m-1}| + H_{m-1}}{\alpha_n - \beta_{m-1} - 1} + H_m \frac{|a_{n-1}| + F_{n-1}}{\beta_m + 1 - \alpha_{n-1}}. \tag{16.3.30}$$

Chapter 17

Alternative expansions of Laplace-type integrals

In this chapter we mention expansions that can be viewed as alternatives or modifications of the expansions obtained by using Watson's lemma. Some of these expansions are convergent and have asymptotic properties.[1]

17.1 Hadamard-type expansions

We have considered the modified Bessel function $I_\nu(z)$ in §9.3 with asymptotic expansion in (9.3.14), and a complete expansion in §9.3.1. The divergence of the asymptotic expansion in (9.3.14) is due to expanding $(1-t)^{\nu-\frac{1}{2}}$ in powers of t, which expansion is valid for $|t|<1$, and replacing the integrals $\int_0^1 e^{-2zt}t^k\,dt$ by $\int_0^\infty e^{-2zt}t^k\,dt$. If, after substituting the power series, we keep integrating the integrals $\int_0^1 e^{-2zt}t^k\,dt$ over $[0,1]$, we obtain a convergent expansion (for some values of ν). The new expansion is called a *Hadamard expansion*, see Paris (2011), and is of the form

$$I_\nu(z) = \frac{e^z}{\sqrt{2\pi z}}\sum_{k=0}^{\infty}\frac{b_k(\nu)}{(2z)^k}, \quad b_k(\nu) = \frac{(\frac{1}{2}-\nu)_k(\frac{1}{2}+\nu)_k}{k!}\,P\left(\tfrac{1}{2}+\nu+k, 2z\right), \quad (17.1.1)$$

and $P(a,z)=\gamma(a,z)/\Gamma(a)$ is the normalized incomplete gamma function, see Chapter 7,

$$P(a,z) = \frac{1}{\Gamma(a)}\int_0^z e^{-t}t^{a-1}\,dt, \quad \Re a > 0. \quad (17.1.2)$$

The incomplete gamma functions can be computed by using the recurrence relation

$$P(a+1,z) = P(a,z) - \frac{z^a e^{-z}}{\Gamma(a+1)}, \quad (17.1.3)$$

[1]This chapter is based on Gil *et al.* (2007, §2.4.4).

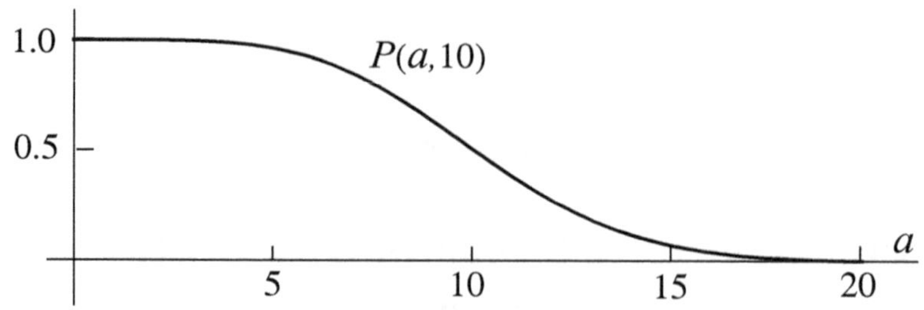

Fig. 17.1 Graph of the incomplete gamma function $P(a, x)$ as a function of a for $x = 10$.

but computation in the forward direction is not stable; see Gil *et al.* (2012) for computing a sequence of these functions by using a backward recursion scheme.

For a graph of $P(a, z)$ with $z = x = 10$ and $0 \le a \le 20$, see Figure 17.1. For fixed values of z this function has the asymptotic behavior

$$P(a, z) = \frac{e^{-z} z^a}{\Gamma(a+1)} \left(1 + \mathcal{O}(a^{-1})\right), \quad a \to \infty, \tag{17.1.4}$$

which follows from the representation in (7.3.13). Hence, the terms in (17.1.1) behave like $\mathcal{O}(k^{-\nu-3/2})$, and convergence of (17.1.1) is guaranteed if $\Re\nu > -\frac{1}{2}$.

The presence of the term $P\left(\frac{1}{2} + \nu + k, 2z\right)$ in (17.1.1) acts as a "smoothing" factor on the coefficients $b_k(\nu)$, since the behavior of $P(a, z)$ when z is large is characterized by a transition point at $z = a$. When the variables are positive, the asymptotic behavior of $P(a, z)$ changes from approximately unity when $a \le z$ to a rapid decay to zero when a becomes larger than z.

To be more specific, we have the asymptotic behavior as given in (17.1.4) for $a \gg z$ and

$$P(a, z) = 1 - \frac{e^{-z} z^{a-1}}{\Gamma(a)} \left(1 + \mathcal{O}(z^{-1})\right), \quad z \to \infty, \tag{17.1.5}$$

which follows from $P(a, z) = 1 - \Gamma(a, z)/\Gamma(a)$ and the result (7.2.9) of $\Gamma(a, z)$ in §7.2.

So, summing n terms in the Hadamard expansion (17.1.1), where $n \sim z$ as z is large, gives nearly the same result as summing n terms in the asymptotic expansion (9.3.14). But including more terms in the Hadamard expansion, trying to benefit from the fact that it is convergent, does not help very much because of the poor convergence of the Hadamard series (recall that convergence of (17.1.1) is guaranteed if $\Re\nu > -\frac{1}{2}$, and that the rate of convergence is controlled by $k^{-\nu-3/2}$).

In Paris (2004a,b) modifications of the Hadamard expansions are discussed from which much faster convergence can be obtained. The method is also used for infinite Laplace-type integrals.

17.2 An expansion in terms of Kummer functions

Next we consider an expansion of the modified Bessel function $K_\nu(z)$. We recall from (9.1.1)

$$K_\nu(z) = \frac{\sqrt{\pi}(2z)^\nu e^{-z}}{\Gamma(\nu+\frac12)} \int_0^\infty e^{-2zt} t^{\nu-\frac12}(1+t)^{\nu-\frac12}\, dt. \qquad (17.2.6)$$

We expand this function in terms of the Kummer function $U(a,c,z)$, with integral representation given in (10.1.5). By expanding in (17.2.6) the factor $(1+t)^{\nu-\frac12}$ in powers of t we get the well-known expansion as given in (9.1.3). As an alternative, we expand

$$(1+t)^{\nu-\frac12} = \sum_{k=0}^\infty c_k \left(\frac{t}{1+t}\right)^k, \quad c_k = (-1)^k \binom{\frac12-\nu}{k} = \frac{\left(\nu-\frac12\right)_k}{k!}. \qquad (17.2.7)$$

This gives a convergent expansion in terms of confluent hypergeometric functions

$$K_\nu(z) = \sqrt{\frac{\pi}{2z}} e^{-z} \sum_{k=0}^\infty \frac{(\nu+\frac12)_k(\nu-\frac12)_k}{k!} U\left(k, \tfrac12-\nu, 2z\right). \qquad (17.2.8)$$

We have

$$\begin{aligned} &U\left(0,\tfrac12-\nu,2z\right) = 1,\\ &U\left(1,\tfrac12-\nu,2z\right) = (2z)^{\nu+\frac12}\, e^{2z}\,\Gamma\left(-\tfrac12-\nu,2z\right), \end{aligned} \qquad (17.2.9)$$

again an incomplete gamma function. Other U-functions can be obtained by the recurrence relation (see Olde Daalhuis (2010a, Eq. 13.3.7))

$$U(a-1,b,z) + (b-2a-z)\,U(a,b,z) + a(a-b+1)\,U(a+1,b,z) = 0. \qquad (17.2.10)$$

For large k and $z \neq 0$ the U-function behaves like

$$k!\,U\left(k,\tfrac12-\nu,2z\right) = \mathcal{O}\left(k^\alpha\, e^{-2\sqrt{2kz}}\right), \qquad (17.2.11)$$

where α is some constant. This follows from the expansions given in §10.3.2 by assuming that z is bounded away from the origin. We see that the convergence is much better than for the Hadamard expansion of §17.1.

The confluent hypergeometric functions that occur in (17.2.8) can be computed by applying the recursion in (17.2.10) in the backward direction; see Gil *et al.* (2007, Chapter 4).

With a minor modification we can expand the Kummer function $U(a,c,z)$ in a similar series as in (17.2.8). When we take the integral in (10.1.5) and expand

$$(1+t)^{c-a-1} = \sum_{k=0}^\infty b_k \left(\frac{t}{1+t}\right)^k, \qquad (17.2.12)$$

where

$$b_k = (-1)^k \binom{a+1-c}{k} = \frac{(c-a-1)_k}{k!}, \qquad (17.2.13)$$

we obtain

$$U(a,c,z) = \sum_{k=0}^{\infty} (a)_k b_k U(a+k, a+1, z). \tag{17.2.14}$$

Using (10.1.10) we can write this in the form

$$U(a,c,z) = z^{-a} \sum_{k=0}^{\infty} \frac{(a)_k (c-a-1)_k}{k!} U(k, 1-a, z), \tag{17.2.15}$$

which may be seen as a modification of the expansion (10.2.18).

For more details on expanding Laplace-type integral in terms of Kummer U-functions we refer to Deaño and Temme (2009).

17.3 An expansion in terms of factorial series

A *factorial series* is a series of the form

$$F(z) = \sum_{n=0}^{\infty} \frac{a_n\, n!}{z(z+1)\cdots(z+n)}, \tag{17.3.16}$$

where a_n are real or complex numbers that do not depend on z. The domain of convergence is usually a half-plane $\Re z > z_0$, and coincides with the domain of convergence of the Dirichlet series $\sum_{n=1}^{\infty} a_n n^{-z}$, except for the points $0, -1, -2, \ldots$. For a proof we refer to Knopp (1964, §258).

There is a formal method to transform an asymptotic expansion

$$F(z) \sim \frac{b_1}{z} + \frac{b_2}{z^2} + \frac{b_3}{z^3} + \cdots, \quad z \to \infty, \tag{17.3.17}$$

into a factorial series. This can be done by substituting

$$\frac{1}{z^m} = \sum_{n=m}^{\infty} \frac{S_n^{n-m}}{z(z+1)\cdots(z+n)}, \tag{17.3.18}$$

and rearranging terms. Here, S_n^m is the *Stirling number of the second kind* (see Chapter 34).

This method can be used for functions defined in terms of a Laplace integral, that is,

$$F(z) = \int_0^{\infty} f(t) e^{-zt}\, dt. \tag{17.3.19}$$

When f satisfies the conditions of Watson's lemma, we can obtain an asymptotic expansion as in (17.3.17), and obtain a factorial series for $F(z)$ in a formal way.

On the other hand, we can take the new variable of integration by writing $w = e^{-t}$. This gives

$$F(z) = \int_0^1 w^{z-1} \varphi(w)\, dw, \quad \varphi(w) = f(-\ln w). \tag{17.3.20}$$

Expanding

$$\varphi(w) = \sum_{n=0}^{\infty} a_n(1-w)^n, \tag{17.3.21}$$

and substituting this series into (17.3.20), using the beta integral in (6.5.67), we again obtain (17.3.16) in a formal way.

We summarize necessary and sufficient conditions from Nielsen (1906) such that the function $F(z)$ with integral in (17.3.20) can be written as a factorial series.

The function $\varphi(w)$ should be analytic at $w = 1$ and the Maclaurin series at $w = 1$ should be convergent inside the disk $|w-1| < 1$. Next, if $w = 0$ is a singular point of φ, and $\varphi^{(k)}(w)$ is the first of all derivatives of φ that becomes infinitely large at $w = 0^+$, then there should be a real number λ such that

$$\lim_{w\downarrow 0} \left|w^{z+\lambda}\varphi^{(k)}(w)\right| = \begin{cases} 0 & \text{if } \Re z > \lambda, \\ \infty & \text{if } \Re z < \lambda. \end{cases} \tag{17.3.22}$$

In addition, there is a condition for the case that φ has other singularities on the circle $|w-1| = 1$.

Example 17.1 (Exponential integral $E_1(z)$). Consider the exponential integral in the form

$$e^z E_1(z) = \int_0^\infty \frac{e^{-zt}}{1+t}\,dt = \int_0^1 w^{z-1}\varphi(w)\,dw, \quad \varphi(w) = \frac{1}{1-\ln w}. \tag{17.3.23}$$

The function $\varphi(w)$ is analytic at $w = 1$ and the Maclaurin series at $w = 1$ converges inside the disk $|w-1| < 1$. Next, $\varphi(0)$ is bounded, and

$$\varphi'(w) = \frac{1}{w(1-\ln w)^2}, \tag{17.3.24}$$

which is infinite at $w = 0$. It is clear that the number λ of (17.3.22) equals 1. The function $\varphi(w)$ has no other singular points on the circle $|w-1| = 1$, and we conclude that we have the convergent expansion

$$e^z E_1(z) = \sum_{n=0}^{\infty} \frac{a_n\, n!}{z(z+1)\cdots(z+n)}, \tag{17.3.25}$$

where the coefficients a_n follow from the expansion (17.3.21). The first few coefficients are

$$a_0 = 1,\ a_1 = -1,\ a_2 = \tfrac{1}{2},\ a_3 = -\tfrac{1}{3},\ a_4 = \tfrac{1}{6},\ a_5 = -\tfrac{7}{60},\ a_6 = \tfrac{19}{360}. \tag{17.3.26}$$

Higher coefficients follow easily from a recurrence relation

$$a_n = -\sum_{m=0}^{n-1} \frac{a_m}{n-m}, \quad n = 1, 2, \ldots. \tag{17.3.27}$$

Table 17.1 Evaluations of the factorial series (17.3.25) with n terms to compute $ze^zE_1(z)$, for several values of n and z.

n	$z=2$	$z=5$	$z=10$	$z=25$	$z=50$
5	0.60e-03	0.71e-04	0.49e-05	0.62e-07	0.15e-08
10	0.44e-03	0.39e-05	0.34e-07	0.16e-10	0.21e-13
15	0.17e-03	0.53e-06	0.12e-08	0.37e-13	0.34e-17
20	0.91e-04	0.15e-06	0.12e-09	0.45e-15	0.44e-20
25	0.56e-04	0.51e-07	0.19e-10	0.11e-16	0.13e-22
30	0.38e-04	0.21e-07	0.39e-11	0.41e-18	0.71e-25
35	0.27e-04	0.10e-07	0.99e-12	0.23e-19	0.67e-27
40	0.20e-04	0.52e-08	0.30e-12	0.18e-20	0.97e-29
45	0.16e-04	0.29e-08	0.10e-12	0.18e-21	0.20e-30
50	0.12e-04	0.17e-08	0.38e-13	0.21e-22	0.54e-32

In Table 17.1 we give the results of computing $ze^zE_1(z)$ for several values of z by using the factorial series in (17.3.25). We give the relative errors, where we have used the Maple code (with Digits = 50) of the Kummer U-function and the relation $E_1(z) = e^{-z}U(1,1,z)$. We see that convergence for $z=2$ is too slow for practical purposes. For $z=10$ we see ten correct digits for $n=25$. As explained in §1.6 and §2.6, with the asymptotic expansion of this function we can obtain a precision of about $\exp(-10) \doteq 0.000045$. ♢

This method can also be used for other special functions, for example, for the function defined by the integral

$$F_\nu(z) = \int_0^\infty \frac{e^{-zt}}{(1+t)^\nu}\,dt = z^{\nu-1}e^z\Gamma(1-\nu,z), \tag{17.3.28}$$

an incomplete gamma function.

In the literature (see, for example, Dunster and Lutz (1991)), convergent factorial series for Bessel functions are considered as alternatives for their uniform asymptotic expansions. In Weniger (2010) transformations into factorial series are discussed, together with elements of nonlinear sequence transformations. These are also used for summing Poincaré-type asymptotic expansions. For a short introduction to sequence transformations we refer to Gil *et al.* (2007, §9.3).

17.4 The Franklin–Friedman expansion

We consider the integral

$$F_\lambda(z) = \frac{1}{\Gamma(\lambda)} \int_0^\infty t^{\lambda-1} e^{-zt} f(t)\,dt, \quad \lambda > 0, \tag{17.4.29}$$

for large values of z with suitable assumptions on f, which will be given later.

In Watson's lemma, the function f is expanded at $t = 0$; in the present method a type of interpolation process is applied to f. The first interpolation point is obtained by considering (we assume that z is positive) the t-value for which $t^\lambda e^{-zt}$ attains its maximum, that is, the point $t_0 = \lambda/z$, and we obtain

$$F_\lambda(z) = f(t_0) z^{-\lambda} + \frac{1}{\Gamma(\lambda)} \int_0^\infty t^{\lambda-1} e^{-zt} \left(f(t) - f(t_0)\right) dt. \tag{17.4.30}$$

Next, observing that

$$\frac{d}{dt} t^\lambda e^{-zt} = -\frac{z}{t}(t - t_0) t^\lambda e^{-zt}, \tag{17.4.31}$$

we obtain

$$F_\lambda(z) = f(t_0) z^{-\lambda} - \frac{1}{z}\frac{1}{\Gamma(\lambda)} \int_0^\infty \frac{f(t) - f(t_0)}{t - t_0}\, d\left(t^\lambda e^{-zt}\right). \tag{17.4.32}$$

Now, integrating by parts, observing that the integrated terms will vanish, we obtain

$$F_\lambda(z) = f(t_0) z^{-\lambda} + \frac{1}{z}\frac{1}{\Gamma(\lambda)} \int_0^\infty t^\lambda e^{-zt} f_1(t)\,dt, \tag{17.4.33}$$

where

$$f_1(t) = \frac{d}{dt}\frac{f(t) - f(t_0)}{t - t_0}. \tag{17.4.34}$$

The new integral in (17.4.33) is of the same form as the integral in (17.4.29), with λ replaced by $\lambda + 1$ and f by f_1. The new interpolation point is $t_1 = (\lambda+1)/z$ and we obtain

$$F_\lambda(z) = (t_0) z^{-\lambda} + \frac{1}{z}\left(f_1(t_1)\lambda z^{-\lambda-1} + \frac{1}{z}\frac{1}{\Gamma(\lambda)} \int_0^\infty t^{\lambda+1} e^{-zt} f_2(t)\,dt\right), \tag{17.4.35}$$

where

$$f_2(t) = \frac{d}{dt}\frac{f_1(t) - f_1(t_1)}{t - t_1}. \tag{17.4.36}$$

This gives by further iteration

$$F_\lambda(z) = z^{-\lambda} \sum_{k=0}^{n-1} f_k\left(t_k\right) \frac{(\lambda)_k}{z^{2k}} + \frac{1}{z^n}\frac{1}{\Gamma(\lambda)} \int_0^\infty t^{\lambda+n-1} e^{-zt} f_n(t)\,dt, \tag{17.4.37}$$

where $n = 0, 1, 2, \ldots$, $f_0(t) = f(t)$, and

$$f_{k+1}(t) = \frac{d}{dt}\left(\frac{f_k(t) - f_k(t_k)}{t - t_k}\right), \quad t_k = \frac{\lambda + k}{z}, \quad k = 0, 1, 2, \ldots. \tag{17.4.38}$$

To obtain the functions f_k and the coefficients in (17.4.37), we require $f \in C^{2n}[0,\infty)$. In addition, we assume for the derivatives

$$|f^{(m)}(t)| \le Me^{\sigma t}, \quad t>0, \quad m=0,1,\ldots,2n, \tag{17.4.39}$$

for suitable M and σ. With these conditions expansion (17.4.37) has an asymptotic property and the remainder is of order $\mathcal{O}(z^{-2n-\lambda})$ as $z\to\infty$. Under further mild conditions on f it can be shown that the remainder tends to zero as $n\to\infty$, and that, hence, the expansion is convergent. For proofs we refer to Franklin and Friedman (1957).

Example 17.2 (The incomplete gamma function). For some special functions we can obtain the coefficients $f_k(t_k)$ easily (by using computer algebra). For example, when $f(t)=1/(1+t)$ we have the incomplete gamma function (see (7.1.6)),

$$F_\lambda(z)=\frac{1}{\Gamma(\lambda)}\int_0^\infty t^{\lambda-1}e^{-zt}\,\frac{dt}{1+t}=e^z\Gamma(1-\lambda,z). \tag{17.4.40}$$

We write $\zeta=z+\lambda$ and obtain the expansion

$$e^z\Gamma(1-\lambda,z)\sim z^{-\lambda}\sum_{k=0}^{\infty}(\lambda)_k\, f_k\, z^{-2k}, \tag{17.4.41}$$

where we write $f_k=f_k(t_k)$, and find the coefficients

$$\begin{aligned}
f_0&=\frac{z}{\zeta},\\
f_1&=\frac{z^3}{\zeta(\zeta+1)^2},\\
f_2&=\frac{z^5(3\zeta+4)}{\zeta(\zeta+1)^2(\zeta+2)^3},\\
f_3&=\frac{z^7\left(15\zeta^3+90\zeta^2+175\zeta+108\right)}{\zeta(\zeta+1)^2(\zeta+2)^3(\zeta+3)^4},\\
f_4&=\frac{z^9\left(105\zeta^6+1680\zeta^5+11025\zeta^4+37870\zeta^3+71540\zeta^2+70120\zeta+27648\right)}{\zeta(\zeta+1)^2(\zeta+2)^3(\zeta+3)^4(\zeta+4)^5}.
\end{aligned} \tag{17.4.42}$$

We note that these coefficients satisfy the order estimate $f_k=\mathcal{O}(1)$ as $z\to\infty$. In addition, the coefficients become small when λ is large. Combining the coefficients with the Pochhammer symbols in (17.4.37), we conclude that

$$(\lambda)_k f_k\,(t_k)=\mathcal{O}\left(\lambda^{-k-1}\right),\quad \lambda\to\infty. \tag{17.4.43}$$

In fact, the expansion holds for $z\to\infty$, uniformly with respect to $\lambda\ge 0$, and for complex values of z and λ as well. In Watson's lemma the asymptotic property of the expansion disappears when λ and z are both large. The same occurs when in (7.2.9) a and z are of the same size.

For a numerical verification of the method for the case $f(t)=1/(1+t)$, we have verified the relation

$$z^\lambda F_\lambda(z)+\frac{\lambda}{z}+z^{\lambda+1}F_{\lambda+1}(z)-1=0, \tag{17.4.44}$$

Table 17.2 Verification of the relation in (17.4.44) for some values of z and λ by using (17.4.37) with $n = 5$ and the coefficients given in (17.4.42).

z	λ	$z^\lambda F_\lambda(z)$	left-hand side of (17.4.44)
10	1	0.9156330438	−0.38e−06
10	10	0.5121792821	−0.53e−05
20	1	0.9543709085	−0.17e−08
20	10	0.6738352273	−0.14e−06
20	20	0.5061709401	−0.22e−06
50	1	0.9807554965	−0.52e−12
50	25	0.6695906479	−0.16e−08
50	50	0.5024874402	−0.26e−08

which follows from the recurrence relation for the incomplete gamma function

$$\Gamma(a+1, z) = a\Gamma(a, z) + z^a e^{-z}. \tag{17.4.45}$$

In Table 17.2 we give the values of $z^\lambda F_\lambda(z)$ together with computed values of the left-hand side of relation (17.4.44) for a few values of z and λ. We have used expansion (17.4.41) with five terms and the coefficients given in (17.4.42). ◇

Remark 17.1. A modification of the expansion in (17.4.37) will be given in §25.1, where the coefficients can be computed from the Maclaurin coefficients of f at the fixed interpolation point $t_0 = \lambda/z$. The aim of that method is to obtain an expansion for large z, and holding uniformly with respect to $\lambda \geq 0$. Convergence of the expansion will not be discussed. △

Chapter 18

Two-point Taylor expansions

In the derivation of uniform asymptotic expansions of integrals we often encounter the problem of expanding a function in two points.[1] It seems that in Chester *et al.* (1957) this type of expansion has been used for the first time in asymptotic analysis. It was used to derive Airy-type expansions for integrals that have two nearby (or coalescing) saddle points. This reference does not give further details about two-point Taylor expansions.

To explain the use of two-point Taylor expansions in asymptotic analysis we consider the integral

$$F_b(z) = \frac{1}{2\pi i}\int_{\mathcal{C}} e^{z\left(\frac{1}{3}t^3-b^2t\right)} f(t)\,dt, \tag{18.0.1}$$

where z is a large parameter and b is a parameter that may assume small values. The contour starts at $\infty e^{-\frac{1}{3}\pi i}$ and terminates at $\infty e^{\frac{1}{3}\pi i}$, and lies in a domain Ω where the function f is analytic. In particular, the saddle points $\pm b$ of the exponential function in (18.0.1) belong to Ω. When $f(t) = 1$ the function $F_b(z)$ becomes an Airy function, see Chapters 8 and 23.

The expansion used in Chester *et al.* (1957) is of the form

$$f(t) = \sum_{n=0}^{\infty} A_n\left(t^2-b^2\right)^n + t\sum_{n=0}^{\infty} B_n\left(t^2-b^2\right)^n. \tag{18.0.2}$$

As will be explained in §23.3, one method for obtaining an asymptotic expansion of $F_b(z)$ that holds uniformly for small values of b is based on substituting this expansion into (18.0.1). When interchanging summation and integration, the result is a formal expansion in two series in terms of functions related to Airy functions. Rearrangement of these series gives the canonical expansion in terms of the Airy function and its derivative, and that expansion can be obtained directly by using an integration by parts procedure introduced in Bleistein (1966).

In §18.4 we give details on manipulating two-point Taylor expansions, and in §23.5 on Airy-type expansion of the Bessel function $J_\nu(z)$. An application to Airy-type expansions of parabolic cylinder functions, is given in Vidunas and Temme (2002).

[1]This chapter is based on López and Temme (2002, 2004).

Fig. 18.1 Left: Contour $\mathcal{C}$ in the integrals (18.1.3)–(18.1.5). Right: For $z \in O_{z_1,z_2}$, we can take a contour $\mathcal{C}$ in Ω which contains O_{z_1,z_2} in its interior and therefore, $|(z-z_1)(z-z_2)| < |(w-z_1)(w-z_2)|$ for all $w \in \mathcal{C}$.

18.1 The expansions

We consider the expansion in a more symmetric form and give information on the coefficients and the remainder in the expansion.

Theorem 18.1 (Two-point expansion). *Let f be an analytic function in an open set $\Omega \subset \mathbb{C}$ and $z_1,\ z_2 \in \Omega$ with $z_1 \neq z_2$. Then, f can be expanded in a two-point Taylor expansion*

$$f(z) = \sum_{n=0}^{N-1} \left(a_n(z_1,z_2)\zeta_1 + a_n(z_2,z_1)\zeta_2\right) \zeta_1^n \zeta_2^n + r_N(z_1,z_2;z), \tag{18.1.3}$$

where $\zeta_j = z - z_j$, $j = 1,2$. The coefficients $a_n(z_1,z_2)$ and $a_n(z_2,z_1)$ of the expansion are given by the Cauchy integral

$$a_n(z_1,z_2) = \frac{1}{2\pi i(z_2-z_1)} \int_{\mathcal{C}} \frac{f(w)\,dw}{(w-z_1)^n(w-z_2)^{n+1}}. \tag{18.1.4}$$

The remainder term $r_N(z_1,z_2;z)$ is given by the Cauchy integral

$$r_N(z_1,z_2;z) = \frac{\zeta_1^N\zeta_2^N}{2\pi i} \int_{\mathcal{C}} \frac{f(w)\,dw}{(w-z_1)^N(w-z_2)^N(w-z)}. \tag{18.1.5}$$

The contour of integration $\mathcal{C}$ is a simple closed loop which encircles the points z_1 and z_2 (for a_n) and z, z_1 and z_2 (for r_N) anti-clockwise and is contained in Ω (see Figure 18.1 (Left)).

When $N \to \infty$, the expansion (18.1.3) is convergent for z inside the Cassini oval (see Figure 18.2)

$$O_{z_1,z_2} = \{z \in \Omega,\ \ |(z-z_1)(z-z_2)| = r\}, \tag{18.1.6}$$

where

$$r = \mathrm{Inf}_{w \in \mathbb{C}\setminus\Omega}\left(|(w-z_1)(w-z_2)|\right). \tag{18.1.7}$$

In particular, if f is an analytic function ($\Omega = \mathbb{C}$), then the expansion (18.1.3) converges for all $z \in \mathbb{C}$.

Remark 18.1. In the following we identify the Cassini oval defined in (18.1.6) with the open set inside this curve. △

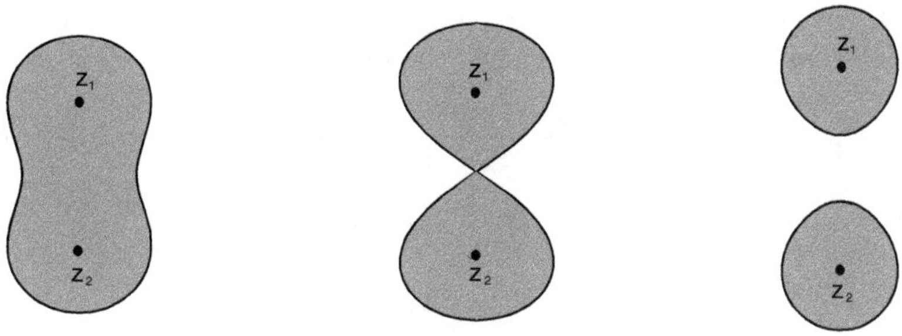

Fig. 18.2 Cassini ovals used in the proof of Theorem 18.1 for three possible values of r (see (18.1.7)). Left: $4r > |z_1 - z_2|^2$. Middle: $4r = |z_1 - z_2|^2$. Right: $4r < |z_1 - z_2|^2$.

Proof. For the proof of Theorem 18.1 we write, by Cauchy's theorem,

$$f(z) = \frac{1}{2\pi i}\int_{\mathcal{C}} f(w)\,\frac{dw}{w-z}, \tag{18.1.8}$$

where $\mathcal{C}$ is the contour defined above (Figure 18.1 (Left)). We write

$$\frac{1}{w-z} = \frac{z+w-z_1-z_2}{(w-z_1)(w-z_2)}\,\frac{1}{1-u}, \tag{18.1.9}$$

where

$$u = \frac{(z-z_1)(z-z_2)}{(w-z_1)(w-z_2)}. \tag{18.1.10}$$

Next we use the expansion

$$\frac{1}{1-u} = \sum_{n=0}^{N-1} u^n + \frac{u^N}{1-u}, \tag{18.1.11}$$

and substitute the form of $1/(w-z)$ in (18.1.8). After straightforward calculations we obtain (18.1.3)–(18.1.5).

For every $z \in O_{z_1,z_2}$, we can take a contour $\mathcal{C}$ in Ω such that $|(z-z_1)(z-z_2)| < |(w-z_1)(w-z_2)|$ for all $w \in \mathcal{C}$ (see Figure 18.1 (Right)). On this contour $|f(w)|$ is bounded by some constant C: $|f(w)| \le C$. Introducing these two bounds in (18.1.5) we see that $\lim_{N\to\infty} r_N(z_1,z_2;z) = 0$ and the proof follows. □

18.2 An alternative form of the expansion

The present expansion of $f(z)$ in the form (18.1.3) stresses the symmetry of the expansion with respect to z_1 and z_2. In this representation it is not possible, however, to let z_1 and z_2 coincide, which causes a little inconvenience (the coefficients $a_n(z_1,z_2)$ become infinitely large as $z_1 \to z_2$; the remainder $r_N(z_1,z_2;z)$ remains well defined).

Furthermore, the sum $a_n(z_1,z_2)\zeta_1 + a_n(z_2,z_1)\zeta_2$ remains finite when $z_1 \to z_2$. This follows from collecting from (18.1.4) the relevant terms:

$$\frac{w-z_2}{z_2-z_1} + \frac{w-z_1}{z_1-z_2} = \frac{1}{(w-z_1)(w-z_2)}. \tag{18.2.12}$$

An alternative way is the representation (cf. (18.0.2)),

$$f(z) = \sum_{n=0}^{\infty} \left(A_n(z_1,z_2) + z\, B_n(z_1,z_2)\right)(z-z_1)^n(z-z_2)^n. \tag{18.2.13}$$

Now we have the relations

$$\begin{aligned} A_n(z_1,z_2) &= -z_1 a_n(z_1,z_2) - z_2 a_n(z_2,z_1),\\ B_n(z_1,z_2) &= a_n(z_1,z_2) + a_n(z_2,z_1), \end{aligned} \tag{18.2.14}$$

which are regular when $z_1 \to z_2$. In fact we have

$$\begin{aligned} A_n(z_1,z_2) &= \frac{1}{2\pi i}\int_{\mathcal{C}} \frac{w-z_1-z_2}{((w-z_1)(w-z_2))^{n+1}} f(w)\,dw,\\ B_n(z_1,z_2) &= \frac{1}{2\pi i}\int_{\mathcal{C}} \frac{1}{((w-z_1)(w-z_2))^{n+1}} f(w)\,dw. \end{aligned} \tag{18.2.15}$$

Letting $z_1 \to 0$ and $z_2 \to 0$, we obtain the standard Maclaurin series of $f(z)$ with even part (the A_n series) and odd part (the B_n series).

18.3 Explicit forms of the coefficients

Definition (18.1.4) is not appropriate for numerical computations. A more practical formula to compute the coefficients of the above two-point Taylor expansion is given in the following proposition.

Proposition 18.1. *The coefficients $a_n(z_1,z_2)$ in the expansion* (18.1.3) *are also given by the formulas*

$$a_0(z_1,z_2) = \frac{f(z_2)}{z_2-z_1}, \tag{18.3.16}$$

and for $n = 1,2,3,\ldots$

$$a_n(z_1,z_2) = \sum_{k=0}^{n} \frac{(n+k-1)!}{k!(n-k)!} \frac{(-1)^{n+1} n f^{(n-k)}(z_2) + (-1)^k k f^{(n-k)}(z_1)}{n!(z_1-z_2)^{n+k+1}}. \tag{18.3.17}$$

Proof. We deform the contour of integration $\mathcal{C}$ in equation (18.1.4) into any contour of the form $\mathcal{C}_1 \cup \mathcal{C}_2$ that is also contained in Ω, where $\mathcal{C}_1$ ($\mathcal{C}_2$) is a simple closed

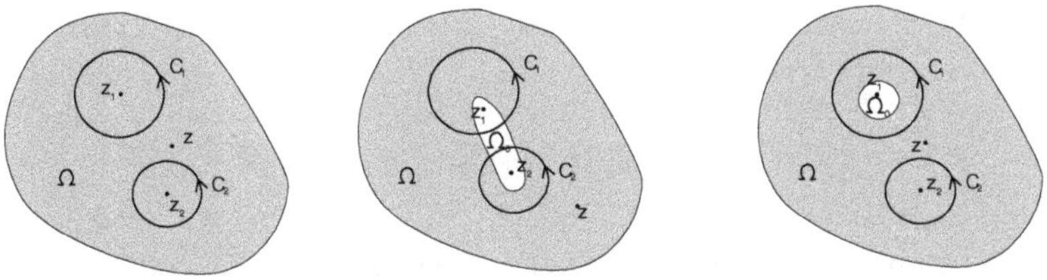

Fig. 18.3 Left: The function $(w-z_2)^{-n-1}f(w)$ is analytic inside $\mathcal{C}_1$, whereas $(w-z_1)^{-n}f(w)$ is analytic inside $\mathcal{C}_2$. Middle: The function $(w-z_2)^{-n-1}g_1(w)$ is analytic inside $\mathcal{C}_1$, whereas $(w-z_1)^{-n}g_2(w)$ is analytic inside $\mathcal{C}_2$. Right: The function $(w-z_2)^{-n-1}g(w)$ is analytic inside $\mathcal{C}_1$, whereas $(w-z_1)^{-n}f(w)$ is analytic inside $\mathcal{C}_2$.

loop which encircles the point z_1 (z_2) anti-clockwise and does not contain the point z_2 (z_1) inside (see Figure 18.3 (Left)). Then,

$$\begin{array}{l}(z_2-z_1)a_n(z_1,z_2) = \\ \displaystyle\frac{1}{2\pi i}\int_{\mathcal{C}_1}\frac{f(w)}{(w-z_2)^{n+1}}\frac{dw}{(w-z_1)^n}+\frac{1}{2\pi i}\int_{\mathcal{C}_2}\frac{f(w)}{(w-z_1)^n}\frac{dw}{(w-z_2)^{n+1}} = \\ \displaystyle\frac{1}{(n-1)!}\frac{d^{n-1}}{dw^{n-1}}\left.\frac{f(w)}{(w-z_2)^{n+1}}\right|_{w=z_1}+\frac{1}{n!}\frac{d^n}{dw^n}\left.\frac{f(w)}{(w-z_1)^n}\right|_{w=z_2}.\end{array} \tag{18.3.18}$$

From here, equations (18.3.16)–(18.3.17) follow straightforwardly. □

18.4 Manipulations with two-point Taylor expansions

To obtain the coefficients in asymptotic expansions we usually derive a recursive scheme of functions and coefficients. For example, in §14.6.2 we use the scheme

$$\begin{aligned} f_n(t) &= a_n + b_n(t-\alpha) + (t-\alpha)(\beta-t)g_n(t), \\ f_{n+1}(t) &= -(t-\alpha)^{1-\lambda}(\beta-t)^{1-\mu}\frac{d}{dt}\left((t-\alpha)^\lambda(\beta-t)^\mu g_n(t)\right), \end{aligned} \tag{18.4.19}$$

with $f_0 = f$, a given function. The coefficients a_n and b_n show up in the asymptotic expansion in §14.6.2 and follow from this scheme:

$$a_n = f_n(\alpha), \quad b_n = \frac{f_n(\beta)-f_n(\alpha)}{\beta-\alpha}. \tag{18.4.20}$$

Because the problem of finding a_n and b_n occurs frequently in deriving uniform expansions we give a few steps for handling a scheme like the one in (18.4.19). Observe first that the second line can be written in the form

$$f_{n+1}(t) = (pt+q)g_n(t) + rw\, g_n'(t), \tag{18.4.21}$$

where

$$p = \mu-\lambda, \quad q = \lambda\beta-\mu\alpha, \quad r = 1, \quad w = (t-\alpha)(t-\beta). \tag{18.4.22}$$

A systematic way is to write the functions f_n as a two-point Taylor expansion. In the present case we write

$$f_n(t) = \sum_{k=0}^{\infty} \left(c_k^n(t-\alpha) + d_k^n(t-\beta)\right)w^k. \tag{18.4.23}$$

By using the results of §18.2 and §18.3 the values c_k^0, d_k^0 can be obtained from the derivatives of f at α and β. The coefficients a_n and b_n defined in (18.4.20) are given by

$$a_n = (\alpha-\beta)d_0^n, \quad b_n = c_0^n + d_0^n. \tag{18.4.24}$$

When we use these a_n and b_n, it follows that

$$a_n + b_n(t-\alpha) = c_0^n(t-\alpha) + d_0^n(t-\beta), \tag{18.4.25}$$

and that the function g_n shown in the first line in (18.4.19) has the expansion

$$g_n(t) = \sum_{k=0}^{\infty} \left(c_{k+1}^n(t-\alpha) + d_{k+1}^n(t-\beta)\right)w^k. \tag{18.4.26}$$

The coefficients of the two-point Taylor expansion of f_{n+1} follow from (18.4.21), and we need to perform a few manipulations.

We use the symmetric form given in (18.1.3), and it is easy to translate the results into expansions of another form by using the results of §18.2. We have the following lemmas.

Lemma 18.1. *Let f have the expansion*

$$f(z) = \sum_{k=0}^{\infty} \left(a_k(z-z_1) + b_k(z-z_2)\right)w^k, \quad w = (z-z_1)(z-z_2). \tag{18.4.27}$$

Then the coefficients of the expansion

$$zf(z) = \sum_{k=0}^{\infty} \left(c_k(z-z_1) + d_k(z-z_2)\right)w^k \tag{18.4.28}$$

are given by $c_0 = z_2a_0$, $d_0 = z_1b_0$, and

$$c_k = z_2a_k + \frac{a_{k-1}+b_{k-1}}{z_2-z_1}, \quad d_k = z_1b_k - \frac{a_{k-1}+b_{k-1}}{z_2-z_1}, \quad k \ge 1. \tag{18.4.29}$$

Proof. We write

$$a_kz(z-z_1)+b_kz(z-z_2) = p_k(z-z_1)w+q_k(z-z_2)w+r_k(z-z_1)+s_k(z-z_2). \tag{18.4.30}$$

Then it is easily verified that

$$p_k = \frac{a_k+b_k}{z_2-z_1}, \quad q_k = -p_k, \quad r_k = z_2a_k, \quad s_k = z_1b_k, \tag{18.4.31}$$

which values give the coefficients in (18.4.29). □

Lemma 18.2. *Let f have the expansion given in* (18.4.27). *Then the coefficients of the expansion*

$$f'(z) = \sum_{k=0}^{\infty} \left(c_k(z-z_1) + d_k(z-z_2)\right) w^k \tag{18.4.32}$$

are given by $c_0 = r_0$, $d_0 = s_0$, and for $k \ge 1$ by

$$c_k = r_k + p_{k-1}, \quad d_k = s_k + q_{k-1}, \tag{18.4.33}$$

where

$$\begin{aligned} p_k &= \frac{2(k+1)(a_{k+1}+b_{k+1})}{z_2 - z_1}, \quad q_k = -p_k, \\ r_k &= \frac{a_k + b_k + (k+1)(z_2-z_1)^2 a_{k+1}}{z_2 - z_1}, \\ s_k &= \frac{a_k + b_k + (k+1)(z_2-z_1)^2 b_{k+1}}{z_1 - z_2}. \end{aligned} \tag{18.4.34}$$

Proof. After differentiating (18.4.27) we write

$$a_k + b_k + (k+1)(2z - z_1 - z_2)\left(a_{k+1}(z-z_1) + b_{k+1}(z-z_2)\right) \tag{18.4.35}$$

in the form of the right-hand side of (18.4.30). This gives the results for the coefficients p_k, q_k, r_k and s_k. □

When we apply the lemmas to obtain the coefficients of the function f_{n+1} from the relation in (18.4.21), with f_n and g_n expanded as in (18.4.23) and (18.4.26), respectively, and $z_1 = \alpha$ and $z_2 = \beta$, we obtain

$$\begin{aligned} c_0^{n+1} &= (p\beta + q)c_1^n, \quad d_0^{n+1} = (p\alpha + q)d_1^n, \\ c_k^{n+1} &= (p\beta + q + rk(\beta - \alpha))c_{k+1}^n + \frac{(p + (2k-1)r)\left(c_k^n + d_k^n\right)}{\beta - \alpha}, \\ d_k^{n+1} &= (p\alpha + q - rk(\beta - \alpha))d_{k+1}^n - \frac{(p + (2k-1)r)\left(c_k^n + d_k^n\right)}{\beta - \alpha}. \end{aligned} \tag{18.4.36}$$

Remark 18.2. When we write the coefficients a_k and b_k of the expansion in (18.4.27) in the form $a_k(z_1, z_2)$ and $b_k(z_1, z_2)$ we have the symmetry rule $a_k(z_1, z_2) = b_k(z_2, z_1)$, just as in the expansion in (18.1.3). Similar relations are valid for the coefficients c_k and d_k in both lemmas. △

Chapter 19

Hermite polynomials as limits of other classical orthogonal polynomials

Hermite polynomials show up in several problems of asymptotic analysis.[1] We consider three different instances where these classical orthogonal polynomials can be used as main approximants.

(1) As limits of other polynomials such as Laguerre and Jacobi orthogonal polynomials, but of generalized Bernoulli polynomials as well.
(2) In turning point problems for second-order linear differential equations when two nearby turning points are present.
(3) For functions defined as an integral in which the saddle points follow a certain pattern for certain values of the parameters (see Chapter 24).

In all three cases, asymptotic representations of polynomials are considered in terms of Hermite polynomials. In the second and third case the Hermite polynomials arise as special cases of another set of special functions, the parabolic cylinder functions, which can be used in similar and more general problems of asymptotic analysis.

In this chapter we concentrate on the first topic; in particular we give examples from the Askey scheme of hypergeometric polynomials. We give examples in which Hermite polynomials are used in asymptotic approximations, and we also give approximations in terms of other polynomials.

19.1 Limits between orthogonal polynomials

The Hermite polynomials have the explicit representation

$$H_n(x) = n! \sum_{k=0}^{\lfloor n/2 \rfloor} \frac{(-1)^k}{k!\,(n-2k)!}(2x)^{n-2k}, \quad n = 0, 1, 2, \ldots. \tag{19.1.1}$$

We recall a few limits between the classical orthogonal polynomials.

[1]This chapter is based on López and Temme (1999a,b).

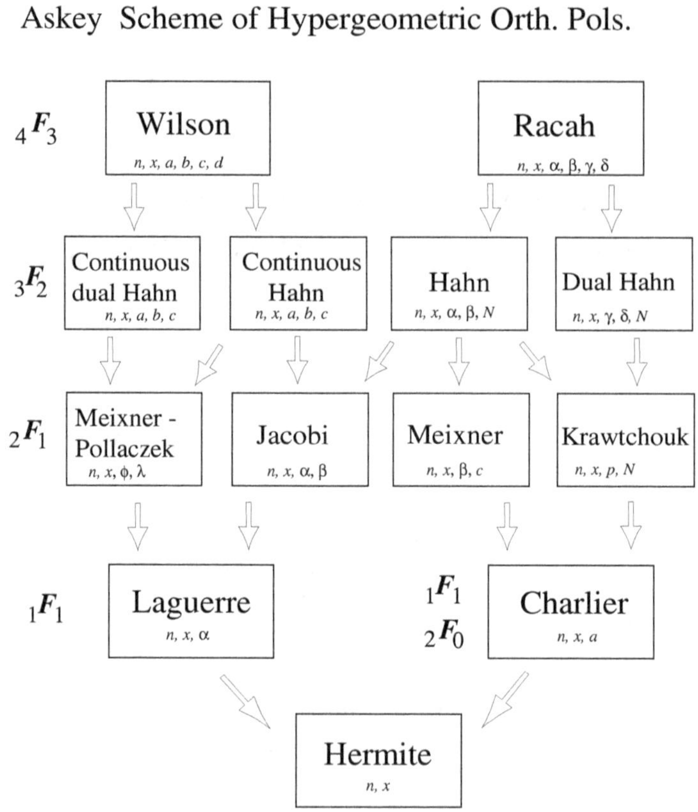

Fig. 19.1 The Askey scheme for hypergeometric orthogonal polynomials, with indicated limit relations between the polynomials.

For the Gegenbauer (ultraspherical) polynomials $C_n^\gamma(x)$ (see §15.3), we have

$$\lim_{\gamma\to\infty} \frac{C_n^\gamma(x)}{C_n^\gamma(1)} = x^n, \quad \lim_{\gamma\to\infty} \gamma^{-n/2} C_n^\gamma(x/\sqrt{\gamma}) = \frac{1}{n!} H_n(x). \tag{19.1.2}$$

The first limit shows that the zeros of $C_n^\gamma(x)$ tend to the origin if the order γ tends to infinity. The second limit is more interesting; it gives the relation to the Hermite polynomials if the order becomes large and the argument x is properly scaled.

For the Laguerre polynomials similar results are

$$\begin{aligned} &\lim_{\alpha\to\infty} \alpha^{-n} L_n^{(\alpha)}(\alpha x) = \frac{(1-x)^n}{n!}, \\ &\lim_{\alpha\to\infty} \alpha^{-n/2} L_n^{(\alpha)}\left(x\sqrt{\alpha}+\alpha\right) = \frac{(-1)^n\, 2^{-n/2}}{n!}\, H_n\left(x/\sqrt{2}\right). \end{aligned} \tag{19.1.3}$$

This again gives insight in the location of the zeros for large values α, and the relation to the Hermite polynomials if α becomes large and x is properly scaled.

Many methods are available to prove these and other limits. In this chapter we concentrate on asymptotic relations between the polynomials, from which the limits follow as special cases.

19.2 The Askey scheme of orthogonal polynomials

In Koekoek *et al.* (2010) many relations are given for hypergeometric orthogonal polynomials and their q-analogues, including limit relations between many polynomials.

In Figure 19.1 we show the *Askey scheme* with examples of limit relations between neighboring polynomials, but many other limit relations are mentioned in Godoy *et al.* (1998); Koekoek and Swarttouw (1998); Koekoek *et al.* (2010); Ronveaux *et al.* (1998).

In Ferreira *et al.* (2003) and López and Temme (1999a,b) several asymptotic relations between polynomials and Hermite polynomials are given by using the method described in this chapter. In these references the following polynomials are considered in this way:

Gegenbauer	Laguerre
Jacobi	Charlier
Meixner–Pollaczek	Jacobi, Meixner
Krawtchouk	Tricomi–Carlitz
generalized Bernoulli	generalized Euler
generalized Bessel	Buchholz

In Ferreira *et al.* (2003, 2008a,b) several other asymptotic expansions are obtained for other elements of the Askey scheme by using the same method. We observe that the method also works for polynomials outside the class of hypergeometric polynomials. The method is different from the one described in Godoy *et al.* (1998), where moreover extra terms in the limit relation are constructed in order to obtain more insight in the limiting process.

19.3 Asymptotic representations

The starting point is a generating series

$$F(x,w)=\sum_{n=0}^{\infty} p_n(x)\,w^n, \tag{19.3.4}$$

F is a given function, which is analytic with respect to w at $w=0$, and p_n is independent of w.

The relation (19.3.4) gives for p_n the Cauchy-type integral

$$p_n(x)=\frac{1}{2\pi i}\int_{\mathcal{C}} F(x,w)\,\frac{dw}{w^{n+1}}, \tag{19.3.5}$$

where $\mathcal{C}$ is a circle around the origin inside the domain where F is analytic (as a function of w).

We write

$$F(x,w) = e^{Aw-Bw^2} f(x,w), \tag{19.3.6}$$

where A and B do not depend on w. This gives

$$p_n(x) = \frac{1}{2\pi i}\int_{\mathcal{C}} e^{Aw-Bw^2} f(x,w)\,\frac{dw}{w^{n+1}}. \tag{19.3.7}$$

Because f is analytic as well (as a function of w), we can expand

$$f(x,w) = e^{-Aw+Bw^2} F(x,w) = \sum_{k=0}^{\infty} c_k w^k, \tag{19.3.8}$$

that is,

$$f(x,w) = 1 + (p_1(x) - A)w + (p_2(x) - Ap_1(x) + B + \tfrac12 A^2)w^2 + \ldots \tag{19.3.9}$$

if we assume that $p_0(x) = 1$ (which implies $c_0 = 1$).

We substitute the expansion (19.3.8) into (19.3.7). The Hermite polynomials have the generating function

$$e^{2xw-w^2} = \sum_{n=0}^{\infty} \frac{H_n(x)}{n!} w^n, \quad x, w \in \mathbb{C}, \tag{19.3.10}$$

which gives the Cauchy-type integral

$$H_n(x) = \frac{n!}{2\pi i}\int_{\mathcal{C}} e^{2xz-z^2} z^{-n-1}\,dz, \tag{19.3.11}$$

where $\mathcal{C}$ is a circle around the origin and the integration is in the positive direction. The result is the finite expansion

$$p_n(x) = z^n \sum_{k=0}^{n} \frac{c_k}{z^k}\frac{H_{n-k}(\xi)}{(n-k)!}, \quad z = \sqrt{B}, \quad \xi = \frac{A}{2\sqrt{B}}, \tag{19.3.12}$$

because terms of the expansion in (19.3.8) with $k > n$ do not contribute to the integral in (19.3.7).

To give the representation of $p_n(x)$ in (19.3.12) an asymptotic property, we take A and B such that $c_1 = c_2 = 0$. This happens if we take

$$A = p_1(x), \quad B = \tfrac12 p_1^2(x) - p_2(x). \tag{19.3.13}$$

As we will show, the asymptotic property follows from the behavior of the coefficients c_k if a parameter (different from the degree k) of the polynomial $p_k(x)$ becomes large. We use the following lemma, and explain what happens by considering a few examples.

Lemma 19.1. *Let $\phi(w)$ be analytic at $w = 0$, with Maclaurin expansion of the form*

$$\phi(w) = \mu w^n (a_0 + a_1 w + a_2 w^2 + \ldots), \tag{19.3.14}$$

where n is a positive integer and a_k are complex numbers that do not depend on the complex number μ, $a_0 \neq 0$. Let c_k denote the coefficients of the power series for $f(w) = e^{\phi(w)}$, that is,

$$f(w) = e^{\phi(w)} = \sum_{k=0}^{\infty} c_k w^k. \tag{19.3.15}$$

Then $c_0 = 1, c_k = 0, k = 1, 2 \ldots, n-1$ and, for fixed k,

$$c_k = \mathcal{O}\left(|\mu|^{\lfloor k/n \rfloor}\right), \quad \mu \to \infty. \tag{19.3.16}$$

Proof. The proof follows from expanding

$$\begin{aligned} \sum_{k=0}^{\infty} c_k w^k &= e^{\phi(w)} = \sum_{k=0}^{\infty} \frac{(\phi(w))^k}{k!} \\ &= \sum_{k=0}^{\infty} \frac{\mu^k w^{kn}}{k!} (a_0 + a_1 w + a_2 w^2 + \ldots)^k, \end{aligned} \tag{19.3.17}$$

and comparing equal powers of w. □

19.4 Gegenbauer polynomials

The generating function is

$$F(x, w) = \left(1 - 2xw + w^2\right)^{-\gamma} = \sum_{n=0}^{\infty} C_n^{\gamma}(x)\, w^n, \tag{19.4.18}$$

with

$$C_0^{\gamma}(x) = 1, \quad C_1^{\gamma}(x) = 2\gamma x, \quad C_2^{\gamma}(x) = 2\gamma(\gamma+1)x^2 - \gamma. \tag{19.4.19}$$

Hence,

$$A = C_1^{\gamma}(x) = 2x\gamma, \quad B = \tfrac{1}{2}\left(C_1^{\gamma}(x)\right)^2 - C_2^{\gamma}(x) = \gamma(1 - 2x^2), \tag{19.4.20}$$

and we can write

$$C_n^{\gamma}(x) = z^n \sum_{k=0}^{n} \frac{c_k}{z^k} \frac{H_{n-k}(\xi)}{(n-k)!}, \tag{19.4.21}$$

where

$$z = \sqrt{\gamma(1 - 2x^2)}, \quad \xi = \frac{x\gamma}{z}. \tag{19.4.22}$$

We have

$$c_0 = 1, \quad c_1 = c_2 = 0, \quad c_3 = \tfrac{2}{3}\gamma x(4x^2 - 3), \quad c_4 = \tfrac{1}{2}\gamma\left(1 + 8x^2(x^2 - 1)\right). \tag{19.4.23}$$

Higher coefficients follow from the recursion relation

$$kc_k = 2x(k-1)c_{k-1} - (k-2)c_{k-2} + 2\gamma x(4x^2 - 3)c_{k-3} + 2\gamma(1 - 2x^2)c_{k-4}. \tag{19.4.24}$$

This relation follows from substituting the Maclaurin series of f given by

$$f(x,w) = e^{-2x\gamma w+\gamma((1-2x^2)} \left(1-2xw+w^2\right)^{-\gamma} = \sum_{k=0}^{\infty} c_k w^k \qquad (19.4.25)$$

into the differential equation

$$(1-2xw+w^2)\frac{df}{dw} = 2\gamma\left(-3x+4x^3+w-2x^2w\right)w^2 f. \qquad (19.4.26)$$

The function $f(x,w)$ of (19.3.8) has the form $f(x,w) = e^{\phi(x,w)}$, where $\phi(x,w) = \gamma w^3(a_0+a_1w+a_2w^2+\ldots)$. By using Lemma 19.1 and $\xi = \mathcal{O}(\sqrt{\gamma})$ we conclude that the sequence $\{\Phi_k\}$ with $\Phi_k = c_k/z^k H_{n-k}(\xi)$ has the following asymptotic structure:

$$\Phi_k = \mathcal{O}\left(\gamma^{n/2+\lfloor k/3\rfloor-k}\right), \quad \gamma\to\infty, \qquad (19.4.27)$$

with fixed $k = 0,1,2,\ldots$. This explains the asymptotic nature of the representation in (19.4.21) for large values of γ, with x and n fixed.

To verify the limits given in (19.1.2), we first write x in terms of ξ:

$$x = \frac{\xi}{\sqrt{\gamma+2\xi^2}}. \qquad (19.4.28)$$

With this value of x we can verify that

$$c_k/z^k = o(1), \quad \gamma\to\infty, \quad k>0, \qquad (19.4.29)$$

and in fact we have the limit

$$\lim_{\gamma\to\infty} \frac{\gamma^n}{(\gamma+2x^2)^{n/2}}\, C_n^\gamma\left(\frac{x}{\sqrt{\gamma+2x^2}}\right) = \frac{1}{n!}H_n(x). \qquad (19.4.30)$$

Remark 19.1. In Elbert and Laforgia (1992) the following expansion is given:

$$\gamma^{-n/2}C_n^\gamma\left(x/\sqrt{\gamma}\right) = \sum_{k=0}^{n-1} Q_j(x)\gamma^{-j}, \qquad (19.4.31)$$

where $Q_j(x)$ are polynomials in x and explicitly given in terms of Stirling numbers of the first kind; see §34.1. The first one is $Q_0(x) = H_n(x)/n!$. This expansion has the elegant property of a representation in negative powers of γ. As our expansion in (19.4.21) it is finite, but our coefficients c_k follow easily from the recurrence relation given in (19.4.24). △

19.5 Laguerre polynomials

We take as generating function (see (15.0.2))

$$F(x,w) = (1+w)^{-\alpha-1}e^{wx/(1+w)} = \sum_{n=0}^{\infty}(-1)^n L_n^{(\alpha)}(x)\, w^n. \qquad (19.5.32)$$

We have

$$\begin{aligned} &L_0^{(\alpha)}(x) = 1, \quad L_1^{(\alpha)}(x) = \alpha + 1 - x, \\ &L_2^{(\alpha)}(x) = \tfrac{1}{2}((\alpha+1)(\alpha+2) - 2(\alpha+2)x + x^2), \end{aligned} \tag{19.5.33}$$

which gives

$$A = x - \alpha - 1, \quad B = x - \tfrac{1}{2}(\alpha+1), \tag{19.5.34}$$

and we obtain

$$L_n^{(\alpha)}(x) = (-1)^n\, z^n \sum_{k=0}^{n} \frac{c_k}{z^k} \frac{H_{n-k}(\xi)}{(n-k)!}, \tag{19.5.35}$$

where

$$z = \sqrt{x - \tfrac{1}{2}(\alpha+1)}, \quad \xi = \frac{x-\alpha-1}{2z}. \tag{19.5.36}$$

We have

$$c_0 = 1, \quad c_1 = c_2 = 0, \quad c_3 = \tfrac{1}{3}(3x - \alpha - 1). \tag{19.5.37}$$

Higher coefficients follow from a recurrence relation, which can be obtained as in §19.4. The representation in (19.5.35) has an asymptotic character for large values of $|\alpha| + |x|$.

To verify that the limits given in (19.1.3) follow from (19.5.35), we need to invert the relation between x and ξ:

$$x = 2\xi^2 + \alpha + 1 + \xi\sqrt{2\,(2\xi^2 + \alpha + 1)}, \tag{19.5.38}$$

which gives

$$x = \alpha + \xi\sqrt{2\alpha} + 2\xi^2 + 1 + \mathcal{O}\left(1/\sqrt{\alpha}\right), \quad \alpha \to \infty. \tag{19.5.39}$$

With this estimate we find for z given in (19.5.36) $z = \sqrt{\alpha/2} + \xi + \mathcal{O}\,(1/\sqrt{\alpha})$, and the limit in (19.1.3) easily follows.

19.6 Generalized Bessel polynomials

For notation, properties and further details on these polynomials we refer to Chapter 33, where the large degree asymptotics is considered.

We have

$$\begin{aligned} &A = Y_1^{\mu}(z) = 1 + \tfrac{1}{2}(\mu+2)z, \\ &B = \tfrac{1}{2}A^2(z) - \tfrac{1}{2}Y_2^{\mu}(z) = -\tfrac{1}{8}z\,(4 + (3\mu+8)z)\,, \end{aligned} \tag{19.6.40}$$

and the expansion

$$Y_n^{\mu}(z) = n!\, B^{\frac{1}{2}n} \sum_{k=0}^{n} \frac{c_k(z,\mu)}{B^{\frac{1}{2}k}} \frac{H_{n-k}(\zeta)}{(n-k)!}, \quad \zeta = \frac{A}{2\sqrt{B}}, \tag{19.6.41}$$

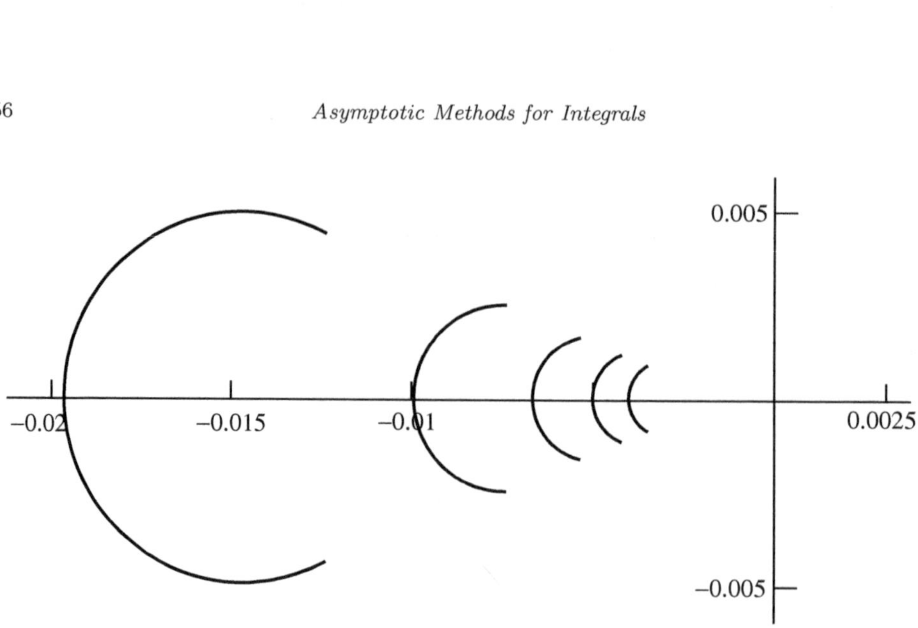

Fig. 19.2 The curves in the z-plane under the mapping $\zeta \mapsto z(\zeta)$ (see (19.6.44)) are the images of the ζ-intervals $[-\sqrt{2n+1}, \sqrt{2n+1}]$ where the zeros of the Hermite polynomial $H_n(\zeta)$ occur. We take $n = 10$ and show the curves (from left to right) for $\mu = 100, 200, \ldots, 500$.

where $c_0(z, \mu) = 1$, $c_1(z, \mu) = c_2(z, \mu) = 0$, and

$$c_3(z, \mu) = \tfrac{1}{12} z^2 \left((5\mu + 16)z + 6\right). \tag{19.6.42}$$

In López and Temme (1999b, Lemma 5.2) it is shown that the representation in (19.6.41) is an asymptotic expansion for $|\mu| \to \infty$, and holds for fixed values of z and n. In this reference a recurrence relation for the coefficients $c_k(z, \mu)$ is derived.

We are especially interested in the location of the zeros of $Y_n^\mu(z)$, which are complex, in contrast to those of the classical orthogonal polynomials, which are real and inside the domain of orthogonality.

In Figure 19.2 we show the curves $z(\zeta)$ for $\zeta \in [-\sqrt{2n+1}, \sqrt{2n+1}]$, in which interval the zeros $h_{n,m}$ of the Hermite polynomial $H_n(\zeta)$ occur. The curves cut the negative z-axis for $\zeta = 0$ at $z = -2/(\mu + 2)$, see (19.6.44).

Let $y_{n,m}$ and $h_{n,m}$ be the mth zero of $Y_n^\mu(z)$ and $H_n(z)$, respectively, $m = 1, 2, \ldots, n$. By using the expansion (19.6.41), we can compute a first approximation of $y_{n,m}$ for given μ and n. Inverting the relation for ζ given in (19.6.41), we obtain for z the equation $pz^2 + qz + 1 = 0$, where

$$p = \tfrac{1}{4}\left(\mu^2 + 4\mu + 4 + 2\zeta^2(3\mu + 8)\right), \quad q = \mu + 2 + 2\zeta^2, \tag{19.6.43}$$

which gives

$$z(\zeta) = \frac{-q + i\zeta\sqrt{2(\mu + 4 - 2\zeta^2)}}{2p}. \tag{19.6.44}$$

Using this with $\zeta = h_{n,m}$ we obtain a first approximation of $z = y_{n,m}$. For $\mu = 500$ and $n = 10$ the maximal relative error in the approximations of the zeros is 0.035.

19.7 Meixner–Pollaczek polynomials into Laguerre polynomials

We give an example on how to use Laguerre polynomials for approximating other polynomials.

Lemma 19.2. *Let the polynomials $p_n(x)$ be defined by the generating function*

$$F(x,w) = \sum_{n=0}^{\infty} p_n(x)\, w^n, \tag{19.7.45}$$

where $F(x,w)$ is analytic in $w=0$ and $F(x,0)=1$. Let the coefficients $c_k(x)$ be defined by the expansion

$$e^{-Aw/(Bw-1)}(1-Bw)^{C+1}\, F(x,w) = \sum_{k=0}^{\infty} c_k(x) w^k, \quad c_0 = 1, \tag{19.7.46}$$

where A, B and C do not depend on w. Then $p_n(x)$ can be represented as the finite sum

$$p_n(x) = B^{n/2} \sum_{k=0}^{n} \frac{c_k(x)}{B^{k/2}}\, L_{n-k}^{(C)}(\xi), \quad \xi = \frac{A}{B}, \tag{19.7.47}$$

where $L_n^{(\alpha)}(x)$ are the Laguerre polynomials. Moreover, A, B and C can be chosen such that $c_1 = 0,\ c_2 = 0,\ c_3 = 0$.

Proof. Use the Cauchy integrals of $p_n(x)$ and the Laguerre polynomials that follow from (19.7.45) and (15.0.2), respectively. □

For the Meixner–Pollaczek polynomials we have the generating function

$$F(w) = \left(1 - e^{i\phi} w\right)^{-\lambda+ix} \left(1 - e^{-i\phi} w\right)^{-\lambda-ix} = \sum_{n=0}^{\infty} P_n^{(\lambda)}(x;\phi) w^n. \tag{19.7.48}$$

From (15.0.2) it follows that

$$G(w) = e^{Aw/(Bw-1)} \left(1 - Bw\right)^{-C-1} = \sum_{n=0}^{\infty} L_n^{(C)}(\xi) B^n w^n, \tag{19.7.49}$$

where $\xi = A/B$. We define c_k by $f(w) = F(w)/G(w) = \sum_{k=0}^{\infty} c_k w^k$. Then the expansion of the Meixner–Pollaczek polynomials reads

$$P_n^{(\lambda)}(x;\phi) = \sum_{k=0}^{n} B^{n-k} c_k L_{n-k}^{(C)}(\xi), \quad \xi = A/B. \tag{19.7.50}$$

We write $x + i\lambda = re^{i\theta}, \theta \in (0,\pi), r \ge 0$, and consider $r \to \infty$; the asymptotic results hold uniformly with respect to θ. We consider two cases.

(1) First we consider a simple case by taking $B = 1$ and $C = \alpha$, and solve $c_1 = 0$ for A. This gives

$$A = \alpha + 1 - 2\lambda\cos\phi - 2x\sin\phi. \tag{19.7.51}$$

The first coefficients c_k are given by

$$c_0 = 1, \quad c_1 = 0, \quad c_2 = x\sin 2\phi + \lambda\cos 2\phi - 2(x\sin\phi + \lambda\cos\phi) + \tfrac{1}{2}\alpha. \tag{19.7.52}$$

The first-term approximation can be written as

$$P_n^{(\lambda)}(x;\phi) = L_n^{(\alpha)}(\xi) + \mathcal{O}\left(r^{n-1}\right), \quad \xi = A. \tag{19.7.53}$$

In this case a limit can be obtained by putting $\lambda = (\alpha+1)/2$. Then we have $c_k = \mathcal{O}(\phi^2)$ as $\phi \to 0$ for $k \ge 2$, and we obtain

$$\lim_{\phi\to 0} P_n^{(\alpha+1)/2}\left((\alpha+1)(1-\cos\phi) - \xi)/(2\sin\phi);\phi\right) = L_n^{(\alpha)}(\xi). \tag{19.7.54}$$

This includes the limit of the Askey scheme

$$\lim_{\phi\to 0} P_n^{(\alpha+1)/2}(-\xi/(2\phi);\phi) = L_n^{(\alpha)}(\xi). \tag{19.7.55}$$

(2) Next we solve $c_1 = 0, c_2 = 0$ for A and C, with $B = 1$. This gives

$$\begin{aligned} A &= 2\bigl(x(\sin\phi - \sin 2\phi) + \lambda(\cos\phi - \cos 2\phi)\bigr),\\ C &= 2\bigl(x(2\sin\phi - \sin 2\phi) + \lambda(2\cos\phi - \cos 2\phi)\bigr) - 1, \end{aligned} \tag{19.7.56}$$

and the first-term approximation can be written as

$$P_n^{(\lambda)}(x;\phi) = L_n^{(\alpha)}(\xi) + \mathcal{O}\left(r^{n-2}\right), \quad r \to \infty, \tag{19.7.57}$$

uniformly with respect to θ, where $\xi = A$ and $\alpha = C$.
Solving $A = \xi, C = \alpha$ for x and λ, we obtain

$$\begin{aligned} \lambda &= (1-\cos\phi)\xi + \tfrac{1}{2}(\alpha+1)(2\cos\phi - 1),\\ x &= \frac{2(\xi-\alpha-1)\cos^2\phi + (\alpha+1-2\xi)\cos\phi + \alpha + 1 - \xi}{2\sin\phi}. \end{aligned} \tag{19.7.58}$$

Then $c_3 = \frac{2}{3}(\alpha+1-2\xi)(1-\cos\phi)$ and $c_k = \mathcal{O}(\phi^2)$ as $\phi \to 0$ for $k \ge 3$. We obtain the limit

$$\lim_{\phi\to 0} P_n^{(\lambda)}(x;\phi) = L_n^{(\alpha)}(\xi). \tag{19.7.59}$$

PART 4

Uniform Methods for Integrals

Chapter 20

An overview of standard forms

It is clear from the examples in earlier chapters that an essential step in the saddle point method is to perform one or several substitutions after which the integral

$$\int_{\mathcal{C}} e^{-\omega\phi(z)}\psi(z)\,dz \tag{20.0.1}$$

can be written in the form

$$\int_{-\infty}^{\infty} e^{-\omega t^2}\, f(t)\,dt, \tag{20.0.2}$$

or as a sum of such integrals when more saddle points have to be taken into account.

The Gaussian dominant exponential term is quite common in many problems and examples. However, when in the immediate vicinity of one of the saddle points the second derivative of ϕ vanishes as well, we need other forms than shown in the exponential function in (20.0.2). We should not transform ϕ into a pure quadratic term, but, for example, into a third degree polynomial of the form $\frac{1}{3}t^3 - \alpha t$.

Indeed, when two saddle points are proximate or coalescing (in which case the second derivative of ϕ vanishes), a better approach is to transform into an integral of the type

$$\int_{\mathcal{C}} e^{\omega\left(\frac{1}{3}t^3-\alpha t\right)}\, f(t)\,dt, \tag{20.0.3}$$

where $\mathcal{C}$ is a contour in the complex plane and α may assume small values. We will see in Chapter 23 how this can be done in the case of Bessel functions, orthogonal polynomials, and so on.

Furthermore, complications will arise in (20.0.2) when f has a pole or singular point close to the origin, or that move to the origin under the influence of extra parameters. We will see this in §21.1, where a simple first-order pole is in close proximity to the saddle point at the origin, or may even coalesce with the origin.

In a number of chapters we will explain the many aspects of uniform asymptotic expansions, usually with cases that are relevant in the asymptotic behavior of special functions, but we also give examples from singular boundary value problems.

Table 20.1 An overview of standard forms.

Case	Standard Form	Approximant	Critical points	§
1	$\displaystyle\int_0^\infty e^{-zt}\frac{f(t)}{t+\alpha}\,dt$	Exponential integral	$0, -\alpha$	§2.7.1
2	$\displaystyle\int_{-\infty}^\infty e^{-zt^2}\frac{f(t)}{t-i\alpha}\,dt$	Error function	$0, i\alpha$	§21.1
3	$\displaystyle\int_{-\infty}^\alpha e^{-zt^2}f(t)\,dt$	Error function	$0, \alpha$	§36.1
4	$\displaystyle\int_0^\infty t^{\beta-1}e^{-z(\frac12 t^2-\alpha t)}f(t)\,dt$	Par. cylinder function	$0, \alpha$	§22.1
5	$\displaystyle\int_{c-i\infty}^{c+i\infty} t^{-\beta-1}e^{z(\frac12 t^2-\alpha t)}f(t)\,dt$	Par. cylinder function	$0, \alpha$	§22.4
6	$\displaystyle\int_{\mathcal{L}} e^{z(\frac13 t^3-\alpha t)}f(t)\,dt$	Airy function	$\pm\sqrt{\alpha}$	§23.2
7	$\displaystyle\int_0^\infty t^{\lambda-1}e^{-zt}f(t)\,dt$	Gamma function	$0, \lambda/z$	§25.1
8	$\displaystyle\int_\alpha^\infty t^{\beta-1}e^{-zt}f(t)\,dt$	Inc. gamma function	$0, \alpha, \beta/z$	§26.1
9	$\displaystyle\int_{-\infty}^{(0+)} t^{-\beta-1}e^{z(t+\alpha/t)}f(t)\,dt$	Bessel I function	$0, \pm\sqrt{\alpha}$	§27.3
10	$\displaystyle\int_0^\infty t^{\beta-1}e^{-z(t+\alpha/t)}f(t)\,dt$	Bessel K function	$0, \pm\sqrt{\alpha}$	§27.3
11	$\displaystyle\int_0^\infty t^{\lambda-1}(t+\alpha)^{-\mu}e^{-zt}f(t)\,dt$	Kummer U function	$0, -\alpha$	§28.1
12	$\displaystyle\int_{-\alpha}^\alpha e^{-zt}(\alpha^2-t^2)^\mu f(t)\,dt$	Bessel I function	$0, \pm\alpha$	§29.3
13	$\displaystyle\int_\alpha^\infty e^{-zt}(t^2-\alpha^2)^\mu f(t)\,dt$	Bessel K function	$0, \pm\alpha$	§29.3
14	$\displaystyle\int_0^\infty \frac{\sin z(t-\alpha)}{t-\alpha}f(t)\,dt$	Sine integral	$0, \alpha$	§28.5
15	$\displaystyle\int_0^\alpha e^{\frac12 zt^2}f(t)\,dt$	Dawson's integral	$0, \alpha$	§36.5
16	$\displaystyle\int_{-\infty}^{(0+)} e^{z(2\xi t-\rho^2\ln t-\frac12 t^2)}f(t)\,\frac{dt}{t}$	Par. cylinder function	0, 2 saddles	§24.1

In Table 20.1 we give an overview of standard forms considered in later chapters and in the literature. Usually these forms arise in the asymptotic analysis of special functions, but in all cases special functions are used as leading terms in approximations. In recent survey papers (Temme, 2012, 2013b) we have discussed similar overviews, together with other aspects of asymptotics for integrals.

The critical points mentioned in the table are the points in the interval of integration (or close to this interval) where the main contributions to the approximation can be obtained. When these points coalesce uniform methods are needed.

20.1 Comments on the table

We give a few comments on the cases in Table 20.1.

Case 1 This can be done as in §21.1. When $f = 1$ we have

$$e^{\alpha z} E_1(\alpha z) = \int_0^\infty \frac{e^{-zt}}{t+\alpha}\, dt. \tag{20.1.4}$$

For details and an application we refer to §2.7.1.

Case 2 This will be considered in §21.1. When $f = 1$ we have the complementary error function:

$$\int_{-\infty}^{\infty} e^{-zt^2} \frac{dt}{t - i\alpha} = \pi i \operatorname{erfc}\left(\alpha\sqrt{z}\right), \quad \Re\alpha > 0. \tag{20.1.5}$$

Case 3 Again we can use the complementary error function for this case. When $f = 1$ we have

$$\tfrac{1}{2}\sqrt{\frac{\pi}{z}} \operatorname{erfc}\left(-\alpha\sqrt{z}\right) = \int_{-\infty}^{\alpha} e^{-zt^2}\, dt. \tag{20.1.6}$$

This standard form has important applications for cumulative distribution functions of probability theory. As we will show in §36.1, we can transform the well-known gamma and beta distributions, and several other ones, into this standard form.

Case 4 When $f = 1$ the integral becomes the parabolic cylinder function $U(a, z)$ with integral representation

$$U\left(\beta - \tfrac{1}{2}, -\alpha\sqrt{z}\right) = \frac{z^{\frac{1}{2}\beta} e^{-\frac{1}{4}\alpha^2 z}}{\Gamma(\beta)} \int_0^\infty t^{\beta-1} e^{-z\left(\frac{1}{2}t^2 - \alpha t\right)}\, dt, \tag{20.1.7}$$

$\Re z > 0$ and $\Re\beta > 0$. When $\beta = 1$, this is related to Case 3. Case 4 has been considered in detail in the literature, for the first time in Bleistein (1966). In that paper an integration-by-parts method is given that can be used in many other uniform expansions.

Case 5 We use the integral representation of the parabolic cylinder function

$$U\left(\beta - \tfrac{1}{2}, \alpha\sqrt{z}\right) = i\sqrt{2\pi}\, z^{-\frac{1}{2}\beta} e^{-\frac{1}{4}\alpha^2 z} \int_{c-i\infty}^{c+i\infty} t^{-\beta-1} e^{z\left(\frac{1}{2}t^2 - \alpha t\right)}\, dt, \tag{20.1.8}$$

where $\Re z > 0$ and $c > 0$. When $\beta = 0$ this gives Case 2, after some modifications.

Case 6 For $f = 1$ we have

$$z^{-\frac{1}{3}} \operatorname{Ai}\left(\alpha z^{\frac{2}{3}}\right) = \frac{1}{2\pi i} \int_{\mathcal{C}_1} e^{z\left(\frac{1}{3}t^3 - \alpha t\right)}\, dt, \tag{20.1.9}$$

where $\mathcal{C}_1$ is shown in Figure 8.2. It starts at $\infty e^{-\frac{1}{3}\pi i}$ and ends at $\infty e^{\frac{1}{3}\pi i}$. This case will be considered in Chapter 23.

Case 7 In the integral

$$F_\lambda(z) = \frac{1}{\Gamma(\lambda)} \int_0^\infty t^{\lambda-1} e^{-z\,t} f(t)\, dt, \quad \lambda > 0, \tag{20.1.10}$$

we assume that z is large and that λ may be large as well. This is different from Watson's lemma considered in Chapter 2, where we have assumed that λ is fixed. There is a saddle point at $\mu = \lambda/z$. A special integration by parts procedure will give a uniform expansion which allows small and large values of λ.

Case 8 This case extends the previous one with an extra parameter:

$$F_\lambda(z, \alpha) = \frac{1}{\Gamma(\lambda)} \int_\alpha^\infty t^{\lambda-1} e^{-z\,t} f(t)\, dt, \tag{20.1.11}$$

with $\lambda \geq 0$, $\alpha \geq 0$ and z large. As in the previous case, λ may also be large; α may be large as well, even larger than λ. When $f = 1$ this integral becomes an incomplete gamma function. In §26.3 we will give an application to the incomplete beta function.

Case 9 When $f = 1$ and $\alpha > 0$, the contour integral

$$F_\beta(z, \alpha) = \int_{-\infty}^{(0+)} t^{-\beta-1} e^{z(t+\alpha/t)} f(t)\, dt \tag{20.1.12}$$

reduces to a modified I-Bessel function, and when $\alpha < 0$ to a J-Bessel function. See §10.3.1 for applications to the Kummer functions. For an application to Laguerre polynomials, we refer to §32.3.2.

Case 10 Related to the previous case is the real integral

$$F_\lambda(z, \alpha) = \int_0^\infty t^{\lambda-1} e^{-z\,t-\alpha/t} f(t)\, dt, \tag{20.1.13}$$

which reduces to the modified K-Bessel function when f is a constant. We have

$$2(\alpha/z)^{\frac{1}{2}\lambda} K_\lambda\left(2\sqrt{\alpha z}\right) = \int_0^\infty t^{\lambda-1} e^{-z\,t-\alpha/t}\, dt. \tag{20.1.14}$$

We give an asymptotic expansion of the function $F_\lambda(z, \alpha)$ for large values of z, which is uniformly valid for $\lambda \geq 0$, $\alpha \geq 0$. For more details we refer to Chapter 27, where an application is given to confluent hypergeometric functions.

Case 11 This is more general than Case 1. See Oberhettinger (1959) for more details.

Case 12 and Case 13 The integrals are of the type

$$\begin{aligned} F(\omega, \alpha, \mu) &= \int_{-\alpha}^{\alpha} e^{-\omega t} f(t) \left(\alpha^2 - t^2\right)^\mu\, dt, \\ G(\omega, \alpha, \mu) &= \int_{\alpha}^{\infty} e^{-\omega t} f(t) \left(t^2 - \alpha^2\right)^\mu\, dt, \end{aligned} \tag{20.1.15}$$

where $\omega > 0$ (and large), $\Re\mu > -1$ (fixed), and $\alpha \ge 0$. When $f(t) = 1$, these integrals reduce to modified Bessel functions:

$$\begin{aligned} F_0(\omega, \alpha, \mu) &= \sqrt{\pi}\Gamma(\mu+1)\left(\frac{2\alpha}{\omega}\right)^{\mu+\frac{1}{2}} I_{\mu+\frac{1}{2}}(\alpha\omega), \\ G_0(\omega, \alpha, \mu) &= \frac{1}{\sqrt{\pi}}\Gamma(\mu+1)\left(\frac{2\alpha}{\omega}\right)^{\mu+\frac{1}{2}} K_{\mu+\frac{1}{2}}(\alpha\omega). \end{aligned} \tag{20.1.16}$$

When α is bounded away from zero asymptotic expansions for large values of ω can be obtained by using Watson's lemma. However, when α is allowed to become small, or even 0, uniform expansions in terms of the Bessel functions can be given. In Chapter 29 we give applications to Legendre functions.

Case 14 The integral

$$D(x, \lambda) = \frac{1}{\pi}\int_0^\infty f(t)\frac{\sin(\lambda(t-x))}{t-x}\, dt \tag{20.1.17}$$

is considered for $x \ge 0$ and large positive values of λ. The special function that takes care of the smooth transition from $x = 0$ to $x > 0$ is the sine integral. We have

$$D_0(x, \lambda) = \int_0^\infty \frac{\sin(\lambda(t-x))}{t-x}\, dt = \pi + \mathrm{si}(\lambda x) = \tfrac{1}{2}\pi + \mathrm{Si}(\lambda x). \tag{20.1.18}$$

For more details we refer to §28.5.

Case 15 This is related to Case 3. When $f = 1$ we have Dawson's integral, a special case of the error function. When $\alpha \to 0$ the critical points at 0 (saddle point) and α (endpoint) coalesce. The special feature is that the saddle point does not give the main contribution. In §36.5 we give an application to the Von Mises distribution.

Case 16 This is related to Case 5. Now we let β of that case depend on the large parameter (see also §22.3). When $f = 1$ we have a parabolic cylinder function. In §24.1 we consider cases in which the parabolic cylinder functions become Hermite polynomials when $z\rho^2 = n$, a nonnegative integer. In Chapter 24 we give applications to Gegenbauer and Tricomi–Carlitz polynomials, and in §32.4 to Laguerre polynomials.

Chapter 21

A saddle point near a pole

When the function f in Laplace's method, see Chapter 3 and (3.0.1), has a singularity near the origin, the straightforward method, which is based on expanding this function in a power series, may fail. The coefficients of this expansion will show the effect of this singularity. In particular, when the singular point approaches the origin under the influence of a parameter, we need other methods. In this chapter we discuss the simple case of a single pole near the saddle point.

21.1 A saddle point near a pole: Van der Waerden's method

The standard form for this kind of problem is

$$F_\alpha(\omega) = \frac{1}{2\pi i}\int_{-\infty}^{\infty} e^{-\omega t^2}\frac{f(t)}{t - i\alpha}\,dt, \quad \Re\alpha > 0, \tag{21.1.1}$$

where f is analytic in a domain $\mathcal{D}$ containing the real line. We also assume that $i\alpha \in \mathcal{D}$. When $f = 1$ this integral can be expressed in terms of the complementary error function defined in §3.4.

When $\Re\alpha > 0$ we have

$$\frac{e^{-\omega\alpha^2}}{2\pi i}\int_{-\infty}^{\infty} e^{-\omega t^2}\frac{dt}{t - i\alpha} = \tfrac{1}{2}\operatorname{erfc}\left(\alpha\sqrt{\omega}\right). \tag{21.1.2}$$

To prove this representation, observe that this integral vanishes as $\omega \to +\infty$ (assuming that α is fixed with $\Re\alpha > 0$). Then, differentiation of the left-hand side with respect to ω gives the result. The restriction $\Re\alpha > 0$ can be dropped: α may assume all complex values; when $\Re\alpha < 0$ the path should be indented to avoid the pole. When we pick up the residue of the pole, we can write (using the third relation in (3.4.28))

$$\frac{e^{-\omega\alpha^2}}{2\pi i}\int_{-\infty}^{\infty} e^{-\omega t^2}\frac{dt}{t - i\alpha} = -\tfrac{1}{2}\operatorname{erfc}\left(-\alpha\sqrt{\omega}\right), \quad \Re\alpha < 0. \tag{21.1.3}$$

This also simply follows from (21.1.2) by changing $t \to -t$.

We also have the representation given in (3.4.29), which easily follows from (21.1.2).

When $\alpha > 0$ and ω is large, the right-hand side of (21.1.2) quickly approaches $\frac{1}{2}$; see Figure 3.1. When α becomes negative the right-hand side quickly approaches unity, as follows from $\operatorname{erfc}(-z) = 2 - \operatorname{erfc} z$. Many probability distribution functions show this kind of behavior.

A standard approach for obtaining an asymptotic expansion that holds uniformly with respect to α in a domain that contains the origin is to split off the pole by writing

$$f(t) = (f(t) - f(i\alpha)) + f(i\alpha). \tag{21.1.4}$$

Then the integral in (21.1.1) can be written in the form

$$F_\alpha(\omega) = \tfrac{1}{2} f(i\alpha) e^{\omega\alpha^2} \operatorname{erfc}\left(\alpha\sqrt{\omega}\right) + \frac{1}{2\pi i}\int_{-\infty}^{\infty} e^{-\omega t^2} g(t)\, dt, \tag{21.1.5}$$

where

$$g(t) = \frac{f(t) - f(i\alpha)}{t - i\alpha}. \tag{21.1.6}$$

This function is analytic in the same domain $\mathcal{D}$ where f is analytic. When we substitute $g(t) = \sum_{n=0}^{\infty} c_n(\alpha) t^n$ into (21.1.5), we obtain the large-ω asymptotic representation of $F_\alpha(\omega)$ in the form

$$F_\alpha(\omega) \sim \tfrac{1}{2} f(i\alpha) e^{\omega\alpha^2} \operatorname{erfc}\left(\alpha\sqrt{\omega}\right) + \frac{1}{2i\sqrt{\pi\omega}} \sum_{n=0}^{\infty} c_{2n}(\alpha) \frac{\left(\frac{1}{2}\right)_n}{\omega^n}. \tag{21.1.7}$$

The validity of this expansion follows from Theorem 3.1 when we can verify if g satisfies the conditions of this theorem.

In this chapter we give examples of certain K-Bessel function integrals, which can be used to obtain asymptotic approximations in a singular perturbation problem. In that case the error function plays the part of a boundary layer function that describes the fast transition at a part of the domain boundary. In Chapter 37 we demonstrate this method for contour integrals that represent the incomplete gamma functions.

The method of splitting off the pole was considered for the first time by Van der Waerden (1951) in a problem of Sommerfeld concerning the propagation of radio waves over a plane earth. Van der Waerden provided a simpler uniform expansion than the one given by Ott (1943), who obtained a uniform expansion in which each term is an incomplete gamma function. For such expansions we refer to §21.2. For a different approach to obtain coefficients in uniform expansions as in (21.1.7), as well as for more general cases, we refer to López and Pagola (2011b).

The uniformity aspect of the expansion in (21.1.7) is that we may assume $i\alpha \in \mathcal{D}$, in particular that α takes values in a neighborhood of the origin. Although the method is rather simple, it will become clear that the numerical evaluation of the coefficients of the asymptotic expansions may need some extra attention. Usually, the function f in (21.1.1) has been obtained in the saddle point method by using some transformations. Often, for numerical applications, the coefficients $c_n(\alpha)$ of the expansion of the function g defined in (21.1.6) have to be expanded in powers of α when this parameter is small.

The role of the error function in uniform expansions will also be considered as a special case in Chapter 22, where we consider a saddle point coalescing with an endpoint of the interval of integration (where in the general case an algebraic singularity may be present).

21.2 An alternative expansion

If we expand $f(t) = \sum_{n=0}^{\infty} a_n t^n$ we find an alternative expansion of (21.1.1) in the form

$$F_\alpha(\omega) \sim \sum_{n=0}^{\infty} a_n \Phi_n, \tag{21.2.8}$$

where initially $\Re\alpha > 0$ and

$$\Phi_n = \frac{1}{2\pi i}\int_{-\infty}^{\infty} e^{-\omega t^2}\frac{t^n}{t-i\alpha}\,dt, \quad n = 0, 1, 2, \ldots. \tag{21.2.9}$$

These functions can be expressed in terms of incomplete gamma functions when we use the integral representation in (7.1.6). We have

$$\begin{aligned}\Phi_{2n} &= \frac{1}{2\pi}\Gamma\left(\tfrac12 + n\right)\alpha^{2n} e^{\alpha^2\omega}\Gamma\left(\tfrac12 - n, \alpha^2\omega\right),\\ \Phi_{2n+1} &= \frac{1}{2\pi i}\Gamma\left(\tfrac32 + n\right)\alpha^{2n+1} e^{\alpha^2\omega}\Gamma\left(-\tfrac12 - n, \alpha^2\omega\right).\end{aligned} \tag{21.2.10}$$

By using the starting value

$$\Phi_0 = \tfrac12 e^{\alpha^2\omega}\operatorname{erfc}\left(\alpha\sqrt{\omega}\right), \tag{21.2.11}$$

it is possible to obtain the Φ_n for $n \ge 1$ by using the recurrence relation

$$\Phi_{n+1} = i\alpha\Phi_n + \frac{(-1)^n + 1}{4\pi i\,\omega^{n+\frac12}}\Gamma\left(\tfrac12 n + \tfrac12\right), \quad n = 0, 1, 2, \ldots. \tag{21.2.12}$$

It can be concluded from (21.2.10) and this recursion that the Φ_n are defined for all complex α, and are in fact analytic functions of α. The expansion in (21.2.8) is also defined for all $\alpha \in \mathbb{C}$.

From (21.2.10) and (7.2.9) we see for fixed n the following asymptotic behavior of Φ_n:

$$\begin{aligned}\Phi_{2n} &\sim \frac{1}{2\pi\alpha}\frac{\Gamma\left(\frac12 + n\right)}{\omega^{\frac12 + n}},\\ \Phi_{2n+1} &\sim \frac{1}{2\pi i\alpha}\frac{\Gamma\left(\frac32 + n\right)}{\omega^{\frac32 + n}},\end{aligned} \tag{21.2.13}$$

as $\alpha^2\omega \to \infty$. When $\alpha^2\omega = \mathcal{O}(1)$, then $\alpha \to 0$ (we always assume $\omega \to \infty$) and $\Phi_n = \mathcal{O}(\alpha^n)$. Hence, the sequence $\{\Phi_n\}$ is an asymptotic sequence as $\omega \to \infty$, uniformly for all α.

Each term in the expansion in (21.2.8) contains an incomplete gamma function, and the expansion may be considered as less efficient than the one given in (21.1.7) by splitting off the pole as in (21.1.4). However, the Φ_n follow from the simple recursion in (21.2.12) (although it is not stable, see Temme (1986a)), they constitute a proper asymptotic scale, and the coefficients a_n in (21.2.8) do not depend on α, whereas the coefficients $c_{2n}(\alpha)$ in (21.1.7) need to be expanded for small values of α.

In Clemmow (1950) it has been shown that the theorems for Watson's lemma apply for the expansion in (21.2.8), and that the error bounds can be given in expansions with a finite number of terms and a remainder.

21.3 An example from De Bruijn

In De Bruijn's book (De Bruijn, 1958, §5.12) the influence of poles near the saddle point is considered by studying the example

$$F_\alpha(z) = \beta^2 \int_{-\infty}^{\infty} e^{-\omega t^2} \frac{f(t)}{\beta^2 + t^2}\, dt, \quad \beta = \omega^{-\frac{1}{2}\alpha}, \quad \alpha > 0, \tag{21.3.14}$$

where ω is a positive large parameter. Observe that for all α the parameter β is small.

Three separate cases are distinguished: $0 < \alpha < 1$, $\alpha = 1$, and $\alpha > 1$, for the special choice $f(t) = e^t$. For each case an asymptotic expansion is given. These expansions are really different in the sense that they do not pass into each other when α passes unity. Van der Waerden's method is not used.

We can use partial fraction decomposition to get two integrals with a single pole, but we can also write

$$f(t) = a_0 + b_0 t + \left(\beta^2 + t^2\right) g(t), \tag{21.3.15}$$

where we assume that g is regular at the points $\pm i\beta$. This gives

$$a_0 = \frac{f(i\beta) + f(-i\beta)}{2}, \quad b_0 = \frac{f(i\beta) - f(-i\beta)}{2i}, \tag{21.3.16}$$

and where $g(t)$ follows with these values from (21.3.15).

Hence,

$$F_\alpha(z) = a_0\beta^2 \int_{-\infty}^{\infty} e^{-\omega t^2} \frac{dt}{\beta^2 + t^2}\, dt + \beta^2 \int_{-\infty}^{\infty} e^{-\omega t^2} g(t)\, dt, \tag{21.3.17}$$

where the first integral can be written in terms of the complementary error function (see (3.4.29)):

$$F_\alpha(z) = a_0 \beta\pi e^{\omega^{1-\alpha}} \operatorname{erfc}\left(\omega^{\frac{1}{2}(1-\alpha)}\right) + \beta^2 \int_{-\infty}^{\infty} e^{-\omega t^2} g(t)\, dt. \tag{21.3.18}$$

By expanding $g(t) = \sum_{k=0}^{\infty} g_k(\beta)t^k$ the asymptotic representation follows:

$$F_\alpha(z) \sim a_0\beta\pi e^{\omega^{1-\alpha}} \operatorname{erfc}\left(\omega^{\frac{1}{2}(1-\alpha)}\right) + \beta^2\sqrt{\frac{\pi}{\omega}}\sum_{k=0}^{\infty} c_k(\beta)\frac{1}{\omega^k}, \tag{21.3.19}$$

where β is defined in (21.3.14) and

$$c_k(\beta) = g_{2k}(\beta)\left(\tfrac{1}{2}\right)_k. \tag{21.3.20}$$

These and all other coefficients are regular when $\beta \to 0$. The expansion in (21.3.19) is valid for all $\alpha \geq 0$.

We give a few coefficients for $f(t) = e^t$, the example used in De Bruijn (1958). We have

$$\begin{aligned} c_0(\beta) &= \frac{1-\cos(\beta)}{\beta^2}, \\ c_1(\beta) &= \frac{\beta^2 - 2 + 2\cos(\beta)}{4\beta^4}, \\ c_2(\beta) &= \frac{\beta^4 - 12\beta^2 + 24 - 24\cos(\beta)}{32\beta^6}. \end{aligned} \tag{21.3.21}$$

From the representation in (21.3.18) we see at once the special value $\alpha = 1$: the complementary error function can be expanded by using the asymptotic expansion given in (3.4.31) when $0 < \alpha < 1$, and when $\alpha > 1$ it can be expanded by using the convergent power series expansion of $\operatorname{erf} z = 1 - \operatorname{erfc} z$. When $\alpha = 1$, De Bruijn gives an expansion in terms of functions related to the complementary error function. His first term is $\beta\,\pi\,e\operatorname{erfc}(1)$, which corresponds to our term $\beta\cos(\beta)\,\pi\,e\operatorname{erfc}(1)$.

It should be noted that De Bruijn is not aiming at a uniform expansion with respect to α. His discussion concerns the role of an extra parameter which causes poles in the neighborhood of the saddle point. But using the uniform method gives a very short explanation of De Bruijn's lengthy discussion.

In addition, Van der Waerden's paper that introduced splitting off the pole was published in 1951. The first edition of De Bruijn's book (De Bruijn, 1958) was published in 1957, and was based on lectures in 1954/1955 at the Mathematical Center in Amsterdam and in 1956/1957 in Eindhoven.

21.4 A pole near a double saddle point

Boersma (1991) has discussed a problem where the exponential function has a double saddle point at the origin. The integral has the form (after some transformations)

$$F_\alpha(\omega) = \frac{1}{2\pi i}\int_{\infty\exp\left(-\frac{1}{3}\pi i\right)}^{\infty\exp\left(\frac{1}{3}\pi i\right)} e^{\omega t^3}\,\frac{f(t)}{t-\alpha}\,dt, \tag{21.4.22}$$

where α is possibly small.

This type of integral arises in the asymptotic analysis of the series

$$S(k,\theta) = J_k(k) + 2\sum_{n=1}^{\infty} J_{n+k}(k)\cos(2n\theta), \quad 0 \le \theta \le \pi, \tag{21.4.23}$$

and $\Im k \to +\infty$. By using a contour integral of the Bessel function the integral is brought into the form in (21.4.22), and to achieve a uniform expansion for all $\theta \in [0,\pi]$ the function

$$G(z) = \frac{3}{2\pi i}\int_{\infty \exp\left(-\frac{1}{3}\pi i\right)}^{\infty \exp\left(\frac{1}{3}\pi i\right)} e^{t^3}\,\frac{dt}{t-z} \tag{21.4.24}$$

has been used. Here the integration path passes to the right of $t = z$. This function is an analytic function of z and can be considered as the analogue of the complementary error function. Boersma has given several properties of this function, and he has expressed it in terms of known special functions.

21.5 A singular perturbation problem and K-Bessel integrals

In this section we consider a singular perturbation problem in a quarter plane (see §21.5.3), and we express the solution of this problem in terms of K-Bessel integrals.

21.5.1 *A Bessel K_0-integral*

Consider the integral

$$K_0(x,y) = \int_x^{\infty} K_0\left(\sqrt{\xi^2+y^2}\right)d\xi, \tag{21.5.25}$$

where $K_0(z)$ is the modified Bessel function and x and y are real parameters. In the analysis we assume that $y \ge 0$ (from the definition we see that $K_0(x,y)$ is an even function of y).

Integrals of this type are used in Boersma *et al.* (1984) in the study of a diffusion problem in semiconductor technology. The results for $K_0(x,y)$ will also be used in §21.5.3 in the asymptotic analysis of a singular perturbation problem.

We want to know the asymptotic behavior of $K_0(x,y)$ for large values of x and y. In particular, we want to know the behavior for $r \to \infty$, uniformly with respect to $\theta \in [0, \frac{1}{2}\pi]$, where

$$x = r\cos\theta, \quad y = r\sin\theta. \tag{21.5.26}$$

We use the cosine transform (see Oberhettinger (1957, p. 13))

$$K_0\left(\sqrt{\xi^2+y^2}\right) = \int_0^{\infty} \frac{e^{-\xi\sqrt{t^2+1}}}{\sqrt{t^2+1}}\cos(yt)\,dt, \tag{21.5.27}$$

which is valid for $\xi \ge 0$. This gives

$$K_0(x,y) = \int_0^\infty \frac{e^{-x\sqrt{t^2+1}}}{t^2+1}\cos(yt)\,dt, \quad x \ge 0. \tag{21.5.28}$$

Observe that

$$K_0(0,y) = \int_0^\infty \frac{\cos(yt)}{t^2+1}\,dt = \tfrac{1}{2}\pi e^{-|y|}, \tag{21.5.29}$$

which is a well-known cosine transform. From this result we can derive the symmetry relation

$$K_0(x,y) = \pi e^{-|y|} - K_0(-x,y), \tag{21.5.30}$$

which can be used to obtain results for $x < 0$.

By writing $t = \sinh u$ in (21.5.28) we obtain

$$K_0(x,y) = \tfrac{1}{2}\int_{-\infty}^\infty e^{-x\cosh u + iy\sinh u}\,\frac{du}{\cosh u}, \quad x \ge 0, \tag{21.5.31}$$

and using polar coordinates introduced in (21.5.26) gives

$$K_0(x,y) = \tfrac{1}{2}\int_{-\infty}^\infty e^{-r\cosh(u-i\theta)}\,\frac{du}{\cosh u}. \tag{21.5.32}$$

If $0 \le \theta < \frac{1}{2}\pi$ we can shift the path of integration upwards, through the saddle point at $u = \theta$, and obtain

$$K_0(x,y) = \tfrac{1}{2}\int_{-\infty}^\infty e^{-r\cosh u}\,\frac{du}{\cosh(u+i\theta)}. \tag{21.5.33}$$

The poles are at $u_k = i(\frac{1}{2}\pi - \theta) + ik\pi, k \in \mathbb{Z}$, and u_0 is close to the origin (the saddle point) if $\theta \sim \frac{1}{2}\pi$. The other poles are not close to the origin for the values of θ considered.

Taking $v = \sinh(u/2)$ as the new variable of integration, we transform the integral in (21.5.33) into the standard form (21.1.1). First we write

$$\frac{1}{\cosh(u+i\theta)} = \frac{\cos\theta\cosh u}{\cosh^2 u - \sin^2\theta} - i\frac{\sin\theta\sinh u}{\cosh^2 u - \sin^2\theta}. \tag{21.5.34}$$

The imaginary part is an odd function of u and can be neglected. We obtain

$$K_0(x,y) = \tfrac{1}{2}\cos\theta\, e^{-r}\int_{-\infty}^\infty e^{-2rv^2} f(v)\,dv, \tag{21.5.35}$$

where

$$f(v) = \frac{v^2+\frac{1}{2}}{(v^2+\alpha^2)(v^2+\beta^2)\sqrt{v^2+1}}, \tag{21.5.36}$$

with

$$\begin{aligned} \alpha &= \sqrt{\tfrac{1}{2}(1-\sin\theta)} = \sin\left(\tfrac{1}{4}\pi - \tfrac{1}{2}\theta\right), \\ \beta &= \sqrt{\tfrac{1}{2}(1+\sin\theta)} = \cos\left(\tfrac{1}{4}\pi - \tfrac{1}{2}\theta\right). \end{aligned} \tag{21.5.37}$$

We split off the poles that are close to the origin:

$$\frac{v^2+\frac{1}{2}}{(v^2+\alpha^2)(v^2+\beta^2)\sqrt{v^2+1}} = \frac{A}{v^2+\alpha^2} + g(v), \tag{21.5.38}$$

where

$$g(v) = \frac{v^2+\frac{1}{2}}{(v^2+\alpha^2)(v^2+\beta^2)\sqrt{v^2+1}} - \frac{A}{v^2+\alpha^2}. \tag{21.5.39}$$

If we take $A = \frac{1}{2\beta}$ this function is regular at $v = \pm i\alpha$. Hence, when we use (3.4.29), we obtain

$$\begin{aligned} K_0(x,y) = {} & \tfrac{1}{2}\pi e^{-y}\operatorname{erfc}\left(\sqrt{2r}\sin\left(\tfrac{1}{4}\pi - \tfrac{1}{2}\theta\right)\right) + \\ & \tfrac{1}{2}\cos\theta\, e^{-r}\int_{-\infty}^{\infty} e^{-2rv^2} g(v)\,dv. \end{aligned} \tag{21.5.40}$$

Observe that this reduces to $K_0(0,y)$ in (21.5.29) when we take $\theta = \frac{1}{2}\pi$.

Expanding the function $g(v) = \sum_{k=0}^{\infty} g_{2k}v^{2k}$ and substituting this expansion we obtain the asymptotic expansion of the integral in (21.5.40)

$$\tfrac{1}{2}\cos\theta\, e^{-r}\int_{-\infty}^{\infty} e^{-2rv^2} g(v)\,dv \sim \tfrac{1}{2}\cos\theta\, e^{-r}\sqrt{\frac{\pi}{2r}}\sum_{k=0}^{\infty}\frac{g_{2k}\left(\frac{1}{2}\right)_k}{(2r)^k}, \tag{21.5.41}$$

as $r \to \infty$, uniformly with respect to $\theta \in [0, \frac{1}{2}\pi]$. The first few coefficients are

$$\begin{aligned} g_0 &= \frac{1}{2\beta^2(\beta+1)}, \quad g_2 = -\frac{3\beta^2+4\beta+2}{4\beta^4(\beta+1)^2}, \\ g_4 &= \frac{5\beta^5+15\beta^4+20\beta^3+28\beta^2+24\beta+8}{16\beta^6(\beta+1)^3}, \end{aligned} \tag{21.5.42}$$

where β is defined in (21.5.37).

We see that these coefficients remain regular when $\theta \in [0, \frac{1}{2}\pi]$, in particular when $\theta \to \frac{1}{2}\pi$.

21.5.2 *A similar Bessel K_1-integral*

In this case the integral is

$$K_1(x,y) = \int_x^{\infty} \frac{K_1\left(\sqrt{\xi^2+y^2}\right)}{\sqrt{\xi^2+y^2}}\,d\xi, \tag{21.5.43}$$

which has been studied in Boersma *et al.* (1984) and for which first-term asymptotic approximations have been given in terms of the complementary error function.

This time we use the sine transform (see Oberhettinger (1957, p. 13))

$$\frac{K_1\left(\sqrt{\xi^2+y^2}\right)}{\sqrt{\xi^2+y^2}} = \frac{1}{y}\int_0^{\infty} \frac{te^{-\xi\sqrt{t^2+1}}}{\sqrt{t^2+1}}\sin(yt)\,dt, \tag{21.5.44}$$

which is valid for $\xi \ge 0$. This gives

$$K_1(x,y) = \frac{1}{y}\int_0^\infty \frac{te^{-x\sqrt{t^2+1}}}{t^2+1}\sin(yt)\,dt, \quad x \ge 0. \tag{21.5.45}$$

By writing $t = \sinh u$, and using polar coordinates given in (21.5.26), we obtain

$$K_1(x,y) = \frac{1}{2iy}\int_{-\infty}^\infty \tanh u\, e^{-r\cosh(u-i\theta)}\,du. \tag{21.5.46}$$

Shifting the path of integration upwards in the complex plane we find

$$K_1(x,y) = \frac{1}{2iy}\int_{-\infty}^\infty \tanh(u+i\theta)\, e^{-r\cosh u}\,du, \quad 0 \le \theta < \tfrac{1}{2}\pi. \tag{21.5.47}$$

This can be written as (see (21.5.33))

$$\begin{aligned} K_1(x,y) &= \frac{1}{y}K_0(x,y) + \frac{1}{2iy}\int_{-\infty}^\infty e^{-r\cosh u}\,\frac{\sinh(u+i\theta)-i}{\cosh(u+i\theta)}\,du \\ &= \frac{1}{y}K_0(x,y) - \frac{\cos\theta}{2y}\int_{-\infty}^\infty e^{-r\cosh u}\,\frac{du}{\cosh u + \sin\theta} \\ &= \frac{1}{y}K_0(x,y) - \frac{\cos\theta\, e^{-r}}{2y}\int_{-\infty}^\infty e^{-2rv^2}\,h(v)\,dv, \end{aligned} \tag{21.5.48}$$

where

$$h(v) = \frac{1}{(v^2+\beta^2)\sqrt{v^2+1}}, \tag{21.5.49}$$

with β defined in (21.5.37). This function does not have poles in the neighborhood of the saddle point and the final integral in (21.5.48) can be expanded in a standard way.

21.5.3 *A singular perturbation problem*

The results of the Bessel function integrals of the previous sections can be used in a singular perturbation problem. Consider the elliptic partial differential equation

$$\varepsilon\Delta\Phi(x,y) - \frac{\partial\Phi}{\partial y}(x,y) = 0, \tag{21.5.50}$$

where $\varepsilon > 0$, in the quarter plane

$$D = \{x > 0, \quad y > 0\} \tag{21.5.51}$$

with boundary conditions

$$\Phi(x,0) = 0, \quad \Phi(0,y) = 1. \tag{21.5.52}$$

The problem is to find the asymptotic behavior of Φ as $\varepsilon \to 0$, in particular near the boundary $x = 0$, where a boundary layer occurs. Observe that we accept a corner discontinuity in the boundary values of Φ.

First we derive an integral representation of Φ. Put

$$\Phi(x,y) = e^{\omega y}F(x,y), \quad \omega = \frac{1}{2\varepsilon}, \tag{21.5.53}$$

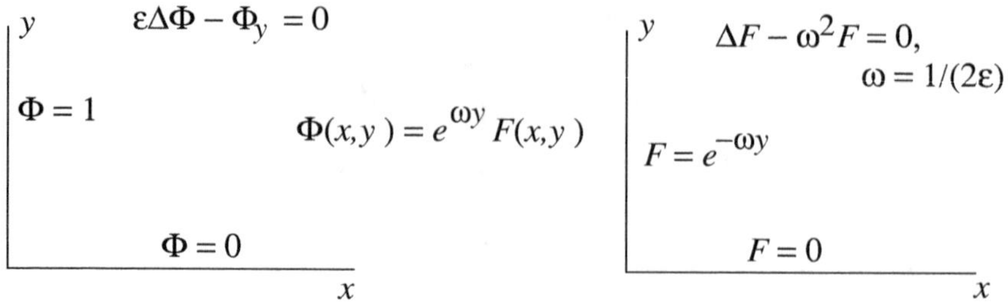

Fig. 21.1 Boundary values of Φ and F in the boundary value problems.

then F satisfies the Helmholtz equation

$$\Delta F(x,y) - \omega^2 F(x,y) = 0, \tag{21.5.54}$$

with boundary conditions (see Figure 21.1)

$$F(0,y) = e^{-\omega y}, \quad F(x,0) = 0. \tag{21.5.55}$$

Separating the variables we obtain

$$F(x,y) = \frac{1}{\pi i}\int_{-\infty}^{\infty} e^{i\lambda y - x\sqrt{\lambda^2+\omega^2}}\,\frac{\lambda\,d\lambda}{\lambda^2+\omega^2}. \tag{21.5.56}$$

Put $\lambda = \omega\sinh u$ and use the polar coordinates

$$x = r\cos\theta, \quad y = r\sin\theta, \quad 0 \le \theta \le \tfrac{1}{2}\pi. \tag{21.5.57}$$

Then

$$F(x,y) = \frac{1}{\pi i}\int_{-\infty}^{\infty} e^{-\omega r\cosh(u-i\theta)}\,\frac{\sinh u}{\cosh u}\,du, \tag{21.5.58}$$

and it follows that

$$F(x,y) = \frac{2y}{\pi}K_1(\omega x, \omega y), \tag{21.5.59}$$

the function that we have considered in §21.5.2. We have from (21.5.48)

$$F(x,y) = \frac{2}{\pi}K_0(\omega x,\omega y) - \frac{\cos\theta\, e^{-r}}{\pi}\int_{-\infty}^{\infty} e^{-2r\omega v^2}\,h(v)\,dv, \tag{21.5.60}$$

where $h(v)$ is given in (21.5.49).

Hence, by using the results for $K_0(x,y)$ from §21.5.1 we can obtain the asymptotic expansion of $F(x,y)$ for $\omega r \to \infty$, uniformly with respect to $\theta \in [0, \frac{1}{2}\pi]$.

The expansion obtained in this way is not valid for small values of r. We would like to find the value of $F(0,0)$ from the representations found so far.

First observe that $K_0(0,0) = \frac{1}{2}\pi$ (see (21.5.29)). It follows that

$$\begin{aligned} F(0,0) &= 1 - \frac{\cos\theta}{\pi}\int_{-\infty}^{\infty} h(v)\,dv, \\ &= 1 - \frac{\cos\theta}{\pi}\int_0^{\infty} \frac{dt}{\sqrt{t(1+t)}(t+\beta^2)}. \end{aligned} \tag{21.5.61}$$

This integral can be written in terms of the Gauss hypergeometric function. To verify this, substitute $t = s/(1-s)$ and compare with (12.0.3). It follows that

$$F(0,0) = 1 - \frac{2\cos\theta}{\pi\beta^2}\,{}_2F_1\left(\begin{matrix}1, \frac{1}{2}\\ \frac{3}{2}\end{matrix}; -z^2\right), \quad z = \tan\left(\tfrac{1}{4}\pi - \tfrac{1}{2}\theta\right). \tag{21.5.62}$$

Because (see Olde Daalhuis (2010b, Eq. 15.4.4))

$${}_2F_1\left(\begin{matrix}1, \frac{1}{2}\\ \frac{3}{2}\end{matrix}; -z^2\right) = \frac{\arctan z}{z}, \tag{21.5.63}$$

we obtain finally $F(0,0) = \dfrac{2\theta}{\pi}$, which shows a smooth transition of the boundary values from 0 to 1.

21.6 A double integral with poles near saddle points

In this section we consider the following double integral as the two-dimensional analogue of the complementary error function, that is, the analogue of the integral representation given in (21.1.2):

$$W(\alpha,\beta) = \int_{-\infty}^{\infty}\int_{-\infty}^{\infty} \frac{e^{-\omega(s^2+t^2)}}{(t+ps-i\alpha)(t+qs-i\beta)}\, ds\, dt, \tag{21.6.64}$$

where p and q are fixed real numbers, and α and β are possibly small real numbers. Initially we take $\alpha > 0$ and $\beta > 0$; ω is a positive large parameter.

The integral in (21.6.64) cannot simply be written as a sum of two integrals, because a partial fraction splitting gives less attractive integrals.

The poles are located on certain lines in the complex two-dimensional (s,t)-plane, and the lines will pass through the origin when α and β become 0. We transform (21.6.64) into two one-dimensional integrals that can be viewed as standard forms, and as generalizations of the complementary error function.

In Jones (1970/71) the two-dimensional integral

$$I(\alpha,\beta) = \int_{-\infty}^{\infty}\int_{-\infty}^{\infty} \frac{e^{-ia\cosh x - ib\cosh y}}{\sin\frac{1}{2}(ix+\alpha)\sin\frac{1}{2}(ix-iy+\beta)}\, dx\, dy \tag{21.6.65}$$

is considered with similar phenomena when α and β tend to zero. Jones considered this integral as a prototype and he introduced the function

$$G(z,\zeta) = \zeta e^{iz^2}\int_z^{\infty} \frac{e^{-it^2}}{t^2+\zeta^2}\, dt, \tag{21.6.66}$$

that can be used to describe the uniform asymptotic phenomena of (21.6.65).

The function $G(z,\zeta)$ cannot be expressed in terms of a familiar special function. It can be viewed as a generalization of the Fresnel integral (see §14.5.1).

Inspired by Jones' approach we write the integral (21.6.64) as a sum of two functions of the form

$$F(\omega,u,v) = \int_0^{\infty} \frac{re^{-\omega r^2}\, dr}{\sqrt{r^2+u^2(r^2+v^2)}}, \quad u \ge 0, \quad v > 0. \tag{21.6.67}$$

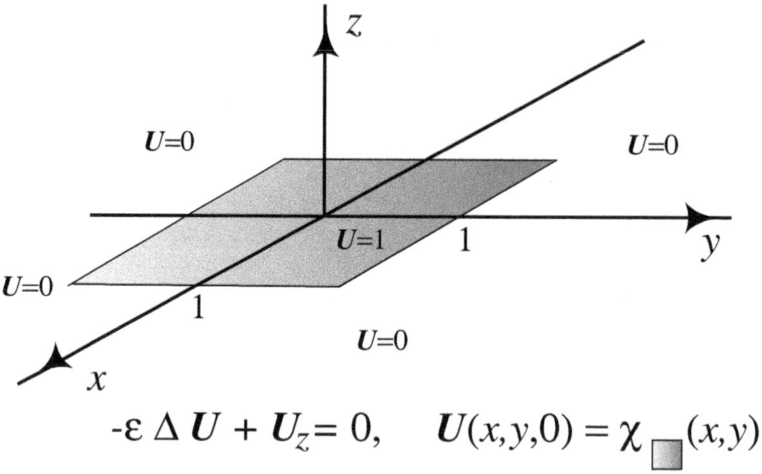

Fig. 21.2 Boundary values of $U(x, y, z)$ in the boundary value problem (21.6.70).

$F(\omega, u, v)$ can be viewed as a generalization of the complementary error function. For $u = 0$ it becomes, see (3.4.29),

$$F(\omega, 0, v) = \frac{\pi}{2v} e^{\omega v^2} \operatorname{erfc}\left(\sqrt{\omega}\, v\right). \tag{21.6.68}$$

When we change the variable of integration by writing $r^2 + u^2 = s^2$, we obtain

$$F(\omega, u, v) = e^{\omega u^2} \int_u^\infty \frac{e^{-\omega s^2}\, ds}{s^2 + v^2 - u^2}. \tag{21.6.69}$$

Observe that F and G are related by $\zeta F(i, z, \zeta) = G(z, \zeta)$, and that $F(\omega, u, v)$ is a function of two variables, because $F(\omega, u, v) = \sqrt{\omega} F(1, u\sqrt{\omega}, v\sqrt{\omega})$.

When ω is large and $v \to 0$, the poles at $r = \pm iv$ of the integral in (21.6.67) coalesce with the saddle point at the origin. If, in addition, $u \to 0$, these points also coalesce with two algebraic singularities.

In (21.6.69) the saddle point at $s = 0$ is located outside the domain of integration, and when $u \to 0$ the saddle point coalesces with an endpoint. If, in addition, $v \to 0$, the saddle point coalesces also with two poles.

21.6.1 *Application to a singular perturbation problem*

We have used the integral in (21.6.64) to describe the asymptotic behavior of the solution of the following singularly perturbed convection-diffusion problem defined in the positive half-space: $\Omega = (-\infty, \infty) \times (-\infty, \infty) \times (0, \infty)$, with a "square shaped source of contamination" located in the boundary plane $z = 0$ (see Figure 21.2):

$$\begin{cases} -\varepsilon \Delta U + U_z = 0, & \text{if } (x, y, z) \in \Omega, \\ U(x, y, 0) = \chi_\square(x, y), & \text{for } -\infty \le x, y \le \infty, \end{cases} \tag{21.6.70}$$

where ε is a small positive parameter and $\chi_\square(x,y)$ is the characteristic function of the unit square:

$$\chi_\square(x,y) \equiv \begin{cases} 1 & \text{if} \quad x \in (-1,1) \wedge y \in (-1,1), \\ 0 & \text{if} \quad x \notin (-1,1) \vee y \notin (-1,1). \end{cases} \tag{21.6.71}$$

Observe that the Dirichlet data are discontinuous at the boundary of the unit square in the plane $z = 0$. Due to these discontinuities boundary layers occur for $z > 0$ if $x \sim \pm 1$ and $y \sim \pm 1$.

The solution of this problem can be derived by using Fourier transforms with respect to x and y, and solving the resulting equation by separating the variables. We obtain

$$U(x,y,z) = \frac{e^{\omega z}}{\pi^2} \int_{-\infty}^{\infty} \int_{-\infty}^{\infty} \frac{\sin(\omega t)}{t} \frac{\sin(\omega s)}{s} e^{-\omega \phi(s,t)}\, ds\, dt, \tag{21.6.72}$$

where

$$\phi(s,t) = -ixt - iys + z\sqrt{1+t^2+s^2}, \quad \omega = \frac{1}{2\varepsilon}. \tag{21.6.73}$$

This problem has been discussed in detail in López *et al.* (2006) (see also López *et al.* (2007) for a different approach), where we have also shown that the above given solution $U(x,y,z)$ is the unique solution of the problem.

After shifting the paths of integration upwards or downwards into the complex s and t planes, and by writing the trigonometric functions in terms of exponential functions, the saddle points can be determined. After a few transformations, we can derive four integrals with paths through the saddle points. It turns out that a first approximation of these integrals is of the type shown in (21.6.64). In $W(\alpha,\beta)$ small values of α and β correspond to values of x and y near ± 1. For x and y near the sides of the unit square complementary error functions can be used to describe the behavior of $U(x,y,z)$. However, for x and y near the corners of the unit square we need the functions $F(\omega,u,v)$ as given in (21.6.67).

In the next section we show how the integral of $W(\alpha,\beta)$ can be written in terms of the function $F(\omega,u,v)$ defined in (21.6.67).

21.6.1.1 *Transformations into a standard form*

We introduce polar coordinates

$$s = r\sin\theta, \quad t = r\cos\theta, \quad 0 \le \theta \le 2\pi, \tag{21.6.74}$$

and write the double integral in (21.6.64) in the form

$$W(\alpha,\beta) = \int_0^{\infty} e^{-\omega r^2} f(r)\, r\, dr, \tag{21.6.75}$$

where

$$f(r) = \int_0^{2\pi} \frac{d\theta}{(r\cos\theta + pr\sin\theta - i\alpha)(r\cos\theta + qr\sin\theta - i\beta)}. \tag{21.6.76}$$

We evaluate this integral by putting $u = e^{i\theta}$ and integrating around the unit circle in the complex u-plane. This gives

$$f(r) = \frac{-4i}{P_1P_2}\int_{|u|=1}\frac{u\,du}{(u^2+Q_1u+R_1)(u^2+Q_2u+R_2)}, \tag{21.6.77}$$

where

$$\begin{aligned} P_1 &= -r(1-ip), & P_2 &= -r(1-iq), \\ Q_1 &= \frac{2i\alpha}{P_1}, & Q_2 &= \frac{2i\beta}{P_2}, \\ R_1 &= \frac{1+ip}{1-ip}, & R_2 &= \frac{1+iq}{1-iq}. \end{aligned} \tag{21.6.78}$$

The zeros of the quadratic factors in the denominator of the integrand in (21.6.77) are

$$\begin{aligned} u_1^{(\pm)} &= \frac{p-i}{r\,(p^2+1)}\left(-\alpha \pm \sqrt{\alpha^2+r^2\,(p^2+1)}\right), \\ u_2^{(\pm)} &= \frac{q-i}{r\,(q^2+1)}\left(-\beta \pm \sqrt{\beta^2+r^2\,(q^2+1)}\right). \end{aligned} \tag{21.6.79}$$

Observe that $\left|u_1^{(+)}u_1^{(-)}\right| = 1$ and $\left|u_2^{(+)}u_2^{(-)}\right| = 1$. Because $\alpha > 0$ and $\beta > 0$, the zeros $u_1^{(+)}$ and $u_2^{(+)}$ are inside the unit circle, and can be used for evaluating the integral by using residues.

First we write

$$\begin{aligned} &\frac{u}{(u^2+Q_1u+R_1)(u^2+Q_2u+R_2)} \\ &\qquad = \frac{a_1u+b_1}{u^2+Q_1u+R_1} + \frac{a_2u+b_2}{u^2+Q_2u+R_2}. \end{aligned} \tag{21.6.80}$$

It is straightforward to verify that

$$\begin{aligned} a_1 &= -a_2 = -\frac{i(p-q)(p+i)(q+i)r^2}{2T}, \\ b_1 &= -\frac{(p-i)(q+i)\left(\alpha(q+i)-\beta(p+i)\right)r}{2T}, \\ b_2 &= \frac{(p+i)(q-i)\left(\alpha(q+i)-\beta(p+i)\right)r}{2T}, \end{aligned} \tag{21.6.81}$$

where

$$T = (\alpha-\beta)^2 + (p-q)^2r^2 + (\alpha q - \beta p)^2. \tag{21.6.82}$$

Calculation of the two residues in the integral in (21.6.77) gives

$$f(r) = \frac{8\pi}{P_1P_2}\left(\frac{a_1u_1^{(+)}+b_1}{u_1^{(+)}-u_1^{(-)}} + \frac{a_2u_2^{(+)}+b_2}{u_2^{(+)}-u_2^{(-)}}\right), \tag{21.6.83}$$

which can be evaluated in the form

$$f(r) = \frac{2\pi}{T}\left(\frac{\alpha-\beta+p(\alpha q-\beta p)}{\sqrt{\alpha^2+r^2(1+p^2)}} - \frac{\alpha-\beta+q(\alpha q-\beta p)}{\sqrt{\beta^2+r^2(1+q^2)}}\right). \tag{21.6.84}$$

It is easily seen that with T defined in (21.6.82) and the representation in (21.6.75), the function $W(\alpha,\beta)$ can be written as a sum of two functions $F(\omega,u,v)$ defined in (21.6.67).

When $\alpha=\beta$ the function $f(r)$ reduces to

$$f(r) = -\frac{2\pi\alpha}{(r^2+\alpha^2)\sqrt{(1+p^2)\,r^2+\alpha^2}}. \tag{21.6.85}$$

21.7 The Fermi–Dirac integral

We define the Fermi–Dirac integral by

$$\mathcal{F}_q(z) = \frac{1}{\Gamma(q+1)}\int_0^\infty \frac{t^q}{1+e^{t-z}}\,dt, \quad \Re q > -1, \quad z\in\mathbb{C}. \tag{21.7.86}$$

This function has been considered in detail in Dingle (1973), where several relations are given for this function and asymptotic expansions are derived for large values of z with q fixed. First we give the convergent representation

$$\mathcal{F}_q(-z) = \sum_{n=1}^{\infty}\frac{(-1)^{n-1}e^{-nz}}{n^{q+1}}, \quad |\operatorname{ph} z| < \tfrac{1}{2}\pi, \tag{21.7.87}$$

which easily follows from (21.7.86) by expanding $1/\left(1+e^{t-z}\right)$.

A result from Dingle (1973, p. 20) reads

$$\mathcal{F}_q(z) \sim 2z^{q+1}\sum_{n=0}^{\infty}\frac{t_{2n}}{\Gamma(q+2-2n)\,z^{2n}} + \cos(\pi q)\mathcal{F}_q(-z), \tag{21.7.88}$$

as $z\to\infty$, $|\operatorname{ph} z| < \frac{1}{2}\pi$. Here,

$$t_{2n} = \sum_{m=1}^{\infty}\frac{(-1)^{m-1}}{m^{2n}} = \left(1-2^{1-2n}\right)\zeta(2n) = \left(1-2^{1-2n}\right)\frac{(2\pi)^{2n}}{2(2n)!}B_{2n}, \tag{21.7.89}$$

for $n\ge 1$, where B_n are the Bernoulli numbers. Dingle does not give this explicit form of t_{2n}.

The expansion in (21.7.88) has been studied in detail in Garoni *et al.* (2001), where in particular the role of term $\cos(\pi q)\mathcal{F}_q(-z)$ has been explained. Observe that this term vanishes if $q=\frac{1}{2},\frac{3}{2},\ldots$, and that, when q assumes other values, it gives an exponentially small contribution compared to terms of the series in (21.7.88). As an example of the importance of this term, the authors show that for a 2-dimensional nonrelativistic ideal Fermi gas, the subdominant exponentially small term becomes dominant.

In Schell (1987) expansions are given for large q and real $z = x$ in the three cases:

(1) $q \sim x$, in terms of $\mathcal{F}_q(-x)$ and the incomplete gamma function $\Gamma(q+2, x)$, which in turn is approximated by using the complementary error function;
(2) $q = ax,\ 0 < a < 2$, again $\mathcal{F}_q(-x)$ and the incomplete gamma function are used;
(3) $\sqrt{q} = o(q - x)$, which implies $x = o(q)$; the expansion is in terms of elementary functions and is obtained by using the saddle points method.

In this section we derive an asymptotic expansion of $\mathcal{F}_q(x)$ for large values of x, which is uniformly valid with respect to $q \in [0, \infty)$. We summarize the main points of Temme and Olde Daalhuis (1990).

We have the representation

$$\mathcal{F}_q(z) = \frac{1}{2i} \int_{\mathcal{L}} \frac{e^{zs}}{s^{q+1} \sin \pi s}\, ds, \tag{21.7.90}$$

where, initially, $\mathcal{L}$ is a vertical contour that cuts the real s-axis between 0 and 1. The integral represents an analytic function of z (we can change the direction of the path at infinity when z assumes complex values) and we obtain the representation in (21.7.87) by shifting the path across the poles on the left. Because $\mathcal{F}_q(z)$ is an analytic function of z, and by invoking the principle of analytic continuation, it follows that we can use (21.7.90) for all $z \in \mathbb{C}$.

We proceed with positive values of $z = x$. The saddle point of $e^{xs}s^{-q} = \exp(xs - q \ln s)$ is located at $s_0 = q/x$. We shift the contour to the point s_0, and pick up the residues. Assuming $s_0 \neq 1, 2, 3, \ldots$, we obtain

$$\mathcal{F}_q(x) = \sum_{n=1}^{N-1} \frac{(-1)^{n-1} e^{xn}}{n^{q+1}} + \frac{1}{2i} \int_{\mathcal{L}} \frac{e^{xs}}{s^{q+1} \sin \pi s}\, ds, \tag{21.7.91}$$

where N is the integer satisfying $N - 1 < s_0 < N$ and $\mathcal{L}$ cuts the real positive axis at s_0, the saddle point. When s_0 assumes an integer value the saddle point coincides with a pole of $1/\sin \pi s$. We can apply Van der Waerden's method, which gives an error function as main approximant. However, in the present case, the analysis is much easier by introducing an incomplete gamma function.

We split off the pole near the saddle point by writing

$$h_N(s) = \frac{\pi}{\sin \pi s} - \frac{(-1)^N}{s - N}, \quad N = 0, 1, 2, \ldots. \tag{21.7.92}$$

Then, by using the loop integral of the incomplete gamma function $Q(a, z)$ in (37.2.15),

$$\begin{aligned}
\mathcal{F}_q(x) &= \sum_{n=1}^{N-1} \frac{(-1)^{n-1} e^{xn}}{n^{q+1}} + G_q(x) + H_q(x), \\
G_q(x) &= \frac{(-1)^N}{2\pi i} \int_{\mathcal{L}} \frac{e^{xs}}{s^{q+1}(s - N)}\, ds = \frac{(-1)^N e^{xN}}{N^{q+1}} Q(q+1, xN), \\
H_q(x) &= \frac{1}{2\pi i} \int_{\mathcal{L}} \frac{e^{xs}}{s^{q+1}} h_N(s)\, ds.
\end{aligned} \tag{21.7.93}$$

The absolute values of the terms in the finite series in the first line of (21.7.93) decrease as n increases. This follows from looking at the graph of $\exp(xs - q\ln s)$ for $0 < s < s_0$. Therefore, the main contribution to $\mathcal{F}_q(x)$ comes from the term with $n = 1$, and the integral in (21.7.91) does not dominate any term of the finite sum.

In addition, when the sum contains more than one term, that is, if $N > 2$, then the terms generate an asymptotic scale (see §1.4). This follows from

$$\frac{e^{x(n+1)}}{(n+1)^q} = R\,\frac{e^{xn}}{n^q}, \qquad R = e^{x+q\ln\frac{n}{n+1}}. \tag{21.7.94}$$

We estimate R for the case that $n+1 < s_0 = q/x$, and we have

$$x + q\ln\frac{n}{n+1} = x\left(1 + s_0\ln\frac{n}{n+1}\right) \le -\frac{x}{2(n+1)}. \tag{21.7.95}$$

Hence, $R = o(1)$ as $q, x \to \infty$.

As an example, take $x = 50$, $q = 205$. Then $s_0 = 4.1$ and $N = 5$, and the four terms in the expansion are

$$\begin{aligned} n = 1:&\quad 5.1847\text{e+}21,\\ n = 2:&\quad 2.6138\text{e–}19,\\ n = 3:&\quad 7.1976\text{e–}34,\\ n = 4:&\quad 6.8319\text{e–}38. \end{aligned} \tag{21.7.96}$$

We see that the terms decrease very fast, and that only one term is needed for computations. In addition, in this example, we can forget about the terms $G_q(x)$ and $H_q(x)$ in (21.7.93).

However, these terms are important when N or s_0 are relatively small, say $1 \le N \le 3$. In that case, the asymptotic expansion of $H_q(x)$ can be obtained by using the saddle point method, and we can use the method of §25.3. By expanding $h_N(s)$ at the saddle point

$$h_N(s) = \sum_{n=0}^{\infty} c_n(s_0)(s - s_0)^n, \tag{21.7.97}$$

we obtain

$$H_q(x) \sim \frac{x^q}{\Gamma(q+1)} \sum_{n=0}^{\infty} c_n(s_0)\Phi_n(q)\frac{1}{x^n}, \qquad x \to \infty, \tag{21.7.98}$$

where $q = \mathcal{O}(x)$ and the Φ_n are simple polynomials, defined by

$$\Phi_n(q) = \frac{\Gamma(q+1)}{2\pi i}\int_{\mathcal{L}} \frac{e^t(t-q)^n}{t^{q+1}}\,dt. \tag{21.7.99}$$

We have

$$\Phi_0(q) = 1, \qquad \Phi_1(q) = 0, \tag{21.7.100}$$

and other ones follow from the recursion

$$\Phi_{n+1}(q) = -n\Phi_n(q) - nq\Phi_{n-1}(q). \tag{21.7.101}$$

Chapter 22

Saddle point near algebraic singularity

In this chapter we consider Case 4 of Table 20.1 in which the exponential function $\exp\left(-z\left(\frac{1}{2}t^2-\alpha t\right)\right)$ has a saddle point at $t=\alpha$. When $\alpha \le \alpha_0 < 0$ (with α_0 fixed) we can use Laplace's method as we have explained in §10.4.2. When $\alpha \ge \alpha_1 > 0$ we can use Watson's lemma. An extra point of attention is the factor $t^{\beta-1}$, an algebraic singularity at the origin.

In Chapter 11 we have introduced the parabolic cylinder function, where we considered the large argument behavior, and in Chapter 30 we give the large parameter case. In this chapter we explain how this function can be used as a main approximant when in a semi-infinite integral a saddle point coalesces with the finite endpoint. In this way, we obtain an expansion in which the saddle point $\alpha \in [\alpha_0, \alpha_1]$, an interval that contains the origin as an interior point. We also consider contour integrals with the same phenomena.

We give examples for the Kummer functions $U(a,c,z)$ and ${}_1F_1(a;c;z)$.

22.1 A saddle point near an endpoint of the interval

The standard form is

$$F_\beta(z,\eta) = \frac{1}{\Gamma(\beta)} \int_0^\infty t^{\beta-1} e^{-z\left(\frac{1}{2}t^2-\eta t\right)} f(t)\, dt, \tag{22.1.1}$$

where $z \to +\infty$. We assume that β is fixed and positive. Because of the gamma function in front of the integral we may relax the condition $\beta > 0$ by taking a loop integral or by integrating by parts. We assume that f is an analytic function in a domain $\mathcal{D}$ containing $[0,\infty)$, and $\eta \in \mathcal{D}_0$, with $\mathcal{D}_0$ properly inside $\mathcal{D}$. For example, when $f(t) = 1/(1+t)$ and we take real values of η, we assume that $\eta \ge \eta_0 > -1$. Extension to complex variables is possible, of course.

When f is a constant function the integral in (22.1.1) reduces to a parabolic cylinder function. We use the notation

$$W_\beta(z) = \frac{1}{\Gamma(\beta)} \int_0^\infty t^{\beta-1} e^{-\frac{1}{2}t^2+zt}\, dt, \tag{22.1.2}$$

and using (11.2.8) we have

$$W_\beta(z) = e^{\frac14 z^2} U\left(\beta+\tfrac12, -z\right). \tag{22.1.3}$$

When $\beta = 1$ we have an important case that will be considered in a somewhat different presentation, see §36.1. In that section we consider cumulative distribution functions, and the main approximant in the uniform expansion is the error function.

The saddle point in (22.1.1) is located at $t_0 = \eta$, and we allow η to be positive, zero, or negative. This causes the saddle point to coalesce with the endpoint of integration, and when η changes sign the asymptotic behavior for large z will change drastically. By using the parabolic cylinder function an approximation can be obtained that is uniformly valid with respect to η in an interval containing the origin as an internal point, the precise domain following from the analytical properties of the function f in (22.1.1)

22.2 The Bleistein expansion

We construct the asymptotic approximation by using an integration-by-parts method due to Bleistein (1966). This method can be applied in slightly different versions in many other cases, for example in Airy-type expansions in Chapter 23.

In Bleistein's method the function f is written in the form

$$f(t) = A_0 + B_0 t + t(t-\eta)g(t), \tag{22.2.4}$$

where A_0, B_0 follow from substituting $t = 0, t = \eta$, which gives

$$A_0 = f(0), \quad B_0 = \frac{f(\eta) - f(0)}{\eta}. \tag{22.2.5}$$

With these values of A_0, B_0 the function g is analytic in $\mathcal{D}$. An elegant proof follows from verifying that, if f is written as the Cauchy integral

$$f(t) = \frac{1}{2\pi i}\int_{\mathcal{C}} \frac{f(\tau)}{\tau - t}\,d\tau, \tag{22.2.6}$$

then the function g can be written as

$$g(t) = \frac{1}{2\pi i}\int_{\mathcal{C}} \frac{f(\tau)}{\tau(\tau-t)(\tau-\eta)}\,d\tau, \tag{22.2.7}$$

where $\mathcal{C}$ is a contour in $\mathcal{D}$ containing the points $0, t, \eta$ in its interior.

It is easily verified that

$$\begin{aligned}\frac{1}{\Gamma(\beta)}\int_0^\infty t^{\beta-1} e^{-z\left(\frac12 t^2 - \eta t\right)}\,dt &= z^{-\frac12\beta} W_\beta\left(\eta\sqrt{z}\right),\\ \frac{1}{\Gamma(\beta)}\int_0^\infty t^{\beta} e^{-z\left(\frac12 t^2 - \eta t\right)}\,dt &= z^{-\frac12(\beta+1)} W_\beta'\left(\eta\sqrt{z}\right),\end{aligned} \tag{22.2.8}$$

and that, by integrating by parts,

$$\begin{aligned}\frac{1}{\Gamma(\beta)}\int_0^\infty t^{\beta-1}e^{-z\left(\frac12 t^2-\eta t\right)}t(t-\eta)g(t)\,dt &=\\ \frac{1}{z\Gamma(\beta)}\int_0^\infty t^{\beta-1}e^{-z\left(\frac12 t^2-\eta t\right)}f_1(t)\,dt,&\end{aligned} \tag{22.2.9}$$

where

$$f_1(t)=t^{1-\beta}\frac{d}{dt}\left(t^\beta g(t)\right)=\beta g(t)+tg'(t). \tag{22.2.10}$$

After these preparations it is not difficult to verify that we can construct the compound expansion

$$F_\beta(z,\eta)\sim z^{-\frac12\beta}W_\beta\left(\eta\sqrt{z}\right)\sum_{n=0}^\infty\frac{A_n}{z^n}+z^{-\frac12(\beta+1)}W_\beta'\left(\eta\sqrt{z}\right)\sum_{n=0}^\infty\frac{B_n}{z^n}, \tag{22.2.11}$$

as $z\to\infty$. The quantities A_n, B_n are given by

$$A_n=f_n(0),\quad B_n=\frac{f_n(\eta)-f_n(0)}{\eta}, \tag{22.2.12}$$

where the functions f_n are defined recursively by the scheme

$$f_n(t)=A_n+B_nt+t(t-\eta)g_n(t),\quad f_{n+1}(t)=t^{1-\beta}\frac{d}{dt}\left(t^\beta g_n(t)\right), \tag{22.2.13}$$

$n=0,1,2,\ldots$, with $f_0(t)=f(t)$.

An exact form of (22.2.11) including a remainder reads as follows

$$\begin{aligned}F_\beta(z,\eta)&=z^{-\frac12\beta}W_\beta\left(\eta\sqrt{z}\right)\sum_{k=0}^{n-1}\frac{A_k}{z^k}+\\ &\qquad z^{-\frac12(\beta+1)}W_\beta'\left(\eta\sqrt{z}\right)\sum_{k=0}^{n-1}\frac{B_k}{z^k}+E_n(z,\eta,\beta),\\ E_n(z,\eta,\beta)&=\frac{1}{z^n\,\Gamma(\beta)}\int_0^\infty t^{\beta-1}e^{-z\left(\frac12 t^2-\eta t\right)}f_n(t)\,dt.\end{aligned} \tag{22.2.14}$$

For real parameters we construct a bound for the remainder. Assume that we can find positive numbers M_n and σ_n that do not depend on η or t, such that

$$|f_n(t)|\le M_ne^{\sigma_n\left(\frac12 t^2-\eta t\right)},\quad t\ge0, \tag{22.2.15}$$

and real values of η in $\mathcal{D}_0$. Then, if $z>\sigma_n$,

$$|E_n(z,\eta,\beta)|\le\frac{M_n}{z^n(z-\sigma_n)^{\frac12\beta}}W_\beta\left(\eta\sqrt{z-\sigma_n}\right), \tag{22.2.16}$$

and we see that

$$E_n(z,\eta,\beta)=\mathcal{O}\left(z^{-n-\frac12\beta}W_\beta\left(\eta\sqrt{z}\right)\right), \tag{22.2.17}$$

as $z\to\infty$, uniformly with respect to real values of η in $\mathcal{D}_0$.

The functions f_n defined in (22.2.13) are all analytic in the same domain $\mathcal{D}$ as the source function f, and can be represented in the form of Cauchy-type integrals. We have the following theorem.

Theorem 22.1. *Let the rational functions $R_n(s,t,\eta)$ be defined by*

$$R_0(s,t,\eta) = \frac{1}{s-t}, \quad R_{n+1}(s,t,\eta) = -\frac{s^{\beta-1}}{s-\eta}\frac{d}{ds}\left(s^{1-\beta}R_n(s,t,\eta)\right), \tag{22.2.18}$$

where $n = 0,1,2,\ldots$, $s,t,\eta \in \mathbb{C}, s \neq 0,t,\eta$. Let $f_n(t)$ be defined by the recursive scheme (22.2.13), *where f_0 is a given analytic function in a domain $\mathcal{D}$. Then we have*

$$f_n(t) = \frac{1}{2\pi i}\int_{\mathcal{C}} R_n(s,t,\eta) f_0(s)\,ds, \tag{22.2.19}$$

where $\mathcal{C}$ is a simple closed contour in $\mathcal{D}$ that encircles the points t,η and the origin. Similar integrals for the coefficients A_n, B_n follow from (22.2.12) *and* (22.2.19).

Proof. In the proof of Theorem 25.1 we explain in detail a related case; the present proof is similar, and is left as an exercise. □

Remark 22.1. In Soni and Sleeman (1987) a set of polynomials $\{P_n\}$ is introduced and the function f in (22.1.1) is expanded in the form $f(t) = \sum_{n=0}^{\infty} C_n P_n(t)$. The coefficients C_n are closely related to the coefficients A_n, B_n of the expansion given in (22.2.11). This gives a variant of Bleistein's method. △

Remark 22.2. In Olver (1997, p. 344) the integral (22.1.1) is expanded by substituting a power series expansion of the function f at the saddle point:

$$f(t) = \sum_{n=0}^{\infty} c_n (t-\eta)^n, \tag{22.2.20}$$

which gives the expansion

$$F_\beta(z,\eta) \sim \frac{1}{\Gamma(\beta)}\sum_{n=0}^{\infty} c_n \int_0^{\infty} t^{\beta-1}(t-\eta)^n e^{-z(\frac{1}{2}t^2-\eta t)}\,dt. \tag{22.2.21}$$

The integrals can be expressed in terms of the functions $W_\beta(x)$ used in (22.2.11), and recursions can be derived for them.

When $\eta < 0$ it seems to be better to expand f at the origin, the point of the interval of integration where the exponential function of the integrand assumes its maximal value. In addition, the coefficients may be simpler than those of an expansion at $t = \eta$. △

Problem 22.1. *Compute by using computer algebra* $R_n(s,t,\eta)$, $n=0,1,2$. *Take* $f(t)=1/(1+t^2)$ *and compute by using residue calculus the functions* $f_n(t)$ *from* (22.2.19). *Verify that*

$$
\begin{aligned}
A_0 &= 1,\\
B_0 &= -\frac{\eta}{1+\eta^2},\\
A_1 &= -\frac{\beta}{1+\eta^2},\\
B_1 &= \frac{\eta\left(3-\eta^2+2\beta\eta^2+2\beta\right)}{\left(1+\eta^2\right)^3},\\
A_2 &= \frac{\beta\left(2-3\eta^2-\eta^4+\beta-\beta\eta^4\right)}{\left(1+\eta^2\right)^3}.
\end{aligned}
\tag{22.2.22}
$$

♡

22.3 Extending the role of the parameter β

Considering the expansions in (22.2.11), we observe that the main approximant, the parabolic cylinder function $W_\beta(z)$, is a function of two variables, and that the parameter β is assumed to be fixed.

However, there is an easy modification by assuming that β is not necessarily fixed (in Chapter 25 we consider Laplace-type integrals with a similar phenomenon). We take the same conditions as for the integral in (22.1.1), and write it in the form

$$F_\beta(z,\eta)=\frac{1}{\Gamma(\beta)}\int_0^\infty e^{-z\phi(t)}f(t)\,\frac{dt}{t}, \tag{22.3.23}$$

where

$$\phi(t)=-\gamma\ln t+\tfrac{1}{2}t^2-\eta t,\quad \gamma=\frac{\beta}{z}. \tag{22.3.24}$$

The saddle points are given by

$$t_\pm=\frac{\eta\pm\sqrt{\eta^2+4\gamma}}{2}. \tag{22.3.25}$$

We can repeat the recursive scheme shown in (22.2.13), now with interpolating points the two saddle points $t_\pm$. That is, we write

$$f(t)=C_0+D_0t+(t-t_-)(t-t_+)g_0(t), \tag{22.3.26}$$

where

$$C_0=\frac{t_+f(t_-)-t_-f(t_+)}{t_+-t_-},\quad D_0=\frac{f(t_+)-f(t_-)}{t_+-t_-}. \tag{22.3.27}$$

After substituting $f(t)$ of (22.3.26) into (22.3.23) contributions with parabolic cylinder functions are obtained, and these are the same as those given in (22.2.8).

Integration by parts now runs as follows:

$$\begin{aligned}&\frac{1}{\Gamma(\beta)}\int_0^\infty e^{-z\phi(t)}(t-t_-)(t-t_+)g_0(t)\,\frac{dt}{t}=\\&\frac{-1}{z\Gamma(\beta)}\int_0^\infty g_0(t)\,de^{-z\phi(t)}=\frac{1}{z\Gamma(\beta)}\int_0^\infty e^{-z\phi(t)}f_1(t)\,\frac{dt}{t},\end{aligned}\tag{22.3.28}$$

where

$$f_1(t)=t\frac{d}{dt}g(t).\tag{22.3.29}$$

By repeating this we obtain the same type of expansion as in (22.2.11):

$$F_\beta(z,\eta)\sim z^{-\frac12\beta}W_\beta\left(\eta\sqrt{z}\right)\sum_{n=0}^\infty\frac{C_n}{z^n}+z^{-\frac12(\beta+1)}W'_\beta\left(\eta\sqrt{z}\right)\sum_{n=0}^\infty\frac{D_n}{z^n},\tag{22.3.30}$$

or in the form with a remainder as in (22.2.14), which now has the form

$$E_n(z,\eta,\beta)=\frac{1}{z^n\,\Gamma(\beta)}\int_0^\infty e^{-z\phi(t)}f_n(t)\,\frac{dt}{t}.\tag{22.3.31}$$

Assume that we can find positive M_n and σ_n that do not depend on t,η,γ, such that

$$|f_n(t)|\le M_ne^{\sigma_n\phi(t)},\quad t\ge 0.\tag{22.3.32}$$

Then, if $z>\sigma_n$,

$$|E_n(z,\eta,\beta)|\le M_n\frac{\Gamma(\beta(1-\sigma_n/z))}{\Gamma(\beta)}\frac{W_{\beta(1-\sigma_n/z)}\left(\eta\sqrt{z-\sigma_n}\right)}{z^n(z-\sigma_n)^{\frac12\beta(1-\sigma_n/z)}},\tag{22.3.33}$$

and again we have an order estimate as in (22.2.17), which holds as $z\to\infty$, uniformly with respect to real values of η in $\mathcal{D}_0$ and $\beta\ge 0$.

Remark 22.3. In the construction of the expansion in this section we use the two saddle points given in (22.3.25), and these saddle points may coalesce at zero when both parameters γ and η tend to zero, with γ always being nonnegative.

As we will see in Chapter 23, a uniform expansion for the case of two coalescing saddle points usually involves Airy functions. In that case the phase function behaves like a cubic polynomial in a neighborhood of the coalescing saddle points.

The present case is different, however. The phase function in (22.3.24) has a logarithmic singularity at the origin, which disappears when $\gamma=0$. At the same instance one of the saddle points disappears (and we have the same case as in §22.2), in a similar way as we will see in Chapter 25. △

Remark 22.4. We have extended Bleistein's original method for the real integral in (22.1.1) by including the parameter β that defines the order of the parabolic cylinder functions. It seems that this has not been been discussed earlier in the literature. There are examples for contour integrals, however, with applications to orthogonal polynomials. For details we refer to Chapter 24, where we consider orthogonal polynomials (Gegenbauer, Tricomi–Carlitz) and to §32.4 (Laguerre). In these cases the parabolic cylinder functions reduce to Hermite polynomials. We mention other examples in §24.4. △

Problem 22.2. *Verify that in the present case the functions f_n can be represented in the form*

$$f_n(t) = \frac{1}{2\pi i}\int_{\mathcal{C}} S_n(s,t,\eta) f_0(s)\,ds, \tag{22.3.34}$$

where $\mathcal{C}$ is a simple closed contour in $\mathcal{D}$ (the domain where f is analytic) that encircles the points t,η and the origin. The functions S_n are defined by

$$\begin{aligned} S_0(s,t,\eta) &= \frac{1}{s-t}, \\ S_{n+1}(s,t,\eta) &= -\frac{1}{(s-t_-)(s-t_+)}\frac{d}{ds}\left(sS_n(s,t,\eta)\right), \end{aligned} \tag{22.3.35}$$

where $n = 0,1,2,\ldots,\ s,t,\eta \in \mathbb{C},\ s \ne 0,t,\eta$. Verify that

$$S_1(s,t,\eta) = \frac{t}{(s-t_-)(s-t_+)(s-t)^2}. \tag{22.3.36}$$

Take $f(t) = 1/(1+t^2)$ and compute by using residue calculus the function $f_1(t)$ from (22.3.34) and (22.3.36). With this function f, verify that

$$\begin{aligned} C_0 &= \frac{\eta^2+\gamma+1}{\eta^2+(\gamma+1)^2}, \\ D_0 &= -\frac{\eta}{\eta^2+(\gamma+1)^2}, \\ C_1 &= \frac{\gamma(2-3\eta^2+6\gamma-\eta^4+3\eta^2\gamma^2+6\gamma^2+2\gamma^3)}{(\eta^2+(\gamma+1)^2)^3}, \\ D_1 &= \frac{\eta(1-\gamma)(3\gamma^2+6\gamma+3-\eta^2)}{(\eta^2+(\gamma+1)^2)^3}. \end{aligned} \tag{22.3.37}$$

Observe that these coefficients are bounded functions of real η and γ. ♡

22.4 Contour integrals

The same asymptotic phenomenon (saddle point close to a singularity) occurs in integrals of the type

$$G_\beta(z,\eta) = \frac{\Gamma(1-\beta)}{2\pi i}\int_{\mathcal{L}} t^{\beta-1} e^{-z\left(\frac12 t^2-\eta t\right)} f(t)\,dt, \quad -\pi < \operatorname{ph} t < \pi, \tag{22.4.38}$$

where $\beta \ne 1,2,3,\ldots$ and $\mathcal{L}$ is a Hankel contour as shown in Figure 2.1. We assume that the singularities of f are "inside" the loop $\mathcal{L}$.

When f is a constant $G_\beta(z,\eta)$ can be expressed in terms of the parabolic cylinder function $U(a,z)$. This follows from the integral representation (see §30.4.3)

$$U(a,z) = \frac{\Gamma\left(\frac12-a\right)}{2\pi i} e^{-\frac14 z^2} \int_{-\infty}^{(0+)} e^{zt-\frac12 t^2} t^{a-\frac12}\,dt, \tag{22.4.39}$$

where $a \ne \frac12, \frac32, \frac52, \ldots$ and $-\pi < \operatorname{ph} t < \pi$.

We define

$$Z_\beta(z) = \frac{\Gamma(1-\beta)}{2\pi i} \int_{\mathcal{L}} t^{\beta-1} e^{-\frac{1}{2}t^2+zt}\, dt, \tag{22.4.40}$$

and have

$$Z_\beta(z) = e^{\frac{1}{4}z^2} U\left(\beta - \tfrac{1}{2}, z\right). \tag{22.4.41}$$

We use the integration by parts procedure as in §22.2 and obtain

$$\begin{aligned} G_\beta(z,\eta) &= z^{-\frac{1}{2}\beta} Z_\beta\left(\eta\sqrt{z}\right) \sum_{k=0}^{n-1} \frac{A_k}{z^k} + \\ &\quad z^{-\frac{1}{2}(\beta+1)} Z'_\beta\left(\eta\sqrt{z}\right) \sum_{k=0}^{n-1} \frac{B_k}{z^k} + E_n(z,\eta,\beta), \\ E_n(z,\eta,\beta) &= \frac{\Gamma(1-\beta)}{2\pi i} \int_{\mathcal{L}} t^{\beta-1} e^{-z\left(\frac{1}{2}t^2 - \eta t_n\right)} f_n(t)\, dt, \end{aligned} \tag{22.4.42}$$

where the coefficients A_n, B_n and the functions f_n are obtained from the scheme as explained in (22.2.12) and (22.2.13).

It will be clear that the same type of expansion can be obtained for the integral

$$G_\beta(z,\eta) = \frac{1}{2\pi i} \int_{c-i\infty}^{c+i\infty} t^{-\beta-1} e^{z\left(\frac{1}{2}t^2 - \eta t\right)} f(t)\, dt, \quad -\tfrac{1}{2}\pi < \operatorname{ph} t < \tfrac{1}{2}\pi, \tag{22.4.43}$$

with c so large that the singularities of the integrand are on the left of the contour.

For constant f this again becomes a parabolic cylinder function, see §30.4.2,

$$U(a,z) = \frac{e^{\frac{1}{4}z^2}}{i\sqrt{2\pi}} \int_{c-i\infty}^{c+i\infty} t^{-a-\frac{1}{2}} e^{\frac{1}{2}t^2 - zt}\, dt, \tag{22.4.44}$$

with $c > 0$.

In the next section we show how to obtain expansions of the Kummer U-function when starting from an integral of the form (22.1.1) or of ${}_1F_1(a;c;z)$ when starting from (22.4.38), after some transformations.

22.5 Kummer functions in terms of parabolic cylinder functions

In Chapter 10 we have considered Kummer functions ${}_1F_1(a;c;z)$ and $U(a,c,z)$ for large values of the parameters. These expansions are only valid for bounded values of $|z|$. For example, when $z \sim c$ the coefficients in the expansion in (10.4.90) tend to infinity. In this section we give expansions in which large values of z are allowed. The applications to Kummer functions are earlier considered in Temme (1978).

When $a = 1$ the Kummer functions reduce to incomplete gamma functions. We have

$$\begin{aligned} \gamma(a,z) &= a^{-1} z^a e^{-z} {}_1F_1\left({1 \atop a+1}; z\right) = a^{-1} z^a {}_1F_1\left({a \atop a+1}; -z\right), \\ \Gamma(a,z) &= z^a e^{-z} U(1, a+1, z) = z^a e^z U(1-a, 1-a, z). \end{aligned} \tag{22.5.45}$$

When $a = 1$ the parabolic cylinder functions in the expansions become complementary error functions. The asymptotic results are then comparable with those for incomplete gamma functions given in Chapter 37.

Remark 22.5. We give the results for large c; however, we can include a as a large parameter when we use the method of §22.3. △

22.5.1 *Uniform expansion of $U(a,c,z)$, $c \to +\infty$*

We use the integral representation given in (10.4.83). We use the notation

$$U(a,c,c\lambda) = \frac{1}{\Gamma(a)} \int_0^\infty g(t) e^{-c\phi(t)}\, dt, \quad \lambda = \frac{z}{c}, \tag{22.5.46}$$

where

$$g(t) = t^{a-1}(1+t)^{-a-1}, \quad \phi(t) = \lambda t - \ln(1+t). \tag{22.5.47}$$

We assume that λ may be any positive number. Hence, we assume that the saddle point at $t_0 = 1/\lambda - 1$ runs through the interval $[-1, \infty)$.

An expansion of ϕ at the origin reads

$$\phi(t) = (\lambda - 1)t + \tfrac{1}{2}t^2 + \mathcal{O}\left(t^3\right), \tag{22.5.48}$$

and we notice the typical behavior as in the exponential function of the standard form in (22.1.1). This suggests the transformation

$$\phi(t) = \tfrac{1}{2}w^2 + \eta w, \tag{22.5.49}$$

where we assume that the saddle point at $t = t_0$ corresponds to $w = -\eta$ (the saddle point in the w-domain), and that $\operatorname{sign}(t - t_0) = \operatorname{sign}(w + \eta)$. Using this transformation in (22.5.46) we obtain

$$U(a,c,c\lambda) = \frac{1}{\Gamma(a)} \int_0^\infty w^{a-1} e^{-c\left(\frac{1}{2}w^2 + \eta w\right)} f(w)\, dw, \tag{22.5.50}$$

where

$$f(w) = \left(\frac{t}{w}\right)^{a-1} (1+t)^{-a-1} \frac{dt}{dw}. \tag{22.5.51}$$

For η we have the relation

$$\tfrac{1}{2}\eta^2 = \lambda - 1 - \ln\lambda, \tag{22.5.52}$$

where we have to determine the sign when taking the square root.

When $\lambda > 1$ we have $t_0 < 0$, and in this case the saddle point in the t-integral is outside the interval of integration. We need the same for the w-integral with the saddle point $-\eta$, that is, η should have the sign of $\lambda - 1$. This gives

$$\eta = \operatorname{sign}(\lambda - 1)\sqrt{2(\lambda - 1 - \ln\lambda)}. \tag{22.5.53}$$

This corresponds to the condition for the transformation in (22.5.49).

An expansion for small values of $|\lambda - 1|$ reads

$$\eta = (\lambda - 1) - \tfrac{1}{3}(\lambda - 1)^2 + \tfrac{7}{36}(\lambda - 1)^3 + \mathcal{O}\left((\lambda - 1)^4\right), \quad \lambda \to 1, \tag{22.5.54}$$

which can be inverted into

$$\lambda = 1 + \eta + \tfrac{1}{3}\eta^2 + \tfrac{1}{36}\eta^3 + \mathcal{O}\left(\eta^4\right), \quad \eta \to 0. \tag{22.5.55}$$

We have met a relation like the one in (22.5.52) several times, see, for example, §6.2. In Figure 6.2 we have shown in detail the relation for complex parameters.

We return to (22.5.50) which is of the standard form shown in (22.1.1). The change of sign in η is only for convenience, because we use the same relation (that is, the one in (22.5.53)) in other chapters. We can construct an expansion as in (22.2.14) with $\beta = a$, z replaced by c and η by $-\eta$. In this way,

$$\begin{aligned} U(a,c,c\lambda) &= c^{-\frac{1}{2}a} W_a\left(-\eta\sqrt{c}\right) \sum_{k=0}^{n-1} \frac{A_k}{c^k} + \\ &\quad c^{-\frac{1}{2}(a+1)} W_a'\left(-\eta\sqrt{c}\right) \sum_{k=0}^{n-1} \frac{B_k}{c^k} + E_n(c,\eta,a), \\ E_n(c,\eta,a) &= \frac{1}{c^n\,\Gamma(a)} \int_0^\infty w^{a-1} e^{-z(\frac{1}{2}w^2+\eta w)} f_n(w)\,dw, \end{aligned} \tag{22.5.56}$$

where the coefficients A_n, B_n and the functions f_n are obtained from the scheme as explained in (22.2.12) and (22.2.13), with starting function $f = f_0$ defined in (22.5.51).

We compute the first coefficients. From (22.5.48) and (22.5.49) it is clear that $t/w \sim \eta/(\lambda - 1)$ as $w \to 0$, and this relation is well defined when $\lambda \to 1$. This gives

$$f(0) = \left(\frac{\eta}{\lambda - 1}\right)^a. \tag{22.5.57}$$

To determine $f(-\eta)$ we write (22.5.49) in the form

$$\tfrac{1}{2}(w+\eta)^2 = \tfrac{1}{2}\eta^2 + \phi(t) = \tfrac{1}{2}\phi''(t_0)\,(t - t_0)^2 + \mathcal{O}\left((t - t_0)^3\right), \tag{22.5.58}$$

as $t \to t_0$, which gives $w + \eta \sim \lambda(t - t_0)$ (where the correct sign when taking the square root follows from the condition on the transformation in (22.5.49)), and, hence, $dt/dw \sim 1/\lambda$ as $t \to t_0$. This gives

$$f(-\eta) = \left(\frac{t_0}{-\eta}\right)^{a-1} \lambda^{a+1} \frac{1}{\lambda} = \lambda\left(\frac{\lambda - 1}{\eta}\right)^{a-1}. \tag{22.5.59}$$

In this way we obtain the first approximation

$$U(a,c,c\lambda) \sim c^{-\frac{1}{2}a} A_0 W_a\left(-\eta\sqrt{c}\right) + c^{-\frac{1}{2}(a+1)} B_0 W_a'\left(-\eta\sqrt{c}\right), \tag{22.5.60}$$

where A_0, B_0 follow from (22.2.5) with η replaced by $-\eta$. That is,

$$U(a,c,c\lambda) \sim c^{-\frac{1}{2}a} \left(\frac{\eta}{\lambda - 1}\right)^a \left(W_a\left(-\eta\sqrt{c}\right) + \frac{C_0}{\sqrt{c}} W_a'\left(-\eta\sqrt{c}\right)\right), \tag{22.5.61}$$

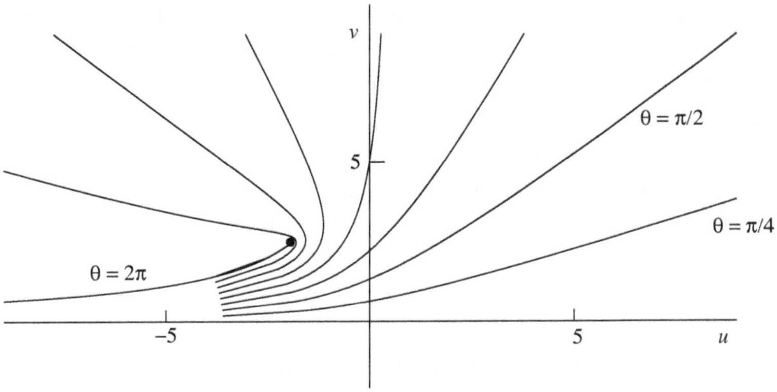

Fig. 22.1 Images in the w-plane of the half-lines $t = -1 + re^{i\theta}$, $r > 0$, where $\theta = \pm\frac{1}{4}\pi k$, $0 \le k \le 8$ for $\lambda = \frac{1}{2}$.

where

$$\eta C_0 = 1 - \lambda \left(\frac{\lambda - 1}{\eta}\right)^{2a-1}. \tag{22.5.62}$$

For small values of η we have

$$C_0 = -\tfrac{2}{3}(a+1) - \tfrac{1}{36}\left(8a^2 + 14a + 3\right)\eta + \mathcal{O}\left(\eta^2\right). \tag{22.5.63}$$

In these approximations the W-function is defined in terms of the parabolic cylinder function $U(a,z)$, see (22.1.3).

More terms A_n, B_n in the expansion in (22.2.11) can be obtained by expanding the function f defined in (22.5.51) at the points $w = 0$ and $w = -\eta$.

The function f is analytic in a domain around the real positive axis. This follows from locating the singularities of this function, which can be found by considering those of the transformation in (22.5.49). We have $dt/dw = (w+\eta)/\phi'(t)$.

The function $\phi'(t)$ vanishes at $t = t_0$, but at this point dt/dw is well defined, and non-vanishing. Because of the logarithmic function in $\phi(t)$ we should also consider the points $t_k = t_0 e^{2\pi k i}$, for $k = \pm1, \pm2, \ldots$. In the w-plane we identify the singular points w_k defined by $(w_k + \eta)^2 = 4k\pi i$, $k \ne 0$.

Taking the square root should be done by considering details of the transformation in (22.5.49), and we conclude that the singularities nearest to the real axis are (if $\lambda > 0$)

$$w_\pm = -\eta + 2\sqrt{\pi}e^{\pm\frac{3}{4}\pi i}. \tag{22.5.64}$$

In Figure 22.1 we give details of the mapping of part of the t-plane to the w-plane. We show the images in the w-plane of the half-lines $t = -1 + re^{i\theta}$, $r > 0$, where $\theta = \frac{1}{4}\pi k$, $1 \le k \le 8$ for $\lambda = \frac{1}{2}$. The black dot indicates the singular point at w_+. The images fold around the singular point w_+ when θ approaches 2π. A branch cut for the mapping runs from w_+ to $-\infty$. A similar picture follows by conjugation for negative values of θ.

From the location of the singular points $w_\pm$ of $f(w)$ we conclude the following. Let $\alpha = \operatorname{ph} w_-$ and $\beta = \operatorname{ph} w_+$. Then $f(w)$ is analytic inside the sector $\alpha < \operatorname{ph} w < \beta$. It also follows from the transformation and the definition of f that inside this sector $f(w) \sim 4\lambda w^{-a-2}$ as $w \to \infty$.

Because the iterated functions $f_n(w)$, which are defined by a similar scheme as in §22.2, are generated as linear combinations of derivatives of $f(w)$, it also follows by invoking Theorem 1.1, that we can assume that an estimate as in (22.2.15) is valid, that is,

$$|f_n(w)| \le M_n e^{\sigma_n(\frac{1}{2}w^2+\eta w)}, \quad w \ge 0, \tag{22.5.65}$$

where M_n and σ_n are some positive numbers independent of w and η. From this estimate we can construct bounds of the remainder in the asymptotic expansion as in (22.3.33).

Remark 22.6. The mapping in (22.5.49) may be compared with the one considered in §10.4.4 for the expansion of $U(a,-c,z)$ for large positive values of c. See also Figure 10.1, where the half-lines are only for $-\pi \le \theta \le \pi$. △

22.5.2 *Uniform expansion of ${}_1F_1(a;c;z)$, $c \to +\infty$*

We use the representation given in (10.1.6), and we write it in the form

$${}_1F_1\left(\begin{matrix} a+c \\ c \end{matrix}; -z\right) = \frac{\Gamma(c)\Gamma(1+a)}{2\pi i\,\Gamma(a+c)} \int_0^{(1+)} e^{-zt} t^{a+c-1}(t-1)^{-a-1}\,dt, \tag{22.5.66}$$

where $\Re(a+c) > 0$ and $-a \ne 1,2,3,\ldots$. Next we use Kummer's transformation (10.1.9), shift the contour one unit to the left, and change a by $-a$. This gives

$${}_1F_1\left(\begin{matrix} a \\ c \end{matrix}; z\right) = \frac{\Gamma(c)\Gamma(1-a)}{2\pi i\,\Gamma(c-a)} \int_{-1}^{(0+)} e^{-zt} t^{a-1} t(t+1)^{c-a-1}\,dt, \tag{22.5.67}$$

where $\Re(c-a) > 0$ and $a \ne 1,2,3,\ldots$. We write this in the form

$${}_1F_1\left(\begin{matrix} a \\ c \end{matrix}; c\lambda\right) = \frac{\Gamma(c)\Gamma(1-a)}{2\pi i\,\Gamma(c-a)} \int_{-1}^{(0+)} e^{-c\phi(t)} g(t)\,dt, \tag{22.5.68}$$

where $\phi(t) = \lambda t - \ln(1+t)$ and $g(t) = t^{a-1}(1+t)^{-a-1}$ are the same as in (22.5.47).

We use the substitution $\phi(t) = \frac{1}{2}w^2 + \eta w$ introduced in (22.5.49), again with the condition $\operatorname{sign}(t-t_0) = \operatorname{sign}(w+\eta)$ for $t > -1$, and obtain

$${}_1F_1\left(\begin{matrix} a \\ c \end{matrix}; c\lambda\right) = \frac{\Gamma(c)\Gamma(1-a)}{2\pi i\,\Gamma(c-a)} \int_{\mathcal{L}} w^{a-1} e^{-c(\frac{1}{2}w^2+\eta w)} f(w)\,dw, \tag{22.5.69}$$

where $\mathcal{L}$ is a Hankel contour as shown in Figure 2.1 and (cf. (22.5.51))

$$f(w) = \left(\frac{t}{w}\right)^{a-1} (1+t)^{-a-1} \frac{dt}{dw}. \tag{22.5.70}$$

The quantity η in (22.5.69) is defined in (22.5.53). From the previous section we know that the relevant singularities are $w_\pm$ given in (22.5.64), and we assume that these points are inside the Hankel contour.

By using the expansion given in (22.4.42) we find

$$\begin{aligned}\frac{\Gamma(c-a)}{\Gamma(c)}\,{}_1F_1\left({a \atop c};\, c\lambda\right) = c^{-\frac{1}{2}a} Z_a\left(-\eta\sqrt{c}\right) \sum_{k=0}^{n-1} \frac{A_k}{c^k} + \\ c^{-\frac{1}{2}(a+1)} Z_a'\left(-\eta\sqrt{c}\right) \sum_{k=0}^{n-1} \frac{B_k}{c^k} + E_n(c,\eta,a),\end{aligned} \tag{22.5.71}$$

where

$$E_n(c,\eta,a) = \frac{\Gamma(1-a)}{2\pi i} \int_{\mathcal{L}} w^{a-1} e^{-a\left(\frac{1}{2}w^2+\eta w\right)} f_n(w)\, dw, \tag{22.5.72}$$

and the coefficients A_n, B_n and the functions f_n are the same as in (22.5.56).

Chapter 23

Two coalescing saddle points: Airy-type expansions

In this chapter we consider Case 6 of Table 20.1 in which the exponential function $\exp\left(z\left(\frac{1}{3}t^3 - \eta t\right)\right)$ has two saddle points at $t = \pm\sqrt{\eta}$. When $\eta \neq 0$ and fixed, we can use Laplace's method for one or both saddle points. When $\eta \to 0$ we should take into account the contributions from both saddle points in one expansion. As we will see, we can use Airy functions to handle this case.

23.1 The standard form

The standard form of this chapter is

$$F_\eta(\nu) = \frac{1}{2\pi i} \int_{\mathcal{C}} e^{\nu\left(\frac{1}{3}t^3 - \eta t\right)} f(t)\, dt, \tag{23.1.1}$$

where the contour is one of the contours $\mathcal{C}_i$ of Figure 8.2. When we take the contour as for the Airy function in (8.1.4), and replace f by unity, we obtain

$$\frac{1}{2\pi i} \int_{\mathcal{C}} e^{\nu\left(\frac{1}{3}t^3 - \eta t\right)}\, dt = \nu^{-\frac{1}{3}} \operatorname{Ai}\left(\eta \nu^{\frac{2}{3}}\right). \tag{23.1.2}$$

For the general case the asymptotic expansion of $F_\eta(\nu)$ can be given in terms of this Airy function. The asymptotic feature of this type of integral is that the phase function $\phi(t) = \frac{1}{3}t^3 - \eta t$ has two saddle points at $\pm\sqrt{\eta}$ that coalesce when $\eta \to 0$, and it is not possible to describe the asymptotic behavior of $F_\eta(\nu)$ in terms of simple functions when η is small. When the parameter η is positive and bounded away from 0, one can perform a saddle point analysis on (23.1.1) and use a transformation $\phi(t) - \phi(\sqrt{\eta}) = \frac{1}{2}u^2$ with the condition $\operatorname{sign}(t - \sqrt{\eta}) = \operatorname{sign}(u)$ when $t > 0$. We obtain

$$F_\eta(\nu) = \frac{e^{\nu\phi(\sqrt{\eta})}}{2\pi i} \int_{-i\infty}^{i\infty} e^{\frac{1}{2}\nu u^2} g(u)\, du, \tag{23.1.3}$$

where $g(u) = f(t)\, dt/du$, with $dt/du = u/(t^2 - \eta)$, which is regular at the positive saddle point, but not at the negative saddle point. It follows that, when η becomes small, a singularity due to dt/du in the u-plane approaches the origin, and an expansion of dt/du at $u = 0$ will have coefficients that become infinite as $\eta \to 0$.

Hence, by using the standard saddle point method we obtain an expansion that is not uniformly valid as $\eta \to 0$.

In this chapter we give details on constructing Airy-type expansions, and how the coefficients can be represented as Cauchy-type contour integrals. We also give details for evaluating the coefficients in the expansion from the values of the function f in (23.1.1) and its derivatives at the saddle points $\pm\eta$.

An important step in this method is the transformation of a phase function exhibiting the same asymptotic features as above into the standard form (23.1.1). This transformation was used for the first time in Chester *et al.* (1957) with a detailed discussion of the regularity of the transformation. In that paper a uniform asymptotic expansion was obtained that is different from the one we will give in (23.2.9), although it can be transformed into this canonical form by using Bleistein's method (see §22.2).

In §23.4 we explain how an integral for the Hermite polynomial with two coalescing saddle points can be transformed into the standard form (23.1.1). In §23.5 we do the same for the J-Bessel function. In §23.6 we consider an infinite series in terms of modified Bessel functions, which is transformed into an integral along a semi-infinite interval. We use an incomplete Scorer function to describe the asymptotic behavior. Other examples will be considered in §30.5 (for parabolic cylinder functions) and in §31.7 (for Coulomb wave functions).

23.2 An integration by parts method

We describe the integration by parts procedure which is a variant of Bleistein's method given in §22.2. We assume that f is an analytic function in a certain domain $\mathcal{D}$ and write

$$f(t) = A_0(\eta) + B_0(\eta)t + (t^2 - \eta)g(t), \tag{23.2.4}$$

where

$$\begin{aligned} A_0(\eta) &= \tfrac{1}{2}\left(f(\sqrt{\eta}) + f(-\sqrt{\eta})\right), \\ B_0(\eta) &= \tfrac{1}{2}\frac{f(\sqrt{\eta}) - f(-\sqrt{\eta})}{\sqrt{\eta}}. \end{aligned} \tag{23.2.5}$$

Clearly $A_0(\eta) \to f(0), B_0(\eta) \to f'(0)$ as $\eta \to 0$ and the following Cauchy integral representations hold

$$\begin{aligned} f(t) &= \frac{1}{2\pi i}\int \frac{f(\tau)}{\tau - t}\,d\tau, & g(t) &= \frac{1}{2\pi i}\int \frac{f(\tau)}{(\tau - t)(\tau^2 - \eta)}\,d\tau, \\ A_0(\eta) &= \frac{1}{2\pi i}\int \frac{\tau f(\tau)}{\tau^2 - \eta}\,d\tau, & B_0(\eta) &= \frac{1}{2\pi i}\int \frac{f(\tau)}{\tau^2 - \eta}\,d\tau, \end{aligned} \tag{23.2.6}$$

for suitable contours of integration (see Theorem 23.1). Upon substituting (23.2.4) into (23.1.1), we obtain

$$
\begin{aligned}
F_\eta(\nu) = {} & \nu^{-\frac13}\operatorname{Ai}\left(\eta\nu^{\frac23}\right)A_0(\eta) - \nu^{-\frac23}\operatorname{Ai}'\left(\eta\nu^{\frac23}\right)B_0(\eta) + \\
& \frac{1}{2\pi i}\int_{\mathcal{C}} e^{\nu(\frac13 t^3-\eta t)}(t^2-\eta)g(t)\,dt.
\end{aligned}
\tag{23.2.7}
$$

An integration by parts gives

$$
\begin{aligned}
F_\eta(\nu) = {} & \nu^{-\frac13}\operatorname{Ai}\left(\eta\nu^{\frac23}\right)A_0(\eta) - \nu^{-\frac23}\operatorname{Ai}'\left(\eta\nu^{\frac23}\right)B_0(\eta) - \\
& \frac{1}{2\pi i\nu}\int_{\mathcal{C}} e^{\nu(\frac13 t^3-\eta t)}f_1(t)\,dt,
\end{aligned}
\tag{23.2.8}
$$

where $f_1(t) = g'(t)$. Repeating this procedure we obtain the compound expansion

$$
F_\eta(\nu) \sim \nu^{-\frac13}\operatorname{Ai}\left(\eta\nu^{\frac23}\right)\sum_{n=0}^{\infty}(-1)^n\frac{A_n(\eta)}{\nu^n} - \nu^{-\frac23}\operatorname{Ai}'\left(\eta\nu^{\frac23}\right)\sum_{n=0}^{\infty}(-1)^n\frac{B_n(\eta)}{\nu^n}, \tag{23.2.9}
$$

where the coefficients A_n, B_n are defined as (cf. (23.2.5))

$$
\begin{aligned}
A_n(\eta) &= \tfrac12\left(f_n(\sqrt{\eta}) + f_n(-\sqrt{\eta})\right), \\
B_n(\eta) &= \tfrac12\frac{f_n(\sqrt{\eta}) - f_n(-\sqrt{\eta})}{\sqrt{\eta}}.
\end{aligned}
\tag{23.2.10}
$$

The functions f_n are defined by the scheme

$$
f_{n+1}(t) = g_n'(t), \quad f_n(t) = A_n(\eta) + B_n(\eta)t + (t^2-\eta)g_n(t), \tag{23.2.11}
$$

with $n = 0, 1, 2, \ldots$ and $f_0(t) = f(t)$.

The expansion in (23.2.9) is valid for large values of ν and holds uniformly with respect to η in a neighborhood of the origin. A more precise formulation can be given when more information on the function f is available. When $\eta\nu^{\frac23}$ is large and positive the Airy function and its derivative are exponentially small; when the argument becomes negative, the Airy functions start to oscillate. When $\eta < 0$ the saddle points are purely imaginary.

The functions $f_n(t)$ defined in (23.2.11) can be represented in the form of Cauchy-type integrals. We have the following theorem.

Theorem 23.1. *Let the rational functions $R_n(s,t,\eta)$ be defined by*

$$
R_0(s,t,\eta) = \frac{1}{s-t}, \quad R_{n+1}(s,t,\eta) = \frac{-1}{s^2-\eta}\frac{d}{ds}R_n(s,t,\eta), \tag{23.2.12}
$$

for $n = 0, 1, 2, \ldots$, where $s, t, \eta \in \mathbb{C}, s \neq t, s^2 \neq \eta$. Let $f_n(t)$ be defined by the recursive scheme (23.2.11)*, where f_0 is a given analytic function in a domain G. Then we have*

$$
f_n(t) = \frac{1}{2\pi i}\int_{\mathcal{C}} R_n(s,t,\eta)f_0(s)\,ds, \tag{23.2.13}
$$

where $\mathcal{C}$ is a simple closed contour in G that encircles the points t and $\pm\sqrt{\eta}$.

Proof. In the proof of Theorem 25.1 we explain in detail a related case; the present proof is similar, and is left as an exercise. □

Representations for the coefficients $A_n(\eta)$ and $B_n(\eta)$ follow from (23.2.10), and are of the form

$$\begin{aligned} A_n(\eta) &= \frac{1}{2\pi i}\int_{\mathcal{C}} R_n^A(s,\eta) f_0(s)\,ds, \\ B_n(\eta) &= \frac{1}{2\pi i}\int_{\mathcal{C}} R_n^B(s,\eta) f_0(s)\,ds, \end{aligned} \tag{23.2.14}$$

where $\mathcal{C}$ is now a simple contour around $s = \pm\sqrt{\eta}$, and may, for example, consist of two small circles around these points. The new rational functions are given by

$$\begin{aligned} R_n^A(s,\eta) &= \tfrac{1}{2}\left(R_n(s,\sqrt{\eta},\eta) + R_n(s,-\sqrt{\eta},\eta)\right), \\ R_n^B(s,\eta) &= \tfrac{1}{2}\frac{R_n(s,\sqrt{\eta},\eta) - R_n(s,-\sqrt{\eta},\eta)}{\sqrt{\eta}}. \end{aligned} \tag{23.2.15}$$

The first ones of these are

$$\begin{array}{ll} R_0^A(s,\eta) = \dfrac{s}{s^2-\eta}, & R_0^B(s,\eta) = \dfrac{1}{s^2-\eta}, \\ R_1^A(s,\eta) = \dfrac{s^2+\eta}{(s^2-\eta)^3}, & R_1^B(s,\eta) = \dfrac{2s}{(s^2-\eta)^3}, \\ R_2^A(s,\eta) = \dfrac{4s(s^2+2\eta)}{(s^2-\eta)^5}, & R_2^B(s,\eta) = \dfrac{2(5s^2+\eta)}{(s^2-\eta)^5}, \\ R_3^A(s,\eta) = \dfrac{4(7s^4+21\eta s^2+2\eta^2)}{(s^2-\eta)^7}, & R_3^B(s,\eta) = \dfrac{40s(2s^2+\eta)}{(s^2-\eta)^7}. \end{array} \tag{23.2.16}$$

Remark 23.1. The aim of constructing Airy-type expansions is to obtain expansions valid for coalescing saddle points, in this case for small values of η. However, we observe that for large values of η the sequences $\{A_n(\eta)\}$ and $\{B_n(\eta)\}$ may constitute asymptotic sequences, and this may give the expansion in (23.2.9) a double asymptotic property: valid for large values of ν as well as for large values of η. Such a property will depend on the analytic nature of the function f_0, in particular on the distance of its singularities from the saddle points $\pm\eta$. In Olde Daalhuis and Temme (1994) details of these conditions are discussed for allowing the parameter η to range through an unbounded interval, with an application to Laguerre polynomials. For an application of this approach with rational functions to Airy-type expansions of Hermite polynomials, see Shi (2008). In Berry and Howls (1993) new representations are given for the coefficients for contour integrals involving a cluster of coalescing saddles. Moreover, the leading behavior of the late coefficients in the expansions is studied. △

23.3 How to compute the coefficients

From the representations in (23.2.14) we can obtain representations of the coefficients A_n, B_n of (23.2.9). As we see from the first rational functions given in

(23.2.16), we can use partial fraction decomposition of the rational functions, and express A_n, B_n in terms of the derivatives of the function f_0 at $\pm\sqrt{\eta}$.

Here we explain how the coefficients A_n, B_n of (23.2.9) can be represented in such forms by using a more systematic approach. We use the notation $b = \sqrt{\eta}$, and write (23.2.11) in the form

$$\begin{aligned} f_0(t) &= f(t), \quad f_{n+1}(t) = g_n'(t), \\ f_n(t) &= A_n + B_n t + (t^2 - b^2) g_n(t), \end{aligned} \tag{23.3.17}$$

for $n = 0, 1, 2, \ldots$.

First we introduce a different type of expansion:

$$f_n(t) = \sum_{k=0}^{\infty} C_k^{(n)} (t^2 - b^2)^k + t \sum_{k=0}^{\infty} D_k^{(n)} (t^2 - b^2)^k. \tag{23.3.18}$$

When we know the coefficients $C_k^{(n)}, D_k^{(n)}$, the coefficients A_n and B_n of (23.2.9) follow easily from

$$A_n = C_0^{(n)}, \quad B_n = D_0^{(n)}, \quad n \geq 0. \tag{23.3.19}$$

In addition, when we know the coefficients in (23.3.18) for f_n, the coefficients for f_{n+1} follow from the recursion

$$\begin{aligned} C_k^{(n+1)} &= (2k+1) D_{k+1}^{(n)} + 2b^2 (k+1) D_{k+2}^{(n)}, \\ D_k^{(n+1)} &= 2(k+1) C_{k+2}^{(n)}, \end{aligned} \tag{23.3.20}$$

for $n, k = 0, 1, 2, \ldots$.

So, we concentrate on the computation of the coefficients C_k, D_k in the expansion

$$f(t) = \sum_{k=0}^{\infty} C_k (t^2 - b^2)^k + t \sum_{k=0}^{\infty} D_k (t^2 - b^2)^k. \tag{23.3.21}$$

This is a two-point Taylor expansion, and for this topic we refer to Chapter 18. To use the results of that chapter we write the expansion in the symmetric form

$$f(t) = \sum_{k=0}^{\infty} \left(a_k(t_1, t_2) \tau_1 + a_k(t_2, t_1) \tau_2 \right) \tau_1^k \tau_2^k, \tag{23.3.22}$$

see (18.1.3), where

$$t_1 = -b, \quad t_2 = b, \quad \tau_1 = t - t_1, \quad \tau_2 = t - t_2. \tag{23.3.23}$$

Then, see (18.2.14),

$$\begin{aligned} C_k &= -t_1 a_k(t_1, t_2) - t_2 a_k(t_2, t_1), \\ D_k &= a_k(t_1, t_2) + a_k(t_2, t_1). \end{aligned} \tag{23.3.24}$$

For $a_k(t_1, t_2)$ we have, cf. (18.3.16) and (18.3.17),

$$a_0(t_1, t_2) = \frac{f(b)}{2b}, \quad a_0(t_2, t_1) = -\frac{f(-b)}{2b}, \tag{23.3.25}$$

and, for $k = 1, 2, 3, \ldots,$

$$a_k(t_1,t_2) = \sum_{j=0}^{k} \frac{(k+j-1)!}{j!(k-j)!} \frac{(-1)^{k+1}kf^{(k-j)}(b) + (-1)^j jf^{(k-j)}(-b)}{k!(-2b)^{k+j+1}}; \tag{23.3.26}$$

$a_k(t_2,t_1)$ follows from $a_k(t_1,t_2)$ by replacing b by $-b$. The first few values are

$$\begin{aligned}
C_0 &= u_0, \qquad\qquad D_0 = \frac{1}{b}v_0,\\
C_1 &= \frac{1}{2b}v_1, \qquad\quad D_1 = \frac{1}{2b^3}(u_1 b - v_0),\\
C_2 &= \frac{1}{8b^3}(u_2 b - v_1), \quad D_2 = \frac{1}{8b^5}\left(v_2 b^2 - 3bu_1 + 3v_0\right),\\
C_3 &= \frac{1}{48b^5}\left(v_3 b^2 - 3bu_2 + 3v_1\right),\\
D_3 &= \frac{1}{48b^7}\left(u_3 b^3 - 6b^2 v_2 + 15bu_1 - 15v_0\right),\\
C_4 &= \frac{1}{384b^7}\left(u_4 b^3 - 6b^2 v_3 + 15bu_2 - 15v_1\right),\\
D_4 &= \frac{1}{384b^9}\left(v_4 b^4 - 10b^3 u_3 + 45b^2 v_2 - 105bu_1 + 105v_0\right),
\end{aligned} \tag{23.3.27}$$

where the coefficients u_k, v_k are defined by

$$u_k = \tfrac{1}{2}\left(f^{(k)}(b) + f^{(k)}(-b)\right), \quad v_k = \tfrac{1}{2}\left(f^{(k)}(b) - f^{(k)}(-b)\right). \tag{23.3.28}$$

Looking at the form of the given coefficients C_k, D_k it seems not to be difficult to write down the values of $C_5, D_5, C_6, D_6, \ldots$.

These values of $C_k = C_k^{(0)}, D_k = D_k^{(0)}$ are the starting values of the recursions in (23.3.20), and the first A_n, B_n are (see (23.3.19))

$$\begin{aligned}
A_0 &= u_0, \quad B_0 = \frac{1}{b}v_0,\\
A_1 &= \frac{1}{4b^3}\left(b^2 v_2 - bu_1 + v_0\right), \quad B_1 = \frac{1}{4b^3}(bu_2 - v_1),\\
A_2 &= \frac{1}{96b^5}\left(3b^3 u_4 - 10b^2 v_3 + 21bu_2 - 21v_1\right),\\
B_2 &= \frac{1}{96b^7}\left(3b^4 v_4 - 10b^3 u_3 + 15b^2 v_2 - 15bu_1 + 15v_0\right),\\
A_3 &= \frac{1}{384b^9}\left(b^6 v_6 - 7b^5 u_5 + 28b^4 v_4 - 70b^3 u_3 + 105b^2 v_2 - 105bu_1 + 105v_0\right),\\
B_3 &= \frac{1}{384b^9}\left(b^5 u_6 - 7b^4 v_5 + 25b^3 u_4 - 60b^2 v_3 + 105bu_2 - 105v_1\right).
\end{aligned} \tag{23.3.29}$$

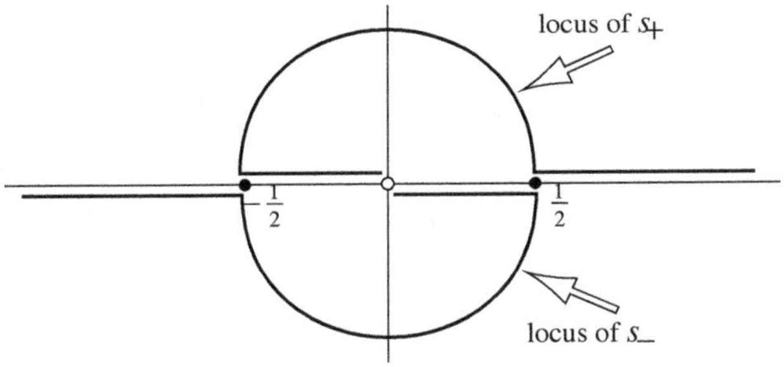

Fig. 23.1 The location of the two saddle points $s_\pm$ defined in (23.4.34) for $\xi \in \mathbb{R}$.

23.4 An Airy-type expansion of the Hermite polynomial

The Hermite polynomials follow from the generating function

$$e^{2xz-z^2} = \sum_{n=0}^{\infty} \frac{H_n(x)}{n!} z^n, \quad x, z \in \mathbb{C}, \tag{23.4.30}$$

which gives the Cauchy-type integral

$$H_n(x) = \frac{n!}{2\pi i} \int_{\mathcal{C}} e^{2xz-z^2} z^{-n-1}\, dz, \tag{23.4.31}$$

where $\mathcal{C}$ is a circle around the origin and the integration is in the positive direction. By substituting $z = Ns$, $x = N\xi$, $N = \sqrt{2n+1}$, it follows that the Hermite polynomials can be represented in the form

$$H_n(x) = \frac{n!}{N^n} \frac{1}{2\pi i} \int_{\mathcal{C}} e^{N^2\phi(s)} \frac{ds}{\sqrt{s}}, \tag{23.4.32}$$

where

$$\phi(s) = 2\xi s - \tfrac{1}{2} \ln s - s^2, \tag{23.4.33}$$

and $\mathcal{C}$ is a path that runs from $-\infty$ (with $\operatorname{ph} s = -\pi$), encircles the origin in positive direction, and returns to $-\infty$, now with $\operatorname{ph} s = +\pi$. In fact, this is a Hankel contour as shown in Figure 2.1. If we wish we can extend the contour to $+\infty$, which we will do in the oscillatory case.

The saddle points of the integral are defined by the equation $\phi'(s) = 2\xi - 1/(2s) - 2s = 0$ and are given by

$$s_\pm = \tfrac{1}{2}\left(\xi \pm \sqrt{\xi^2 - 1}\right). \tag{23.4.34}$$

In Figure 23.1 we show the location of $s_\pm$ for all $\xi \in \mathbb{R}$.

When $\xi = \pm 1$ the saddle points coalesce at $\pm\frac{1}{2}$, and when $\xi \sim 1$ uniform Airy-type expansions can be derived. When $-1 < \xi < 1$ the saddle points are complex (on the circle around the origin with radius $\frac{1}{2}$); for these values of ξ, that is, if $-\sqrt{2n+1} < x < \sqrt{2n+1}$, zeros occur. For a proof of this property we refer to

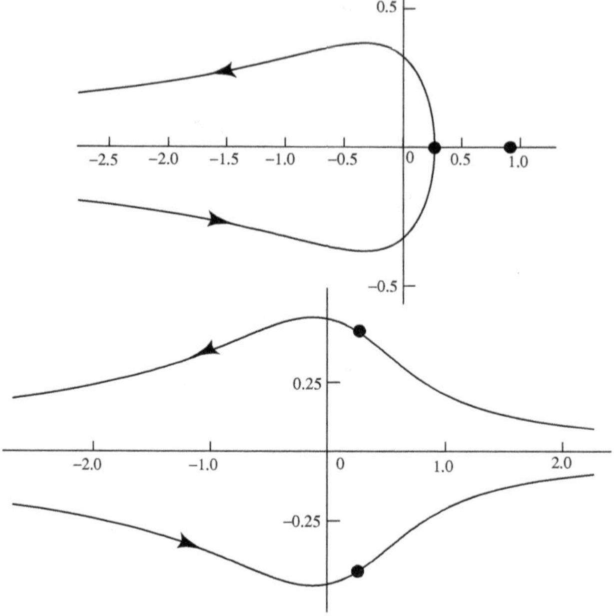

Fig. 23.2 Saddle point contours for $\xi = 1.2$ (upper figure) and $\xi = 0.5$ (lower figure).

Temme (1996a, Exercise 6.9). When $\xi > 1$ or $\xi < -1$ the saddle points are real, and the Hermite polynomials are non-oscillating.

In Figure 23.2 we show the saddle point contours of the integral in (23.4.32) for $\xi = 1.2$ (upper figure) and $\xi = 0.5$ (lower figure). The saddle points are indicated by black dots.

23.4.1 *The cubic transformation*

We give the details on the asymptotic behavior for real x, and concentrate on $x \ge 0$, because for $x < 0$ we can use the symmetry of the Hermite polynomials $H_n(-x) = (-1)^n H_n(x)$.

The cubic transformation to obtain the standard form in (23.1.1) is given by

$$\phi(s) = \psi(t), \quad \psi(t) = \tfrac{1}{3}t^3 - \eta t + A, \tag{23.4.35}$$

where η and A are functions of ξ to be determined. These values follow from substituting corresponding saddle points in the s-plane (see (23.4.34)) and those in the t-plane ($t_\pm = \pm\sqrt{\eta}$).

This correspondence easily follows from considering the relations in (23.4.35) for $s > 0$ and $t \in \mathbb{R}$. These intervals contain the saddle points, and on these intervals we have $\Im\phi(s) = \Im\psi(t) = 0$, when the values of η and A are real. Although only parts of these intervals can be used for saddle point contours (because remaining parts will be steepest ascent paths), it will be instructive to see how these complete intervals will act in the cubic transformation.

Assume that $\xi > 1$. In that case the saddle points in (23.4.34) are real and $\eta > 0$, as we will see. The graphs of $\phi(s), s > 0$ and $\psi(t), t \in \mathbb{R}$ easily show that the saddle points $s_\pm$ should correspond to the saddle points $t_\mp$.

For $\xi > 1$ the saddle point contour is given by

$$r = \frac{\xi - \sqrt{\xi^2 - \theta \cot\theta}}{2\cos\theta}, \qquad -\pi < \theta < \pi, \tag{23.4.36}$$

where we have used polar coordinates $s = re^{i\theta}$. This saddle point contour through s_- for the integral in (23.4.32) will be transformed into the saddle point contour in the t-plane through t_+.

For $\xi \in (0,1)$ the contour splits up into two parts, running from $-\infty$ through s_- in the lower half-plane to $+\infty$, and from $+\infty$ through s_+ in the upper half-plane back to $-\infty$.

For η and A we have the relations

$$\begin{aligned} \tfrac{4}{3}\eta^{\frac{3}{2}} &= \xi\sqrt{\xi^2-1} - \operatorname{arccosh}\xi, \quad \xi \ge 1, \\ \tfrac{4}{3}(-\eta)^{\frac{3}{2}} &= \arccos\xi - \xi\sqrt{1-\xi^2}, \quad\ \ 0 \le \xi \le 1, \\ A &= \tfrac{1}{2}\xi^2 + \tfrac{1}{4} + \tfrac{1}{2}\ln 2. \end{aligned} \tag{23.4.37}$$

The quantity η is analytic in a neighborhood of $\xi = 1$. We have the differential equation

$$\eta\left(\frac{d\eta}{d\xi}\right)^2 = \xi^2 - 1, \tag{23.4.38}$$

and for small values of $|\xi - 1|$ the expansion

$$\eta = 2^{\frac{1}{3}}(\xi-1)\left(1 + \tfrac{1}{10}(\xi-1) - \tfrac{2}{175}(\xi-1)^2 + \mathcal{O}\left((\xi-1)^3\right)\right). \tag{23.4.39}$$

The cubic transformation in (23.4.35) gives the representation

$$H_n(x) = \frac{n!}{N^n}e^{N^2A}F_\eta(\nu), \quad F_\eta(\nu) = \frac{1}{2\pi i}\int_{\mathcal{L}} e^{\nu\left(\frac{1}{3}t^3 - \eta t\right)} f(t)\,dt, \tag{23.4.40}$$

where $\nu = N^2$ and

$$f(t) = -\frac{1}{\sqrt{s}}\frac{ds}{dt} = -\frac{1}{\sqrt{s}}\frac{\psi'(t)}{\phi'(s)} = 2\sqrt{s}\frac{t^2 - \eta}{4s^2 - 4\xi s + 1}. \tag{23.4.41}$$

For $\xi > 1$ the contour $\mathcal{L}$ in the t-plane is as the contour C_1 for the Airy function shown in Figure 8.2, and in (23.4.40) we use the same direction of integration as in the figure. The choice of this contour follows from the earlier observation that the saddle point contour through s_- for the integral in (23.4.32) should be mapped to the saddle point contour through t_+.

With the representation in (23.4.40) we can start an integration by parts method as in §23.2, and in this way we obtain an expansion as in (23.2.9).

23.4.2 *Details on the coefficients*

We give a few details on the coefficients. As explained in §23.3 we need the derivatives of f at the saddle point $t_\pm$. For this, we first try to find the coefficients in the expansions

$$s = s_- + \sum_{k=1}^{\infty} p_k (t-b)^k, \quad s = s_+ + \sum_{k=1}^{\infty} q_k (t+b)^k, \quad b = \sqrt{\eta}. \tag{23.4.42}$$

By using (23.4.33) or the relations for ds/dt that follow from (23.4.41) we obtain

$$p_1 = -\sqrt{s_-}\,\chi(\eta), \quad q_1 = -\sqrt{s_+}\,\chi(\eta), \quad \chi(\eta) = \left(\frac{\eta}{\xi^2-1}\right)^{\frac14}, \tag{23.4.43}$$

and for the higher coefficients we have

$$\begin{aligned}
p_2 &= -\frac{p_1^3 + 2s_-^3}{3s_-p_1(4s_-^2-1)},\\
p_3 &= -\frac{8s_-^6 - 16p_1^3s_-^3 - 36s_-^2p_1^6 - p_1^6}{36p_1^3s_-^2(4s_-^2-1)^2},\\
p_4 &= -\frac{40s_-^9 + 540s_-^5p_1^6 + 15p_1^6s_-^3 + 864s_-^4p_1^9 + 108p_1^9s_-^2 - p_1^9}{270s_-^3p_1^5(4s_-^2-1)^3}.
\end{aligned} \tag{23.4.44}$$

For q_k we can use the same expressions with s_- replaced by s_+ and p_1 by q_1.

When the coefficients p_k and q_k are available we can obtain the coefficients r_k and s_k in the expansions

$$f(t) = \chi(\eta)\sum_{k=0}^{\infty} r_k(t-b)^k, \quad f(t) = \chi(\eta)\sum_{k=0}^{\infty} s_k(t+b)^k. \tag{23.4.45}$$

We have

$$r_0 = s_0 = 1, \tag{23.4.46}$$

and

$$\begin{aligned}
r_1 &= -\frac{8s_-^3 + (12s_-^2+1)p_1^3}{6s_-p_1^2(4s_-^2-1)},\\
r_2 &= -\frac{16s_-^6 - (96s_-^5 + 8s_-^3)p_1^3 + (-144s_-^4 - 48s_-^2 + 1)p_1^6}{24p_1^4s_-^2(4s_-^2-1)^2},\\
r_3 &= -\big(1280s_-^9 + (34560s_-^7 + 11520s_-^5 - 240s_-^3)p_1^6 +\\
&\quad (43200s_-^6 + 29808s_-^4 + 36s_-^2 + 13)p_1^9\big) \,/\, \big(2160s_-^3p_1^6(4s_-^2-1)^3\big),\\
r_4 &= -\big(6400s_-^{12} + (-69120s_-^{10} - 23040s_-^8 + 480s_-^6)p_1^6 -\\
&\quad (691200s_-^9 + 476928s_-^7 + 576s_-^5 + 208s_-^3)p_1^9 -\\
&\quad (725760s_-^8 + 815616s_-^6 + 47520s_-^4 - 288s_-^2 + 3)p_1^{12}\big) \,/\\
&\quad \big(10368s_-^4p_1^8(4s_-^2-1)^4\big).
\end{aligned} \tag{23.4.47}$$

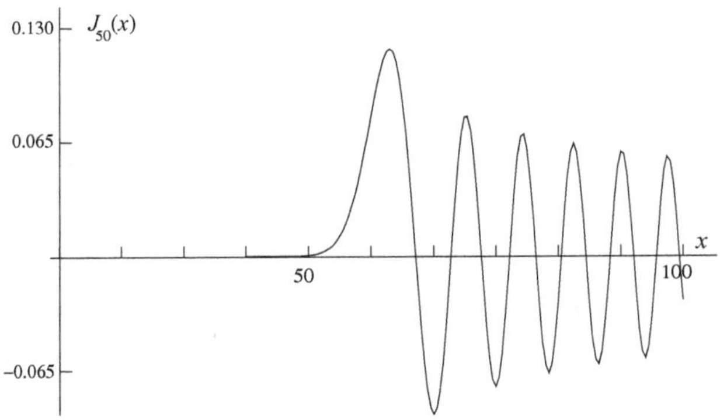

Fig. 23.3 The Bessel function $J_{50}(x), 0 \le x \le 100$.

For s_k we can use the same expressions with s_- replaced by s_+ and p_1 by q_1.

For the Hermite polynomials with the representation in (23.4.40) we have the expansion

$$
\begin{aligned}
H_n(x) \sim \frac{n!}{\nu^{\frac12 n}} e^{\nu A}\chi(\eta)\Big(\nu^{-\frac13}\operatorname{Ai}\left(\eta\nu^{\frac23}\right)\sum_{n=0}^{\infty}(-1)^n\frac{A_n(\eta)}{\nu^n} - \\
\nu^{-\frac23}\operatorname{Ai}'\left(\eta\nu^{\frac23}\right)\sum_{n=0}^{\infty}(-1)^n\frac{B_n(\eta)}{\nu^n}\Big),
\end{aligned}
\tag{23.4.48}
$$

where $\nu = 2n+1$, A and η are given in (23.4.37), with $\xi = x/\sqrt{\nu}$, and $\chi(\eta)$ is defined in (23.4.43). For the coefficients we have $A_0(\eta) = 1$, $B_0(\eta) = 0$, and a few other coefficients follow from (23.3.29) with u_k and v_k given by (cf. (23.3.28))

$$
u_k = \tfrac12 k!\,(r_k + s_k)\,, \quad v_k = \tfrac12 k!\,(r_k - s_k)\,. \tag{23.4.49}
$$

For further details on this expansion we refer to Shi (2008), where the singularities of the function $f(t)$ are discussed. It is shown that the expansion is also valid for unbounded values of ξ. For earlier results, see Skovgaard (1959).

23.5 An Airy-type expansion of the Bessel function $J_\nu(z)$

We consider positive ν and z and use the integral representation for the ordinary Bessel function (see (9.4.39))

$$
J_\nu(z) = \frac{1}{2\pi i}\int_{\mathcal{C}} e^{z\sinh s - \nu s}\,ds, \tag{23.5.50}
$$

where the contour $\mathcal{C}$ starts at $\infty - \pi i$ and terminates at $\infty + \pi i$.

From graphs of the Bessel function of high positive order and positive z, see Figure 23.3, it can be seen that $J_\nu(z)$ starts oscillating when z exceeds the value ν. We concentrate on the transition area $z \sim \nu$.

We write the integral in the form

$$J_\nu(\nu z) = \frac{1}{2\pi i}\int_{\mathcal{C}} e^{\nu\phi(s)}\,ds, \quad \phi(s) = z\sinh s - s. \tag{23.5.51}$$

When $z \sim 1$ there are two small real saddle points $s_\pm$ defined by the equation $\cosh s = 1/z$. For $0 < z \le 1$ the saddle points are real, and we have in that case

$$s_\pm = \pm\theta, \quad \text{where} \quad z = 1/\cosh\theta, \quad \theta \ge 0. \tag{23.5.52}$$

We have $\phi(s_\pm) = \pm(\tanh\theta - \theta)$. When $\nu \sim z$, the parameter θ becomes small, and the two saddle points coalesce when $\nu = z$. When $1 < z$ the saddle points leave the real axis and are located on the imaginary axis. A local expansion at $s = 0$, that is,

$$\phi(s) = (z-1)s + \tfrac{1}{6}zs^3 + \ldots \tag{23.5.53}$$

shows that when $z \to 1$ $(\theta \to 0)$ the first two derivatives of ϕ vanish at $s = 0$, and that ϕ behaves like a cubic polynomial. This suggests the transformation

$$\phi(s) = \tfrac{1}{3}t^3 - \eta t + A, \tag{23.5.54}$$

where η and A have to be determined. By choosing $t(s_\pm) = \pm\sqrt{\eta}$, it follows that $A = 0$. For $0 < z \le 1$ we have $\eta \ge 0$ and

$$\tfrac{2}{3}\eta^{\frac{3}{2}} = \theta - \tanh\theta, \quad \sqrt{\eta} = 2^{-1/3}\theta\left(1 + \mathcal{O}(\theta^2)\right), \quad \theta \to 0. \tag{23.5.55}$$

In terms of z this definition of η can be written as

$$\tfrac{2}{3}\eta^{\frac{3}{2}} = \operatorname{arccosh}(1/z) - \sqrt{1-z^2}, \quad 0 < z \le 1. \tag{23.5.56}$$

For $z \ge 1$ we have $\eta \le 0$ and

$$\tfrac{2}{3}(-\eta)^{\frac{3}{2}} = \sqrt{z^2-1} - \arccos(1/z), \quad z \ge 1. \tag{23.5.57}$$

Upon inverting one of these relations we have

$$z(\eta) = 1 - \chi + \tfrac{3}{10}\chi^2 + \tfrac{1}{350}\chi^3 - \tfrac{479}{63000}\chi^4 + \ldots, \quad \chi = 2^{-\frac{1}{3}}\eta. \tag{23.5.58}$$

The mapping in (23.5.54) transforms the integral in (23.5.51) into the standard form

$$J_\nu(\nu z) = \frac{1}{2\pi i}\int_{\mathcal{C}} e^{\nu\left(\frac{1}{3}t^3 - \eta t\right)} f(t)\,dt, \quad f(t) = \frac{ds}{dt}. \tag{23.5.59}$$

The function f is analytic in both parameters t and η; the specification of the domains of analyticity is not given here. After having obtained the standard form we can derive the expansion (23.2.9). We can compute the coefficients $A_n(\eta), B_n(\eta)$ using the scheme given in §23.3.

We need the derivatives of f at $\pm\sqrt{\eta}$. For the special functions that are solutions of second-order linear differential equations, as the Bessel functions, recurrence relations for the coefficients A_n, B_n can be derived. For integrals they follow from a recursive scheme (see (23.2.11)), and we show which steps are needed in the case of the Bessel function.

The function f defined in (23.5.59) has to be expanded in Taylor series at the points $\pm b$, where $b = \sqrt{\eta}$ (see (23.5.55)).

We have from (23.5.54) and (23.5.59)

$$f(t) = \frac{ds}{dt} = \frac{t^2 - b^2}{z \cosh s - 1}, \tag{23.5.60}$$

where $s = \pm\theta$ corresponds to $t = \pm b$ (see the condition given before (23.5.55)). We expand

$$s = \sum_{k=0}^{\infty} a_k (t-b)^k, \quad a_0 = \theta. \tag{23.5.61}$$

By substituting this expansion into (23.5.54) we can find the coefficients a_k and the derivatives

$$f^{(k)}(b) = (k+1)!\, a_{k+1}, \quad k = 0, 1, 2, \ldots. \tag{23.5.62}$$

The first coefficients are

$$a_1 = \left(\frac{4\eta}{1-z^2}\right)^{\frac14}, \quad a_2 = \frac{2 - a_1^3}{6a_1\sqrt{1-z^2}}, \quad a_3 = \frac{(3z^2+2)a_1^6 - 8a_1^3 - 4}{72a_1^3(1-z^2)}. \tag{23.5.63}$$

The coefficient a_1 is analytic at $\eta = 0$. By using the expansion in (23.5.58) we can find the expansion

$$a_1 = 2^{\frac13}\left(1 + \tfrac15\chi + \tfrac{9}{350}\chi^2 - \tfrac{89}{15750}\chi^3 - \tfrac{4547}{1155000}\chi^4 + \ldots\right), \quad \chi = 2^{-\frac13}\eta. \tag{23.5.64}$$

For the derivatives of f at $-b$ we observe (see (23.5.54) with $A = 0$) that changing the sign of t is equivalent to changing the sign of s. When we change the signs of t and s in (23.5.61) we obtain

$$s = -\theta + \sum_{k=1}^{\infty} (-1)^{k+1} a_k (t+b)^k, \tag{23.5.65}$$

and

$$\frac{ds}{dt} = \sum_{k=0}^{\infty} (-1)^k (k+1) a_{k+1} (t+b)^k, \tag{23.5.66}$$

which gives

$$f^{(k)}(-b) = (-1)^k (k+1)!\, a_{k+1}, \quad k = 0, 1, 2, \ldots. \tag{23.5.67}$$

It follows that (see (23.3.28))

$$\begin{aligned} u_{2k+1} &= 0, & v_{2k} &= 0, \\ u_{2k} &= (2k+1)!\, a_{2k+1}, & v_{2k+1} &= (2k+2)!\, a_{2k+2}, \end{aligned} \tag{23.5.68}$$

for $k = 0, 1, 2, \ldots$. From the given coefficients C_n, D_n in (23.3.27) we see that $D_n = 0$ and that

$$\begin{aligned} &C_0 = a_1, \quad C_1 = \frac{1}{b} v_1, \quad C_2 = \frac{1}{4b^3}(3ba_3 - a_2), \\ &C_3 = \frac{1}{8b^5}(4b^2 a_4 - 3ba_3 + a_2), \\ &C_4 = \frac{1}{64b^7}(20b^3 a_5 - 24b^2 a_4 + 15ba_3 - 5a_2). \end{aligned} \tag{23.5.69}$$

For the coefficients A_n, B_n we have $A_{2n+1} = 0$, $B_{2n} = 0$ and

$$\begin{aligned} &A_0 = a_1, \quad B_1 = \frac{1}{2b^3}(3ba_3 - a_2), \\ &A_2 = \frac{1}{16b^5}(60b^3 a_5 - 40b^2 a_4 + 21ba_3 - 7a_2), \\ &B_3 = \frac{5}{64b^9}\left(168b^5 a_7 - 168b^4 a_6 + 100b^3 a_5 - 48b^2 a_4 + 21ba_3 - 7a_2\right). \end{aligned} \tag{23.5.70}$$

Because of the vanishing coefficients, and because the series in (23.2.9) are usually written with the first coefficient of the first series equal to unity, we write the expansion of the Bessel function in the form

$$J_\nu(\nu z) \sim a_1 \left(\frac{\mathrm{Ai}\left(\eta \nu^{\frac{2}{3}}\right)}{\nu^{\frac{1}{3}}} \sum_{n=0}^{\infty} \frac{\widetilde{A}_n(\eta)}{\nu^{2n}} + \frac{\mathrm{Ai}'\left(\eta \nu^{\frac{2}{3}}\right)}{\nu^{\frac{5}{3}}} \sum_{n=0}^{\infty} \frac{\widetilde{B}_n(\eta)}{\nu^{2n}} \right), \tag{23.5.71}$$

where η is given in (23.5.56) and (23.5.57), a_1 in (23.5.63) and

$$\widetilde{A}_0(\eta) = 1, \quad \widetilde{A}_n(\eta) = \frac{A_{2n}}{a_1}, \quad \widetilde{B}_n(\eta) = \frac{B_{2n+1}}{a_1}, \tag{23.5.72}$$

with A_n, B_n given in (23.3.29) with $b^2 = \eta$.

Further details on evaluating the coefficients $\widetilde{A}_n(\eta), \widetilde{B}_n(\eta)$ and those of the derivative $J'_\nu(z)$ can be found in Gil *et al.* (2007, §8.4) and Temme (1997). For details on the validity of the expansion in (23.5.71) we refer to Olver and Maximon (2010), where, moreover, Airy-type expansions are given for the Bessel function $Y_\nu(z)$, the Hankel functions, and the modified Bessel functions. In addition, explicit forms of the coefficients of expansion (23.5.71) in terms of the coefficients of the Debye-type expansions are given.

Problem 23.1. *Verify that the coefficient*

$$\widetilde{B}_0(\eta) = \frac{3ba_3 - a_2}{2b^3 a_1} \tag{23.5.73}$$

can be written as

$$\widetilde{B}_0(\eta) = \frac{(3z^2 + 2)a_1^6 - 20}{48\eta a_1^4 (1 - z^2)}. \tag{23.5.74}$$

♡

23.6 A semi-infinite interval: Incomplete Scorer function

We give an example in which an incomplete Scorer function can be used when two saddle points coalesce with a finite endpoint of the interval of integration. We have mentioned the Scorer function in §4.3.2 with definition in (4.3.22). The incomplete Scorer function can be defined by

$$\mathrm{Hi}_\sigma(z) = \frac{1}{\pi}\int_\sigma^\infty e^{zt-\frac{1}{3}t^3}\,dt, \tag{23.6.75}$$

where $z \in \mathbb{C}$ and $\sigma \in \mathbb{C}$.

Observe that when z is small, two saddle points at $\pm\sqrt{z}$ are close to the origin, which may coalesce with the endpoint σ when $\sigma \to 0$.

We can use this function for describing the asymptotic behavior of the function defined by the series

$$S(r,\theta,\omega) = \sum_{n=-\infty}^{\infty} \frac{I_n(\omega r)}{n^2+\omega^2}\cos n\phi, \quad \phi = \tfrac{1}{2}\pi + \theta, \tag{23.6.76}$$

for large values of ω. Functions of this type arise in a singular perturbation problem of which the details are given in §23.6.1. In particular we want to know the behavior uniformly with respect to r and ϕ for certain values of these parameters.

We transform this series into an integral by using

$$\frac{\omega}{n^2+\omega^2} = \int_0^\infty e^{-\omega t}\cos(nt)\,dt. \tag{23.6.77}$$

This gives

$$S(r,\theta,\omega) = \frac{1}{\omega}\int_0^\infty e^{-\omega t} \sum_{n=-\infty}^{\infty} I_n(\omega r)\cos n\phi \cos nt\,dt. \tag{23.6.78}$$

The series can be evaluated by using the well-known generating function of the Bessel coefficients

$$e^{z\cos\alpha} = \sum_{n=-\infty}^{\infty} I_n(z)\cos n\alpha, \tag{23.6.79}$$

and we obtain

$$S(r,\theta,\omega) = \frac{1}{2\omega}\left(T(r,\theta,\omega) + T(r,-\pi-\theta,\omega)\right), \tag{23.6.80}$$

where

$$T(r,\theta,\omega) = \int_0^\infty e^{-\omega\psi(t)}\,dt, \quad \psi(t) = t - r\sin(t-\theta). \tag{23.6.81}$$

It is clear that for large positive z, when $\cos\alpha > 0$, the series in (23.6.79) is exponentially large, and when $\cos\alpha < 0$, it is exponentially small. The same happens for the series in (23.6.76): the values $\phi = \pm\frac{1}{2}\pi$ are transition points, and to describe this change in behavior we need more than a simple exponential function as in (23.6.79). In addition, the parameter r plays an important role, in particular when $r \sim 1$. We will see that the integral in (23.6.81) and the incomplete Scorer function defined in (23.6.75) can be used to describe the peculiar behavior when ω is large, $r \sim 1$, and $\theta \sim 0$ or $\theta \sim \pi$.

The integrand in (23.6.81) has saddle points at

$$t^{\pm} = \theta \pm i\rho, \quad \rho = \operatorname{arccosh} \frac{1}{r} \tag{23.6.82}$$

(and at $t^{\pm} + 2k\pi, k = \pm 1, \pm 2, \ldots$, but these points are not of interest in our case). We assume that $r \in [0, 1]$ and $\theta \in [-\pi, \pi]$.

When $r \to 1$ the two saddle points coalesce at θ, which is in or outside the domain of integration, or at the endpoint $t = 0$. So, on the one hand we have a typical Airy-type asymptotic problem (coalescing saddle points), but we cannot expect that the standard Airy functions can describe the asymptotics because the saddle points coalesce near the finite endpoint of a semi-infinite interval.

As in earlier sections of this chapter, a uniform expansion of $T(r, \theta, \omega)$ can be obtained by using a cubic transformation $t \to s(t)$ of the form

$$\psi(t) = \tfrac{1}{3} s^3 + \zeta s + A, \tag{23.6.83}$$

where ψ is given in (23.6.81) and ζ and A have to be determined. These quantities follow from prescribing that the saddle points $t^{\pm}$ correspond to the saddle points $s^{\pm} = \pm i\sqrt{\zeta}$ in the s-plane. This gives the equations

$$\begin{aligned} \theta + i\rho - ir\sinh\rho &= \tfrac{2}{3} i\zeta^{\frac{3}{2}} + A, \\ \theta - i\rho + ir\sinh\rho &= -\tfrac{2}{3} i\zeta^{\frac{3}{2}} + A. \end{aligned} \tag{23.6.84}$$

Hence,

$$A = \theta, \quad \tfrac{2}{3}\zeta^{\frac{3}{2}} = \rho - \tanh\rho = \operatorname{arccosh}\frac{1}{r} - \sqrt{1 - r^2}. \tag{23.6.85}$$

We see that for every $r \in (0, 1)$ we can find a positive ζ, and $\zeta = 0$ if $r = 1$. In addition we define σ to be the point in the s-domain that corresponds to $t = 0$. Because $\psi(0) = r\sin\theta$, it follows from (23.6.83) and $A = \theta$, that for all considered r and θ we can find a real number σ defined by the equation

$$\tfrac{1}{3}\sigma^3 + \zeta\sigma = r\sin\theta - \theta. \tag{23.6.86}$$

We have $\operatorname{sign}(\sigma) = \operatorname{sign}(-\theta)$.

After these preparations we can write

$$T(r, \theta, \omega) = e^{-\omega\theta} \int_{\sigma}^{\infty} e^{-\omega(\frac{1}{3}s^3 + \zeta s)}\, h(s)\, ds, \tag{23.6.87}$$

where

$$h(s) = \frac{dt}{ds} = \frac{s^2 + \zeta}{1 - r\cos(t - \theta)}. \tag{23.6.88}$$

A first approximation follows from replacing $h(s)$ by

$$h_0 = \tfrac{1}{2}\left(h(s^+) + h(s^-)\right) = \left(\frac{2\zeta}{1-r^2}\right)^{\frac{1}{4}}. \tag{23.6.89}$$

This gives

$$T(r,\theta,\omega) \sim h_0\, e^{-\omega\theta} \int_\sigma^\infty e^{-\omega(\frac{1}{3}s^3+\zeta s)}\, ds, \tag{23.6.90}$$

or, in terms of the incomplete Scorer function defined in (23.6.75),

$$T(r,\theta,\omega) \sim \pi h_0\, e^{-\omega\theta}\, \omega^{-\frac{1}{3}} H_{\sigma\omega^{\frac{1}{3}}}\left(-\omega^{\frac{2}{3}}\zeta\right). \tag{23.6.91}$$

For all values of θ and r considered, the endpoint gives the dominant contribution. At the endpoint we have

$$h_0\, e^{-\omega\theta}\, e^{-\omega(\frac{1}{3}\sigma^3+\zeta\sigma)} = h_0\, e^{-\omega r \sin\theta}. \tag{23.6.92}$$

When $r \to 1$ we have $\zeta \to 0$, and we have

$$T(1,\theta,\omega) \sim h_0\, e^{-\omega\theta} \int_\sigma^\infty e^{-\frac{1}{3}\omega s^3}\, ds, \tag{23.6.93}$$

where in this case (see (23.6.86))

$$h_0 = 2^{\frac{1}{3}}, \quad \sigma = (3(\sin\theta - \theta))^{\frac{1}{3}}. \tag{23.6.94}$$

If, in addition, $\theta \to 0$ we obtain

$$T(1,0,\omega) \sim h_0 \int_0^\infty e^{-\frac{1}{3}\omega s^3}\, ds = \left(\frac{2}{9\omega}\right)^{\frac{1}{3}} \Gamma\left(\tfrac{1}{3}\right). \tag{23.6.95}$$

When $1-r$ and θ are not small, the saddle points in the integral in (23.6.81) are bounded away from the dominant point $t=0$. Substituting $t - r\sin(t-\theta) = u - r\sin\theta$ and applying Watson's lemma, we find

$$T(r,\theta,\omega) = \frac{e^{-\omega r\sin\theta}}{\omega(1-r\cos\theta)}\left(1+\mathcal{O}(1/\omega)\right), \quad \omega\to\infty. \tag{23.6.96}$$

The term $T(r,-\pi-\theta,\omega)$ in (23.6.80) can be treated in a similar way, with special care for the case that $1-r$ and $-\pi-\theta$ are small.

23.6.1 *A singular perturbation problem inside a circle*

The series considered in (23.6.76) arises in the asymptotic analysis of the following elliptic partial differential equation

$$\varepsilon\Delta\Phi(x,y) - \frac{\partial\Phi}{\partial y}(x,y) = 1, \quad x^2+y^2<1, \tag{23.6.97}$$

where $\varepsilon > 0$. The boundary condition reads

$$\Phi(\cos\theta, \sin\theta) = 0 \tag{23.6.98}$$

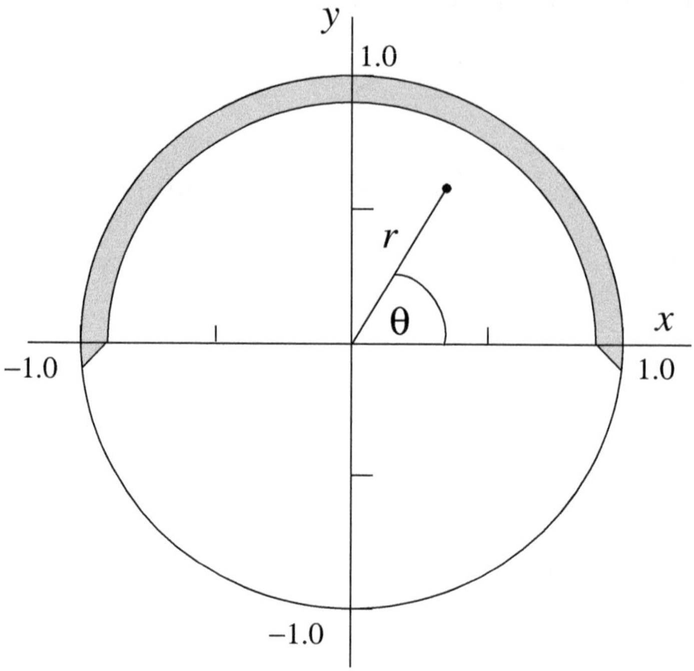

Fig. 23.4 Boundary layer inside the circle along the upper boundary $r = 1, y > 0$, and near the points $(\pm 1, 0)$.

on the boundary of the circle $r = 1$, where we have introduced the polar coordinates

$$x = r\cos\theta, \quad y = r\sin\theta, \quad 0 \le r \le 1, \quad -\pi < \theta \le \pi. \tag{23.6.99}$$

The problem is to find the asymptotic behavior of Φ as $\varepsilon \to 0$. The solution to equation (23.6.97) has a boundary layer along the upper part of the unit disk. In particular it is of interest to find the behavior of Φ in small neighborhoods of the points $(x, y) = (\pm 1, 0)$, the places of birth of the boundary layer; see Figure 23.4. In Temme (2007) we have described several methods for obtaining asymptotic approximations that are valid outside the boundary layer and in the boundary layer near the upper part of the circle.

When $\varepsilon \to 0$ the second-order elliptic operator in (23.6.97) reduces (in the limit $\varepsilon = 0$) to a first-order operator. The solution of the reduced equation cannot satisfy the boundary condition on the whole circle. The capricious behavior of the solution occurs in the boundary layer, whereas below the boundary layer the solution behaves very regularly; see Eckhaus and de Jager (1966, Theorem IV).

By using the expansion in (23.6.79) it is not difficult to verify that the exact solution of the boundary value problem can be written as

$$\Phi(x, y) = -y - e^{\omega y} F(x, y), \quad \omega = \frac{1}{2\varepsilon}, \tag{23.6.100}$$

where $F(x, y)$ can be expanded in the form of a Fourier series

$$F(x,y) = 2 \sum_{n=0}^{\infty}{}' \frac{I_n'(\omega)}{I_n(\omega)} I_n(\omega r) \cos n \left(\theta + \tfrac{1}{2}\pi\right). \tag{23.6.101}$$

The prime in the summation symbol means that the first term of the sum is to be halved.

We replace the ratios $I_n'(\omega)/I_n(\omega)$ by approximations that hold for large values of ω, uniformly with respect to n. From the Debye expansion given in §9.5 it follows after a few manipulations that

$$\frac{I_n'(\omega)}{I_n(\omega)} \sim \sum_{m=0}^{\infty} d_m \left(\frac{\omega}{\sqrt{n^2+\omega^2}}\right)^{m-1}, \tag{23.6.102}$$

where

$$\begin{aligned} &d_0 = 1, \quad d_1 = d_2 = 0, \quad d_3 = -\frac{1}{2\omega}, \quad d_4 = \frac{1}{2\omega^2}, \quad d_5 = -\frac{1}{2\omega^3}, \\ &d_6 = \frac{1}{2\omega^4} - \frac{5}{8\omega^2}, \quad d_7 = \frac{9}{4\omega^3} - \frac{1}{2\omega^5}, \quad d_8 = \frac{1}{2\omega^6} - \frac{47}{8\omega^4}. \end{aligned} \tag{23.6.103}$$

When we substitute (23.6.102) into (23.6.101) and rearrange, it follows that we have to evaluate the series

$$S_\mu(r,\theta,\omega) = \sum_{n=-\infty}^{\infty} \frac{\omega^{2\mu}}{(n^2+\omega^2)^\mu} I_n(\omega r) \cos n \left(\theta + \tfrac{1}{2}\pi\right), \tag{23.6.104}$$

where $\mu = -\frac{1}{2}, \frac{1}{2}, 1, \frac{3}{2}, \ldots$. With this notation the representation in (23.6.102) can be replaced by the asymptotic relation

$$F(x,y) \sim \sum_{m=0}^{\infty} d_m\, S_{(m-1)/2}(r,\theta,\omega). \tag{23.6.105}$$

The function $S_1(r, \theta, \omega)$ appears in (23.6.76). The terms in (23.6.105) with odd values of m follow from similar methods, whereas the terms with even m are essentially more difficult to evaluate.

Chapter 24

Hermite-type expansions of integrals

In this chapter we consider what happens when for certain polynomials the degree n is large, together with other parameters. We will see that Hermite polynomials can be used in uniform expansions. The main examples considered in this chapter are $C_n^\gamma(x)$, the Gegenbauer or ultraspherical polynomials (with as special case the relativistic Hermite polynomials) and $f_n^{(\alpha)}(x)$, the Tricomi–Carlitz polynomials for large values of the degree n and the orders γ or α. The Laguerre polynomials for large α are considered in §32.4. For all these cases we show how to derive approximations of the zeros of the polynomials in terms of the zeros of the Hermite polynomials, and we compare these with numerical values. In §24.4 we mention several other examples considered in the literature.

Our starting point is an integral representation of the polynomials and we use the saddle point method. This approach can also be used for other special functions, although in that case the large parameter corresponding to the degree n in the Hermite case may not be necessarily an integer, and we need parabolic cylinder functions as main approximants; see Case 16 of Table 20.1.

The parabolic cylinder functions occur in similar expansions when starting from a second-order linear differential equation. In that case two turning points coalesce. For an application to Whittaker functions with both parameters large we refer to Olver (1980). In §24.2 we will explain the coalescence of turning points for the Gegenbauer polynomials. In Temme (1990a,b) we have discussed approximations for the classical orthogonal polynomials that can be obtained by using methods based on differential equations. In particular we derived Hermite-type expansions for the Gegenbauer and Laguerre polynomials.

In Chapter 19 we have considered certain limits of a number of orthogonal polynomials, and we showed how the limits can be written in terms of Hermite polynomials. In all the cases of that chapter the degree of the polynomials is fixed.

24.1 An expansion in terms of Hermite polynomials

We consider integrals of the form

$$F_\nu(\xi) = \frac{1}{2\pi i}\int_{\mathcal{C}} e^{\nu\Psi(t)}\, f(t)\,\frac{dt}{\sqrt{t}}, \tag{24.1.1}$$

where

$$\Psi(t) = 2\xi t - \rho^2 \ln t - \tfrac{1}{2}t^2. \tag{24.1.2}$$

We assume that ν is a positive large parameter and that ρ is positive. The logarithmic function in Ψ and $\sqrt{t}$ assume their principal values, which are real for positive values of t. The contour is a Hankel loop shown in Figure 2.1, and runs from $t = -\infty$, $\operatorname{ph} t = -\pi$, encircles the origin in positive direction, and terminates at $-\infty$, now with $\operatorname{ph} t = +\pi$.

The saddle points $t_{1,2}$ follow from solving $\Psi'(t) = 0$, that is, from

$$t^2 - 2\xi t + \rho^2 = 0, \tag{24.1.3}$$

and are given by

$$t_{1,2} = \xi \mp \sqrt{\xi^2 - \rho^2}. \tag{24.1.4}$$

For large values of ν the function $F_\nu(\xi)$ defined in (24.1.1) can be expanded in terms of parabolic cylinder functions. This asymptotic expansion holds uniformly with respect to $\xi \in \mathbb{R}$ and $\rho \in [0,\infty)$. For certain values of ν and ρ the parabolic cylinder functions reduce to Hermite polynomials.

The procedure for obtaining the Hermite-type expansions runs as follows. We assume for constructing the expansion that f is an analytic function in a domain Ω of the complex plane that contains the saddle points and the contour $\mathcal{C}$.

We proceed as in a number of earlier examples, where we have used Bleistein's method. We write

$$f(t) = \alpha_0 + \beta_0 t + (t - t_1)(t - t_2) g_0(t), \tag{24.1.5}$$

where α_0, β_0 follow from substituting $t = t_{1,2}$. That is,

$$\alpha_0 = \frac{t_2 f(t_1) - t_1 f(t_2)}{t_2 - t_1}, \quad \beta_0 = \frac{f(t_2) - f(t_1)}{t_2 - t_1}. \tag{24.1.6}$$

We obtain, on substituting (24.1.5) into (24.1.1) and integrating by parts the integral containing $g_0(t)$,

$$F_\nu(\xi) = \alpha_0\, \mathcal{H}(\xi,\nu,\rho) + \frac{\beta_0}{2\nu}\,\mathcal{H}'(\xi,\nu,\rho) + \frac{1}{\nu}\frac{1}{2\pi i}\int_{\mathcal{C}} e^{\nu\Psi(t)}\, f_1(t)\,\frac{dt}{\sqrt{t}}, \tag{24.1.7}$$

where

$$f_1(t) = \sqrt{t}\frac{d}{dt}\left(\sqrt{t}\, g_0(t)\right) = t g_0'(t) + \tfrac{1}{2} g_0(t), \tag{24.1.8}$$

and $\mathcal{H}(\xi,\nu,\rho)$ equals $F_\nu(\xi)$ with $f(t)$ replaced by unity, that is,

$$\mathcal{H}(\xi,\nu,\rho) = \frac{1}{2\pi i}\int_{\mathcal{C}} e^{\nu\Psi(t)}\,\frac{dt}{\sqrt{t}}. \tag{24.1.9}$$

The prime in $\mathcal{H}'(\xi,\nu,\rho)$ denotes the derivative with respect to ξ.[1] Repeating this procedure, we obtain for $N = 0, 1, 2, \ldots$

$$F_\nu(\xi) = \mathcal{H}(\xi,\nu,\rho) \sum_{s=0}^{N-1} \frac{\alpha_s}{\nu^s} + \frac{\mathcal{H}'(\xi,\nu,\rho)}{2\nu} \sum_{s=0}^{N-1} \frac{\beta_s}{\nu^s} + \frac{1}{\nu^N} R_N(\xi,\nu), \tag{24.1.10}$$

where

$$R_N(\xi,\nu) = \frac{1}{2\pi i} \int_{\mathcal{C}} e^{\nu\Psi(t)}\, f_N(t)\, \frac{dt}{\sqrt{t}}, \tag{24.1.11}$$

and $\alpha_s, \beta_s, f_s(t)$ follow from the recursive scheme

$$\begin{aligned} f_s(t) &= \alpha_s + \beta_s t + (t-t_1)(t-t_2) g_s(t), \quad f_{s+1}(t) = t g_s'(t) + \tfrac{1}{2} g_s(t), \\ \alpha_s &= \frac{t_2 f_s(t_1) - t_1 f_s(t_2)}{t_2 - t_1}, \quad \beta_s = \frac{f_s(t_2) - f_s(t_1)}{t_2 - t_1}, \end{aligned} \tag{24.1.12}$$

for $s = 0, 1, 2, \ldots$, with $f_0(t) = f(t)$.

The function $\mathcal{H}(\xi,\nu,\rho)$ is a parabolic cylinder function. From (30.4.68) we obtain

$$\mathcal{H}(\xi,\nu,\rho) = \frac{\nu^{\frac{1}{2}\nu\rho^2 - \frac{1}{4}}\, e^{\nu\xi^2}}{\Gamma\left(\frac{1}{2} + \nu\rho^2\right)} U\left(-\nu\rho^2, 2\xi\sqrt{\nu}\right). \tag{24.1.13}$$

If $\nu\rho^2 = n + \frac{1}{2}$ $(n = 0, 1, 2, \ldots)$, then $\mathcal{H}(\xi,\nu,\rho)$ becomes a Hermite polynomial (see (11.1.3)):

$$\mathcal{H}(\xi,\nu,\rho) = \frac{1}{n!} \left(\tfrac{1}{2}\nu\right)^{\frac{1}{2}n} H_n\left(\xi\sqrt{2\nu}\right), \quad \nu\rho^2 = n + \tfrac{1}{2}. \tag{24.1.14}$$

All coefficients α_s, β_s are well defined and the functions f_s are analytic in the domain where $f_0 = f$ is analytic.

Remark 24.1. The factor $\sqrt{t}$ in the integral in (24.1.1) (and later) could have been replaced by any other power of t, and this would have given a different condition for $\nu\rho^2$ for the Hermite polynomial case. The present choice is convenient for the applications given later in this chapter. See also Remark 24.4. △

24.1.1 *Cauchy-type integrals for the coefficients*

As in other uniform expansions (see §25.2.1 and Chapter 23), we can give Cauchy-type integral representations of α_s, β_s and f_s. We introduce a set of rational functions $R_s(\tau, t), s = 0, 1, 2, \ldots$, by writing

$$\begin{aligned} R_0(\tau,t) &= \frac{1}{\tau - t}, \\ R_{s+1}(\tau,t) &= -\frac{1}{(\tau^2 - 2\xi\tau + \rho^2)} \left(\tfrac{1}{2} R_s(\tau,t) + \tau \frac{d}{d\tau} R_s(\tau,t)\right), \end{aligned} \tag{24.1.15}$$

[1]For convenience, we indicate three variables in $\mathcal{H}(\xi,\nu,\rho)$, although it is a function of two variables.

where t_1, t_2 are the saddle points defined in (24.1.4). That is, $(\tau - t_1)(\tau - t_2) = \tau^2 - 2\xi\tau + \rho^2$. The next few R_s are

$$R_1(\tau,t) = \frac{\tau + t}{2(\tau^2 - 2\xi\tau + \rho^2)(\tau - t)^2},$$
$$R_2(\tau,t) = \Bigl((5\tau^2 - 6\tau\xi + \rho^2)\tau^2 + 6(\rho^2 + \tau^2 - 2\tau\xi)\tau t + (\rho^2 - 3\tau^2 + 2\tau\xi)t^2\Bigr) \Big/ \Bigl(4(\tau^2 - 2\xi\tau + \rho^2)^3 (\tau - t)^3\Bigr). \tag{24.1.16}$$

Lemma 24.1. *Let $\mathcal{L}$ be a simple closed contour in Ω and let $\mathcal{L}$ contain the points t, t_1 and t_2 in its interior. Then,*

$$f_s(t) = \frac{1}{2\pi i}\int_{\mathcal{L}} R_s(\tau,t) f(\tau)\, d\tau, \quad s = 0,1,2,\ldots. \tag{24.1.17}$$

Proof. In the proof of Theorem 25.1 we explain in detail a related case; the present proof is similar, and is left as an exercise. □

For the coefficients α_s, β_s we have the following representations.

Lemma 24.2. *Let $A_s(\tau), B_s(\tau)$ $(s = 0,1,2,\ldots)$ be defined by*

$$A_s(\tau) = \frac{t_2 R_s(\tau,t_1) - t_1 R_s(\tau,t_2)}{t_2 - t_1}, \quad B_s(\tau) = \frac{R_s(\tau,t_2) - R_s(\tau,t_1)}{t_2 - t_1}. \tag{24.1.18}$$

Then

$$\alpha_s = \frac{1}{2\pi i}\int_{\mathcal{L}} A_s(\tau) f(\tau)\, d\tau, \quad \beta_s = \frac{1}{2\pi i}\int_{\mathcal{L}} B_s(\tau) f(\tau)\, d\tau. \tag{24.1.19}$$

Proof The proof easily follows from the previous lemma and the representations in (24.1.12). □

Observe that the representations of α_s, β_s and f_s obtained by using the recursive scheme in (24.1.12) contain derivatives of the function f, and that the integrals in (24.1.17) and (24.1.19) are in terms of f itself.

The first few A_s, B_s are:

$$A_0(\tau) = \frac{\tau - 2\xi}{\tau^2 - 2\xi\tau + \rho^2}, \quad B_0(\tau) = \frac{1}{\tau^2 - 2\xi\tau + \rho^2},$$
$$A_1(\tau) = \frac{4\tau\xi^2 - 3\tau\rho^2 + 2\xi\rho^2 - 4\tau^2\xi + \tau^3}{2(\tau^2 - 2\xi\tau + \rho^2)^3}, \quad B_1(\tau) = \frac{3\tau^2 - 2\tau\xi - \rho^2}{2(\tau^2 - 2\xi\tau + \rho^2)^3},$$
$$A_2(\tau) = \Bigl(5\tau^5 - 24\tau^2\xi^3 - 2\xi\rho^4 - 32\tau\xi^2\rho^2 + 9\tau\rho^4 + 44\tau^3\xi^2 + 60\tau^2\xi\rho^2 - 34\tau^3\rho^2 - 26\tau^4\xi\Bigr) \Big/ \Bigl(4(\tau^2 - 2\xi\tau + \rho^2)^5\Bigr),$$
$$B_2(\tau) = \frac{16\tau\xi\rho^2 + 12\tau^2\xi^2 - 26\tau^2\rho^2 + 21\tau^4 - 24\tau^3\xi + \rho^4}{4(\tau^2 - 2\xi\tau + \rho^2)^5}. \tag{24.1.20}$$

These coefficients do also satisfy the recurrence relation in (24.1.15) with initial conditions given in (24.1.20).

24.2 Gegenbauer polynomials

We consider the Gegenbauer polynomials $C_n^\gamma(x)$ for large values of γ, while n may be of the same order or smaller. For expansions for large n we refer to §15.3.

The Gegenbauer polynomials, or ultraspherical polynomials, can be defined by the generating function (15.3.26). An explicit representation is (see Koornwinder *et al.* (2010, Eq. 18.5.10))

$$C_n^\gamma(x) = \frac{1}{\Gamma(\gamma)} \sum_{k=0}^{\lfloor n/2 \rfloor} \frac{(-1)^k \Gamma(\gamma+n-k)}{k!(n-2k)!}(2x)^{n-2k}, \tag{24.2.21}$$

where $\lfloor m \rfloor$ is the integer number satisfying $\lfloor m \rfloor \le m < \lfloor m \rfloor + 1$, with $m \in \mathbb{R}$. Special values are

$$C_{2m}^\gamma(0) = (-1)^m \frac{\Gamma(\gamma+m)}{\Gamma(\gamma)\, m!}, \quad C_n^\gamma(1) = \frac{1}{n!}\frac{\Gamma(2\gamma+n)}{\Gamma(2\gamma)}. \tag{24.2.22}$$

These polynomials satisfy the differential equation

$$(1-x^2)y'' - (2\gamma+1)xy' + n(n+2\gamma)y = 0. \tag{24.2.23}$$

A simple transformation

$$W(x) = (1-x^2)^{(2\gamma+1)/4} C_n^\gamma(x) \tag{24.2.24}$$

gives for W the equation

$$(1-x^2)W'' + \left((n+\gamma)^2 + \frac{2+4\gamma-4\gamma^2+x^2}{4(1-x^2)}\right)W = 0. \tag{24.2.25}$$

This can be written in the form

$$W'' = \nu^2 \frac{x^2-x_0^2}{(1-x^2)^2} W - \frac{x^2+3}{4(1-x^2)^2} W, \tag{24.2.26}$$

with

$$\nu = n+\gamma, \quad x_0 = \frac{\sqrt{\nu^2 - (\gamma-\frac{1}{2})^2}}{\nu}. \tag{24.2.27}$$

We assume that $\gamma > \frac{1}{2}$ and $n = 0, 1, 2, \ldots$. We have $x_0 \in [0,1]$. For the asymptotic problem we assume that ν is large.

Especially interesting is the behavior of the polynomials when x crosses the values $\pm x_0$ (turning points of (24.2.26)) and ± 1 (singular points of the differential equation, although not of the Gegenbauer polynomials). When $\gamma \gg n$ the turning points (and all zeros of the Gegenbauer polynomial) tend to zero and coalesce. As mentioned in the beginning of this chapter, parabolic cylinder functions can be used for describing the asymptotics. In this case, for the Gegenbauer polynomials, Hermite polynomials can be used.

When $n \gg \gamma$ the turning points coalesce with the singularities at ± 1, and for the transition near $x = \pm 1$ we need a J-Bessel function as approximant. In Temme (1990b) we have described these cases in more detail.

We are interested in the asymptotic behavior of $C_n^\gamma(x)$ for large values of γ; n may be large as well. We will give an asymptotic expansion that holds uniformly with respect to $x \in (-1,1)$ and with respect to n such that n/γ remains bounded.

24.2.1 *Preliminary steps*

We recall the Cauchy-type integral given in (15.3.27):

$$C_n^\gamma(x) = \frac{1}{2\pi i}\int_{\mathcal{C}} \frac{dw}{(1-2xw+w^2)^\gamma\, w^{n+1}}. \tag{24.2.28}$$

We assume that $x \in (-1,1)$ and that γ is a positive number. The contour $\mathcal{C}$ is a simple closed contour around the origin; the singular points $w_\pm = x \pm i\sqrt{1-x^2}$ are in the exterior of the contour $\mathcal{C}$. From the singular points $w_\pm$ branch cuts run to ∞. We assume that $(1-2xw+w^2)^\gamma$ is positive for real values of w.

We use the notation (see also (24.2.27))

$$\widehat{\gamma} = \gamma - \tfrac{1}{2}, \quad \nu = \gamma + n, \quad x_0 = \frac{\sqrt{\nu^2 - \widehat{\gamma}^2}}{\nu}, \quad \rho = \sqrt{\frac{n+\frac{1}{2}}{\nu}}, \quad \sigma = \frac{\widehat{\gamma}}{\nu}. \tag{24.2.29}$$

We have $x_0 \in [0,1]$ and the interrelations

$$\sigma = 1 - \rho^2 = \sqrt{1 - x_0^2}. \tag{24.2.30}$$

For the asymptotic problem we assume that ν is large.

We write

$$C_n^\gamma(x) = \frac{1}{2\pi i}\int_{\mathcal{C}} e^{\nu\Phi(w)}\, \frac{dw}{\sqrt{w(1-2xw+w^2)}}, \tag{24.2.31}$$

where

$$\Phi(w) = -\rho^2 \ln w - \sigma \ln(1-2xw+w^2). \tag{24.2.32}$$

The saddle points of $\Phi(w)$ follow from the equation $\Phi'(w) = 0$, which gives the equation

$$(\nu + \widehat{\gamma})w^2 - 2x\nu w + (\nu - \widehat{\gamma}) = 0. \tag{24.2.33}$$

Hence, the saddle points are given by

$$\begin{aligned} w_1 &= \frac{\nu}{\nu+\widehat{\gamma}}\left(x - \sqrt{x^2 - x_0^2}\right), \\ w_2 &= \frac{\nu}{\nu+\widehat{\gamma}}\left(x + \sqrt{x^2 - x_0^2}\right). \end{aligned} \tag{24.2.34}$$

The saddle points coalesce if $x = -x_0$ or $x = x_0$, the corresponding turning points of the differential equation in (24.2.26). When $x_0 \le x \le 1$ the saddle points are real. When x crosses the point x_0 from above, the real saddle points collide and become complex at

$$w_1 = w_2 = w_0 = \sqrt{\frac{\nu - \widehat{\gamma}}{\nu + \widehat{\gamma}}}. \tag{24.2.35}$$

When $x \in [-x_0, x_0]$ the saddle points are located on the circle around the origin with radius w_0. They collide again (when $x = -x_0$) at $w = -w_0$, and they become real again if $-1 < x < -x_0$. When $x = 1$, we have $w_2 = 1$ and $w_1 = w_0^2$; when

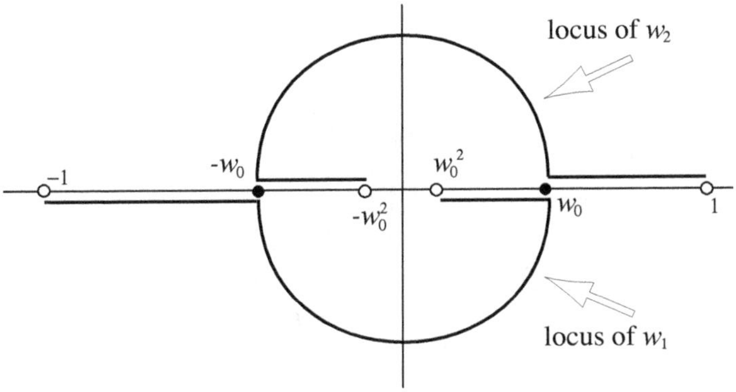

Fig. 24.1 The location of the two saddle points $w_{1,2}$ defined in (24.2.34) for x-values in the interval $(-1, 1)$. When $x \in [-x_0, x_0]$, $w_{1,2}$ are located on the circle around the origin with radius w_0.

$x = -1$, we have $w_1 = -1$ and $w_2 = -w_0^2$. In Figure 24.1 the location of the saddle points is visualized.

Remark 24.2. The location of the saddle points if $|x| \ge 1$ is not considered here. In that case the singularities $w_\pm = x \pm \sqrt{x^2-1}$ of the function $(1-2xw+w^2)^\gamma$ become real, and the asymptotics is quite different outside the interval of orthogonality. Outside $[-1,1]$, the Hermite-type expansions are not valid. As remarked earlier, if ν is large and x is close to the points ± 1, Bessel functions are needed to describe the asymptotics of the Gegenbauer polynomials. △

Remark 24.3. If $x_0 \to 1$ (which means that γ/ν tends to zero), the singularity w_+ and the saddle point w_2 coalesce (w_- and w_1 too). Hence, n/γ should be bounded. That is, we restrict x_0 to a compact subinterval of $(-1,1)$ and $x \in (-1,1)$. If $n \gg \gamma$ the asymptotics can be described in terms of elementary functions (if $|x| \le 1-\delta$, see §15.3) or in terms of Bessel functions (if x is close to unity). △

24.2.2 *A first approximation*

If $\gamma \gg n$, the saddle points coalesce in the neighborhood of the origin $w = 0$. Near the origin the phase function $\Phi(w)$ can be expanded in the form

$$\Phi(w) = -\rho^2 \ln w + 2x\sigma w - \sigma(1-2x^2)w^2 + \mathcal{O}\left(w^3\right), \qquad (24.2.36)$$

and using this approximation up to the quadratic term we obtain by (23.4.31)

$$C_n^\gamma(x) = \frac{1}{n!}\,\gamma^{n/2}\,(1-2x^2)^{n/2}\left(H_n\left(x\sqrt{\gamma}/\sqrt{1-2x^2}\right) + \varepsilon_n^\gamma(x)\right), \qquad (24.2.37)$$

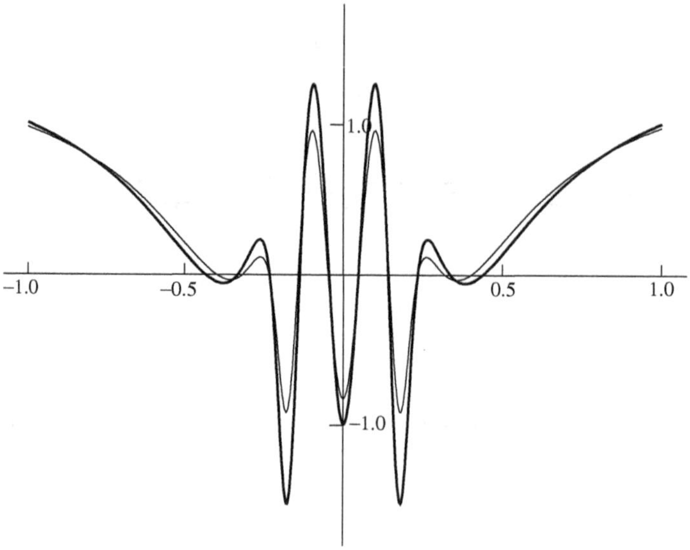

Fig. 24.2 $C_n^\gamma(x)$ (bold curve) and its approximation in (24.2.37) for $-1 \le x \le 1$, $n = 10$, and $\gamma = 50$, properly scaled.

where $\varepsilon_n^\gamma(x)$ is expected to be small if $\gamma \to \infty, \gamma \gg n$ and $x_0 = \mathcal{O}(n/\gamma)$ (so, x_0 is small as well).

It is not difficult to verify that this approximation is well defined when $x = \pm 1/\sqrt{2}$ and that the limits given in (19.1.2) follow from this first approximation, if indeed $\varepsilon_n^\gamma(x)$ vanishes in the limits.

In Figure 24.2 we show the graphs of the Gegenbauer polynomial and its approximation (thin line) in (24.2.37) for $-1 \le x \le 1$, $n = 10$, and $\gamma = 50$ (scaled by using the values of the Gegenbauer polynomial at the origin and at $x = 1$, see (24.2.22)).

24.2.3 *Transformation to the standard form*

We observe that the pattern of the saddle points $w_{1,2}$ is quite similar to that of the integral that defines the Hermite polynomial; cf. Figure 23.1 and Figure 24.1, we use the transformation

$$\Phi(w) = \Psi(t) + A, \tag{24.2.38}$$

where A does not depend on t and $\Psi(t)$ is given in (24.1.2). We assume the following conditions:

(1) The mapping is analytic at $w = 0$ (with $t(0) = 0$).
(2) The saddle points $w_{1,2}$ correspond to $t_{1,2}$ given in (24.1.4).

These two conditions determine the parameters ξ and ρ in $\Psi(t)$ and the quantity A. The parameter ν in (24.1.1) will be taken equal to the present value $\nu = n + \gamma$.

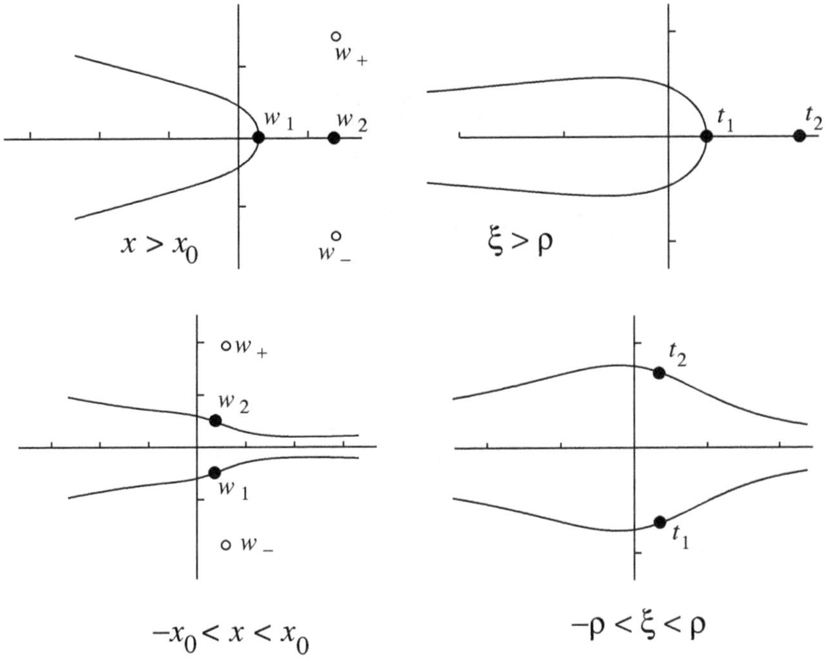

Fig. 24.3 Saddle point contours in the w-plane (left) and the t-plane. The pictures at the top correspond to $x = 0.7$, those at the bottom to $x = 0.3$. The other parameters are: $\gamma = 500, n = 100$.

From the first condition it follows that we have to take the value of ρ in (24.1.2) as given in (24.2.29), that is, $\rho = \sqrt{(n+\frac{1}{2})/\nu}$. This means, the logarithmic singularities in $\Psi(t)$ and $\Phi(w)$ at the origin should be the of the same nature.

The second condition gives the relations

$$\Phi(w_1) = \Psi(t_1) + A, \quad \Phi(w_2) = \Psi(t_2) + A, \tag{24.2.39}$$

from which we can eliminate A. Hence,

$$\Phi(w_1) - \Phi(w_2) = \Psi(t_1) - \Psi(t_2). \tag{24.2.40}$$

This equation defines the parameter ξ. More details about ξ will be given later. When we know ξ, the quantity A follows from either relation in (24.2.39).

In Figure 24.3 we show the saddle point contours for special choices of the parameters. We take $n = 100, \gamma = 500$ and in the pictures at the top $x = 0.7$. From these values the other parameters can be computed:

$$\nu = 600, \quad x_0 = 0.554\ldots, \quad \rho = 0.409\ldots, \quad \xi = 0.554\ldots. \tag{24.2.41}$$

For the graphs at the bottom of Figure 24.3 we have $x = 0.3, \xi = 0.207\ldots$; in this case the asymptotic behavior of the Gegenbauer polynomial is determined by the contributions of the saddle point contours through both saddle points (oscillatory

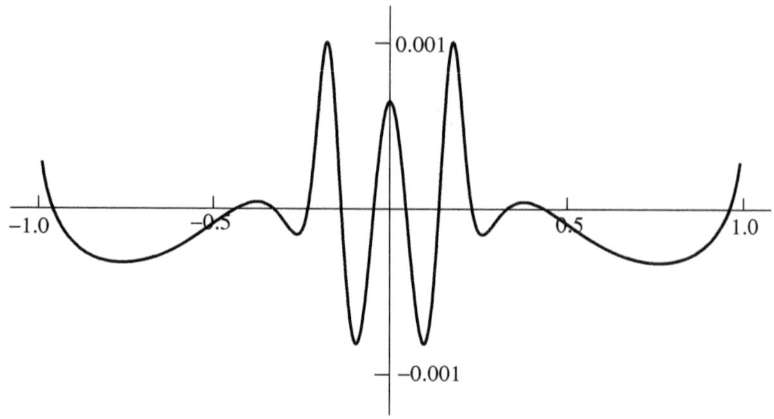

Fig. 24.4 The difference of the left-hand side and the first-term approximation of the expansion given in (24.2.45), for $-1 \le x \le 1$, $n = 10$, and $\gamma = 50$, properly scaled.

case). The points $w_\pm$ are the zeros of $1 - 2xw + w^2$, the singular points of the function $\Phi(w)$, that is, $w_\pm = x \pm i\sqrt{1-x^2}$.

Using the transformation (24.2.38) we obtain

$$C_n^\gamma(x) = \frac{e^{\nu A}}{2\pi i} \int_{\mathcal{C}} e^{\nu\Psi(t)} f(t) \frac{dt}{\sqrt{t}}, \tag{24.2.42}$$

where

$$f(t) = \frac{\sqrt{t}}{\sqrt{w(1-2xw+w^2)}} \frac{dw}{dt}, \tag{24.2.43}$$

and using (24.2.38),

$$\frac{dw}{dt} = \nu \frac{w(1-2xw+w^2)\,(t^2 - 2\xi t + \rho^2)}{t\left((\nu+\widehat{\gamma})w^2 - 2x\nu w + \nu - \widehat{\gamma}\right)}. \tag{24.2.44}$$

By combining (24.1.1), (24.1.10), (24.1.14), and (24.2.42) we find

$$C_n^\gamma(x) \sim \frac{e^{\nu A}}{n!} \left(\tfrac{1}{2}\nu\right)^{n/2} \left(H_n\left(\xi\sqrt{2\nu}\right) \sum_{s=0}^{\infty} \frac{\alpha_s}{\nu^s} + \frac{H_n'\left(\xi\sqrt{2\nu}\right)}{\sqrt{2\nu}} \sum_{s=0}^{\infty} \frac{\beta_s}{\nu^s} \right). \tag{24.2.45}$$

For $e^{\nu A}$ we refer to (24.2.58). If we wish we can write the expansion in terms of finite series with a remainder, as we have done in (24.1.10).

In Figure 24.4 we show the difference of the left-hand side and the first-term approximation of the expansion given in (24.2.45), for $-1 \le x \le 1$, $n = 10$, and $\gamma = 50$, the same as in Figure 24.2 with the same scaling of the polynomials. Showing graphs of both scaled polynomials in one picture does not give noticeable separate graphs.

From (24.2.44) it follows that the value of dw/dt at the points $t = t_j$ is given by (we need to apply l'Hôpital's rule)

$$\left.\frac{dw}{dt}\right|_{t=t_j} = \frac{\sqrt{w_j(1-2xw_j+w_j^2)}}{\sqrt{t_j}} \left(\frac{\xi^2-\rho^2}{x^2-x_0^2}\right)^{\frac{1}{4}}, \tag{24.2.46}$$

and from (24.2.43) it follows that

$$\alpha_0 = f(t_j) = \left(\frac{\xi^2-\rho^2}{x^2-x_0^2}\right)^{\frac14}. \tag{24.2.47}$$

Remark 24.4. From (24.1.6) it follows that $\alpha_0 = f(t_1) = f(t_2)$ and $\beta_0 = 0$. We have achieved this thanks to the choice of the factor $\sqrt{t}$ in (24.1.1) and the square root in (24.2.31). △

24.2.3.1 *Evaluating* ξ

To find a relation for ξ we use the relations $t_j^2 = 2\xi t_j - \rho^2$ and $w_j^2 = (2x\nu w_j - (\nu-\widehat{\gamma}))/(\nu+\widehat{\gamma})$, $j = 1, 2$, which follow from (24.1.3) and (24.2.33). We also use the explicit forms of the saddle points $t_{1,2}$ and $w_{1,2}$ (see (24.1.4) and (24.2.34). Then it follows from (24.2.40), by using (24.1.2), (24.2.30) and (24.2.32), after straightforward manipulations,

$$\rho^2 \operatorname{arccosh}\frac{x}{x_0} - \sigma \operatorname{arctanh}\frac{x\sqrt{x^2-x_0^2}}{1-x^2+\sigma} = \rho^2 \operatorname{arccosh}\frac{\xi}{\rho} - \xi\sqrt{\xi^2-\rho^2}, \tag{24.2.48}$$

where $x_0 \le x < 1$ and $\xi \ge \rho$. (The left-hand side equals $\frac12(\Phi(w_1) - \Phi(w_2))$, the right-hand side equals $\frac12(\Psi(t_1) - \Psi(t_2))$.) For $-x_0 \le x \le x_0$ and $-\rho \le \xi \le \rho$ we have

$$\rho^2 \arcsin\frac{x}{x_0} + \sigma \arctan\frac{x\sqrt{x_0^2-x^2}}{1-x^2+\sigma} = \rho^2 \arcsin\frac{\xi}{\rho} + \xi\sqrt{\rho^2-\xi^2}. \tag{24.2.49}$$

For $-1 < x < x_0$ we can use (24.2.48) with x and ξ replaced by $-x$ and $-\xi$.

To compute ξ from (24.2.48) the numerical inversion can be started by writing $\xi = \rho\cosh\theta$, $\theta \ge 0$. The right-hand side becomes $\rho^2(\theta - \sinh\theta\cosh\theta)$, and a simple Newton process can be used to find θ. Similar for (24.2.49) by using $\xi = \rho\sin\theta$, $-\frac12\pi \le \theta \le \frac12\pi$.

If $x_0 \to 0$, which implies $\rho \to 0$, (24.2.48) reduces to

$$\xi^2 = -\tfrac12 \ln\left(1-x^2\right), \quad \operatorname{sign}(\xi) = \operatorname{sign}(x), \quad -1 < x < 1. \tag{24.2.50}$$

From (24.2.29) we see that x_0 tends to zero when n is fixed

By differentiating the left-hand side of (24.2.48) with respect to x and the right-hand side with respect to ξ, we find that both sides monotonically increase functions on $(x_0, 1)$ and (ρ, ∞), respectively. Hence, for each value of $x \in (x_0, 1)$ we can find one (and only one) value of $\xi > \rho$, such that (24.2.48) is satisfied. The same holds for the other intervals.

From (24.2.40) we obtain

$$\frac{d\xi}{dx} = \frac{1}{2(1-x^2)}\sqrt{\frac{x_0^2-x^2}{\rho^2-\xi^2}}, \tag{24.2.51}$$

which holds for $x \in (-1, 1)$ and $\xi \in \mathbb{R}$.

It follows that ξ is an increasing function of x with $\lim_{x\to\pm1}\xi(x)=\pm\infty$. The function $\xi(x)$ is defined for all $x\in(-1,1)$, and is analytic in this interval (and in a large domain of the complex plane), in particular at the points $\pm x_0$. It can be shown that ξ defined by (24.2.40) is analytic in both variables x and x_0.

The analytical properties of the function $\xi(x,x_0)$ follow from methods discussed in Qu and Wong (1990). In Temme (1990c) a proof is given for a different mapping, but the same methods can be used for the present case.

We give a few coefficients in Maclaurin expansions of ξ as a function of x. It easily follows from (24.2.49) and (24.2.51) that $\xi(x)$ is an odd function of x. We have

$$x(\xi)=c_1\xi+c_3\xi^3+\ldots, \tag{24.2.52}$$

where

$$\begin{aligned} c_1&=\frac{2\rho}{x_0}=2\sqrt{\frac{\nu}{\nu+\widehat{\gamma}}},\\ c_3&=-\frac{x_0^4+8\rho^4x_0^2-4\rho^4}{3\rho x_0^5}=-\frac{\nu^{3/2}(7\widehat{\gamma}+5\nu)}{3(\nu+\widehat{\gamma})^{5/2}}. \end{aligned} \tag{24.2.53}$$

An expansion at the point $\xi=\rho$ has the form

$$x(\xi)=x_0+d_1(\xi-\rho)+d_2(\xi-\rho)^2+d_3(\xi-\rho)^3+\ldots, \tag{24.2.54}$$

where

$$\begin{aligned} d_1^3&=\frac{4\rho(1-x_0^2)^2}{x_0},\quad (d_1>0),\\ d_2&=\frac{-d_1^4+4(1-x_0^2)^2-32d_1\rho x_0(1-x_0^2)}{10x_0d_1^2},\\ d_3&=[-64(1-x_0^2)^4+704\rho x_0(1-x_0^2)^3d_1-1536\rho^2x_0^2(1-x_0^2)^2d_1^2-\\ &\qquad 4(93x_0^2+7)(1-x_0^2)d_1^4+32\rho x_0(28x_0^2-3)d_1^5+11d_1^8]/(350x_0^2d_1^5), \end{aligned} \tag{24.2.55}$$

Because $\rho\in(0,1)$ and $x_0\in(0,1)$, we see that the shown coefficients are bounded if $\nu\to\infty$. If $x_0\to0$ (observe that $\rho/x_0=\sqrt{\nu/(\nu+\widehat{\gamma})}$) the coefficients d_j tend to c_j.

24.2.3.2 *Evaluating A*

The quantity A can be obtained by using (24.2.39). It is convenient to observe that

$$A=\tfrac12\left(\Phi(w_1)+\Phi(w_2)\right)-\tfrac12\left(\Psi(t_1)+\Psi(t_2)\right). \tag{24.2.56}$$

We also use

$$(1-2xw_1+w_1^2)(1-2xw_2+w_2^2)=\frac{4\widehat{\gamma}^2(1-x^2)}{(\nu+\widehat{\gamma})^2}. \tag{24.2.57}$$

The result can be written in the form

$$e^{\nu A}=\nu^{-\frac12n-\frac14}\left(2\widehat{\gamma}\right)^{-\widehat{\gamma}}\left(1-x^2\right)^{-\frac12\widehat{\gamma}}\left(\nu+\widehat{\gamma}\right)^{\frac12\nu+\frac12\widehat{\gamma}}\,e^{-\frac12n-\frac14-\nu\xi^2}; \tag{24.2.58}$$

see (24.2.29) for the notation.

24.2.4 *Special cases of the expansion*

We consider some special cases of the expansion given in (24.2.45).

(1) If $x = 0$, or equivalently $\xi = 0$, we can simplify many quantities. We take n even, $n = 2m$, and use

$$\begin{aligned} C_{2m}^{\gamma}(0) &= (-1)^m \frac{\Gamma(\gamma+m)}{\Gamma(\gamma)\, m!}, \\ H_{2m}(0) &= (-1)^m \frac{(2m)!}{m!}, \quad H_{2m}'(0) = 0. \end{aligned} \tag{24.2.59}$$

Then (see (24.2.47))

$$\alpha_0 = \left(\frac{\nu}{2\gamma + 2m - \frac{1}{2}}\right)^{\frac{1}{4}} \tag{24.2.60}$$

and (24.2.45) reduces to

$$\frac{\Gamma(\gamma+m)}{\Gamma(\gamma)} \sim e^{-m-\frac{1}{4}\widehat{\gamma}-\widehat{\gamma}} \left(\gamma + m - \tfrac{1}{4}\right)^{\gamma+m-\frac{1}{2}} \sum_{s=0}^{\infty} \frac{\alpha_s/\alpha_0}{\nu^s}. \tag{24.2.61}$$

This corresponds to expansions for the ratio of gamma functions; see §25.6.1. The factor in front of the series gives the correct first-order approximation of the left-hand side for large values of γ.

(2) If $x \to 1$ ($\xi \to \infty$) we can also verify some limits. The relation in (24.2.48) gives:

$$\begin{aligned} -\xi^2 + \rho^2 \ln \xi + \tfrac{1}{2}\rho^2 + \rho^2 \ln \frac{2}{\rho} + \mathcal{O}(\xi^{-2}) &= \\ \rho^2 \operatorname{arccosh} \frac{1}{x_0} + \tfrac{1}{2}\sigma \ln \frac{(\sigma+1)^2(1-x)}{2\sigma^2} &+ \mathcal{O}(1-x). \end{aligned} \tag{24.2.62}$$

By using

$$C_n^{\gamma}(1) = \frac{1}{n!} \frac{\Gamma(2\gamma+n)}{\Gamma(2\gamma)}, \quad H_n(z) \sim (2z)^n, \quad z \to \infty, \tag{24.2.63}$$

and $\alpha_0 \sim \sqrt{\xi/\sigma}$, we conclude that, for $x = 1$, the expansion in (24.2.45) reduces to

$$\frac{\Gamma(2\gamma+n)}{\Gamma(2\gamma)} \sim \frac{(\nu+\widehat{\gamma})^{n+\frac{1}{2}+\frac{1}{2}\nu+\frac{3}{2}\widehat{\gamma}}}{(2\widehat{\gamma})^{2\widehat{\gamma}+\frac{1}{2}}} \left(\frac{\rho}{x_0 e \sqrt{\nu}}\right)^{n+\frac{1}{2}} \sum_{s=0}^{\infty} \frac{\alpha_s/\alpha_0}{\nu^s}. \tag{24.2.64}$$

Again, this corresponds to expansions for the ratio of gamma functions. The factor in front of the series gives the correct first-order approximation of the left-hand side for large values of γ.

Table 24.1 Correct decimal digits in the approximations of zeros of $G_{10}^{\gamma}(x)$.

m \ γ	1.0	2.5	5.0	10	25	50	75	100
1	3.5	3.7	4.0	4.3	4.8	5.3	5.6	5.9
2	3.4	3.6	3.9	4.2	4.8	5.3	5.6	5.9
3	3.3	3.6	3.8	4.2	4.8	5.3	5.6	5.9
4	3.3	3.5	3.8	4.2	4.8	5.3	5.6	5.8
5	3.3	3.5	3.8	4.1	4.8	5.3	5.6	5.8
6	3.3	3.5	3.8	4.1	4.8	5.3	5.6	5.8
7	3.3	3.5	3.8	4.2	4.8	5.3	5.6	5.8
8	3.3	3.6	3.8	4.2	4.8	5.3	5.6	5.9
9	3.4	3.6	3.9	4.2	4.8	5.3	5.6	5.9
10	3.5	3.7	4.0	4.3	4.8	5.3	5.6	5.9

24.2.5 *Approximating the zeros*

When computing approximations of the zeros of the Gegenbauer polynomials, we start with the zeros of the Hermite polynomial in (24.2.45).

Let $g_{n,m}^{\gamma}, h_{n,m}$ be the mth zero of $C_n^{\gamma}(x)$, $H_n(x)$, respectively, $m = 1, 2, \ldots, n$. For given γ and n we compute

$$\xi_{n,m} = \frac{h_{n,m}}{\sqrt{2\nu}}, \qquad m = 1, 2, \ldots, n. \tag{24.2.65}$$

Upon inverting with these $\xi_{n,m}$ equation (24.2.49), we compute $x_{n,m}$, giving the approximations $\widetilde{g}_{n,m}^{\gamma}$ for the zeros of $C_n^{\gamma}(x)$:

$$\widetilde{g}_{n,m}^{\gamma} = x_{n,m}, \qquad m = 1, 2, \ldots, n. \tag{24.2.66}$$

From asymptotic expansions of Hermite polynomials (see, for instance, Skovgaard (1959)) it follows that $h_{n,0}, h_{n,n}$ have the asymptotic estimate

$$-h_{n,0} = h_{n,n} = \sqrt{2n+1} - \varepsilon(n), \tag{24.2.67}$$

where $\varepsilon(n)$ is a positive function of n, with the estimate $\varepsilon(n) = \mathcal{O}(n^{-1/6})$, as $n \to \infty$. It follows that the numbers $\xi_{n,m}$ (if n is large) belong to the interval $[-\rho, \rho]$.

In Table 24.1 we show for $n = 10$ the "correct number of decimal digits" in the approximation of $g_{10,m}^{\gamma}$ by $\widetilde{g}_{10,m}^{\gamma}$. That is, we show

$$-{}^{10}\log\left|\frac{g_{10,m}^{\gamma} - \widetilde{g}_{10,m}^{\gamma}}{g_{10,m}^{\gamma}}\right|, \qquad m = 1, 2, \ldots, 10, \tag{24.2.68}$$

where $\widetilde{g}_{10,m}^{\gamma}$ is the approximation obtained by the procedure just described.

We conclude from the table that the approximations are quite uniform with respect to m and that the zeros near the endpoints of the interval $[-1, 1]$ are slightly better approximated than the zeros near the origin. Furthermore, small values of the large parameter $\nu = n + \gamma$ give already at least 3 significant digits.

24.2.6 The relativistic Hermite polynomials

The relativistic Hermite polynomials $H_n^N(z)$ can be defined by the generating function

$$\sum_{n=0}^{\infty} \frac{H_n^N(z)}{n!} t^n = \left(\left(1-\frac{tz}{N}\right)^2 + \frac{t^2}{N}\right)^{-N}, \quad |t| < \frac{N}{\sqrt{N+z^2}}, \tag{24.2.69}$$

where N and z are real numbers, $N > 0$ (the polynomials and series are also defined for complex N and z, with a different region of convergence).

These polynomials have been introduced for the investigation of the harmonic oscillator in the frame of relativistic quantum theory (Aldaya *et al.*, 1991). They are extensions of the Hermite polynomials in the sense that

$$\lim_{N\to\infty} H_n^N(z) = H_n(z). \tag{24.2.70}$$

They can be expressed in terms of Gegenbauer polynomials (Nagel, 1994):

$$H_n^N\left(z\sqrt{N}\right) = \frac{n!}{N^{\frac{1}{2}n}} \left(1+z^2\right)^{\frac{1}{2}n} C_n^N\left(\frac{z}{\sqrt{1+z^2}}\right). \tag{24.2.71}$$

Applying the results given for the Gegenbauer polynomials is straightforward, and we omit further details.

Because the zeros of $C_n^\gamma(x)$ occur inside the interval $[-x_0, x_0]$, where x_0 is given in (24.2.29), it follows that the zeros of $H_n^N\left(z\sqrt{N}\right)$ occur for large $n+N$ inside the interval $[-z_0, z_0]$, where

$$z_0 = \frac{\sqrt{n(n+2N)}}{N}, \tag{24.2.72}$$

and that, hence, the zeros of $H_n^N(x)$ occur inside the interval

$$\left(-\sqrt{\frac{n(n+2N)}{N}}, \sqrt{\frac{n(n+2N)}{N}}\right). \tag{24.2.73}$$

In Gawronski and Van Assche (2003) asymptotic approximations are given for the polynomials $H_n^N\left(z\sqrt{N}\right)$ inside an interval similar to $[-z_0, z_0]$ (the oscillatory case), and outside this interval in the complex z-plane. For many other details about these polynomials we refer to Vignat (2011).

24.3 Tricomi–Carlitz polynomials

The Tricomi–Carlitz polynomials are defined by

$$t_n^{(\alpha)}(x) = \sum_{k=0}^{n} (-1)^k \binom{x-\alpha}{k} \frac{x^{n-k}}{(n-k)!}. \tag{24.3.74}$$

We obtain from (15.5.36) the relation to the Laguerre polynomials:

$$t_n^{(\alpha)}(x) = (-1)^n L_n^{(x-\alpha-n)}(x), \tag{24.3.75}$$

and we observe that the class of polynomials $\{P_n(\lambda)\}$ introduced in Chapter 25 (see (25.1.9)) follows from the present set by putting $\alpha = 0$. The polynomials are related to random walks on the positive integers, see Karlin and McGregor (1959).

The Tricomi–Carlitz polynomials satisfy the recurrence relation

$$(n+1)t_{n+1}^{(\alpha)}(x) - (n+\alpha)\,t_n^{(\alpha)}(x) + x\,t_{n-1}^{(\alpha)}(x) = 0, \quad n \ge 1, \tag{24.3.76}$$

with initial values $t_0^{(\alpha)}(x) = 1,\ t_1^{(\alpha)}(x) = \alpha$. A few other values are

$$\begin{aligned} t_2^{(\alpha)}(x) &= \tfrac{1}{2}\left(\alpha + \alpha^2 - x\right), \\ t_3^{(\alpha)}(x) &= \tfrac{1}{6}\left(2\alpha + 3\alpha^2 + \alpha^3 - 2x - 3x\alpha\right). \end{aligned} \tag{24.3.77}$$

Tricomi (1948) introduced these polynomials. He observed that $\{t_n^{(\alpha)}(x)\}$ is not a system of orthogonal polynomials, the recurrence relations failing to have the required form (cf. Szegő (1975, p. 43)). However, Carlitz (1958) discovered that some scaling gives the desired property.

If we take

$$f_n^{(\alpha)}(x) = x^n\, t_n^{(\alpha)}\left(x^{-2}\right), \tag{24.3.78}$$

then $\{f_n^{(\alpha)}(x)\}$ satisfies

$$(n+1)f_{n+1}^{(\alpha)}(x) - (n+\alpha)\,x\,f_n^{(\alpha)}(x) + f_{n-1}^{(\alpha)}(x) = 0, \quad n \ge 1, \tag{24.3.79}$$

with initial values $f_0^{(\alpha)}(x) = 1,\ f_1^{(\alpha)}(x) = \alpha\,x$. A few other values are

$$\begin{aligned} f_2^{(\alpha)}(x) &= \tfrac{1}{2}\left(\alpha(1+\alpha)x^2 - 1\right), \\ f_3^{(\alpha)}(x) &= \tfrac{1}{6}\,x\left(-2 + 2\alpha x^2 - 3\alpha + 3\alpha^2 x^2 + \alpha^3 x^2\right). \end{aligned} \tag{24.3.80}$$

There is a generating function for $f_n^{(\alpha)}(x)$:

$$e^{w/x + (1-\alpha x^2)/x^2\,\ln(1-xw)} = \sum_{n=0}^{\infty} f_n^{(\alpha)}(x) w^n, \quad |wx| < 1, \tag{24.3.81}$$

and it follows that

$$f_n^{(\alpha)}(-x) = (-1)^n f_n^{(\alpha)}(x), \quad n = 0, 1, 2, \ldots. \tag{24.3.82}$$

Hence, we can concentrate on $x \ge 0$ (when we consider real x, as we do).

If $x = 0$ the generating function reduces to

$$e^{-\frac{1}{2}w^2} = \sum_{n=0}^{\infty} f_{2n}^{(\alpha)}(0)\, w^{2n}, \tag{24.3.83}$$

giving

$$f_{2n}^{(\alpha)}(0) = (-1)^n\, 2^{-n}/n! \quad f_{2n+1}^{(\alpha)}(0) = 0, \quad n = 0, 1, 2, \ldots. \tag{24.3.84}$$

Carlitz proved that for $\alpha > 0$, the polynomials $f_n^{(\alpha)}(x)$ satisfy the orthogonality relation

$$\int_{-\infty}^{\infty} f_m^{(\alpha)}(x)\, f_n^{(\alpha)}(x)\, d\Psi^{(\alpha)}(x) = \frac{2\,e^{\alpha}}{(n+\alpha)\,n!}\,\delta_{mn}, \tag{24.3.85}$$

where $\Psi^{(\alpha)}(x)$ is the step function whose jumps are

$$d\,\Psi^{(\alpha)}(x) = \frac{(k+\alpha)^{k-1}\,e^{-k}}{k!} \quad \text{at} \quad x = x_k = \pm\frac{1}{\sqrt{k+\alpha}}, \tag{24.3.86}$$

$k = 0, 1, 2, \ldots$. The values x_k play a special role in the generating function, because for these x-values we have

$$e^{w/x_k}\,(1 - x_k w)^k = \sum_{n=0}^{\infty} f_n^{(\alpha)}(x_k)\,w^n, \tag{24.3.87}$$

and now the series converges for all values of w.

Goh and Wimp (1994) and Goh and Wimp (1997) establish the asymptotic behavior of the Tricomi–Carlitz polynomials and discuss their zero distribution. They observe that the polynomials $f_n^{(\alpha)}(x/\sqrt{\alpha})$ have all zeros in the interval $[-1, 1]$.

In this section we describe how to obtain an asymptotic representation of the Tricomi–Carlitz polynomials in terms of the Hermite polynomial. We concentrate on large values of the parameter α and n is allowed to be large as well.

The asymptotic distribution of the zeros of $f_n^{(\alpha)}(x)$ can be obtained by using the zeros of the Hermite polynomials. The role of the Hermite polynomials follows from the saddle point analysis, but also by observing that

$$\lim_{\alpha\to\infty} f_n^{(\alpha)}\left(\frac{x\sqrt{2}}{\alpha}\right) = \frac{2^{-n/2}}{n!}\,H_n(x). \tag{24.3.88}$$

This follows from the generating function given in (24.3.81). Replacing x on the left-hand side by $x\sqrt{2}/\alpha$ yields, if $\alpha \to \infty$, $\exp(x\sqrt{2}\,w - \frac{1}{2}w^2)$. This is, up to some scaling, the generating function for the Hermite polynomials given in (15.0.6). From the relation in (24.3.88) we can obtain information on the location of the zeros for large values of α; for more details on these zeros we refer to §24.3.4.

Although the Tricomi–Carlitz polynomials can be expressed in terms of the Laguerre polynomials (see (24.3.75)), it is not possible to use the standard results on Laguerre polynomials from the literature to describe the asymptotics of $f_n^{(\alpha)}(x)$; this is due to the peculiar role and position of the parameters n and x in (24.3.75). For an overview of large-n asymptotics of $L_n^{(\alpha)}(x)$ we refer to Chapter 32. In Lee and Wong (2010) asymptotic results of the Tricomi–Carlitz polynomials have been obtained by using the recurrence relation. In Lee and Wong (2014) the integral representation given in (24.3.89) is used, and large degree expansions in terms of Airy functions are derived.

24.3.1 *Contour integral and saddle points*

We take the generating function (24.3.81) as starting point, and use the Cauchy-type integral:

$$f_n^{(\alpha)}(x) = \frac{1}{2\pi i}\int_{\mathcal{C}} e^{w/x + (1-\alpha x^2)/x^2\,\ln(1-xw)}\,\frac{dw}{w^{n+1}}. \tag{24.3.89}$$

The contour $\mathcal{C}$ is a circle around the origin with radius less than $1/|x|, x \neq 0$.

Rescaling the parameters in (24.3.89) by writing

$$\widehat{\alpha} = \alpha - \tfrac{1}{2}, \quad \nu = \left(n + \tfrac{1}{2}\right)/\widehat{\alpha}, \quad x \to x/\sqrt{\widehat{\alpha}}, \quad w \to w\sqrt{\widehat{\alpha}}, \tag{24.3.90}$$

we obtain

$$f_n^{(\alpha)}\left(x/\sqrt{\widehat{\alpha}}\right) = \frac{\widehat{\alpha}^{-n/2}}{2\pi i} \int_{\mathcal{C}} e^{\widehat{\alpha}\Phi(w)} \frac{dw}{\sqrt{w(1-xw)}}, \tag{24.3.91}$$

where

$$\Phi(w) = \frac{w}{x} + \frac{1-x^2}{x^2} \ln(1-xw) - \nu \ln w, \tag{24.3.92}$$

with

$$\Phi'(w) = -\frac{w^2 - x(\nu+1)w + \nu}{(1-xw)w}. \tag{24.3.93}$$

The saddle points are given by

$$w_1 = \tfrac{1}{2}\left(x(\nu+1) - \sqrt{W}\right), \quad w_2 = \tfrac{1}{2}\left(x(\nu+1) + \sqrt{W}\right), \tag{24.3.94}$$

where

$$W = x^2(\nu+1)^2 - 4\nu. \tag{24.3.95}$$

If

$$-\frac{2\sqrt{\nu}}{\nu+1} < x < \frac{2\sqrt{\nu}}{\nu+1} \tag{24.3.96}$$

the saddle points are complex, and for these values of x the zeros of $f_n^{(\alpha)}\left(x/\sqrt{\alpha}\right)$ occur. In that case the saddle points are located on the circle with radius $\sqrt{\nu}$.

Observe that for $\nu = 1$ this interval becomes $-1 < x < 1$, and that for other positive values of ν the bounds of the interval are less than unity. This confirms the result from Goh and Wimp that the zeros of $f_n^{(\alpha)}\left(x/\sqrt{\alpha}\right)$ are located inside the x-interval $(-1, 1)$. However, from the location of the saddle points it follows that the asymptotic distribution of the zeros for large α and/or n is inside the interval shown in (24.3.96), which will be much smaller as $\nu \to 0$ or $\nu \to \infty$.

We see from (24.3.88) that the Hermite polynomials will arise in the asymptotic behavior when α is much larger than n (in a similar manner with the corresponding parameters in the Gegenbauer and Laguerre cases). That is why we concentrate on the case $\nu \leq \nu_0 < 1$, where ν_0 is a fixed number.

Furthermore, we have to take into account the role of the logarithmic singularity of the function $\Phi(w)$ at $w = 1/x$, which disappears when $x = 1$. For $x \sim 1$ and for $x > 0$ the asymptotic behavior is quite different from what we consider in the present chapter, and other methods from uniform asymptotic analysis are needed. Therefore we assume that $0 \leq x \leq x_0 < 1$, where x_0 is a fixed number.

When we take these assumptions on x and ν, we see that the location of the saddle points is the same as described for the Hermite polynomials in §24.1.

24.3.2 A first approximation

Before giving details of the saddle point analysis we give a first result. If n is fixed and α is large, the complex saddle points given in (24.3.94) are close to the origin. For small values of w the phase function $\Phi(w)$ can be approximated by

$$\Phi_0(w) = xw - \tfrac{1}{2}\,(1-x^2)\,w^2 - \nu \ln w. \tag{24.3.97}$$

Taking this approximation for $\Phi(w)$ in (24.3.91) and using (23.4.31) we obtain the approximation

$$f_n^{(\alpha)}(x) = \left(\frac{1-\alpha x^2}{2}\right)^{n/2} \frac{1}{n!}\left(H_n\left(\frac{\alpha x}{\sqrt{2(1-\alpha x^2)}}\right) + \varepsilon_n^{(\alpha)}(x)\right), \tag{24.3.98}$$

where we expect that $|\varepsilon_n^{(\alpha)}(x)|$ is small if $\alpha \gg n$.

This approximation is well defined when $x = \pm 1/\sqrt{\alpha}$. Observe that the limit in (24.3.88) follows from (24.3.98) (if indeed $\lim\limits_{\alpha\to\infty} \varepsilon_n^{(\alpha)}(x) = 0$).

Computation of the zeros of $f_n^{(\alpha)}(x)$ for $n = 10, \alpha = 50$ with the help of (24.3.98) and the zeros of $H_{10}(x)$ gives a maximal absolute error of 0.0054 for the zeros of $f_{10}^{(50)}(x)$ and a relative error of about 5%.

24.3.3 Transformation to the standard form

We transform

$$\Phi(w) = \Psi(t) + A, \tag{24.3.99}$$

where $\Phi(w)$ is given in (24.3.92) and

$$\Psi(t) = 2\xi t - \rho^2 \ln t - \tfrac{1}{2}t^2, \quad \rho = \sqrt{\nu}. \tag{24.3.100}$$

The quantities A and ξ follow from the condition that the saddle points in the w-plane (see (24.3.94)) correspond to the saddle points

$$t_1 = \xi - \sqrt{\xi^2-\rho^2}, \quad t_2 = \xi + \sqrt{\xi^2-\rho^2} \tag{24.3.101}$$

in the t-plane.

Using the transformation (24.3.99), we obtain from (24.3.91) the representation

$$f_n^{(\alpha)}\left(x/\sqrt{\widehat{\alpha}}\right) = \frac{\widehat{\alpha}^{-n/2}\, e^{\widehat{\alpha}A}}{2\pi i}\int_{\mathcal{C}} e^{\widehat{\alpha}\,\Psi(t)}\, f(t)\,\frac{dt}{\sqrt{t}}, \tag{24.3.102}$$

where

$$f(t) = \frac{\sqrt{t}}{\sqrt{w(1-xw)}}\,\frac{dw}{dt}, \tag{24.3.103}$$

and

$$\frac{dw}{dt} = \frac{(1-xw)w}{t}\,\frac{(t-t_1)(t-t_2)}{(w-w_1)(w-w_2)}. \tag{24.3.104}$$

At the turning points we have

$$\left.\frac{dw}{dt}\right|_{t=t_j} = \sqrt{\frac{2(1-xw_j)w_j}{t_j}}\left(\frac{\xi^2-\rho^2}{x^2(\nu+1)^2-4\nu}\right)^{\frac14}, \quad j=1,2 \tag{24.3.105}$$

and

$$f(t_1)=f(t_2)=\sqrt{2}\left(\frac{\xi^2-\rho^2}{x^2(\nu+1)^2-4\nu}\right)^{\frac14}. \tag{24.3.106}$$

Evaluating the equation $\Psi(t_1)-\Psi(t_2)=\Phi(w_1)-\Phi(w_2)$, which defines the quantity ξ, we obtain

$$\begin{aligned}&2\rho^2\operatorname{arccosh}\frac{\xi}{\rho}-2\xi\sqrt{\xi^2-\rho^2}=\\&\quad-\frac{\sqrt{W}}{x}+2\frac{1-x^2}{x^2}\operatorname{arcsinh}\frac{x\sqrt{W}}{2\sqrt{1-x^2}}+2\nu\operatorname{arccosh}\frac{x(\nu+1)}{2\sqrt{\nu}},\end{aligned} \tag{24.3.107}$$

where W is defined in (24.3.95). The relation in (24.3.107) holds for $2\sqrt{\nu}/(\nu+1)\le x<1$, which correspond to ξ-values in $[\sqrt{\nu},\xi_0)$, where ξ_0 is the value that corresponds to $x=1$.

For $-2\sqrt{\nu}/(\nu+1)\le x\le 2\sqrt{\nu}/(\nu+1)$, the ξ-interval becomes $[-\sqrt{\nu},\sqrt{\nu}]$. In that case W is negative, and it is better to write

$$w_1=\tfrac12\left(x(\nu+1)-i\sqrt{-W}\right), \quad w_2=\tfrac12\left(x(\nu+1)+i\sqrt{-W}\right), \tag{24.3.108}$$

and

$$t_1=\xi-i\sqrt{\nu-\xi^2}, \quad t_2=\xi+i\sqrt{\nu-\xi^2}. \tag{24.3.109}$$

In this case the equation $\Psi(t_1)-\Psi(t_2)=\Phi(w_1)-\Phi(w_2)$ gives

$$\begin{aligned}&2\xi\sqrt{\rho^2-\xi^2}+2\rho^2\arcsin\frac{\xi}{\rho}=\\&\quad\frac{\sqrt{-W}}{x}-2\frac{1-x^2}{x^2}\arcsin\frac{x\sqrt{-W}}{2\sqrt{1-x^2}}+2\nu\arcsin\frac{x(\nu+1)}{2\sqrt{\nu}}.\end{aligned} \tag{24.3.110}$$

The function ξ is an odd function of x. The first few coefficients in the Maclaurin expansion are given:

$$\begin{aligned}\xi=&\frac{\nu+3}{6}x+\frac{8\nu^2+45\nu+135}{1620}x^3+\\&\frac{166\nu^3+1302\nu^2+4977\nu+14175}{408240}x^5+\ldots.\end{aligned} \tag{24.3.111}$$

When $\nu\to0$, we have $\rho\to0$, $W\to x$, and the relation in (24.3.107) becomes

$$-2\xi^2=-1+2\frac{1-x^2}{x^2}\operatorname{arcsinh}\frac{x^2}{2\sqrt{1-x^2}}, \quad \operatorname{sign}(\xi)=\operatorname{sign}(x). \tag{24.3.112}$$

In the case of the Gegenbauer and Laguerre polynomials, differential equations for the relation between ξ and x are available, see (24.2.51) and (32.4.88). These equations also arise in the Liouville–Green approach when using differential equations. In the present case, a differential equation for the Tricomi–Carlitz equation does not exist, but we can find the following relation:

$$\frac{d\xi}{dx} = -\frac{1}{2x^3\sqrt{\xi^2-\rho^2}}\left(x\sqrt{W} + \ln\frac{1-\frac{1}{2}x(x(\nu+1)+\sqrt{W})}{1-\frac{1}{2}x(x(\nu+1)-\sqrt{W})}\right). \tag{24.3.113}$$

When ξ is available, A follows by straightforward calculations from

$$A = \Phi(w_1) - \Psi(t_1) = \Phi(w_2) - \Psi(t_2), \tag{24.3.114}$$

or from the more symmetric form

$$A = \tfrac{1}{2}\left(\Phi(w_1)+\Phi(w_2)\right) - \tfrac{1}{2}\left(\Psi(t_1)+\Psi(t_2)\right). \tag{24.3.115}$$

The result is

$$A = \tfrac{1}{2} - \xi^2 + \frac{1-x^2}{2x^2}\ln\left(1-x^2\right). \tag{24.3.116}$$

We have the expansion

$$f_n^{(\alpha)}\left(x/\sqrt{\widehat{\alpha}}\right) \sim \frac{e^{\widehat{\alpha}A}}{2^{n/2}n!}\left(H_n\left(\xi\sqrt{2\widehat{\alpha}}\right)\sum_{s=0}^{\infty}\frac{\alpha_s}{\widehat{\alpha}^s} + \frac{H_n'\left(\xi\sqrt{2\widehat{\alpha}}\right)}{\sqrt{2\widehat{\alpha}}}\sum_{s=0}^{\infty}\frac{\beta_s}{\widehat{\alpha}^s}\right), \tag{24.3.117}$$

where $\widehat{\alpha}\to\infty$, $x\in[-1,1]$. The coefficients can be obtained as in the scheme given in (24.1.12), with f_0 given in (24.3.103). The first coefficients are

$$\alpha_0 = \sqrt{2}\left(\frac{\xi^2-\rho^2}{x^2(\nu+1)^2-4\nu}\right)^{\frac{1}{4}}, \quad \beta_0 = 0. \tag{24.3.118}$$

For the first-order approximation we obtain

$$f_n^{(\alpha)}\left(x/\sqrt{\widehat{\alpha}}\right) = \alpha_0\,\frac{e^{\widehat{\alpha}A}}{2^{n/2}n!}\left(H_n\left(\xi\sqrt{2\alpha-1}\right) + \varepsilon_n^{(\alpha)}(x)\right). \tag{24.3.119}$$

24.3.4 *Approximating the zeros*

In Table 24.2 we give the zeros x_k of $f_n^{(\alpha)}\left(x/\sqrt{\widehat{\alpha}}\right)$ for $n=10, \alpha=50$ and compare the zeros with approximations x_k^a obtained from this asymptotic formula. That is, let (for $k=1,2,\ldots,10$) h_k be the zeros of $H_{10}(x)$. Define $\xi_k = h_k/\sqrt{2\alpha-1}$, and invert the relation in (24.3.110) to obtain x_k. Then the approximations of the zeros are given by $x_k^a = x_k/\sqrt{\widehat{\alpha}}$; ε_k and δ_k are the absolute and relative errors, respectively. We observe that the approximations for these values of n and α are quite satisfactory; at least 5 significant decimal digits can be obtained in this way.

Table 24.2 Comparison of the zeros of the function $f_n^{(\alpha)}\left(x/\sqrt{\tilde{\alpha}}\right)$ for $n = 10, \alpha = 50$ with approximations based on the zeros of $H_n(x)$. We show $x_k, k = 1, 2, \ldots, 10$ (the zeros of $f_{10}^{(50)}(x)$) with their approximations x_k^a, and the absolute and relative errors.

k	x_k	x_k^a	ε_k	δ_k
1	−0.0855233907	−0.0855230252	0.36e-06	0.42e-05
2	−0.0650754635	−0.0650753259	0.13e-06	0.21e-05
3	−0.0460298897	−0.0460298453	0.44e-07	0.96e-06
4	−0.0274857009	−0.0274856920	0.89e-08	0.32e-06
5	−0.0091433976	−0.0091433973	0.32e-09	0.35e-07
6	0.0091433976	0.0091433973	0.32e-09	0.35e-07
7	0.0274857009	0.0274856920	0.89e-08	0.32e-06
8	0.0460298897	0.0460298454	0.44e-07	0.96e-06
9	0.0650754635	0.0650753259	0.13e-06	0.21e-05
10	0.0855233907	0.0855230252	0.36e-06	0.42e-05

24.4 More examples

We mention other cases for Hermite-type expansions for polynomials. First we observe that Gegenbauer polynomials are special cases of the Jacobi polynomials $P_n^{(\alpha,\beta)}(x)$ with equal α and β. We have

$$C_n^\gamma(x) = \frac{(2\gamma)_n}{\left(\gamma+\frac{1}{2}\right)_n} P_n^{(\gamma-\frac{1}{2},\gamma-\frac{1}{2})}(x). \tag{24.4.120}$$

It will be clear that the Jacobi polynomials with slightly different large α and β have a similar asymptotic behavior as the Gegenbauer polynomials, and that Hermite-type asymptotic expansions can be derived for that case. It is even possible to expand the Jacobi polynomials in terms of the Gegenbauer polynomials, which may yield a larger domain of validity for the parameters. In Temme (1990b) we have shown the first steps of a Liouville–Green approach by using the differential equation of the Jacobi polynomials with large parameter $\kappa = n+\frac{1}{2}(\alpha+\beta+1)$. When the parameters satisfy $\alpha \gg \beta$ or $\beta \gg \alpha$ (the "skew" case), the Jacobi polynomials can be approximated in terms of Laguerre polynomials.

In a series of papers, Roderick Wong and his co-workers have derived uniform asymptotic approximations for polynomials for which integral methods are used. The main approximants are, for certain domains of the parameters, parabolic cylinder functions, which can be written as Hermite polynomials. See Jin and Wong (1998); Li and Wong (2000, 2001), where Meixner, Meixner–Pollaczek, and Krawtchouk polynomials, respectively, are considered.

PART 5

Uniform Methods for Laplace-Type Integrals

Chapter 25

The vanishing saddle point

We consider Laplace-type integrals of the form[1]

$$F_\lambda(z) = \frac{1}{\Gamma(\lambda)} \int_0^\infty t^{\lambda-1} e^{-z\,t} f(t)\,dt, \tag{25.0.1}$$

with $\Re\lambda > 0$ and z large, in which λ may also be large. In Watson's lemma for this type of integral, see §2.1, z is the large parameter, and λ is assumed to be fixed. When λ is not fixed (say, λ depends on z) the expansion in (2.1.4) may become invalid.

When z and λ are large, the dominant part of the integral in (25.0.1) is

$$t^\lambda\, e^{-zt} = e^{-z\phi(t)}, \quad \phi(t) = t - \mu \ln t, \quad \mu = \frac{\lambda}{z}. \tag{25.0.2}$$

The function ϕ has a saddle point at $t = \mu$. When z is large and λ is fixed μ tends to zero, and the saddle point vanishes. See Figure 25.1.

When μ is bounded away from zero, we can transform the integral by using Laplace's method, substituting

$$\phi(t) - \phi(\mu) = \tfrac{1}{2}w^2, \quad \operatorname{sign}(t-\mu) = \operatorname{sign}(w). \tag{25.0.3}$$

This gives

$$F_\lambda(z) = \frac{e^{-z\phi(\mu)}}{\Gamma(\lambda)} \int_{-\infty}^\infty e^{-\frac{1}{2}zw^2} g(w)\,dw, \quad g(w) = \frac{1}{t}\frac{dt}{dw}. \tag{25.0.4}$$

We have

$$g(0) = \frac{1}{\mu}\left.\frac{dt}{dw}\right|_{w=0} = \frac{1}{\sqrt{\mu}}, \tag{25.0.5}$$

and we see that $g(0)$ becomes unbounded when $\mu \to 0$. Furthermore, all higher coefficients in the asymptotic expansion obtained from the representation in (25.0.4) by expanding $g(w)$ in powers of w, become unbounded when $\mu \to 0$.

The phase function ϕ has a quadratic shape when μ is bounded away from zero, but not if $\mu \to 0$. In fact, in the limit $\mu = 0$, ϕ becomes a linear function: the saddle point disappears when it approaches the endpoint of the integration interval.

[1] This chapter is based on Temme (1983, 1985).

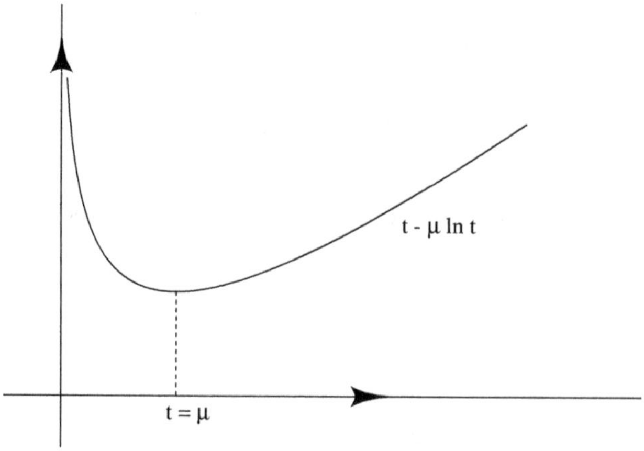

Fig. 25.1 The function $e^{-z\phi(t)}$ has a saddle point at $t = \mu$, which vanishes when $\mu = 0$.

In the following sections we describe two methods that yield expansions in which small and other values of μ are allowed. These approaches are modifications of Laplace's method: we expand at the saddle point, but we do not transform the phase function into a quadratic function.

25.1 Expanding at the saddle point

To cover small and other values of μ, we first expand f at $t = \mu$:

$$f(t) = \sum_{n=0}^{\infty} a_n(\mu)\ (t-\mu)^n. \tag{25.1.6}$$

Upon substituting this into (25.0.1) we obtain the formal expansion

$$F_\lambda(z) \sim z^{-\lambda} \sum_{n=0}^{\infty} a_n(\mu)\ P_n(\lambda)\ z^{-n}, \tag{25.1.7}$$

where

$$\begin{aligned} P_n(\lambda) &= \frac{z^{n+\lambda}}{\Gamma(\lambda)} \int_0^\infty t^{\lambda-1} e^{-z\,t} (t-\mu)^n\, dt \\ &= \frac{1}{\Gamma(\lambda)} \int_0^\infty x^{\lambda-1} e^{-x} (x-\lambda)^n\, dt. \end{aligned} \tag{25.1.8}$$

These functions are polynomials and an explicit representation easily follows:

$$\begin{aligned} P_n(\lambda) &= \sum_{k=0}^{n} \binom{n}{k} (\lambda)_k (-\lambda)^{n-k} \\ &= (-1)^n n! \sum_{k=0}^{n} \binom{-\lambda}{k} (\lambda)_k \frac{\lambda^{n-k}}{(n-k)!} \\ &= n!\, L_n^{(-n-\lambda)}(-\lambda), \end{aligned} \tag{25.1.9}$$

where we have used the explicit representation of the Laguerre polynomials given in (15.5.36). $P_n(\lambda)$ is a special case of the Tricomi–Carlitz polynomial, see (24.3.75).

The first few polynomials are

$$P_0(\lambda) = 1, \quad P_1(\lambda) = 0, \quad P_2(\lambda) = \lambda, \quad P_3(\lambda) = 2\lambda, \tag{25.1.10}$$

with recursion relation

$$P_{n+1}(\lambda) = n\left(P_n(\lambda) + \lambda P_{n-1}(\lambda)\right), \quad n = 1, 2, 3, \ldots. \tag{25.1.11}$$

By using induction it is easily seen that

$$P_n(\lambda) = \mathcal{O}\left(\lambda^{\lfloor n/2 \rfloor}\right), \quad \lambda \to \infty. \tag{25.1.12}$$

Remark 25.1. Under mild conditions on $a_n(\mu)$, that is, on f, this expansion is uniformly valid with respect to $\lambda \in [0, \infty)$, and in a larger domain of the complex plane. The main condition on f is that its singularities are not too close to the point $t = \mu$. Let R_μ denote the radius of convergence of the Taylor expansion of f at $t = \mu$. Then we require

$$R_\mu^{-1} = \mathcal{O}\left((1+\mu)^{-\kappa}\right), \quad \mu \geq 0, \quad \kappa \geq \tfrac{1}{2}, \quad \kappa \text{ fixed}. \tag{25.1.13}$$

For details we refer to Temme (1983, 1985). △

Example 25.1 (Incomplete gamma function). When we take $f(t) = 1/(1+t)$ the function $F_\lambda(z)$ can be written in terms of the incomplete gamma function (see (7.1.6)) or generalized exponential integral:

$$F_\lambda(z) = e^z \Gamma(1-\lambda, z) = z^{1-\lambda} e^z E_\lambda(z), \tag{25.1.14}$$

where

$$E_\lambda(z) = \int_1^\infty t^{-\lambda} e^{-zt}\, dt. \tag{25.1.15}$$

We obtain the expansion

$$\Gamma(1-\lambda, z) \sim z^{-\lambda} e^{-z} \sum_{n=0}^{\infty} a_n(\mu) P_n(\lambda) z^{-n}, \quad \mu = \frac{\lambda}{z}, \tag{25.1.16}$$

where

$$a_n(\mu) = \frac{(-1)^n}{(1+\mu)^{n+1}}. \tag{25.1.17}$$

We clearly see why the method works: the terms of the expansion in (25.1.16), that is,

$$a_n(\mu) P_n(\lambda) z^{-n}, \tag{25.1.18}$$

constitute an asymptotic scale as $z \to \infty$, uniformly with respect to λ or μ in the interval $[0, \infty)$. In fact, large values of λ improve the asymptotic convergence of the expansion. ♢

25.2 An integration by parts method

Consider (25.0.1) and write $f(t) = \bigl(f(t) - f(\mu)\bigr) + f(\mu)$. Then we have

$$\begin{aligned} F_\lambda(z) &= z^{-\lambda} f(\mu) - \frac{1}{z\Gamma(\lambda)} \int_0^\infty \frac{f(t)-f(\mu)}{t-\mu}\, de^{-z\phi(t)} \\ &= z^{-\lambda} f(\mu) + \frac{1}{z\Gamma(\lambda)} \int_0^\infty t^{\lambda-1} e^{-zt} f_1(t)\, dt, \end{aligned} \tag{25.2.19}$$

where

$$f_1(t) = t\frac{d}{dt}\frac{f(t)-f(\mu)}{t-\mu}. \tag{25.2.20}$$

Continuing this procedure we obtain for $K = 0, 1, 2, \ldots$

$$\begin{aligned} z^\lambda F_\lambda(z) &= \sum_{k=0}^{K-1} \frac{f_k(\mu)}{z^k} + \frac{1}{z^K} E_K(z,\mu), \\ f_k(t) &= t\frac{d}{dt}\frac{f_{k-1}(t) - f_{k-1}(\mu)}{t-\mu}, \quad k = 1, 2, \ldots, \quad f_0(t) = f(t), \\ E_K(t,\mu) &= \frac{1}{\Gamma(\lambda)} \int_0^\infty t^{\lambda-1} e^{-zt} f_K(t)\, dt. \end{aligned} \tag{25.2.21}$$

Eventually we obtain the complete asymptotic expansion

$$F_\lambda(z) \sim z^{-\lambda} \sum_{n=0}^\infty \frac{f_n(\mu)}{z^n}. \tag{25.2.22}$$

This expansion is valid under the same conditions as given in Remark 25.1, and it has a more canonical form: the large parameter z and the uniformity parameter μ are separated in the terms of the expansion. Another reason for giving this procedure is that similar methods are used in obtaining uniform expansions in more complicated cases, as we demonstrate in many examples in other chapters.

The coefficients $f_n(\mu)$ can be expressed in terms of the coefficients $a_n(\mu)$. To verify this we write

$$f_n(t) = \sum_{m=0}^\infty c_m^{(n)} (t-\mu)^m. \tag{25.2.23}$$

Then $a_m(\mu) = c_m^{(0)}$, $f_n(\mu) = c_0^{(n)}$ and we have from (25.2.21)

$$f_{n+1}(t) = \sum_{m=0}^\infty c_m^{(n+1)} (t-\mu)^m = t \sum_{m=1}^\infty c_m^{(n)} (-1)(t-\mu)^{m-2}. \tag{25.2.24}$$

This gives the recursion

$$c_m^{(n+1)} = m c_{m+1}^{(n)} + \mu(m+1) c_{m+2}^{(n)}, \quad m, n = 0, 1, 2, \ldots, \tag{25.2.25}$$

and the few first relations are

$$
\begin{aligned}
f_0(\mu) &= a_0(\mu),\\
f_1(\mu) &= \mu a_2(\mu),\\
f_2(\mu) &= \mu(2a_3(\mu) + 3\mu a_4(\mu)),\\
f_3(\mu) &= \mu(6a_4(\mu) + 20\mu a_5(\mu) + 15\mu^2 a_6(\mu)),\\
f_4(\mu) &= \mu(24a_5(\mu) + 130\mu a_6(\mu) + 210\mu^2 a_7(\mu) + 105\mu^3 a_8(\mu)).
\end{aligned}
\tag{25.2.26}
$$

Example 25.2. The first coefficients $f_n(\mu)$ for the expansion of the incomplete gamma function (Example 25.1), that is, with $f(t) = 1/(1+t)$, are given by

$$
\begin{aligned}
f_0(\mu) &= 1/(1+\mu),\\
f_1(\mu) &= \mu/(1+\mu)^3,\\
f_2(\mu) &= \mu(\mu-2)/(1+\mu)^5,\\
f_3(\mu) &= \mu(\mu^2 - 8\mu + 6)/(1+\mu)^7,\\
f_4(\mu) &= \mu(\mu^3 - 22\mu^2 + 58\mu - 24)/(1+\mu)^9.
\end{aligned}
\tag{25.2.27}
$$

In this way, the following expansion can be obtained

$$
\Gamma(1-\lambda, z) \sim z^{-\lambda} e^{-z} \sum_{n=0}^{\infty} \frac{f_n(\mu)}{z^n}. \tag{25.2.28}
$$

This expansion is the same as the one derived in §7.4, see in particular (7.4.41), where an expansion is given of $\Gamma(a,x)$ with $a < x$. In fact, it can be shown that the expansion in (25.2.28) holds for large positive z, uniformly with respect to $\mu \ge \mu_0 > -1$, where μ_0 is fixed. ◊

25.2.1 *Representing coefficients as a Cauchy-type integral*

The functions $f_k(t)$ introduced in (25.2.21) can be represented in the form of Cauchy-type integrals. We have the following theorem.

Theorem 25.1. *Let the rational functions $R_k(s,t,\mu)$ be defined by*

$$
R_0(s,t,\mu) = \frac{1}{s-t}, \quad R_{k+1}(s,t,\mu) = \frac{-1}{s-\mu}\frac{d}{ds}\left(sR_k(s,t,\mu)\right), \quad k \ge 0, \tag{25.2.29}
$$

where $s,t,\mu \in \mathbb{C}, s \ne t, s \ne \mu$. Let $f_k(t)$ be defined by the recursive scheme

$$
f_k(t) = t\frac{d}{dt}\frac{f_{k-1}(t) - f_{k-1}(\mu)}{t-\mu}, \tag{25.2.30}
$$

where $f_0(t)$ is a given analytic function in a domain $\mathcal{D}$. Then we have

$$
f_k(t) = \frac{1}{2\pi i}\int_{\mathcal{C}} R_k(s,t,\mu) f_0(s)\, ds, \tag{25.2.31}
$$

where $\mathcal{C}$ is a simple closed contour in $\mathcal{D}$ that encircles the points t and μ.

Proof.

$$
\begin{aligned}
f_k(t) &= \frac{1}{2\pi i}\int_{\mathcal{C}} R_0(s,t,\mu)\, f_k(s)\, ds \\
&= \frac{1}{2\pi i}\int_{\mathcal{C}} sR_0(s,t,\mu)\frac{d}{ds}\frac{f_{k-1}(s)-f_{k-1}(\mu)}{s-\mu}\, ds \\
&= \frac{1}{2\pi i}\int_{\mathcal{C}} sR_0(s,t,\mu)\left(\frac{d}{ds}\frac{f_{k-1}(s)}{s-\mu}+\frac{f_{k-1}(\mu)}{(s-\mu)^2}\right) ds \\
&= \frac{1}{2\pi i}\int_{\mathcal{C}} sR_0(s,t,\mu)\, d\frac{f_{k-1}(s)}{s-\mu} \\
&= \frac{-1}{2\pi i}\int_{\mathcal{C}} \frac{f_{k-1}(s)}{s-\mu}\frac{d}{ds}\left(sR_0(s,t,\mu)\right) ds \\
&= \frac{1}{2\pi i}\int_{\mathcal{C}} R_1(s,t,\mu) f_{k-1}(s)\, ds \\
&\ \ \vdots \\
&= \frac{1}{2\pi i}\int_{\mathcal{C}} R_k(s,t,\mu) f_0(s)\, ds.
\end{aligned}
\tag{25.2.32}
$$

The term containing $f_{k-1}(\mu)$ in the third line does not contribute in the integration because the rational function $sR_0(s,t,\mu)/(s-\mu)^2 = \mathcal{O}(s^{-2})$ as $s\to\infty$, and all poles of this function are inside $\mathcal{C}$. Hence, the integral of this function vanishes. □

Remark 25.2. Observe that the rational functions R_k introduced in (25.2.29) are independent of the function f_0 and that the representation of $f_k(t)$ given in (25.2.31) can be considered as the analogue of the Cauchy integral defining the remainder of a Maclaurin series. An estimate of $f_k(t)$ can be obtained as in Cauchy's inequality if bounds for R_k are available. In this way bounds for the remainder $E_n(t,\mu)$ of the expansion in (25.2.21) can be obtained. △

25.3 Expansions for loop integrals

We can use the same methods for loop integrals. In this way, we modify Watson's lemma for loop integrals, see §2.2. The starting point is the integral

$$
G_\lambda(z) = \frac{\Gamma(\lambda+1)}{2\pi i}\int_{-\infty}^{(0+)} s^{-\lambda-1} f(s) e^{zs}\, ds, \tag{25.3.33}
$$

where we assume that $z>0$. Along the lower side of the negative real axis we have $\operatorname{ph} s = -\pi$, along the upper side $\operatorname{ph} s = \pi$.

The function $s^{-\lambda}e^{zs}$ has a saddle point at $t=\mu$, where again $\mu=\lambda/z$. We expand f as in (25.1.6) and obtain

$$
G_\lambda(z) \sim z^\lambda \sum_{n=0}^{\infty} a_n(\mu) Q_n(\lambda) z^{-n}, \tag{25.3.34}
$$

where

$$Q_n(\lambda) = \frac{z^{n-\lambda}\Gamma(\lambda+1)}{2\pi i}\int_{-\infty}^{(0+)} s^{-\lambda-1}(s-\mu)^n e^{zs}\,ds. \tag{25.3.35}$$

The $Q_n(\lambda)$ are, again, polynomials in λ. It is easily verified that

$$Q_n(\lambda) = (-1)^n P_n(-\lambda), \quad n = 0, 1, 2, \ldots. \tag{25.3.36}$$

Integration by parts in (25.3.33) gives the expansion

$$G_\lambda(z) \sim z^\lambda \sum_{n=0}^{\infty} (-1)^n \frac{f_n(\mu)}{z^n}, \tag{25.3.37}$$

where the coefficients $f_n(\mu)$ are the same as in (25.2.22).

Example 25.3 (Incomplete gamma function). As in Example 25.1, the special case $f(s) = 1/(1+s)$, gives an incomplete gamma function. Consider

$$G_\lambda(z) = \frac{\Gamma(\lambda+1)}{2\pi i}\int_{-\infty}^{(0+)} s^{-\lambda-1} e^{zs}\,\frac{ds}{1+s}, \quad z \ge 0. \tag{25.3.38}$$

When $z = 0$ and $\lambda > 0$ this function vanishes, as follows from deforming the contour into a vertical line and shifting it to $+\infty$. Using the loop integral for the reciprocal gamma function in (6.0.3), we find

$$e^{-z}\frac{d}{dz}\left(e^z\, G_\lambda(z)\right) = z^\lambda. \tag{25.3.39}$$

Hence,

$$G_\lambda(z) = e^{-z}\int_0^z e^t t^\lambda\,dt = z^{\lambda+1}e^{-z}\Gamma(\lambda)\gamma^*(\lambda, -z), \tag{25.3.40}$$

where $\gamma^*(a, z)$ can be expressed in terms of the standard incomplete gamma function $\gamma(a, z)$, see (7.1.4).

The function $\gamma^*(a, z)$ is an analytic function of z and a, and can be defined for negative values of z, and we obtain the expansion

$$\gamma^*(\lambda, -z) \sim \frac{e^z}{z\Gamma(\lambda)}\sum_{n=0}^{\infty} a_n(\mu) Q_n(\lambda) z^{-n}, \tag{25.3.41}$$

where $a_n(\mu)$ are given in (25.1.17). Similarly,

$$\gamma^*(\lambda, -z) \sim \frac{e^z}{z\Gamma(\lambda)}\sum_{n=0}^{\infty} (-1)^n \frac{f_n(\mu)}{z^n}, \quad \mu = \lambda/z, \tag{25.3.42}$$

uniformly for $\mu \ge \mu_0 > -1$. The coefficients $f_n(\mu)$ follow from the method described in §25.2, and the first few are given in (25.2.27).

This expansion can be compared with the one for $\gamma(a, x)$ in (7.3.18) for large a, $a > x$, where the coefficient $a_k(\lambda)$ are not the same as the coefficients in (25.3.41). In fact, the expansions are the same, after rewriting the terms.

The contour integral in (25.3.38) is related to other contour integrals for the incomplete gamma functions $\gamma(a, z)$ and $\Gamma(a, z)$ as given in §37.2. These are used in that section for obtaining uniform expansions that are valid near the transition point of $\gamma(a, z)$ and $\Gamma(a, z)$ at $z = a$. ♢

25.4 Kummer functions

For a more general case we can consider $f(t) = (1+t)^\sigma$, $\sigma \in \mathbb{C}$, which gives expansions of the Kummer (or confluent hypergeometric functions); see Chapter 10. These functions can also be expressed in terms of Whittaker functions.

We take the integral representation of the U-function in (10.1.5) and write

$$F_\lambda(z) = \frac{1}{\Gamma(\lambda)} \int_0^\infty t^{\lambda-1} e^{-zt} (1+t)^\sigma \, dt. \tag{25.4.43}$$

This gives the relations

$$F_\lambda(z) = U(\lambda, \sigma+\lambda+1, z) = z^{-\sigma-\lambda} U(-\sigma, 1-\lambda-\sigma, z), \tag{25.4.44}$$

where we used a well-known functional relation of the U-function given in (10.1.10).

It follows that we can obtain the asymptotic expansion of the function $U(a,c,z)$ for $z \to \infty$, uniformly valid with respect to a and c such that $a-c$ belongs to a bounded set. The second U-function in (25.4.44) reveals that it is possible to obtain an asymptotic expansion of $U(a,c,z)$ for $z \to \infty$, uniformly valid with respect to $c \in (-\infty, 1+a]$, where a belongs to a bounded set.

The coefficients $a_n(\mu)$ for the expansion (25.1.7) of the function $F_\lambda(z)$ in (25.4.44) are given by

$$a_n(\mu) = (1+\mu)^{\sigma-n} \binom{\sigma}{n}, \quad n = 0, 1, 2, \ldots. \tag{25.4.45}$$

With $f(s) = (1+s)^\sigma$, the function $G_\lambda(z)$ of (25.3.33) becomes the other Kummer function. We have (see (10.1.7) and (10.1.9))

$$\begin{aligned} G_\lambda(z) &= \frac{z^{\lambda-\sigma}\Gamma(\lambda+1)}{\Gamma(\lambda+1-\sigma}\, {}_1F_1\left(\begin{matrix}\lambda+1\\ \lambda+1-\sigma\end{matrix}; z\right) \\ &= \frac{z^{\lambda-\sigma}\Gamma(\lambda+1)e^z}{\Gamma(\lambda+1-\sigma}\, {}_1F_1\left(\begin{matrix}-\sigma\\ \lambda+1-\sigma\end{matrix}; -z\right). \end{aligned} \tag{25.4.46}$$

The expansions of $G_\lambda(z)$ are given in (25.3.34) and (25.3.37) with coefficients $a_n(\mu)$ as in (25.4.45); the first coefficients $f_n(\mu)$ follow from (25.2.26).

The expansions of the Kummer functions obtained in this way are more powerful than those obtained in Chapter 10, where in §10.4 we have given expansions for $c \to \infty$ with z bounded.

25.5 Generalized zeta function

The generalized zeta function is defined by

$$\zeta(\lambda, z) = \sum_{n=0}^{\infty} (n+z)^{-\lambda}, \quad z \neq 0, -1, -2, \ldots, \quad \Re\lambda > 1. \tag{25.5.47}$$

This function is also known as the Hurwitz zeta function and reduces to the more familiar Riemann zeta function $\zeta(\lambda)$ for $z = 1$. It can be expressed as an integral by using

$$F_\lambda(z) = \frac{1}{\Gamma(\lambda)} \int_0^\infty t^{\lambda-1} e^{-zt} f(t)\, dt, \quad f(t) = \frac{1}{e^t - 1} - \frac{1}{t}, \tag{25.5.48}$$

and the relation is

$$\zeta(\lambda, z) = F_\lambda(z) + z^{-\lambda} - \frac{z^{1-\lambda}}{1-\lambda}. \tag{25.5.49}$$

Because $F_\lambda(z)$ is an analytic function of λ (see Remark 2.2), this gives the analytic continuation of (25.5.47) with respect to λ.

For a representation as a loop integral, cf. (25.3.33), we have

$$G_\lambda(z) = \frac{\Gamma(\lambda+1)}{2\pi i} \int_{-\infty}^{(0+)} s^{-\lambda-1} f(s) e^{zs}\, ds, \tag{25.5.50}$$

with the same function f as in (25.5.48), in which case the relation to the generalized zeta function is

$$\zeta(-\lambda, z) = -G_\lambda(z) - \frac{z^{\lambda+1}}{\lambda+1}. \tag{25.5.51}$$

Hence, for $z \to \infty$ we can obtain for $\zeta(\lambda, z)$ an asymptotic expansion, which is uniformly valid with respect to $\lambda \in [0, \infty)$ (by using (25.1.7) or (25.2.22)), and an expansion which is uniformly valid with respect to $\lambda \in (-\infty, 0]$ (by using (25.3.34) and (25.3.37)).

An interesting application is found in considering the expansion of the sum

$$S_n(s) = \sum_{j=1}^{n} j^s, \tag{25.5.52}$$

for $n \to \infty$, uniformly with respect to s. This sum can be expressed in terms of the generalized zeta function. The relation is

$$S_n(s) = n^s + \zeta(-s) - \zeta(-s, n). \tag{25.5.53}$$

25.6 Transforming to the standard form

We give examples that show the same asymptotic features as our standard form (25.0.1), and for which a transformation is needed to obtain this standard form. For example we have the integrals and their role for representing certain special functions:

$$\begin{aligned}
&\int_0^\infty e^{-zw} \left(w(w+1)\right)^{\lambda-1} dw, && \text{Modified Bessel function,} \\
&\int_0^\infty e^{-zw} \left(w(w+1)^\nu\right)^{\lambda-1} dw, && \text{Kummer function,} \\
&\int_0^\infty e^{-z(w+aw^2)} w^{\lambda-1}\, dw, && \text{Parabolic cylinder function,} \\
&\int_0^\infty e^{-zw} \left(1-e^{-w}\right)^{\lambda-1} dw, && \text{Ratio of gamma functions.}
\end{aligned} \tag{25.6.54}$$

The nature of the asymptotic expansion of these integrals for $z \to \infty$ with $\mu = \lambda/z$ as a uniformity parameter in $[0, \infty)$ is the same as that of our standard form (25.0.1). We give details of the transformation of the fourth example to the standard form, and a few details for the parabolic cylinder function. All four examples have alternative loop integrals, which represent functions from the same group of special functions as those following from the real integrals. For details we refer to Temme (1985).

25.6.1 *The ratio of two gamma functions*

We have given expansions of $\Gamma(a+z)/\Gamma(b+z)$ in §6.5, where we have introduced the normalized gamma function $\Gamma^*(z)$ by writing

$$\Gamma^*(z) = \sqrt{z/2\pi}e^z z^{-z}\Gamma(z). \tag{25.6.55}$$

Hence, this function is $\mathcal{O}(1)$ as for large values of z, as can be seen from the expansion in (6.1.15). By dividing that expansion by a similar one with z replaced by $z + \lambda = z(1+\mu)$, we obtain

$$\frac{\Gamma^*(z)}{\Gamma^*(z+\lambda)} \sim \sum_{k=0}^{\infty} d_k(\mu) z^{-k}, \quad z \to \infty, \tag{25.6.56}$$

where

$$\begin{aligned}
d_0(\mu) &= 1, \\
d_1(\mu) &= \frac{\mu}{12(\mu+1)}, \\
d_2(\mu) &= \frac{\mu^2}{288(\mu+1)^2}, \\
d_3(\mu) &= -\frac{\mu(139\mu^2 + 432\mu + 432)}{51840(\mu+1)^3}, \\
d_4(\mu) &= -\frac{\mu^2(571\mu^2 + 1728\mu + 1728)}{2488320(\mu+1)^4}.
\end{aligned} \tag{25.6.57}$$

We see that all these coefficients are of order $\mathcal{O}(1)$ as $\mu \to \infty$, and it is expected that the expansion in (25.6.56) is uniformly valid with respect to $\mu \in [0, \infty)$ or $\lambda \in [0, \infty)$. The expansion of the ratio of gamma functions given in (6.5.72) is not uniformly valid with respect to large a and/or b. We will derive the same expansion as in (25.6.56) by using the beta integral.

A simple transformation in the beta integral (see (6.5.67)) gives

$$F_\lambda(z) = \frac{\Gamma(z)}{\Gamma(z+\lambda)} = \frac{1}{\Gamma(\lambda)} \int_0^\infty e^{-zw} \left(1 - e^{-w}\right)^{\lambda-1} dw. \tag{25.6.58}$$

Comparing this with (25.0.1), we observe that it has the standard form when we take $f(w) = ((1-e^{-w})/w)^{\lambda-1}$. However, this choice of f will not give a uniform expansion for the λ-interval $[0, \infty)$. The main reason is that f depends on λ and that large λ-values will have much influence on the coefficients of the expansion obtained in this way. Instead we include possible influences of λ in a transformation to the standard form.

The saddle point of $e^{-zw}(1-e^{-w})^\lambda$ is

$$w_0 = \ln(1+\mu), \quad \mu = \lambda/z. \tag{25.6.59}$$

We transform the integral in (25.6.58) into the standard form (25.0.1) by using the mapping $w \mapsto t(w)$ given by

$$w - \mu \ln\left(1-e^{-w}\right) = t - \mu \ln t + A(\mu), \tag{25.6.60}$$

where $A(\mu)$ is a function to be determined. The right-hand side has a saddle point at $t=\mu$, and we prescribe that the value $w = w_0$ should correspond to $t=\mu$. This gives

$$A(\mu) = (\mu+1)\ln(\mu+1) - \mu. \tag{25.6.61}$$

In addition, we prescribe $\operatorname{sign}(w-w_0) = \operatorname{sign}(t-\mu)$.

The transformation gives

$$F_\lambda(z) = \frac{e^{-zA(\mu)}}{\Gamma(\lambda} \int_0^\infty t^{\lambda-1} e^{-zt} f(t)\,dt, \tag{25.6.62}$$

where

$$f(t) = \frac{t}{1-e^{-w}}\frac{dw}{dt} = \frac{t-\mu}{1-(1+\mu)e^{-w}}. \tag{25.6.63}$$

We expand $f(t) = \sum_{n=0}^\infty a_n(\mu)(t-\mu)^n$ and the first coefficients are

$$\begin{aligned}
a_0(\mu) &= \sqrt{\mu+1},\\
a_1(\mu) &= \frac{\mu-1+a_0(\mu)}{3\mu},\\
a_2(\mu) &= \frac{1}{12a_0(\mu)},\\
a_3(\mu) &= \frac{8\mu^3+12\mu^2-12\mu-8+(8+8\mu-15\mu^2)a_0(\mu)}{540(\mu+1)\mu^3},\\
a_4(\mu) &= \frac{\mu a_2(\mu) - 48(\mu+1)a_3(\mu)}{72(\mu+1)\mu}.
\end{aligned} \tag{25.6.64}$$

By computing a few more coefficients, and using (25.2.26), we can compute the first coefficients $f_0(\mu), \cdots, f_5(\mu)$. It turns out that $f_n(\mu) = \sqrt{\mu+1}\,d_n(\mu)$ (see (25.6.56)). In this way we obtain, using (25.2.22), (25.6.58), (25.6.61), and (25.6.62),

$$\frac{\Gamma(z)}{\Gamma(z+\lambda)} \sim z^{-\lambda} e^{-zA(\mu)} \sqrt{\mu+1} \sum_{n=0}^\infty d_n(\mu) z^{-n}, \tag{25.6.65}$$

or, using the normalized gamma function $\Gamma^*(z)$, the expansion in (25.6.56).

25.6.2 *Parabolic cylinder functions*

We have the following integral representation (see (11.2.8))

$$U(a,z) = \frac{e^{-\frac{1}{4}z^2}}{\Gamma\left(a+\frac{1}{2}\right)} \int_0^\infty w^{a-\frac{1}{2}} e^{-\frac{1}{2}w^2 - zw}\, dw, \quad \Re a > -\tfrac{1}{2}. \tag{25.6.66}$$

In Chapter 11 we have given the expansions for large complex z with a fixed, and in Chapter 30 we will give expansions in which a may be large as well. In the present section we consider representations that give some of the expansions in Chapter 30.

The representation in (25.6.66) can be written in the form

$$U\left(\lambda - \tfrac{1}{2}, z\right) = \frac{z^\lambda e^{-\frac{1}{4}z^2}}{\Gamma(\lambda)} \int_0^\infty w^{\lambda-1} e^{-z^2\left(\frac{1}{2}w^2+w\right)}\, dw, \tag{25.6.67}$$

and the saddle points of $t^\lambda e^{-z^2\left(\frac{1}{2}t^2+t\right)}$ are located at

$$w_\pm = \tfrac{1}{2}\left(-1 \pm \sqrt{1+4\mu}\right), \quad \mu = \lambda/z^2. \tag{25.6.68}$$

Observe that the saddle point w_- is bounded away from $[0,\infty)$ for all $\mu \ge 0$.

The transformation to the standard form reads

$$\tfrac{1}{2}w^2 + w - \mu \ln w = t - \mu \ln t + A(\mu), \tag{25.6.69}$$

where $A(\mu)$ follows from substituting $w = w_+$ and $t = \mu$. Moreover, we prescribe $\operatorname{sign}(w - w_+) = \operatorname{sign}(t - \mu)$. This gives

$$U\left(\lambda - \tfrac{1}{2}, z\right) = \frac{z^\lambda e^{-\frac{1}{4}z^2 - z^2 A(\mu)}}{\Gamma(\lambda)} \int_0^\infty t^{\lambda-1} e^{-z^2 t} f(t)\, dt, \tag{25.6.70}$$

where

$$f(t) = \frac{t}{w}\frac{dw}{dt}. \tag{25.6.71}$$

From (25.6.70) the asymptotic expansion for $z \to \infty$ can be derived by expanding f at $t = \mu$. It is uniformly valid with respect $\lambda \in [0,\infty)$.

For the corresponding loop integral we can take (see Temme (2010a, Eq. 12.5.6))

$$U(a,z) = \frac{e^{\frac{1}{4}z^2}}{i\sqrt{2\pi}} \int_{c-i\infty}^{c+i\infty} e^{-zw+\frac{1}{2}w^2} w^{-a-\frac{1}{2}}\, dw, \quad -\tfrac{1}{2}\pi < \operatorname{ph} w < \tfrac{1}{2}\pi, \tag{25.6.72}$$

where c is a positive number. With this representation we can obtain an expansion of $U(a,-z)$ by using the saddle point w_+ given in (25.6.68) and the same transformation as in (25.6.69).

Chapter 26

A moving endpoint: Incomplete Laplace integrals

In this chapter we consider Case 8 of Table 20.1.[1] The integrand has the form $t^{\lambda-1}e^{-zt}f(t)$, which is of a simple Laplace-type. However, we consider an incomplete Laplace integral with interval $[\alpha,\infty)$, where $\alpha \ge 0$, and we consider both α and λ as uniformity parameters that can range through all nonnegative values, whether or not they are large in comparison with z.

We consider a corresponding loop integral and discuss an application to the incomplete beta function for both types of integral.

26.1 The standard form

The standard form in this case is

$$F_\lambda(z,\alpha) = \frac{1}{\Gamma(\lambda)}\int_\alpha^\infty t^{\lambda-1}e^{-zt}f(t)\,dt, \tag{26.1.1}$$

with $\lambda \ge 0$, $\alpha \ge 0$ and z large. As in Chapter 25, λ may also be large. So, introducing the extra parameter α gives a generalization of the case considered in that chapter. As we will see in §26.3, this generalization is of interest for describing the asymptotic behavior of the incomplete beta function $I_x(p,q)$ for a wide range of all parameters.

An interpretation of $F_\lambda(z,\alpha)$ for $\lambda = 0$ follows from

$$F_0(z,\alpha) = \begin{cases} 0, & \text{if } \alpha > 0,\\ f(0), & \text{if } \alpha = 0. \end{cases} \tag{26.1.2}$$

The second case follows from integration by parts.

We are interested in the asymptotic expansion of $F_\lambda(z,\alpha)$ as $z \to \infty$, which is uniformly valid with respect to both λ and α in $[0,\infty)$. The parameters λ and α may be coupled with the large parameter z, or they may vary independently through the uniformity interval.

For a description of the various asymptotic features, four different cases can be distinguished.

---

[1]This chapter is based on Temme (1987).

(1) λ and α both fixed. For this classical case Watson's lemma gives an expansion. When $\alpha = 0$, then $f(t)$ is expanded in powers of t, when $\alpha > 0$, then $t^{\lambda-1}f(t)$ is expanded in powers of $(t-\alpha)$.
(2) λ fixed and $\alpha \ge 0$. An incomplete gamma function (that is, (26.1.1) with $f = 1$) is needed to describe the uniform transition of $\alpha = 0$ to $\alpha > 0$; Erdélyi (1974), Temme (1976), Zil'bergleĭt (1976) and Soni (1983) give more details on this case. The asymptotic feature is the possible coalescence of two critical points: $t = 0$ (an algebraic singularity) and $t = \alpha$ (endpoint of integration).
(3) $\lambda \ge 0$ and $\alpha = 0$. This case is studied in Chapter 25.
(4) $\lambda \ge 0$ and $\alpha > 0$ (fixed). When $\mu = \lambda/z$ is larger than α, the saddle point at $t = \mu$ is inside the interval of integration, and otherwise it is outside. This transition follows as a special case from Chapter 22, in which the main approximant is a parabolic cylinder function. Here we need the complementary error function, which is a special case of this function.

These four cases are combined in the present chapter, where $\lambda \ge 0$ and $\alpha \ge 0$. As in the second case, the basic approximant is an incomplete gamma function. However, in that case (with fixed λ), the full ranges of both arguments of the incomplete gamma function are not completely exploited.

26.2 Constructing the expansion

We consider (26.1.1) and take $\mu = \lambda/z$. We can repeat the integration by parts method used in Chapter 25, see §25.2, and we have contributions from the endpoint α as well. We recall the notation of the incomplete gamma function ratios considered in Chapter 7. We have

$$P(a,x) = \frac{1}{\Gamma(a)}\int_0^x t^{a-1}e^{-t}\,dt, \quad Q(a,x) = \frac{1}{\Gamma(a)}\int_x^\infty t^{a-1}e^{-t}\,dt. \tag{26.2.3}$$

It is not difficult to see that the integration by parts method of §25.2 now gives the expansion

$$F_\lambda(z,\alpha) = z^{-\lambda}Q(\lambda,\alpha z)\sum_{k=0}^{\infty}\frac{f_k(\mu)}{z^k} + \frac{\alpha^\lambda e^{-\alpha z}}{z\Gamma(\lambda)}\sum_{k=0}^{\infty}\frac{B_k(\alpha,\mu)}{z^k}, \tag{26.2.4}$$

where the functions f_k follow from the same scheme as in §25.2, that is,

$$f_k(t) = t\frac{d}{dt}\frac{f_{k-1}(t) - f_{k-1}(\mu)}{t-\mu}, \quad k = 1,2,\ldots, \tag{26.2.5}$$

with $f_0(t) = f(t)$. The coefficients $B_k(\alpha,\mu)$ are defined by

$$B_k(\alpha,\mu) = \frac{f_k(\alpha) - f_k(\mu)}{\alpha-\mu}, \quad k = 0,1,2,\ldots. \tag{26.2.6}$$

As in §25.2 we can write the expansion with a finite number of terms and a remainder.

Observe that the functions f_k, and, hence, the coefficients of the first series in (26.2.4) do not depend on α; in fact we recognize the expansion given in (25.2.22). Furthermore, the integrated terms at $t = \alpha$, which generate the second series in (26.2.4), all vanish when $\alpha \to 0$.

These observations lead us to write

$$F_\lambda(z,\alpha) = Q(\lambda,\alpha z)F_\lambda(z) + \frac{\alpha^\lambda e^{-\alpha z}}{z\Gamma(\lambda)} B_\lambda(z,\alpha), \tag{26.2.7}$$

where the functions $F_\lambda(z)$ and $B_\lambda(z,\alpha)$ have the asymptotic expansions following from (26.2.4). Letting $\alpha \to 0$ in (26.2.7) and using $Q(\lambda, 0) = 1$, we see indeed that the right-hand side of (26.2.7) reduces to $F_\lambda(z,0) = F_\lambda(z)$, and in fact, (26.2.7) gives an exact identity, defining a new function $B_\lambda(z,\alpha)$ for which the asymptotic expansion follows from the second series in (26.2.4).

Alternatively, we can obtain the expansion of $B_\lambda(z,\alpha)$ by using a differential equation for this function. By differentiating (26.2.7) with respect to α, we find

$$\frac{\alpha}{z} B'_\lambda(z,\alpha) + (\mu - \alpha) B_\lambda(z,\alpha) = z^\lambda F_\lambda(z) - f(\alpha). \tag{26.2.8}$$

Substitution of

$$F_\lambda(z) \sim z^{-\lambda} \sum_{k=0}^{\infty} \frac{f_k(\mu)}{z^k}, \quad B_\lambda(z,\alpha) \sim \sum_{k=0}^{\infty} \frac{B_k(\alpha,\mu)}{z^k} \tag{26.2.9}$$

into (26.2.8) shows that this equation is formally satisfied if

$$\begin{aligned} (\mu - \alpha) B_k(\alpha,\mu) &= f_k(\mu) - \alpha B'_{k-1}(\alpha), \quad k = 1,2,3,\ldots, \\ B_0(\alpha,\mu) &= \frac{f(\alpha) - f(\mu)}{\alpha - \mu}. \end{aligned} \tag{26.2.10}$$

Again, the prime denotes differentiation with respect to α. It easily follows that (26.2.10) generates the same coefficients $B_k(\alpha,\mu)$ as those defined in (26.2.6). Therefore, by using (26.2.7), we again arrive at (26.2.4).

26.2.1 *The complementary function*

The following complementary function

$$E_\lambda(z,\alpha) = \frac{1}{\Gamma(\lambda)} \int_0^\alpha t^{\lambda-1} e^{-zt} f(t)\, dt \tag{26.2.11}$$

is strongly related to (26.1.1). We have

$$E_\lambda(z,\alpha) + F_\lambda(z,\alpha) = F_\lambda(z), \tag{26.2.12}$$

which follows from the remarks below (26.2.7).

The function $E_\lambda(z,\alpha)$ has a similar representation as in (26.2.7). When we use the complementary relation in (26.2.12), and the one for the incomplete gamma function ratios, we obtain

$$E_\lambda(z,\alpha) = P(\lambda,\alpha z)F_\lambda(z) - \frac{\alpha^\lambda e^{-\alpha z}}{z\Gamma(\lambda)} B_\lambda(z,\alpha). \tag{26.2.13}$$

It follows that, when we have the expansions of the functions $F_\lambda(z)$ and $B_\lambda(z,\alpha)$, the results can be used for both integrals (26.1.1) and (26.2.11).

The relation in (26.2.13) can be used for $E_\lambda(z,\alpha)$ when $\frac{1}{2} < F_\lambda(z,\alpha)/F_\lambda(z) < 1$, according to the principle "compute the smallest one first".

26.3 Application to the incomplete beta function

The incomplete beta function is defined by

$$I_x(p,q) = \frac{1}{B(p,q)} \int_0^x \tau^{p-1}(1-\tau)^{q-1}\, d\tau, \quad p>0, \quad q>0, \tag{26.3.14}$$

where $B(p,q)$ is Euler's (complete) beta integral

$$B(p,q) = \int_0^1 \tau^{p-1}(1-\tau)^{q-1}\, d\tau = \frac{\Gamma(p)\Gamma(q)}{\Gamma(p+q)}. \tag{26.3.15}$$

The asymptotic problem for $I_x(p,q)$ is to give an expansion with p as the large parameter and $x \in [0,1]$ and $\mu = p/q$ as uniformity parameters. As explained in §26.2.1, it is also convenient to have an expansion of the complementary function. We can use

$$J_x(p,q) = \frac{1}{B(p,q)} \int_x^1 \tau^{p-1}(1-\tau)^{q-1}\, d\tau = 1 - I_x(p,q) = I_{1-x}(q,p), \tag{26.3.16}$$

but the result for $J_x(p,q)$ also follows from (26.2.13).

The saddle point of the function $\tau^p(1-\tau)^q$ is

$$\tau_0 = \frac{p}{p+q}. \tag{26.3.17}$$

If $p+q$ is large, the value of $I_x(p,q)$ is very small when $x < \tau_0$, and it is close to unity when $x > \tau_0$. When τ_0 is restricted to a compact interval of $(0,1)$, this transition can properly be described by an error function; when $\tau_0 \to 1$ the basic approximant is an incomplete gamma function. We will use this function to cover the complete uniformity domain of q, that is, $[0,\infty)$.

It is essential to transform (26.3.14) into the standard form (26.1.1). First we write (26.3.14) in the form

$$I_x(p,q) = \frac{1}{B(p,q)} \int_{-\ln x}^{\infty} \left(1-e^{-w}\right)^{q-1} e^{-pw}\, dw. \tag{26.3.18}$$

The transformation to the standard form is similar to the one used in (25.6.60) for the complete beta integral. We repeat a few steps. We have

$$w - \mu \ln\left(1-e^{-w}\right) = t - \mu \ln t + A(\mu), \quad \mu = \frac{q}{p}, \tag{26.3.19}$$

where $A(\mu)$ does not depend on t. The left-hand side has a vanishing derivative at

$$w_0 = \ln(\mu+1) = -\ln \tau_0, \tag{26.3.20}$$

see (26.3.17), and the right-hand side at $t = \mu$. To make the mapping properly defined we require the correspondences

$$w = 0 \Longleftrightarrow t = 0, \quad w = w_0 \Longleftrightarrow t = \mu, \quad w = +\infty \Longleftrightarrow t = +\infty. \tag{26.3.21}$$

The middle one gives

$$A(\mu) = (1+\mu)\ln(1+\mu) - \mu. \tag{26.3.22}$$

The point $w = -\ln x$, the lower endpoint of integration in (26.3.18), is mapped to a point in the t-domain that we call α. This number follows from the relation in (26.3.19), and we find the implicit relation

$$-\ln x - \mu\ln(1-x) = \alpha - \mu\ln\alpha + A(\mu), \tag{26.3.23}$$

with corresponding points

$$x = 0 \Longleftrightarrow \alpha = +\infty, \quad x = \tau_0 \Longleftrightarrow \alpha = \mu, \quad x = 1 \Longleftrightarrow \alpha = 0. \tag{26.3.24}$$

Hence, $\operatorname{sign}(x - \tau_0) = -\operatorname{sign}(\alpha - \mu)$. The transformations in (26.3.19) and (26.3.23) are the same, up to parametrization.

The transformed version of (26.3.18) is

$$I_x(p,q) = \frac{e^{-pA(\mu)}}{B(p,q)}\int_\alpha^\infty t^{q-1}e^{-pt}f(t,\mu)\,dt, \tag{26.3.25}$$

where

$$f(t,\mu) = \frac{t}{1-e^{-w}}\frac{dw}{dt} = \frac{t-\mu}{1-(1+\mu)e^{-w}}. \tag{26.3.26}$$

The function $f(t,\mu)$ is a positive function of t on $[0,\infty)$; $f(0,\mu) = 1$, $f(\mu,\mu) = \sqrt{1+\mu}$, and $f(t,\mu) \sim t$ as $t \to +\infty$.

We write (26.3.25) in the standard form (26.1.1):

$$e^{pA(\mu)}\frac{\Gamma(p)}{\Gamma(p+q)}I_x(p,q) = F_q(p,\alpha) = \frac{1}{\Gamma(q)}\int_\alpha^\infty t^{q-1}e^{-pt}f(t,\mu)\,dt, \tag{26.3.27}$$

and obtain (see (26.2.7))

$$e^{pA(\mu)}\frac{\Gamma(p)}{\Gamma(p+q)}I_x(p,q) = Q(q,\alpha p)F_q(p) + \frac{\alpha^q e^{-\alpha p}}{p\Gamma(q)}B_q(p,\alpha), \tag{26.3.28}$$

where $F_q(p) = F_q(p,0)$ and with expansions of $F_q(p)$ and $B_q(p,\alpha)$ as in (26.2.9).

However, in the present case, the complete integral $F_q(p)$ is known explicitly, because $I_x(p,q) = 1$ for $x = 1$ ($\alpha = 0$). That is, we have

$$F_q(p) = e^{pA(\mu)}\frac{\Gamma(p)}{\Gamma(p+q)}. \tag{26.3.29}$$

This gives

$$I_x(p,q) = Q(q,\alpha p) + \frac{\alpha^q e^{-\alpha p}e^{-pA(\mu)}}{pB(p,q)}B_q(p,\alpha), \tag{26.3.30}$$

or, by using (26.3.22) and (26.3.23),

$$I_x(p,q) = Q(q,\alpha p) + \frac{\alpha^q e^{-\alpha p+q}}{pB(p,q)}\left(\frac{p}{p+q}\right)^{p+q}B_q(p,\alpha). \tag{26.3.31}$$

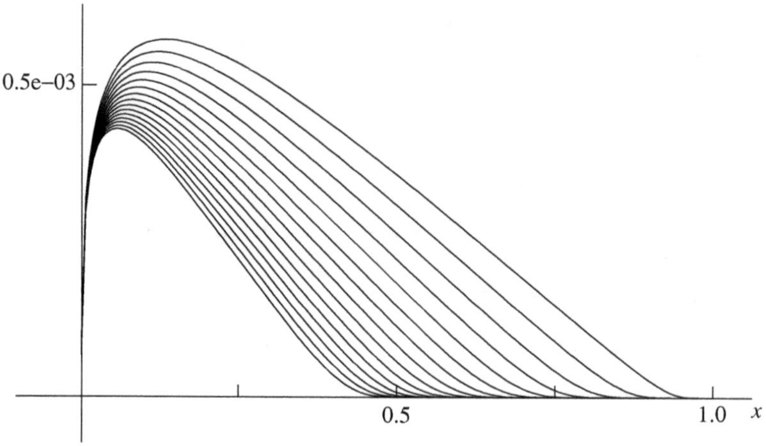

Fig. 26.1 The relative error in the approximation in (26.3.32) with only the term $B_0(\alpha,\mu)$ of the series. We take $p = 100$ and $q = 10, 20, \ldots, 150$. The curves become lower as q increases.

As an alternative, by using (26.3.23), we can write (26.3.30) in the form

$$
\begin{aligned}
I_x(p,q) &= Q(q,\alpha p) + \frac{x^p(1-x)^q}{pB(p,q)} B_q(p,\alpha), \\
B_q(p,\alpha) &\sim \sum_{k=0}^{\infty} \frac{B_k(\alpha,\mu)}{p^k}, \quad B_k(\alpha,\mu) = \frac{f_k(\alpha,\mu) - f_k(\mu,\mu)}{\alpha-\mu},
\end{aligned}
\tag{26.3.32}
$$

as $p \to \infty$, uniformly with respect to $q \in [0,\infty)$ and $x \in [0,1]$.

Observe that only one asymptotic series is left in the representation of $I_x(p,q)$, the expansion of $B_q(p,\alpha)$. The first coefficient is

$$
B_0(\alpha,\mu) = \frac{f(\alpha,\mu) - f(\mu,\mu)}{\alpha-\mu}, \tag{26.3.33}
$$

with

$$
f(\alpha,\mu) = \frac{\alpha-\mu}{1-(1+\mu)x}, \quad f(\mu,\mu) = \sqrt{1+\mu}. \tag{26.3.34}
$$

Special values are

$$
B_0(0,\mu) = \frac{\sqrt{1+\mu}-1}{\mu}, \quad B_0(\mu,\mu) = \frac{\mu - 1 + \sqrt{1+\mu}}{3\mu}, \tag{26.3.35}
$$

and $B_0(\alpha,\mu) \to 1$, as $\alpha \to \infty$. They satisfy

$$
0 \le B_0(0,\mu) \le B_0(\mu,\mu) \le B_0(\infty,\mu), \quad \mu \ge 0. \tag{26.3.36}
$$

The functions $f_k(t,\mu)$ used in the other coefficients $B_k(\alpha,\mu)$ can be generated by the scheme given in (26.2.5) with starting function $f_0(t,\mu) = f(t,\mu)$ given in (26.3.26).

In §42.3 we will use the asymptotic representation of $I_x(p,q)$, together with simpler representations in terms of the error function, to obtain an asymptotic expansion of the value x that solves the equation $I_x(p,q) = \rho$, $0 < \rho < 1$, for large values of $p + q$.

For further details on the transformation (26.3.19), the analytic properties of $f(t,\mu)$, and the nature of the expansion, we refer to Temme (1987).

In Figure 26.1 we show the relative errors on the x-interval $(0,1)$ for the approximation in (26.3.32) with only the term $B_0(\alpha,\mu)$ of the series. We take $p = 100$ and $q = 10, 20, \ldots, 150$. The curves become lower as q increases. On the right of the transition point $\tau_0 = p/(p+q)$ (see (26.3.17)), the values of $I_x(p,q)$ become almost equal to 1, and for $x > \tau_0$ the relative errors are much smaller than on the left of τ_0. In fact, for $x > \tau_0$ it is better to check relative errors by using the complementary function $J_x(p,q)$ defined in (26.3.16); see also Remark 26.2.

Remark 26.1. Because (26.3.29) can be written in the form

$$F_q(p) = \sqrt{\mu+1}\, p^{-q} \frac{\Gamma^*(p)}{\Gamma^*(p+q)}, \tag{26.3.37}$$

where $\Gamma^*(z)$ is defined in (25.6.55), it follows that the first six coefficients $f_k(\mu,\mu)$ needed in (26.3.32) follow from the $d_k(\mu)$ given in (25.6.57), the relation being $f_k(\mu,\mu) = \sqrt{\mu+1}\, d_k(\mu)$. See also (25.6.56) and (25.6.65). △

Remark 26.2. For the complementary function $J_x(p,q)$ defined in (26.3.16) it will be convenient to use

$$J_x(p,q) = P(q,\alpha p) - \frac{x^p(1-x)^q}{pB(p,q)} B_q(p,\alpha), \tag{26.3.38}$$

where α and $B_q(p,\alpha)$ are the same as in (26.3.32). △

26.3.1 *Expansions of the coefficients*

We give a few details for the evaluation of $B_k(\alpha,\mu)$ when $\alpha \sim \mu$. First we need to expand $x = \sum_{k=0}^{\infty} x_k(\mu)(\alpha-\mu)^k$ and we can obtain the coefficients from (26.3.23). We know that $x_0(\mu) = 1/(\mu+1)$. To avoid the logarithmic terms in (26.3.23) it is easier to use the derivative:

$$\frac{dx}{d\alpha} = \frac{x_0(\mu)x(1-x)(\alpha-\mu)}{\alpha(x - x_0(\mu))}. \tag{26.3.39}$$

For $x_1(\mu)$ we find

$$x_1(\mu)^2 = (\mu+1)^{-3} \quad \Longrightarrow \quad x_1(\mu) = -(\mu+1)^{-3/2}, \tag{26.3.40}$$

where the minus sign follows from the relations given in (26.3.24). To avoid the square roots, we write $\rho = \sqrt{\mu+1}$. Then,

$$\begin{aligned}
x_0(\mu) &= \frac{1}{\rho^2}, \quad x_1(\mu) = -\frac{1}{\rho^3}, \quad x_2(\mu) = \frac{\rho+2}{3\rho^4(\rho+1)},\\
x_3(\mu) &= -\frac{\rho^2+10\rho+13}{36\rho^5(\rho+1)^2},\\
x_4(\mu) &= -\frac{2\rho^3-9\rho^2-81\rho-92}{540\rho^6(\rho+1)^3},\\
x_5(\mu) &= -\frac{3\rho^4-52\rho^3+26\rho^2+812\rho+939}{12960\rho^7(\rho+1)^4}.
\end{aligned} \tag{26.3.41}$$

When $\mu = 0$ ($\rho = 1$) the equation in (26.3.23) reduces to $x = e^{-\alpha}$, hence, the limiting values are $x_k(0) = (-1)^k/k!$.

The expansion of x can be used to obtain an expansion of $f(\alpha,\mu)$ given in (26.3.34), and of the functions $f_k(\alpha,\mu)$, that is,

$$f_k(\alpha,\mu) = \sum_{n=0}^{\infty} c_n^{(k)}(\mu)(\alpha-\mu)^k, \tag{26.3.42}$$

as we did in (25.2.23) with recursion of the coefficients given in (25.2.25). Then the coefficients of the expansion

$$B_k(\alpha,\mu) = \sum_{n=0}^{\infty} b_n^{(k)}(\mu)(\alpha-\mu)^k \tag{26.3.43}$$

follow from $b_n^{(k)}(\mu) = c_{n+1}^{(k)}(\mu)$.

The first coefficients of the function $f_0(\alpha,\mu)$ defined in (26.3.34) are

$$\begin{aligned}
c_0^{(0)}(\mu) &= \rho, \quad c_1^{(0)}(\mu) = \frac{\rho+2}{3(\rho+1)}, \quad c_2^{(0)}(\mu) = \frac{1}{12\rho},\\
c_3^{(0)}(\mu) &= \frac{(\rho-1)(8\rho^2+17\rho+8)}{540\rho^2(\rho+1)^3},\\
c_4^{(0)}(\mu) &= \frac{15\rho^4-68\rho^3-182\rho^2-68\rho+15}{12960\rho^3(\rho+1)^4},\\
c_5^{(0)}(\mu) &= -\frac{(\rho-1)(32\rho^4+297\rho^3+550\rho^2+297\rho+32)}{90720\rho^4(\rho+1)^5},
\end{aligned} \tag{26.3.44}$$

again with $\rho = \sqrt{\mu+1}$.

26.4 A corresponding loop integral

As in many other examples we can consider a loop integral with similar asymptotic phenomena. In §25.3 we have considered the integral

$$G_\lambda(z) = \frac{\Gamma(\lambda+1)}{2\pi i}\int_{-\infty}^{(0+)} s^{-\lambda-1} f(s) e^{zs}\, ds, \tag{26.4.45}$$

as the corresponding loop integral for $F_\lambda(z,\alpha)$ of (26.1.1) with $\alpha=0$. In the present case we can take

$$G_\lambda(z,\alpha)=\frac{\Gamma(\lambda)}{2\pi i}\int_{c-i\infty}^{c+i\infty} s^{-\lambda}f(s)e^{zs}\,\frac{ds}{s-\alpha},\quad 0<c<\alpha, \tag{26.4.46}$$

where f is analytic in the domain $\Re s>c$. If we wish, and if f allows, we can deform the contour into the form of a loop as in (26.4.45).

To construct an asymptotic expansion we first split off the pole by writing

$$\frac{f(s)}{s-\alpha}=\frac{f(\alpha)}{s-\alpha}+g(s),\quad g(s)=\frac{f(s)-f(\alpha)}{s-\alpha}. \tag{26.4.47}$$

This gives, by using the loop integral of the incomplete gamma function given in §37.2,

$$\begin{aligned}G_\lambda(z,\alpha)&=-f(\alpha)\Gamma(\lambda)\alpha^{-\lambda}e^{\alpha z}Q(\lambda,\alpha z)+R_\lambda(z,\alpha),\\ R_\lambda(z,\alpha)&=\frac{\Gamma(\lambda)}{2\pi i}\int_{c-i\infty}^{c+i\infty} s^{-\lambda}g(s)e^{zs}\,ds.\end{aligned} \tag{26.4.48}$$

This integral is of the same type as in §25.3. We write $g(s)=g(\mu)+(g(s)-g(\mu))$, where $\mu=\lambda/z$. Then, integrating by parts and using the loop integral of the reciprocal gamma function in (2.2.17), gives the expansion

$$R_\lambda(z,\alpha)\sim z^{\lambda-1}\sum_{n=0}^{\infty}(-1)^n\frac{g_n(\mu)}{z^n},\quad z\to\infty, \tag{26.4.49}$$

where the functions g_n are defined by the scheme

$$g_{n+1}(s)=\frac{d}{ds}\left(s\frac{g_n(s)-g_n(\mu)}{s-\mu}\right),\quad n=0,1,2,\ldots, \tag{26.4.50}$$

and $g_0(s)=g(s)$.

If we wish we can write the expansion in (26.4.49) with a finite number of terms and a remainder.

26.4.1 *Application to the incomplete beta function*

The incomplete beta function considered in (26.3.14) has the integral representation (see Temme (1996a, §11.3.2))

$$I_x(p,q)=\frac{x^p(1-x)^q}{2\pi i}\int_{c-i\infty}^{c+i\infty}\sigma^{-p}(1-\sigma)^{-q}\frac{d\sigma}{\sigma-x}, \tag{26.4.51}$$

where $0<x<c<1$, $p>0$, $q>0$, and the multivalued functions σ^{-p} and $(1-\sigma)^{-q}$ assume their principal values.

To obtain the standard form in (26.4.46) we first take $\sigma=e^{-w}$. The contour in the w-plane becomes a Hankel contour integral shown in Figure 2.1, and we obtain

$$I_x(p,q)=\frac{x^p(1-x)^q}{2\pi i}\int_{-\infty}^{(0+)} e^{pw}\left(1-e^{-w}\right)^{-q}\frac{e^{-w}\,dw}{e^{-w}-x}, \tag{26.4.52}$$

where the contour cuts the real axis at a point w_0 satisfying $0 < w_0 < -\ln x$.

The saddle point analysis and the transformation into the standard form can be done as in §26.3. By using (26.3.19) we obtain

$$I_x(p,q) = \frac{x^p(1-x)^q}{2\pi i} e^{pA(\mu)} \int_{-\infty}^{(0+)} e^{ps} t^{-q} h(t,\alpha,\mu) \frac{dt}{t-\alpha}, \tag{26.4.53}$$

where α satisfies equation (26.3.23), $A(\mu)$ is given in (26.3.22), the contour cuts the positive real axis between 0 and α, and

$$\begin{aligned} h(t,\alpha,\mu) &= \frac{e^{-w}}{e^{-w}-x}(t-\alpha)\frac{dw}{dt} \\ &= \frac{t-\mu}{1-(1+\mu)e^{-w}} \frac{1-e^{-w}}{t} \frac{t-\alpha}{1-xe^{w}}. \end{aligned} \tag{26.4.54}$$

From the first relation for h it is not difficult to verify, observing that $t = \alpha$ corresponds to $w = -\ln x$ and using l'Hôpital's rule, that $h(\alpha,\alpha,\mu) = -1$. This gives, using (26.3.23),

$$I_x(p,q) = Q(q,\alpha p) + \frac{x^p(1-x)^q}{2\pi i} e^{pA(\mu)} \int_{-\infty}^{(0+)} e^{ps} t^{-q} g(t,\alpha,\mu)\, dt, \tag{26.4.55}$$

with g defined by

$$g(t,\alpha,\mu) = \frac{h(t,\alpha,\mu) - h(\alpha,\alpha,\mu)}{t-\alpha}, \tag{26.4.56}$$

as in (26.4.47). The integral can be expanded as the function $R_\lambda(z,\alpha)$ in (26.4.49). This gives

$$I_x(p,q) = Q(q,\alpha p) + \frac{x^p(1-x)^q}{\Gamma(q)} e^{pA(\mu)} p^{q-1} S_q(p,\alpha), \tag{26.4.57}$$

where

$$S_q(p,\alpha) \sim \sum_{n=0}^{\infty} (-1)^n \frac{g_n(\alpha,\mu)}{p^n}, \quad p \to \infty. \tag{26.4.58}$$

Remark 26.3. The factor in front of $B_q(p,\alpha)$ in (26.3.32) is not the same as the one in front of $S_q(p,\alpha)$ in (26.4.57). However, the ratio of these factors, that is,

$$\frac{x^p(1-x)^q}{\Gamma(q)} e^{pA(\mu)} p^{q-1} \Big/ \frac{x^p(1-x)^q}{pB(p,q)} = e^{pA(\mu)} p^q \frac{\Gamma(p)}{\Gamma(p+q)}, \tag{26.4.59}$$

can be expanded as in (25.6.65). We have

$$e^{pA(\mu)} p^q \frac{\Gamma(p)}{\Gamma(p+q)} \sim \sqrt{1+\mu} \sum_{n=0}^{\infty} \frac{d_n(\mu)}{p^n}, \tag{26.4.60}$$

where the first coefficients $d_n(\mu)$ are given in (25.6.57). The expansions of $B_q(p,\alpha)$ and $S_q(p,\alpha)$ satisfy

$$\sum_{n=0}^{\infty} \frac{B_n(\alpha,\mu)}{p^n} \sim \sqrt{1+\mu} \sum_{n=0}^{\infty} \frac{d_n(\mu)}{p^n} \sum_{n=0}^{\infty} (-1)^n \frac{g_n(\alpha,\mu)}{p^n}. \tag{26.4.61}$$

From (26.3.33) and (26.4.56) it is not difficult to verify that $B_0(\alpha,\mu) = \sqrt{1+\mu}\, g_0(\alpha,\mu)$. More relations between the coefficients follow from (26.4.61).

△

Chapter 27

An essential singularity: Bessel-type expansions

In this chapter we consider Laplace-type integrals in which the exponential function $\exp(-z(t+\beta^2/t)$ has an essential singularity at the origin and two saddle points at $\pm\beta$ that approach the origin as $\beta \to 0$. As indicated in Case 9 and Case 10 of Table 20.1, we can use modified Bessel functions to handle this asymptotic feature.[1]

We have considered similar integrals in Chapter 10 to describe the asymptotic behavior of the Kummer functions for large values of the a-parameter. The expansions so obtained are valid for bounded values of the argument of the Kummer functions, and in this chapter we can accept unbounded values.

27.1 Expansions in terms of modified Bessel functions

We write the standard form for this case as

$$F_\lambda(z,\beta) = \int_0^\infty t^{\lambda-1} e^{-z\left(t+\beta^2/t\right)} f(t)\, dt. \tag{27.1.1}$$

To define this integral for complex values of the parameters we refer to Remark 10.2, and we need to know the domain of analyticity of the function $f(t)$. For the asymptotic analysis we assume that $z \to \infty$ through positive real values, and that $\beta \neq 0$. Hence we can take for λ any complex value. It will appear, however, that the coefficients of the expansion to be derived are well defined at $\beta = 0$ when f is analytic at the origin. This remains to be the case when we apply the results to the Kummer U-function; see §27.4.1, where we also consider complex values of the large a-parameter.

When f is a constant we can express this integral in terms of the modified Bessel function; see (10.3.35). We introduce a function $A_\lambda(z,\beta)$ writing

$$A_\lambda(z,\beta) = \int_0^\infty t^{\lambda-1} e^{-z\left(t+\beta^2/t\right)}\, dt = 2\beta^\lambda K_\lambda(2\beta z). \tag{27.1.2}$$

We have used this Bessel function in §10.3.2, where we have expanded the function f in (27.1.1) at the origin, with applications to the Kummer U-function. The expansion obtained in this way is only valid for bounded values of β. In the present chapter

[1]This chapter is based on Temme (1990c).

we expand at the two saddle points, again with an application to the Kummer U-function, and we consider contour integrals with an application to the Kummer F-function.

The saddle points occur at $\pm\beta$, and when β is bounded away from zero, we can concentrate on the positive saddle point β and use Laplace's method. When, however, $\beta \to 0$, the internal saddle point coalesces with the point $t = 0$, where the argument of the exponential function has a pole. In addition, there is an algebraic singularity, but the influence of the essential singularity due to the pole is more significant. Observe that in the limit $\beta \to 0$, the pole in the argument of the exponential function disappears; furthermore, both saddle points coalesce with the pole at the origin. These asymptotic features are also shown in the integral in (27.1.2), and the modified Bessel function serves as a basic approximant in the uniform asymptotic expansion of the function $F_\lambda(z,\beta)$ in (27.1.1).

The procedure of this section takes into account both saddle points, although $-\beta$ lies outside the interval of integration. For this reason we assume that f is also defined at negative values of its argument. In §27.3 we consider an expansion by using only the internal saddle point β.

The first step is the representation

$$f(t) = a_0 + b_0 t + \left(t - \beta^2/t\right) g(t), \tag{27.1.3}$$

where a_0, b_0 follow from substitution of $t = \pm\beta$. We have

$$a_0 = \frac{f(\beta) + f(-\beta)}{2}, \quad b_0 = \frac{f(\beta) - f(-\beta)}{2\beta}. \tag{27.1.4}$$

Inserting (27.1.3) into (27.1.1) we obtain

$$F_\lambda(z,\beta) = a_0 A_\lambda(z,\beta) + b_0 A_{\lambda+1}(z,\beta) + F_\lambda^{(1)}(z,\beta), \tag{27.1.5}$$

where $A_\lambda(z,\beta)$ is defined in (27.1.2). An integration by parts gives

$$\begin{aligned} F_\lambda^{(1)}(z,\beta) &= -\frac{1}{z}\int_0^\infty t^\lambda g(t)\, d\left(e^{-z(t+\beta^2/t)}\right) \\ &= \frac{1}{z}\int_0^\infty t^{\lambda-1} f_1(t) e^{-z(t+\beta^2/t)}\, dt, \end{aligned} \tag{27.1.6}$$

where

$$f_1(t) = t^{1-\lambda}\frac{d}{dt}\left(t^\lambda g(t)\right) = \lambda g(t) + t g'(t). \tag{27.1.7}$$

We see that $F_\lambda^{(1)}(z,\beta)$ is of the same form as $F_\lambda(z,\beta)$, and the procedure can now be applied to $F_\lambda^{(1)}(z,\beta)$. In this way, we obtain for (27.1.1) the expansion

$$F_\lambda(z,\beta) \sim A_\lambda(z,\beta)\sum_{k=0}^\infty \frac{a_k}{z^k} + A_{\lambda+1}(z,\beta)\sum_{k=0}^\infty \frac{b_k}{z^k}, \quad z \to \infty, \tag{27.1.8}$$

where we define inductively $f_0 = f$, $g_0 = g$, and

$$\begin{aligned} f_k(t) &= t^{1-\lambda}\frac{d}{dt}\left(t^\lambda g_{k-1}(t)\right) = a_k + b_k t + \left(t - \beta^2/t\right) g_k(t), \quad k \ge 1, \\ a_k &= \frac{f_k(\beta) + f_k(-\beta)}{2}, \quad b_k = \frac{f_k(\beta) - f_k(-\beta)}{2\beta}, \quad k \ge 0. \end{aligned} \tag{27.1.9}$$

We can introduce a remainder for the expansion in (27.1.8) by writing

$$F_\lambda(z,\beta) = A_\lambda(z,\beta)\sum_{k=0}^{K-1}\frac{a_k}{z^k} + A_{\lambda+1}(z,\beta)\sum_{k=0}^{K-1}\frac{b_k}{z^k} + z^{-K}R_K, \qquad (27.1.10)$$

where $K = 0,1,2,\ldots$. When $K=0$ the sums are empty and $R_0 = F_\lambda(z,\beta)$. The integration by parts procedure yields for R_K the representation

$$R_K = \int_0^\infty t^{\lambda-1} f_K(t) e^{-z(t+\beta^2/t)}\, dt, \qquad (27.1.11)$$

where f_K is defined by the scheme in (27.1.9).

When a bound for $|f_K(t)|$ is available, say,

$$|f_K(t)| \le M_K, \quad t \ge 0, \quad K = 0,1,2,\ldots, \qquad (27.1.12)$$

then a bound for R_K reads

$$|R_K| \le M_K A_\lambda(z,\beta). \qquad (27.1.13)$$

Since f_K depends on β, the quantity M_K may also depend on β. It follows that for bounded values of β, say $\beta \in [0,\beta_0]$, β_0 fixed and finite, the estimate (27.1.13) of the remainder R_K, shows the asymptotic nature of the expansion (27.1.10), provided that (27.1.12) is satisfied.

Observe that the estimate in (27.1.12) is rather global, since it takes into account values of f_K, in the complete interval $[0,\infty)$. A sharper and more realistic bound for R_K may be obtained as follows. Let

$$w_\sigma(t) = e^{\sigma(t+\beta^2/t-2\beta)}, \quad t>0, \quad \sigma \ge 0. \qquad (27.1.14)$$

Observe that $w_\sigma(\beta) = 1$ and that when $\sigma > 0$

$$\lim_{t\to 0} w_\sigma(t) = \lim_{t\to\infty} w_\sigma(t) = \infty. \qquad (27.1.15)$$

We assume that we can assign quantities σ_K and M_K, which may depend on β and which satisfy

$$\sigma_K \ge 0, \quad M_K \ge 1 + \varepsilon_K, \qquad (27.1.16)$$

where ε_K is fixed and positive, such that for all $t > 0$ we have

$$|f_K(t)| \le M_K |f_K(\beta)| w_{\sigma_K}(t). \qquad (27.1.17)$$

Then instead of (27.1.13) we obtain

$$|R_K| \le M_K |f_K(\beta)| \widetilde{A}_\lambda(z,\beta), \quad z \ge \sigma_K, \qquad (27.1.18)$$

where

$$\widetilde{A}_\lambda(z,\beta) = A_\lambda(z-\sigma_K,\beta)\, e^{-2\beta\sigma_K}. \qquad (27.1.19)$$

When $f_K(\beta) = 0$ a slight modification is needed. The idea about this approach is that in (27.1.17) function values outside a neighborhood of $t=\beta$ may be estimated very roughly, and that the integral, which results after inserting the right-hand side of (27.1.17) into (27.1.11), can be written in terms of one of the approximants in front of the series in (27.1.10).

Observe that the function defined in (27.1.19) satisfies

$$\widetilde{A}_\lambda(z,\beta)/A_\lambda(z,\beta) = 1 + o(1), \quad z \to \infty, \qquad (27.1.20)$$

uniformly with respect to $\beta \in [0,\infty)$. This follows from (27.1.2) and well-known asymptotic relations for the modified Bessel function.

In §27.4.1 we give an application to the Kummer U-function.

27.2 A corresponding loop integral

For a corresponding loop integral we take the form

$$G_\lambda(z,\beta) = \frac{1}{2\pi i}\int_{\mathcal{L}} e^{z(t+\beta^2/t)} t^{\lambda-1} f(t)\, dt, \qquad (27.2.21)$$

where $\mathcal{L}$ is a vertical line in the half-plane $\Re t > 0$, such that the singularities of f are on the left of this line. For convergence we may deform $\mathcal{L}$ into a loop such that $\Re t \to -\infty$ on $\mathcal{L}$.

We start an integration by parts procedure as in §27.1 and obtain

$$G_\lambda(z,\beta) \sim B_\lambda(z,\beta)\sum_{k=0}^{\infty}(-1)^k\frac{a_k}{z^k} + B_{\lambda+1}(z,\beta)\sum_{k=0}^{\infty}(-1)^k\frac{b_k}{z^k}, \qquad (27.2.22)$$

as $z\to\infty$, where (see (10.3.49))

$$B_\lambda(z,\beta) = \beta^\lambda I_{-\lambda}(2\beta z). \qquad (27.2.23)$$

The coefficients can be generated from the scheme as given in (27.1.9).

In §27.4.3 we give an application to the Kummer F-function.

27.3 Expansion at the internal saddle point

In the expansion (27.1.10) we have used function values of f at the negative saddle point $-\beta$. These values appear in the coefficients a_k, b_k of the expansion. The form of the expansion is very attractive, since only two special functions arise, and furthermore since the parameters β and z are nicely separated in both series. Although the expansion (27.1.10) has a canonical form, there remains the drawback that the function f must be defined on $(-\infty,0)$ in order to obtain for β a uniformity domain $[0,\infty)$. For example, it is not possible to obtain such a uniformity domain when $f(t)=1/(t+1)$. In this section we only expand the function f at the internal saddle point.

We expand f in the form

$$f(t) = \sum_{k=0}^{\infty} a_k(\beta)(t-\beta)^k. \qquad (27.3.24)$$

Substituting (27.3.24) in (27.1.1), we obtain after interchanging the order of summation and integration the formal result

$$F_\lambda(z,\beta) \sim z^{-\lambda}\sum_{k=0}^{\infty} a_k(\beta) Q_k(\zeta) z^{-k}, \quad z\to\infty, \qquad (27.3.25)$$

where

$$Q_k(\zeta) = \zeta^{\lambda+k}\int_0^\infty t^{\lambda-1}(t-1)^k e^{-\zeta(t+1/t)}\, dt, \quad \zeta=\beta z. \qquad (27.3.26)$$

The functions $Q_k(\zeta)$ can be expressed in terms of the modified Bessel functions defined in (27.1.2). Via the binomial expansion we obtain

$$Q_k(\zeta) = 2\zeta^{\lambda+k} \sum_{m=0}^{k} (-1)^{k-m} \binom{k}{m} K_{\lambda+m}(2\zeta). \tag{27.3.27}$$

On the other hand, integrating by parts in (27.3.26), gives the recurrence relation

$$Q_{k+2}(\zeta) = (k+\lambda+1-2\zeta)Q_{k+1}(\zeta) + \zeta(2k+\lambda+1)Q_k(\zeta) + k\zeta^2 Q_{k-1}(\zeta), \tag{27.3.28}$$

which can be used for $k = 0, 1, 2, \ldots$, with starting values that follow from (27.3.27).

For further details on this method we refer to Temme (1990c).

27.4 Application to Kummer functions

Kummer functions are introduced in Chapter 10, where we have given expansions of these functions for large values of a with bounded values of $|z|$. In this section we show how to extend the domain for z. The expansions derived for large a and unbounded z may be compared with the one given in Olver (1997, p. 447) for the Whittaker function $W_{-\kappa,\mu}(z)$ for large values of κ. For the relations between Kummer and Whittaker functions, see (10.1.14).

27.4.1 *Expansion of* $U(a,c,z)$, $a \to +\infty$, $z > 0$

For the U-function we start with the integral representation given in (10.1.5) for which we use a simple transformation $t/(t+1) = e^{-w}$. We write the result in the form

$$\Gamma(a)U(a, 1-\lambda, az) = \int_0^\infty w^{\lambda-1} e^{-a\phi(w)} g(w)\, dw, \tag{27.4.29}$$

where

$$\phi(w) = w + \frac{z}{e^w - 1}, \quad g(w) = \left(\frac{1-e^{-w}}{w}\right)^{\lambda-1}. \tag{27.4.30}$$

We assume that $\Re(az) > 0$. For convenience we may assume that $\Re\lambda \ge 0$; observe that the convergence at the origin is controlled by the condition on az. On the other hand, we can use the relation in (10.1.10), now in the form

$$U(a, 1-\lambda, az) = (az)^\lambda U(a+\lambda, 1+\lambda, az). \tag{27.4.31}$$

In the following analysis we assume that $a > 0$, $z > 0$, and $\lambda \ge 0$.

With the transformation[2]

$$\phi(w) = t + \frac{\beta^2}{t} + A, \tag{27.4.32}$$

[2] A transformation of this type seems to have been proposed for the first time in the paper Handelsman and Bleistein (1969).

we transform (27.4.29) into the standard form (27.1.1). The saddle points of $\phi(w)$ in (27.4.32) must correspond to those of the function on the right-hand side. Saddle points are $\pm w_0$ and $\pm t_0$, where $t_0 = \beta$ and

$$w_0 = 2\,\mathrm{arcsinh}\left(\tfrac{1}{2}\sqrt{z}\right). \tag{27.4.33}$$

It follows that $A = -\frac{1}{2}z$ and that the function $t(w)$ defined in (27.4.32) is an odd function of w. This can be concluded from rewriting (27.4.32) in the form

$$\phi(w) + \tfrac{1}{2}z = w + \tfrac{1}{2}z\coth\left(\tfrac{1}{2}w\right) = t + \frac{\beta^2}{t}. \tag{27.4.34}$$

The quantity β follows from

$$2\beta = \tfrac{1}{2}z + w_0 + \frac{z}{e^{w_0} - 1}, \tag{27.4.35}$$

which gives

$$\beta = \tfrac{1}{2}\left(w_0 + \sinh(w_0)\right). \tag{27.4.36}$$

Observe that both functions of t and w in (27.4.34) are convex on the positive real axis, and with the extra condition $\mathrm{sign}(w - w_0) = \mathrm{sign}(t - \beta)$ we can assign for every positive w a unique positive value of t.

With the obtained values of A and β the mapping $w \mapsto t$ is analytic at $w = \pm w_0$ and at $w = 0$. In fact it is analytic in $\mathbb{R}$ and as a conformal mapping in a large domain Ω of the complex plane. We have the correspondences

$$t(\pm\infty) = \pm\infty, \quad t(\pm w_0) = \pm\beta, \quad t(0) = 0. \tag{27.4.37}$$

Using transformation (27.4.32) in (27.4.29), we arrive at the standard form (cf. (27.1.1))

$$\Gamma(a)e^{-\frac{1}{2}az}U(a, 1-\lambda, az) = \int_0^\infty t^{\lambda-1}e^{-a(t+\beta^2/t)}f(t)\,dt, \tag{27.4.38}$$

where

$$f(t) = \left(\frac{w}{t}\right)^{\lambda-1} g(w)\frac{dw}{dt} = \left(\frac{1-e^{-w}}{t}\right)^{\lambda-1}\frac{dw}{dt}, \tag{27.4.39}$$

and

$$\frac{dw}{dt} = \frac{t^2-\beta^2}{t^2\phi'(w)} = \left(\frac{1-e^{-w}}{t}\right)^2 \frac{t^2-\beta^2}{\left(1-e^{-w}\right)^2 - ze^{-w}}. \tag{27.4.40}$$

This gives

$$f(t) = \left(\frac{1-e^{-w}}{t}\right)^{\lambda+1} \frac{t^2-\beta^2}{\left(1-e^{-w}\right)^2 - ze^{-w}}. \tag{27.4.41}$$

After these preparations the expansions (27.1.8) and (27.3.25) can be constructed. We have

$$U(a, 1-\lambda, az) \sim \frac{e^{\frac{1}{2}az}}{\Gamma(a)}\left(A_\lambda(a,\beta)\sum_{k=0}^{\infty}\frac{a_k}{a^k} + A_{\lambda+1}(a,\beta)\sum_{k=0}^{\infty}\frac{b_k}{a^k}\right), \tag{27.4.42}$$

as $a \to \infty$. $A_\lambda(a,\beta)$ is given in (27.1.2) and the coefficients can be generated by the scheme given in (27.1.9), with f given in (27.4.41).

When we write $c = 1-\lambda$, use the recurrence relation in (10.3.38) and the symmetry rule $K_\nu(z) = K_{-\nu}(z)$, we can rearrange the expansion in (27.4.42) to obtain an expansion as in (10.3.39) with two Bessel functions $K_{c-1}(2\beta a)$ and $K_c(2\beta a)$. The result is

$$U(a,c,az) \sim 2\beta^{1-c}\frac{e^{\frac{1}{2}az}}{\Gamma(a)}\left(K_{c-1}(2\beta a)\sum_{k=0}^{\infty}\frac{\widetilde{a}_k}{a^k} + \beta K_c(2\beta a)\sum_{k=0}^{\infty}\frac{b_k}{a^k}\right), \tag{27.4.43}$$

where

$$\widetilde{a}_0 = a_0, \quad \widetilde{a}_k = a_k - (c-1)b_{k-1}, \quad k \ge 1. \tag{27.4.44}$$

The coefficients a_k and b_k follow from the scheme in (27.1.9), with f defined in (27.4.39), with $\lambda = 1-c$.

The expansion holds for bounded values of c and is uniformly valid with respect to $\beta \in [0,\infty)$; that is, uniformly valid with respect to $z \in [0,\infty)$. For details we refer to Temme (1990c).

We give the first coefficient $\widetilde{a}_0$ and b_0 of (27.4.43). A few calculations based on (27.4.40) and l'Hôpital's rule yield

$$\left.\frac{dw}{dt}\right|_{t=\pm\beta} = \sqrt{\frac{2\tanh\left(\frac{1}{2}w_0\right)}{\beta}}. \tag{27.4.45}$$

So we obtain

$$\begin{aligned} \widetilde{a}_0 &= \sqrt{\frac{2\tanh\left(\frac{1}{2}w_0\right)}{\beta}}\left(\frac{2\sinh\left(\frac{1}{2}w_0\right)}{\beta}\right)^{-c}\cosh\left(\tfrac{1}{2}cw_0\right), \\ b_0 &= \sqrt{\frac{2\tanh\left(\frac{1}{2}w_0\right)}{\beta}}\left(\frac{2\sinh\left(\frac{1}{2}w_0\right)}{\beta}\right)^{-c}\frac{\sinh\left(\frac{1}{2}cw_0\right)}{\beta}, \end{aligned} \tag{27.4.46}$$

with w_0 and β defined in (27.4.33) and (27.4.36). The coefficients are well defined as $\beta \to 0$, but it will be clear that we need expansions in that case.

The expansion in (27.4.43) may be compared with the one in (10.3.39), which is valid for bounded values of z.

Remark 27.1. When we need the expansion of $U(a+\alpha,c,az)$ with α fixed, we can obtain similar results; see Remark 10.3. For example, we may consider the large-a expansion with the U-function in the form (cf. (27.4.29))

$$\Gamma\left(\tfrac{1}{2}-\tfrac{1}{2}\lambda+a\right)U\left(\tfrac{1}{2}-\tfrac{1}{2}\lambda+a,1-\lambda,az\right) = \int_0^\infty w^{\lambda-1}e^{-a\phi(w)}h(w)\,dw, \tag{27.4.47}$$

where $\phi(w)$ is defined in (27.4.30) and

$$h(w) = e^{-\frac{1}{2}(1-\lambda)w}g(w) = \left(\frac{\sinh\left(\frac{1}{2}w\right)}{\frac{1}{2}w}\right)^{\lambda-1}. \tag{27.4.48}$$

Then the transformation in (27.4.32) gives

$$\Gamma\left(\tfrac{1}{2}-\tfrac{1}{2}\lambda+a\right)U\left(\tfrac{1}{2}-\tfrac{1}{2}\lambda+a,1-\lambda,az\right)= \\ e^{\frac{1}{2}az}\int_0^\infty t^{\lambda-1}e^{-a(t+\beta^2/t)}F(t)\,dt, \tag{27.4.49}$$

with

$$F(t)=\left(\frac{w}{t}\right)^{\lambda-1}h(w)\frac{dw}{dt}. \tag{27.4.50}$$

The function $F(t)$ is even, and in that case the functions $F_k(t)$ that follow from the recursive scheme in (27.1.9) (with starting function $f(t)$ replaced by $F(t)$) are even (odd) when n is even (odd). This will have favorable consequences for the expansion in (27.4.43), because the corresponding coefficients b_{2k} and $\widetilde{a}_{2k+1}$ will vanish.

This occurs when we consider expansions for the Coulomb wave functions in Chapter 31. In particular, see §31.6. △

27.4.2 *Auxiliary expansions and further details*

We see from (27.4.34) that β should vanish when $z\to 0$ (because $t\to 0$ if $w\to 0$). From the simple differential equation

$$\frac{d\beta}{dz}=\tfrac{1}{4}\sqrt{(z+4)/z},\quad \beta(0)=0, \tag{27.4.51}$$

and a Taylor expansion of the right-hand side, it follows that β^2 of (27.4.32) is an analytic function of z, at least in the disk $|z|<4$. We have

$$\beta^2=z+\tfrac{1}{12}z^2-\tfrac{1}{720}z^3+\tfrac{1}{6720}z^4+\ldots. \tag{27.4.52}$$

Conversely, z is an analytic function of β^2 in some neighborhood of the origin. These domains can be extended to domains containing $[0,\infty)$. We have

$$z=\beta^2-\tfrac{1}{12}\beta^4+\tfrac{11}{720}\beta^6-\tfrac{73}{20160}\beta^8+\ldots. \tag{27.4.53}$$

From (27.4.34) the relation between w and t follows. As we have observed earlier, w is an odd function of t, and

$$w=\frac{z}{\beta^2}t-\frac{z(12\beta^2-12z-z^2)}{12\beta^6}t^3+\mathcal{O}\left(t^5\right),\quad t\to 0. \tag{27.4.54}$$

From the expansion in (27.4.53) we can verify that the first coefficients in this expansion are well defined when $z\to 0$.

The singular points of the transformation in (27.4.34) follow from the zeros of $\phi'(t)$, see (27.4.40). The zeros $\pm w_0$ defined in (27.4.33) are the saddle points, which correspond to $\pm\beta$. At these points the mapping is analytic. To find other zeros we observe that the equation for the saddle points is given by

$$e^{2w}-(z+2)e^w+1=0, \tag{27.4.55}$$

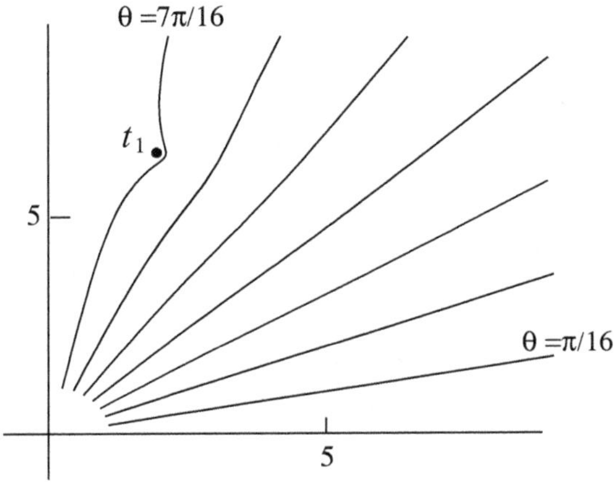

Fig. 27.1 Images in the t-plane of the half-lines $w = re^{i\theta}$, $r > 0$, where $\theta = j\pi/16$, $0 \le j \le 7$ for $z = 1$; t_1 is a singular point of the transformation in (27.4.34).

and that except for the saddle points $\pm w_0$ other solutions are $\pm w_k$ where $w_k = w_0 + 2\pi k i$, $k = \pm1, \pm2, \pm3, \ldots$. The points w_k correspond in the t-plane to the points t_k given by (see also (27.4.34))

$$t_k + \frac{\beta^2}{t_k} = 2\beta + 2\pi k i. \tag{27.4.56}$$

This gives

$$t_k = \beta + \pi k i + \sqrt{(\beta + \pi k i)^2 - \beta^2}, \quad k = \pm1, \pm2, \pm3, \ldots. \tag{27.4.57}$$

In Figure 27.1 we give details of the mapping of part of the w-plane to the t-plane. We show the images in the t-plane of the half-lines $w = re^{i\theta}$, $r \ge 1$, where $\theta = j\pi/16$, $1 \le j \le 7$ for $z = 1$. The black dot indicates the singular point at t_1 defined in (27.4.57). The images fold around the singular point t_1 as θ approaches the value $\operatorname{ph} w_1$. A branch cut for the mapping runs from t_1 to $-\infty$. A similar figure follows by conjugation for negative values of θ.

For the figure we have used $z = 1$. From this value the other values follow: $w_0 = 0.96$, $\beta = 1.04$, $w_1 = w_0 + 2\pi i$, $t_1 = 2.03 + 6.44i$, $\operatorname{ph} w_1 = 1.42$. This last value is somewhat larger than $7\pi/16 \doteq 1.37$.

From these details we conclude that the function $f(t)$ defined in (27.4.39) is analytic inside a sector $|\operatorname{ph} t| < \operatorname{ph} t_1$, with vertex at the origin. As in Watson's lemma, this sector determines the sector for complex values of a (see (2.1.6)):

$$-\operatorname{ph} t_1 - \tfrac{1}{2}\pi + \delta \le \operatorname{ph} a \le \operatorname{ph} t_1 + \tfrac{1}{2}\pi - \delta. \tag{27.4.58}$$

In fact this can also be used for complex values of z.

It is also easily verified that $f(t) \sim t^{1-\lambda}$ as $t \to \infty$ inside this sector. Because the iterated functions $f_k(t)$, which are defined by a similar scheme as in §27.1, are generated as linear combinations of derivatives of $f(t)$, it also follows by invoking Theorem 1.1, that we can assume that an estimate as in (27.1.17) is valid, that is,

$$|f_k(t)| \le M_k e^{\sigma_k (t + \beta^2/t - 2\beta)}, \quad t \ge 0, \tag{27.4.59}$$

where M_k and σ_k are some positive numbers independent of t. From this estimate we can construct bounds of the remainder in the asymptotic expansion as in (27.1.18).

27.4.3 *Expansion of* ${}_1F_1(a;c;z)$, $a \to +\infty$, $z \geq 0$

For the F-function we take (10.3.46) in the form

$$ {}_1F_1\left(\begin{matrix} a \\ 1-\lambda \end{matrix}; az\right) = \frac{\Gamma(1-\lambda)\Gamma(a+\lambda)\, e^{az}}{\Gamma(a)\; 2\pi i} \int_{\mathcal{L}} e^{a\phi(w)} w^{\lambda-1} g(-w)\, dw, \tag{27.4.60} $$

where the functions ϕ and g are defined in (27.4.30).

We use the transformation (27.4.32) and obtain

$$ \begin{aligned} &{}_1F_1\left(\begin{matrix} a \\ 1-\lambda \end{matrix}; az\right) = \\ &\quad \frac{\Gamma(1-\lambda)\Gamma(a+\lambda)\, e^{\frac{1}{2}az}}{\Gamma(a)\; 2\pi i} \int_{\mathcal{L}} e^{a(t+\beta^2/t)} t^{\lambda-1} F(t)\, dt, \end{aligned} \tag{27.4.61} $$

where β is defined in (27.4.36) and $F(t) = f(-t)$, with f defined in (27.4.41) (observe that t is an odd function of w).

The expansion in (27.2.22) gives

$$ \begin{aligned} &{}_1F_1\left(\begin{matrix} a \\ 1-\lambda \end{matrix}; az\right) \sim \frac{\Gamma(1-\lambda)\Gamma(a+\lambda)\, e^{\frac{1}{2}az}}{\Gamma(a)} \times \\ &\left(B_\lambda(a,\beta) \sum_{k=0}^{\infty} (-1)^k \frac{c_k}{a^k} + B_{\lambda+1}(a,\beta) \sum_{k=0}^{\infty} (-1)^k \frac{d_k}{a^k}\right), \end{aligned} \tag{27.4.62} $$

as $a \to \infty$, where $B_\lambda(a,\beta)$ is the modified Bessel function defined in (27.2.23).

The coefficients can be generated from a scheme as given in (27.1.9), with $f(t)$ replaced by $F(t) = f(-t)$. A further analysis shows a correspondence with the coefficients in (27.4.42). We find

$$ \begin{aligned} &{}_1F_1\left(\begin{matrix} a \\ 1-\lambda \end{matrix}; az\right) \sim \beta^\lambda \frac{\Gamma(1-\lambda)\Gamma(a+\lambda)\, e^{\frac{1}{2}az}}{\Gamma(a)} \times \\ &\quad \left(I_{-\lambda}(2\beta a) \sum_{k=0}^{\infty} \frac{a_k}{a^k} - \beta I_{-\lambda-1}(2\beta a) \sum_{k=0}^{\infty} \frac{b_k}{a^k}\right), \end{aligned} \tag{27.4.63} $$

where the coefficients a_k and b_k are the same as in the expansion in (27.4.42).

When we write $c = 1 - \lambda$, using the recurrence relation in (10.3.52), we can rearrange the expansion in (27.4.63) to obtain an expansion as in (10.3.53) with two Bessel functions $I_{c-1}(2\beta a)$ and $I_c(2\beta a)$. The result is

$$ \begin{aligned} &\frac{1}{\Gamma(c)} {}_1F_1\left(\begin{matrix} a \\ c \end{matrix}; az\right) \sim \beta^{1-c} \frac{\Gamma(1+a-c) e^{\frac{1}{2}az}}{\Gamma(a)} \times \\ &\quad \left(I_{c-1}(2\beta a) \sum_{k=0}^{\infty} \frac{\widetilde{a}_k}{a^k} - \beta I_c(2\beta a) \sum_{k=0}^{\infty} \frac{b_k}{a^k}\right), \end{aligned} \tag{27.4.64} $$

with $\widetilde{a}_k$ as in (27.4.44). The first coefficients are given in (27.4.46).

For a large-a expansion of ${}_1F_1\left(\frac{1}{2}c+a;c,z\right)$ in which the corresponding coefficients $\widetilde{a}_{2k+1}$ and b_{2k} will vanish, we refer to Remark 27.1.

Remark 27.2. This correspondence between the coefficients of the expansions of the U- and F-functions also occurs in the simpler expansions in (10.3.39) and (10.3.53). The expansion in (27.4.64) is derived for $a \to \infty$, and is a uniform expansion with respect to $z \in [0,\infty)$, with c a bounded parameter. For $c = 0, -1, -2, \ldots$ we have the usual exception, although the left-hand side has a proper interpretation when we use the limit in (10.1.3). Also, all quantities on the right-hand side are well defined for these c-values. △

Remark 27.3. We can write $\beta^{1-c} I_{c-1}(2\beta a) = a^{c-1} E_{c-1}(-2\beta a)$, where $E_\nu(z)$ is Tricomi's function; see (10.3.23). The Kummer function ${}_1F_1(a;c;z)$ is an entire function of z, same as Tricomi's function, and we see that all quantities in the expansion (27.4.64) are analytic functions at $\beta = 0$ ($z = 0$). This also happens for the other expansions of the F-function in this chapter; see (27.4.71) and (27.5.112). △

27.4.4 *Expansion of* ${}_1F_1(a;c;z)$, $a \to -\infty$, $0 \le z < -4a$

We start as in (10.3.54), and have

$$\begin{aligned} {}_1F_1\left(\begin{matrix} -a \\ c \end{matrix}; az\right) &= e^{az}\, {}_1F_1\left(\begin{matrix} c+a \\ c \end{matrix}; -az\right) \\ &= \frac{\Gamma(c)\Gamma(1+a)}{2\pi i\,\Gamma(a+c)} \int_{\mathcal{L}} e^{a\psi(w)} w^{-c} g(w)\, dw, \end{aligned} \tag{27.4.65}$$

where $g(w)$ is the same as in (27.4.30) and $\psi(w)$ follows from $\phi(w)$ by replacing z by $-z$. That is,

$$\psi(w) = w - \frac{z}{e^w - 1}, \quad g(w) = \left(\frac{1-e^{-w}}{w}\right)^{-c}. \tag{27.4.66}$$

The saddle points are $\pm w_0 = \pm 2i \arcsin\left(\frac{1}{2}\sqrt{z}\right)$, and the transformation reads

$$\psi(w) - \tfrac{1}{2}z = w - \tfrac{1}{2}z \coth\left(\tfrac{1}{2}w\right) = t - \frac{\gamma^2}{t}, \tag{27.4.67}$$

where γ is given by

$$\gamma = \tfrac{1}{2}\left(2\theta + \sin(2\theta)\right), \quad \theta = -\tfrac{1}{2} i w_0 = \arcsin\left(\tfrac{1}{2}\sqrt{z}\right). \tag{27.4.68}$$

Compare these quantities with those given in (27.4.33)–(27.4.36).

The transformation gives

$${}_1F_1\left(\begin{matrix} -a \\ c \end{matrix}; az\right) = \frac{\Gamma(c)\Gamma(1+a)}{2\pi i\,\Gamma(a+c)} e^{\frac{1}{2}az} \int_{\mathcal{L}} e^{a(t-\gamma^2/t)} t^{-c} p(t)\, dt, \tag{27.4.69}$$

where

$$p(t) = \left(\frac{w}{t}\right)^{-c} g(w)\frac{dw}{dt}. \qquad (27.4.70)$$

Using the integration by parts procedure we obtain the expansion

$$\begin{aligned} {}_1F_1\left(\begin{matrix}-a\\ c\end{matrix};\, az\right) \sim \gamma^{1-c}\frac{\Gamma(c)\Gamma(a+1)\,e^{\frac12 az}}{\Gamma(a+c)} \times \\ \left(J_{c-1}(2\gamma a)\sum_{k=0}^{\infty}\frac{A_k}{a^k} + \gamma J_{c-2}(2\gamma a)\sum_{k=0}^{\infty}\frac{B_k}{a^k}\right), \end{aligned} \qquad (27.4.71)$$

where the coefficients A_k and B_k follow from the scheme

$$\begin{aligned} p_k(t) &= -t^c\frac{d}{dt}\left(t^{1-c}q_{k-1}(t)\right) = A_k + B_kt + \left(t+\gamma^2/t\right)q_k(t),\\ A_k &= \frac{p_k(i\gamma)+p_k(-i\gamma)}{2}, \qquad B_k = \frac{p_k(i\gamma)-p_k(-i\gamma)}{2i\gamma}, \end{aligned} \qquad (27.4.72)$$

where $p_0(t) = p(t)$ is defined in (27.4.70).

The first coefficients follow from

$$p(\pm i\gamma) = \left(\frac{\gamma}{2\sin\theta}\right)^c e^{\pm ic\theta}\sqrt{\frac{2}{\gamma}\tan\theta}, \qquad (27.4.73)$$

and are given by

$$\begin{aligned} A_0 &= \left(\frac{\gamma}{2\sin\theta}\right)^c\sqrt{\frac{2}{\gamma}\tan\theta}\,\cos(c\theta),\\ B_0 &= \left(\frac{\gamma}{2\sin\theta}\right)^c\sqrt{\frac{2}{\gamma}\tan\theta}\,\frac{\sin(c\theta)}{\gamma}. \end{aligned} \qquad (27.4.74)$$

Remark 27.4. The saddle points $\pm w_0$ are defined by

$$e^{\pm w_0} = \tfrac12\left(2 - z \pm i\sqrt{z(4-z)}\right) = e^{\pm 2i\theta}. \qquad (27.4.75)$$

When $z = 4$, the saddle points coincide and the saddle points satisfy $e^{\pm w_0} = -1$. For the Kummer function ${}_1F_1(-a; c; az)$ we need Airy functions to describe the asymptotic behavior for large positive a and $z \sim 4$; see Dunster (1989). The expansion in (27.4.71) is uniformly valid with respect to $z \in [0, z_0]$, where z_0 is a fixed number in $(0, 4)$. For negative z the Bessel functions can be written in terms of modified Bessel functions. In that case θ and γ become purely imaginary, and the expansion remains valid. △

Remark 27.5. When $a = n = 0, 1, 2, \ldots$ and $c = \alpha + 1$, the Kummer function in (27.4.71) is a Laguerre polynomial $L_n^{(\alpha)}(x)$, see (10.1.17). In that case the expansion corresponds to the one given in §32.3.2, where the expansion has $\kappa = n + \frac12(\alpha+1)$ as the large parameter. This can be achieved in our case when we write (27.4.65) in the form

$${}_1F_1\left(\begin{matrix}-a\\ c\end{matrix};\, \kappa z\right) = \frac{\Gamma(c)\Gamma(1+a)}{2\pi i\,\Gamma(a+c)}\int_{\mathcal{L}} e^{\kappa\psi(w)}w^{-c}G(w)\,dw, \qquad (27.4.76)$$

where $\kappa = a + \frac{1}{2}c$ and

$$G(w) = \left(\frac{\sinh\left(\frac{1}{2}w\right)}{\frac{1}{2}w}\right)^{-c}. \tag{27.4.77}$$

The expansion is as in (27.4.71):

$$\begin{aligned}{}_1F_1\left(\begin{matrix}-a\\ c\end{matrix};\kappa z\right) \sim \gamma^{1-c}\frac{\Gamma(c)\Gamma(a+1)\,e^{\frac{1}{2}\kappa z}}{\Gamma(a+c)} \times \\ \left(J_{c-1}(2\gamma\kappa)\sum_{k=0}^{\infty}\frac{C_k}{\kappa^k} + \gamma J_{c-2}(2\gamma\kappa)\sum_{k=0}^{\infty}\frac{D_k}{\kappa^k}\right),\end{aligned} \tag{27.4.78}$$

where C_k and D_k follow from the scheme in (27.4.72) with $p(t)$ replaced by

$$P(t) = \left(\frac{w}{t}\right)^{-c} G(w)\frac{dw}{dt}. \tag{27.4.79}$$

Because $G(w)$ is an even function of w, and w is an odd function of t (see (27.4.67)), the function $P(t)$ is even, and $D_0 = 0$. In addition, $C_{2k+1} = D_{2k} = 0$, $k = 0, 1, 2, \ldots$. This makes the expansion more efficient than the one given in (27.4.71). △

27.4.5 *Expansion of* $U(a, c, z)$, $a \to -\infty$, $0 < z < -4a$

For this case we can use the method described in §10.3.5, and we start with the connection formula given in (10.3.61), that is,

$$\frac{1}{\Gamma(a+c)}U(-a,c,z) = \frac{e^{\mp\pi ia}}{\Gamma(c)}\,{}_1F_1\left(\begin{matrix}-a\\ c\end{matrix};z\right) - \frac{e^{z\pm\pi ic}}{\Gamma(-a)}U\left(a+c,c,ze^{\pm\pi i}\right). \tag{27.4.80}$$

For the F-function we use the expansion given in (27.4.71) and for the U-function we use the representation (see also (10.3.62))

$$U(a+c,c,az) = \frac{1}{\Gamma(a+c)}\int_0^{\infty} w^{-c}e^{-a\phi(w)}g(-w)\,dw, \tag{27.4.81}$$

where $\phi(w)$ and $g(w)$ are defined in (27.4.30), with $\lambda = 1-c$. We use the transformation given in (27.4.34) and obtain

$$U(a+c,c,az) = \frac{e^{\frac{1}{2}az}}{\Gamma(a+c)}\int_0^{\infty} t^{-c}e^{-a(t+\beta^2/t)}f(-t)\,dt, \tag{27.4.82}$$

where $f(t)$ is given in (27.4.41). This gives the expansion (cf. (27.4.42))

$$\begin{aligned}U(a+c,\,c,\,az) \sim 2\beta^{1-c}\frac{e^{\frac{1}{2}az}}{\Gamma(a+c)} \times \\ \left(K_{1-c}(2\beta a)\sum_{k=0}^{\infty}(-1)^k\frac{a_k}{a^k} - \beta K_{2-c}(2\beta a)\sum_{k=0}^{\infty}(-1)^k\frac{b_k}{a^k}\right),\end{aligned} \tag{27.4.83}$$

where β is defined in (27.4.36). The coefficients follow from the recursive scheme given in (27.1.9) with starting function $f(-t)$ and $\lambda = 1-c$. Indeed, we obtain the same coefficients, up to a change of sign in the odd numbered terms.

Next we replace z by $ze^{\pm\pi i}$. From (27.4.33), (27.4.36) and (27.4.68) it follows that β should be replaced by γ. We obtain

$$
\begin{aligned}
&U\left(a+c,c,aze^{\pm\pi i}\right) \sim 2\gamma^{1-c}e^{\pm\frac{1}{2}(1-c)\pi i}\frac{e^{-\frac{1}{2}az}}{\Gamma(a+c)} \times \\
&\left(K_{1-c}\left(\zeta e^{\pm\frac{1}{2}\pi i}\right)\sum_{k=0}^{\infty}\frac{A_k}{a^k} \mp i\gamma K_{2-c}\left(\zeta e^{\pm\frac{1}{2}\pi i}\right)\sum_{k=0}^{\infty}\frac{B_k}{a^k}\right),
\end{aligned}
\tag{27.4.84}
$$

where $\zeta = 2\gamma a$, A_k, B_k follow from a_k, b_k by replacing z by $ze^{\pm\pi i}$, and are the same as those in (27.4.71).

We use $K_\nu(z) = K_{-\nu}(z)$ and (10.3.64) to write the K-Bessel functions in terms of ordinary Bessel functions. Then, by using the expansions in (27.4.71) and (27.4.84), the connection formula in (27.4.80) gives

$$
\begin{aligned}
U(-a,c,az) &\sim \gamma^{1-c}\Gamma(a+1)e^{\frac{1}{2}az} \times \\
&\left(C_{c-1}(\zeta)\sum_{k=0}^{\infty}\frac{A_k}{a^k} + \gamma C_{c-2}(\zeta)\sum_{k=0}^{\infty}\frac{B_k}{a^k}\right),
\end{aligned}
\tag{27.4.85}
$$

where

$$
C_\nu(\zeta) = \cos(\pi a)J_\nu(\zeta) + \sin(\pi a)Y_\nu(\zeta), \quad \zeta = 2\gamma a, \tag{27.4.86}
$$

and γ is defined in (27.4.68).

The expansion in (27.4.85) is uniformly valid with respect to $z \in [0, z_0]$, where z_0 is a fixed number in $(0, 4)$. The first coefficients are given in (27.4.74). As for the F-function we need Airy functions to describe the asymptotic behavior of $U(-a,c,az)$ for large positive a and $z \sim 4$; see Dunster (1989).

For $a = n$ (a nonnegative integer) the expansion can be used for the Laguerre polynomials. See (10.1.17) and Remark 27.5, where we also describe how to obtain a more efficient expansion.

27.5 A second uniformity parameter

In this section we consider (27.1.1) with λ as a second uniformity parameter in $[0, \infty)$. Thus we take further advantage of the fact that the modified Bessel function is a function of two variables. Again we take the standard form

$$
\begin{aligned}
F_\lambda(z,\beta) &= \int_0^\infty t^{\lambda-1}e^{-z\left(t+\beta^2/t\right)}f(t)\,dt \\
&= \int_0^\infty t^{-1}e^{-z\left(t+\beta^2/t-\mu\ln t\right)}f(t)\,dt, \quad \lambda = \mu z.
\end{aligned}
\tag{27.5.87}
$$

In Chapter 25 we considered (27.5.87) with $\beta = 0$, again with $z \to \infty$ and λ as a uniformity parameter in $[0, \infty)$.

The saddle points $t_\pm$ of the integrand are now defined as the points where the derivative of $t + \beta^2/t - \mu \ln t$ vanishes. We have

$$t_\pm = \tfrac{1}{2}(\mu \pm T), \quad T = \sqrt{\mu^2 + 4\beta^2}. \tag{27.5.88}$$

As in the previous case, one of the real saddle points is outside the interval of integration and now the "phase function" used to compute the saddle points has a logarithmic singularity at $t = 0$. The two saddle points coincide with this singularity when β and μ both vanish. At the same moment, however, the logarithmic singularity disappears.

First we construct an expansion by using the integration by parts procedure of §27.1. The modification of (27.1.3) is

$$f(t) = c_0 + d_0 t + \left(t - \mu - \beta^2/t\right) h(t). \tag{27.5.89}$$

Using this in (27.5.87) we obtain, after repeating the procedure,

$$F_\lambda(z,\beta) \sim A_\lambda(z,\beta) \sum_{k=0}^{\infty} \frac{c_k}{z^k} + A_{\lambda+1}(z,\beta) \sum_{k=0}^{\infty} \frac{d_k}{z^k}, \tag{27.5.90}$$

as $z \to \infty$, where A_λ is the modified K-Bessel function, see (27.1.2). The coefficients c_k, d_k follow from the recursive scheme

$$\begin{aligned} f_0(t) &= f(t), \\ f_k(t) &= t\frac{d}{dt} h_{k-1}(t) = c_k + d_k t + \left(t - \mu - \beta^2/t\right) h_k(t), \\ c_k &= \frac{t_+ f_k(t_-) - t_- f_k(t_+)}{t_+ - t_-}, \quad d_k = \frac{f_k(t_+) - f_k(t_-)}{t_+ - t_-}. \end{aligned} \tag{27.5.91}$$

In a similar way, we can obtain the analogue of expansion (27.3.25) by expanding f at the point t_+. We use

$$f(t) = \sum_{k=0}^{\infty} c_k(\beta,\mu) \left(t - t_+\right)^k, \tag{27.5.92}$$

and upon substituting we obtain

$$F_\lambda(z,\beta) \sim z^{-\lambda} \sum_{k=0}^{\infty} c_k(\beta,\mu) \Phi_k(z) z^{-k}, \tag{27.5.93}$$

where

$$\begin{aligned} \Phi_k(z) &= z^{\lambda+k} \int_0^\infty t^{\lambda-1} \left(t - t_+\right)^k e^{-z(t+\beta^2/t)}\, dt \\ &= 2z^{\lambda+k} \beta^\lambda \sum_{m=0}^{k} \binom{k}{m} \left(-t_+\right)^{k-m} \beta^m K_{\lambda+m}(2\beta z). \end{aligned} \tag{27.5.94}$$

A recurrence relation for $\Phi_k(z)$ follows from the above integral representation, similar as in (27.3.28).

27.5.1 Expansion of $U(a,c,z)$, $a \to \infty$, $z > 0$, $c \le 1$

Our starting point is (cf. (27.4.29))

$$\Gamma(a)U(a, 1-\lambda, az) = \int_0^\infty e^{-a\psi(w)} \frac{dw}{1-e^{-w}}, \tag{27.5.95}$$

where

$$\psi(w) = -\mu \ln(e^w - 1) + (\mu+1)w + \frac{z}{e^w - 1}, \quad \mu = \frac{\lambda}{a}. \tag{27.5.96}$$

The real saddle points of the function ψ are

$$w_\pm = \ln\left(1 + \tfrac{1}{2}(\mu + z \pm W)\right), \quad W = \sqrt{(\mu+z)^2 + 4z}. \tag{27.5.97}$$

The transformation to the standard form (27.5.87) reads

$$\psi(w) = t + \frac{\beta^2}{t} - \mu \ln t + A, \tag{27.5.98}$$

and we use the corresponding saddle points $w_\pm$ and $t_\pm$ (see (27.5.88)) to compute A and β. We assume that $\operatorname{sign}(w - w_+) = \operatorname{sign}(t - t_+)$.

Elimination of A from the two equations that follow from (27.5.98) (with $w = w_\pm, t = t_\pm$) gives a relation for the unknown parameter β in terms of μ and z:

$$(\mu+1)\ln\frac{2+\mu+z+W}{2+\mu+z-W} - \mu\ln\frac{W+\mu+z}{W-\mu-z} + W = 2T - \mu\ln\frac{T+\mu}{T-\mu}, \tag{27.5.99}$$

where T is defined in (27.5.88).

By considering $\mu \in [0,\infty)$ as a fixed parameter, we obtain a more transparent relation for $\beta(z)$ in the form of a differential equation:

$$\frac{d\beta(z)}{dz} = \frac{\beta W}{2zT}, \quad \beta(0) = 0. \tag{27.5.100}$$

Observe that $\beta = 0$ is a solution, but we need the one that satisfies

$$\beta^2(z) \sim (1+\mu)^{(1+\mu)/\mu}\frac{z}{e}, \quad z \downarrow 0, \tag{27.5.101}$$

which follows from (27.5.99). The function $y(z) = \beta^2(z)/z$ is analytic at the origin and satisfies

$$\frac{dy(z)}{dz} = \frac{W-T}{zT}y(z), \quad y(0) = (1+\mu)^{(1+\mu)/\mu}e^{-1}. \tag{27.5.102}$$

The value of A follows from (27.5.98) by substituting $w = w_+, t = t_+$. We have

$$A = (\mu+1)w_+ - \mu\ln\frac{\mu+z+W}{\mu+T} - \tfrac{1}{2}(\mu+z-W) - T. \tag{27.5.103}$$

Using (27.5.99), we can eliminate $\frac{1}{2}W - T$ and we obtain

$$A = \tfrac{1}{2}\left((\mu+1)\ln(\mu+1) + \mu\ln\left(\beta^2/z\right) - \mu - z\right). \tag{27.5.104}$$

Using the transformation (27.5.98) in (27.5.95) we obtain the standard form (27.5.87):

$$e^{aA}\Gamma(a)\,U(a,1-\lambda,az)=\int_0^\infty t^{\lambda-1}e^{-a(t+\beta^2/t)}\,f(t)\,dt, \tag{27.5.105}$$

where β^2 follows from (27.5.99) with $\mu=\lambda/a$. Furthermore,

$$\begin{aligned} f(t)&=\frac{t}{1-e^{-w}}\frac{dw}{dt}\\ &=\frac{e^w(e^w-1)}{t}\frac{t^2-\mu t-\beta^2}{(e^w-1)^2-(\mu+z)(e^w-1)-z}. \end{aligned} \tag{27.5.106}$$

Using (27.5.90) we obtain for the U-function

$$U(a,c,az)\sim 2\beta^{1-c}\frac{e^{-aA}}{\Gamma(a)}\left(K_{c-1}(\zeta)\sum_{k=0}^{\infty}\frac{c_k}{a^k}+\beta K_{c-2}(\zeta)\sum_{k=0}^{\infty}\frac{d_k}{a^k}\right), \tag{27.5.107}$$

where $\zeta=2\beta a$, A is defined in (27.5.104), β follows from (27.5.99), $\mu=(1-c)/a$, and the coefficients c_k, d_k follow from the scheme given in (27.5.91) with f given in (27.5.106).

This expansion is derived for $c\le 1$ and $z\ge 0$, both on unbounded intervals. By using the transformation formula in (10.1.10) it follows that the expansion in (27.5.107) can also be used for $U(1+a-c,2-c,z)$, $c\ge 1$.

A few computations give

$$f\left(t_\pm\right)=e^{\frac12 w_\pm}\sqrt{T/W}=e^{\frac12 w_\pm}\left(\frac{\mu^2+4\beta^2}{(\mu+z)^2+4z}\right)^{\frac14}, \tag{27.5.108}$$

and by using (27.5.91) the first coefficients c_0 and d_0 can be computed with these values.

By using the recurrence relation

$$U(a,c,z)=aU(a+1,c,z)+U(a,c-1,z), \tag{27.5.109}$$

we have numerically verified the first-order approximation by taking in the series in (27.5.107) the first terms c_0 and d_0. We took $a=100$, $c=-10,-20,\ldots,-200$ and $z=10,20,\ldots,200$ and found a maximal relative error 0.47e-6 at $z=10$, $c=-170$.

27.5.2 *Expansion of* ${}_1F_1(a;c;z)$, $a\to\infty$, $z\ge 0$, $c\ge 1$

For the F-function we start with (10.3.54), and write it in the form

$$e^{-az}{}_1F_1\left({a+c\atop c};az\right)=\frac{\Gamma(c)\Gamma(1+a)}{2\pi i\,\Gamma(a+c)}\int_{\mathcal{L}}e^{a\psi(w)}\frac{dw}{1-e^{-w}}, \tag{27.5.110}$$

where $\psi(w)$ is defined in (27.5.96), now with $\mu=(c-1)/a$.

The transformation in (27.5.98) gives

$$\begin{aligned} {}_1F_1\left({a+c\atop c};az\right)&=\frac{\Gamma(c)\Gamma(1+a)e^{az+aA}}{2\pi i\,\Gamma(a+c)}\times\\ &\qquad\int_{\mathcal{L}}e^{a(t+\beta^2/t)}t^{-\lambda-1}f(t)\,dt, \end{aligned} \tag{27.5.111}$$

where $f(t)$ is given in (27.5.106), and β can be obtained from (27.5.99).

Integrating by parts, using the scheme given in (27.5.91), we obtain

$$
\begin{aligned}
{}_1F_1\left({a+c \atop c}; az\right) \sim \beta^{1-c}\frac{\Gamma(c)\Gamma(1+a)e^{az+aA}}{\Gamma(a+c)} \times \\
\left(I_{c-1}(\zeta)\sum_{k=0}^{\infty}(-1)^k\frac{c_k}{a^k} + \beta I_{c-2}(\zeta)\sum_{k=0}^{\infty}(-1)^k\frac{d_k}{a^k}\right),
\end{aligned}
\tag{27.5.112}
$$

where A, ζ and the coefficients are as in (27.5.107), but $\mu = (c-1)/a$.

This expansion is derived for $a \to \infty$ with $c \geq 1$ and $z \geq 0$, both on unbounded intervals. Because of the transformation formula (10.1.9) written in the form

$$
e^{az}{}_1F_1\left({-a \atop c}; -az\right) = {}_1F_1\left({a+c \atop c}; az\right),
\tag{27.5.113}
$$

it follows that we can use the expansion in (27.5.112) also for the left-hand side of this relation.

For a numerical verification we have used the recurrence relation

$$
a\,{}_1F_1\left({a+1 \atop c}; z\right) = (1+a-c)\,{}_1F_1\left({a \atop c}; z\right) + (c-1)\,{}_1F_1\left({a \atop c-1}; z\right),
\tag{27.5.114}
$$

and by taking the series in (27.5.112) with the first terms c_0 and d_0 for $a = 100$, $c, z = 10, 20, \ldots, 200$ and found a maximal relative error 0.48e-6 at $z = 10$, $c = 170$.

Chapter 28

Expansions in terms of Kummer functions

The Kummer function ${}_1F_1$, or confluent hypergeometric function (see Chapter 10), can be defined as the limit

$$\lim_{b\to+\infty} {}_2F_1\left(\begin{matrix}a,b\\c\end{matrix};\frac{z}{b}\right) = {}_1F_1\left(\begin{matrix}a\\c\end{matrix};z\right). \tag{28.0.1}$$

In this chapter we give two examples where the Kummer functions ${}_1F_1(a;c;z)$ and $U(a,c,z)$ can be used as main terms in certain uniform expansions of the Gauss hypergeometric function. In Chapter 12 we have given asymptotic expansions of the ${}_2F_1$-function for a single large parameter b or c in terms of elementary functions. In the present cases the variable z plays an extra role.

28.1 Approximation in terms of the Kummer U-function

In Case 11 of Table 20.1 and in §2.7.2 we have mentioned a Laplace-type integral of the form

$$F_{\lambda,\nu}(\alpha,\omega) = \frac{1}{\Gamma(\lambda)}\int_0^\infty t^{\lambda-1}e^{-\omega t}f(t)\,\frac{dt}{(t+\alpha)^\nu}, \tag{28.1.2}$$

where $\Re\omega > 0$, $\Re\lambda > 0$, $|\mathrm{ph}\,\alpha| < \pi$. In this chapter we derive asymptotic expansions for large values of ω that are uniform with respect to small values of α. The special functions used in these expansions are Kummer U-functions.

As a corresponding loop integral we consider

$$G_{\lambda,\nu}(\alpha,\omega) = \frac{1}{2\pi i}\int_{-\infty}^{(-\alpha+,0+)} t^{\lambda-1}e^{\omega t}f(t)\,\frac{dt}{(t+\alpha)^\nu}, \quad \Re\omega > 0, \tag{28.1.3}$$

where we assume that the contour encloses the point $t = -\alpha$ and no singularities of f. Complex values of α can be considered as well, and for α the domain of interest is around the origin. We take the branch cuts from 0 to $-\infty$ and from $-\alpha$ to $-\infty$. The multivalued functions assume their principal values. When f is a constant, the loop integral becomes a ${}_1F_1$-function, see (10.1.7) and (28.1.26).

Similar asymptotic problems have been considered in Zil′bergleĭt (1976) for the following integrals

$$
\begin{aligned}
I(x,\lambda) &= \int_a^\infty g(t)(t-x)^{-\nu} e^{-\lambda t}\,dt, \quad x\in[\alpha,a],\\
K(x,\lambda) &= \int_a^x f(t)(x-t)^{-\nu} e^{-\lambda t}\,dt, \quad x\in[a,\beta],
\end{aligned}
\tag{28.1.4}
$$

and some of their generalizations. The first integral is expanded in terms of the Kummer U-function, and in the second one the ${}_1F_1$-function is used. The expansions are for large values of λ, $\Re\lambda \ge 0$, and uniform with respect to x in the indicated intervals.

28.1.1 *Constructing the expansions*

The easiest approach for (28.1.2) is to expand

$$
f(t) = \sum_{n=0}^\infty f_n t^n \quad \Longrightarrow \quad F_{\lambda,\nu}(\alpha,\omega) \sim \sum_{n=0}^\infty f_n \Psi_n,
\tag{28.1.5}
$$

where

$$
\Psi_n = \frac{1}{\Gamma(\lambda)} \int_0^\infty t^{n+\lambda-1}(t+\alpha)^{-\nu} e^{-\omega t}\,dt.
\tag{28.1.6}
$$

The functions Ψ_n can be expressed in terms of the Kummer U-function, see Chapter 10, with integral representation (10.1.5).

We have, using these relation in (10.1.10),

$$
\begin{aligned}
\Psi_n &= (\lambda)_n \alpha^{n+\lambda-\nu} U(n+\lambda, n+\lambda+1-\nu, \alpha\omega)\\
&= (\lambda)_n \omega^{\nu-n-\lambda} U(\nu, \nu+1-n-\lambda, \alpha\omega).
\end{aligned}
\tag{28.1.7}
$$

The functions Ψ_n can be obtained by using recurrence relations of the U-function or by integrating by parts. We have

$$
\omega\Psi_{n+1} = (n+\lambda-\nu-\alpha\omega)\Psi_n + \alpha(\lambda+n-1)\Psi_{n-1}.
\tag{28.1.8}
$$

For numerical aspects of such recursions, we refer to Gil *et al.* (2007, §4.5.1).

The expansion in (28.1.5) has an asymptotic character for large values of ω, uniformly with respect to α. The second line of (28.1.7) gives the integral representation

$$
\Psi_n = \frac{(\lambda)_n}{\omega^{n+\lambda-\nu}\Gamma(\nu)} \int_0^\infty t^{\nu-1}(1+t)^{-n-\lambda} e^{-\alpha\omega t}\,dt,
\tag{28.1.9}
$$

which shows that

$$
\Psi_n = \mathcal{O}\left(\omega^{-n-\lambda+\nu}\right), \quad \omega\to\infty,
\tag{28.1.10}
$$

uniformly with respect to $\alpha \ge 0$. This is true when n is large enough, that is, if $n > \Re(\nu-\lambda)$. Then convergence at infinity of the integral is guaranteed if $\alpha = 0$.

The approach described in (28.1.5) can also be used for the simpler cases when $\nu = 0, 1$ and/or $\lambda = 1$, which we have considered in §2.7.

Using integration by parts we can obtain an expansion in which only the functions Ψ_0 and Ψ_1 are present. The first step is writing

$$f(t) = a_0 + b_0 t + t(t+\alpha)g_0(t), \tag{28.1.11}$$

as we did on several other occasions for obtaining uniform expansions. In this way,

$$F_{\lambda,\nu}(\alpha,\omega) = a_0\Psi_0 + b_0\Psi_1 + \frac{1}{\Gamma(\lambda)}\int_0^\infty t^\lambda (t+\alpha)^{1-\nu} e^{-\omega t} g_0(t)\,dt, \tag{28.1.12}$$

Integration by parts gives

$$F_{\lambda,\nu}(\alpha,\omega) = a_0\Psi_0 + b_0\Psi_1 + \frac{1}{\omega\Gamma(\lambda)}\int_0^\infty t^{\lambda-1} (t+\alpha)^{-\nu} e^{-\omega t} f_1(t)\,dt, \tag{28.1.13}$$

where

$$f_1(t) = t(t+\alpha)g_0'(t) + (\alpha\lambda + (1-\nu+\lambda)t)g_0(t). \tag{28.1.14}$$

By repeating the steps, it follows that

$$F_{\lambda,\nu}(\alpha,\omega) \sim \Psi_0 \sum_{n=0}^\infty \frac{a_n}{\omega^n} + \Psi_1 \sum_{n=0}^\infty \frac{b_n}{\omega^n}, \tag{28.1.15}$$

where the coefficients a_n and b_n follow from the recursive scheme

$$\begin{aligned} f_n(t) &= a_n + b_n t + t(t+\alpha)g_n(t), \\ a_n &= f_n(0), \quad b_n = \frac{f_n(-\alpha) - f_n(0)}{-\alpha}, \\ f_{n+1}(t) &= t(t+\alpha)\,g_n'(t) + (\alpha\lambda + (1-\nu+\lambda)t)\,g_n(t), \end{aligned} \tag{28.1.16}$$

where $n = 0, 1, 2, \ldots$ and $f_0 = f$.

We can write the expansion in (28.1.15) with a remainder:

$$F_{\lambda,\nu}(\alpha,\omega) \sim \Psi_0 \sum_{n=0}^{N-1} \frac{a_n}{\omega^n} + \Psi_1 \sum_{n=0}^{N-1} \frac{b_n}{\omega^n} + \frac{1}{\omega^N} R_N, \quad N = 0, 1, 2, \ldots, \tag{28.1.17}$$

where

$$R_N = \frac{1}{\Gamma(\lambda)}\int_0^\infty t^{\lambda-1}(t+\alpha)^{-\nu} e^{-\omega t} f_N(t)\,dt. \tag{28.1.18}$$

When estimates are available in the form $|f_N(t)| \le M_N e^{\sigma t}$, σ and M_N not depending on α in some domain, it is possible to prove the asymptotic nature of the expansion in (28.1.15).

Remark 28.1. By applying the recurrence relation in (28.1.8), the expansion in (28.1.5) can be rearranged to obtain an expansion of the form shown in (28.1.15). However, in that way, a different expansion arises, because the coefficients a_n and b_n in (28.1.15) depend on function values of f and its derivatives at $t = 0$ and $t = -\alpha$. The coefficients f_n in (28.1.5) are obtained from $t = 0$. △

Remark 28.2. For the construction of the expansion in (28.1.17) we need conditions on the function f. For example, $f \in C^N[\beta, \infty)$, where $\beta < -\alpha$. For applications to special functions, for example to the Gauss hypergeometric functions in later sections, we assume that f is analytic in a domain $\mathcal{D}$ that contains the interval $[\beta, \infty)$, and that $|f(t)| \le Me^{\sigma t}$, $t \ge 0$, for some M and σ. △

Remark 28.3. The functions f_n defined by the scheme in (28.1.16) can be represented in terms of two-point Taylor expansions, and the coefficients a_n and b_n follow from these representations. For details we refer to Chapter 18, in particular to §18.4. We represent $f_0 = f$ in the form

$$f_0(t) = \sum_{k=0}^{\infty} \left(c_k^0 t + d_k^0 (t+\alpha)\right) w^k, \quad w = t(t+\alpha), \tag{28.1.19}$$

where the coefficients can be expressed in terms of the derivatives of the function f at $t = 0$ and $t = -\alpha$; see §18.3. The function $g_0 = g$ defined in (28.1.11) has the expansion

$$g_0(t) = \sum_{k=0}^{\infty} \left(c_{k+1}^0 t + d_{k+1}^0 (t+\alpha)\right) w^k. \tag{28.1.20}$$

By applying Lemmas 18.1 and 18.2, it follows that f_1 given in (28.1.14) satisfies

$$f_1(t) = \sum_{k=0}^{\infty} \left(c_k^1 t + d_k^1 (t+\alpha)\right) w^k, \tag{28.1.21}$$

where $c_0^1 = (q - p\alpha)c_1^0$, $d_0^1 = qd_1^0$, $p = 1 - \nu + \lambda$, $q = \alpha\lambda$, and for $k \ge 1$

$$\begin{aligned} c_k^1 &= (q - p\alpha)c_{k+1}^0 - \frac{(p + 2k - 1)\left(c_k^0 + d_k^0\right) + k\alpha^2 c_{k+1}^0}{\alpha}, \\ d_k^1 &= qd_{k+1}^0 + \frac{(p + 2k - 1)\left(c_k^0 + d_k^0\right) + k\alpha^2 d_{k+1}^0}{\alpha}. \end{aligned} \tag{28.1.22}$$

The coefficients for

$$f_n(t) = \sum_{k=0}^{\infty} \left(c_k^n t + d_k^n (t+\alpha)\right) w^k, \quad n \ge 2, \tag{28.1.23}$$

follow from iterating the formulas in (28.1.22). The coefficients a_n and b_n of the expansion in (28.1.17) follow from (28.1.16), and are given by

$$a_n = \alpha d_0^n, \quad b_n = c_0^n + d_0^n, \quad n = 0, 1, 2, \ldots. \tag{28.1.24}$$

△

28.1.2 Expansion for the loop integral

For the loop integral in (28.1.3) we define the functions

$$\Phi_n = \frac{1}{2\pi i}\int_{-\infty}^{(0+)} t^{n+\lambda-1} e^{\omega t}\,\frac{dt}{(t+\alpha)^{\nu}}, \quad n = 0, 1. \tag{28.1.25}$$

These functions can be written as (see (10.1.7))

$$\begin{aligned}\Phi_n &= \frac{\omega^{\nu-\lambda-n}}{\Gamma(\nu+1-\lambda-n)}\,{}_1F_1\left(\begin{matrix}\nu\\ \nu+1-\lambda-n\end{matrix};\,-\alpha\omega\right)\\ &= \frac{\omega^{\nu-\lambda-n}e^{-\alpha\omega}}{\Gamma(\nu+1-\lambda-n)}\,{}_1F_1\left(\begin{matrix}1-\lambda-n\\ \nu+1-\lambda-n\end{matrix};\,\alpha\omega\right),\end{aligned} \tag{28.1.26}$$

where for the second relation we have used (10.1.9).

Again, an expansion of the form (28.1.15) can be obtained by integrating by parts. This time we obtain

$$G_{\lambda,\nu}(\alpha,\omega) \sim \Phi_0 \sum_{n=0}^{\infty} (-1)^n \frac{a_n}{\omega^n} + \Phi_1 \sum_{n=0}^{\infty} (-1)^n \frac{b_n}{\omega^n}, \tag{28.1.27}$$

where a_n and b_n are the same as in (28.1.15).

28.2 The ${}_2F_1$-function, large c, in terms of U

We take the integral representation given in (12.1.8) of §12.1.1, and write it in the standard form of this chapter:

$$\begin{aligned}&(1+z)^a \frac{\Gamma(c+\omega-b)}{\Gamma(c+\omega)}\;{}_2F_1\left(\begin{matrix}a, b\\ c+\omega\end{matrix};\,-z\right) =\\ &\qquad \frac{1}{\Gamma(b)}\int_0^{\infty} t^{b-1} f(t) e^{-\omega t}\,\frac{dt}{(t+\alpha)^a},\end{aligned} \tag{28.2.28}$$

where

$$f(t) = \left(\frac{e^t-1}{t}\right)^{b-1} e^{(1+a-c)t}\left(\frac{e^t-e^{-\alpha}}{t+\alpha}\right)^{-a}, \quad \alpha = \ln\frac{z+1}{z}. \tag{28.2.29}$$

Constructing the expansions derived in §28.1 is straightforward. The function f is analytic inside the strip $|\Im t| < 2\pi$ (when $\alpha \geq 0$). For complex α the width of the strip depends on $\Im\alpha$, because of the singularities at $-\alpha \pm 2\pi i$. The function f and its successive descendants f_n can be bounded on $[0,\infty)$ as indicated below (28.1.18) with $\sigma > b - a$, uniformly with respect to $\alpha \geq 0$.

It follows that the expansion holds for $\omega \to \infty$, uniformly with respect to $\alpha \geq 0$, with bounded values of a, b and c. Of course, when α is bounded from 0 (that is, when z is bounded), there is no need for the uniform expansion, and we can use the results of §12.1.1.

The complete expansion reads (see (28.1.15))

$$(1+z)^a \frac{\Gamma(c+\omega-b)}{\Gamma(c+\omega)} {}_2F_1\left({a,b \atop c+\omega};-z\right) \sim \Psi_0 \sum_{n=0}^{\infty} \frac{a_n}{\omega^n} + \Psi_1 \sum_{n=0}^{\infty} \frac{b_n}{\omega^n}, \qquad (28.2.30)$$

where Ψ_n is given in (28.1.7), and the coefficients a_n and b_n can be obtained from the recursive scheme shown in (28.1.16), with f defined in (28.2.29). In Ψ_n and in the scheme we use $\lambda = b$, $\nu = a$, and $\alpha = \ln(1+1/z)$.

The first coefficients a_0 and b_0 follow from

$$f(0) = \left(\frac{1-e^{-\alpha}}{\alpha}\right)^{-a}, \quad f(-\alpha) = \left(\frac{1-e^{-\alpha}}{\alpha}\right)^{b-1} e^{(c-1)\alpha}, \qquad (28.2.31)$$

and the first-order approximation reads

$$\begin{aligned} &(1+z)^a \frac{\Gamma(c+\omega-b)}{\Gamma(c+\omega)} {}_2F_1\left({a,b \atop c+\omega};-z\right) \sim \\ &\quad \alpha^{b-a}\left(a_0\, U(b,b+1-a,\alpha\omega) + \alpha b\, b_0\, U(b+1,b+2-a,\alpha\omega)\right). \end{aligned} \qquad (28.2.32)$$

When we have a set of derivatives of the function f at $t=0$ and $t=-\alpha$, we can generate more coefficients; see Remark 28.3.

Remark 28.4. This case has been considered in Farid Khwaja and Olde Daalhuis (2013, 2014), where an expansion is given for complex ω inside the sector $|\mathrm{ph}\,\omega| \le \frac{1}{2}\pi$, uniformly for large $|z|$, provided with estimates for the remainder. The main terms in the approximation are the same Kummer functions as in our (28.2.32). $\triangle$

28.2.1 *Legendre polynomials: Uniform expansions*

In §15.2 we have considered the large-n expansion of the Legendre polynomial, and we have observed that for small values of $1-x^2$ we need a uniform expansion. We see in (15.2.20) a ${}_2F_1$-function with the large parameter n at the right place for the present analysis. But we return to the representation in (15.2.14), now with $n+\frac{1}{2}$ as the large parameter. This gives simpler forms and expansions related to those in the literature.

We write

$$P_n^+(\cos\theta) = \sqrt{1-e^{-2i\theta}}\, \frac{e^{-i\theta\omega+\frac{1}{4}\pi i}}{\pi\sqrt{2\sin\theta}} \int_0^{\infty} e^{-\omega s} f(s)\, \frac{ds}{\sqrt{s(s+2i\theta)}}, \qquad (28.2.33)$$

where

$$\omega = n+\tfrac{1}{2}, \quad f(s) = e^{\frac{1}{2}s}\sqrt{\frac{s}{e^s-1}\,\frac{s+2i\theta}{e^s-e^{-2i\theta}}}. \qquad (28.2.34)$$

We recall that $P_n(\cos\theta) = 2\Re P_n^+(\cos\theta)$, and we concentrate on $\theta \in [0, \pi-\theta_0]$, where θ_0 is a fixed positive small number. For values of θ near π we can use the relation $P_n(\cos(\pi-\theta)) = (-1)^n P_n(\cos\theta)$, that is, $P_n(-x) = (-1)^n P_n(x)$.

Using the standard form in (28.1.2) and the expansion in (28.1.15), we obtain

$$P_n^+(\cos\theta) \sim \frac{i\,e^{-(n+1)i\theta}}{\sqrt{\pi}}\left(\Psi_0\sum_{k=0}^{\infty}\frac{a_k}{\omega^k}+\Psi_1\sum_{k=0}^{\infty}\frac{b_k}{\omega^k}\right), \tag{28.2.35}$$

where (see (28.1.7), (10.1.13) and the connection formulas for the Bessel functions in §9.2)

$$\begin{aligned}\Psi_0 &= U\left(\tfrac{1}{2},1,2i\theta\omega\right) = -\tfrac{1}{2}i\sqrt{\pi}\,e^{i\theta\omega}\left(J_0\left(\theta\omega\right)-iY_0\left(\theta\omega\right)\right),\\ \Psi_1 &= -\frac{d}{d\omega}\Psi_0 = -i\theta\Psi_0+\tfrac{1}{2}i\theta\sqrt{\pi}\,e^{i\theta\omega}\left(J_0'\left(\theta\omega\right)-iY_0'\left(\theta\omega\right)\right),\end{aligned} \tag{28.2.36}$$

and a_k and b_k follow from the scheme given in (28.1.16).

We have

$$a_0 = e^{\frac{1}{2}i\theta}\sqrt{\frac{\theta}{\sin\theta}},\quad b_0 = 0. \tag{28.2.37}$$

This gives the first-order approximation, called Hilb's formula, see Szegő (1975, §8.21),

$$P_n(\cos\theta) = \sqrt{\frac{\theta}{\sin\theta}}\,J_0(\theta\omega)+\mathcal{O}\left(n^{-1}\right),\quad n\to\infty, \tag{28.2.38}$$

uniformly for $\theta\in[0,\pi-\theta_0]$.

Remark 28.5. After computing the coefficients a_k and b_k and taking twice the real part of the expansion in (28.2.35), an expansion of the Legendre polynomials can be obtained in which only the J-Bessel functions are present. In fact, a more general result is available (see Koornwinder *et al.* (2010, Eq. 18.15.6)) for the Jacobi polynomials. △

Remark 28.6. In Chapter 29 we will give results for the Legendre functions $P_\nu^\mu(z)$, from which expansions of the Legendre polynomials $P_n(z)$ follow as a special case. We will use the method of §28.3, which gives expansions in terms of the ${}_1F_1$-functions, and which immediately yield expansions in terms of J-Bessel functions. △

28.3 The ${}_2F_1$-function, large b, in terms of ${}_1F_1$

In §12.2.2 we have seen how the limit in (28.0.1) follows from the large b asymptotic approximation with $z=\mathcal{O}(1/b)$. In this section we explain two methods for obtaining uniform expansions that are valid for large b with z in a domain containing the origin as an interior point. We exclude a fixed neighborhood of the point $z=1$; that is, we assume $|1-z|\ge z_0>0$.

28.3.1 *Using a real integral*

We give an expansion of ${}_2F_1(a,b;c,z)$ for $b \to -\infty$.[1] By using the first connection formula in (12.0.6) we can obtain a similar expansion for $b \to +\infty$.

We use the integral representation given in (12.0.3) in the form

$$ {}_2F_1\left({a, -b \atop c} ; z \right) = \frac{\Gamma(c)}{\Gamma(a)\,\Gamma(c-a)} \int_0^1 t^{a-1}(1-t)^{c-a-1}(1-zt)^b\, dt. \qquad (28.3.39)$$

We take a new variable of integration s defined by

$$1 - zt = (1-z)^s = e^{s\ln(1-z)}, \quad \text{or} \quad s = \frac{\ln(1-zt)}{\ln(1-z)}, \qquad (28.3.40)$$

with $s = t$ if $z = 0$, and

$$\frac{dt}{ds} = \frac{\ln(1-z)}{-z}(1-z)^s. \qquad (28.3.41)$$

Here, $\ln(1-z)$ has the principal branch, which is real for $z < 1$.

Using the transformation defined in (28.3.40) we obtain

$$ {}_2F_1\left({a, -b \atop c} ; z \right) = \frac{\Gamma(c)}{\Gamma(a)\,\Gamma(c-a)} \int_0^1 F(s) s^{a-1}(1-s)^{c-a-1}(1-z)^{(b+1)s}\, ds, \qquad (28.3.42)$$

where

$$F(s) = \frac{\ln(1-z)}{-z}\left(\frac{t}{s}\right)^{a-1}\left(\frac{1-t}{1-s}\right)^{c-a-1}. \qquad (28.3.43)$$

We have

$$\frac{t}{s} = \frac{\ln(1-z)}{-z}\,\frac{e^{s\ln(1-z)}-1}{s\ln(1-z)}, \qquad (28.3.44)$$

$$\frac{1-t}{1-s} = \frac{(1-z)\ln(1-z)}{-z}\,\frac{e^{(s-1)\ln(1-z)}-1}{(s-1)\ln(1-z)}. \qquad (28.3.45)$$

It follows that

$$F(s) = (1-z)^{c-a-1}\left(\frac{\ln(1-z)}{-z}\right)^{c-1}\phi(\alpha s)\psi\left(\alpha(s-1)\right), \qquad (28.3.46)$$

where

$$\phi(w) = \left(\frac{e^w-1}{w}\right)^{a-1}, \quad \psi(w) = \left(\frac{e^w-1}{w}\right)^{c-a-1}, \quad \alpha = \ln(1-z). \qquad (28.3.47)$$

The functions ϕ and ψ are singular at the points $w_k = 2k\pi i$, $k \in \mathbb{Z}\setminus\{0\}$. This gives for $F(s)$ the singularities

$$\frac{2k\pi i}{\ln(1-z)}, \quad 1 + \frac{2\ell\pi i}{\ln(1-z)}, \quad k,\ell \in \mathbb{Z}\setminus\{0\}. \qquad (28.3.48)$$

---

[1]This section is based on Temme (1986c).

Hence, when z ranges through compact subsets of $\mathbb{C}\setminus\{0\}$, the singularities of F are bounded away from the interval $[0,1]$.

We write

$$F(s) = (1-z)^{c-a-1}\left(\frac{\ln(1-z)}{-z}\right)^{c-1} f(s), \tag{28.3.49}$$

$$f(s) = \phi(\alpha s)\psi\left(\alpha(s-1)\right), \tag{28.3.50}$$

and

$$ {}_2F_1\left({a,-b \atop c};z\right) = (1-z)^{c-a-1}\left(\frac{\ln(1-z)}{-z}\right)^{c-1} F_b(a,c,\omega), \tag{28.3.51}$$

with

$$F_b(a,c,\omega) = \frac{\Gamma(c)}{\Gamma(a)\Gamma(c-a)}\int_0^1 f(s)s^{a-1}(1-s)^{c-a-1}e^{\omega s}\,ds, \tag{28.3.52}$$

and

$$\omega = (b+1)\ln(1-z). \tag{28.3.53}$$

28.3.1.1 *The uniformity aspects*

Before constructing the uniform asymptotic expansion, we first remark that from (28.3.52) simpler non-uniform expansions can be obtained for two separate cases:

(*i*) $\Re\omega \to -\infty$.
The dominant point in the integral in (28.3.52) is $s=0$, and an expansion follows by expanding

$$(1-s)^{c-a-1}f(s) = \sum_{n=0}^{\infty} c_n s^n. \tag{28.3.54}$$

The result is

$$F_b(a,c,\omega) \sim \frac{\Gamma(c)}{\Gamma(c-a)}\sum_{n=0}^{\infty} c_n \frac{(a)_n}{(-\omega)^{n+a}}, \quad \Re(c-a) > 0. \tag{28.3.55}$$

(*ii*) $\Re\omega \to +\infty$.
The dominant point in the integral in (28.3.52) is $s=1$, and an expansion follows by expanding

$$s^{a-1}f(s) = \sum_{n=0}^{\infty} d_n(s-1)^n. \tag{28.3.56}$$

The result is

$$F_b(a,c,\omega) \sim \frac{\Gamma(c)}{\Gamma(a)}e^{\omega}\sum_{n=0}^{\infty}(-1)^n d_n \frac{(c-a)_n}{\omega^{n+c-a}}, \quad \Re a > 0. \tag{28.3.57}$$

In the uniform expansion, contributions from both $s=0$ and $s=1$ will be taken into account. In this way we can allow $\Re\omega$ to be negative as well as positive; even $\omega = 0$ is accepted.

Observe that $\Re\omega > 0$, $\Re\omega < 0$ is equivalent to $|1-z| > 1$, $|1-z| < 1$, respectively, so that in fact all points on the unit circle around $z=1$ are "transition" points; that is, points for which the asymptotic behavior of the Gauss function will change drastically as $b \to -\infty$.

28.3.1.2 *Constructing the expansion*

We use an expansion of $f(s)$ with interpolating points $s = 0$ and $s = 1$ by writing

$$f(s) = a_0 + b_0 s + s(1-s)g_0(s), \tag{28.3.58}$$

with

$$a_0 = f(0), \quad b_0 = f(1) - f(0). \tag{28.3.59}$$

Then we obtain

$$F_b(a, c, \omega) = a_0\, {}_1F_1\left(\begin{matrix} a \\ c \end{matrix}; \omega\right) + b_0 \frac{d}{d\omega} {}_1F_1\left(\begin{matrix} a \\ c \end{matrix}; \omega\right) + R_1. \tag{28.3.60}$$

If $\omega \neq 0$, integration by parts gives for R_1

$$\begin{aligned} R_1 &= \frac{\Gamma(c)}{\omega\Gamma(a)\Gamma(c-a)} \int_0^1 g_0(s) s^a (1-s)^{c-a}\, d\, e^{\omega s} \\ &= \frac{\Gamma(c)}{\omega\Gamma(a)\Gamma(c-a)} \int_0^1 f_1(s) s^{a-1} (1-s)^{c-a-1} e^{\omega s}\, ds, \end{aligned} \tag{28.3.61}$$

with

$$f_1(s) = (cs - a)g_0(s) + (s-1)s g_0'(s). \tag{28.3.62}$$

The procedure can be continued and we obtain

$$F_b(a, c, \omega) = {}_1F_1\left(\begin{matrix} a \\ c \end{matrix}; \omega\right) \sum_{m=0}^{M-1} \frac{a_m}{\omega^m} + \frac{d}{d\omega} {}_1F_1\left(\begin{matrix} a \\ c \end{matrix}; \omega\right) \sum_{m=0}^{M-1} \frac{b_m}{\omega^m} + R_M, \tag{28.3.63}$$

with $M \geq 0$, $R_0 = F_b(a, c, \omega)$,

$$a_m = f_m(0), \quad b_m = f_m(1) - f_m(0), \quad f_0 = f, \tag{28.3.64}$$

$$\begin{aligned} f_m(s) &= (cs - a)g_{m-1}(s) + (s-1)s g_{m-1}'(s) \\ &= a_m + b_m s + s(1-s)g_m(s), \end{aligned} \tag{28.3.65}$$

$m = 0, 1, 2, \ldots, M-1$, and

$$R_M = \frac{\Gamma(c)}{\omega^M \Gamma(a)\Gamma(c-a)} \int_0^1 f_M(s) s^{a-1} (1-s)^{c-a-1} e^{\omega s}\, ds. \tag{28.3.66}$$

28.3.1.3 *On the regularity of the coefficients at $z = 0$*

When starting the integration by parts procedure in (28.3.61) we have assumed $\omega \neq 0$. However, we can show that the coefficients a_m and b_m in (28.3.63) and the functions $f_m(s)$ generated by the scheme in (28.3.65) can be rescaled in the form

$$A_m = \frac{a_m}{\alpha^m}, \quad B_m = \frac{b_m}{\alpha^m}, \quad F_m(s) = \frac{f_m(s)}{\alpha^m}, \tag{28.3.67}$$

with α as in (28.3.47), and $A_m, B_m, F_m(s)$ are analytic functions at $\alpha = 0$.

This follows from writing in (28.3.52) $s = u/\alpha$ (initially assume $\alpha > 0$). This gives

$$F_b(a,c,\omega) = \frac{\alpha^{1-c}\Gamma(c)}{\Gamma(a)\Gamma(c-a)} \int_0^\alpha \chi(u) u^{a-1} (\alpha - u)^{c-a-1} e^{(b+1)u}\, du, \tag{28.3.68}$$

where

$$\chi(u) = \phi(u)\psi(u-\alpha), \tag{28.3.69}$$

with ϕ, ψ defined in (28.3.47). We can repeat the integration by parts procedure, now by using interpolating points $u = 0$ and $u = \alpha$. This gives an expansion in negative powers of $(b+1)$.

For complex values of α with $\Re\alpha > 0$ this method can also be used; when $\Re\alpha < 0$ we can substitute in (28.3.52) $s = -v/\alpha$.

28.3.1.4 *A class of polynomials biorthogonal on the unit circle*

In Temme (1986c) the uniform asymptotic expansion of polynomials of the type

$$\begin{aligned} P_n(z;\alpha,\beta) &= {}_2F_1\left({-n, \alpha+\beta+1 \atop 2\alpha+1} ; 1-z \right), \\ Q_n(z;\alpha,\beta) &= P_n(z;\alpha,-\beta) \end{aligned} \tag{28.3.70}$$

has been considered.[2] These polynomials constitute a biorthogonal set on the unit circle. This means that there is a weight function, in this case

$$w(\theta) = \left(1 - e^{i\theta}\right)^{\alpha+\beta} \left(1 - e^{-i\theta}\right)^{\alpha-\beta}, \tag{28.3.71}$$

such that for $m, n \in \mathbb{N}_0$

$$\begin{aligned} &\frac{1}{2\pi} \int_{-\pi}^{\pi} P_m\left(e^{i\theta};\alpha,\beta\right) Q_n\left(e^{-i\theta};\alpha,\beta\right) w(\theta)\, d\theta = \\ &\qquad \frac{\Gamma(2\alpha+1)}{\Gamma(\alpha+\beta+1)\Gamma(\alpha-\beta+1)} \frac{n!}{(2\alpha+1)_n} \delta_{m,n}. \end{aligned} \tag{28.3.72}$$

Orthogonal polynomials on the unit circle (also called *Szegő polynomials*) are discussed in Ismail (2005, Chapter 8). In Temme (1986c) the relation to *Heisenberg polynomials* is explained. See also Liu *et al.* (2013), where a uniform asymptotic expansion of the Heisenberg polynomials is obtained by applying a uniform treatment of Darboux's method.

Applying the results of the previous sections to $P_n(z;\alpha,\beta)$ we now have $b = n$, $a = \alpha+\beta+1$, $c = 2\alpha+1$, and $\omega = (n+1)\ln z$. Thus in our uniform asymptotic expansion the Kummer function becomes ${}_1F_1(\alpha+\beta+1; 2\alpha+1; \omega)$. For the special case $\beta = -\frac{1}{2}$ the Kummer function is a Bessel function, see (10.1.13). In addition, the function $P_n\left(z;\alpha,\frac{1}{2}\right)$ is a Gegenbauer polynomial (see Koornwinder *et al.* (2010, Eq. 18.5.11))

$$P_n\left(e^{i\theta};\alpha,-\tfrac{1}{2}\right) = \frac{n!}{(2\alpha+1)_n} e^{\frac{1}{2}in\theta} C_n^{(\alpha+\frac{1}{2})}\left(\cos\tfrac{1}{2}\theta\right). \tag{28.3.73}$$

[2]This problem was suggested by Richard Askey.

In this way, the familiar large-n asymptotics of Gegenbauer polynomials in terms of Bessel functions follows, which holds uniformly in a neighborhood of $\theta = 0$, the endpoint of the interval of orthogonality of these polynomials. In fact, the result for the Gegenbauer polynomials is a special case of the result given for the general Gauss hypergeometric function.

Remark 28.7. Gegenbauer polynomials are special cases of the Jacobi polynomials. In Remark 28.5 we have given references for these polynomials. △

28.3.2 *Using a loop integral*

We use the loop integral given in (12.0.4) and use transformations as described in §10.3.3 for the ${}_1F_1$-function. This gives for $\Re b > 0$

$$\begin{aligned} {}_2F_1\left(\begin{matrix} a,b \\ c \end{matrix}; z\right) = \frac{\Gamma(c)\Gamma(1+b-c)}{2\pi i\, \Gamma(b)} \times \\ \int_{\mathcal{L}} e^{bt}\left(e^t-1\right)^{-c}\left(1-\frac{ze^t}{e^t-1}\right)^{-a} dt, \end{aligned} \tag{28.3.74}$$

where $\mathcal{L}$ is a Hankel contour as shown in Figure 2.1; it also encloses the point $t = -\ln(1-z)$. We write this in the form

$${}_2F_1\left(\begin{matrix} a,b \\ c \end{matrix}; z\right) = (1-z)^{-a}\frac{\Gamma(c)\Gamma(1+b-c)}{\Gamma(b)} G_{\lambda,\nu}(\alpha,\omega), \tag{28.3.75}$$

where $G_{\lambda,\nu}(\alpha,\omega)$ is the loop integral defined in (28.1.3), with $\lambda = a-c+1$, $\nu = a$, $\alpha = \ln(1-z)$, $\omega = b$, and

$$f(t) = \left(\frac{e^t-1}{t}\right)^{a-c}\left(\frac{e^t-e^{-\alpha}}{t+\alpha}\right)^{-a}. \tag{28.3.76}$$

In this case, the functions Φ_0, Φ_1 in the expansion (28.1.27) are given by

$$\begin{aligned} \Phi_n &= \frac{b^{c-1-n}e^{-\alpha b}}{\Gamma(c-n)}\, {}_1F_1\left(\begin{matrix} c-a-n \\ c-n \end{matrix}; \alpha b\right) \\ &= \frac{b^{c-1-n}}{\Gamma(c-n)}\, {}_1F_1\left(\begin{matrix} a \\ c-n \end{matrix}; -\alpha b\right). \end{aligned} \tag{28.3.77}$$

The coefficients follow from (28.1.16); a_0 and b_0 follow from

$$f(0) = \left(\frac{1-e^{-\alpha}}{\alpha}\right)^{-a}, \quad f(-\alpha) = e^{a\alpha}\left(\frac{1-e^{-\alpha}}{\alpha}\right)^{a-c}. \tag{28.3.78}$$

28.4 Jacobi polynomials of large degree: Laguerre-type expansion

Usually, the large degree expansions of Jacobi polynomials that are uniformly valid near $x = 1$ are given in terms of the J-Bessel functions; see, for example, Wong and Zhao (2003), with references to Szegő's work and with error bounds.

In this section we derive an expansion by using a transformation mentioned in Farid Khwaja and Olde Daalhuis (2014, §3.1.2), where the hypergeometric function ${}_2F_1\left(a+\lambda, a-\lambda; c; \frac{1}{2}-\frac{1}{2}z\right)$ is considered for large values of λ, with special attention to the behavior for $z \sim 1$. In general that transformation gives an expansion in terms of the F-Kummer functions, which for the special case considered can be approximated in terms of I-Bessel functions. The authors use a different transformation that leads to a Bessel-type expansion in one step.

From the representation of the Jacobi polynomial in (15.4.34), together with the relation $P_n^{(\beta,\alpha)}(-x) = (-1)^n P_n^{(\alpha,\beta)}(x)$, it follows that the large λ-problem is related to the large degree asymptotics of $P_n^{(\alpha,\beta)}(x)$, with special attention to $x \sim 1$. We show how the transformation suggested by Farid Khwaja and Olde Daalhuis can be used for the Jacobi polynomials, and we obtain an expansion in terms of Kummer functions in the form of a Laguerre polynomials.

The representation in (15.4.33) can be written in the form

$$P_n^{(\alpha,\beta)}(x) = \frac{(-1)^n}{2\pi i} \frac{2^{-n}}{(1-x)^\alpha(1+x)^\beta} \int_{\mathcal{C}} (1-z)^\alpha(1+z)^\beta e^{n\phi(z)} \frac{dz}{z-x}, \qquad (28.4.79)$$

where $\mathcal{C}$ is a simple closed contour encircling the point $z = x$, but not the points $z = \pm 1$, and

$$\phi(z) = \ln \frac{1-z^2}{z-x}. \qquad (28.4.80)$$

The saddle points are

$$z_\pm = x \pm i\sqrt{1-x^2} = e^{\pm i\theta}, \quad x = \cos\theta, \qquad (28.4.81)$$

and by using these values it follows that

$$\phi(z_\pm) = \ln 2 \pm i(\theta - \pi), \quad \phi''(z_\pm) = \frac{e^{\mp(\frac{1}{2}\pi+\theta)i}}{\sin\theta}. \qquad (28.4.82)$$

The function

$$\psi(w) = \ln \frac{1-w}{w-\xi} + w \qquad (28.4.83)$$

has saddle points at

$$w_\pm = \frac{1+\xi}{2} \pm i\sqrt{1-\left(\frac{1+\xi}{2}\right)^2} = e^{\pm i\sigma}, \quad \xi = 2\cos\sigma - 1, \qquad (28.4.84)$$

and by using these values it follows that

$$\psi(w_\pm) = \pm i(\sigma - \pi) + e^{\pm i\sigma}, \quad \psi''(w_\pm) = \mp i \cot\left(\tfrac{1}{2}\sigma\right). \qquad (28.4.85)$$

Observe that when $\xi \to 1$, $\sigma \to 0$, and that the saddle points in (28.4.84) tend to 1, the same as for the saddle points in (28.4.81) as $x \to 1$.

We transform

$$\phi(z) = \psi(w) + A, \qquad (28.4.86)$$

and A and ξ follow from the corresponding saddle points $z_\pm$ and $w_\pm$. This gives

$$A = \ln 2 - \cos\sigma, \quad \theta = \sigma + \sin\sigma, \tag{28.4.87}$$

and the latter relation defines $\xi = 2\cos\sigma - 1$.

The transformation in (28.4.86) leads to

$$P_n^{(\alpha,\beta)}(x) = \frac{(-1)^n}{2\pi i}\frac{e^{-n\cos\sigma}}{(1-x)^\alpha(1+x)^\beta}\int_{\mathcal{C}}(1-w)^{n+\alpha}e^{nw}f(w)\,\frac{dw}{(w-\xi)^{n+1}}, \tag{28.4.88}$$

where $\mathcal{C}$ is a small circle around $w = \xi$ and

$$f(w) = (1+z)^\beta\left(\frac{1-z}{1-w}\right)^\alpha \frac{w-\xi}{z-x}\frac{dz}{dw}. \tag{28.4.89}$$

We use Bleistein's procedure by writing

$$f(w) = a_0 + b_0(1-w) + (w-w_-)(w-w_+)g(w), \tag{28.4.90}$$

where $w_\pm$ are the saddle points defined in (28.4.84), and obtain after a few steps

$$P_n^{(\alpha,\beta)}(x) \sim \frac{(-1)^n\, e^{-n\cos\sigma}}{(1-x)^\alpha(1+x)^\beta}\left(\Phi_n^{(\alpha)}(\xi)\sum_{k=0}^{\infty}\frac{a_k}{n^k} + \Phi_n^{(\alpha+1)}(\xi)\sum_{k=0}^{\infty}\frac{b_k}{n^k}\right), \tag{28.4.91}$$

where

$$\Phi_n^{(\alpha)}(\xi) = \frac{1}{2\pi i}\int_{\mathcal{C}}(1-w)^{n+\alpha}e^{nw}\,\frac{dw}{(w-\xi)^{n+1}}, \tag{28.4.92}$$

and the coefficients follow from the recursive scheme

$$\begin{aligned}
&f_k(w) = a_k + b_k(1-w) + (w-w_-)(w-w_+)g_k(w),\\
&a_k = \frac{(1-w_-)f(w_+) - (1-w_+)f(w_-)}{w_+ - w_-}, \quad b_k = \frac{f(w_+) - f(w_-)}{w_- - w_+},\\
&f_{k+1}(w) = (w-\xi)(1-w)^{-\alpha}\frac{d}{dw}\left((1-w)^{\alpha+1}g_k(w)\right).
\end{aligned} \tag{28.4.93}$$

The function $\Phi_n^\alpha(\xi)$ can be written in terms of a Laguerre polynomial, which follows from the Rodrigues formula

$$L_n^{(\alpha)}(z) = \frac{z^{-\alpha}\,e^z}{n!}\frac{d^n}{dz^n}\left(e^{-z}z^{n+\alpha}\right). \tag{28.4.94}$$

This gives

$$L_n^{(\alpha)}(z) = \frac{z^{-\alpha}\,e^z}{2\pi i}\int_{\mathcal{C}} e^{-t}t^{n+\alpha}\frac{dt}{(t-z)^{n+1}}, \tag{28.4.95}$$

where $\mathcal{C}$ is a small circle around $t = z$. A few manipulations give

$$\Phi_n^\alpha(\xi) = (-1)^n e^{n\xi}(1-\xi)^\alpha L_n^{(\alpha)}\left(n(1-\xi)\right), \quad n = 0, 1, 2, \ldots, \tag{28.4.96}$$

and using the relation for ξ in (28.4.84) it follows that

$$\begin{aligned}
P_n^{(\alpha,\beta)}(x) \sim\ & \frac{e^{-\frac12 n(1-\xi)}}{(1+x)^\beta}\frac{(1-\xi)^\alpha}{(1-x)^\alpha}\times\\
&\left(L_n^{(\alpha)}((1-\xi)n)\sum_{k=0}^{\infty}\frac{a_k}{n^k} + (1-\xi)L_n^{(\alpha+1)}((1-\xi)n)\sum_{k=0}^{\infty}\frac{b_k}{n^k}\right).
\end{aligned} \tag{28.4.97}$$

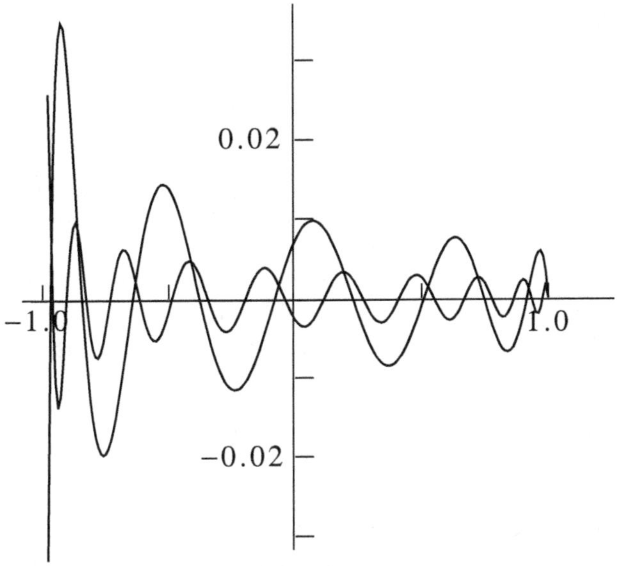

Fig. 28.1 Error curves for the approximation of $P_n^{(\alpha,\beta)}(x)$ by using (28.4.97) with the terms $k = 0$; $x \in (-0.98, 1]$, $\alpha = \frac{1}{3}$, $\beta = \frac{1}{5}$ and $n = 10$ (with the larger errors) and $n = 20$.

Here, $1 - \xi = 4\sin^2\left(\frac{1}{2}\sigma\right)$ with σ defined through the relation in (28.4.87).

In Figure 28.1 we show the error curves of the approximation of $P_n^{(\alpha,\beta)}(x)$ by using (28.4.102) with one term in both series. We took $x \in (-0.98, 1]$, $\alpha = \frac{1}{3}$, $\beta = \frac{1}{5}$ and $n = 10$ (with the larger errors) and $n = 20$.

We give some details for the evaluation of the first coefficients a_0, b_0. We need $f(w_\pm)$ and have (see (28.4.84) and (28.4.86))

$$\left.\frac{dz}{dw}\right|_{w=w_\pm} = \sqrt{\frac{\psi''(w_\pm)}{\phi''(z_\pm)}} = \sqrt{\sin\theta\cot\left(\tfrac{1}{2}\sigma\right)e^{\pm\frac{1}{2}i\theta}}. \tag{28.4.98}$$

This gives

$$f(w_\pm) = Re^{\pm i\chi}, \tag{28.4.99}$$

where

$$R = 2^{\beta+\frac{1}{2}}\cos^\beta\left(\tfrac{1}{2}\theta\right)\left(\frac{\sin\frac{1}{2}\theta}{\sin\frac{1}{2}\sigma}\right)^\alpha\sqrt{\frac{\sin\sigma}{\sin\theta}}, \quad \chi = \tfrac{1}{2}\beta\theta + \tfrac{1}{2}(\alpha+1)\sin\sigma. \tag{28.4.100}$$

The relation between θ and σ follows from (28.4.87). For the first coefficients we have

$$a_0 = \frac{\cos\left(\chi - \frac{1}{2}\sigma\right)}{\cos\left(\frac{1}{2}\sigma\right)}R, \quad b_0 = -\frac{\sin\chi}{\sin\sigma}R. \tag{28.4.101}$$

We combine R with the front factor in (28.4.97), introduce the notation $a_k^* = a_k/R$, $b_k^* = b_k/R$, and obtain

$$\begin{aligned} P_n^{(\alpha,\beta)}(x) \sim \frac{2^{\alpha+\frac12}e^{-\frac12 n(1-\xi)}}{\cos^\beta\left(\frac12\theta\right)}\left(\frac{\sin\frac12\sigma}{\sin\frac12\theta}\right)^\alpha\sqrt{\frac{\sin\sigma}{\sin\theta}}\times \\ \left(L_n^{(\alpha)}((1-\xi)n)\sum_{k=0}^{\infty}\frac{a_k^*}{n^k} + (1-\xi)L_n^{(\alpha+1)}((1-\xi)n)\sum_{k=0}^{\infty}\frac{b_k^*}{n^k}\right). \end{aligned} \tag{28.4.102}$$

Remark 28.8. Large-degree expansions of the Laguerre polynomials follow from those of the Kummer functions; see (10.1.17) and (27.4.71). The argument of the Bessel function $J_{c-1}(2\gamma a)$ in that result becomes $2\gamma n$, and using the notation of (27.4.68), we have

$$\gamma = \arcsin\left(\tfrac12\sqrt{z}\right) + \tfrac12\sqrt{z}\sqrt{1-z/4}, \quad z = 1-\xi = 4\sin^2\left(\tfrac12\sigma\right). \tag{28.4.103}$$

This gives $\gamma = \frac12(\sigma+\sin\sigma) = \frac12\theta$, see (28.4.87). So, the first Bessel function in (27.4.71) becomes $J_\alpha(n\gamma)$, as we usually see in approximations of the Jacobi polynomials for $x \sim 1$. A similar correspondence follows from the Frenzen–Wong expansion of $L_n^{(\alpha)}(x)$ given in §32.3.2, with a slightly different choice of the large parameter. △

28.4.1 *Laguerre-type expansion for large values of β*

A known limit is

$$\lim_{\beta\to\infty} P_n^{(\alpha,\beta)}\left(1-\frac{2x}{\beta}\right) = L_n^{(\alpha)}(x), \tag{28.4.104}$$

which shows the role of the Laguerre polynomial as approximant of the Jacobi polynomial for $x \sim 1$ as β becomes large. This limit does not follow from our expansion, because we concentrate on large n, albeit with special attention to $x \sim 1$.

The role of β can be included in the asymptotic analysis by modifying (28.4.79):

$$P_n^{(\alpha,\beta)}(x) = \frac{(-1)^n}{2\pi i}\frac{2^{-n}}{(1-x)^\alpha(1+x)^\beta}\int_{\mathcal{C}}(1-z)^\alpha e^{n\widetilde{\phi}(z)}\,\frac{dz}{z-x}, \tag{28.4.105}$$

where

$$\widetilde{\phi}(z) = (1+b)\ln(1+z) + \ln(1-z) - \ln(z-x), \quad \beta = bn. \tag{28.4.106}$$

This time the saddle points are

$$z_\pm = \frac{b+(b+2)x \pm i(b+2)\sqrt{(1-x)(x-x_0)}}{2(b+1)}, \quad x_0 = \frac{b^2-4b-4}{(b+2)^2}, \tag{28.4.107}$$

where we assume that $x_0 \le x \le 1$; x_0 is an increasing function from -1 when $b=0$ (as earlier) and $x_0 \to 1$ as $b \to +\infty$. Hence, for increasing b the turning points x_0 and 1 coalesce, and when we keep $x \in (x_0, 1)$ (the domain of the zeros of the Jacobi polynomial), the saddle points coalesce at 1. In Chapter 24 we have seen the role of the Hermite polynomials when turning points coalesce, but the present case is different, because in (28.4.105) the integrand is singular at $z = 1$, and Laguerre polynomials should be used. See also Temme (1990b, §4.2), where the first steps are given for this case when using the differential equation.

It is rather straightforward to repeat the previous approach with the new integral and function $\widetilde{\phi}(z)$ to obtain a large-n Laguerre-type expansion in which we can take $\beta \in [0, \infty)$. For $x \in [x_0, 1]$ it is convenient to write

$$x = \tfrac{1}{2}(1 - x_0)\cos\theta + \tfrac{1}{2}(1 + x_0), \quad \theta \in [0, \pi]. \tag{28.4.108}$$

Then we have (cf. (28.4.82))

$$z_\pm = \frac{b + 2e^{\pm i\theta}}{b+2}, \quad \widetilde{\phi}\,(z_\pm) = \ln 2 \pm i(\theta - \pi) + b\ln(1 + z_\pm), \tag{28.4.109}$$

and

$$\widehat{\phi}''\,(z_\pm) = Te^{\mp(\frac{1}{2}\pi+\lambda)i}, \tag{28.4.110}$$

where

$$\begin{aligned} T &= \frac{(b+1)(b+2)}{2(1+x)}\cot\tfrac{1}{2}\theta, \\ \lambda &= \arctan\frac{2\sin\theta(b+\cos\theta+1)}{\left((b+\cos\theta+1)^2 - \sin^2\theta\right)}. \end{aligned} \tag{28.4.111}$$

When the argument of the arctan–function is negative, we need to add π to the right-hand side.

We use the transformation as in (28.4.86) with the same $\psi(w)$ given in (28.4.83). The saddle points are given in (28.4.84) with the same notation $\xi = 2\cos\sigma - 1$. This time A and σ are given by

$$\begin{aligned} A &= \ln 2 - \cos\sigma + \tfrac{1}{2}b\ln(2(1+x)), \\ \theta + b\arctan&\frac{\sin\theta}{b+\cos\theta+1} = \sigma + \sin\sigma. \end{aligned} \tag{28.4.112}$$

When b is large, we have

$$\theta + b\arctan\frac{\sin\theta}{b+\cos\theta+1} = \theta + \sin\theta - \frac{\sin\theta(\cos\theta+1)}{b} + \mathcal{O}\left(b^{-2}\right), \tag{28.4.113}$$

and it follows that

$$\sigma = \theta - \frac{\sin\theta}{b} + \mathcal{O}\left(b^{-2}\right), \quad b \to \infty. \tag{28.4.114}$$

The transformation gives for (28.4.105)

$$P_n^{(\alpha,\beta)}(x) = \frac{(-1)^n}{2\pi i}\frac{2^{-n}e^{nA}}{(1-x)^\alpha(1+x)^\beta}\int_{\mathcal{C}}(1-w)^{n+\alpha}e^{nw}h(w)\,\frac{dw}{(w-\xi)^{n+1}}, \tag{28.4.115}$$

where

$$h(w) = \left(\frac{1-z}{1-w}\right)^\alpha \frac{w-\xi}{z-x}\frac{dz}{dw}. \tag{28.4.116}$$

Table 28.1 Relative errors in the approximation using the expansion in (28.4.117) with two terms c_0 and d_0.

n \ β	100	200	300	400	500
50	0.17e-3	0.32e-2	0.81e-3	0.56e-4	0.37e-4
100	0.12e-2	0.48e-3	0.71e-4	0.71e-4	0.14e-4
150	0.21e-3	0.55e-3	0.46e-3	0.17e-3	0.70e-4
200	0.12e-2	0.87e-3	0.14e-4	0.15e-3	0.14e-4
250	0.91e-3	0.40e-4	0.43e-3	0.31e-3	0.47e-4

We obtain the expansion

$$\begin{aligned} P_n^{(\alpha,\beta)}(x) \sim \frac{2^{\frac12\beta}e^{-\frac12 n(1-\xi)}}{(1+x)^{\frac12\beta}}\frac{(1-\xi)^\alpha}{(1-x)^\alpha} \times \\ \left(L_n^{(\alpha)}((1-\xi)n)\sum_{k=0}^{\infty}\frac{c_k}{n^k} + (1-\xi)L_n^{(\alpha+1)}((1-\xi)n)\sum_{k=0}^{\infty}\frac{d_k}{n^k}\right), \end{aligned} \tag{28.4.117}$$

where c_k and d_k follow from the recursive scheme (28.4.93) with $f(w)$ replaced by $h(w)$.

For the coefficients we need

$$\left.\frac{dz}{dw}\right|_{w=w_\pm} = \sqrt{\frac{\cot\left(\frac12\sigma\right)}{T}}\,e^{\pm\frac12 i\lambda}, \tag{28.4.118}$$

and the other components of $h\left(w_\pm\right)$ are

$$\begin{aligned} \frac{1-z_\pm}{1-w_\pm} &= \frac{2\sin\frac12\theta}{(b+2)\sin\frac12\sigma}e^{\pm\frac12 i(\theta-\sigma)}, \\ \frac{w_\pm-\xi}{z_\pm - x} &= \frac{b+2}{\sqrt{2(1+x)}}\frac{\sin\frac12\sigma}{\sin\frac12\theta}e^{\pm i\left(\arctan\left(\frac{b}{b+2}\tan\frac12\theta\right)-\frac12\sigma\right)}. \end{aligned} \tag{28.4.119}$$

This gives

$$h(w_\pm) = Re^{\pm i\chi}, \tag{28.4.120}$$

where

$$\begin{aligned} R &= \left(\frac{2\sin\frac12\theta}{(b+2)\sin\frac12\sigma}\right)^{\alpha-1}\frac{2}{\sqrt{2(1+x)}}\sqrt{\frac{\cot\left(\frac12\sigma\right)}{T}} \\ &= 2\left(\frac{2\sin\frac12\theta}{(b+2)\sin\frac12\sigma}\right)^{\alpha-1}\sqrt{\frac{\tan\frac12\theta\cot\frac12\sigma}{(b+1)(b+2)}}, \\ \chi &= \tfrac12\alpha(\theta-\sigma)+\tfrac12\lambda+\arctan\left(b\tan\tfrac12\theta\right)-\tfrac12\sigma. \end{aligned} \tag{28.4.121}$$

From these quantities we can compute the first coefficients (compare (28.4.101))

$$c_0 = \frac{\cos\left(\chi-\frac12\sigma\right)}{\cos\left(\frac12\sigma\right)}R, \quad d_0 = -\frac{\sin\chi}{\sin\sigma}R. \tag{28.4.122}$$

The relation between x and θ is given in (28.4.108).

In Table 28.1 we give the absolute relative errors in the computation of $P_n^{(\alpha,\beta)}(x)$ using the approximation (28.4.117) with two terms c_0 and d_0, and $\alpha = \frac{1}{3}$, $x = \frac{1}{2}(x_0+1)$, that is, $\theta = \frac{1}{2}\pi$.

28.5 Expansion of a Dirichlet-type integral

We consider the integral

$$D(x,\lambda) = \frac{1}{\pi}\int_0^\infty f(t)\frac{\sin(\lambda(t-x))}{t-x}\,dt, \quad x \ge 0, \tag{28.5.123}$$

for large positive values of λ and for continuous functions f that are absolutely integrable on $[0,\infty)$. This case is related to the problems considered in this chapter, although the form of the integral is different. The main approximant is the sine integral, which is a special case of the Kummer functions.

This integral has been considered in Zil′bergleĭt (1977), and we summarize the main result of this paper.

For fixed values of x we have the limits

$$\lim_{\lambda\to\infty} D(x,\lambda) = \begin{cases} f(x), & \text{if} \quad x > 0, \\ \frac{1}{2}f(+0), & \text{if} \quad x = 0. \end{cases} \tag{28.5.124}$$

This non-uniform result can be extended by giving a complete asymptotic expansion of $D(x,\lambda)$, valid uniformly in $x \ge 0$.

The special function that takes care of the smooth transition $x = 0$ to $x > 0$ is the sine integral defined in (1.3.20) and (1.3.27). We have

$$D_0(x,\lambda) = \int_0^\infty \frac{\sin(\lambda(t-x))}{t-x}\,dt = \pi + \mathrm{si}(\lambda x) = \tfrac{1}{2}\pi + \mathrm{Si}(\lambda x). \tag{28.5.125}$$

We write

$$D(x,\lambda) = \frac{1}{\pi}f(x)D_0(x,\lambda) + \frac{1}{\pi}\int_0^\infty g(x,t)\sin(\lambda(t-x))\,dt, \tag{28.5.126}$$

where

$$g(x,t) = \frac{f(t)-f(x)}{t-x}. \tag{28.5.127}$$

The integral in (28.5.126) can be integrated by parts:

$$\begin{aligned} &\frac{1}{\pi}\int_0^\infty g(x,t)\sin(\lambda(t-x))\,dt = \\ &\quad \frac{\cos(\lambda x)}{\pi\lambda}\sum_{n=0}^{\lfloor\frac{1}{2}(N-1)\rfloor}\frac{(-1)^n g_{2n}(x,0)}{\lambda^{2n}} + \\ &\quad \frac{\sin(\lambda x)}{\pi\lambda^2}\sum_{n=0}^{\lfloor\frac{1}{2}(N-2)\rfloor}\frac{(-1)^n g_{2n+1}(x,0)}{\lambda^{2n}} + \frac{(-1)^{\lfloor\frac{1}{2}N\rfloor}}{\pi\lambda^N}R_N(x,\lambda), \end{aligned} \tag{28.5.128}$$

where $\lfloor\nu\rfloor$ is the integral part of the real number ν,

$$g_n(x,t) = \frac{d^n}{dt^n}g(x,t), \quad R_N(x,\lambda) = \int_0^\infty g_n(x,t)\,p_n\bigl(\lambda(t-x)\bigr)\,dt, \tag{28.5.129}$$

and

$$p_n(z) = \begin{cases} \sin z, & \text{if } n \text{ is even,} \\ \cos z, & \text{if } n \text{ is odd.} \end{cases} \tag{28.5.130}$$

In the integration by parts procedure it was assumed that for every fixed $x \ge 0$ and every $n = 0,1,2,\ldots,N-1$, we have

$$\lim_{t\to\infty} g_n(x,t) = 0, \quad \text{and} \quad f \in C^{N+1}[0,\infty). \tag{28.5.131}$$

Collecting the results, we have

$$\begin{aligned} D(x,\lambda) = \frac{1}{\pi}f(x)D_0(x,\lambda) + \frac{\cos(\lambda x)}{\pi\lambda}\sum_{n=0}^{\lfloor\frac12(N-1)\rfloor}\frac{(-1)^n g_{2n}(x,0)}{\lambda^{2n}} + \\ \frac{\sin(\lambda x)}{\pi\lambda^2}\sum_{n=0}^{\lfloor\frac12(N-2)\rfloor}\frac{(-1)^n g_{2n+1}(x,0)}{\lambda^{2n}} + \frac{(-1)^{\lfloor\frac12 N\rfloor}}{\pi\lambda^N}R_N(x,\lambda). \end{aligned} \tag{28.5.132}$$

Observe that the limit in (28.5.124) follows from this result, since

$$\mathrm{Si}(0) = 0, \quad \lim_{z\to+\infty}\mathrm{Si}(z) = \tfrac12\pi. \tag{28.5.133}$$

Regarding the asymptotic representation in (28.5.132) we have the following theorem.

Theorem 28.1. *Let* $f \in C^{N+1}[0,\infty)$, $N \ge 1$, *let the remainders* $R_N(x.\lambda)$ *in* (28.5.129) *exist, and let the following conditions be satisfied:*

- $\lim_{t\to\infty} t^{-1}f^{(n)}(t) = 0$, $n = 0,1,2,\ldots,N$.
- $\int_0^\infty F(t)\,dt < \infty$, *where* $F(t) = \sup_{x\ge0}|g_n(x,t)|$.

Then, as $\lambda \to \infty$, *the asymptotic formula in* (28.5.132) *holds and* $R_N(x,\lambda) = o(1)$, *uniformly for all* $x \ge 0$.

Proof. For the proof we refer to Zil′bergleĭt (1977). □

PART 6

Uniform Examples for Special Functions

Chapter 29

Legendre functions

Legendre functions are special cases of the Gauss hypergeometric functions, and we can use the many connection formulas for these functions. For certain combinations of the parameters the hypergeometric functions satisfy quadratic transformations, and Legendre functions are those special cases of the hypergeometric functions for which a quadratic transformation exists. This property gives special relations that we will use for asymptotic representations.

We consider the associated Legendre functions $P_\nu^\mu(z)$ and $Q_\nu^\mu(z)$, which are solutions of the differential equation

$$\left(1-z^2\right)\frac{d^2w}{dz^2}-2z\frac{dw}{dz}+\left(\nu(\nu+1)-\frac{\mu^2}{1-z^2}\right)w=0. \tag{29.0.1}$$

Examples of the relations with the hypergeometric functions are

$$P_\nu^\mu(z)=\frac{1}{\Gamma(1-\mu)}\left(\frac{z+1}{z-1}\right)^{\frac12\mu}{}_2F_1\left({-\nu,\nu+1 \atop 1-\mu};\tfrac12-\tfrac12 z\right), \tag{29.0.2}$$

and

$$\begin{aligned}Q_\nu^\mu(z)=&\frac{\sqrt{\pi}\,e^{\mu\pi i}\,\Gamma(\nu+\mu+1)\,\left(z^2-1\right)^{\frac12\mu}}{2^{\nu+1}\,z^{\nu+\mu+1}\,\Gamma\left(\nu+\frac32\right)}\times\\ &{}_2F_1\left({\frac12\nu+\frac12\mu+1,\frac12\nu+\frac12\mu+\frac12 \atop \nu+\frac32};z^{-2}\right).\end{aligned} \tag{29.0.3}$$

We concentrate on $z>1$, and the results can be extended into the complex plane as long as we take $|\mathrm{ph}(z-1)|\le\pi-\delta$. On the other hand, connection formulas can be used to write $P_\nu^\mu(-z)$ and $Q_\nu^\mu(-z)$ in terms of $P_\nu^\mu(z)$ and $Q_\nu^\mu(z)$.

In addition we consider the Legendre functions on the interval $(-1,1)$, and we cannot (as a rule) use the results for $P_\nu^\mu(z)$ and $Q_\nu^\mu(z)$. We write the functions on $(-1,1)$ as $\mathsf{P}_\nu^\mu(x)$ and $\mathsf{Q}_\nu^\mu(x)$, and, again, real representations are available in terms of hypergeometric functions. For example,

$$\mathsf{P}_\nu^\mu(x)=\frac{1}{\Gamma(1-\mu)}\left(\frac{1+x}{1-x}\right)^{\frac12\mu}{}_2F_1\left({-\nu,\nu+1 \atop 1-\mu};\tfrac12-\tfrac12 x\right),\quad -1<x<1. \tag{29.0.4}$$

The functions $\mathsf{P}_\nu^\mu(x)$ and $\mathsf{Q}_\nu^\mu(x)$ are called *Ferrers functions*. The definitions are

$$\begin{aligned}
\mathsf{P}_\nu^\mu(x) &= e^{\frac12\mu\pi i} P_\nu^\mu(x+i0) = e^{-\frac12\mu\pi i} P_\nu^\mu(x-i0),\\
\mathsf{P}_\nu^\mu(x) &= \tfrac12\left(e^{\frac12\mu\pi i} P_\nu^\mu(x+i0) + e^{-\frac12\mu\pi i} P_\nu^\mu(x-i0)\right),\\
\mathsf{P}_\nu^\mu(x) &= \frac{i}{\pi} e^{-\mu\pi i}\left(e^{-\frac12\mu\pi i} Q_\nu^\mu(x+i0) - e^{\frac12\mu\pi i} Q_\nu^\mu(x-i0)\right),\\
\mathsf{Q}_\nu^\mu(x) &= \tfrac12 e^{-\mu\pi i}\left(e^{-\frac12\mu\pi i} Q_\nu^\mu(x+i0) + e^{\frac12\mu\pi i} Q_\nu^\mu(x-i0)\right),
\end{aligned} \tag{29.0.5}$$

where $-1 < x < 1$ and $f(x \pm i0)$ is defined as $\lim_{\varepsilon\to 0} f(x \pm i\varepsilon)$.

For more detailed information on the Legendre functions we refer to Dunster (2010), also for uniform expansions for large parameters with references to methods based on differential equations.

29.1 Expansions of $P_\nu^\mu(z), Q_\nu^\mu(z)$; $\nu \to \infty$, $z \ge 1$

We consider two cases, one with $z \in [z_0, \infty)$ with $z_0 > 1$, a fixed number. For this case we use a simple approach by using suitable representations in terms of hypergeometric functions. In the other case we use uniform expansions in terms of Bessel functions, and these expansions correspond to the results given by Olver in Olver (1997, Chapter 12).

We concentrate on $\nu \to +\infty$, because the asymptotic behavior for $\nu \to -\infty$ can be obtained from

$$\begin{aligned}
P_{-\nu-1}^\mu(z) &= P_\nu^\mu(z),\\
Q_{-\nu-1}^\mu(z) &= Q_\nu^\mu(z) + e^{\pi i\mu}\cos(\nu\pi)\Gamma(\nu+\mu+1)\Gamma(\mu-\nu)P_\nu^{-\mu}(z).
\end{aligned} \tag{29.1.6}$$

29.1.1 *Expansions for $z \ge z_0 > 1$*

For $P_\nu^\mu(z)$ we use the representation (see Erdélyi *et al.* (1953–1955, Vol. I, p. 142, (21)))

$$\begin{aligned}
&\sqrt{2\pi}\,(z^2-1)^{\frac14}\frac{\Gamma\left(\nu+\frac32\right)}{\Gamma(\nu+\mu+1)} P_\nu^\mu(z) = \\
&\qquad\left(z+\sqrt{z^2-1}\right)^{\nu+\frac12} {}_2F_1\left(\begin{matrix}\frac12+\mu,\ \frac12-\mu\\ \nu+\frac32\end{matrix};\zeta_1\right) + \\
&\qquad ie^{-\mu\pi i}\left(z-\sqrt{z^2-1}\right)^{\nu+\frac12} {}_2F_1\left(\begin{matrix}\frac12+\mu,\ \frac12-\mu\\ \nu+\frac32\end{matrix};\zeta_2\right),
\end{aligned} \tag{29.1.7}$$

where

$$\zeta_1 = \frac{z+\sqrt{z^2-1}}{2\sqrt{z^2-1}}, \quad \zeta_2 = \frac{-z+\sqrt{z^2-1}}{2\sqrt{z^2-1}}. \tag{29.1.8}$$

For $Q_\nu^\mu(z)$ we use (see Erdélyi *et al.* (1953–1955, Vol. I, p. 137, (44)))

$$e^{-\pi i\mu} Q_\nu^\mu(z) = A_3\, {}_2F_1\left(\begin{matrix} \frac12+\mu, \frac12-\mu \\ \nu+\frac32 \end{matrix} ; \zeta_2\right), \tag{29.1.9}$$

where

$$A_3 = \sqrt{\tfrac12\pi}\,\left(z^2-1\right)^{-\frac14}\left(z+\sqrt{z^2-1}\right)^{-\nu-\frac12}\frac{\Gamma\left(\mu+\nu+1\right)}{\Gamma\left(\nu+\frac32\right)}. \tag{29.1.10}$$

With ν as the large parameter, and using the power series of the hypergeometric functions, we obtain asymptotic expansions, whether or not the series are convergent. Observe that the series with ζ_1 does not converge for $z>1$, whereas the one with ζ_2 converges if $z>\frac34\sqrt2$. We can also use the method based on Watson's lemma; see §12.1.1.

The expansions obtained from these representations are also valid for complex values of ν; see §12.1.1. Complex values of the fixed parameter μ and variable z (with z bounded away from 1); for proper conditions see Wagner (1988) or Olde Daalhuis (2010b, §15.12(ii)).

29.1.2 *Expansion in terms of modified Bessel functions*

We use the methods of Chapter 28 that we have applied to the Gauss hypergeometric functions. The expansions of this section are especially of interest when $z\downarrow 1$, although they can be used for all $z>1$.

29.1.2.1 *Expansion of $Q_\nu^\mu(z)$ in terms of K-Bessel functions*

We use the method of §28.1.1. We take the representation given in (29.1.9), and replace ζ_2 by $1/\eta$. We use the standard integral representation of the hypergeometric function, and obtain

$$\begin{aligned} e^{-\pi i\mu} Q_\nu^\mu(z) = {} & \sqrt{\tfrac12\pi}\,\left(z^2-1\right)^{-\frac14}\left(z+\sqrt{z^2-1}\right)^{-\nu-\frac12}\times \\ & \frac{1}{\Gamma\left(\frac12-\mu\right)}\int_0^1 t^{-\mu-\frac12}(1-t)^{\nu+\mu}(1-t/\eta)^{-\mu-\frac12}\,dt, \end{aligned} \tag{29.1.11}$$

where $\Re\mu<\frac12$ and

$$\eta = -2\sqrt{z^2-1}\left(z+\sqrt{z^2-1}\right). \tag{29.1.12}$$

After transforming $1-t\to e^{-t}$ and some extra manipulations, we obtain the representation

$$\begin{aligned} e^{-\pi i\mu} Q_\nu^\mu(z) = {} & \sqrt{\tfrac12\pi}\,\alpha^{\mu+\frac12}\left(z^2-1\right)^{-\frac14}\left(z+\sqrt{z^2-1}\right)^{-\nu-\frac12}\times \\ & \frac{1}{\Gamma\left(\frac12-\mu\right)}\int_0^\infty t^{-\mu-\frac12} f(t) e^{-\omega t}\,\frac{dt}{(t+\alpha)^{\frac12+\mu}}, \end{aligned} \tag{29.1.13}$$

where $\omega = \nu + \frac{1}{2}$, $\alpha = \ln(1-\eta)$, and

$$f(t) = \left(\frac{\sinh\left(\frac{1}{2}t\right)}{\frac{1}{2}t} \; \frac{\sinh\left(\frac{1}{2}(t+\alpha)\right)}{\frac{1}{2}(t+\alpha)} \; \frac{\frac{1}{2}\alpha}{\sinh\left(\frac{1}{2}\alpha\right)} \right)^{-\mu-\frac{1}{2}}. \tag{29.1.14}$$

This choice of ω and f gives a convenient form of the asymptotic expansion. We have $f(0) = f(-\alpha) = 1$, and for the derivatives $f^{(n)}(-\alpha) = (-1)^n f^{(n)}(0)$, $n = 1, 2, 3, \ldots$.

First we expand $f(t) = \sum_{k=0}^{\infty} p_k t^k$, and obtain the expansion

$$e^{-\pi i\mu} Q_\nu^\mu(z) \sim \sqrt{\tfrac{1}{2}\pi}\, \alpha^{\mu+\frac{1}{2}} \left(z^2-1\right)^{-\frac{1}{4}} \left(z + \sqrt{z^2-1}\right)^{-\nu-\frac{1}{2}} \sum_{n=0}^{\infty} p_k \Psi_k, \tag{29.1.15}$$

where

$$\Psi_k = \left(\tfrac{1}{2} - \mu\right)_k \omega^{2\mu-k} U\left(\tfrac{1}{2} + \mu, 1 + 2\mu - k, \alpha\omega\right). \tag{29.1.16}$$

The first coefficients are

$$p_0 = 1, \quad p_1 = \frac{1}{4\alpha\eta}(2\mu+1)(2\alpha - \alpha\eta + 2\eta). \tag{29.1.17}$$

The functions Ψ_k can be written in terms of the modified Bessel functions. The first one is

$$\Psi_0 = \frac{1}{\sqrt{\pi}} (\omega/\alpha)^\mu e^{\frac{1}{2}\alpha\omega} K_\mu\left(\tfrac{1}{2}\alpha\omega\right). \tag{29.1.18}$$

For Ψ_1 we have

$$\begin{aligned} \Psi_1 = -\frac{d}{d\omega}\Psi_0 &= -\frac{1}{\sqrt{\pi}}(\omega/\alpha)^\mu e^{\frac{1}{2}\alpha\omega} \times \\ &\quad \left(\left(\mu/\omega + \tfrac{1}{2}\alpha\right) K_\mu\left(\tfrac{1}{2}\alpha\omega\right) + \tfrac{1}{2}\alpha K_\mu'\left(\tfrac{1}{2}\alpha\omega\right) \right) \\ &= -\frac{\alpha}{2\sqrt{\pi}} (\omega/\alpha)^\mu e^{\frac{1}{2}\alpha\omega} \left(K_\mu\left(\tfrac{1}{2}\alpha\omega\right) - K_{\mu-1}\left(\tfrac{1}{2}\alpha\omega\right) \right). \end{aligned} \tag{29.1.19}$$

To evaluate other Ψ-functions we can use the recurrence relation given in (28.1.8).

An expansion with only two Bessel functions can be derived as in (28.2.30). In the present case we have

$$\begin{aligned} e^{-\pi i\mu} Q_\nu^\mu(z) &\sim \sqrt{\alpha/2}\, \left(z^2-1\right)^{-\frac{1}{4}} \left(z + \sqrt{z^2-1}\right)^{-\nu-\frac{1}{2}} \omega^\mu e^{\frac{1}{2}\alpha\omega} \times \\ &\quad \left(K_\mu\left(\tfrac{1}{2}\alpha\omega\right) \sum_{k=0}^{\infty} \frac{A_{2k}}{\omega^{2k}} + K_{\mu-1}\left(\tfrac{1}{2}\alpha\omega\right) \sum_{k=0}^{\infty} \frac{B_{2k+1}}{\omega^{2k+1}} \right), \end{aligned} \tag{29.1.20}$$

with ω and α as in (29.1.13). The coefficients of this expansion follow from a recursive scheme given in (28.1.16) with $\lambda = \frac{1}{2} - \mu$, $\nu = \frac{1}{2} + \mu$, and $\alpha = \ln(1-\eta)$. This gives an expansion of the form (28.1.17) with coefficients a_k, b_k and a remainder given in (28.1.18). By using Ψ_1 given in (29.1.19), we obtain the expansion in (29.1.20) where $A_k = a_k - \frac{1}{2}\alpha b_k$, $B_k = \frac{1}{2}\alpha b_k$, that is,

$$A_k = \tfrac{1}{2}\left(f_k(0) + f_k(-\alpha)\right), \quad B_k = \tfrac{1}{2}\left(f_k(0) - f_k(-\alpha)\right). \tag{29.1.21}$$

Because of the special choice of ω and f in (29.1.13) and (29.1.14), we have indeed $A_{2k+1} = B_{2k} = 0$, $k = 0, 1, 2, \ldots$. See Remark 28.3 for the procedure to evaluate the coefficients by using two-point Taylor expansions.

The first coefficients are

$$\begin{aligned} A_0 &= 1, \quad B_1 = \frac{(4\mu^2-1)\left(\eta(\alpha-2)-2\alpha\right)}{8\alpha\eta}, \\ A_2 &= \frac{(4\mu^2-1)(2\mu-3)\left(p_2\eta^2+p_1\eta+p_0\right)}{384\alpha^2\eta^2}, \end{aligned} \tag{29.1.22}$$

where

$$\begin{aligned} p_0 &= 12\alpha^2(3+2\mu), \\ p_1 &= -12\alpha(3\alpha+2\mu\alpha-4\mu+2), \\ p_2 &= \alpha^2-60+6\mu\alpha^2-24\mu\alpha+24\mu+12\alpha. \end{aligned} \tag{29.1.23}$$

These and the higher coefficients are analytic at $\eta = 0$ (where also $\alpha = \ln(1-\eta)$ vanishes), which means that the coefficients are analytic at $z = 1$. We have

$$\begin{aligned} B_1 &= -\frac{(4\mu^2-1)}{96}\eta\left(1+2\eta+\mathcal{O}\left(\eta^2\right)\right), \\ A_2 &= \frac{(4\mu^2-1)(2\mu-3)}{23040}\eta^2\left(1+(10\mu+7)\eta+\mathcal{O}\left(\eta^2\right)\right), \end{aligned} \tag{29.1.24}$$

as $\eta \to 0$. In addition, the coefficients are bounded functions of z for $z \ge 1$.

The restriction $\Re\mu < \frac{1}{2}$ given after (29.1.11) is not needed in the expansion in (29.1.20). This follows from a similar discussion as given for Watson's lemma; see Remark 2.2.

29.1.2.2 *Expansion of $P_\nu^\mu(z)$ in terms of I-Bessel functions*

We derive an expansion valid for $z > 1$ using the methods of §28.3. We take the representation (see Erdélyi *et al.* (1953–1955, Vol. I, p. 131, (29)))

$$P_\nu^\mu(z) = A\,{}_2F_1\left({\frac{1}{2}-\mu, -\nu-\mu \atop 1-2\mu} ; \eta\right), \tag{29.1.25}$$

where η is defined in (29.1.12) and

$$A = 2^\mu (z^2-1)^{-\frac{1}{2}\mu}\left(z+\sqrt{z^2-1}\right)^{-\nu-\mu}\frac{1}{\Gamma(1-\mu)}. \tag{29.1.26}$$

We have the following expansion:

$$\begin{aligned} P_\nu^\mu(z) \sim \sqrt{\frac{\alpha}{-\eta}}\left(z+\sqrt{z^2-1}\right)^{-\nu}\omega^\mu e^{\frac{1}{2}\alpha\omega} \times \\ \left(I_{-\mu}\left(\tfrac{1}{2}\alpha\omega\right)\sum_{k=0}^\infty \frac{A_{2k}}{\omega^{2k}} - I_{1-\mu}\left(\tfrac{1}{2}\alpha\omega\right)\sum_{k=0}^\infty \frac{B_{2k+1}}{\omega^{2k+1}}\right), \end{aligned} \tag{29.1.27}$$

where α, ω, η, and the coefficients are the same as in (29.1.20).

This expansion can be derived by using the method of §28.3.1. The representation in (28.3.42) becomes (after minor modifications)

$$\begin{aligned} {}_2F_1\left({\tfrac12-\mu,-\nu-\mu \atop 1-2\mu};\eta\right) &= \left(\frac{\alpha}{-\eta}\right)^{\frac12-\mu}\times \\ &\frac{\Gamma(1-2\mu)}{\Gamma(\frac12-\mu)\,\Gamma(\frac12-\mu)}\int_0^1 \widetilde{f}(s)s^{-\frac12-\mu}(1-s)^{-\frac12-\mu}e^{\widetilde{\omega}s}\,ds, \end{aligned} \tag{29.1.28}$$

where $\Re\mu<\frac12$, $\widetilde{\omega}=\alpha\left(\nu+\frac12\right)$, $\alpha=\ln(1-\eta)$, and

$$\widetilde{f}(s)=\left(\frac{\sinh\left(\frac12\alpha s\right)}{\frac12\alpha s}\ \frac{\sinh\left(\frac12\alpha(s-1)\right)}{\frac12\alpha(s-1)}\ \frac{\frac12\alpha}{\sinh\left(\frac12\alpha\right)}\right)^{-\mu-\frac12}. \tag{29.1.29}$$

This function is the same as the one in (29.1.14), after some scaling. We have $\widetilde{f}(0)=\widetilde{f}(1)=1$. Furthermore, for the derivatives we have $\widetilde{f}^{(n)}(1)=(-1)^n\widetilde{f}^{(n)}(0)$, $n=0,1,2,\ldots$.

The expansion of the hypergeometric function becomes

$$\begin{aligned} {}_2F_1\left({\tfrac12-\mu,-\nu-\mu \atop 1-2\mu};\eta\right) &= \left(\frac{\alpha}{-\eta}\right)^{\frac12-\mu}\times \\ \left({}_1F_1\left({\tfrac12-\mu \atop 1-2\mu};\widetilde{\omega}\right)\sum_{m=0}^{M-1}\frac{\alpha_m}{\widetilde{\omega}^m}\right. &+ \left.\frac{d}{d\widetilde{\omega}}{}_1F_1\left({\tfrac12-\mu \atop 1-2\mu};\widetilde{\omega}\right)\sum_{m=0}^{M-1}\frac{\beta_m}{\widetilde{\omega}^m}+R_M\right), \end{aligned} \tag{29.1.30}$$

where

$$\begin{aligned} {}_1F_1\left({\tfrac12-\mu \atop 1-2\mu};\widetilde{\omega}\right) &= B\,I_{-\mu}\left(\tfrac12\widetilde{\omega}\right),\quad B=\Gamma(1-\mu)\left(\tfrac14\widetilde{\omega}\right)^{\mu}e^{\frac12\widetilde{\omega}}, \\ \frac{d}{d\widetilde{\omega}}{}_1F_1\left({\tfrac12-\mu \atop 1-2\mu};\widetilde{\omega}\right) &= \tfrac12 B\left(I_{-\mu}\left(\tfrac12\widetilde{\omega}\right)+I_{1-\mu}\left(\tfrac12\widetilde{\omega}\right)\right). \end{aligned} \tag{29.1.31}$$

This gives

$$\begin{aligned} {}_2F_1\left({\tfrac12-\mu,-\nu-\mu \atop 1-2\mu};\eta\right) &= \left(\frac{\alpha}{-\eta}\right)^{\frac12-\mu}\Gamma(1-\mu)\left(\tfrac14\widetilde{\omega}\right)^{\mu}e^{\frac12\widetilde{\omega}}\times \\ &\left(I_{-\mu}\left(\tfrac12\widetilde{\omega}\right)\sum_{m=0}^{M-1}\frac{a_m}{\widetilde{\omega}^m}+I_{1-\mu}\left(\tfrac12\widetilde{\omega}\right)\sum_{m=0}^{M-1}\frac{b_m}{\widetilde{\omega}^m}+R_M\right), \end{aligned} \tag{29.1.32}$$

where $a_m=\alpha_m+\frac12\beta_m$ and $b_m=\frac12\beta_m$.

The coefficients α_m and β_m follow from the recursive scheme given in (28.3.64)–(28.3.65), with $a=\frac12-\mu$, $c=1-2\mu$, and with f the function given in (29.1.29). The remainder R_M is as in (28.3.66).

After replacing in the expansion in (29.1.32) $\widetilde{\omega}$ by $\alpha\omega$, verifying that $a_{2m+1}=b_{2m}=0$, and that $a_{2m}/\alpha^{2m}=A_{2m}$, $b_{2m+1}/\alpha^{2m+1}=-B_{2m+1}$, we find the expansion in (29.1.27). For the regularity of the coefficients A_m and B_m near $\alpha=0$, we refer to §28.3.1.3. The first coefficients are given in (29.1.22). The restriction $\Re\mu<\frac12$ given after (29.1.28) is not needed in the expansion in (29.1.27).

29.1.3 Expansions of $\mathsf{P}_\nu^\mu(x)$ and $\mathsf{Q}_\nu^\mu(x)$ in terms of Bessel functions

We give expansions valid for $-1+\delta \le x < 1$ (with δ a small positive fixed number). First we use the result of §29.1.2.2. Taking the first relation in the first line of (29.0.5) and writing $x = \cos\theta$, $\theta \in [0,\pi]$, we obtain the representation

$$\mathsf{P}_\nu^\mu(\cos\theta) = A\,{}_2F_1\left({\frac12-\mu, -\nu-\mu \atop 1-2\mu} ; \eta\right), \tag{29.1.33}$$

where

$$\eta = 1-e^{2i\theta} = -2i\sin\theta e^{i\theta}, \quad A = 2^\mu (\sin\theta)^{-\mu} e^{-i(\nu+\mu)\theta}\frac{1}{\Gamma(1-\mu)}. \tag{29.1.34}$$

With this form of η we have $\alpha = \ln(1-\eta) = 2i\theta$. The right-hand side of (29.1.33) is real, which follows from applying the relation in the second line of (12.0.6), and from observing that we obtain the representation as above with i replaced by $-i$ in A and η.

Next we modify the expansion of $e^{\frac12\mu\pi i}P_\nu^\mu(x+i\varepsilon)$ (see the first line of (29.0.5)) by using the new relations for η and α. This gives the expansion

$$\mathsf{P}_\nu^\mu(\cos\theta) \sim \sqrt{\frac{\theta}{\sin\theta}}\omega^\mu \left(J_{-\mu}(\theta\omega)\sum_{k=0}^\infty \frac{A_{2k}}{\omega^{2k}} + J_{1-\mu}(\theta\omega)\sum_{k=0}^\infty \frac{\widetilde{B}_{2k+1}}{\omega^{2k+1}}\right), \tag{29.1.35}$$

where $\omega = \nu+\frac12$ and $\widetilde{B}_{2k+1} = -iB_{2k+1}$. The coefficients A_{2k} and $\widetilde{B}_{2k+1}$ are real for these values of η and α.

In a similar way we obtain in terms of the Y-Bessel function

$$\mathsf{Q}_\nu^\mu(\cos\theta) \sim -\tfrac12\pi\sqrt{\frac{\theta}{\sin\theta}}\omega^\mu \left(Y_{-\mu}(\theta\omega)\sum_{k=0}^\infty \frac{A_{2k}}{\omega^{2k}} + Y_{1-\mu}(\theta\omega)\sum_{k=0}^\infty \frac{\widetilde{B}_{2k+1}}{\omega^{2k+1}}\right), \tag{29.1.36}$$

where we have used several relations between the modified K-Bessel functions, the Hankel functions, and the ordinary Bessel functions (see §9.2).

In the present case the first coefficients are in real form

$$\begin{aligned} A_0 &= 1, \quad \widetilde{B}_1 = \frac{(4\mu^2-1)(\sin\theta-\theta\cos\theta)}{8\theta\sin\theta}, \\ A_2 &= -\frac{(4\mu^2-1)(2\mu-3)}{384\,\theta^2\sin^2\theta}\left(6\mu\left(\sin^2\theta+\theta^2\cos^2\theta\right) + \right. \\ &\qquad \left. \theta^2\left(8+\cos^2\theta\right) - 15\sin^2\theta + 3\theta\sin(2\theta)(1-2\mu)\right). \end{aligned} \tag{29.1.37}$$

The expansions on the interval $(-1,1)$ can also be derived by applying directly the methods of §28.2 and §28.3.1.

The expansion of $\mathsf{P}_\nu^\mu(x)$ corresponds to the one given in Olver (1997, Chapter 12, §12.4), where also an error bound for the remainder is given. Expansions of $\mathsf{P}_\nu^{-\mu}(x)$ and $\mathsf{Q}_\nu^{-\mu}(x)$ with both parameters μ and ν large are given in Olver (1975); the approximations are given in terms of parabolic cylinder functions. These results are mentioned in Dunster (2010, §14.15), together with more uniform estimates of the Ferrers functions.

29.2 Expansions of $P_\nu^\mu(z), Q_\nu^\mu(z)$; $\mu \to \infty$, $z > 1$

Again, we consider two cases, one with bounded values of z, where we use power series of the hypergeometric functions, and one with $z > 1$. For the first case we use power series of the hypergeometric functions, and in the other case we obtain uniform expansions in terms of modified Bessel functions.

29.2.1 *Expansions for bounded z*

We use Erdélyi *et al.* (1953–1955, Vol. I, p. 141, (16) and (17)):

$$\frac{2e^{-\pi i\mu}\Gamma(\mu+1)}{\Gamma(\mu+\nu+1)\Gamma(\mu-\nu)}\left(\frac{z-1}{z+1}\right)^{\frac12\mu} Q_\nu^\mu(z) = {}_2F_1\left({-\nu,\nu+1 \atop \mu+1}; \tfrac12+\tfrac12 z\right) - e^{\mp\pi i\nu}\left(\frac{z-1}{z+1}\right)^{\mu} {}_2F_1\left({-\nu,\nu+1 \atop \mu+1}; \tfrac12-\tfrac12 z\right), \tag{29.2.38}$$

$$\frac{\pi\Gamma(\mu+1)}{\Gamma(\mu+\nu+1)\Gamma(\mu-\nu)}\left(\frac{z-1}{z+1}\right)^{\frac12\mu} P_\nu^\mu(z) = \sin(\mu\pi)\, {}_2F_1\left({-\nu,\nu+1 \atop \mu+1}; \tfrac12+\tfrac12 z\right) - \sin(\nu\pi)e^{\mp\pi i\mu}\left(\frac{z-1}{z+1}\right)^{\mu} {}_2F_1\left({-\nu,\nu+1 \atop \mu+1}; \tfrac12-\tfrac12 z\right), \tag{29.2.39}$$

with the upper or lower sign chosen accordingly as $\Im z \gtrless 0$.

For $\mu \to -\infty$ we can use a similar formula for the Q-function, which follows from the symmetry relation

$$e^{\pi i\mu}\Gamma(\nu+\mu+1)Q_\nu^{-\mu}(z) = e^{-\pi i\mu}\Gamma(\nu-\mu+1)Q_\nu^\mu(z). \tag{29.2.40}$$

For $P_\nu^{-\mu}(z)$ we can use the representation in (29.0.2). For the case of complex z and μ we refer to Wagner (1988) or Olde Daalhuis (2010b, §15.12(ii)).

29.2.2 *Expansions in terms of modified Bessel functions*

The expansions of this section are especially of interest when $z \to \infty$, although they can be used for all $z > 1$.

For the uniform expansions with $z > 1$ we use the results of §29.1.2.1 and §29.1.2.2 by applying Whipple's formula, which gives two equivalent relations:

$$\begin{aligned} e^{-\pi i\mu}Q_\nu^\mu(z) &= \sqrt{\tfrac12\pi}\,(z^2-1)^{-\frac14}\,\Gamma(\nu+\mu+1)P_{-\mu-\frac12}^{-\nu-\frac12}\left(\frac{z}{\sqrt{z^2-1}}\right), \\ \Gamma(-\nu-\mu)P_\nu^\mu(z) &= \sqrt{2/\pi}\,(z^2-1)^{-\frac14}\, e^{\pi i(\nu+\frac12)}Q_{-\mu-\frac12}^{-\nu-\frac12}\left(\frac{z}{\sqrt{z^2-1}}\right). \end{aligned} \tag{29.2.41}$$

Using (29.1.27), the first relation in (29.2.41), and the first relation in (29.1.6), we obtain

$$\begin{aligned} &e^{-\pi\mu i}Q_\nu^\mu(z) \sim \tfrac{1}{2}\sqrt{\pi\alpha}\,\mu^{-\nu-\frac{1}{2}}\Gamma(\nu+\mu+1)\,\times \\ &\left(I_{\nu+\frac{1}{2}}\left(\tfrac{1}{2}\alpha\mu\right)\sum_{k=0}^{\infty}\frac{A_{2k}^*}{\mu^{2k}} - I_{\nu+\frac{3}{2}}\left(\tfrac{1}{2}\alpha\mu\right)\sum_{k=0}^{\infty}\frac{B_{2k+1}^*}{\mu^{2k+1}}\right), \end{aligned} \tag{29.2.42}$$

as $\mu \to \infty$, $z > 1$. The coefficients follow from those in (29.1.20) and (29.1.27), the first few following from (29.1.22). When we write the coefficients in the earlier expansions as $A_k(\alpha,\mu)$, $B_k(\alpha,\mu)$, the relations are $A_k^* = A_k\left(\alpha, -\nu-\frac{1}{2}\right)$, $B_k^* = B_k\left(\alpha,-\nu-\frac{1}{2}\right)$. In the present case, $\alpha = \ln(1-\eta)$, $\eta = 2/(1-z)$.

For $\mu \to -\infty$ we can use the relation in (29.2.40).

For $P_\nu^\mu(z)$ we first give the expansion for $\mu \to -\infty$. From the second relation in Whipple's relation in (29.2.41) we have

$$P_\nu^{-\mu}(z) = \frac{\sqrt{2/\pi}\,\left(z^2-1\right)^{-\frac{1}{4}}}{\Gamma(-\nu+\mu)}e^{\pi i(\nu+\frac{1}{2})}Q_{\mu-\frac{1}{2}}^{-\nu-\frac{1}{2}}\left(\frac{z}{\sqrt{z^2-1}}\right). \tag{29.2.43}$$

Then, (29.1.20) gives

$$\begin{aligned} &P_\nu^{-\mu}(z) \sim \frac{\sqrt{\alpha/\pi}\,\mu^{-\nu-\frac{1}{2}}}{\Gamma(\mu-\nu)}\times \\ &\left(K_{\nu+\frac{1}{2}}\left(\tfrac{1}{2}\alpha\mu\right)\sum_{k=0}^{\infty}\frac{A_{2k}^*}{\mu^{2k}} + K_{\nu+\frac{3}{2}}\left(\tfrac{1}{2}\alpha\mu\right)\sum_{k=0}^{\infty}\frac{B_{2k+1}^*}{\mu^{2k+1}}\right), \end{aligned} \tag{29.2.44}$$

where α and the coefficients are the same as in (29.2.42).

Finally, for $P_\nu^\mu(z)$ we can use

$$P_\nu^\mu(z) = \frac{\Gamma(\nu+\mu+1)}{\Gamma(\nu-\mu+1)}P_\nu^{-\mu}(z) + \frac{2}{\pi}\sin(\mu\pi)e^{-\pi i\mu}Q_\nu^\mu(z). \tag{29.2.45}$$

29.2.3 *Expansions of* $\mathsf{P}_\nu^\mu(x)$ *and* $\mathsf{Q}_\nu^\mu(x)$

The expansions for large μ follow from the modifications of the representations in (29.2.38) and (29.2.39) for $x \in (-1,1)$. We have

$$\begin{aligned} &\frac{2\Gamma(\mu+1)}{\Gamma(\mu+\nu+1)\Gamma(\mu-\nu)}\left(\frac{1-x}{1+x}\right)^{\frac{1}{2}\mu}\mathsf{Q}_\nu^\mu(x) = \\ &\qquad\cos(\mu\pi)\;{}_2F_1\!\left({-\nu,\nu+1 \atop \mu+1};\tfrac{1}{2}+\tfrac{1}{2}x\right) - \\ &\qquad\cos(\nu\pi)\left(\frac{1-x}{1+x}\right)^{\mu}{}_2F_1\!\left({-\nu,\nu+1 \atop \mu+1};\tfrac{1}{2}-\tfrac{1}{2}x\right), \end{aligned} \tag{29.2.46}$$

$$\frac{\pi\Gamma(\mu+1)}{\Gamma(\mu+\nu+1)\Gamma(\mu-\nu)}\left(\frac{1-x}{1+x}\right)^{\frac12\mu}\mathsf{P}_\nu^\mu(x) = \\ \sin(\mu\pi)\,{}_2F_1\left({-\nu,\nu+1 \atop \mu+1};\tfrac12+\tfrac12 x\right) - \\ \sin(\nu\pi)\left(\frac{1-x}{1+x}\right)^{\mu}\,{}_2F_1\left({-\nu,\nu+1 \atop \mu+1};\tfrac12-\tfrac12 x\right). \tag{29.2.47}$$

For both representations we can use the standard power series of the ${}_2F_1$-functions to obtain asymptotic expansions for large positive values of μ, uniformly for $x\in[-1,1]$. For $\mu\to-\infty$ we can use (29.0.4) for $\mathsf{P}_\nu^\mu(x)$, and the analogue of (29.2.40) for $\mathsf{Q}_\nu^\mu(x)$, that is,

$$\Gamma(\nu+\mu+1)\mathsf{Q}_\nu^{-\mu}(x) = \Gamma(\nu-\mu+1)\left(\cos(\mu\pi)\mathsf{Q}_\nu^\mu(x)+\tfrac12\pi\sin(\mu\pi)\mathsf{P}_\nu^\mu(x)\right). \tag{29.2.48}$$

29.3 Integrals with nearly coincident branch points

In Table 20.1, Cases 12 and 13, we have shown integrals of the type

$$\begin{aligned} F(\omega,\alpha,\mu) &= \int_{-\alpha}^{\alpha} e^{-\omega t} f(t)\left(\alpha^2-t^2\right)^\mu\,dt,\\ G(\omega,\alpha,\mu) &= \int_{\alpha}^{\infty} e^{-\omega t} f(t)\left(t^2-\alpha^2\right)^\mu\,dt, \end{aligned} \tag{29.3.49}$$

where $\omega>0$ (and large), $\Re\mu>-1$, and $\alpha\ge0$. The asymptotic problem is to give expansions that remain valid as $\alpha\downarrow0$.

When $f(t)=1$, these integrals reduce to modified Bessel functions (see Olver and Maximon (2010, Eqs. 10.32.2, 10.32.8))

$$\begin{aligned} F_0(\omega,\alpha,\mu) &= \sqrt{\pi}\,\Gamma(\mu+1)\left(\frac{2\alpha}{\omega}\right)^{\mu+\frac12} I_{\mu+\frac12}(\alpha\omega),\\ G_0(\omega,\alpha,\mu) &= \frac{1}{\sqrt{\pi}}\Gamma(\mu+1)\left(\frac{2\alpha}{\omega}\right)^{\mu+\frac12} K_{\mu+\frac12}(\alpha\omega). \end{aligned} \tag{29.3.50}$$

When α is bounded away from zero it is an easy matter to obtain expansions for large values of ω. However, when α is allowed to become small, or even 0, uniform expansions in terms of the Bessel functions can be given.

Integrals with this uniformity feature are considered by F. Ursell in Ursell (1984) (with applications to Legendre functions). In Ursell (2007) a contour integral is considered, with an application to Gegenbauer polynomials. In that case the expansion is in terms of the J-Bessel function.

We give from Ursell (1984) the results for the Legendre functions. We have

$$\begin{aligned} P_n^{-m}(\cosh z) &= \frac{(2\sinh z)^{-m}}{\sqrt{\pi}\,\Gamma\left(m+\frac12\right)}F(\omega,z,\mu),\\ Q_n^{-m}(\cosh z) &= e^{-m\pi i}\frac{\sqrt{\pi}\,(2\sinh z)^{-m}}{\Gamma\left(m+\frac12\right)}G(\omega,z,\mu), \end{aligned} \tag{29.3.51}$$

where $\omega = n + \frac{1}{2}$, $\mu = m - \frac{1}{2}$ and the function f shown in (29.3.49) is defined by

$$f(t) = \left(\frac{2\cosh z - 2\cosh t}{z^2 - t^2}\right)^{\mu}. \tag{29.3.52}$$

29.3.1 *Ursell's expansions of Legendre functions*

The procedure for obtaining asymptotic expansions, in which the modified Bessel functions are the leading term approximations, runs as in other cases, namely by using integration by parts. For $F(\omega, \alpha, \mu)$ in (29.3.49) we write

$$f(t) = a_0 + b_0 t + \left(\alpha^2 - t^2\right) g_0(t), \tag{29.3.53}$$

and a_0 and b_0 follow from substitution of $t = \pm\alpha$. For the representations of the Legendre functions the function f in (29.3.52) is even in t (and in z), and, hence, $b_0 = 0$. Therefore, in the case of $P_n^{-m}(\cosh z)$, we write

$$f(t) = a_0 + b_0\left(z^2 - t^2\right) + \left(z^2 - t^2\right)^2 g_0(t), \tag{29.3.54}$$

with

$$a_0 = f(z), \quad b_0 = -\frac{1}{2z} f'(z). \tag{29.3.55}$$

When we use (29.3.54) for $F(\omega, z, \mu)$ in (29.3.51) we obtain, after integrating by parts,

$$\begin{aligned} F(\omega, z, \mu) = {} & a_0 F_0(\omega, z, \mu) + b_0 F_0(\omega, z, \mu + 1) + \\ & \frac{1}{\omega}\int_{-z}^{z} e^{-\omega t} \frac{d}{dt}\left(\left(z^2 - t^2\right)^{\mu+2} g_0(t)\right) dt. \end{aligned} \tag{29.3.56}$$

Observe that, after differentiating in the integrand, the integrated terms vanish. Another integration by parts gives

$$\begin{aligned} F(\omega, z, \mu) = {} & a_0 F_0(\omega, z, \mu) + b_0 F_0(\omega, z, \mu + 1) + \\ & \frac{1}{\omega^2}\int_{-z}^{z} e^{-\omega t}\left(z^2 - t^2\right)^{\mu} f_1(t)\, dt, \end{aligned} \tag{29.3.57}$$

where

$$f_1(t) = \left(z^2 - t^2\right)^{-\mu} \frac{d^2}{dt^2}\left(\left(z^2 - t^2\right)^{\mu+2} g_0(t)\right). \tag{29.3.58}$$

By repeating this procedure it follows that for $p = 0, 1, 2, \ldots$,

$$\begin{aligned} F(\omega, z, \mu) = {} & F_0(\omega, z, \mu)\sum_{k=0}^{p-1}\frac{a_k}{\omega^{2k}} + F_0(\omega, z, \mu + 1)\sum_{k=0}^{p-1}\frac{b_k}{\omega^{2k}} + \\ & \frac{1}{\omega^{2p}}\int_{-z}^{z} e^{-\omega t}\left(z^2 - t^2\right)^{\mu} f_p(t)\, dt, \end{aligned} \tag{29.3.59}$$

where

$$\begin{aligned} f_{p+1}(t) &= \left(z^2 - t^2\right)^{-\mu} \frac{d^2}{dt^2}\left(\left(z^2 - t^2\right)^{\mu+2} g_p(t)\right), \\ f_p(t) &= a_p + b_p\left(z^2 - t^2\right) + \left(z^2 - t^2\right)^2 g_p(t), \end{aligned} \tag{29.3.60}$$

and

$$a_p = f_p(z), \quad b_p = -\frac{1}{2z} f_p'(z). \tag{29.3.61}$$

For the function $G(\omega, z, \mu)$ in (29.3.51) we obtain in a similar way

$$\begin{aligned} G(\omega, z, \mu) = G_0(\omega, z, \mu) \sum_{k=0}^{p-1} \frac{a_k}{\omega^{2k}} - G_0(\omega, z, \mu+1) \sum_{k=0}^{p-1} \frac{b_k}{\omega^{2k}} + \\ \frac{1}{\omega^{2p}} \int_z^\infty e^{-\omega t} \left(t^2 - z^2\right)^\mu f_p(t)\, dt, \end{aligned} \tag{29.3.62}$$

where a_k, b_k, and f_p are the same as for $F(\omega, z, \mu)$.

For the Legendre functions the asymptotic representations are

$$\begin{aligned} P_n^{-m}\ (\cosh z) &= \left(\frac{z}{\omega \sinh z}\right)^m \times \\ &\left(I_m(\omega z) \mathfrak{A}^{-m}(z, \omega) + \frac{z(2m+1)}{\omega} I_{m+1}(\omega z) \mathfrak{B}^{-m}(z, \omega)\right), \\ Q_n^{-m}\ (\cosh z) &\sim \left(\frac{z e^{-\pi i}}{\omega \sinh z}\right)^m \times \\ &\left(K_m(\omega z) \mathfrak{A}^{-m}(z, \omega) - \frac{z(2m+1)}{\omega} K_{m+1}(\omega z) \mathfrak{B}^{-m}(z, \omega)\right), \end{aligned} \tag{29.3.63}$$

where the functions $\mathfrak{A}^{-m}(z, \omega)$ and $\mathfrak{B}^{-m}(z, \omega)$ have the asymptotic expansions

$$\mathfrak{A}^{-m}(z, \omega) \sim \sum_{k=0}^{\infty} \frac{a_k}{\omega^{2k}}, \quad \mathfrak{B}^{-m}(z, \omega) \sim \sum_{k=0}^{\infty} \frac{b_k}{\omega^{2k}}. \tag{29.3.64}$$

This expansion is valid for bounded m and z. For estimates of the remainder terms in (29.3.59) and (29.3.62) we refer to Ursell (1984), where it is also shown that the functions $\mathfrak{A}^{-m}(z, \omega)$, $\mathfrak{B}^{-m}(z, \omega)$ and related functions are analytic functions of z^2 at $z = 0$. Observe that in the expansions derived in §29.1 we allow unbounded z. For an alternative expansion, of $P_n^{-m}\,(\cosh z)$, also valid for unbounded z and with error bounds, see §29.3.3.

29.3.2 *Coefficients of the expansion*

Ursell has not given the first coefficients a_k and b_k of the expansion. The best way to obtain these is based on using the expansion

$$f_p(t) = c_0(z) \sum_{k=0}^{\infty} c_k^p \left(z^2 - t^2\right)^k, \quad c_0(z) = \lim_{t \to z} f(t) = \left(\frac{\sinh z}{z}\right)^\mu, \tag{29.3.65}$$

where the coefficients c_k^0 can be obtained from function values of f and its derivatives at $t = z$. This is a two-point Taylor series as considered in Chapter 18. Because the function $f(t)$ used for the Legendre function is even, only one series is needed.

By using the recursion scheme given in (29.3.60) we can obtain recurrence relations in the form

$$c_k^p = \gamma_k c_{k+1}^{p-1} + \delta_k c_{k+2}^{p-1}, \quad k \ge 0, \quad p \ge 1, \tag{29.3.66}$$

where $\gamma_0 = 0$ and

$$\begin{aligned} \gamma_k &= -2(2k+1)(k+1) - 2(4k+3)\mu - 4\mu^2, \qquad k \ge 1, \\ \delta_k &= 4z^2\left((k+1)(k+2) + (2k+3)\mu + \mu^2\right), \qquad k \ge 0. \end{aligned} \tag{29.3.67}$$

Then,

$$a_p = c_0^p = f_p(z), \quad b_p = -\frac{1}{2z} f_p'(z), \quad p = 0, 1, 2, \ldots. \tag{29.3.68}$$

The coefficients of $f_0 = f$ defined in (29.3.52) can be obtained by using $w = z^2 - t^2$ and expanding

$$f_0(t) = \left(2\frac{\cosh z - \cosh\left(\sqrt{z^2 - w}\right)}{w}\right)^{\mu} = c_0(z) \sum_{k=0}^{\infty} c_k^0 w^k, \tag{29.3.69}$$

where $c_0(z)$ is defined in (29.3.65). The first coefficients are

$$\begin{aligned} c_0^0 &= 1, \\ c_1^0 &= \mu\frac{\sinh z - z\cosh z}{4z^2 \sinh z}, \\ c_2^0 &= \mu\frac{(9 + 4z^2 + 3\mu)\sinh^2 z - 3(1+\mu) z \sinh 2z + 3z^2(\mu - 1)\cosh^2 z}{96 z^4 \sinh^2 z}. \end{aligned} \tag{29.3.70}$$

29.3.3 *An alternative expansion of* $P_n^{-m}(\cosh z)$

The expansion given in (29.3.69) can be used for $F(\omega, \alpha, \mu)$ of (29.3.51) and gives a series of modified Bessel functions. For the function $P_n^{-m}(\cosh z)$ the result is

$$P_n^{-m}(\cosh z) = \frac{1}{\omega^m}\sqrt{\frac{z}{\sinh z}} \sum_{k=0}^{\infty} c_k^0 \left(m + \tfrac{1}{2}\right)_k I_{m+k}(\omega z) \left(\frac{2z}{\omega}\right)^k, \tag{29.3.71}$$

where $\omega = n + \frac{1}{2}$ and the first few c_k^0 are given in (29.3.70); $\mu = m - \frac{1}{2}$.

The expansion in (29.3.71) is considered in Shivakumar and Wong (1988), where a remainder is provided. The authors showed that the result is uniformly valid for $z \ge 0$ and they derived realistic and computable bounds for the remainder, of which the first few are given explicitly. These bounds are more useful than Ursell's bounds mentioned after (29.3.64).

The function $f(t)$ defined in (29.3.52) is singular at $t = \pm(z - 2\pi i)$. Hence, the expansion in (29.3.69) is convergent for $|w| < 2\pi\left(1 + z^2/\pi^2\right)^{\frac{1}{4}}$. It follows that the expansion converges uniformly for all $t \in [-z, z]$, and hence, the expansion in (29.3.71) is convergent for all finite positive z. This follows also from

$$I_\nu(z) \sim \frac{\left(\frac{1}{2}z\right)^\nu}{\Gamma(\nu+1)}, \quad \nu \to \infty, \quad z \ne 0, \tag{29.3.72}$$

with z fixed.

29.3.4 A related integral with nearly coincident branch points

In Schmidt (1978) an integral of the type

$$I(q,z,\omega)=\int_q^z \frac{f\left(\sqrt{t^2-q^2}\right)}{\sqrt{t^2-q^2}}\sin(\omega t)\,dt \tag{29.3.73}$$

has been discussed for large values of ω and possibly small values of q. That is, an approximation is given valid for $\omega q\ge 0$ with z such that $\omega z\to\infty$. Besides, an order estimate of the remainder in the expansion is given.

In this case the Bessel function $J_0(z)$ can be used, which has the integral representation (see Olver and Maximon (2010, Eq. 10.9.12))

$$J_0(qx)=\frac{2}{\pi}\int_q^\infty \frac{\sin(xt)}{\sqrt{t^2-q^2}}\,dt. \tag{29.3.74}$$

The first-term approximation is obtained by writing $f(t)=f(0)+tg(t)$ and integrating by parts. In this way, higher-order terms can be obtained as well.

According to Schmidt, integrals of the type (29.3.73) arise in membrane research in biology and in small angle x-ray scattering problems.

29.4 Toroidal harmonics and conical functions

Toroidal harmonics are a subclass of the Legendre functions with $\mu=m$, $m=0,1,2,\ldots$. For these functions the connection formula in (29.2.45) becomes

$$P_\nu^m(z)=\frac{\Gamma(1+\nu+m)}{\Gamma(1+\nu-m)}P_\nu^{-m}(z). \tag{29.4.75}$$

In Gil *et al.* (2000) we have given algorithms for the computation of the toroidal functions $P_\nu^m(z)$ for large values of m, with $z>1$, possibly large. These values can be obtained from the expansions in §29.2 and were needed for obtaining initial values in a backward recursion algorithm.

Conical functions are associated Legendre functions of the form $P^\mu_{-\frac12+i\tau}(z)$ and $Q^\mu_{-\frac12+i\tau}(z)$. They show up in a large number of applications in engineering, applied physics, quantum physics (related to the amplitude for Yukawa potential scattering), or cosmology, among others. Also, they are the kernel of the Mehler–Fock transform, which has numerous applications.

Expansions for large values of τ can be obtained from the results in §29.1. In the expansion in (29.1.20) we have $\omega=\nu+\frac12=i\tau$, and the K-Bessel functions become Hankel functions, which can be expressed in terms of ordinary J- and Y-Bessel functions (see §9.2). In the same manner, the I-Bessel functions in the expansion given in (29.1.27) become J-Bessel functions.

For expansions of the conical functions with both parameters large we refer to Dunster (1991). Numerical applications of these expansions are given in Gil *et al.* (2009). In Dunster *et al.* (2014) the expansion of the Q-function in (29.1.15) is used for the computation of $Q^\mu_{-\frac12+i\tau}(z)$ for large values of τ and $z>1$.

Chapter 30

Parabolic cylinder functions: Large parameter

In Chapter 11 we have given a few properties of parabolic cylinder functions and we have derived the expansions for large argument z.[1] When the parameter is large, we need expansions in the form of uniform expansions, which are in terms of elementary functions and in terms of Airy functions, and have been obtained in Olver (1959) by using the governing differential equation. In Temme (2000) we have indicated that some of Olver's expansions can be modified in the sense that they become valid with a double asymptotic property: valid when either the argument or the order, or both, are large. Jones (2006) has also given similar expansions, by using the differential equation in the complex plane, and he provided error bounds for remainders.

In this chapter we give an overview of the modified expansions for real argument and order by using several integral representations.

30.1 Notation for uniform asymptotic expansions

In the asymptotic expansions the following notations will be used:

$$a = \tfrac{1}{2}\mu^2, \quad z = \mu t\sqrt{2}, \tag{30.1.1}$$

$$h(\mu) = 2^{-\frac{1}{4}\mu^2-\frac{1}{4}} e^{-\frac{1}{4}\mu^2} \mu^{\frac{1}{2}\mu^2-\frac{1}{2}}, \tag{30.1.2}$$

$$\xi = \tfrac{1}{2}t\sqrt{t^2-1} - \tfrac{1}{2}\ln\left(t+\sqrt{t^2-1}\right), \tag{30.1.3}$$

$$\widetilde{\xi} = \tfrac{1}{2}t\sqrt{t^2+1} + \tfrac{1}{2}\ln\left(t+\sqrt{t^2+1}\right), \tag{30.1.4}$$

$$\tau = \tfrac{1}{2}\left(\frac{t}{\sqrt{t^2-1}}-1\right), \quad \widetilde{\tau} = \tfrac{1}{2}\left(\frac{t}{\sqrt{t^2+1}}-1\right), \tag{30.1.5}$$

$$\eta(t) = \tfrac{1}{2}\arccos t - \tfrac{1}{2}t\sqrt{1-t^2}, \quad \sigma = \frac{t}{\sqrt{1-t^2}}. \tag{30.1.6}$$

[1]This chapter is based on Temme (2000).

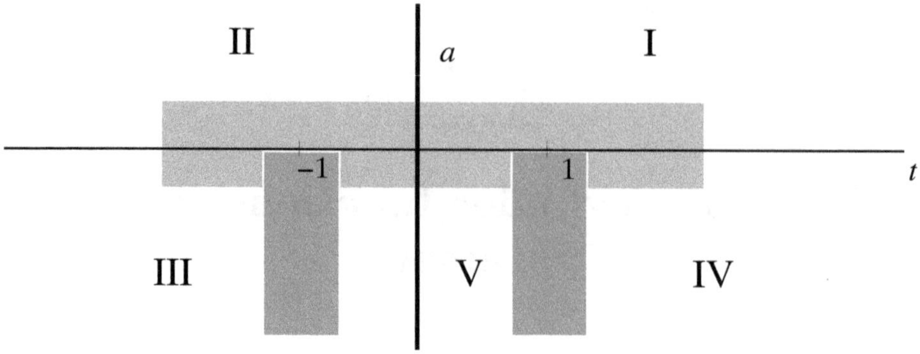

Fig. 30.1 Five regions for the asymptotic expansions of $U(a,z)$ and $V(a,z)$ and their derivatives given in §30.2 and §30.3. In the strips below $t = \pm 1$ Airy-type expansions should be used (which are valid in much larger domains than indicated in the figure).

By using the notation in (30.1.1), the differential equation in (11.0.2) can be written in the form

$$\frac{d^2w}{dt^2} - \mu^4\left(t^2+1\right)w = 0, \tag{30.1.7}$$

with solutions $U\left(\frac{1}{2}\mu^2, \mu t\sqrt{2}\right)$ and $V\left(\frac{1}{2}\mu^2, \mu t\sqrt{2}\right)$.

On the other hand, the functions $U\left(-\frac{1}{2}\mu^2, \mu t\sqrt{2}\right)$ and $V\left(-\frac{1}{2}\mu^2, \mu t\sqrt{2}\right)$ satisfy the equation

$$\frac{d^2w}{dt^2} - \mu^4\left(t^2-1\right)w = 0, \tag{30.1.8}$$

which has turning points at $t = \pm 1$. The functions $U(a,z)$ and $V(a,z)$ are oscillatory in the interval $(-2\sqrt{-a}, 2\sqrt{-a})$, $(a < 0)$. The Hermite case in (11.1.3) is a well-known example.

In Figure 30.1 we show the five domains in the (t,a)-plane where the various expansions in terms of elementary functions of the parabolic cylinder functions of the following sections are valid.

I	§30.3.1:	$a > 0,$	$z \ge 0,$	$a + z \to \infty;$
II	§30.3.2:	$a > 0,$	$z \le 0,$	$a - z \to \infty;$
III	§30.2.2:	$a < 0,$	$z < -2\sqrt{-a},$	$-a - z \to \infty;$
IV	§30.2.1:	$a < 0,$	$z > 2\sqrt{-a},$	$-a + z \to \infty;$
V	§30.2.3:	$a < 0,$	$-2\sqrt{-a} < z < 2\sqrt{-a},$	$-a \to \infty.$

In §30.5 the Airy-type expansion for $a < 0$ and $t \sim 1$ is given.

30.2 The case $a < 0$

For this case we need three subcases, two for outside and one for inside the oscillatory interval.

30.2.1 *The case* $z > 2\sqrt{-a},\ -a + z \to \infty$

We have the representation

$$U\left(-\tfrac{1}{2}\mu^2, \mu t\sqrt{2}\right) = \frac{h(\mu)\, e^{-\mu^2\xi}}{(t^2-1)^{\frac{1}{4}}} F_\mu(t), \tag{30.2.9}$$

with expansion

$$F_\mu(t) \sim \sum_{s=0}^{\infty} \frac{\phi_s(\tau)}{\mu^{2s}}. \tag{30.2.10}$$

The coefficients $\phi_s(\tau)$ are polynomials in τ, $\phi_0(\tau) = 1$, and are given by the recursion

$$\phi_{s+1}(\tau) = -4\tau^2(\tau+1)^2 \frac{d}{d\tau}\phi_s(\tau) - \tfrac{1}{4}\int_0^\tau \left(20u^2 + 20u + 3\right) \phi_s(u)\, du, \tag{30.2.11}$$

$s = 0, 1, 2, \ldots$. This recursion follows from writing the differential equation in (11.0.2) for $F_\mu(t)$ and substituting $t = (\tau + \frac{1}{2})/\sqrt{\tau(\tau+1)}$, which is the inverse of the relation in (30.1.5).

The first coefficients ϕ_s are given by

$$\begin{aligned}
\phi_0(\tau) &= 1,\\
\phi_1(\tau) &= -\frac{\tau}{12}\left(20\tau^2 + 30\tau + 9\right),\\
\phi_2(\tau) &= \frac{\tau^2}{288}\left(6160\tau^4 + 18480\tau^3 + 19404\tau^2 + 8028\tau + 945\right).
\end{aligned} \tag{30.2.12}$$

Observe that $\lim\limits_{t\to\infty} \tau(t) = 0$ and that all coefficients $\phi_s(\tau)$ vanish at infinity for $s > 0$, as follows easily from induction.

The expansion in (30.2.10) holds uniformly with respect to $t \geq 1 + \delta$. Moreover, it is valid as $t \to \infty$, uniformly with respect to $\mu \geq \delta$. In this sense, the expansion has a double asymptotic property: it holds if one or both parameters t and μ are large, and not only if μ is large.

For the function $V(a, z)$ we have the representation

$$V\left(-\tfrac{1}{2}\mu^2, \mu t\sqrt{2}\right) = \frac{e^{\mu^2\xi}}{\mu\sqrt{\pi}\, h(\mu)\, (t^2-1)^{\frac{1}{4}}} P_\mu(t), \tag{30.2.13}$$

with expansion

$$P_\mu(t) \sim \sum_{s=0}^{\infty} (-1)^s \frac{\phi_s(\tau)}{\mu^{2s}}, \tag{30.2.14}$$

where the $\phi_s(\tau)$ are the same as in (30.2.10).

For the derivatives we can use the identities

$$\begin{aligned}\frac{d}{dt}\frac{e^{-\mu^2\xi}}{(t^2-1)^{\frac14}}F_\mu(t) &= -\mu^2\left(t^2-1\right)^{\frac14}e^{-\mu^2\xi}\,G_\mu(t),\\ \frac{d}{dt}\frac{e^{+\mu^2\xi}}{(t^2-1)^{\frac14}}P_\mu(t) &= +\mu^2\left(t^2-1\right)^{\frac14}e^{+\mu^2\xi}\,Q_\mu(t),\end{aligned} \tag{30.2.15}$$

with expansions

$$G_\mu(t)\sim\sum_{s=0}^{\infty}\frac{\psi_s(\tau)}{\mu^{2s}},\quad Q_\mu(t)\sim\sum_{s=0}^{\infty}(-1)^s\frac{\psi_s(\tau)}{\mu^{2s}}. \tag{30.2.16}$$

The coefficients ψ_s can be obtained from the relation

$$\psi_s(\tau)=\phi_s(\tau)+2\tau(\tau+1)(2\tau+1)\phi_{s-1}(\tau)+8\tau^2(\tau+1)^2\frac{d\phi_{s-1}(\tau)}{d\tau}, \tag{30.2.17}$$

$s=0,1,2,\ldots$. The first few are

$$\begin{aligned}\psi_0(\tau)&=1,\\ \psi_1(\tau)&=\frac{\tau}{12}\left(28\tau^2+42\tau+15\right),\\ \psi_2(\tau)&=-\frac{\tau^2}{288}\left(7280\tau^4+21840\tau^3+23028\tau^2+9684\tau+1215\right).\end{aligned} \tag{30.2.18}$$

This gives the representations

$$\begin{aligned}U'\left(-\tfrac12\mu^2,\mu t\sqrt2\right)&=-\frac{\mu}{\sqrt2}h(\mu)\left(t^2-1\right)^{\frac14}e^{-\mu^2\xi}\,G_\mu(t),\\ V'\left(-\tfrac12\mu^2,\mu t\sqrt2\right)&=\frac{\left(t^2-1\right)^{\frac14}e^{\mu^2\xi}}{\sqrt{2\pi}\,h(\mu)}\,Q_\mu(t),\end{aligned} \tag{30.2.19}$$

with expansions given in (30.2.16).

Remark 30.1. The functions $F_\mu(t), G_\mu(t), P_\mu(t)$ and $Q_\mu(t)$ introduced in the asymptotic representations satisfy the following exact relation:

$$F_\mu(t)\,Q_\mu(t)+G_\mu(t)\,P_\mu(t)=2. \tag{30.2.20}$$

This follows from the first Wronskian relation in (11.1.7). △

30.2.2 *The case* $z<-2\sqrt{-a},\ -a-z\to\infty$

For this case we have for $t\ge 1+\delta$ the representations

$$\begin{aligned}U\left(-\tfrac12\mu^2,-\mu t\sqrt2\right)&=\frac{h(\mu)}{(t^2-1)^{\frac14}}\quad\times\\ &\left(\sin\left(\tfrac12\pi\mu^2\right)e^{-\mu^2\xi}\,F_\mu(t)+\frac{\Gamma\left(\frac12+\frac12\mu^2\right)\cos\left(\frac12\pi\mu^2\right)}{\mu\sqrt{\pi}\,h^2(\mu)}e^{\mu^2\xi}\,P_\mu(t)\right),\end{aligned} \tag{30.2.21}$$

where $F_\mu(t)$ and $P_\mu(t)$ have the expansions given in (30.2.10) and (30.2.14), respectively. An expansion of $V(-\frac{1}{2}\mu^2, -\mu t\sqrt{2})$ follows from (11.1.5), (30.2.9) and (30.2.21). A few manipulations give

$$
\begin{aligned}
V\left(-\tfrac{1}{2}\mu^2, -\mu t\sqrt{2}\right) &= \frac{h(\mu)}{(t^2-1)^{\frac{1}{4}}\,\Gamma\left(\frac{1}{2}+\frac{1}{2}\mu^2\right)} \quad \times \\
&\left(\cos\left(\tfrac{1}{2}\pi\mu^2\right) e^{-\mu^2\xi}\, F_\mu(t) - \frac{\Gamma\left(\frac{1}{2}+\frac{1}{2}\mu^2\right)\sin\left(\frac{1}{2}\pi\mu^2\right)}{\mu\sqrt{\pi}\,h^2(\mu)} e^{\mu^2\xi}\, P_\mu(t)\right).
\end{aligned} \tag{30.2.22}
$$

If $a = -\frac{1}{2}\mu^2 = -n - \frac{1}{2}$, $n = 0, 1, 2, \ldots$, the cosine in (30.2.21) vanishes, and, hence, the dominant part vanishes. This is the Hermite case, cf. (11.1.3).

Representations for the derivatives follow from the relations in (30.2.15). We have

$$
\begin{aligned}
U'\left(-\tfrac{1}{2}\mu^2, -\mu t\sqrt{2}\right) &= \frac{\mu h(\mu)}{\sqrt{2}}\,(t^2-1)^{\frac{1}{4}} \quad \times \\
&\left(\sin\left(\tfrac{1}{2}\pi\mu^2\right) e^{-\mu^2\xi}\, G_\mu(t) - \frac{\Gamma\left(\frac{1}{2}+\frac{1}{2}\mu^2\right)\cos\left(\frac{1}{2}\pi\mu^2\right)}{\mu\sqrt{\pi}\,h^2(\mu)} e^{\mu^2\xi}\, Q_\mu(t)\right),
\end{aligned} \tag{30.2.23}
$$

and

$$
\begin{aligned}
V'\left(-\tfrac{1}{2}\mu^2, -\mu t\sqrt{2}\right) &= \frac{\mu h(\mu)\,(t^2-1)^{\frac{1}{4}}}{\sqrt{2}\,\Gamma\left(\frac{1}{2}+\frac{1}{2}\mu^2\right)} \quad \times \\
&\left(\cos\left(\tfrac{1}{2}\pi\mu^2\right) e^{-\mu^2\xi}\, G_\mu(t) + \frac{\Gamma\left(\frac{1}{2}+\frac{1}{2}\mu^2\right)\sin\left(\frac{1}{2}\pi\mu^2\right)}{\mu\sqrt{\pi}\,h^2(\mu)} e^{\mu^2\xi}\, Q_\mu(t)\right).
\end{aligned} \tag{30.2.24}
$$

30.2.3 *The case* $-2\sqrt{-a} < z < 2\sqrt{-a}$

For negative a and $-1 < t < 1$ the expansions are essentially different, because now oscillations with respect to t occur. We use Olver's result, see Olver (1959) and Temme (2010a, §12.10.iv). We can give modified versions based on our earlier modifications, but because in the present case t belongs to a finite interval, the modified expansions do not have the double asymptotic property.

In Olver's expansions coefficients $u_s(t)$ and $v_s(t)$ are used. The first few are

$$
\begin{aligned}
u_0(t) &= 1, \quad u_1(t) = \frac{t(t^2-6)}{24}, \quad u_2(t) = \frac{-9t^4+249t^2+145}{1152},\\
v_0(t) &= 1, \quad v_1(t) = \frac{t(t^2+6)}{24}, \quad v_2(t) = \frac{15t^4-327t^2-143}{1152},
\end{aligned} \tag{30.2.25}
$$

and the higher coefficients follow from the recurrence relations

$$
\begin{aligned}
&(t^2-1)u_s'(t) - 3stu_s(t) = r_{s-1}(t),\\
&8r_s(t) = (3t^2+2)u_s(t) - 12(s+1)tr_{s-1}(t) + 4(t^2-1)r_{s-1}'(t),\\
&v_s(t) = u_s(t) + \tfrac{1}{2}tu_{s-1}(t) - r_{s-2}(t).
\end{aligned} \tag{30.2.26}
$$

We introduce the expansions

$$\begin{aligned}\mathcal{A}_\mu(t) &\sim \sum_{s=0}^{\infty} \frac{(-1)^s u_{2s}(t)}{(1-t^2)^{3s}\mu^{4s}}, & \mathcal{B}_\mu(t) &\sim \sum_{s=0}^{\infty} \frac{(-1)^s u_{2s+1}(t)}{(1-t^2)^{3s+\frac{3}{2}}\mu^{4s+2}},\\ \mathcal{C}_\mu(t) &\sim \sum_{s=0}^{\infty} \frac{(-1)^s v_{2s}(t)}{(1-t^2)^{3s}\mu^{4s}}, & \mathcal{D}_\mu(t) &\sim \sum_{s=0}^{\infty} \frac{(-1)^s v_{2s+1}(t)}{(1-t^2)^{3s+\frac{3}{2}}\mu^{4s+2}}.\end{aligned} \tag{30.2.27}$$

Then we have

$$\begin{aligned}U\left(-\tfrac{1}{2}\mu^2, \mu t\sqrt{2}\right) &\sim \frac{2g(\mu)}{(1-t^2)^{\frac{1}{4}}} \times\\ &\left(\cos\left(\mu^2\eta - \tfrac{1}{4}\pi\right)\mathcal{A}_\mu(t) - \sin\left(\mu^2\eta - \tfrac{1}{4}\pi\right)\mathcal{B}_\mu(t)\right),\end{aligned} \tag{30.2.28}$$

$$\begin{aligned}U'\left(-\tfrac{1}{2}\mu^2, \mu t\sqrt{2}\right) &\sim \mu\sqrt{2}g(\mu)(1-t^2)^{\frac{1}{4}} \times\\ &\left(\sin\left(\mu^2\eta - \tfrac{1}{4}\pi\right)\mathcal{C}_\mu(t) + \cos\left(\mu^2\eta - \tfrac{1}{4}\pi\right)\mathcal{D}_\mu(t)\right),\end{aligned} \tag{30.2.29}$$

$$\begin{aligned}V\left(-\tfrac{1}{2}\mu^2, \mu t\sqrt{2}\right) &\sim \frac{2g(\mu)}{\Gamma(\frac{1}{2}+\frac{1}{2}\mu^2)(1-t^2)^{\frac{1}{4}}} \times\\ &\left(\cos\left(\mu^2\eta + \tfrac{1}{4}\pi\right)\mathcal{A}_\mu(t) - \sin\left(\mu^2\eta + \tfrac{1}{4}\pi\right)\mathcal{B}_\mu(t)\right),\end{aligned} \tag{30.2.30}$$

$$\begin{aligned}V'\left(-\tfrac{1}{2}\mu^2, \mu t\sqrt{2}\right) &\sim \frac{\mu\sqrt{2}g(\mu)(1-t^2)^{\frac{1}{4}}}{\Gamma(\frac{1}{2}+\frac{1}{2}\mu^2)} \times\\ &\left(\sin\left(\mu^2\eta + \tfrac{1}{4}\pi\right)\mathcal{C}_\mu(t) + \cos\left(\mu^2\eta + \tfrac{1}{4}\pi\right)\mathcal{D}_\mu(t)\right),\end{aligned} \tag{30.2.31}$$

as $\mu \to \infty$, uniformly with respect to $|t| \le 1-\delta$; η is defined in (30.1.6).

The quantity $g(\mu)$ is only known in the form of an asymptotic expansion

$$g(\mu) \sim h(\mu)\left(\sum_{k=0}^{\infty} \frac{g_k}{\mu^{2k}}\right)^{-1}, \tag{30.2.32}$$

where

$$g_0 = 1, \quad g_1 = \tfrac{1}{24}, \quad g_3(0) = -\tfrac{2021}{207360}, \quad g_{2k} = 0, \ (k = 1, 2, 3, \ldots). \tag{30.2.33}$$

The coefficients g_k are certain limits of coefficients used in Olver's expansions. For $h(\mu)$ we refer to (30.1.2).

By using the Wronskian relation (11.1.7) it follows that we have the following asymptotic identity

$$\mathcal{A}_\mu(t)\mathcal{C}_\mu(t) + \mathcal{B}_\mu(t)\mathcal{D}_\mu(t) \sim \frac{\Gamma(\frac{1}{2}+\frac{1}{2}\mu^2)}{2\mu\sqrt{\pi}g^2(\mu)} \sim 1 - \frac{1}{576\mu^4} + \frac{2021}{2488320\mu^8} + \ldots. \tag{30.2.34}$$

30.3 The case $a > 0$

In this case the differential equation in (11.0.2) has no real turning points, see also (30.1.7), and we give expansions for $z > 0$ and $z < 0$. These expansions are quite similar and in Olver (1959) one expansion has been given for both cases (and for complex values of z). We prefer two expansions in order to obtain the double asymptotic property.

30.3.1 *The case* $z \ge 0,\ a + z \to \infty$

The results are

$$\begin{aligned} U\left(\tfrac{1}{2}\mu^2, \mu t\sqrt{2}\right) &= \frac{e^{-\mu^2\widetilde{\xi}}}{\mu\sqrt{2}\,h(\mu)\,(t^2+1)^{\frac{1}{4}}}\,\widetilde{F}_\mu(t), \\ U'\left(\tfrac{1}{2}\mu^2, \mu t\sqrt{2}\right) &= -\frac{(t^2+1)^{\frac{1}{4}}}{2h(\mu)}\,e^{-\mu^2\widetilde{\xi}}\widetilde{G}_\mu(t), \end{aligned} \tag{30.3.35}$$

with expansions

$$\widetilde{F}_\mu(t) \sim \sum_{s=0}^{\infty} (-1)^s\,\frac{\phi_s(\widetilde{\tau})}{\mu^{2s}}, \quad \widetilde{G}_\mu(t) \sim \sum_{s=0}^{\infty} (-1)^s\,\frac{\psi_s(\widetilde{\tau})}{\mu^{2s}}. \tag{30.3.36}$$

The coefficients ϕ_s and ψ_s in (30.3.36) are the same as in (30.2.10) and (30.2.16), respectively; $\widetilde{\tau}$ is defined in (30.1.5).

30.3.2 *The case* $z \le 0,\ a - z \to \infty$

We have the representations

$$\begin{aligned} U\left(\tfrac{1}{2}\mu^2, -\mu t\sqrt{2}\right) &= \frac{\sqrt{2\pi}\,h(\mu)}{\Gamma\left(\frac{1}{2}+\frac{1}{2}\mu^2\right)}\frac{e^{\mu^2\widetilde{\xi}}}{(t^2+1)^{\frac{1}{4}}}\,\widetilde{P}_\mu(t), \\ U'\left(\tfrac{1}{2}\mu^2, -\mu t\sqrt{2}\right) &= -\frac{\mu\sqrt{\pi}\,h(\mu)}{\Gamma\left(\frac{1}{2}+\frac{1}{2}\mu^2\right)}e^{\mu^2\widetilde{\xi}}\,(t^2+1)^{\frac{1}{4}}\,\widetilde{Q}_\mu(t), \end{aligned} \tag{30.3.37}$$

with expansions

$$\widetilde{P}_\mu(t) \sim \sum_{s=0}^{\infty} \frac{\phi_s(\widetilde{\tau})}{\mu^{2s}}, \quad \widetilde{Q}_\mu(t) \sim \sum_{s=0}^{\infty} \frac{\psi_s(\widetilde{\tau})}{\mu^{2s}}. \tag{30.3.38}$$

Remark 30.2. By using the relation in (11.1.5), the representations for $V(a, z)$ and $V'(a, z)$ for positive a can be obtained from the results for $U(a, z)$ and $U'(a, z)$ in (30.3.35), (30.3.36) and (30.3.37). △

Remark 30.3. The functions $\widetilde{F}_\mu(t), \widetilde{G}_\mu(t), \widetilde{P}_\mu(t)$ and $\widetilde{Q}_\mu(t)$ introduced in (30.3.35) and (30.3.37) satisfy the following exact relation

$$\widetilde{F}_\mu(t)\widetilde{Q}_\mu(t) + \widetilde{G}_\mu(t)\widetilde{P}_\mu(t) = 2. \tag{30.3.39}$$

This follows from the second Wronskian relation in (11.1.7). See also Remark 30.1. △

Remark 30.4. The expansions in (30.3.36) and (30.3.38) have the double asymptotic property: they are valid if $a + |z| \to \infty$. △

30.4 Expansions from integral representations

In this section we explain how some of the expansions given in §§30.2–30.3 can be obtained from the integrals that define $U(a,z)$ and $V(a,z)$. Again we consider real values of the parameters, and we use the notation given in §30.1.

30.4.1 *The case* $a > 0,\ z \ge 0;\ a+z\to\infty$

We start with the integral representation (see (11.2.8))

$$U(a,z)=\frac{e^{-\frac14 z^2}}{\Gamma\left(\frac12+a\right)}\int_0^\infty w^{a-\frac12}e^{-\frac12 w^2-zw}\,dw,\quad a>-\tfrac12, \tag{30.4.40}$$

which we write in the form

$$U(a,z)=\frac{z^{a+\frac12}\,e^{-\frac14 z^2}}{\Gamma\left(\frac12+a\right)}\int_0^\infty w^{-\frac12}e^{-z^2\phi(w)}\,dw, \tag{30.4.41}$$

where

$$\phi(w)=w+\tfrac12 w^2-\lambda\ln w,\quad \lambda=\frac{a}{z^2}. \tag{30.4.42}$$

Initially, we consider z as the large parameter.

The positive saddle point w_0 of the integrand in (30.4.42) is computed from

$$\frac{d\phi(w)}{dw}=\frac{w^2+w-\lambda}{w}=0, \tag{30.4.43}$$

giving

$$w_0=\tfrac12\left(\sqrt{1+4\lambda}-1\right). \tag{30.4.44}$$

When λ is bounded away from 0 we can use Laplace's method (see Chapter 3). When a and z are such that $\lambda\to 0$, Laplace's method cannot be applied. However, we can use the method described in Chapter 25 that gives an expansion that holds uniformly with respect to $\lambda\ge 0$.

To obtain a standard form for this Laplace-type integral, we transform $w\to t$ (see §25.6) by writing

$$\phi(w)=t-\lambda\ln t+A, \tag{30.4.45}$$

where A does not depend on t or w, and we prescribe that $w=0$ should correspond to $t=0$ and $w=w_0$ to $t=\lambda$, the saddle point in the t-plane.

This gives

$$U(a,z)=\frac{z^{a+\frac12}\,e^{-\frac14 z^2-Az^2}}{(1+4\lambda)^{\frac14}\Gamma\left(\frac12+a\right)}\int_0^\infty t^{a-\frac12}e^{-z^2t}f(t)\,dt, \tag{30.4.46}$$

where

$$f(t)=(1+4\lambda)^{\frac14}\sqrt{\frac{t}{w}}\,\frac{dw}{dt}=(1+4\lambda)^{\frac14}\sqrt{\frac{w}{t}}\,\frac{t-\lambda}{w^2+w-\lambda}. \tag{30.4.47}$$

By normalizing with the quantity $(1+4\lambda)^{\frac{1}{4}}$ we obtain $f(\lambda)=1$, as can be verified from (30.4.47) and a limiting process (using l'Hôpital's rule). The quantity A is given by

$$A=\tfrac{1}{2}w_0^2+w_0-\lambda\ln w_0-\lambda+\lambda\ln\lambda. \tag{30.4.48}$$

Using the method described in §25.2 we can obtain the expansion

$$U(a,z)\sim\frac{e^{-\frac{1}{4}z^2-Az^2}}{z^{a+\frac{1}{2}}\,(1+4\lambda)^{\frac{1}{4}}}\sum_{k=0}^{\infty}\frac{f_k(\lambda)}{z^{2k}}. \tag{30.4.49}$$

The coefficients $f_k(\lambda)$ can be obtained by using an integration by parts method, which in the present case runs as follows.

Consider the integral

$$F_a(z)=\frac{1}{\Gamma\left(\frac{1}{2}+a\right)}\int_0^\infty t^{a-\frac{1}{2}}e^{-z^2t}f(t)\,dt. \tag{30.4.50}$$

We have with $\lambda=a/z^2$ and writing $f(t)=f(\lambda)+(f(t)-f(\lambda))$,

$$\begin{aligned}F_a(z)&=z^{-2a-1}f(\lambda)-\frac{1}{z^2\Gamma\left(\frac{1}{2}+a\right)}\int_0^\infty t^{\frac{1}{2}}\frac{(f(t)-f(\lambda))}{t-\lambda}\,de^{-z^2(t-\lambda\ln t)}\\&=z^{-2a-1}f(\lambda)+\frac{1}{z^2\Gamma\left(\frac{1}{2}+a\right)}\int_0^\infty t^{a-\frac{1}{2}}e^{-z^2t}f_1(t)\,dt.\end{aligned} \tag{30.4.51}$$

Repeating this procedure we obtain (30.4.49), where the functions $\{f_k\}$ are defined by

$$f_{k+1}(t)=\sqrt{t}\,\frac{d}{dt}\left(\sqrt{t}\frac{f_k(t)-f_k(\lambda)}{t-\lambda}\right),\quad k=0,1,2,\ldots, \tag{30.4.52}$$

with $f_0(t)=f(t)$.

The coefficients $f_k(\lambda)$ can be expressed in terms of the Taylor coefficients of $f(t)$ at $t=\lambda$. Let these coefficients be defined by $a_n(\lambda)$, that is,

$$f(t)=\sum_{n=0}^{\infty}a_n(\lambda)(t-\lambda)^n. \tag{30.4.53}$$

Then we write for the iterated functions

$$f_k(t)=\sum_{n=0}^{\infty}a_n^{(k)}(\lambda)(t-\lambda)^n, \tag{30.4.54}$$

and by substituting this in (30.4.52) it follows that

$$a_n^{(k+1)}(\lambda)=\lambda(n+1)a_{n+2}^{(k)}(\lambda)+\left(n+\tfrac{1}{2}\right)a_{n+1}^{(k)}(\lambda),\quad k\ge 0,\quad n\ge 0. \tag{30.4.55}$$

Hence, the coefficients $f_k(\lambda)$ of (30.4.49) are given by

$$f_k(\lambda)=a_0^{(k)}(\lambda),\quad k\ge 0. \tag{30.4.56}$$

We have

$$
\begin{aligned}
f_0(\lambda) &= 1, \\
f_1(\lambda) &= \tfrac{1}{2}\bigl(a_1(\lambda) + 2\lambda a_2(\lambda)\bigr), \\
f_2(\lambda) &= \tfrac{1}{4}\bigl(12\lambda^2 a_4(\lambda) + 14\lambda a_3(\lambda) + 3a_2(\lambda)\bigr).
\end{aligned}
\tag{30.4.57}
$$

Explicitly,

$$
\begin{aligned}
f_0(\lambda) &= 1, \\
f_1(\lambda) &= -\frac{(1+2\widetilde{\tau})^2}{24(1+\widetilde{\tau})}\left(9 + 30\widetilde{\tau} + 20\widetilde{\tau}^2\right), \\
f_2(\lambda) &= \frac{(1+2\widetilde{\tau})^4}{1152(1+\widetilde{\tau})^2}\left(945 + 8028\widetilde{\tau} + 19404\widetilde{\tau}^2 + 18480\widetilde{\tau}^3 + 6160\widetilde{\tau}^4\right),
\end{aligned}
\tag{30.4.58}
$$

where $\widetilde{\tau}$ is defined in (30.1.5), assuming that we use the notation for a and z in (30.1.1).

After a few manipulations it follows that we can write (30.4.49) in the form of the expansion in (30.3.35), because

$$
\phi_k(\widetilde{\tau}) = \frac{(-1)^k}{(2t^2)^k} f_k(\lambda), \quad k = 0, 1, 2, \ldots. \tag{30.4.59}
$$

30.4.2 *The case $a > 0$, $z \le 0$; $a - z \to \infty$*

In this case we can use again the integral in (30.4.40), writing $U(a, -z)$ with $z > 0$. We have

$$
U(a, -z) = \frac{z^{a+\frac{1}{2}}\, e^{-\frac{1}{4}z^2}}{\Gamma\left(\frac{1}{2} + a\right)} \int_0^\infty w^{-\frac{1}{2}} e^{-z^2\phi(w)}\, dw, \tag{30.4.60}
$$

where

$$
\phi(w) = -w + \tfrac{1}{2}w^2 - \lambda \ln w, \quad \lambda = \frac{a}{z^2}, \tag{30.4.61}
$$

with positive saddle point

$$
w_0 = \tfrac{1}{2}\left(\sqrt{1+4\lambda} + 1\right). \tag{30.4.62}
$$

In this case the positive saddle point w_0 is, for all $\lambda \ge 0$, bounded away from 0, and we can use Laplace's method after transforming the integral by substituting

$$
\phi(w) - \phi(w_0) = \tfrac{1}{2}\phi''(w_0)\, s^2, \quad \operatorname{sign}(s) = \operatorname{sign}(w - w_0). \tag{30.4.63}
$$

Remark 30.5. To derive the expansions for $a > 0$ we can again use the contour integral (see Temme (2010a, Eq. 12.5.6))

$$
U(a, z) = \frac{e^{\frac{1}{4}z^2}}{i\sqrt{2\pi}} \int_{\mathcal{C}} e^{-zs+\frac{1}{2}s^2} s^{-a-\frac{1}{2}}\, ds, \tag{30.4.64}
$$

where $\mathcal{C}$ is a vertical line in the half-plane $\Re s > 0$. After scaling the parameters, we have

$$U\left(\tfrac{1}{2}\mu^2, \mu t\sqrt{2}\right) = \frac{e^{\frac{1}{4}z^2}(\mu\sqrt{2})^{\frac{1}{2}-\frac{1}{2}\mu^2}}{i\sqrt{2\pi}} \int_{\mathcal{C}} e^{\mu^2\phi(w)} \frac{dw}{\sqrt{w}}, \tag{30.4.65}$$

where

$$\phi(w) = w^2 - 2tw - \tfrac{1}{2}\ln w, \tag{30.4.66}$$

with positive saddle point

$$w_0 = \tfrac{1}{2}\left(t + \sqrt{1+t^2}\right). \tag{30.4.67}$$

For $z \ge 0$, that is, $t \ge 0$ we can use Laplace's method because for all $t \ge 0$ the saddle point is bounded away from the singular point $w = 0$. For $t \to -\infty$ the saddle point approaches the origin, and to obtain an expansion that is uniformly valid with respect to $t \le 0$ we can use the method described in §25.3 for loop integrals. We need an integration by parts procedure as in §30.4.1.

In both cases we obtain the same expansions as in §30.4.1 and as indicated in §30.4.2. △

30.4.3 *The case $a < 0$, $|z| > 2\sqrt{-a}$; $-a + |z| \to \infty$*

For $a < 0$ we take as starting point the integral (see Temme (2010a, Eq. 12.5.5))

$$U(-a, z) = \frac{\Gamma(\frac{1}{2}+a)}{2\pi i} e^{-\frac{1}{4}z^2} \int_{\mathcal{L}} e^{zs - \frac{1}{2}s^2}\, s^{-a-\frac{1}{2}}\, ds, \tag{30.4.68}$$

where $\mathcal{L}$ is a contour that encircles the negative s-axis in positive direction, starting at $-\infty$ with $\operatorname{ph} s = -\pi$ and terminating at $-\infty$ with $\operatorname{ph} s = +\pi$.

Using the scaled parameters, we can write the integral in the form

$$U\left(-\tfrac{1}{2}\mu^2, \mu t\sqrt{2}\right) = \frac{\Gamma(\frac{1}{2}+a)\,(\mu\sqrt{2})^{\frac{1}{2}-\frac{1}{2}\mu^2}\, e^{-\frac{1}{4}z^2}}{2\pi i} \int_{\mathcal{L}} e^{-\mu^2\phi(w)} \frac{dw}{\sqrt{w}}, \tag{30.4.69}$$

where

$$\phi(w) = w^2 - 2tw + \tfrac{1}{2}\ln w, \tag{30.4.70}$$

with positive saddle point

$$w_0 = \tfrac{1}{2}\left(t + \sqrt{t^2-1}\right). \tag{30.4.71}$$

Again, for $t \ge 1+\delta$ we can use Laplace's method. For $t \to -\infty$ the saddle point approaches the origin and for $t \le -1-\delta$ we can use the method described in §25.3 for loop integrals and an integration by parts procedure as described in §30.4.1.

When we have the expansions of $U(-a, \pm z)$ we can use the relation in (11.1.5). However, the gamma function with negative argument is not attractive to give a solid representation. To obtain a result for $V(-a,z)$ a suitable integral representation is (see Temme (2010a, Eq. 12.5.7))

$$V(a,z) = \frac{e^{-\frac14 z^2}}{2\pi}\int_{\mathcal{L}_+\cup\mathcal{L}_-} e^{-\frac12 s^2+zs}\, s^{a-\frac12}\, ds, \tag{30.4.72}$$

where $\mathcal{L}_+$ and $\mathcal{L}_-$ are two horizontal lines, $\mathcal{L}_+$ in the upper half-plane $\Im s>0$ and $\mathcal{L}_-$ in the lower half-plane $\Im s<0$. On both paths the integration is from $\Re s=-\infty$ to $\Re s=+\infty$.

In this case, the path of steepest descent is as follows. On $\mathcal{L}_+$ and $\mathcal{L}_-$ it runs from $-\infty$ to the saddle point w_0 along the path of steepest descent, and then from w_0 to $+\infty$ on the real axis (which is also a path of steepest descent for the integrand). The first contributions are complex conjugates, the contributions from the real interval are the same. For $t\ge 1+\delta$ we can use Laplace's method on semi-infinite intervals for both contributions.

When we have the contributions for $U(-a,z)$ and $V(-a,z)$ (which will be the same as in (30.2.9) and (30.2.13)) for $t\ge 1+\delta$, we can use the relation in (11.1.4) written in the form

$$V(-a,-z) = \frac{\cos\pi a}{\Gamma(\frac12+a)}U(-a,z) - \sin\pi a V(-a,z). \tag{30.4.73}$$

This will give the result as in (30.2.22).

Observe that when we integrate on $\mathcal{L}_+$ in the other direction (from $\Re s=+\infty$ to $\Re s=-\infty$) the contour $(-\mathcal{L}_+)\cup\mathcal{L}_-$ can be deformed into $\mathcal{C}$ of (30.4.68), and the integral defines $U(a,z)$, up to a factor. That is,

$$U(-a,z) = \frac{\Gamma(\frac12+a)}{2\pi i}e^{-\frac14 z^2}\int_{(-\mathcal{L}_+)\cup\mathcal{L}_-} e^{zs-\frac12 s^2}\, s^{-a-\frac12}\, ds. \tag{30.4.74}$$

30.5 Airy-type expansions

We start with the representation in (30.4.64). After scaling the parameters, we have

$$U\left(-\tfrac12\mu^2,\mu t\sqrt{2}\right) = \frac{e^{\frac14 z^2}(\mu\sqrt{2})^{\frac12+\frac12\mu^2}}{i\sqrt{2\pi}}\int_{\mathcal{C}} e^{\mu^2\phi(w)}\frac{dw}{\sqrt{w}}, \tag{30.5.75}$$

where

$$\phi(w) = w^2 - 2tw + \tfrac12\ln w. \tag{30.5.76}$$

We assume that $t\ge 1$. The real saddle points are

$$w_\pm = \tfrac12\left(t\pm\sqrt{t^2-1}\right). \tag{30.5.77}$$

When $t\to 1$ the saddle points in (30.5.77) coalesce. We take for the contour $\mathcal{C}$ in (30.5.75) vertical line trough w_+, or the saddle point contour through this point.

Following the method described in §23.5, we use the transformation in (23.5.54):

$$\phi(w) = \tfrac13 s^3 - \zeta s + A, \tag{30.5.78}$$

where s should be real and positive if $w > w_+$, with $\frac{1}{3}s^3 \sim w^2$ as $w \to +\infty$. This determines the branch of the three possible ones.

The values of ζ and A have to be determined. These quantities follow from substituting $w = w_\pm$ and the corresponding saddle points $\pm\sqrt{\zeta}$ in the s-plane, and the equations for A and ζ become

$$\tfrac{2}{3}\zeta^{\frac{3}{2}} = \tfrac{1}{2}t\sqrt{t^2-1} - \tfrac{1}{2}\operatorname{arccosh} t = \xi, \quad A = \tfrac{1}{2}\ln\tfrac{1}{2} - \tfrac{1}{4} - \tfrac{1}{2}t^2, \tag{30.5.79}$$

where ξ is defined in (30.1.3).

This gives for the representation in (30.5.75)

$$U\left(-\tfrac{1}{2}\mu^2, \mu t\sqrt{2}\right) = 2\sqrt{\pi}\,\mu h(\mu) F_\zeta(\mu), \tag{30.5.80}$$

where $h(\mu)$ is defined in (30.1.2), and (see (23.1.1))

$$F_\zeta(\mu) = \frac{1}{2\pi i}\int_{\mathcal{C}_1} e^{\mu^2\left(\frac{1}{3}s^3 - \zeta s\right)} f(s)\,ds, \tag{30.5.81}$$

with

$$f(s) = \frac{1}{\sqrt{w}}\frac{dw}{ds} = 2\sqrt{w}\frac{s^2-\zeta}{4w^2-4wt+1}. \tag{30.5.82}$$

The contour $\mathcal{C}_1$ is the one shown in Figure 8.2.

We can construct an asymptotic expansion by using the integration by parts method described in §23.2, and we obtain (cf. (23.2.9))

$$F_\zeta(\mu) \sim \frac{\operatorname{Ai}\left(\zeta\mu^{\frac{4}{3}}\right)}{\mu^{\frac{2}{3}}}\sum_{n=0}^{\infty}(-1)^n\frac{A_n(\zeta)}{\mu^{2n}} - \frac{\operatorname{Ai}'\left(\zeta\mu^{\frac{4}{3}}\right)}{\mu^{\frac{4}{3}}}\sum_{n=0}^{\infty}(-1)^n\frac{B_n(\zeta)}{\mu^{2n}}, \tag{30.5.83}$$

where the coefficients A_n, B_n are defined as in (23.2.10), with the function f_n defined in the scheme (23.2.11) and $f_0 = f$ defined in (30.5.82).

By using l'Hôpital's rule, it follows from (30.5.82) that

$$f(\sqrt{\zeta}) = f(-\sqrt{\zeta}) = \left(\frac{\zeta}{t^2-1}\right)^{\frac{1}{4}}, \tag{30.5.84}$$

and that, hence, $A_0(\zeta) = f(\sqrt{\zeta})$ and $B_0(\zeta) = 0$.

As explained in §23.3, the coefficients A_n and B_n can be expressed in terms of the derivatives of f at $\pm\sqrt{\zeta}$. For the first relations, see (23.3.28) and (23.3.29), where $b = \sqrt{\zeta}$.

To obtain these derivatives we need the coefficients in the expansions

$$w = w_+ + \sum_{k=1}^{\infty} c_k^+\left(s - \sqrt{\zeta}\right)^k, \quad w = w_- + \sum_{k=1}^{\infty} c_k^-\left(s + \sqrt{\zeta}\right)^k, \tag{30.5.85}$$

which can be obtained from (30.5.78) or from the relation for $\dfrac{dw}{ds}$ that follows from (30.5.82). From (30.5.82) and (30.5.84) we obtain

$$c_1^\pm = \sqrt{w_\pm}\left(\frac{\zeta}{t^2-1}\right)^{\frac{1}{4}} \tag{30.5.86}$$

and a few other coefficients are given by

$$
\begin{aligned}
c_2^+ &= \frac{2w_+^3 - (c_1^+)^3}{3w_+c_1^+(4w_+^2-1)},\\
c_3^+ &= \frac{-8w_+^6 - 16w_+^3(c_1^+)^3 + 36(c_1^+)^6w_+^2 + (c_1^+)^6}{36w_+^2(c_1^+)^3(4w_+^2-1)^2}.
\end{aligned}
\tag{30.5.87}
$$

It follows that

$$
\begin{aligned}
f'\left(\sqrt{\zeta}\right) &= \frac{4w_+c_2^+ - (c_1^+)^2}{2w_+^{\frac{3}{2}}},\\
f''\left(\sqrt{\zeta}\right) &= \frac{3\left(-4c_2^+c_1^+w_+ + (c_1^+)^3 + 8c_3^+w_+^2\right)}{4w_+^{\frac{5}{2}}},
\end{aligned}
\tag{30.5.88}
$$

and the same for the derivatives at $-\sqrt{\zeta}$ with w_+, c_j^+ replaced by $w_-, c_j^-, (j = 1, 2, 3)$.

With these quantities the coefficients A_1, B_1 can be evaluated, by using (23.3.28) and (23.3.29), where $b = \sqrt{\zeta}$.

At the transition point $t = 1$ ($\zeta = 0$) the coefficients $c_k^\pm$ are analytic functions of ζ, however, their numerical evaluation needs extra attention for small values of ζ (observe that $w_\pm \to \frac{1}{2}$, as $\zeta \to 0$). It is possible to give Maclaurin expansions of the coefficients. A different numerical approach is described in Temme (2000, §3.4). The same method works for the Airy-type expansions of the Bessel functions, and this can be found with more details in Temme (1997); see also Gil *et al.* (2007, §8.4).

Remark 30.6. For the expansions in terms of Airy functions, we have assumed $t \geq 1$, but the results are valid for $t \geq -1 + \delta$ (and for complex t-values). For $-1 < t < 1$, ζ is negative and we can replace (30.5.79) by

$$
\tfrac{2}{3}(-\zeta)^{\frac{3}{2}} = -\tfrac{1}{2}\sqrt{1-t^2} + \tfrac{1}{2}\arccos t, \quad -1 < t < 1. \tag{30.5.89}
$$

We can expand

$$
\zeta = 2^{\frac{1}{3}}(t-1)\left(1 + \tfrac{1}{10}(t-1) - \tfrac{2}{173}(t-1)^2 + \mathcal{O}\left((1-t)^3\right)\right), \quad t \to 1. \tag{30.5.90}
$$

△

Chapter 31

Coulomb wave functions

The differential equation

$$w'' + \left(1 - \frac{2\eta}{\rho} - \frac{\lambda(\lambda+1)}{\rho^2}\right) w = 0 \tag{31.0.1}$$

is a special form of Kummer's equation $zw'' + (c-z)w' - aw = 0$. Equation (31.0.1) plays an important role in physics, in particular in quantum mechanics as a form of the Schrödinger equation in a central Coulomb field. The solutions of (31.0.1) are called Coulomb wave functions, and are denoted by $F_\lambda(\eta,\rho)$ and $G_\lambda(\eta,\rho)$. $F_\lambda(\eta,\rho)$ is regular at $\rho = 0$, $G_\lambda(\eta,\rho)$ is irregular. The relation to the Kummer functions is usually written in the form

$$\begin{aligned} F_\lambda(\eta,\rho) &= A\,{}_1F_1\left(\begin{matrix}\lambda+1-i\eta\\ 2\lambda+2\end{matrix};2i\rho\right),\\ G_\lambda(\eta,\rho) &= iF_\lambda(\eta,\rho) + iB\,U(\lambda+1-i\eta, 2\lambda+2, 2i\rho),\\ A &= \frac{|\Gamma(\lambda+1+i\eta)|e^{-\pi\eta/2-i\rho}(2\rho)^{\lambda+1}}{2\Gamma(2\lambda+2)},\\ B &= e^{\pi\eta/2+\lambda\pi i-i\sigma_\lambda(\eta)-i\rho}(2\rho)^{\lambda+1}, \end{aligned} \tag{31.0.2}$$

$$\sigma_\lambda(\eta) = \operatorname{ph}\Gamma(\lambda+1+i\eta) \quad \text{(the Coulomb phase shift).}$$

When $\lambda = L = 0, 1, 2, \ldots$, which is usually the case in physical applications, $G_\lambda(\eta,\rho)$ has a logarithmic singularity at $\rho = 0$. The functions $F_\lambda(\eta,\rho)$ and $G_\lambda(\eta,\rho)$ are real for real values of η, $\rho > 0, \lambda \geq 0$. This certainly does not follow directly from the above definitions, considering the many complex parameters in the definitions. For $F_\lambda(\eta,\rho)$, however, this result follows directly from Kummer's transformation given in (10.1.9).

The Coulomb phase shift $\sigma_\lambda(\eta)$ is zero when $\eta = 0$ and continuous elsewhere. The quantities A and B in the above definitions are chosen such that $F_\lambda(\eta,\rho)$ and $G_\lambda(\eta,\rho)$ behave asymptotically as circular functions,

$$F_\lambda(\eta,\rho) = \sin\theta_\lambda(\eta,\rho) + o(1), \quad G_\lambda(\eta,\rho) = \cos\theta_\lambda(\eta,\rho) + o(1), \quad \rho \to \infty, \tag{31.0.3}$$

where

$$\theta_\lambda(\eta,\rho) = \rho - \eta\ln(2\rho) - \tfrac{1}{2}\lambda\pi + \sigma_\lambda(\eta). \tag{31.0.4}$$

It is possible to introduce representations like those for the Bessel functions in terms of the Hankel functions; see §9.2. For the Coulomb functions we write

$$\begin{aligned} H_\lambda^+(\eta,\rho) &= G_\lambda(\eta,\rho) + iF_\lambda(\eta,\rho),\\ H_\lambda^-(\eta,\rho) &= G_\lambda(\eta,\rho) - iF_\lambda(\eta,\rho). \end{aligned} \tag{31.0.5}$$

In terms of the Kummer functions we have

$$H_\lambda^\pm(\eta,\rho) = (2\rho)^{\lambda+1} e^{\frac12\pi\eta} e^{\pm i\rho \mp \pi i(\lambda+\frac12)\pm i\sigma_\lambda(\eta)} U^\pm, \tag{31.0.6}$$

where

$$U^\pm = U(\lambda+1\pm i\eta, 2\lambda+2, \mp 2i\rho), \tag{31.0.7}$$

with $\sigma_\lambda(\eta)$ defined in (31.0.2). These relations follow from the definitions in (31.0.2) and the connection formulas of the Kummer functions; see (10.1.11).

This gives

$$\begin{aligned} F_\lambda(\eta,\rho) &= \tfrac12 (2\rho)^{\lambda+1} e^{\frac12\pi\eta} \left(e^{i\chi}U^- + e^{-i\chi}U^+\right),\\ G_\lambda(\eta,\rho) &= -\tfrac12 i\, (2\rho)^{\lambda+1} e^{\frac12\pi\eta} \left(e^{i\chi}U^- - e^{-i\chi}U^+\right), \end{aligned} \tag{31.0.8}$$

where

$$\chi = (\lambda+1)\pi - \rho - \sigma_\lambda(\eta). \tag{31.0.9}$$

When $\eta = 0$ the Coulomb functions reduce to Bessel functions, that is,

$$U^- = \tfrac12\sqrt{\pi} e^{-\pi i(\lambda+1)+i\rho} H^{(2)}_{\lambda+\frac12}(\rho), \quad U^+ = \tfrac12\sqrt{\pi} e^{\pi i(\lambda+1)-i\rho} H^{(1)}_{\lambda+\frac12}(\rho), \tag{31.0.10}$$

which give

$$F_\lambda(0,\rho) = \sqrt{\tfrac12\pi\rho}\, J_{\lambda+\frac12}(\rho), \quad G_\lambda(0,\rho) = -\sqrt{\tfrac12\pi\rho}\, Y_{\lambda+\frac12}(\rho). \tag{31.0.11}$$

For more information we refer to Thompson (2010). This reference gives the expansions for large ρ, and some details on Airy-type expansions, which are needed when ρ and η are large, and $\rho \sim 2\eta$.

In this chapter we derive expansions for large positive ρ and for large $\pm\eta$, with λ fixed in all cases.

31.1 Contour integrals for Coulomb functions

We express the Coulomb functions in terms of integrals of the Kummer U-function by using the following integral representations of the U-functions:

$$\begin{aligned} U(a,c,i\zeta) &= \frac{(\frac12\zeta)^{1-c}\Gamma(c-a)e^{\frac12 i\zeta-\pi i a}}{2\pi i}\int_{\mathcal{L}_-} e^{\frac12\zeta w}\frac{(w+i)^a}{(w-i)^a}\frac{dw}{(w+i)^c},\\ U(a,c,-i\zeta) &= \frac{(\frac12\zeta)^{1-c}\Gamma(c-a)e^{-\frac12 i\zeta+\pi i a}}{2\pi i}\int_{\mathcal{L}_+} e^{\frac12\zeta w}\frac{(w-i)^a}{(w+i)^a}\frac{dw}{(w-i)^c}, \end{aligned} \tag{31.1.12}$$

where $\mathcal{L}_\pm$ are Hankel contours around the branch cuts from $\pm i$ to $-\infty$; $(w\pm i)^{a-c} = 1$ and $(w\pm i)^a = 1$ when $w \pm i = 1$. These representations hold when $\Re\zeta > 0$ and $a - c \neq 0, 1, 2, \ldots$, and follow from (10.1.8). A direct proof follows from integrating along the lower and upper sides of the branch cuts, assuming temporarily $\Re(a-c) > -1$, and comparing the result with the standard integral representation in (10.1.5) and using (10.1.10).

By using (10.1.11) with the lower signs and the representations in (31.1.12), we obtain

$$\frac{1}{\Gamma(c)}\,{}_1F_1\left({a \atop c};\, i\zeta\right) = \frac{(\frac12\zeta)^{1-c}e^{\frac12 i\zeta}}{2\pi i}\int_{\mathcal{L}} e^{\frac12\zeta w}\left(\frac{w+i}{w-i}\right)^a \frac{dw}{(w+i)^c}, \tag{31.1.13}$$

where $\mathcal{L}$ is the Hankel contour around the two branch cuts from $\pm i$ to $-\infty$, or $\mathcal{L}$ is the union of two Hankel contours around each branch cut.

For the functions $U^\pm$ we have

$$\begin{aligned} e^{i\chi}U^- &= \frac{\rho^{-2\lambda-1}|\Gamma(\lambda+1+i\eta)|e^{-\pi\eta}}{2\pi i}\int_{\mathcal{L}_-}\left(\frac{w+i}{w-i}\right)^{-i\eta}\frac{e^{\rho w}\,dw}{(w^2+1)^{\lambda+1}},\\ e^{-i\chi}U^+ &= \frac{\rho^{-2\lambda-1}|\Gamma(\lambda+1-i\eta)|e^{-\pi\eta}}{2\pi i}\int_{\mathcal{L}_+}\left(\frac{w-i}{w+i}\right)^{i\eta}\frac{e^{\rho w}\,dw}{(w^2+1)^{\lambda+1}}. \end{aligned} \tag{31.1.14}$$

31.2 Expansions for $\rho \to \infty$ and bounded η

We take the representations in (31.0.7) and (31.0.8). We use (10.2.18) and obtain

$$\begin{aligned} (2\rho)^{\lambda+1}e^{\frac12\pi\eta+i\chi}U^- &= ie^{-i\theta_\lambda(\eta,\rho)}\left(P_\lambda(\eta,\rho)+iQ_\lambda(\eta,\rho)\right),\\ (2\rho)^{\lambda+1}e^{\frac12\pi\eta-i\chi}U^+ &= -ie^{i\theta_\lambda(\eta,\rho)}\left(P_\lambda(\eta,\rho)-iQ_\lambda(\eta,\rho)\right), \end{aligned} \tag{31.2.15}$$

where

$$\begin{aligned} P_\lambda(\eta,\rho)+iQ_\lambda(\eta,\rho) &\sim \sum_{k=0}^\infty \frac{(\lambda+1-i\eta)_k(-\lambda-i\eta)_k}{k!\,(-2i\rho)^k} = \sum_{k=0}^\infty \frac{p_k+iq_k}{(2\rho)^k},\\ P_\lambda(\eta,\rho)-iQ_\lambda(\eta,\rho) &\sim \sum_{k=0}^\infty \frac{(\lambda+1+i\eta)_k(-\lambda+i\eta)_k}{k!\,(2i\rho)^k} = \sum_{k=0}^\infty \frac{p_k-iq_k}{(2\rho)^k}. \end{aligned} \tag{31.2.16}$$

This gives

$$\begin{aligned} F_\lambda(\eta,\rho) &= \sin\theta_\lambda(\eta,\rho)\,P_\lambda(\eta,\rho) - \cos\theta_\lambda(\eta,\rho)\,Q_\lambda(\eta,\rho),\\ G_\lambda(\eta,\rho) &= \cos\theta_\lambda(\eta,\rho)\,P_\lambda(\eta,\rho) + \sin\theta_\lambda(\eta,\rho)\,Q_\lambda(\eta,\rho), \end{aligned} \tag{31.2.17}$$

where $\theta_\lambda(\eta,\rho)$ is defined in (31.0.4).

For p_k, q_k we have $p_0 = 1, q_0 = 0$ and for $k \ge 0$ the recurrence relations

$$\begin{aligned} (k+1)p_{k+1} &= \eta(2k+1)p_k - \left(k+k^2-\lambda^2-\lambda-\eta^2\right)q_k,\\ (k+1)q_{k+1} &= (k+k^2-\lambda^2-\lambda-\eta^2)p_k + \eta(2k+1)q_k. \end{aligned} \tag{31.2.18}$$

For the derivatives we have the representations

$$\begin{aligned} F'_\lambda(\eta,\rho) &= \cos\theta_\lambda(\eta,\rho)\, R_\lambda(\eta,\rho) + \sin\theta_\lambda(\eta,\rho)\, S_\lambda(\eta,\rho), \\ G'_\lambda(\eta,\rho) &= -\sin\theta_\lambda(\eta,\rho)\, R_\lambda(\eta,\rho) + \cos\theta_\lambda(\eta,\rho)\, S_\lambda(\eta,\rho), \end{aligned} \tag{31.2.19}$$

with expansions

$$R_\lambda(\eta,\rho) \sim \sum_{k=0}^{\infty} \frac{r_k}{(2\rho)^k}, \quad S_\lambda(\eta,\rho) \sim \sum_{k=0}^{\infty} \frac{s_k}{(2\rho)^k}, \tag{31.2.20}$$

where $r_0 = p_0$, $s_0 = q_0$, and for $k \ge 1$

$$\begin{aligned} r_k &= p_k + 2(k-1)q_{k-1} - 2\eta p_{k-1}, \\ s_k &= q_k - 2(k-1)p_{k-1} - 2\eta q_{k-1}. \end{aligned} \tag{31.2.21}$$

From the Wronskian

$$G_\lambda(\eta,\rho)F'_\lambda(\eta,\rho) - G'_\lambda(\eta,\rho)F_\lambda(\eta,\rho) = 1, \tag{31.2.22}$$

it follows that

$$P_\lambda(\eta,\rho)R_\lambda(\eta,\rho) + Q_\lambda(\eta,\rho)S_\lambda(\eta,\rho) = 1. \tag{31.2.23}$$

We note that the expansions give the earlier mentioned result in (31.0.3).

Remark 31.1.

In the expansions of this section we need the Coulomb phase shift $\sigma_\lambda(\eta)$, see (31.0.2) and (31.0.4); in later expansions $\sigma_\lambda(\eta)$ does not appear. The computation can be done by using the definition in (31.0.2) or

$$\sigma_\lambda(\eta) = \ln \frac{\Gamma(\lambda+1+i\eta)}{|\Gamma(\lambda+1+i\eta)|}. \tag{31.2.24}$$

However, when we use the standard software for this, assuming that a code for $\Gamma(z)$ (or the logarithm) is available for complex argument, the result of the computation may give a value of $\sigma_\lambda(\eta)$ in $(-\pi,\pi]$, the principal value domain of the logarithm. In particular for large values of η this answer will be correct modulo 2π. From the expansion of the gamma function in Chapter 6, in particular of §6.3, it is not difficult to obtain the expansion (see (6.3.42))

$$\sigma_\lambda(\eta) = \left(x - \tfrac{1}{2}\right)\theta + y\ln(r) - y + S_\lambda(\eta), \tag{31.2.25}$$

where

$$x = \lambda + 1, \quad y = \eta, \quad r = \sqrt{x^2+y^2}, \quad \theta = \arctan(y/x), \tag{31.2.26}$$

and

$$S_\lambda(\eta) \sim -\frac{\sin\theta}{12r} + \frac{\sin(3\theta)}{360r^3} - \frac{\sin(5\theta)}{1260r^5} + \ldots, \quad r \to \infty. \tag{31.2.27}$$

Although we do not assume in this section that η is large, this expansion will give information on the proper choice of the multiple of 2π to be used in calculations.

For example, we take $\lambda = \frac{1}{2}$, $x = \frac{3}{2}$, $y = \eta = 10$. Then, a straightforward computation of $p = \ln\Gamma(x+iy)$ gives $p \doteq -12.48519120 + 1.984487352\,i$. When we take in (31.2.25) $S_\lambda(\eta) = 0$, then $q = \sigma_\lambda(\eta) \doteq 14.55901036$, and $(q - \Im p)/(2\pi) \doteq 2.001297494$. We conclude that we should enhance $\Im p$ by 4π to obtain the correct value of $\sigma_\lambda(\eta)$. In Maple a function $lnGAMMA(z)$ is available, which gives $-12.48519120 + 14.55085797\,i$, and the imaginary part is the correct value of $\sigma_\lambda(\eta)$.

See also Luke (1970), where this choice is based on remainder estimates in the asymptotic expansion of $\ln\Gamma(z)$. △

31.3 Expansions for $\eta \to \infty$ and bounded ρ

We use the results of §10.3.2, where we have given expansions of $U(a,c,z)$ for large a. We slightly modify the approach in order to obtain expansions of $U^\pm = U(\lambda+1+\pm a, 2\lambda+2, \mp 2i\rho)$ with $a = i\eta$ (see (31.0.7)) with large positive η. The expansions in §10.3.2 are valid for complex a and z (see also Remark 10.2).

The result in (10.3.37) now reads

$$U(\lambda+1+a, 2\lambda+2, z) \sim 2\frac{(z/a)^{-\lambda-\frac{1}{2}}e^{\frac{1}{2}z}}{\Gamma(\lambda+1+a)}\sum_{k=0}^{\infty} c_k(\rho)\left(\frac{z}{a}\right)^{\frac{1}{2}k} K_{2\lambda+1-k}(\zeta), \quad (31.3.28)$$

where $\zeta = 2\sqrt{az}$, $z = -2i\rho$, and $c_k(\rho)$ are the coefficients of the expansion $f(s) = \sum_{k=0}^{\infty} c_k(\rho)s^k$, where

$$f(s) = e^{zg(s)}\left(\frac{\frac{1}{2}s}{\sinh\frac{1}{2}s}\right)^{2\lambda+2}, \quad g(s) = \frac{1}{s} - \frac{1}{e^s-1} - \frac{1}{2}. \quad (31.3.29)$$

The first coefficients are

$$\begin{aligned} &c_0(\rho) = 1, \quad c_1(\rho) = \tfrac{1}{6}i\rho, \quad c_2(\rho) = -\tfrac{1}{72}\left(6\lambda+6+\rho^2\right),\\ &c_3(\rho) = -\tfrac{1}{6480}i\rho\left(90\lambda+108+5\rho^2\right), \\ &c_4(\rho) = \tfrac{1}{155520}\left(1188\lambda+540\lambda^2+252\rho^2+180\lambda\rho^2+648+5\rho^4\right). \end{aligned} \quad (31.3.30)$$

Note that $c_{2k}(\rho)$ and $ic_{2k+1}(\rho)$ are real.

As in (10.3.39), we can obtain an expansion with only two Bessel functions. We take $a = i\eta$, $z = -2i\rho$, $z/a = e^{-\pi i}2\rho/\eta$, $\zeta = 2\sqrt{2\rho\eta}$, and obtain

$$U^+ \sim 2\frac{e^{(\lambda+\frac{1}{2})\pi i}(2\rho/\eta)^{-\lambda-\frac{1}{2}}e^{-i\rho}}{\Gamma(\lambda+1+i\eta)}\left(K_{2\lambda+1}(\zeta)\,P + \sqrt{2\rho/\eta}\,K_{2\lambda+2}(\zeta)\,Q\right), \quad (31.3.31)$$

where

$$P \sim \sum_{k=0}^{\infty}\frac{a_k(\rho)}{\eta^k}, \quad Q \sim \sum_{k=0}^{\infty}\frac{b_k(\rho)}{\eta^k}, \quad \eta\to\infty. \quad (31.3.32)$$

This gives, by using (31.0.9),

$$e^{-i\chi}U^{+} \sim -2i\frac{(2\rho/\eta)^{-\lambda-\frac{1}{2}}}{|\Gamma(\lambda+1+i\eta)|}\left(K_{2\lambda+1}(\zeta)\,P+\sqrt{2\rho/\eta}\,K_{2\lambda+2}(\zeta)\,Q\right), \tag{31.3.33}$$

and similarly for $e^{i\chi}U^{-}$ by changing the sign of i.

The first coefficients are

$$\begin{aligned}
&a_0(\rho)=1, \quad b_0(\rho)=-ic_1(\rho)=\tfrac{1}{6}\rho,\\
&a_1(\rho)=2i\lambda c_1(\rho)+ic_1(\rho)-2\rho c_2(\rho), \quad b_1(\rho)=2\lambda c_2(\rho)+2i\rho c_3(\rho),\\
&a_2(\rho)=-4\lambda^2c_2(\rho)-2\lambda c_2(\rho)-8i\lambda\rho c_3(\rho)+4\rho^2c_4(\rho),\\
&b_2(\rho)=4i\lambda^2c_3(\rho)-2i\lambda c_3(\rho)-8\lambda\rho c_4(\rho)+4\rho c_4(\rho)-4i\rho^2c_5(\rho).
\end{aligned} \tag{31.3.34}$$

Note that the coefficients $a_k(\rho)$ and $b_k(\rho)$ are all real.

It follows that the right-hand side of (31.3.31) is purely imaginary and when we use (31.0.8) we obtain a vanishing contribution for $F_\lambda(\eta,\rho)$. For $G_\lambda(\eta,\rho)$ we obtain

$$G_\lambda(\eta,\rho) \sim 2\frac{\sqrt{2\rho}\,\eta^{\lambda+\frac{1}{2}}\,e^{\frac{1}{2}\pi\eta}}{|\Gamma(\lambda+1+i\eta)|}\left(K_{2\lambda+1}(\zeta)\,P+\sqrt{2\rho/\eta}\,K_{2\lambda+2}(\zeta)\,Q\right). \tag{31.3.35}$$

For $F_\lambda(\eta,\rho)$ we need a different approach. The message of the vanishing of the contribution for this function is that it is exponentially small compared with $G_\lambda(\eta,\rho)$ for large positive values of η and bounded ρ. We use the first line of (31.0.2) and the results for the ${}_1F_1$-function given in §10.3.3.

We have (compare (10.3.47))

$$\begin{aligned}
{}_1F_1\left(\begin{matrix}\lambda+1-i\eta\\ 2\lambda+2\end{matrix};\,2i\rho\right) &\sim \frac{\Gamma(2\lambda+2)\Gamma(-\lambda-i\eta)e^{i\rho}}{\Gamma(\lambda+1-i\eta)}\left(\frac{2\rho}{\eta}e^{\pi i}\right)^{-\lambda-\frac{1}{2}}\times\\
&\sum_{k=0}^{\infty}(-i)^k c_k(\rho)\left(\frac{2\rho}{\eta}\right)^{\frac{1}{2}k} I_{2\lambda-k+1}(\zeta),
\end{aligned} \tag{31.3.36}$$

where ζ is the same as in (31.3.31) and the coefficients $c_k(\rho)$ are the same as in (31.3.28).

Rearranging and using (31.0.2), we obtain

$$F_\lambda(\eta,\rho) \sim \tfrac{1}{2}\sqrt{2\rho}\,\eta^{\lambda+\frac{1}{2}}e^{-\frac{1}{2}\pi\eta}R\left(I_{2\lambda+1}(\zeta)\,P-\sqrt{\frac{2\rho}{\eta}}I_{2\lambda+2}(\zeta)\,Q\right), \tag{31.3.37}$$

where P and Q have the expansions given in (31.3.32) (with the same coefficients), and

$$R=\frac{|\Gamma(\lambda+1+i\eta)|\Gamma(-i\eta-\lambda)}{\Gamma(\lambda+1-i\eta)}e^{-(\lambda+\frac{1}{2})\pi i}. \tag{31.3.38}$$

We can write

$$R=\frac{2\pi\,e^{-\pi\eta}}{|\Gamma(\lambda+1+i\eta)|\,(1-e^{2\lambda\pi i-2\pi\eta})}\sim\frac{2\pi\,e^{-\pi\eta}}{|\Gamma(\lambda+1+i\eta)|}, \tag{31.3.39}$$

with an exponentially small error when $\eta\to\infty$. See also (31.4.45).

31.4 Expansions for $\eta \to -\infty$ and bounded ρ

Again we can use the results of §10.3.2, but first we explain how to handle the complex quantities in the expressions for $U^{\pm}$ defined in (31.0.7) and used in (31.0.8). Indeed, this time we can use these relations for both $F_\lambda(\eta,\rho)$ and $G_\lambda(\eta,\rho)$, because now the functions are of the same order.

When $\eta < 0$ we write $\eta = e^{-\pi i}(-\eta)$. This gives $a = i\eta = e^{-\frac{1}{2}\pi i}(-\eta)$. For $z = -2i\rho$ we write $z = 2\rho e^{-\frac{1}{2}\pi i}$. In this way,

$$az = 2\rho(-\eta)e^{-\pi i}, \quad 2\sqrt{az} = -iu, \quad u = 2\sqrt{-2\eta\rho}, \quad w = \sqrt{\frac{z}{a}} = \sqrt{\frac{2\rho}{-\eta}}, \tag{31.4.40}$$

where u and w are real.

The result comparable with (31.3.28) reads for the present case

$$U^+ \sim 2\frac{(-2\rho/\eta)^{-\frac{1}{2}-\lambda}e^{-i\rho}}{\Gamma(\lambda+1+i\eta)}\sum_{k=0}^{\infty} c_k(\rho)\left(\frac{-2\rho}{\eta}\right)^{\frac{1}{2}k} K_{2\lambda+1-k}(-iu), \tag{31.4.41}$$

where $\eta \to -\infty$, and $c_k(\rho)$ are the coefficients as used in (31.3.28).

The expansion with only two Bessel functions is

$$\begin{aligned} U^+ = {} & 2\frac{(-2\rho/\eta)^{-\frac{1}{2}-\lambda}e^{-i\rho}}{\Gamma(\lambda+1+i\eta)} \times \\ & \left(K_{2\lambda+1}(-iu)\,P + i\sqrt{-2\rho/\eta}\,K_{2\lambda+2}(-iu)\,Q\right), \end{aligned} \tag{31.4.42}$$

where P and Q and the coefficients are as in the previous section.

We replace the K-functions by Hankel functions (see the first line in (9.2.5)), and write the Hankel functions in terms of ordinary Bessel functions (see (9.2.8)). This gives

$$\begin{aligned} e^{-i\chi}U^+ = {} & \pi\frac{(-2\rho/\eta)^{-\frac{1}{2}-\lambda}}{|\Gamma(\lambda+1+i\eta)|} \times \\ & \left(\left(J_{2\lambda+1}(u) + iY_{2\lambda+1}(u)\right)P - \sqrt{-2\rho/\eta}\left(J_{2\lambda+2}(u) + iY_{2\lambda+2}(u)\right)Q\right). \end{aligned} \tag{31.4.43}$$

A similar result can be obtained for $e^{i\chi}U^-$ by changing the sign of i throughout.

Using (31.0.8), we obtain for the Coulomb functions

$$\begin{aligned} F_\lambda(\eta,\rho) = {} & \pi\frac{\sqrt{2\rho}(-\eta)^{\lambda+\frac{1}{2}}e^{\frac{1}{2}\pi\eta}}{|\Gamma(\lambda+1+i\eta)|}\left(J_{2\lambda+1}(u)\,P - \sqrt{-2\rho/\eta}\,J_{2\lambda+2}(u)\,Q\right), \\ G_\lambda(\eta,\rho) = {} & -\pi\frac{\sqrt{2\rho}(-\eta)^{\lambda+\frac{1}{2}}e^{\frac{1}{2}\pi\eta}}{|\Gamma(\lambda+1+i\eta)|}\left(Y_{2\lambda+1}(u)\,P - \sqrt{-2\rho/\eta}\,Y_{2\lambda+2}(u)\,Q\right), \end{aligned} \tag{31.4.44}$$

where $u = 2\sqrt{-2\rho\eta}$.

From the expansion of the gamma function given in §6.4, it follows that

$$e^{-\frac{1}{2}\pi\eta}\,|\Gamma(\lambda+1+i\eta)| \sim \sqrt{2\pi}\,(-\eta)^{\lambda+\frac{1}{2}}, \quad \eta \to -\infty. \tag{31.4.45}$$

From this estimate we conclude that the front factors in the representations in (31.4.44) are of order $\mathcal{O}(1)$, as $\eta \to -\infty$, because we have assumed that $\rho = \mathcal{O}(1)$.

31.5 Expansions for $\eta \to -\infty$ and $\rho \ge \rho_0 > 0$

When $\eta \to -\infty$ and $\rho \ge \rho_0 > 0$ we can use the saddle point method, and obtain a result that can be used uniformly for $\rho \ge \rho_0$.

We write

$$\rho = -r\eta, \tag{31.5.46}$$

and using (31.1.14) we have

$$e^{i\chi}U^- = \frac{\rho^{-1-2\lambda}|\Gamma(\lambda+1+i\eta)|e^{-\pi\eta}}{2\pi i}\int_{\mathcal{L}_-}\frac{e^{-\eta\phi(w)}\,dw}{(w^2+1)^{\lambda+1}}, \tag{31.5.47}$$

where

$$\phi(w) = rw + i\ln\frac{w+i}{w-i}. \tag{31.5.48}$$

There are two saddle points:

$$w_\pm = \pm i\sqrt{\frac{r+2}{r}} = \pm i\frac{W}{r}, \quad W = \sqrt{r(r+2)}. \tag{31.5.49}$$

We see that w_- is on the negative imaginary axis below the singularity at $w = -i$; w_+ is above the singularity at $w = i$. It appears that w_- is relevant for the saddle point analysis of U^-.

We have

$$\phi(w_-) = -iW + i\ln\frac{W-r}{W+r}, \quad \phi''(w_-) = irW. \tag{31.5.50}$$

Hence, at the saddle point w_- the saddle point contour makes an angle of $\frac{1}{4}\pi$ with the horizontal direction (see §4.1). In (31.5.47) we use the transformation

$$\phi(w) = \phi\left(w_-\right) + \tfrac{1}{2}\phi''(w_-)s^2, \tag{31.5.51}$$

with $w = w_- + s + \mathcal{O}(s^2)$ as $s \to 0$, and obtain

$$e^{i\chi}U^- = \rho^{-1-2\lambda}|\Gamma(\lambda+1+i\eta)|e^{-\pi\eta+i\eta W-i\eta\ln\frac{W-r}{W+r}}S^-, \tag{31.5.52}$$

where

$$S^- = \frac{1}{2\pi i}\int_{\infty e^{-\frac{3}{4}\pi i}}^{\infty e^{\frac{1}{4}\pi i}} e^{-\frac{1}{2}\eta\phi''(w_-)s^2}f(s)\,ds, \tag{31.5.53}$$

and

$$f(s) = \frac{dw}{ds}\frac{1}{(w^2+1)^{\lambda+1}}, \quad \frac{dw}{ds} = \frac{s\phi''(w_-)}{r+2/(w^2+1)}. \tag{31.5.54}$$

The substitution of the expansion $f(s) = \sum_{k=0}^{\infty} f_k s^k$ gives

$$S^- \sim \sum_{k=0}^{\infty} f_{2k}\frac{1}{2\pi i}\int_{\infty e^{-\frac{3}{4}\pi i}}^{\infty e^{\frac{1}{4}\pi i}} e^{-\frac{1}{2}\eta\phi''(w_-)s^2}s^{2k}\,ds. \tag{31.5.55}$$

Substituting $s = te^{\frac{1}{4}\pi i}$, we obtain

$$S^- \sim \frac{e^{\frac{1}{4}\pi i}}{2\pi i} \sum_{k=0}^{\infty} i^k f_{2k} \int_{-\infty}^{\infty} e^{\frac{1}{2}\eta r W t^2} t^{2k}\, dt, \tag{31.5.56}$$

that is,

$$S^- \sim \left(\tfrac{1}{2}r\right)^{\lambda+1} \frac{e^{-\frac{1}{4}\pi i+(\lambda+1)\pi i}}{\sqrt{-2\pi r W\eta}} \sum_{k=0}^{\infty} i^k \frac{s_k}{(-\eta)^k}, \tag{31.5.57}$$

where

$$s_k = \left(\tfrac{1}{2}\right)_k \left(\frac{2}{rW}\right)^k \frac{f_{2k}}{f(0)}, \quad f(0) = \left(\tfrac{1}{2}re^{\pi i}\right)^{\lambda+1}. \tag{31.5.58}$$

To compute the coefficients we first need the coefficients in the expansion $w = w_- + \sum_{k=0}^{\infty} d_k s^k$. These follow from (31.5.51), and we obtain

$$d_1 = 1, \quad d_2 = \frac{ir(3+2r)}{6W}, \quad d_3 = -\frac{r\left(9+6r+2r^2\right)}{72(r+2)}. \tag{31.5.59}$$

The first s_k are

$$s_0 = 1, \quad s_1 = -\frac{12(r+2)^2\lambda(\lambda+1)+9+6r+2r^2}{24(r+2)W}. \tag{31.5.60}$$

By changing signs of i in (31.3.36) we obtain for U^+:

$$e^{-i\chi}U^+ = \rho^{-1-2\lambda}|\Gamma(\lambda+1-i\eta)|e^{-\pi\eta-i\eta W+i\eta\ln\frac{W-r}{W+r}}S^+, \tag{31.5.61}$$

with expansion (cf. (31.4.41))

$$S^+ \sim \left(\tfrac{1}{2}r\right)^{\lambda+1} \frac{e^{\frac{1}{4}\pi i-(\lambda+1)\pi i}}{\sqrt{-2\pi r\eta W}} \sum_{k=0}^{\infty} (-i)^k \frac{s_k}{(-\eta)^k}. \tag{31.5.62}$$

The real coefficients s_k are the same as those in (31.5.57) for S^-.

These representations in terms of $S^\pm$ give for $F_\lambda(\eta,\rho)$ and $G_\lambda(\eta,\rho)$, by using (31.0.8),

$$\begin{aligned} F_\lambda(\eta,\rho) &= \frac{\sqrt{r}\,(-\eta)^{-\lambda-\frac{1}{2}}e^{-\frac{1}{2}\pi\eta}\,|\Gamma(\lambda+1+i\eta)|}{2\sqrt{2\pi W}}\left(e^{i\tau}T^- + e^{-i\tau}T^+\right), \\ G_\lambda(\eta,\rho) &= \frac{\sqrt{r}\,(-\eta)^{-\lambda-\frac{1}{2}}e^{-\frac{1}{2}\pi\eta}\,|\Gamma(\lambda+1+i\eta)|}{2i\sqrt{2\pi W}}\left(e^{i\tau}T^- - e^{-i\tau}T^+\right), \end{aligned} \tag{31.5.63}$$

where

$$\tau = -\tfrac{1}{4}\pi + (\lambda+1)\pi + \eta W - \eta\ln\frac{W-r}{W+r}, \quad W = \sqrt{r(r+2)}, \quad r = -\rho/\eta, \tag{31.5.64}$$

and

$$T^\pm \sim \sum_{k=0}^{\infty} (\mp i)^k \frac{s_k}{(-\eta)^k}, \quad \eta \to -\infty. \tag{31.5.65}$$

Table 31.1 Relative errors in the computation with the Wronskian relation given in (31.5.66).

η	−100		−1000	
r	ρ	W	ρ	W
1/10	10	0.25e-07	100	0.25e-13
1	100	0.37e-09	1000	0.37e-15
10	1000	0.15e-09	10000	0.15e-15
100	10000	0.18e-09	100000	0.18e-15
1000	100000	0.18e-09	1000000	0.19e-15

From the estimate given in (31.4.45) we conclude that the front factors in the representations in (31.5.63) are of order $\mathcal{O}(1)$, as $\eta \to -\infty$, uniformly with respect to unbounded values of ρ.

We have used the Wronskian relation

$$F_{\lambda-1}(\eta,\rho)G_\lambda(\eta,\rho) - F_\lambda(\eta,\rho)G_{\lambda-1}(\eta,\rho) = \frac{\lambda}{\sqrt{\lambda^2+\eta^2}}, \quad \lambda \ge 1, \tag{31.5.66}$$

to verify the expansions of this section. In Table 31.1 we show the relative errors in the Wronskian relation. We take six terms in the expansions of $T^{\pm}$, $\lambda = 2.5$, $\eta = -100$, $\eta = -1000$, and several values of $r = -\rho/\eta$. We observe that the error remains of the same order as ρ increases.

31.6 Expansions for $\eta \to -\infty$ and $\rho \ge 0$

The expansions in the previous section cannot be used when $r = -\rho/\eta$ tends to zero. When r is bounded from 0, ρ should become large, and we cannot use the results of §31.4. To handle all possible values for $\rho \ge 0$ we need a more powerful result, again in terms of Bessel functions, and this expansion follows from the results for the Kummer U-function given in §27.4.1. Because of the special parameters of the U-function in the case of the Coulomb functions we give a few steps in the derivation of the expansion.

The representation in (27.4.29) becomes

$$\Gamma(\lambda+1+a)U(\lambda+1+a, 2\lambda+2, az) = \int_0^\infty w^{-2\lambda-2}e^{-a\phi(w)}g(w)\,dw, \tag{31.6.67}$$

where

$$\phi(w) = w + \frac{z}{e^w-1}, \quad g(w) = \left(\frac{w/2}{\sinh(w/2)}\right)^{2\lambda+2}. \tag{31.6.68}$$

There are saddle points at $\pm w_0$, with w_0 defined by

$$w_0 = 2\,\mathrm{arcsinh}\left(\tfrac{1}{2}\sqrt{z}\right), \tag{31.6.69}$$

and the transformation

$$\tfrac{1}{2}z + w + \frac{z}{e^w - 1} = t + \frac{\beta^2}{t}, \quad \beta = \tfrac{1}{2}\left(w_0 + \sinh w_0\right), \tag{31.6.70}$$

gives

$$\Gamma(\lambda+1+a)U(\lambda+1+a, 2\lambda+2, az) = e^{\frac{1}{2}za} \int_0^\infty t^{-2\lambda-2} e^{-a(t+\beta^2/t)} f(t)\, dt, \tag{31.6.71}$$

where

$$f(t) = \left(\frac{t}{w}\right)^{2\lambda+2} \left(\frac{w/2}{\sinh(w/2)}\right)^{2\lambda+2} \frac{dw}{dt}. \tag{31.6.72}$$

The analogue of the expansion in (27.4.42) can be obtained by using the scheme given in (27.1.9), with f given in (31.6.72). Observe that the λ in §27.1 has a different meaning compared with the present λ. See the integrals in (27.1.1) and (31.6.71).

Furthermore, the function $f(t)$ is even. In Remark 27.1 we have explained that this is a favorable condition for the expansion, because the coefficients a_{2k+1} and b_{2k} will vanish. We obtain

$$\begin{aligned} U(\lambda+1+a\,,\, 2\lambda+2, az) &\sim \frac{2\beta^{-2\lambda-1} e^{\frac{1}{2}az}}{\Gamma(\lambda+1+a)} \times \\ &\left(K_{2\lambda+1}(2\beta a) \sum_{k=0}^\infty \frac{a_{2k}}{a^{2k}} + \beta K_{2\lambda}(2\beta a) \sum_{k=0}^\infty \frac{b_{2k+1}}{a^{2k+1}}\right), \end{aligned} \tag{31.6.73}$$

as $a \to \infty$.

An expansion with Bessel functions of the same order as in §31.3 and §31.4 reads

$$\begin{aligned} U(\lambda+1+a\,,\, 2\lambda+2, az) &\sim \frac{2\beta^{-2\lambda-1} e^{\frac{1}{2}az}}{\Gamma(\lambda+1+a)} \times \\ &\left(K_{2\lambda+1}(2\beta a) \sum_{k=0}^\infty \frac{\widetilde{a}_{2k}}{a^{2k}} + \beta K_{2\lambda+2}(2\beta a) \sum_{k=0}^\infty \frac{b_{2k+1}}{a^{2k+1}}\right), \end{aligned} \tag{31.6.74}$$

where

$$\widetilde{a}_0 = a_0, \quad \widetilde{a}_{2k} = a_{2k} - (2\lambda+1)b_{2k-1}, \quad k \ge 1. \tag{31.6.75}$$

In the present case we have (see (31.6.72))

$$\widetilde{a}_0 = r_0 \left(\frac{\beta/2}{\sinh(w_0/2)}\right)^{2\lambda+2}, \quad r_0 = \left.\frac{dw}{dt}\right|_{t=\beta} = \sqrt{2\tanh\left(\tfrac{1}{2}w_0\right)/\beta}. \tag{31.6.76}$$

For the function U^+ defined in (31.0.7) we take $a = i\eta$ and $z = -2\rho/\eta$ (with $\eta < 0$). Then the modified Bessel functions in (31.6.74) have argument $2i\eta\beta$, and we replace them by Hankel functions (see the first line in (9.2.5)). Next we write the Hankel function in terms of the Bessel functions. This gives

$$\begin{aligned} F_\lambda(\eta, \rho) &= \pi \frac{(2\rho)^{\lambda+1} \beta^{-2\lambda-1} e^{\frac{1}{2}\pi\eta}}{|\Gamma(\lambda+1+i\eta)|} \left(J_{2\lambda+1}(u)\, P + J_{2\lambda+2}(u)\, Q\right), \\ G_\lambda(\eta, \rho) &= -\pi \frac{(2\rho)^{\lambda+1} \beta^{-2\lambda-1} e^{\frac{1}{2}\pi\eta}}{|\Gamma(\lambda+1+i\eta)|} \left(Y_{2\lambda+1}(u)\, P + Y_{2\lambda+2}(u)\, Q\right), \end{aligned} \tag{31.6.77}$$

where $u = -2\beta\eta$ and

$$P \sim \sum_{k=0}^{\infty} (-1)^k \frac{\widetilde{a}_{2k}}{\eta^{2k}}, \quad Q \sim \sum_{k=0}^{\infty} (-1)^k \frac{b_{2k+1}}{\eta^{2k+1}}, \quad \eta \to -\infty. \tag{31.6.78}$$

To evaluate the coefficients we need those in the expansions

$$w = w_0 + \sum_{k=0}^{\infty} r_k (t-\beta)^{k+1}, \quad f(t) = \sum_{k=0}^{\infty} p_k (t-\beta)^k. \tag{31.6.79}$$

The first coefficients are (r_0 and $p_0 = \widetilde{a}_0$ are defined in (31.6.76))

$$\begin{aligned} r_1 &= r_0 \frac{\beta^2 r_0^3 (z+6) - 6z}{12 z\beta}, \\ r_2 &= r_0 \frac{54z^2 - 12\beta^2 r_0^3 z(z+6) + \beta^4 r_0^6 (z^2+6z+18)}{144\beta^2 z^2}, \\ p_1 &= -a_0 \frac{(\beta^2(3\lambda z + 2z + 12\lambda + 6) r_0^3 - 6z(1+2\lambda)}{6z\beta}. \end{aligned} \tag{31.6.80}$$

Then,

$$\begin{aligned} b_1 &= \frac{-(4\lambda+1)p_1 + 2\beta p_2}{2\beta}, \\ a_2 &= \big(3(4\lambda+3)(4\lambda+1)p_1 + 2\beta(-9+16\lambda(\lambda-1))p_2 + \\ &\qquad 12\beta^2(1-4\lambda)p_3 + 24\beta^3 p_4\big)/(32\beta), \end{aligned} \tag{31.6.81}$$

and straightforward calculations yield

$$\begin{aligned} b_1 = &-\frac{a_0(4\lambda+3)(4\lambda+1)}{16\beta^2} + \\ &\frac{a_0(6z^2\lambda^2 + 6z^2\lambda + z^2 + 48z\lambda^2 + 6z + 48z\lambda + 96\lambda^2 + 96\lambda + 18)}{12\beta(z+4)\sqrt{z(z+4)}}. \end{aligned} \tag{31.6.82}$$

31.7 Expansions for η, $\rho \to \infty$; Airy-type expansions

We write the representation for U^- of (31.1.14) in the form

$$e^{i\chi} U^- = \rho^{-2\lambda-1} |\Gamma(\lambda+1+i\eta)|\, I^-, \quad I^- = \frac{1}{2\pi i}\int_{\mathcal{L}_-} \frac{e^{\rho\phi(w)}\, dw}{(w^2+1)^{\lambda+1}}, \tag{31.7.83}$$

where

$$\phi(w) = w - \xi \arctan w, \quad \xi = \frac{2\eta}{\rho}. \tag{31.7.84}$$

This function has two saddle points $w_\pm = \pm\sqrt{\xi - 1}$, which coalesce when $\xi = 1$ (or $\rho = 2\eta$), and this happens in a neighborhood of the w-plane where the integrand of the integral in (31.7.83) is analytic. To derive Airy-type expansions of the Coulomb functions we can use the method described in Chapter 23. Initially we assume that $\xi \ge 1$, in which case the saddle points are real.

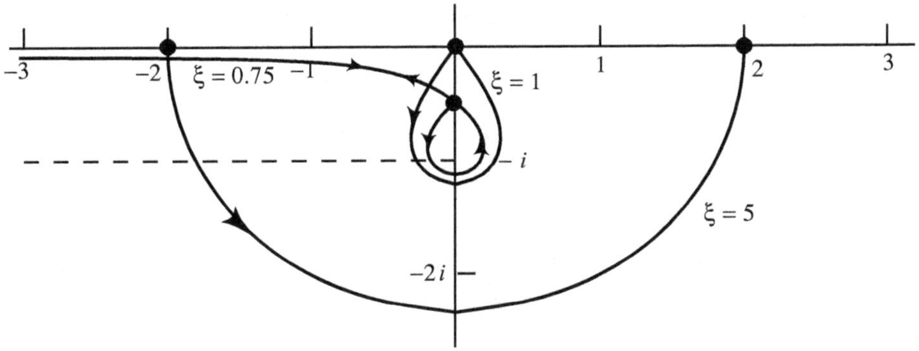

Fig. 31.1 Parts of the saddle point contours of the integral (31.7.83) for $\xi = 1$ and $\xi = 5$. The complete contours start at $-\infty$, make a right turn at w_-, continue along the curve to w_+, and return to $-\infty$. For $\xi = \frac{3}{4}$ the contour $\mathcal{L}_-$ through w_- is shown.

We observe that the real axis is a path through the saddle points satisfying $\Im\phi(w) = 0$, and indeed parts of the real axis are parts of the steepest descent contour. The saddle point contour runs from $-\infty$ to w_-, then along an arc in the lower half-plane to w_+ (encircling the singular point at $w = -i$ anti-clockwise), and returns to $-\infty$. See Figure 31.1, where we have shown these curves for $\xi = 5$ (saddle points at ± 2) and for $\xi = 1$ (saddle points at the origin). We also show the branch cut from $-i$ to $-\infty - i$. Because of the logarithmic singularity at $w = -i$, the phases of the integrand are different on the way out and on the way back along $(-\infty, w_-]$. When $0 < \xi < 1$ the saddle points are $w_\pm = \pm i\sqrt{1-\xi}$; for $\xi = \frac{3}{4}$ the contour $\mathcal{L}_-$ through w_- is shown.

We use the transformation

$$w - \xi \arctan w = \tfrac{1}{3}s^3 - \zeta s, \tag{31.7.85}$$

where ζ follows from substituting the saddle points in the w- and s-variables. This gives

$$\begin{aligned} \tfrac{2}{3}\zeta^{\frac{3}{2}} &= \xi \arctan \sqrt{\xi - 1} - \sqrt{\xi - 1}, \quad \xi \ge 1, \\ \tfrac{2}{3}(-\zeta)^{\frac{3}{2}} &= -\xi \operatorname{arctanh} \sqrt{1-\xi} + \sqrt{1-\xi}, \quad 0 \le \xi \le 1. \end{aligned} \tag{31.7.86}$$

An expansion gives

$$\zeta = (\xi - 1) - \tfrac{2}{15}(\xi-1)^2 + \tfrac{83}{1575}(\xi-1)^3 + \mathcal{O}\left((\xi-1)^4\right), \quad \xi \to 1. \tag{31.7.87}$$

The transformation yields

$$I^- = -\frac{1}{2\pi i}\int_{\mathcal{C}_3} e^{\rho\left(\frac{1}{3}s^3 - \zeta s\right)} f(s)\,ds, \quad f(s) = \frac{dw}{ds}\frac{1}{(w^2+1)^{\lambda+1}}, \tag{31.7.88}$$

where the contour $\mathcal{C}_3$ is as in Figure 8.2 with the same direction of integration.

The expansion follows from the method described in §23.2. We use the notation

$$\begin{aligned}\Phi_0 &= -\frac{1}{2\pi i}\int_{\mathcal{C}_3} e^{\rho\left(\frac13 s^3-\zeta s\right)}\,ds,\\ \Phi_1 &= -\frac{1}{2\pi i}\int_{\mathcal{C}_3} s e^{\rho\left(\frac13 s^3-\zeta s\right)}\,ds,\end{aligned} \tag{31.7.89}$$

and obtain

$$I^- = \Phi_0 P + \Phi_1 Q, \quad P\sim\sum_{k=0}^{\infty}\frac{A_k}{\rho^k}, \quad Q\sim\sum_{k=0}^{\infty}\frac{B_k}{\rho^k}, \tag{31.7.90}$$

where

$$A_k = \frac{f_k\left(\sqrt{\zeta}\right)+f_k\left(-\sqrt{\zeta}\right)}{2}, \quad B_k = \frac{f_k\left(\sqrt{\zeta}\right)-f_k\left(-\sqrt{\zeta}\right)}{2\sqrt{\zeta}}, \tag{31.7.91}$$

and the functions f_k follow from the recursive scheme

$$f_{k+1}(s) = g_k'(s), \quad f_k(s) = A_k + B_k s + \left(s^2-\zeta\right) g_k(s), \quad k=0,1,2,\ldots, \tag{31.7.92}$$

with $f_0 = f$.

We obtain representations of Airy functions by rotating the path of integration in (31.7.89) (by substituting $s = te^{-2\pi i/3}$) and using (8.1.4). It follows that

$$\begin{aligned}\Phi_0 &= -\rho^{-\frac13}e^{-\frac23\pi i}\operatorname{Ai}\left(\zeta\rho^{\frac23}e^{-\frac23\pi i}\right),\\ \Phi_1 &= -\rho^{-\frac23}e^{-\frac43\pi i}\operatorname{Ai}'\left(\zeta\rho^{\frac23}e^{-\frac23\pi i}\right).\end{aligned} \tag{31.7.93}$$

The expansion for $e^{-i\chi}U^+$ follows from (31.7.90) by changing the sign of i in Φ_j, $j=0,1$. For the Coulomb functions with the representations in (31.0.8) we need the sum and difference of the Φ_j. These follow from

$$\begin{aligned}\operatorname{Ai}(z) &= -e^{-\frac23\pi i}\operatorname{Ai}\left(ze^{-\frac23\pi i}\right) - e^{\frac23\pi i}\operatorname{Ai}\left(ze^{\frac23\pi i}\right),\\ \operatorname{Bi}(z) &= e^{-\frac16\pi i}\operatorname{Ai}\left(ze^{-\frac23\pi i}\right) + e^{\frac16\pi i}\operatorname{Ai}\left(ze^{\frac23\pi i}\right).\end{aligned} \tag{31.7.94}$$

This gives

$$\begin{aligned}\Phi_0+\overline{\Phi}_0 &= \rho^{-\frac13}\operatorname{Ai}\left(\zeta\rho^{\frac23}\right), \quad \Phi_0-\overline{\Phi}_0 = i\rho^{-\frac13}\operatorname{Bi}\left(\zeta\rho^{\frac23}\right),\\ \Phi_1+\overline{\Phi}_1 &= \rho^{-\frac23}\operatorname{Ai}'\left(\zeta\rho^{\frac23}\right), \quad \Phi_1-\overline{\Phi}_1 = i\rho^{-\frac23}\operatorname{Bi}'\left(\zeta\rho^{\frac23}\right),\end{aligned} \tag{31.7.95}$$

and by using (31.0.8) the representations

$$\begin{aligned}F_\lambda(\eta,\rho) &= \frac{2^\lambda e^{\frac12\pi\eta}|\Gamma(\lambda+1+i\eta)|}{\rho^{\lambda+\frac13}}\left(\operatorname{Ai}\left(\zeta\rho^{\frac23}\right)P + \rho^{-\frac13}\operatorname{Ai}'\left(\zeta\rho^{\frac23}\right)Q\right),\\ G_\lambda(\eta,\rho) &= \frac{2^\lambda e^{\frac12\pi\eta}|\Gamma(\lambda+1+i\eta)|}{\rho^{\lambda+\frac13}}\left(\operatorname{Bi}\left(\zeta\rho^{\frac23}\right)P + \rho^{-\frac13}\operatorname{Bi}'\left(\zeta\rho^{\frac23}\right)Q\right).\end{aligned} \tag{31.7.96}$$

The expansions of P and Q are given in (31.7.90). Because $w(s)$ is an odd function (see (31.7.85)) it follows that $f(s)$ defined in (31.7.88) is an even function. Hence, considering the scheme in (31.7.92), we conclude that $f_{2k}(s)$ is even and $f_{2k+1}(s)$ is odd, and that $A_{2k+1} = B_{2k} = 0$, $k = 0, 1, 2, \ldots$. We have

$$P \sim \sum_{k=0}^{\infty} \frac{A_{2k}}{\rho^{2k}}, \quad Q \sim \sum_{k=0}^{\infty} \frac{B_{2k+1}}{\rho^{2k+1}}, \quad \rho \to \infty. \tag{31.7.97}$$

The first coefficient follows from

$$\left.\frac{dw}{ds}\right|_{s=\sqrt{\zeta}} = \sqrt{\xi}\left(\frac{\zeta}{\xi - 1}\right)^{\frac{1}{4}}, \tag{31.7.98}$$

and this gives

$$A_0 = \xi^{-\lambda-\frac{1}{2}}\left(\frac{\zeta}{\xi - 1}\right)^{\frac{1}{4}}. \tag{31.7.99}$$

In §23.3 we have explained how to compute the coefficients A_k and B_k. From (23.3.29) we obtain

$$B_1 = \frac{\beta p_2 - p_1}{4\beta^3}, \quad A_2 = \frac{3p_4\beta^3 - 10p_3\beta^2 + 21p_2\beta - 21p_1}{96\beta^5}, \tag{31.7.100}$$

where $\beta = \sqrt{\zeta}$ and $p_k = f^{(k)}(\beta)$. For B_1 we find

$$B_1 = A_0 \frac{\left(8 - 12\xi + 9\xi^2 + 48\lambda\xi(\lambda + 1)(\xi - 2)\right) + 48\lambda(\lambda + 1)\big)\, c_1^6 - 5\xi^4}{48(\xi - 1)^2 c_1^8}, \tag{31.7.101}$$

where c_1 denotes the derivative in (31.7.98).

Chapter 32

Laguerre polynomials: Uniform expansions

Laguerre polynomials $L_n^{(\alpha)}(z)$ have been considered in §15.5, where we have given expansions for large n and bounded values of z and α. In this chapter we give more details on the large degree asymptotics. Frst we consider bounded values of z, with two cases: α bounded and α depending on n. When z is allowed to become large as well, we need other uniform expansions. For this we summarize the results given in Frenzen and Wong (1988), with expansions in terms of Airy functions and Bessel functions, and those from Temme (1986b), with expansions in terms of Hermite polynomials.

In all these approaches the Cauchy-type contour integral that follows from the generating function given in (15.0.2) is the starting point, and we recall

$$L_n^{(\alpha)}(z) = \frac{1}{2\pi i}\int_{\mathcal{C}} (1-t)^{-\alpha-1} e^{-tz/(1-t)} \frac{dt}{t^{n+1}}, \tag{32.0.1}$$

where $\mathcal{C}$ is a circle around the origin with radius less than unity.

32.1 An expansion for bounded z and α

The relation to the Kummer or confluent hypergeometric functions shown in (15.5.37) makes it possible to use the results for these functions for large negative a, see §10.3.4 and §10.3.5. In (10.3.59) we have given a complete expansion in terms of Bessel functions (for the function $E_\nu(z)$ we refer to §10.3.1).

We keep in mind that, usually, for $\Re z \ge 0$ the analytic E-functions are expressed in terms of J-Bessel functions, and for $\Re z \le 0$ in terms of modified I-Bessel functions. The expansion in (10.3.59) is valid as $n \to \infty$ with α and z in bounded domains of the complex plane. When z is bounded away from the origin we can expand the Bessel functions for large values of their argument.

The construction of the coefficients $a_k(z)$ and $b_k(z)$ is explained in §10.3.2, with the first ones shown in terms of the coefficients $c_k(z)$ in (10.3.40). The $c_k(z)$ are used in the expansion of $f(z,s)$ defined in (10.3.28), and we should take $c = \alpha + 1$.

This is an important special case of the Kummer functions, and it is instructive to give an explicit construction of the coefficients $a_k(z)$ and $b_k(z)$ by using an integration by parts method. For the Kummer functions we have used a different method, but the present method can also be used for the expansion in §10.3.

From (10.3.54) and (10.1.17) it follows that

$$e^{\frac12 z}L_n^{(\alpha)}(-z) = \frac{1}{2\pi i}\int_{\mathcal{L}} e^{ns+z/s}s^{-\alpha-1}F(s)\,ds, \tag{32.1.2}$$

where (see (10.3.28))

$$F(s) = f(-z,s) = e^{-zg(s)}\left(\frac{s}{1-e^{-s}}\right)^{\alpha+1}, \quad g(s) = \frac{1}{s} - \frac{1}{e^s-1} - \frac{1}{2}. \tag{32.1.3}$$

We write

$$F(s) = a_0(z) + b_0(z)s + s^2G(s), \tag{32.1.4}$$

where a_0 and b_0 do not depend on s, and a_0 and b_0 follow from (see also (10.3.31))

$$a_0(z) = F(0) = 1, \quad b_0(z) = F'(0) = \tfrac{1}{12}\left(6(\alpha+1)+z\right). \tag{32.1.5}$$

Then, after integrating by parts,

$$e^{\frac12 z}L_n^{(\alpha)}(-z) = a_0(z)\Phi_1(z) + b_0(z)\Phi_1(z) + \frac{1}{2\pi i\,n}\int_{\mathcal{L}} e^{ns+z/s}s^{-\alpha-1}F_1(s)\,ds, \tag{32.1.6}$$

where $\Phi_k(z)$ is defined in (10.3.49) and for the present case is given by

$$\Phi_k(z) = \left(\frac{z}{n}\right)^{\frac12(k-\alpha)} I_{\alpha-k}\left(2\sqrt{nz}\right). \tag{32.1.7}$$

For the function F_1 we have

$$F_1(s) = -s^{\alpha+1}e^{-z/s}\frac{d}{ds}\left(s^{1-\alpha}e^{z/s}G(s)\right). \tag{32.1.8}$$

Continuing this procedure, we obtain

$$\begin{aligned} e^{\frac12 z}L_n^{(\alpha)}(-z) = {} & \Phi_0(z)\sum_{k=0}^{K-1}\frac{a_k(z)}{n^k} + \Phi_1(z)\sum_{k=0}^{K-1}\frac{b_k(z)}{n^k} + \\ & \frac{1}{2\pi i\,n^K}\int_{\mathcal{L}} e^{ns+z/s}s^{-\alpha-1}F_K(s)\,ds, \end{aligned} \tag{32.1.9}$$

where $a_k(z)$, $b_k(z)$ and $F_k(z)$ follow from the recursive scheme, starting with $F_0 = F$,

$$\begin{aligned} F_k(s) &= a_k(z) + b_k(z)s + s^2G_k(s), \\ F_{k+1}(s) &= -s^{\alpha+1}e^{-z/s}\frac{d}{ds}\left(s^{1-\alpha}e^{z/s}G_k(s)\right). \end{aligned} \tag{32.1.10}$$

We can express the coefficients $a_k(z)$ and $b_k(z)$ in terms of the Maclaurin coefficients of $F(s)$. We write

$$F_k(s) = \sum_{j=0}^{\infty} c_j^{(k)}(z)s^j, \quad G_k(s) = \sum_{j=0}^{\infty} c_{j+2}^{(k)}(z)s^j \tag{32.1.11}$$

and after substituting these expansions into

$$F_{k+1}(s) = (\alpha-1)sG_k(s) + zG_k(s) - s^2G_k'(s), \tag{32.1.12}$$

we find for the coefficients $c_j^{(k)}(z)$ the recursion

$$c_0^{(k+1)} = zc_2^{(k)}, \quad c_j^{(k+1)} = zc_{j+2}^{(k)} + (\alpha - j)c_{j+1}^{(k)}, \tag{32.1.13}$$

where $j \ge 1$ and $k \ge 0$.

The first coefficients are

$$\begin{aligned} a_0 &= 1, \quad b_0 = \tfrac{1}{12}(6(\alpha+1)+z), \\ a_1 &= \tfrac{1}{288}z\left(-12\alpha - 12 + 36(\alpha+1)^2 + 12z(\alpha+1) + z^2\right), \\ b_1 &= \tfrac{1}{51840}\left(2160(\alpha^2-1)(3\alpha+2) + 1080(\alpha+1)(\alpha^2+3\alpha-2)z + \right. \\ &\quad \left. 108(10\alpha+1+5\alpha^2)z^2 + 5z^4\right). \end{aligned} \tag{32.1.14}$$

Szegő (1975, §8.22, §8.72, Problem 46) has suggested several methods for Laguerre polynomials of large degree. For a more recent publication, see Borwein *et al.* (2008), where an expansion is given for these polynomials (in negative powers of $(n+1)$. This expansion corresponds to the expansion in (32.1.9) after expanding the modified Bessel functions for large argument, and it is valid for z off the positive axis and bounded away from the origin. A related expansion valid near the positive real axis follows from (10.3.59) after writing the E-functions in terms of J-Bessel functions (see §10.3.1) and by expanding the J-Bessel function.

32.2 An expansion for bounded z; α depends on n

In the previous section we have assumed that α is a fixed or bounded parameter. In the present case we let α depend on n and write the Cauchy integral in (32.0.1) in the form[1]

$$L_n^{(\alpha n)}(z) = \frac{e^z}{2\pi i}\int e^{-z/(1-t)} e^{-n\phi(t)} \frac{dt}{t(1-t)}, \quad \alpha \ge 0, \tag{32.2.15}$$

where the function $\phi(t)$ is given by

$$\phi(t) = \ln t + \alpha \ln(1-t), \quad \phi'(t) = \frac{1-(\alpha+1)t}{t(1-t)}. \tag{32.2.16}$$

The integral is taken over a circle around the origin with radius smaller than 1. We assume that $0 \le \alpha \le \alpha_0$, where α_0 is a fixed positive number; z is bounded as well. Because we consider large values of n, the integral can be taken over a vertical line that cuts the real axis between 0 and 1, where $\mathrm{ph}(1-t) = 0$.

As $\alpha \to 0$ the saddle point $t_0 = 1/(\alpha+1)$ approaches the singular point at $t = 1$. The behavior of the integral at $t = 1$ suggests the transformation

$$\phi(t) = s - 1 + \alpha \ln(1-s) + A, \tag{32.2.17}$$

[1] This case came up in communications with Marcello Seri and Leonid Parnovski (University College London), who studied the integrated density of states of Schrödinger operators with constant magnetic field perturbed by a quasi periodic potential.

where $s \uparrow 1$ as $t \uparrow 1$; A is independent of t or s and follows from substituting $t = t_0$ and $s = s_0 = 1 - \alpha$, the point where the derivative of the s-function vanishes. This gives

$$A = \alpha - (1+\alpha)\log(1+\alpha). \tag{32.2.18}$$

We obtain

$$L_n^{(\alpha n)}(z) = \frac{e^{z+n-nA}}{2\pi i}\int f(s)\frac{e^{-ns}}{(1-s)^{\alpha n+1}}\,ds, \tag{32.2.19}$$

where the new contour comes from $+\infty$, encircles the point $s = 1$ clockwise, and returns to $+\infty$, with $\mathrm{ph}(1-s) = 0$ for $s < 1$. Also,

$$\begin{aligned} \frac{dt}{ds} &= t_0\frac{t(t-1)(s-s_0)}{(s-1)(t-t_0)}, \\ f(s) &= e^{-z/(1-t)}\frac{1-s}{t(1-t)}\frac{dt}{ds} = t_0\frac{s-s_0}{t-t_0}e^{-z/(1-t)}. \end{aligned} \tag{32.2.20}$$

We write

$$f(s) = e^{-z/(c_1(1-s))}g(s), \quad g(s) = t_0\frac{s-s_0}{t-t_0}e^{-z(1/(1-t)-1/(c_1(1-s)))}, \tag{32.2.21}$$

where c_1 is the first coefficient in the expansion

$$t - 1 = c_1(s-1) + c_2(s-1)^2 + \ldots. \tag{32.2.22}$$

This representation makes $g(s)$ defined and analytic at $s = 1$.

By writing (32.2.17) in the form

$$t\left(\frac{t-1}{s-1}\right)^{\alpha} = e^{s-1+A}, \tag{32.2.23}$$

and taking $t = 1$, $s = 1$, it follows that $c_1 = e^{A/\alpha}$.

We continue with

$$L_n^{(\alpha n)}(z) = \frac{e^{z+n-nA}}{2\pi i}\int e^{-z/(c_1(1-s))}g(s)\frac{e^{-ns}}{(1-s)^{\alpha n+1}}\,ds. \tag{32.2.24}$$

We obtain an expansion by interpolating $g(s)$ at the two special points s_0 and 1. The first step is

$$g(s) = a_0 + b_0(s-s_0) + (s-s_0)(1-s)h(s), \tag{32.2.25}$$

where a_0 and b_0 follow from substitution of $s = s_0$ and $s = 1$:

$$a_0 = g(s_0), \quad b_0 = \frac{g(s_0)-g(1)}{s_0-1}. \tag{32.2.26}$$

This gives for (32.2.24):

$$L_n^{(\alpha n)}(z) = e^{z-nA}\bigl(a_0\Phi + b_0\Psi + F_0\bigr), \tag{32.2.27}$$

where, by using the integral representation of the J-Bessel function given in (10.3.24),

$$\begin{aligned}
\Phi &= \left(\frac{nc_1}{z}\right)^{\frac{1}{2}\alpha n} J_{\alpha n}(\zeta), \quad \zeta = 2\sqrt{\frac{nz}{c_1}},\\
\Psi &= \frac{\zeta}{2n}\left(\frac{nc_1}{z}\right)^{\frac{1}{2}\alpha n} J_{\alpha n+1}(\zeta),\\
F_0 &= \frac{e^n}{2\pi i}\int e^{-z/(c_1(1-s))}(s-s_0)(1-s)h(s)\frac{e^{-ns}}{(1-s)^{\alpha n+1}}\,ds.
\end{aligned} \tag{32.2.28}$$

For the evaluation of Ψ we have used the recurrence formula for the Bessel function:

$$J_{\nu+1}(z) + J_{\nu-1}(z) = \frac{2\nu}{z}J_\nu(z). \tag{32.2.29}$$

Integrating by parts follows from writing

$$(s-s_0)\frac{e^{-ns}}{(1-s)^{\alpha n}}\,ds = \frac{1-s}{n}\,de^{-ns-\alpha n\ln(1-s)}, \tag{32.2.30}$$

and we obtain

$$F_0 = \frac{e^n}{2\pi i\, n}\int e^{-z/(c_1(1-s))}g_1(s)\frac{e^{-ns}}{(1-s)^{\alpha n+1}}\,ds, \tag{32.2.31}$$

where

$$g_1(s) = e^{-\frac{z}{c_1(1-s)}}\left(\frac{z}{c_1}h(s) + (1-s)h(s) - (1-s)^2h'(s)\right). \tag{32.2.32}$$

The integral in (32.2.31) has the same form as in (32.2.24), and the procedure can be continued to obtain for $n\to\infty$

$$L_n^{(\alpha n)}(z) \sim \left(\frac{nc_1}{z}\right)^{\frac{1}{2}\alpha n} e^{z-nA}\left(J_{\alpha n}(\zeta)\sum_{k=0}^{\infty}\frac{a_k}{n^k} + \frac{\zeta}{2n}J_{\alpha n+1}(\zeta)\sum_{k=0}^{\infty}\frac{b_k}{n^k}\right). \tag{32.2.33}$$

We evaluate the first coefficients. The value a_0 follows from $g(s)$ given in (32.2.21) with $s=s_0, t=t_0$. We have

$$\left.\frac{dt}{ds}\right|_{s=s_0} = (1+\alpha)^{-\frac{3}{2}}, \tag{32.2.34}$$

and

$$a_0 = g(s_0) = \sqrt{1+\alpha}\,e^{-z(c_1(1+\alpha)-1)/(c_1\alpha)}. \tag{32.2.35}$$

The result for dt/ds at $s=s_0$ follows from (32.2.20) and by applying l'Hôpital's rule. For b_0 we need $g(1)$ and the coefficient c_2 in the expansion given in (32.2.22). We have

$$c_2 = -\frac{c_1(c_1-1)}{\alpha}, \quad g(1) = e^{-zc_2/c_1^2} = e^{z(c_1-1)/(\alpha c_1)}. \tag{32.2.36}$$

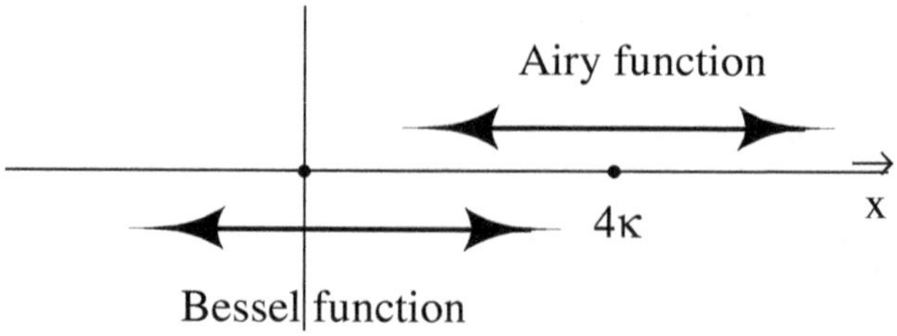

Fig. 32.1 Zeros of $L_n^{(\alpha)}(x)$ occur in the x-interval $(0, 4\kappa)$. The small zeros can be approximated in terms of the zeros of the J-Bessel function, the large zeros in terms of those of the Airy function.

For small α we need expansions:

$$\begin{aligned} A &= -\tfrac{1}{2}\alpha^2 + \tfrac{1}{6}\alpha^3 + \ldots, \\ c_1 &= 1 - \tfrac{1}{2}\alpha + \tfrac{7}{24}\alpha^2 + \ldots, \\ e^{\frac{1}{2}z}a_0 &= 1 + \frac{12-z}{24}\alpha + \frac{z^2-144}{1152}\alpha^2 + \ldots, \\ e^{\frac{1}{2}z}b_0 &= \frac{z-6}{12} + \frac{6-z}{48}\alpha + \frac{5z^3 - 90z^2 + 2016z - 12960}{207360}\alpha^2 + \ldots. \end{aligned} \tag{32.2.37}$$

Remark 32.1. The special feature of the approximation in (32.2.33) is the transition value of α. When the order and the argument of the Bessel function $J_\nu(z)$ are large and $\nu \sim z$, the behavior of this function changes from monotone and very small ($z < \nu$) to oscillatory ($z > 0$). This happens when $\alpha \sim 2\sqrt{z/n}$ (for these values of α we have $c_1 \sim 1$), and in that case the Bessel functions can be approximated by Airy functions; see §23.5. In §32.3.1 we give an expansion of $L_n^{(\alpha)}(z)$ in terms of Airy functions, but there we assume that α is bounded and z may depend on n. △

32.3 Expansions for bounded α; z depends on n

The first large-n asymptotics of these polynomials with z depending on n have been investigated in detail by Tricomi; for a summary of his results see Tricomi (1954). Tricomi's results were later considerably improved by Erdélyi (1960). More precisely, Erdélyi gave two asymptotic formulas for $L_n^{(\alpha)}(4\kappa x)$, where n is large, x is real and

$$\kappa = n + \tfrac{1}{2}(\alpha + 1). \tag{32.3.38}$$

One formula holds uniformly for $-\infty < x \le a$ and the other for $b \le x < \infty$, where a and b are two fixed numbers, $0 < b < a < 1$. See Figure 32.1, where we indicate the Airy and the Bessel function intervals. These two intervals overlap and, moreover, they cover the entire x-axis. Erdélyi used the differential equation.

In Frenzen and Wong (1988) it has been shown that more complete results can be obtained from their integral representations. Frenzen and Wong used special forms of (32.0.1), and applied the saddle point method to obtain expansions in terms of Airy and Bessel functions. In Temme (1990a) we have mentioned several asymptotic forms of the Laguerre polynomials, including forms in which α may be large. These results follow from uniform expansions of Whittaker functions by using differential equations; see Dunster (1989) for Airy-type expansions.

An approach for large α based on the integral representation in (32.0.1) is given in §32.4. The expansion is in terms of Hermite polynomials.

We summarize the results from Frenzen and Wong (1988) (see also Wong (2001, Chapter VII)) on the Bessel and Airy-type expansions of the Laguerre polynomials. The expansions are obtained from the Cauchy integral given in (32.0.1), and are provided with many details on the transformations and with bounds for remainders in the expansions.

32.3.1 *An expansion in terms of Airy functions*

By substituting $t = (s-1)/(s+1)$ in (32.0.1) the following representation is obtained:

$$e^{-\frac{1}{2}z} L_n^{(\alpha)}(z) = \frac{(-1)^n}{2^\alpha\, 2\pi i} \int_{+\infty}^{(1+)} e^{-\frac{1}{2}zs} \left(\frac{1+s}{1-s}\right)^{\kappa} \left(1-s^2\right)^{\frac{1}{2}(\alpha-1)}\, ds, \tag{32.3.39}$$

where κ is defined in (32.3.38). The path of integration encircles the point $s = 1$ in the positive direction; it starts at $+\infty$ and cuts the real axis in the interval $(-1, 1)$, where $\operatorname{ph}(1-s) = 0$.

This representation can be written in the form

$$e^{-\frac{1}{2}\nu x} L_n^{(\alpha)}(\nu x) = \frac{(-1)^n}{2^\alpha} \frac{1}{2\pi i} \int_{+\infty}^{(1+)} e^{\nu f(s,x)} \left(1-s^2\right)^{\frac{1}{2}(\alpha-1)}\, ds, \tag{32.3.40}$$

where $\nu = 4\kappa$ and

$$f(s,x) = \tfrac{1}{2}\operatorname{arctanh} s - \tfrac{1}{2}sx. \tag{32.3.41}$$

The saddle points are given by $s_\pm = \pm\sqrt{1-1/x}$ if $x < 1$, and at $s_\pm = \pm i\sqrt{1/x-1}$ if $0 < x \le 1$. They coalesce at $s = 0$ when $x \to 1$, and $f(x,s)$ and the other factor of the integrand are analytic in a neighborhood of this point. As explained in Chapter 23 we may expect an Airy-type expansion in this case.

To perform the cubic transformation we set

$$f(s,x) = \tfrac{1}{3}u^3 - B^2(x)u + A. \tag{32.3.42}$$

Because f is an odd function of s, $A = 0$. For $B(x)$ we find, after substituting the corresponding saddle points in the s-plane and the u-plane,

$$B(x) = \begin{cases} i\left(\tfrac{3}{2}\beta(x)\right)^{\frac{1}{3}}, & \text{if} \quad 0 < x \le 1, \\[2ex] \left(\tfrac{3}{2}\gamma(x)\right)^{\frac{1}{3}}, & \text{if} \quad x \ge 1, \end{cases} \tag{32.3.43}$$

where

$$\begin{aligned}\beta(x) &= \tfrac{1}{2}\left(\arccos\sqrt{x} - \sqrt{x-x^2}\right),\\ \gamma(x) &= \tfrac{1}{2}\left(\sqrt{x^2-x} - \operatorname{arccosh}\sqrt{x}\right).\end{aligned} \tag{32.3.44}$$

Then, after the integration by parts procedure explained in §23.2, it follows that we can obtain the expansion

$$\begin{aligned}&(-1)^n 2^\alpha e^{-\frac{1}{2}\nu x} L_n^{(\alpha)}(\nu x) \sim \\ &\quad \mathrm{Ai}\left(\nu^{2/3}B^2(x)\right)\sum_{k=0}^{\infty}\frac{\alpha_{2k}}{\nu^{2k+1/3}} - \mathrm{Ai}'\left(\nu^{2/3}B^2(x)\right)\sum_{k=0}^{\infty}\frac{\beta_{2k+1}}{\nu^{2k+5/3}},\end{aligned} \tag{32.3.45}$$

as $n\to\infty$, uniformly for $x \in (b,\infty]$, where $b\in(0,1)$, a fixed number.

The coefficients α_{2k}, β_{2k+1} follow from the recursion

$$h_n(u) = \alpha_n + \beta_n u + \left(u^2 - B^2(x)\right) g_n(u), \quad h_{n+1}(u) = g_n'(u), \tag{32.3.46}$$

with

$$h_0(u) = h(u) = \left(1 - s^2(u)\right)^{(\alpha-1)/2}\frac{ds}{du}, \tag{32.3.47}$$

where the relation between s and u is given by (32.3.42).

The leading term of the expansion is given by

$$\alpha_0 = x^{\frac{1}{2}(1-\alpha)}\left(\frac{4B^2(x)}{(x-1)x^3}\right)^{\frac{1}{4}}. \tag{32.3.48}$$

For small values of $x-1$ we can use

$$\begin{aligned}\alpha_0 &= 2^{\frac{1}{3}} x^{\frac{1}{2}(1-\alpha)} \times \\ &\quad \left(1 - \tfrac{4}{5}(x-1) + \tfrac{5}{7}(x-1)^2 - \tfrac{1042}{1575}(x-1)^3 + \mathcal{O}\left((x-1)^4\right)\right).\end{aligned} \tag{32.3.49}$$

32.3.2 *An expansion in terms of Bessel functions*

In this case we substitute in (32.0.1) $t = e^{-s}$, and obtain

$$e^{-\frac{1}{2}z} L_n^{(\alpha)}(z) = \frac{1}{2\pi i}\int_{-\infty}^{(0+)} \exp\left(ns - \tfrac{1}{2}z\frac{1+e^{-s}}{1-e^{-s}}\right)\frac{ds}{(1-e^{-s})^{\alpha+1}}, \tag{32.3.50}$$

where the path of integration is a Hankel contour, see Figure 2.1. Assume that $\alpha > -1$ and write again $\nu = 4\kappa$ (see (32.3.38)). Then, for $x < 1$,

$$e^{-\frac{1}{2}\nu x} L_n^{(\alpha)}(\nu x) = \frac{2^{-\alpha}}{2\pi i}\int_{-\infty}^{(0+)} e^{\frac{1}{2}\nu f(s,x)}\left(\frac{\sinh s}{s}\right)^{-\alpha-1}\frac{ds}{s^{\alpha+1}}, \tag{32.3.51}$$

where

$$f(s,x) = s - x\coth s. \tag{32.3.52}$$

This function has two saddle points at

$$s_\pm = \pm i \arcsin\sqrt{x}. \tag{32.3.53}$$

As x tends to zero, the saddle points coalesce with each other and moreover with the pole of $f(s,x)$ at $s=0$.

The appropriate transformation is (see also §27.4)

$$f(s,x) = u - \frac{A^2(x)}{u}, \tag{32.3.54}$$

where $A(x)$ can be determined by substituting corresponding saddle points. This gives

$$A(x) = \begin{cases} \frac{1}{2} i \left(\sqrt{x^2 - x} + \operatorname{arcsinh}\sqrt{-x}\right), & \text{if} \quad x \le 0, \\ \frac{1}{2}\left(\sqrt{x - x^2} + \arcsin\sqrt{x}\right), & \text{if} \quad 0 \le x < 1. \end{cases} \tag{32.3.55}$$

With the transformation in (32.3.52) the representation in (32.3.51) becomes

$$e^{-\frac{1}{2}\nu x} L_n^{(\alpha)}(\nu x) = \frac{2^{-\alpha}}{2\pi i} \int_{-\infty}^{(0+)} u^{-\alpha-1} h(u) e^{\frac{1}{2}\nu\left(u - A^2(x)/u\right)}\, du, \tag{32.3.56}$$

where

$$h(u) = \left(\frac{\sinh s}{u}\right)^{-\alpha-1} \frac{ds}{du}, \tag{32.3.57}$$

and the relation between s and u follows from (32.3.54). When $h(u)$ is replaced by a constant and when we use (10.3.24), the integral becomes a J-Bessel function.

The integration by parts procedure starts with writing

$$h(u) = \alpha_0 + \frac{\beta_0}{u} + \left(1 + \frac{A^2(x)}{u^2}\right) g(u), \tag{32.3.58}$$

where α_0 and β_0 follow from substituting the corresponding saddle points in the s- and u-plane. This gives (since $h(u)$ is even)

$$\alpha_0 = h(iA(x)), \quad \beta_0 = 0. \tag{32.3.59}$$

Continuing we obtain the following expansion

$$2^\alpha e^{-\frac{1}{2}\nu x} L_n^{(\alpha)}(\nu x) \sim \\ \frac{J_\alpha(\nu A(x))}{A^\alpha(x)} \sum_{k=0}^{\infty} 2^{2k} \frac{\alpha_{2k}}{\nu^{2k}} - \frac{J_{\alpha+1}(\nu A(x))}{A^{\alpha+1}(x)} \sum_{k=0}^{\infty} 2^{2k+1} \frac{\beta_{2k+1}}{\nu^{2k+1}}, \tag{32.3.60}$$

as $n \to \infty$, uniformly for $x \in (-\infty, a]$, where $a \in (0,1)$; $\nu = 4\kappa$ (see (32.3.38)). For $x < 0$ the quantity $A(x)$ is imaginary, and the Bessel functions become modified Bessel functions.

The coefficients α_{2k}, β_{2k+1} follow from recursions. Define (with $h_0(u) = h(u)$, see (32.3.58)) a set of functions $\{h_n\}$, $\{g_n\}$ and coefficients $\{\alpha_n\}$, $\{\beta_n\}$ by writing

$$h_n(u) = \alpha_n + \beta_n/u + (1 + A^2(x)/u^2) g_n(u), \tag{32.3.61}$$

$$h_{n+1}(u) = g_n'(u) - \frac{\alpha+1}{u} g_n(u). \tag{32.3.62}$$

The first coefficient is

$$\alpha_0 = (1-x)^{-\frac14}\left(\frac{A(x)}{\sqrt{x}}\right)^{\alpha+\frac12}, \quad x<1, \tag{32.3.63}$$

in which $A(x)/\sqrt{x}$ is analytic at $x=0$. For small values of x we can expand

$$\frac{A(x)}{\sqrt{x}} = 1 - \tfrac16 x - \tfrac{1}{40}x^2 - \tfrac{1}{112}x^3 + \mathcal{O}(x^4). \tag{32.3.64}$$

For β_1 we have, again for $x<1$,

$$\beta_1 = \tfrac{1}{48}\alpha_0\left(3(1-4\alpha^2)\left(1-\sqrt{A^2(x)(1-x)/x}\right) + \frac{\sqrt{x}A(x)(6-x)}{(1-x)^{\frac32}}\right), \tag{32.3.65}$$

where we also consider $\sqrt{x}A(x)$ for $x<0$ as analytic continuation of the value in the neighborhood of the origin, see (32.3.64).

Remark 32.2. The expansion of this section corresponds to more general results for the Kummer functions for large negative a. See §27.4.4 and 27.4.5. △

32.4 An expansion in terms of Hermite polynomials; large α

Earlier in this chapter we have given expansions of the Laguerre polynomials $L_n^{(\alpha)}(z)$ with α fixed and for $\alpha = \mathcal{O}(n)$. When α is large (whether or not n is large), we can give approximations in terms of Hermite polynomials, in a similar way as in Chapter 24. Our approach for the Laguerre polynomials has been discussed earlier in Temme (1986b). We summarize the main steps of this reference.

We start with the Cauchy integral representation in (32.0.1), and as earlier in this chapter we use

$$\kappa = n + \tfrac12(\alpha+1). \tag{32.4.66}$$

The substitution $t=(w+1)/(w-1)$ transforms (32.0.1) into

$$L_n^{(\alpha)}(4\kappa x) = \frac{(-1)^n 2^{-\alpha} e^{2\kappa x}}{2\pi i}\int_{-\infty}^{(-1^+)} e^{\kappa\Phi(w)}\frac{dw}{\sqrt{1-w^2}}, \tag{32.4.67}$$

where

$$\Phi(w) = 2xw + (1+\tau)\ln(1-w) - (1-\tau)\ln(1+w), \quad \tau = \frac{\alpha}{2\kappa}. \tag{32.4.68}$$

The contour in (32.4.67) cuts the real axis between $w=-1$ and $w=+1$. We define a branch cut for the many-valued functions in the integrand from -1 to $-\infty$, although $w=-1$ is only a pole of the integrand. When we start the integration at $-\infty$ the phase of $1+w$ equals $-\pi$, and after turning around the point $w=-1$ it equals π.

The saddle points of $\Phi(w)$ are

$$w_1 = \frac{-\frac12\tau - \sqrt{W}}{x}, \quad w_2 = \frac{-\frac12\tau + \sqrt{W}}{x}, \quad W = x^2 - x + \tfrac14\tau^2. \tag{32.4.69}$$

The zeros of W are (observe that $0 \le \tau \le 1$)

$$x_1 = \tfrac{1}{2} - \tfrac{1}{2}\sqrt{1-\tau^2}, \quad x_2 = \tfrac{1}{2} + \tfrac{1}{2}\sqrt{1-\tau^2}. \tag{32.4.70}$$

These are the turning points: for $x \in (x_1, x_2)$ the Laguerre polynomial $L_n^{(\alpha)}(4\kappa x)$ has zeros, outside this interval it is monotonic.

When x is large, we have

$$w_1 = -1 + \frac{1-\tau}{2x} + \mathcal{O}\left(x^{-2}\right), \quad w_2 = 1 - \frac{1+\tau}{2x} + \mathcal{O}\left(x^{-2}\right). \tag{32.4.71}$$

When x crosses the turning point x_2 from above, the saddle points collide at $w_1 = w_2 = -\tau/\left(1+\sqrt{1+\tau^2}\right)$. When x decreases further, passing the turning point x_1 the saddle points collide again at $w_1 = w_2 = -\left(1+\sqrt{1+\tau^2}\right)/\tau$, and become real again. When $x_1 \le x \le x_2$ the saddle points are complex and located on the curves around $w = -1$ described by the equation

$$\tau v = \pm\sqrt{1-\tau^2-(\tau u+1)^2}, \quad w = u + iv, \tag{32.4.72}$$

with

$$-\frac{1+\sqrt{1+\tau^2}}{\tau} \le u \le -\frac{\tau}{1+\sqrt{1+\tau^2}}. \tag{32.4.73}$$

When $x \downarrow 0$, we have $w_1 \to -\infty$ and $w_2 \to -1/\tau$.

Clearly, the location of the saddle points is as described in §24.1 for the Hermite polynomials and as in §24.2.1 for the Gegenbauer polynomials. For $\tau \to 1$ the turning points x_j coalesce, and this happens when $\alpha \gg n$.

32.4.1 *A first approximation*

If $\alpha \gg n$, the saddle points coalesce in the neighborhood of the point $w = -1$. Near this point the phase function $\Phi(w)$ can be expanded:

$$\begin{aligned}\Phi(w) &= (1-\tau)\ln(1-w) - 2x + (\tau+1)\ln 2 + \\ &\left(2x - \tfrac{1}{2}(\tau+1)\right)(w+1) - \tfrac{1}{8}(\tau+1)(w+1)^2 + \mathcal{O}\left((w+1)^3\right),\end{aligned} \tag{32.4.74}$$

and using this approximation up to the quadratic term we obtain by (23.4.31):

$$\begin{aligned}L_n^{(\alpha)}(x) = \frac{(-1)^n 2^n}{n!}\left(\frac{\kappa(\tau+1)}{8}\right)^{n/2} \times \\ \left(H_n\left(\frac{\sqrt{2\kappa}\left(2x - \frac{1}{2}(\tau+1)\right)}{\sqrt{\tau+1}}\right) + \varepsilon_n^\alpha(x)\right),\end{aligned} \tag{32.4.75}$$

where $\varepsilon_n^\alpha(x)$ is expected to be small if $\alpha \to \infty, \alpha \gg n$.

In Figure 32.2 we show the graphs of the Laguerre polynomials and their approximations (thin lines) in (32.4.75) for $\alpha = 20$, $n = 2$, $n = 3$. We have used these rather small values because for larger values of the parameters proper scaling of the functions in the oscillatory interval $[x_1, x_2] \simeq [0.18, 0.82]$ is quite difficult.

It is not difficult to verify, that the limits given in (19.1.3) follow from this first approximation.

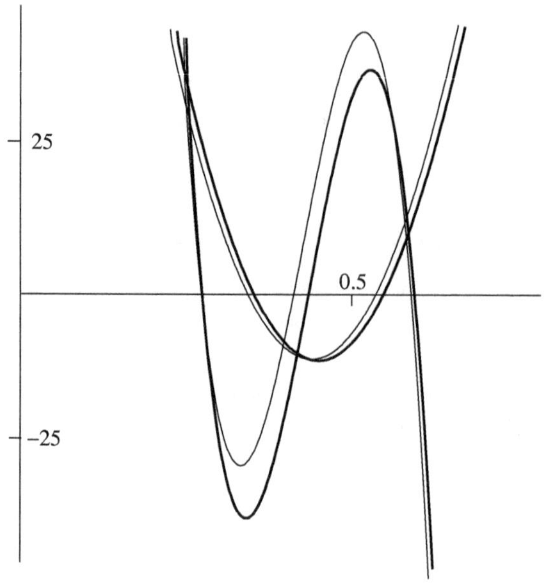

Fig. 32.2 $L_n^{(\alpha)}(x)$ and its approximation (thin lines) in (32.4.75) for $\alpha = 20$ and $n = 2$, $n = 3$.

32.4.2 *Transformation to the standard form*

Next we define the mapping $w \mapsto t(w)$ by writing

$$\Phi(w) = \Psi(t) + A, \tag{32.4.76}$$

where $\Psi(t)$ is given in (24.1.2), with saddle points given in (24.1.4). They correspond to w_j given in (32.4.69) and $t(w_j) = t_j$, $j = 1, 2$. In addition, $t(-1) = 0$. For real values, we have $\operatorname{sign}(w + 1) = \operatorname{sign}(t)$.

The transformation will be analytic at $w = -1$ when we take $\rho^2 = 1 - \tau$. The values of A and ξ (in (32.4.76) and in $\Psi(t)$) follow from the corresponding points $t(w_j) = t_j$.

The result of the transformation is

$$L_n^{(\alpha)}(4\kappa x) = \frac{(-1)^n 2^{-\alpha} e^{2\kappa x + \kappa A}}{2\pi i} \int_{-\infty}^{(0+)} e^{\kappa \Psi(t)} f(t) \, \frac{dt}{\sqrt{t}}, \tag{32.4.77}$$

where

$$f(t) = \frac{\sqrt{t}}{\sqrt{1 - w^2}} \frac{dw}{dt}, \quad \frac{dw}{dt} = \frac{1 - w^2}{2xt} \frac{(t - t_1)(t - t_2)}{(w - w_1)(w - w_2)}. \tag{32.4.78}$$

Using the method of §24.1, we obtain an expansion as in (24.1.10):

$$\begin{aligned} L_n^{(\alpha)}(4\kappa x) \sim \frac{(-1)^n \left(\frac{1}{2}\kappa\right)^{n/2} e^{2\kappa x + \kappa A}}{2^\alpha \, n!} \times \\ \left(H_n\left(\xi\sqrt{2\kappa}\right) \sum_{s=0}^{\infty} \frac{\alpha_s}{\kappa^s} + \frac{H_n'\left(\xi\sqrt{2\kappa}\right)}{\sqrt{2\kappa}} \sum_{s=0}^{\infty} \frac{\beta_s}{\kappa^s} \right). \end{aligned} \tag{32.4.79}$$

We have

$$\left.\frac{dw}{dt}\right|_{t=t_j} = \frac{\sqrt{1-w_j^2}}{\sqrt{2t_j}} \left(\frac{\xi^2-\rho^2}{W}\right)^{\frac{1}{4}}, \tag{32.4.80}$$

which gives

$$f(t_1) = f(t_2) = \frac{1}{\sqrt{2}} \left(\frac{\xi^2-\rho^2}{x^2-x+\frac{1}{4}\tau^2}\right)^{\frac{1}{4}}. \tag{32.4.81}$$

It follows that

$$\alpha_0 = \frac{1}{\sqrt{2}} \left(\frac{\xi^2-\rho^2}{x^2-x+\frac{1}{4}\tau^2}\right)^{\frac{1}{4}}, \quad \beta_0 = 0. \tag{32.4.82}$$

The quantity A can be obtained from an analogue equation in (24.2.56), with the result

$$A = -\tfrac{1}{2}(1+\tau) - \tau\ln x + \tfrac{1}{2}(1+\tau)\ln(1+\tau) - \xi^2. \tag{32.4.83}$$

We give the relation for ξ for three different intervals. We have, using $\Phi(w_2) - \Phi(w_1) = \Psi(t_2) - \Psi(t_1)$,

(1) $-\infty < \xi \le -\rho$, $0 < x \le x_1$:

$$\begin{aligned} &\xi\sqrt{\xi^2-\rho^2} + \rho^2 \operatorname{arccosh}\frac{-\xi}{\rho} = \\ &4\sqrt{W} - 2(1+\tau)\operatorname{arctanh}\frac{\sqrt{W}}{x+\frac{1}{2}\tau} - 2(1-\tau)\operatorname{arctanh}\frac{\sqrt{W}}{x-\frac{1}{2}\tau}, \end{aligned} \tag{32.4.84}$$

(2) $-\rho \le \xi \le \rho$, $x_1 \le x \le x_2$:

$$\begin{aligned} &\xi\sqrt{\rho^2-\xi^2} - \rho^2 \operatorname{atn2}\left(\sqrt{\rho^2-\xi^2}, \xi\right) = \\ &2\sqrt{-W} - (1-\tau)\operatorname{atn2}\left(\sqrt{-W}, x-\tfrac{1}{2}\tau\right) - (1+\tau)\operatorname{atn2}\left(\sqrt{-W}, x+\tfrac{1}{2}\tau\right); \end{aligned} \tag{32.4.85}$$

(3) $\rho \le \xi < \infty$, $x_2 \le x < \infty$:

$$\begin{aligned} &\xi\sqrt{\xi^2-\rho^2} - \rho^2 \operatorname{arccosh}\frac{\xi}{\rho} = \\ &4\sqrt{W} - 2(1+\tau)\operatorname{arctanh}\frac{\sqrt{W}}{x+\frac{1}{2}\tau} - 2(1-\tau)\operatorname{arctanh}\frac{\sqrt{W}}{x-\frac{1}{2}\tau}. \end{aligned} \tag{32.4.86}$$

In (32.4.85) the function $\operatorname{atn2}(y, x)$ denotes the phase in the interval $(-\pi, \pi]$ of the complex number $z = x + iy$; for $x > 0$ we have $\operatorname{atn2}(y, x) = \arctan(y/x)$.

If $\tau = 1$ the relation between x and ξ is given by

$$\tfrac{1}{2}\xi^2 = 2x - \ln(2x) - 1, \quad x > 0, \quad \operatorname{sign}(\xi) = \operatorname{sign}\left(x - \tfrac{1}{2}\right). \tag{32.4.87}$$

The relations that define ξ also follow from the differential equation

$$\left(\xi^2-\rho^2\right)\left(\frac{d\xi}{dx}\right)^2 = \frac{4W}{x^2}, \tag{32.4.88}$$

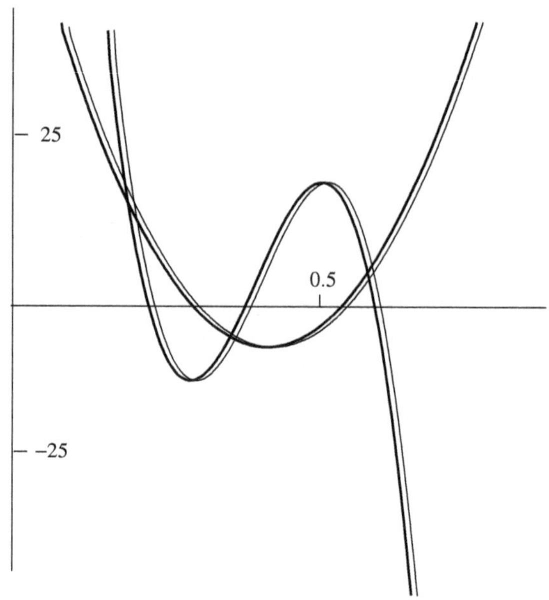

Fig. 32.3 $L_n^{(\alpha)}(x)$ and its approximation (thin lines) in (32.4.79) for $\alpha = 10$ and $n = 2$, $n = 3$; we have only used the term with α_0. For clarity, the graphs of the approximation are shifted 0.01 unit in the x-direction because otherwise the difference would not be discernible.

and the boundary conditions at x_1 and x_2. This equation arises in the Liouville–Green approach to transform the Whittaker differential equation into an equation from which approximations in terms of parabolic cylinder functions can be obtained.

In Figure 32.3 we show the graphs of the Laguerre polynomials and their approximations in (32.4.79) for $\alpha = 10$ and $n = 2$, $n = 3$; we have only used the term with α_0. Compare with Figure 32.2. We have shifted the graphs of the approximations 0.01 unit in the x-direction, because otherwise no distinction without a difference in the graphs could be noticed.

32.4.3 *Approximating the zeros*

As in §24.2.5 we compute approximations of the zeros of the Laguerre polynomials starting with the zeros of the Hermite polynomial in (32.4.79).

Let $\ell_{n,m}^{\alpha}, h_{n,m}$ be the mth zero of $L_n^{(\alpha)}(x)$, $H_n(x)$, respectively, $m = 1, 2, \ldots, n$. For given α and n we compute

$$\xi_{n,m} = \frac{h_{n,m}}{\sqrt{2\kappa}}, \quad m = 1, 2, \ldots, n. \tag{32.4.89}$$

With these $\xi_{n,m}$ we invert equation (32.4.85) to compute $x_{n,m}$ and obtain the approximations $\widetilde{\ell}_{n,m}^{\alpha}$ for the zeros of $L_n^{(\alpha)}(x)$:

$$\widetilde{\ell}_{n,m}^{\alpha} = x_{n,m}, \quad m = 1, 2, \ldots, n. \tag{32.4.90}$$

Table 32.1 Correct decimal digits in the approximations of zeros of $L^{(\alpha)}_{(10)}(x)$.

m \ α	0.0	1.0	5.0	10	25	50	75	100
1	1.7	2.3	3.2	3.7	4.4	5.0	5.3	5.6
2	2.4	2.7	3.4	3.8	4.5	5.0	5.4	5.6
3	2.8	3.0	3.5	3.9	4.5	5.1	5.4	5.6
4	3.0	3.2	3.6	4.0	4.6	5.1	5.4	5.7
5	3.2	3.4	3.8	4.1	4.6	5.1	5.5	5.7
6	3.4	3.5	3.9	4.2	4.7	5.2	5.5	5.7
7	3.5	3.6	4.0	4.2	4.7	5.2	5.5	5.8
8	3.7	3.8	4.1	4.3	4.8	5.3	5.6	5.8
9	3.8	3.9	4.1	4.4	4.9	5.3	5.6	5.8
10	3.9	4.0	4.2	4.5	4.9	5.4	5.6	5.8

In Table 32.1 we show for $n = 10$ the "correct number of decimal digits" in the approximation of $\ell^{\alpha}_{n,m}$ by $\widetilde{\ell}^{\alpha}_{n,m}$. That is, we show

$$-{}^{10}\log\left|\frac{\ell^{\alpha}_{10,m}-\widetilde{\ell}^{\alpha}_{10,m}}{\ell^{\alpha}_{10,m}}\right|, \qquad m = 1, 2, \ldots, 10, \tag{32.4.91}$$

where $\widetilde{\ell}^{\alpha}_{10,m}$ is the approximation obtained by the procedure just described.

It follows that for the smaller values of α the large zeros are better approximated than the small zeros. Furthermore, as expected, large values of α give better approximations, and the approximations become uniform with respect to m.

Chapter 33

Generalized Bessel polynomials

Generalized Bessel polynomials of degree n, complex order μ and complex argument z, denoted by $Y_n^\mu(z)$, have been introduced in Krall and Frink (1949), and can be defined by their generating function.[1] We have (Grosswald, 1978):

$$\left(\frac{2}{1+\sqrt{1-2zw}}\right)^\mu \frac{e^{2w/(1+\sqrt{1-2zw})}}{\sqrt{1-2zw}} = \sum_{n=0}^{\infty} \frac{Y_n^\mu(z)}{n!} w^n, \quad |2zw| < 1, \tag{33.0.1}$$

with special values

$$Y_n^\mu(0) = 1, \quad n = 0, 1, 2, \ldots. \tag{33.0.2}$$

The generalized Bessel polynomials are important in several problems of mathematical physics. For a historical survey and discussion of many interesting properties, we refer to Grosswald's monograph. In Dunster (2001) and Wong and Zhang (1997) detailed contributions on asymptotic approximations are given for these polynomials. Wong and Zhang use integral representations and Dunster's approach is based on a differential equation; in addition, Dunster includes the role of the parameter μ in the asymptotic analysis. Both references first give expansions in terms of elementary functions, which are not uniformly valid with respect to z near the points $\pm i/n$. For neighborhoods of one of these points the cited references give expansions in terms of Airy functions.

In §33.2 we give an expansion in terms of Laguerre polynomials which is valid outside a fixed neighborhood of the origin. In §33.3 we give uniform expansions in terms of elementary functions. Our idea is to stay close to the Debye-type expansions of the modified Bessel functions, and the expansions simply reduce to these expansions when we take $\mu = 0$. In a final section we give the uniform expansion in terms of the modified Bessel functions, and this expansion is valid in both turning points $\pm i/n$. For convenience we start with properties of the generalized Bessel polynomials, in particular with relations to Bessel and Kummer functions.

[1]This chapter is based on López and Temme (2011).

33.1 Relations to Bessel and Kummer functions

For $\mu = 0$ the generalized Bessel polynomials become well-known polynomials that occur in representations of Bessel functions of fractional order. We have in terms of the modified K-Bessel function (see Olver and Maximon (2010, Eq. 10.49.12))

$$Y_n^0(z) = \sqrt{\frac{2}{\pi z}} e^{1/z} K_{n+\frac{1}{2}}(1/z) = \sum_{k=0}^{n} \binom{n}{k} (n+1)_k \left(\tfrac{1}{2}z\right)^k. \tag{33.1.3}$$

For $Y_n^\mu(z)$ an explicit formula reads (Krall and Frink, 1949)

$$Y_n^\mu(z) = \sum_{k=0}^{n} \binom{n}{k} (n+\mu+1)_k \left(\tfrac{1}{2}z\right)^k. \tag{33.1.4}$$

The simple integral representation

$$Y_n^\mu(z) = \frac{1}{\Gamma(n+\mu+1)} \int_0^\infty t^{n+\mu} \left(1+\tfrac{1}{2}zt\right)^n e^{-t}\,dt \tag{33.1.5}$$

gives (33.1.4) by expanding $(1+\frac{1}{2}zt)^n$ in powers of z.

From the generating function in (33.0.1) we have the Cauchy integral representation

$$Y_n^\mu(z) = \frac{n!}{2\pi i} \int_{\mathcal{C}} \left(\frac{2}{1+\sqrt{1-2zw}}\right)^\mu \frac{e^{2w/(1+\sqrt{1-2zw})}}{\sqrt{1-2zw}} \frac{dw}{w^{n+1}}, \tag{33.1.6}$$

where $\mathcal{C}$ is a circle with radius smaller than $1/|2z|$.

When $z = 0$ all polynomials reduce to unity, the first few polynomials being

$$\begin{aligned} &Y_0^\mu(z) = 1, \quad Y_1^\mu(z) = 1 + \tfrac{1}{2}(\mu+2)z, \\ &Y_2^\mu(z) = 1 + (\mu+3)z + \tfrac{1}{4}(\mu+3)(\mu+4)z^2. \end{aligned} \tag{33.1.7}$$

More values can be obtained from the recurrence relation

$$A_n Y_{n+2}^\mu(z) = B_n Y_{n+1}^\mu(z) + C_n Y_n^\mu(z), \tag{33.1.8}$$

where

$$\begin{aligned} A_n &= 2(2n+\mu+2)(n+\mu+2), \\ B_n &= (2n+\mu+3)(2\mu + z(2n+\mu+4)(2n+\mu+2)), \\ C_n &= 2(n+1)(2n+\mu+4). \end{aligned} \tag{33.1.9}$$

Moreover, there is a recursion with respect to μ:

$$(n+\mu+2)Y_n^{\mu+2}(z) = (2n+\mu+2-2/z)Y_n^{\mu+1}(z) + (2/z)Y_n^\mu(z), \tag{33.1.10}$$

and for the derivative we have

$$\frac{d}{dz}Y_n^\mu(z) = \tfrac{1}{2}n(n+\mu+1)Y_{n-1}^{\mu+2}(z) = \frac{n+\mu+1}{z}\left(Y_n^{\mu+1}(z) - Y_n^\mu(z)\right). \tag{33.1.11}$$

These relations are special cases of results for the Kummer functions; see Chapter 10.

The relations to the Kummer functions are

$$\begin{aligned} Y_n^\mu(1/z) &= (2z)^{n+\mu+1}U(n+\mu+1, 2n+\mu+2, 2z) \\ &= (2z)^{-n}\frac{\Gamma(2n+\mu+1)}{\Gamma(n+\mu+1)}\,{}_1F_1\left(\begin{matrix} -n \\ -2n-\mu \end{matrix};\,2z\right). \end{aligned} \tag{33.1.12}$$

For $\Re z < 0$ it is convenient to have the representation

$$Y_n^\mu(-1/z) = F_n^\mu(1/z) + U_n^\mu(1/z), \tag{33.1.13}$$

where

$$\begin{aligned} F_n^\mu(1/z) &= \frac{n!\,(2z)^{n+\mu+1}e^{-2z}}{\Gamma(2n+\mu+2)}\,{}_1F_1\left(\begin{matrix} n+1 \\ 2n+\mu+2 \end{matrix};\,2z\right), \\ U_n^\mu(1/z) &= \frac{(-1)^n n!\,(2z)^{n+\mu+1}e^{-2z}}{\Gamma(n+\mu+1)}U(n+1, 2n+\mu+2, 2z). \end{aligned} \tag{33.1.14}$$

For $\mu = 0$ we have

$$F_n^0(z) = \sqrt{2\pi z}\,e^{-z}I_{n+\frac12}(1/z), \quad U_n^0(z) = (-1)^n\sqrt{\frac{2z}{\pi}}\,e^{-z}K_{n+\frac12}(z), \tag{33.1.15}$$

and this corresponds to the relation

$$Y_n^0(-z) = \sqrt{\frac{2}{\pi z}}\,e^{-1/z}\left((-1)^n K_{n+\frac12}(1/z) + \pi I_{n+\frac12}(1/z)\right). \tag{33.1.16}$$

In §10.1 we have given the relation between the Laguerre polynomials and the Kummer functions:

$$L_n^{(\alpha)}(z) = \frac{(-1)^n}{n!}U(-n, \alpha+1, z) = \binom{n+\alpha}{n}\,{}_1F_1\left(\begin{matrix} -n \\ \alpha+1 \end{matrix};\,z\right), \tag{33.1.17}$$

and from (33.1.12) we obtain for the generalized Bessel polynomials the relation

$$Y_n^\mu(1/z) = n!\,(2z)^{-n}L_n^{(-2n-\mu-1)}(2z). \tag{33.1.18}$$

This relation has been used in Martínez-Finkelshtein *et al.* (2001) in the study of the zeros of both polynomials. For more information on the zeros we refer to §19.6, and for more details to Pasquini (2000) and (Campos and Calderón, 2012).

33.2 An expansion in terms of Laguerre polynomials

The interesting region in the z-plane for uniform asymptotic expansions is a neighborhood of size $\mathcal{O}(1/n)$ of the origin, where the zeros appear. For z outside a fixed neighborhood of the origin a simple expansion will be derived.

First we mention

$$\begin{aligned} Y_n^\mu(z) &= 2^{\mu+\frac12}\left(\frac{2nz}{e}\right)^n e^{1/z} \times \\ &\quad \left(1 - \frac{1+6\mu(\mu+1+2z^{-1})+6z^{-2}}{24n} + \mathcal{O}\left(1/n^2\right)\right). \end{aligned} \tag{33.2.19}$$

This is derived in Dočev (1962), and mentioned in Grosswald (1978, p. 124) and in Wong and Zhang (1997). More terms in this expansion can be obtained, for example by using the Cauchy integral given in (33.1.6).

In this section we derive a simple asymptotic expansion related to the result in (33.2.19) by expanding part of the integrand in (33.1.6) in powers of $W = \sqrt{1-2zw}$. As we have learned in Chapter 15, the main asymptotic contributions in the expansion of the contour integral in (33.1.6) come from the singular point $w = 1/(2z)$, and when $w \sim 1/(2z)$ the quantity W is small.

We have

$$\left(\frac{2}{1+\sqrt{1-2zw}}\right)^{\mu} e^{2w/(1+\sqrt{1-2zw})} = 2^{\mu} e^{1/z}(1+W)^{-\mu} e^{-W/z}, \tag{33.2.20}$$

and we expand for $m = 0, 1, 2, \ldots$

$$(1+W)^{-\mu} e^{-W/z} = \sum_{k=0}^{m-1} L_k^{(-\mu-k)}(1/z) W^k + W^m U_m(W). \tag{33.2.21}$$

The appearance of the Laguerre polynomials becomes clear when expanding both the exponential and binomial and by comparing the coefficients with the representation

$$L_n^{(\alpha)}(x) = \sum_{m=0}^{n} \binom{n+\alpha}{n-m} \frac{(-x)^m}{m!}. \tag{33.2.22}$$

Introducing this expansion in (33.1.6) we find

$$Y_n^{\mu}(z) = n!\, 2^{\mu} e^{1/z} \sum_{k=0}^{m-1} L_k^{(-\mu-k)}(1/z) \Phi_k^{(n)} + n!\, R_m(n), \tag{33.2.23}$$

where

$$\Phi_k^{(n)} = \frac{1}{2\pi i} \int_{\mathcal{C}} (1-2zw)^{(k-1)/2} \frac{dw}{w^{n+1}} = (2z)^n \frac{\left(\frac12 - \frac12 k\right)_n}{n!}, \tag{33.2.24}$$

and

$$R_m(n) = \frac{2^{\mu} e^{1/z}}{2\pi i} \int_{\mathcal{C}} (1-2wz)^{(m-1)/2} U_m(W) \frac{dw}{w^{n+1}}, \tag{33.2.25}$$

with $\mathcal{C}$ a circle with radius less than $1/|2z|$.

After the change of variable $w = (1-t/n)/(2z)$ we have

$$R_m(n) = \frac{(2z)^n 2^{\mu} e^{1/z}}{2\pi i\, n^{(m+1)/2}} \int_{\mathcal{C}} t^{(m-1)/2} U_m\left(\sqrt{t/n}\right) \frac{dt}{(1-t/n)^{n+1}}. \tag{33.2.26}$$

The function $U_m(W)$ is analytic in $|W| < 1$ and $\mathcal{O}(1)$ as $W \to 0$. This means that $|U_m(W)| < C_m$ with C_m a positive constant, on and inside the path $\mathcal{C}$ of integration in (33.2.25). Hence, $|U_m(\sqrt{t/n})| < C_m$ on $\mathcal{C}$ (indeed, the path $\mathcal{C}$ has been modified after the change of variable, but we can set it equal to the previous path). In addition, $(1-t/n)^{-(n+1)}$ is bounded on $\mathcal{C}$ (and converges to e^t as $n \to \infty$). Therefore, the above integral is $\mathcal{O}(1)$ as $n \to \infty$. Hence,

$$R_m(n) = \mathcal{O}\left(\frac{(2z)^n}{n^{(m+1)/2}}\right), \quad n \to \infty,\ z \neq 0, \tag{33.2.27}$$

which is comparable with the large-n behavior of $\Phi_m^{(n)}$.

From the integral in (33.2.24) it easily follows that $\Phi_{2k+1}^{(n)} = 0$ for $k = 0, 1, 2, \ldots, n$, and we see that, when n is large, only the even terms in the series contribute. We notice that the sequence $\{\Phi_{2k}^{(n)}\}$ constitutes an asymptotic sequence. This follows from

$$\frac{\Phi_{2k+2}^{(n)}}{\Phi_{2k}^{(n)}} = \frac{\Gamma(-\frac12 - k + n)}{\Gamma(\frac12 - k + n)} = \frac{1}{n - \frac12 - k} = \mathcal{O}\left(n^{-1}\right), \quad n \to \infty, \tag{33.2.28}$$

with k fixed.

We conclude that we have the asymptotic expansion

$$Y_n^\mu(z) \sim (2z)^n\, 2^\mu e^{1/z} \sum_{k=0}^{\infty} L_k^{(-\mu-k)}(1/z) \left(\tfrac12 - \tfrac12 k\right)_n, \quad n \to \infty, \tag{33.2.29}$$

which holds uniformly for $|z| \geq z_0$, where z_0 is a positive constant.

Remark 33.1. If we wish we can expand the Pochhammer symbols in (33.2.29) for large n and rearrange the series. In that way we can obtain an expansion of $Y_n^\mu(z)$ in negative powers of n, and this expansion is comparable with an expansion of which the first terms are given in (33.2.19). △

Remark 33.2. In (33.2.29) we expand the generalized Bessel polynomials $Y_n^\mu(z)$ in terms of another set of polynomials, the generalized Laguerre polynomials. Because the degree of these polynomials does not depend on the large parameter, they can be evaluated much more easily than the polynomials $Y_n^\mu(z)$. In fact, to compute the Laguerre polynomials we can use a recurrence relation, which follows from differentiating (33.2.21) with respect to W. Let $c_k = L_k^{(-\mu-k)}(1/z)$, then

$$z(k+1)c_{k+1} = -(\mu z + kz + 1)c_k - c_{k-1}, \quad k = 1, 2, 3, \ldots, \tag{33.2.30}$$

with initial values $c_0 = 1$, $c_1 = -(\mu z + 1)/z$. △

Remark 33.3. For general values of μ the expansion in (33.2.29) is not convergent, but for $\mu = 0, -1, -2, \ldots$ it is. For example, for $\mu = 0$ a relation for the K-Bessel function should arise. From (33.2.21) it follows that $L_k^{(-k)}(1/z) = (-1)^k/(k!\, z^k)$, which gives the convergent expansion

$$Y_n^0(z) = (2z)^n e^{1/z} \sum_{k=0}^{\infty} \frac{(-1)^k}{k!\, z^k} \left(\tfrac12 - \tfrac12 k\right)_n, \quad z \neq 0. \tag{33.2.31}$$

Summing the series, separating the terms with even and odd k, we obtain

$$Y_n^0(z) = (-1)^n e^{1/z} \sqrt{\frac{\pi}{2z}} \left(I_{-n-\frac12}(1/z) - I_{n+\frac12}(1/z)\right), \tag{33.2.32}$$

and by using a well-known relation between the modified Bessel functions the representation in (33.1.3) arises. △

Table 33.1 Relative errors δ in the asymptotic expansion in (33.2.29) with terms up and including $k = 20$ for $\mu = 17/4$ and several values of z and n.

	$z = 10^j$			$z = -10^j$		
	j	$Y_n^\mu(z)$	δ	j	$Y_n^\mu(z)$	δ
$n = 50$	-1	0.4232e034	0.16e-03	-1	0.1961e026	0.26e-11
	0	0.1211e081	0.38e-07	0	0.1778e080	0.62e-08
	1	0.5131e130	0.17e-07	1	0.4235e130	0.14e-07
	2	0.4707e180	0.16e-07	2	0.4617e180	0.15e-07
	3	0.4666e230	0.15e-07	3	0.4657e230	0.15e-07
$n = 100$	-1	0.1681e093	0.30e-07	-1	0.5251e084	0.68e-15
	0	0.3190e189	0.10e-10	0	0.4501e188	0.18e-11
	1	0.1325e289	0.47e-11	1	0.1089e289	0.39e-11
	2	0.1213e389	0.43e-11	2	0.1189e389	0.42e-11
	3	0.1202e489	0.43e-11	3	0.1200e489	0.42e-11

In Table 33.1 we give the relative errors δ when we use the expansion in (33.2.29) with terms up and including $k = 20$, for $\mu = 17/4$ and several values of z and n. We see a quite uniform error with respect to z, except when $z = \pm\frac{1}{10}$.

33.3 Expansions in terms of elementary functions

By using the saddle point methods we obtain expansions that hold uniformly inside sectors $|\mathrm{ph}(\pm z)| \le \frac{1}{2}\pi - \delta$, where δ is a fixed small positive number. For $Y_n^\mu(z)$ we take $\nu = n + \frac{1}{2}$ as the large parameter. This gives a suitable identification of the results with those for the Bessel function $K_{n+\frac{1}{2}}(z)$ when $\mu = 0$; see also (33.1.3). In addition we replace the argument z of $Y_n^\mu(z)$ by $1/(\nu z)$ (observe that in Dunster (2001) ν is also the large parameter, and the Bessel polynomial is considered with reversed argument). Because for both cases $|\mathrm{ph}(\pm z)| \le \frac{1}{2}\pi - \delta$ the derivation of the asymptotic expansion is very similar, we first summarize the results in the following two subsections. In a later section we give the details of the analysis. When we take $\mu = 0$, the expansions of $Y_n^\mu(z)$ given in this section reduce to the Debye-type expansions of the modified Bessel functions given in §9.5.

33.3.1 *The case* $|\mathrm{ph}\, z| < \frac{1}{2}\pi$

For large values of n we have the expansion

$$Y_n^\mu(\zeta) \sim \frac{\left(1 - z + \sqrt{1+z^2}\right)^\mu \sqrt{z}}{(1+z^2)^{\frac{1}{4}}} e^{\nu z - \nu\eta} \sum_{k=0}^{\infty} \frac{A_k(\mu, z)}{\nu^k}, \tag{33.3.33}$$

and the expansion holds uniformly inside the sector $|\mathrm{ph}\, z| \le \frac{1}{2}\pi - \delta$. Here, δ is a small positive constant, $A_0(\mu, z) = 1$,

$$A_1(\mu, z) = \frac{t(5t^2 - 3)}{24} - \frac{\mu t^2(z+1)}{4} + \frac{\mu^2(tz-1)}{4}, \tag{33.3.34}$$

and

$$\nu = n + \tfrac{1}{2}, \quad \zeta = \frac{1}{\nu z}, \quad t = \frac{1}{\sqrt{1+z^2}}, \quad \eta = \sqrt{1+z^2} + \ln \frac{z}{1+\sqrt{1+z^2}}. \tag{33.3.35}$$

33.3.2 *The case* $|\mathrm{ph}(-z)| < \frac{1}{2}\pi$

We write (see (33.1.13) and (33.1.14))

$$Y_n^{\mu}(-1/z) = F_n^{\mu}(1/z) + U_n^{\mu}(1/z). \tag{33.3.36}$$

For large values of n we have the expansions

$$U_n^{\mu}(\zeta) \sim (-1)^n \frac{\left(1+z+\sqrt{1+z^2}\right)^{\mu} \sqrt{z}}{(1+z^2)^{\frac{1}{4}}} e^{-\nu z - \nu\eta} \sum_{k=0}^{\infty} \frac{B_k(\mu, z)}{\nu^k}, \tag{33.3.37}$$

$$F_n^{\mu}(\zeta) \sim \frac{\left(1+z-\sqrt{1+z^2}\right)^{\mu} \sqrt{z}}{(1+z^2)^{1/4}} e^{-\nu z + \nu\eta} \sum_{k=0}^{\infty} \frac{C_k(\mu, z)}{\nu^k}, \tag{33.3.38}$$

and the expansions hold uniformly inside the sector $|\mathrm{ph}\, z| \le \frac{1}{2}\pi - \delta$. We have $B_0(\mu, z) = 1$, $C_0(\mu, z) = 1$,

$$\begin{aligned} B_1(\mu, z) &= \frac{t(5t^2-3)}{24} + \frac{\mu t^2(z-1)}{4} - \frac{\mu^2(zt+1)}{4}, \\ C_1(\mu, \zeta) &= -\frac{t(5t^2-3)}{24} + \frac{\mu t^2(z-1)}{4} + \frac{\mu^2(zt-1)}{4}. \end{aligned} \tag{33.3.39}$$

The quantities ν, ζ, t and η are the same as in (33.3.35).

Remark 33.4. For $\mu = 0$ the expansions reduce to those of the modified Bessel functions mentioned in (33.1.15). For example, $A_k(0, z) = (-1)^k u_k(t)$ (see §9.5).

For $\zeta = 0$ all Bessel polynomials $Y_n^{\mu}(\zeta)$ reduce to unity (see (33.0.2)). We have as $\zeta \to 0$:

$$z \to \infty, \; t \to 0, \; zt \to 1, \; z - \eta \to 0. \tag{33.3.40}$$

As a consequence, $A_1(\mu, z) \to 0$ as $\zeta \to 0$. In fact all coefficients $A_k(\mu, z)$ with $k \ge 1$ vanish as $\zeta \to 0$, and both sides of (33.3.33) reduce to unity. △

33.3.3 *Integral representations*

To derive the asymptotic expansions we introduce the integrals

$$\begin{aligned} P_\nu^\mu(z) &= \int_0^\infty p_\mu(s) e^{-\nu\phi(s)}\,ds,\\ Q_\nu^\mu(z) &= \frac{1}{2\pi i}\int_{\mathcal{L}} q_\mu(s) e^{\nu\phi(s)}\,ds, \end{aligned} \tag{33.3.41}$$

where $\nu > 0$ and

$$\phi(s) = 2zs - \ln s - \ln(1+s). \tag{33.3.42}$$

When $z > 0$ the contour $\mathcal{L}$ is a vertical line with $\Re s > 0$; when z is complex we can deform the contour in order to keep convergence. For the same purpose we can rotate the path of integration for $P_\nu^\mu(z)$ in (33.3.41).

For certain choices of $p_\mu(s)$ and $q_\mu(s)$ these integrals give representations of the functions $Y_n^\mu(z)$, $F_n^\mu(z)$, and $U_n^\mu(z)$. We have

$$\begin{aligned} Y_n^\mu(\zeta) &= \frac{(2\nu z)^{n+\mu+1}}{\Gamma(n+\mu+1)} P_\nu^\mu(z), & p_\mu(s) &= \frac{s^\mu}{\sqrt{s(1+s)}},\\ U_n^\mu(\zeta) &= \frac{(-1)^n (2\nu z)^{n+\mu+1} e^{-2\nu z}}{\Gamma(n+\mu+1)} P_\nu^\mu(z), & p_\mu(s) &= \frac{(1+s)^\mu}{\sqrt{s(1+s)}},\\ F_n^\mu(\zeta) &= \frac{n!}{(2\nu z)^n} Q_\nu^\mu(z), & q_\mu(s) &= \frac{(1+s)^{-\mu}}{\sqrt{s(1+s)}}. \end{aligned} \tag{33.3.43}$$

The multivalued functions in $\phi(s)$, $p_\mu(s)$, and $q_\mu(s)$ have their principal branches and are real for $s > 0$.

The first two representations in (33.3.43) follow from (10.1.5), the first line in (33.1.12), and the second line in (33.1.14). For the third one in (33.3.43) we refer to the first line in (33.1.14) and the integral representation of the ${}_1F_1$-function given in (10.1.7).

33.3.4 *Construction of the expansions*

We use the saddle point method to obtain asymptotic expansions of the integrals in (33.3.41). The saddle points follow from the equation $\phi'(s) = 0$, where

$$\phi'(s) = \frac{2zs^2 + 2(z-1)s - 1}{s(1+s)}, \tag{33.3.44}$$

and are given by

$$s_+ = \frac{1 - z + \sqrt{1+z^2}}{2z}, \quad s_- = \frac{1 - z - \sqrt{1+z^2}}{2z}. \tag{33.3.45}$$

When $z > 0$ the saddle points are well separated, with $-1 < s_- < -\frac{1}{2}$ and $s_+ > 0$. We have the following limits: $\lim_{z\to 0} s_+ = +\infty$ and $\lim_{z\to\infty} s_+ = 0$.

Also,

$$s_+(1+s_+)=\frac{1+\sqrt{1+z^2}}{2z^2}, \tag{33.3.46}$$

and

$$\phi(s_+)=1-z+\ln(2z)+\eta,\quad \phi''(s_+)=\frac{4z^2\sqrt{1+z^2}}{1+\sqrt{1+z^2}}, \tag{33.3.47}$$

with η defined in (33.3.35).

We use Laplace's method with the transformation

$$\phi(s)-\phi(s_+)=\tfrac12\phi''(s_+)w^2,\quad \operatorname{sign}(w)=\operatorname{sign}(s-s_+). \tag{33.3.48}$$

We have $s=w+\mathcal{O}(w^2)$ as $w\to0$. The integrals in (33.3.41) become

$$\begin{aligned}P_\nu^\mu(z)&=e^{-\nu\phi(s_+)}\int_{-\infty}^{\infty} f(w)\,e^{-\frac12\nu\phi''(s_+)w^2}\,dw,\qquad f(w)=p_\mu(s)\frac{ds}{dw},\\ Q_\nu^\mu(z)&=\frac{e^{\nu\phi(s_+)}}{2\pi i}\int_{-i\infty}^{i\infty} g(w)\,e^{\frac12\nu\phi''(s_+)w^2}\,dw,\qquad g(w)=q_\mu(s)\frac{ds}{dw}.\end{aligned} \tag{33.3.49}$$

By expanding

$$f(w)=\sum_{k=0}^{\infty}f_kw^k,\quad g(w)=\sum_{k=0}^{\infty}g_kw^k, \tag{33.3.50}$$

we obtain the asymptotic expansions

$$\begin{aligned}P_\nu^\mu(z)&\sim f_0e^{-\nu\phi(s_+)}\sqrt{\frac{2\pi}{\nu\phi''(s_+)}}\sum_{k=0}^{\infty}\frac{F_k(\mu,z)}{\nu^k},\\ Q_\nu^\mu(z)&\sim g_0\frac{e^{\nu\phi(s_+)}}{2\pi}\sqrt{\frac{2\pi}{\nu\phi''(s_+)}}\sum_{k=0}^{\infty}(-1)^k\frac{G_k(\mu,z)}{\nu^k},\end{aligned} \tag{33.3.51}$$

where (see also (33.3.46))

$$\begin{aligned}F_k(\mu,z)&=\frac{(\frac12)_k2^k}{(\phi''(s_+))^k}\frac{f_{2k}}{f_0},\qquad f_0=p_\mu(s_+),\\ G_k(\mu,z)&=\frac{(\frac12)_k2^k}{(\phi''(s_+))^k}\frac{g_{2k}}{g_0},\qquad g_0=q_\mu(s_+),\end{aligned} \tag{33.3.52}$$

because $ds/dw=1$ at $w=0$.

By using (33.3.43) and (33.3.45)–(33.3.47) it follows that

$$Y_n^\mu(\zeta)\sim\frac{\left(1-z+\sqrt{1+z^2}\right)^\mu\sqrt{z}}{(1+z^2)^{\frac14}}\frac{e^{\nu z-\nu\eta}}{\mathcal{G}(\nu+\mu+\frac12)}\sum_{k=0}^{\infty}\frac{F_k^{(1)}(\mu,z)}{\nu^k}, \tag{33.3.53}$$

$$U_n^\mu(\zeta)\sim(-1)^n\frac{\left(1+z+\sqrt{1+z^2}\right)^\mu\sqrt{z}}{(1+z^2)^{\frac14}}\frac{e^{-\nu z-\nu\eta}}{\mathcal{G}(\nu+\mu+\frac12)}\sum_{k=0}^{\infty}\frac{F_k^{(2)}(\mu,z)}{\nu^k}, \tag{33.3.54}$$

$$F_n^\mu(\zeta) \sim \mathcal{G}\left(\nu+\tfrac{1}{2}\right) \frac{\left(1+z-\sqrt{1+z^2}\right)^\mu \sqrt{z}}{(1+z^2)^{1/4}} e^{-\nu z+\nu\eta} \sum_{k=0}^{\infty} (-1)^k \frac{G_k(\mu,z)}{\nu^k}. \tag{33.3.55}$$

The coefficients $F_k^{(1)}(\mu,z)$ are obtained from (33.3.52) and the function $f(w)$ of (33.3.49) with the function $p_\mu(s)$ as given in the first line of (33.3.43), and $F_k^{(2)}(\mu,z)$ follow from taking the function $p_\mu(s)$ as given in the second line. The function $\mathcal{G}(z+a)$ is the slowly varying part of the corresponding gamma function $\Gamma(z+a)$ for large z. That is, see (6.4.58),

$$\mathcal{G}(z+a) = \frac{\Gamma(z+a)}{\sqrt{2\pi}\, z^{z+a-\frac{1}{2}} e^z} \sim \sum_{k=0}^{\infty} \frac{a_k(a)}{z^k}, \qquad z \to \infty. \tag{33.3.56}$$

The first coefficients $a_k(a)$ are given in (6.4.60).

The final form of the expansion of $Y_n^\mu(\zeta)$ given in (33.3.33) can be obtained by dividing the expansion in (33.3.53) by the large-ν expansion of $\mathcal{G}(\nu+\mu+\frac{1}{2})$ given in (33.3.56), and similarly for the other expansions. This gives for $k = 0, 1, 2, \ldots$

$$A_k(\mu,z) = F_k^{(1)}(\mu,z) - \sum_{j=0}^{k-1} A_j(\mu,z) a_{k-j}\left(\mu+\tfrac{1}{2}\right), \tag{33.3.57}$$

$$B_k(\mu,z) = F_k^{(2)}(\mu,z) - \sum_{j=0}^{k-1} B_j(\mu,z) a_{k-j}\left(\mu+\tfrac{1}{2}\right), \tag{33.3.58}$$

$$C_k(\mu,z) = \sum_{j=0}^{k} (-1)^j G_j(\mu,z) a_{k-j}\left(\tfrac{1}{2}\right). \tag{33.3.59}$$

To compute the coefficients $F_k(\mu,z)$ and $A_k(\mu,z)$ we need the coefficients in the expansion

$$s = s_+ + \sum_{k=1}^{\infty} s_k w^k, \tag{33.3.60}$$

which follow from (33.3.48). We write, as in (33.3.35), $t = 1/\sqrt{1+z^2}$, and obtain

$$\begin{aligned} s_1 &= 1, \quad s_2 = \frac{2-t}{6}, \quad s_3 = \frac{(1-t)\left(5t^3-6t^2+2\right)}{18t^2}, \\ s_4 &= -\frac{z(1-t)\left(40t^4-65t^3+24t^2-2t+4\right)}{135t^2}. \end{aligned} \tag{33.3.61}$$

With these coefficients we can compute the coefficients $f(w)$ and $g(w)$ of (33.3.49) and (33.3.50) by choosing the appropriate $p_\mu(s)$ and $q_\mu(s)$.

To obtain the coefficients in (33.3.57)–(33.3.59) we first compute $a_k(\mu+\frac{1}{2})$ that appear in (33.3.56). We have from (6.4.60)

$$\begin{aligned} a_0\left(\mu+\tfrac{1}{2}\right) &= 1, \quad a_1\left(\mu+\tfrac{1}{2}\right) = \tfrac{1}{24}\left(-1+12\mu^2\right), \\ a_2\left(\mu+\tfrac{1}{2}\right) &= \tfrac{1}{1152}\left(1+48\mu-24\mu^2-192\mu^3+144\mu^4\right), \\ a_3\left(\mu+\tfrac{1}{2}\right) &= \tfrac{1}{414720}\left(1003-720\mu-17100\mu^2+11520\mu^3+32400\mu^4-\right. \\ &\qquad \left. 34560\mu^5+8640\mu^6\right). \end{aligned} \tag{33.3.62}$$

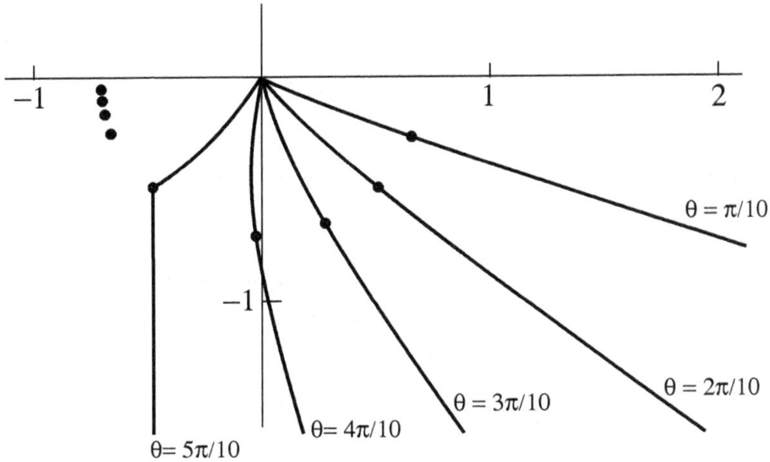

Fig. 33.1 Saddle point contours of the first integral in (33.3.41) for $z = e^{i\theta}$, $\theta = k\pi/10$, $k = 1, 2, 3, 4, 5$.

From Olver (1997, p. 378) it follows that the expansions in §9.5 hold for large values of ν and are uniformly valid for complex values of z inside the sector $|\mathrm{ph}\, z| \le \frac{1}{2}\pi - \delta$ with δ a small positive number.

The singularities of the functions f and g in (33.3.49) follow from those of the mapping in (33.3.48). This transformation does not depend on μ and, hence, Laplace's method remains applicable for all fixed values of μ, and also (33.3.33) is uniformly valid for complex values of z inside the sector $|\mathrm{ph}\, z| \le \frac{1}{2}\pi - \delta$.

For complex z inside the sector $|\mathrm{ph}\, z| < \frac{1}{2}\pi$ the saddle points given in (33.3.45) move into the complex plane, and for all these values of z it is possible to find a single saddle point contour from 0 through s_+ such that $\mathrm{ph}(zs) = 0$ at infinity. If $|\mathrm{ph}\, z| \le \frac{1}{2}\pi - \delta$ the singular points of the transformation in (33.3.48) and of the function f and g in (33.3.49) stay away from the origin.

In Figure 33.1 we show the saddle points s_+ (black dots) and the corresponding saddle point contours of the first integral in (33.3.41) for $z = e^{i\theta}$, $\theta = k\pi/10$, $k = 1, 2, 3, 4, 5$. The black dots at the left are the saddle points s_-. When $z = i$ the saddle points s_+ and s_- coincide at $s = -\frac{1}{2}(1+i)$. The saddle point contours of the second integral in (33.3.41) are the paths of steepest ascent of the first integral.

In Dunster (2001) the expansions of $Y_n^\mu(z)$ are given for large n with possibly large values of μ as well. Starting from an integral we show how to include μ as a second large parameter, leaving out the details. Let $\mu = \alpha\nu$ and write (33.4.68) in the form

$$Y_n^\mu(\zeta) = \frac{(2\nu z)^{n+\mu+1}}{\Gamma(n+\mu+1)} \int_0^\infty \frac{e^{-\nu\psi(s)}}{\sqrt{s(1+s)}}\, ds, \tag{33.3.63}$$

where $\zeta = 1/(\nu z)$, $\nu = n + \frac{1}{2}$, and

$$\psi(s) = 2zs - (1+\alpha)\ln s - \ln(1+s). \tag{33.3.64}$$

Then the saddle point method can be repeated, giving an expansion that holds again in the sector $|\mathrm{ph}\, z| \le \frac{1}{2}\pi - \delta$ and $\alpha \ge -1+\varepsilon$, with δ, ε small positive numbers. In a similar way the expansions for the sector $|\mathrm{ph}(-z)| \le \frac{1}{2}\pi - \delta$ can be modified.

33.4 Expansions in terms of modified Bessel functions

The expansions for $Y_n^\mu(1/(\nu z))$ in the previous section §33.3 become invalid when z approaches the points $\pm i$, because in that case the saddle points coincide. As shown in Dunster (2001) and Wong and Zhang (1997) it is possible to derive uniform expansions in terms of Airy functions, and these expansions are valid in large-z domains.

Because the asymptotic phenomena of the generalized Bessel polynomials $Y_n^\mu(z)$ for large n and fixed μ are the same as those of the polynomial $Y_n^0(z)$, we approach the problem to obtain uniform expansions by expanding the generalized polynomials in terms of the modified Bessel functions $K_\nu(z)$ (with $\nu = n + \frac{1}{2}$), which are the same as the reduced Bessel polynomials $Y_n^0(z)$ (see (33.1.3)). By using the existing results for the Bessel functions a complete description is available in this way.

We summarize the results of this section as follows. For $n \to \infty$ we have the asymptotic expansion

$$Y_n^\mu(\zeta) \sim \frac{(2\nu z)^\mu n!\, e^{\nu z}}{\Gamma(n+\mu+1)}\sqrt{\frac{2\nu z}{\pi}}\left(K_\nu(\nu z)\sum_{k=0}^\infty \frac{C_k}{\nu^k} + K_\nu'(\nu z)\sum_{k=0}^\infty \frac{D_k}{\nu^k}\right). \tag{33.4.65}$$

Here, $\zeta = 1/(\nu z)$, $\nu = n + \frac{1}{2}$, and

$$\begin{aligned} C_0 &= \tfrac{1}{2}U, \quad D_0 = -\frac{zV}{2w}, \\ C_1 &= \frac{1}{8w^3}\Big(\mu w\left(\mu w^2 - 1 - z\right)U + z\left(\mu^2 w^2 - 2z\right)V\Big), \\ D_1 &= \frac{1}{8w^3}\Big(-\mu^2 w z^2 U - \mu z\left(\mu w^2 - 1 - z\right)V\Big), \end{aligned} \tag{33.4.66}$$

where

$$U = s_+^\mu + s_-^\mu, \quad V = s_+^\mu - s_-^\mu, \quad w = \sqrt{1+z^2}, \tag{33.4.67}$$

with saddle points $s_\pm$ given in (33.3.45). Because $s_- < 0$ when $z > 0$, we need an interpretation of s_-^μ; this quantity is written as $e^{\mu\pi i}(-s_-)^\mu$.

33.4.1 *Construction of the expansion*

To start the construction of the expansion we write (33.1.5) in the form

$$Y_n^\mu(\zeta) = \frac{(2\nu z)^{n+\mu+1}}{\Gamma(n+\mu+1)}\int_0^\infty \frac{s^\mu}{\sqrt{s(1+s)}} e^{-\nu\phi(s)}\, ds, \tag{33.4.68}$$

where again

$$\zeta = \frac{1}{\nu z}, \quad \nu = n + \tfrac{1}{2}, \quad \phi(s) = 2zs - \ln s - \ln(1+s). \tag{33.4.69}$$

We write

$$f_0(s) = s^\mu = A_0 + B_0 s + \phi'(s) g_0(s), \tag{33.4.70}$$

and substitute $s = s_+$ and $s = s_-$ to obtain

$$A_0 = \frac{s_+ f_0(s_-) - s_- f_0(s_+)}{s_+ - s_-}, \quad B_0 = \frac{f_0(s_+) - f_0(s_-)}{s_+ - s_-}. \tag{33.4.71}$$

Putting (33.4.70) into (33.4.68) we obtain

$$Y_n^\mu(\zeta) = A_0\Phi_0 + B_0\Phi_1 + \frac{(2\nu z)^{n+\mu+1}}{\Gamma(n+\mu+1)} \int_0^\infty \frac{\phi'(s) g_0(s)}{\sqrt{s(1+s)}} e^{-\nu\phi(s)}\, ds, \tag{33.4.72}$$

where

$$\Phi_0 = \frac{(2\nu z)^\mu n!}{\Gamma(n+\mu+1)} Y_n^0(\zeta), \quad \Phi_1 = \frac{(2\nu z)^{\mu-1}(n+1)!}{\Gamma(n+\mu+1)} Y_n^1(\zeta). \tag{33.4.73}$$

By using (33.1.3) and (33.1.11) it follows that

$$\begin{aligned} \Phi_0 &= \frac{(2\nu z)^\mu n!}{\Gamma(n+\mu+1)} \sqrt{\frac{2\nu z}{\pi}}\, e^{\nu z} K_\nu(\nu z), \\ \Phi_1 &= \frac{(2\nu z)^\mu n!}{2\Gamma(n+\mu+1)} \sqrt{\frac{2\nu z}{\pi}}\, e^{\nu z} \big((1/z - 1)K_\nu(\nu z) - K_\nu'(\nu z)\big). \end{aligned} \tag{33.4.74}$$

In the second line we can also write

$$(1/z - 1)K_\nu(\nu z) - K_\nu'(\nu z) = K_{\nu+1}(\nu z) - K_\nu(\nu z), \tag{33.4.75}$$

but we prefer the notation with the derivative, because the asymptotic expansions of $K_\nu(\nu z)$ and $K_\nu'(\nu z)$ are quite related and usually presented together.

The next step is to use integration by parts in (33.4.72), and this gives

$$Y_n^\mu(\zeta) = A_0\Phi_0 + B_0\Phi_1 + \frac{(2\nu z)^{n+\mu+1}}{\nu\Gamma(n+\mu+1)} \int_0^\infty \frac{f_1(s)}{\sqrt{s(1+s)}} e^{-\nu\phi(s)}\, ds, \tag{33.4.76}$$

where

$$f_1(s) = \sqrt{s(1+s)} \frac{d}{ds} \frac{g_0(s)}{\sqrt{s(1+s)}}. \tag{33.4.77}$$

Repeating this procedure by writing for $k \ge 0$

$$f_k(s) = A_k + B_k s + \phi'(s) g_k(s), \quad f_0(s) = s^\mu, \tag{33.4.78}$$

$$A_k = \frac{s_+ f_k(s_-) - s_- f_k(s_+)}{s_+ - s_-}, \quad B_k = \frac{f_k(s_+) - f_k(s_-)}{s_+ - s_-}, \tag{33.4.79}$$

$$f_{k+1}(s) = \sqrt{s(1+s)} \frac{d}{ds} \frac{g_k(s)}{\sqrt{s(1+s)}} = g_k'(s) - \frac{2s+1}{2s(s+1)} g_k(s), \tag{33.4.80}$$

we obtain for $K \ge 0$

$$\begin{aligned} Y_n^\mu(\zeta) = \Phi_0 \sum_{k=0}^{K-1} \frac{A_k}{\nu^k} + \Phi_1 \sum_{k=0}^{K-1} \frac{B_k}{\nu^k} + \\ \frac{(2\nu z)^{n+\mu+1}}{\nu^M \Gamma(n+\mu+1)} \int_0^\infty \frac{f_K(s)}{\sqrt{s(1+s)}} e^{-\nu\phi(s)}\, ds. \end{aligned} \tag{33.4.81}$$

We rearrange the expansion by using (33.4.74) and writing

$$C_k = A_k + \frac{1-z}{2z} B_k, \quad D_k = -\tfrac{1}{2} B_k, \quad k = 0, 1, 2, \ldots \tag{33.4.82}$$

to obtain the expansion given in (33.4.65).

Remark 33.5. To compute the coefficients A_k, B_k defined in (33.4.79) and the functions $f_k(s)$, say, by using a computer algebra package, it is convenient to write the functions $f_k(s)$ in the form of two-point Taylor expansions at the saddle points s_+ and s_-. More details on this method can be found in Chapter 18. △

Remark 33.6. For integer values of μ we have the following simple cases.

(1) For $\mu = 0, 1, 2, \ldots$ the expansion in (33.4.65) has a finite number of terms which can also be obtained from the recursion in (33.1.10).
(2) For $\mu = -1, -2, -3, \ldots$ we can also obtain an exact result. When $\mu = -1$ we have

$$C_k = \frac{z-1}{2^k}, \quad D_k = -\frac{z}{2^k}, \quad k = 0, 1, 2, \ldots, \tag{33.4.83}$$

and we can sum the convergent series when $2\nu = 2n+1 > 1$. This gives a result that corresponds to the relation in (33.1.10) with $\mu = -1$.

△

Remark 33.7. To avoid the multivalued function s^μ in the computation of the coefficients of the expansion in (33.4.65), we can modify the integral in (33.4.68) by substituting $s = e^t$. What remains is that the coefficients are complex for positive values of z. From numerical computations, we see that when we take a few terms in the asymptotic expansion that the answer has a small imaginary part when $z > 0$. For example, when we take $z = 10$, $n = 1000$, $\mu = \frac{1}{3}$, we have the result from the expansion (using the coefficients as shown in (33.4.66)) in the form

$$Y_n^\mu(\zeta) \doteq 5.16995988669207921 48\text{e}21 + i\, 9.43102070336\text{e}12. \tag{33.4.84}$$

By using the recurrence relation in (33.1.8) we obtain the real result

$$Y_n^\mu(\zeta) \doteq 5.1699599272928779441\text{e}21, \tag{33.4.85}$$

with relative error 7.85e-9. The computations are done in Maple with Digits = 20; when increasing this to 50 we obtain the same result. In addition, for $z = 1/1000$ and $z = 1000$ the relative errors are of the same order. △

Chapter 34

Stirling numbers

We describe a method for obtaining an asymptotic expansion of both Stirling numbers $S_n^{(m)}$ and $\mathcal{S}_n^{(m)}$ for large values of n. The expansions are obtained by using a modification of the saddle point method, and are valid uniformly for $0 \le m \le n$. Short tables are given to show the results for $n = 10$. These experiments confirm the uniform character of our estimates.[1]

34.1 Definitions and integral representations

The Stirling numbers of the first and second kind, denoted by $S_n^{(m)}$ and $\mathcal{S}_n^{(m)}$, respectively, are defined by:

$$x(x-1)\cdots(x-n+1) = \sum_{m=0}^{n} S_n^{(m)} x^m, \tag{34.1.1}$$

$$x^n = \sum_{m=0}^{n} \mathcal{S}_n^{(m)} x(x-1)\cdots(x-m+1). \tag{34.1.2}$$

We shall give asymptotic expansions of the Stirling numbers as $n \to \infty$, and the expansion hold uniformly with respect to $m \in [0, n]$.

Alternative generating function are

$$\frac{\bigl(\ln(x+1)\bigr)^m}{m!} = \sum_{n=m}^{\infty} S_n^{(m)} \frac{x^n}{n!}, \tag{34.1.3}$$

$$\frac{(e^x-1)^m}{m!} = \sum_{n=m}^{\infty} \mathcal{S}_n^{(m)} \frac{x^n}{n!}. \tag{34.1.4}$$

This gives the Cauchy-type integrals

$$\mathcal{S}_n^{(m)} = \frac{n!}{m!}\frac{1}{2\pi i}\oint \frac{(e^x-1)^m}{x^{n+1}}\,dx, \tag{34.1.5}$$

[1]This chapter is based on Temme (1993b).

$$(-1)^{n-m}S_{n+1}^{(m+1)} = \frac{1}{2\pi i}\oint \frac{(x+1)(x+2)\cdots(x+n)}{x^{m+1}}\,dx, \tag{34.1.6}$$

where the contours are small circles around the origin.

We have the following boundary values

$$S_n^{(n)} = \mathcal{S}_n^{(n)} = 1,\ n \ge 0, \quad \text{and} \quad S_n^{(0)} = \mathcal{S}_n^{(0)} = 0,\ n \ge 1. \tag{34.1.7}$$

Furthermore it is convenient to assume that $S_n^{(m)} = \mathcal{S}_n^{(m)} = 0$ if $m > n$.

The Stirling numbers are integers; apart from the above mentioned zero values, the numbers of the second kind are positive; those of the first kind have the sign of $(-1)^{n+m}$.

For the Stirling numbers of the second kind we have a simple explicit representation

$$\mathcal{S}_n^{(m)} = \frac{1}{m!}\sum_{k=0}^{m}(-1)^{m+k}\binom{m}{k}k^n. \tag{34.1.8}$$

The Stirling numbers play an important role in difference calculus, combinatorics, and probability theory. More properties of Stirling numbers can be found in, for example, Jordan (1947, Chapter 4), Comtet (1974), and in Bressoud (2010).

An example from combinatorics is: $\mathcal{S}_n^{(m)}$ is the number of ways of partitioning a set of n elements into m non-empty subsets. We find for $m = 2, n = 4$ the value $\mathcal{S}_4^{(2)} = 7$, since

$$\begin{aligned}\{a,b,c,d\} &= \{a\}\cup\{b,c,d\} = \{b\}\cup\{a,c,d\}\\ &= \{c\}\cup\{a,b,d\} = \{d\}\cup\{a,b,c\}\\ &= \{a,b\}\cup\{c,d\} = \{a,c\}\cup\{b,d\} = \{a,d\}\cup\{b,c\}.\end{aligned} \tag{34.1.9}$$

Asymptotic expansions can be found in several papers. Hsu (1948) has given an expansion for the numbers of the second kind that is useful when $n - m$ is small. Moser and Wyman (1958, 1957) have considered several overlapping domains in the (n, m)-plane with $n \ge m$. In Bleick and Wang (1974) a complete expansion for the numbers of the second kind has been given, which is convergent when $m < (n+1)^{2/3}/[\pi+(n+1)^{-1/3}]$. When divergent, the expansion is still useful if used as an asymptotic series. However, the case $m \sim n$ (both large) is not covered. In Dingle (1973, p. 199) the numbers of both kinds are considered. Again, the approximations are not uniformly valid with respect to m; the results for the numbers of the first kind become worse when $m \ll n$, whereas for the numbers of the second kind the problems arise when $m \sim n$.

In most cases the asymptotic results are obtained by using saddle point techniques for contour integrals. In Knessl and Keller (1991) the asymptotics has been done from a quite different point of view.

Generalizations of Stirling numbers have been studied in several directions. One way is observing that the generating function in (34.1.1) can be written in the form

$$n!\binom{x}{n} = \frac{\Gamma(x+1)}{\Gamma(x-n+1)}, \quad n = 0, 1, 2, \ldots. \tag{34.1.10}$$

When we replace n by α (a general real or complex number) then we can call the numbers $S_\alpha^{(m)}$ in the expansion

$$\frac{\Gamma(x+1)}{\Gamma(x-\alpha+1)} = \sum_{m=0}^{\infty} S_\alpha^{(m)} x^m, \quad |x| < 1, \tag{34.1.11}$$

Stirling functions of the first kind. In the explicit representation (34.1.8) of the numbers of the second kind we can replace n by a number α, and obtain functions $\mathcal{S}_\alpha^{(m)}$. See Butzer and Hauss (1991), also for asymptotic expansions of $\mathcal{S}_{\alpha+m}^{(m)}$ for large m. For Stirling numbers with general complex parameters we refer to Graham *et al.* (1994), Flajolet and Prodinger (1999), and for asymptotic approximations of those to Chelluri *et al.* (2000).

In Broder (1984) so-called r-Stirling numbers are introduced, also with a combinatorial interpretation, by writing the generating identities (34.1.1) and (34.1.4) in the form

$$x^r(x+r)(x+r+1)\cdots(x+n-1) = \sum_{m=r}^{n} \begin{bmatrix} n \\ m \end{bmatrix}_r x^m, \tag{34.1.12}$$

$$\frac{e^{rx}(e^x-1)^m}{m!} = \sum_{n=m}^{\infty} \begin{Bmatrix} n \\ m \end{Bmatrix}_r \frac{x^n}{n!}. \tag{34.1.13}$$

These relations define the r-Stirling numbers of the first and second kind, respectively, for $n \ge r$ and $m \ge 0$ and integer $r \ge 0$. Asymptotic approximations are considered in Corcino *et al.* (1999), and the methods are based on the uniform methods described in this chapter.

34.2 Stirling number of the second kind

A modification of the saddle point method gives a uniform expansion for large n that holds uniformly with respect to $m \in [0, n]$.

We write (34.1.5) in the form:

$$\mathcal{S}_n^{(m)} = \frac{n!}{m!}\frac{1}{2\pi i}\oint e^{\phi(x)}\frac{dx}{x}, \tag{34.2.14}$$

where

$$\phi(x) = -n\ln x + m\ln(e^x - 1). \tag{34.2.15}$$

There is a real positive saddle point x_0 that follows from the equation[2]

$$\mu x = 1 - e^{-x}, \quad \mu = \frac{m}{n}. \tag{34.2.16}$$

The solution $x_0 = 0$ is not of interest, since the contour in (34.2.14) is not allowed to pass through the origin. Moreover, $\phi'(0)$ is undefined.

[2]It is not difficult to verify that complex saddle points occur outside the strip $|\Im z| < 2\pi$.

The limiting behavior of x_0 is

$$m \ll n, \quad \Longrightarrow \quad x_0 \to +\infty, \qquad m \sim n \quad \Longrightarrow \quad x_0 \downarrow 0. \tag{34.2.17}$$

Further details on the saddle point can be found in a later section.

Recall that the saddle point method is based on replacing

$$\phi(x) = -n \ln x + m \ln(e^x - 1) \tag{34.2.18}$$

by a quadratic function

$$\phi(x) - \phi(x_0) = t^2, \quad \text{sign}\,(x - x_0) = \text{sign}(t). \tag{34.2.19}$$

However, this usual approach gives approximations that are not accurate when $n \sim m$. In that case $x_0 \downarrow 0$, and the logarithmic term in $\phi(x)$ vanishes. Clearly when $m \sim n$ and $x \sim 0$, the function $\phi(x)$ should not be transformed into a parabola.

To modify the standard saddle point transformation in (34.2.19) we observe that

$$x \to 0^+ \quad \Longrightarrow \quad \phi(x) \sim (m - n) \ln x, \tag{34.2.20}$$

and

$$x \to \infty \quad \Longrightarrow \quad \phi(x) \sim mx. \tag{34.2.21}$$

This suggests the transformation $x \to t(x)$ defined by

$$\phi(x) = mt + (m - n) \ln t + A, \tag{34.2.22}$$

where A does not depend on t. The derivative of the right-hand side vanishes at $t_0 = (n - m)/m$. We prescribe for the mapping (34.2.22) the corresponding points

$$x = 0 \Longleftrightarrow t = 0, \; x = x_0 \Longleftrightarrow t = t_0, \; x = +\infty \Longleftrightarrow t = +\infty. \tag{34.2.23}$$

The quantity A follows from substitution of $x = x_0$, $t = t_0$ in (34.2.22), which gives

$$A = \phi(x_0) - mt_0 + (n - m) \ln t_0. \tag{34.2.24}$$

For more information on A we refer to a later section.

Transformation (34.2.22) brings (34.2.14) in the form

$$\mathcal{S}_n^{(m)} = \frac{n!}{m!} \frac{e^A}{2\pi i} \oint e^{mt} f(t) \frac{dt}{t^{n-m+1}}, \tag{34.2.25}$$

where

$$f(t) = \frac{t}{x} \frac{dx}{dt} = \frac{m(t - t_0)}{x \phi'(x)}. \tag{34.2.26}$$

Initially, the contour in (34.2.25) is a small circle around $t = 0$, but it can be deformed into a contour through the new saddle point t_0.

A first approximation is obtained by replacing $f(t)$ in (34.2.25) by its value at the saddle point t_0. The result is

$$\mathcal{S}_n^{(m)} \sim e^A m^{n-m} f(t_0) \binom{n}{m}, \quad n \to \infty, \tag{34.2.27}$$

Table 34.1 Approximations of the Stirling numbers of the second kind.

m	$\mathcal{S}_{10}^{(m)}$	(34.2.27)	m	$\mathcal{S}_{10}^{(m)}$	(34.2.27)
1	1	0.9993	6	22827	22922
2	511	512.98	7	5880	5897.7
3	9330	9390.1	8	750	751.45
4	34105	34319	9	45	45.04
5	42525	42750	10	1	1

where

$$f(t_0) = \frac{1}{x_0}\sqrt{\frac{mt_0}{\phi''(x_0)}} = \sqrt{\frac{t_0}{(1+t_0)(x_0-t_0)}}. \tag{34.2.28}$$

A transformation as in (34.2.22) with logarithmic terms is used in earlier chapters, for example in §25.6. The function f is analytic in neighborhoods of the origin and the saddle point, and in a wider domain of the complex plane, including the positive real axis. In particular, when t_0 (that is, x_0) tends to zero, f remains analytic in a neighborhood of the origin.

We consider limiting values of $f(t_0)$. Let $\mu = m/n$. Then

$$f(t_0) = 1 + \mathcal{O}(1-\mu), \quad \mu \to 1, \tag{34.2.29}$$

and

$$f(t_0) = 1 + \mathcal{O}(\mu), \quad \mu \to 0. \tag{34.2.30}$$

Table 34.1 gives exact values $\mathcal{S}_{10}^{(m)}$ $(m = 1, 2, \ldots, 10)$ and the approximations based on (34.2.27). The maximal relative error is 0.0064, and occurs at $m = 3$. Similar computations with $n = 20, n = 30$ show the following: the maximal relative errors are 0.0031, 0.0021, and occur for $m = 7, m = 10$, respectively.

The maximal errors do not occur at boundary values of m, but at about $m = \frac{1}{3}n$. Larger values of n also show the uniform character of the asymptotic estimate (34.2.27).

34.2.1 *Higher-order approximations*

The estimate in (34.2.27) can be supplied with more terms, and eventually written as a complete asymptotic expansion.

Consider the integral in (34.2.25) in the form

$$F_\lambda(m) = \frac{1}{2\pi i}\int e^{mt} f(t)\frac{dt}{t^{\lambda+1}}, \tag{34.2.31}$$

where $\lambda = n - m$. The contour is a contour around the origin, where f is analytic. Writing

$$f(t) = f(t_0) + (t - t_0)g(t), \quad t_0 = \lambda/m, \tag{34.2.32}$$

we obtain

$$F_\lambda(m) = \frac{m^\lambda}{\Gamma(\lambda+1)}\, f(t_0) + \frac{1}{2\pi i}\int (t - t_0)e^{mt} g(t)\frac{dt}{t^{\lambda+1}}. \tag{34.2.33}$$

Integrating by parts, we obtain after several steps, and in the usual way (see §25.2),

$$F_\lambda(m) \sim \frac{m^\lambda}{\Gamma(\lambda+1)}\sum_{k=0}^{\infty}(-1)^k f_k(t_0)m^{-k}, \tag{34.2.34}$$

where the functions $f_k(t)$ are defined by

$$f_{k+1}(t) = t\frac{d}{dt}\frac{f_k(t) - f_k(t_0)}{t - t_0}, \quad k = 0, 1, 2, \ldots, \tag{34.2.35}$$

with $f_0(t) = f(t)$.

This procedure gives for (34.2.25) a complete asymptotic expansion of the form

$$\mathcal{S}_n^{(m)} \sim e^A m^{n-m}\binom{n}{m}\sum_{k=0}^{\infty}(-1)^k f_k(t_0)m^{-k}, \quad n \to \infty, \tag{34.2.36}$$

where $t_0 = (n - m)/m$ and f is given by (34.2.26). The first coefficient of the series is given by (34.2.28), the second one, $f_1(t_0)$, can be obtained as explained in §25.2. The function f is analytic in a domain containing the positive real axis. Consider the Taylor expansion around $t = t_0$ (the point where we expect the main contributions to the integral in (34.2.25)

$$f(t) = \sum_{k=0}^{\infty} a_k(t - t_0)^k, \quad a_k = \frac{1}{k!}\, f^{(k)}(t_0). \tag{34.2.37}$$

Then it follows that

$$f_1(t) = t\sum_{k=0}^{\infty}(k+1)a_{k+2}(t - t_0)^k, \tag{34.2.38}$$

and that

$$\begin{aligned} &f_1(t_0) = a_2 t_0, \quad f_2(t_0) = t_0\,(2a_3 + 3a_4 t_0)\,, \\ &f_3(t_0) = t_0\left(6a_4 + 20a_5 t_0 + 15a_6 t_0^2\right). \end{aligned} \tag{34.2.39}$$

Further coefficients $f_k(t_0)$ easily follow from deriving recurrence relations for the coefficients of Taylor series of $f_k(t)$ around the point $t = t_0$. Let us write $f_k(t) = \sum_{j=0}^{\infty} a_j^{(k)}(t - t_0)^j$, then

$$a_j^{(k+1)} = j a_{j+1}^{(k)} + t_0(j+1)a_{j+2}^{(k)}, \quad a_j^{(0)} = a_j, \quad j, k = 0, 1, 2, \ldots. \tag{34.2.40}$$

Table 34.2 Approximation (34.2.44) of the Stirling numbers of the second kind.

n	maximal relative error	occurs at $m =$
10	0.00047	4
20	0.00012	3
30	0.00006	4
40	0.00003	5
50	0.00002	7

To compute a_k, we need the coefficients x_k in the expansion

$$x = x_0 + x_1(t - t_0) + x_2(t - t_0)^2 + \cdots, \tag{34.2.41}$$

where the relation between x and t is defined in (34.2.22) and x_0 is the solution of (34.2.16). We have $x_1 = a_0 x_0/t_0 = f(t_0)x_0/t_0$. See the first relation in (34.2.26); $f(t_0)$ is given in (34.2.28). With (34.2.26) we can express other values of a_k in terms of x_k. So we obtain by manipulations of power series

$$\begin{aligned} a_1 &= \frac{2t_0x_0x_2 + x_1x_0 - t_0x_1^2}{x_0^2}, \\ a_2 &= \frac{3x_3t_0x_0 + 2x_2x_0^2 - 3t_0x_0x_1x_2 - x_1^2x_0 + t_0x_1^3}{x_0^3}. \end{aligned} \tag{34.2.42}$$

Hence, for $f_1(t_0) = t_0a_2$ we need x_0, x_1, x_2, x_3. To obtain x_2, x_3, we use (34.2.41) and dx/dt that follows from (34.2.26). After several manipulations we finally obtain the coefficient of the second term in (34.2.36), that is

$$\begin{aligned} f_1(t_0) = \big(&-2x_0^3 + 2t_0^5 + 4t_0^3 + 4t_0^4 + 3x_0^2t_0 - 6x_0t_0^4 - 5x_0^2t_0^2 + 2x_0^4t_0 + \\ &x_0^3t_0 - 6x_0^3t_0^2 + 8x_0^2t_0^3\big) \,/\, \big(24a_0(1+t_0)^2(x_0 - t_0)^4\big). \end{aligned} \tag{34.2.43}$$

A further analysis shows that $f_1(t_0)$ is a bounded function of t_0 on $[0,\infty)$. Using the two-term variant of (34.2.27), that is,

$$\mathcal{S}_n^{(m)} \sim e^A m^{n-m}\binom{n}{m}\left(f(t_0) - \frac{f_1(t_0)}{m}\right), \quad n \to \infty, \tag{34.2.44}$$

we obtain the maximal relative errors as shown in Table 34.2.

Remark 34.1. We have obtained an expansion in negative powers of m, whereas we assume that n is the large parameter. When we choose a different scaling of parameters in (34.2.31), for example by using $m = \mu n$, we can obtain an expansion in negative powers of n, and it will be the same expansion. △

34.2.2 *About the positive saddle point*

Consider equation

$$\mu x = 1 - e^{-x}, \tag{34.2.45}$$

that for $\mu = m/n$ defines the real saddle point x_0 for the integral in (34.2.14).

Lemma 34.1. *For the positive solution x_0 of the equation* (34.2.45), *where $0 < \mu < 1$, we have $1/\mu - 1 < x_0 < 1/\mu$.*

Proof. For $x > 0$ we have $x/(1+x) < 1 - e^{-x} < 1$, which gives the bounds for x_0. □

The following lemma gives an expansion of x_0 that is very useful for small values of μ.

Lemma 34.2. *For $\mu \in (0, 1)$ the positive solution x_0 of* (34.2.45) *has the convergent expansion*

$$x_0 = \frac{1}{\mu} - \sum_{k=1}^{\infty} \frac{k^{k-1}}{k!} \delta^k, \quad \delta = \frac{1}{\mu} e^{-\frac{1}{\mu}}. \tag{34.2.46}$$

Proof. Let $x = 1/\mu + y$; then the equation for y reads $ye^y = -\delta$. One solution is $y = -1/\mu$, which corresponds to the trivial solution $x = 0$ of equation (34.2.45). We need the other y-solution that tends to zero if $\delta \to 0$. The equation for y is considered in De Bruijn (1958, p. 23), where the convergent expansion of this lemma is given for $|\delta| < 1/e$. □

Remark 34.2. It may be noted that the solution x_0 can be written in terms of the Lambert W-function (see Roy and Olver (2010, §4.13)) That is, the function $W(x)$ that satisfies the equation $W(x)e^{W(x)} = x$. The equation considered in Lemma 34.2 has the form $ye^y = -\delta$, and because $0 < \delta < 1/e$, the solution is in terms of the Lambert W-function $x = 1/\mu + W(-\delta)$, where the W-function assumes its principal branch. △

The convergence of the expansion in the previous lemma becomes worse if μ is close to unity. For this case we have a different expansion.

Lemma 34.3. *For small values of $|\mu - 1|$ the positive solution x_0 of* (34.2.45) *has the convergent expansion*

$$x_0 = \sum_{k=1}^{\infty} c_k (1-\mu)^k, \tag{34.2.47}$$

where the first few coefficients are given by

$$c_1 = 2, \quad c_2 = \tfrac{4}{3}, \quad c_3 = \tfrac{10}{9}, \quad c_4 = \tfrac{136}{135}. \tag{34.2.48}$$

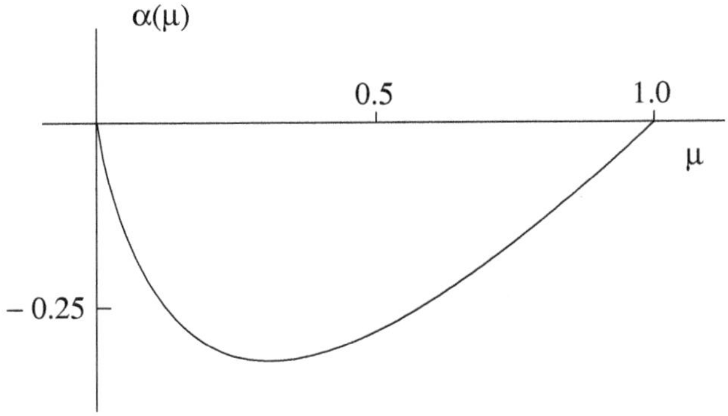

Fig. 34.1 Graph of $\alpha(\mu) = A/n$ defined in (34.2.51) as a function of μ.

Proof. The equation (34.2.45) can be written in the form

$$1 - \mu = \frac{x}{f(x)}, \quad f(x) = \frac{x^2}{e^{-x} - 1 + x}. \tag{34.2.49}$$

The function f is analytic in a neighborhood of the point $x = 0$ with $f(0) = 2$. According to De Bruijn (1958, p. 22), there exist positive numbers a and b such that for $|1 - \mu| < a$ the equation has just one solution x_0 in the domain $|x| < b$, and x_0 has for $|1 - \mu| < a$ the convergent expansion as shown in the lemma, with

$$c_k = \frac{1}{k!} \left(\frac{d}{dx}\right)^{k-1} \left(f(x)\right)^k \bigg|_{x=0}. \tag{34.2.50}$$

□

34.2.3 *About the quantity A*

The quantity A is defined in (34.2.24) where the function ϕ is given in (34.2.15) and x_0 is the positive saddle point. We give a few details about the size of this quantity, which is a function of n and m. It is convenient to define $\alpha(\mu) = A/n, 0 < \mu < 1$, that is,

$$\alpha(\mu) = -\ln x_0 + \mu \ln\left(e^{x_0} - 1\right) - \mu t_0 + (1 - \mu) \ln t_0, \tag{34.2.51}$$

where $\mu = m/n$ and $t_0 = 1/\mu - 1$. The function $\alpha(\mu)$ is only a function of μ, and is negative on the interval $(0, 1)$; see Figure 34.1.

Lemma 34.4. *$\alpha(\mu) \to 0$ if $\mu \to 0$ or $\mu \to 1$.*

Proof. This follows from the behavior of x_0 in these limits. Using the fact that x_0 is a solution of (34.2.45), we write $\alpha(\mu)$ in the form,

$$\alpha(\mu) = (1 - \mu) \ln \frac{t_0}{x_0} + \mu x_0 + \mu \ln \mu + \mu - 1. \tag{34.2.52}$$

If $\mu \to 1$, we use Lemma 34.3. If $\mu \to 0$ we observe that $x_0/t_0 = 1 - \mu\delta + \mathcal{O}\left(\mu\delta^2\right)$ and that $\mu x_0 - 1 = \mathcal{O}(\mu\delta)$, where δ is introduced in Lemma 34.2. □

34.3 Stirling numbers of the first kind

In Dingle (1973, p. 199) the generating function (34.1.3) has been used, which gives the representation

$$S_n^{(m)} = \frac{n!}{m!}\frac{1}{2\pi i}\int \frac{\left(\ln(z+1)\right)^m}{z^{n+1}}\,dz, \tag{34.3.53}$$

where the contour is a small circle around $z=0$. To make the integral representation similar to the previous case, Dingle transformed $1+z=e^x$. This gives

$$S_n^{(m)} = \frac{n!}{m!}\frac{1}{2\pi i}\int \frac{x^m e^x}{(e^x-1)^{n+1}}\,dx, \tag{34.3.54}$$

Integration by parts and changing $x\to -x$ gives

$$S_n^{(m)} = \frac{(n-1)!}{(m-1)!}\frac{(-1)^{n-m}}{2\pi i}\int \frac{x^{m-1}}{(1-e^{-x})^n}\,dx. \tag{34.3.55}$$

Again, the contour is a small circle around $x=0$. This representation has been used by Dingle to apply the saddle point method. We have used the method of the previous section on this integral, but the results were not as satisfactory as in that case.

To try an alternative representation we use (34.1.6), and write it in the form

$$(-1)^{n-m}S_{n+1}^{(m+1)} = \frac{1}{2\pi i}\int e^{\phi(x)}\frac{dx}{x}, \tag{34.3.56}$$

where

$$\phi(x) = \ln\bigl((x+1)(x+2)\cdots(x+n)\bigr) - m\ln x. \tag{34.3.57}$$

The saddle point is the solution of $\phi'(x)=0$. We have the following lemma.

Lemma 34.5. *The function $\phi'(x)$ has one and only one positive zero.*

Proof. Note that $\phi'(x)$ is negative when x is small and that for positive values of x we have $\phi'(x) > n/(x+n) - m/x$. Hence, $\phi'(x)>0$ when $x > nm/(n-m)$. This shows that $\phi'(x)$ has at least one positive zero. Next we observe that $\phi'(x)$ has at least one zero in any of the $n-1$ intervals $(-2,-1),(-3,-2),\ldots,(-n,-n+1)$. Now we reduce the $n+1$ fractions of $\phi'(x)$ to a common denominator. Then the nominator of $\phi'(x)$ is a polynomial of degree n and having at most n zeros. From the distribution of the zeros of $\phi'(x)$ just mentioned we have proved the lemma. □

The present function $\phi(x)$ has the following behavior on the positive real axis:

$$\phi(x)\sim -m\ln x, \quad \text{as } x\to 0, \qquad \phi(x)\sim (n-m)\ln x, \quad \text{as } x\to\infty. \tag{34.3.58}$$

Combining these two limiting cases, we observe that the function $n\ln(x+1) - m\ln x$ has (globally on $(0,\infty)$) the same graph as $\phi(x)$. This suggests the following transformation $x\mapsto t(x)$

$$\phi(x) = n\ln(1+t) - m\ln t + B. \tag{34.3.59}$$

Table 34.3 Approximations of the Stirling numbers of the first kind.

m	$\lvert S_{10}^{(m)}\rvert$	(34.3.65)	m	$\lvert S_{10}^{(m)}\rvert$	(34.3.65)
1	362880	362880	6	63273	63007
2	1026576	1018563	7	9450	9420.8
3	1172700	1163168	8	870	868.2
4	723680	718718	9	45	44.95
5	269325	267855	10	1	1

The derivative of the right-hand side vanishes at $t_0 = m/(n-m)$. We prescribe for the mapping in (34.3.59) the corresponding points

$$x = 0 \Longleftrightarrow t = 0, \quad x = x_0 \Longleftrightarrow t = t_0, \quad x = +\infty \Longleftrightarrow t = +\infty. \tag{34.3.60}$$

The quantity B follows from substitution of $x = x_0$, $t = t_0$ in (34.3.59), which gives

$$B = \phi(x_0) - n\ln(t_0+1) + m\ln t_0. \tag{34.3.61}$$

The transformation (34.3.59) brings (34.3.56) in the form

$$(-1)^{n-m} S_{n+1}^{(m+1)} = \frac{e^B}{2\pi i}\int \frac{(1+t)^n}{t^{m+1}} g(t)\,dt, \tag{34.3.62}$$

where, initially, the contour is a small circle around $t = 0$, and

$$g(t) = \frac{t}{x}\frac{dx}{dt} = \frac{(n-m)t - m}{(t+1)x\phi'(x)}. \tag{34.3.63}$$

A first approximation to $S_{n+1}^{(m+1)}$ is now obtained by replacing $g(t)$ in (34.3.62) by the value of this function at the saddle point t_0. The remaining integral is evaluated by using

$$\frac{1}{2\pi i}\int \frac{(1+t)^n}{t^{m+1}}\,dt = \binom{n}{m}. \tag{34.3.64}$$

This gives the one-term approximation

$$S_{n+1}^{(m+1)} \sim (-1)^{n-m} e^B g(t_0)\binom{n}{m}, \quad n \to \infty. \tag{34.3.65}$$

We finally compute $g(t_0)$. It follows from (34.3.63) that

$$g(t_0) = \frac{t_0}{x_0}\frac{dx}{dt} = \frac{n-m}{(t_0+1)x_0\phi''(x_0)dx/dt}, \tag{34.3.66}$$

where dx/dt is evaluated at $t = t_0$. This gives a relation for dx/dt at $t = t_0$, from which we obtain

$$g(t_0) = \frac{1}{x_0}\sqrt{\frac{m(n-m)}{n\phi''(x_0)}}. \tag{34.3.67}$$

When n and x are large, we may use representations in terms of the logarithm of the gamma function and the derivatives thereof. That is,

$$\begin{aligned}\phi(x) &= \ln\Gamma(x+n+1) - \ln\Gamma(x+1) - m\ln x,\\ \phi'(x) &= \psi(x+n+1) - \psi(x+1) - m/x.\end{aligned} \tag{34.3.68}$$

Asymptotic expansions of these functions are given in Chapter 6; see also (13.2.82). The transformation defined in (34.3.59) is of the same kind as the one in (34.2.22).

In Table 34.3 exact values of $|S_{10}^{(m)}|$, $(m = 1, \ldots, 10)$ are compared with absolute values of the approximations given in (34.3.65) (the values for $m = 1$, $m = n$ are not computed via (34.3.65)). The maximal relative error now occurs at $m = 3$, and is 0.0082. For $n = 20$, $n = 30$, the maximal errors are: 0.0063 and 0.0053, respectively; again they occur at $m = 3$. These experiments confirm the uniform character with respect to m of the result in (34.3.65).

Remark 34.3. A curious finite expansion follows by substituting

$$g(t) = \sum_{k=0}^{\infty} c_k t^k \tag{34.3.69}$$

in (34.3.62), from which we obtain the exact representation

$$(-1)^{n-m} S_{n+1}^{(m+1)} = e^B \sum_{k=0}^{m} c_k \binom{n}{m-k}. \tag{34.3.70}$$

For example, when $m = 0$, we have $c_0 = g(0) = 1$ and $e^B = n!$, which gives $S_{n+1}^{(1)} = (-1)^n n!$. △

Chapter 35

Asymptotics of the integral $\int_0^1 \cos(bx + a/x)\,dx$

We consider the integral

$$P(a,b) = \int_0^1 \cos(bx + a/x)\,dx \tag{35.0.1}$$

for large positive values of a and b. As we will see, the asymptotic behavior of $P(a,b)$ depends on the ratio b/a. Let

$$c = \sqrt{b/a}. \tag{35.0.2}$$

Then we consider the following four cases: $0 < c \le c_0 < 1$, $c = 1$, $c \ge c_1 > 1$, and general $c \ge 0$. Here, c_0, c_1 are constants.

Integrals of this type arose in the study of the so-called Minkowski Question Mark function and Salem's problem; see Alkauskas (2012).

35.1 The case $b < a$

We write $P(a,b)$ in the form

$$P(a,b) = \frac{1}{c}\int_0^c \cos\bigl(\lambda(t + 1/t)\bigr)\,dt, \quad \lambda = \sqrt{ab}, \tag{35.1.3}$$

and introduce a function $u(t)$ by writing

$$t + 1/t = 2(u+1), \quad 0 < t < 1. \tag{35.1.4}$$

Inverting this we have

$$t(u) = u + 1 - \sqrt{(u+1)^2 - 1}, \quad u \ge 0, \tag{35.1.5}$$

where the square root is positive for all $u \ge 0$. We have

$$\frac{dt}{du} = 1 - \frac{u+1}{\sqrt{(u+1)^2 - 1}}. \tag{35.1.6}$$

The integral in (35.1.3) becomes

$$P(a,b) = \frac{1}{c}\int_{u_0}^{\infty} \cos\bigl(2\lambda(u+1)\bigr)f(u)\,du, \tag{35.1.7}$$

where

$$f(u) = -\frac{dt}{du}, \quad u_0 = u(c) = \tfrac{1}{2}(c + 1/c) - 1. \tag{35.1.8}$$

Observe that f and all its derivatives vanish as $u \to +\infty$. We have

$$f^{(k)}(u) = \mathcal{O}\left(u^{-k-2}\right), \quad u \to +\infty. \tag{35.1.9}$$

It follows that we can integrate by parts. After replacing the cosine by exponentials we can use integration by parts as described in §14.2 for the method of stationary phase. In the present case we write

$$P(a,b) = \frac{1}{2\lambda c}\int_{u_0}^{\infty} f(u)\, d\sin\bigl(2\lambda(u+1)\bigr), \tag{35.1.10}$$

and obtain

$$P(a,b) = -\frac{\sin\left(2\lambda(u_0+1)\right)}{2\lambda c} f\left(u_0\right) - \frac{1}{2\lambda c}\int_{u_0}^{\infty} \sin\bigl(2\lambda(u+1)\bigr) f'(u)\, du. \tag{35.1.11}$$

Because

$$2\lambda c = 2b, \quad 2\lambda(u_0+1) = a+b, \quad f\left(u_0\right) = \frac{2c^2}{1-c^2} = \frac{2b}{a-b}, \tag{35.1.12}$$

we can write this in the form

$$P(a,b) = -\frac{\sin\left(a+b\right)}{a-b} - \frac{1}{2\lambda c}\int_{u_0}^{\infty} \sin\bigl(2\lambda(u+1)\bigr) f'(u)\, du, \tag{35.1.13}$$

and we can continue integrating by parts. This gives, eventually, the asymptotic expansion

$$\begin{aligned} P(a,b) \sim & -\frac{\sin\left(a+b\right)}{2\lambda c}\sum_{k=0}^{\infty} \frac{(-1)^k f^{(2k)}(u_0)}{(2\lambda)^{2k}} - \\ & \frac{\cos\left(a+b\right)}{(2\lambda)^2 c}\sum_{k=0}^{\infty} \frac{(-1)^k f^{(2k+1)}(u_0)}{(2\lambda)^{2k}}. \end{aligned} \tag{35.1.14}$$

The derivatives are easily evaluated, and the first few are

$$\begin{aligned} f^{(1)}(u_0) &= -\frac{8c^3}{\left(1-c^2\right)^3}, \\ f^{(2)}(u_0) &= \frac{48\left(c^2+1\right)c^4}{\left(1-c^2\right)^5}, \\ f^{(3)}(u_0) &= -\frac{384\left(c^4+3c^2+1\right)c^5}{\left(1-c^2\right)^7}, \\ f^{(4)}(u_0) &= \frac{3840\left(c^2+1\right)\left(c^4+5c^2+1\right)c^6}{(1-c^2)^9}. \end{aligned} \tag{35.1.15}$$

35.2 The case $a = b$

When $a = b$ we have $c = 1$ and $u_0 = 0$. The integral in (35.1.7) becomes

$$P(a,a) = \int_0^\infty \cos\bigl(2a(1+u)\bigr) f(u)\,du, \tag{35.2.16}$$

where $f(u)$ is defined in (35.1.8) (see also (35.1.6)). In this case we cannot use integration by parts, because f is singular at the origin.

The integral in (35.2.16) can be written in the form

$$P(a,a) = \Re e^{2ia} \int_0^\infty e^{2iau} f(u)\,du. \tag{35.2.17}$$

We rotate the path of integration onto the positive imaginary axis, and obtain (writing $u = iv$)

$$P(a,a) = \Re e^{2ia} \int_0^\infty e^{-2av} g(v)\,dv, \quad g(v) = i\left(\frac{(1+iv)(1-i)}{2\sqrt{v+\frac12 iv^2}} - 1\right). \tag{35.2.18}$$

The asymptotic expansion can be obtained by using Watson's lemma, see §2.1, by expanding $g(v)$ in powers of v at the origin:

$$g(v) = -i + \frac{1}{\sqrt{v}} \sum_{k=0}^\infty (c_k + id_k) v^k, \tag{35.2.19}$$

and interchanging the order of summation and integration. This gives

$$P(a,a) \sim \Re e^{2ia} \left(-\frac{i}{2a} + \sum_{k=0}^\infty (c_k + id_k) \frac{\Gamma\left(k+\frac12\right)}{(2a)^{k+\frac12}}\right), \tag{35.2.20}$$

or

$$P(a,a) \sim \frac{\sin(2a)}{2a} + \sqrt{\frac{\pi}{2a}} \sum_{k=0}^\infty \bigl(\cos(2a)c_k - \sin(2a)d_k\bigr) \frac{\left(\frac12\right)_k}{(2a)^k}. \tag{35.2.21}$$

The first few coefficients are

$$c_0 = \tfrac12, \quad c_1 - \tfrac38, \quad c_2 = \tfrac{5}{64}, \quad c_3 = \tfrac{7}{256}, \tag{35.2.22}$$

and $d_k = (-1)^k c_k$. The general form of the coefficients follows from

$$a_k = \frac{(-1)^{k+1}}{2^{k+1}} \frac{2k+1}{2k-1} \frac{\left(\frac12\right)_k}{k!}, \tag{35.2.23}$$

and

$$c_{2k} = (-1)^k a_{2k}, \quad c_{2k+1} = (-1)^{k+1} a_{2k+1}. \tag{35.2.24}$$

35.3 The case $b > a$

We write (35.0.1) in the form

$$P(a,b) = \Re \frac{1}{c} \int_0^c e^{\lambda\phi(z)}\, dz, \quad \phi(z) = i(z + 1/z), \tag{35.3.25}$$

with $\lambda = \sqrt{ab}$ and $c = \sqrt{b/a}$. In this case we have contributions from the interior stationary point at $z = 1$ and possibly from the endpoints. With this form of the integral we cannot use the method of stationary phase, because there is an essential singularity at the origin. As explained in §4.4 and in Chapter 14 we use the saddle point method.

We replace the interval $[0, c]$ by two contours in the complex plane, and pick up the contributions at $z = 1$ from one that runs through this saddle point. Other contributions come from the other contour that terminates at $z = c$. The point $z = 0$ will not give contributions.

We write

$$z = x + iy = re^{i\theta}, \quad \phi(z) = -(r - 1/r)\sin\theta + i(r + 1/r)\cos\theta. \tag{35.3.26}$$

The saddle point contour through the interior stationary point $z = 1$ is defined by $\Im\phi(z) = \Im\phi(1) = 2$, and is given by

$$(r + 1/r)\cos\theta = 2, \quad \text{or} \quad r = \frac{1 + \sin\theta}{\cos\theta}, \quad -\tfrac{1}{2}\pi < \theta < \tfrac{1}{2}\pi. \tag{35.3.27}$$

In terms of x and y we have $y/\sqrt{x^2 + y^2} = x - 1$, so when $(x, y) \sim (1, 0)$ we have $y \sim x - 1$. Solving for y we have

$$y = \frac{\sqrt{x}(x - 1)}{\sqrt{2 - x}}, \quad 0 < x < 2. \tag{35.3.28}$$

This path starts at $z = 0$, passes the stationary point at $z = 1$ and terminates at $+i\infty$. We call this path $\mathcal{P}_1$.

The second path, starting at $+i\infty$ and terminating at $z = c$, will be called $\mathcal{P}_2$ and is defined by $\Im\phi(z) = \Im\phi(c) = c + 1/c$. This gives for $\mathcal{P}_2$ the representation

$$r = \frac{\frac{1}{2}(c + 1/c) + \sqrt{\frac{1}{4}(c - 1/c)^2 + \sin^2\theta}}{\cos\theta}, \tag{35.3.29}$$

or in terms of x and y:

$$y = \sqrt{\frac{x(x - c)(xc - 1)}{c^2 + 1 - cx}}, \quad c < x < c + 1/c. \tag{35.3.30}$$

In Figure 35.1 we show the paths $\mathcal{P}_1$ and $\mathcal{P}_2$ with indicated directions of integration, for $a = 50$, $b = 150$, $c = 1.73$.

Using these paths $\mathcal{P}_1$ and $\mathcal{P}_2$ we write (35.3.25) in the form

$$P(a,b) = P_1(a,b) + P_2(a,b), \quad P_j(a,b) = \Re \frac{1}{c} \int_{\mathcal{P}_j} e^{\lambda\phi(z)}\, dz, \quad j = 1, 2. \tag{35.3.31}$$

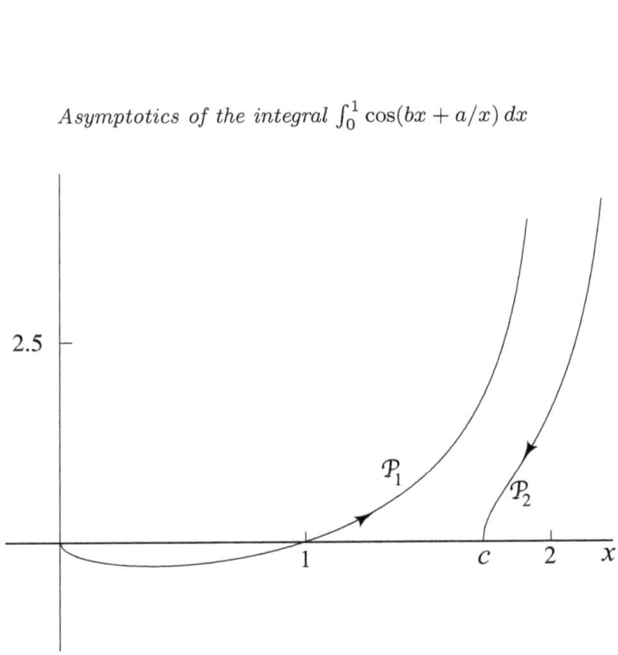

Fig. 35.1 The paths $\mathcal{P}_1$ and $\mathcal{P}_2$.

35.3.1 *The contribution from* $\mathcal{P}_1$

$P_1(a,b)$ can be written in the form

$$P_1(a,b) = \Re \frac{e^{2\lambda i}}{c} \int_{\mathcal{P}_1} e^{\Re\lambda\phi(x)} \left(1 + i\frac{dy}{dx}\right) dx, \quad 0 < x < 2, \tag{35.3.32}$$

where y is given in (35.3.28) and

$$\Re\phi(x) = -\frac{2(1-x)^2}{\sqrt{x(2-x)}}. \tag{35.3.33}$$

We substitute $x = 1 + t/\sqrt{1+t^2}$, which gives

$$P_1(a,b) = \Re \frac{e^{2\lambda i}}{c} \int_{-\infty}^{\infty} e^{-2\lambda t^2} f_1(t)\,dt, \quad f_1(t) = \frac{dx}{dt} + i\frac{dy}{dt}, \tag{35.3.34}$$

where

$$\frac{dx}{dt} = \frac{1}{(1+t^2)^{3/2}}, \quad \frac{dy}{dt} = \frac{(1+t^2)^{3/2} + 2t + t^3}{(1+t^2)^{3/2}}. \tag{35.3.35}$$

The odd parts do not contribute to the integral. We can express the contribution from dx/dt in terms of a special function and the even part of dy/dt gives an elementary integral:

$$P_1(a,b) = \Re \frac{e^{2\lambda i}}{c} \left(\sqrt{\pi}\, U\left(\tfrac{1}{2}, 0, 2\lambda\right) + i\sqrt{\frac{\pi}{2\lambda}}\right), \tag{35.3.36}$$

where we have used the Kummer U-function (see Chapter 10) with integral representation given in (10.1.5). The term with the U-function in (35.3.36) has the asymptotic expansion (see (10.2.18))

$$\sqrt{\pi}\, U\left(\tfrac{1}{2}, 0, 2\lambda\right) \sim \sqrt{\frac{\pi}{2\lambda}} \sum_{k=0}^{\infty} \frac{\left(\frac{1}{2}\right)_k \left(\frac{3}{2}\right)_k}{k!\,(-2\lambda)^k}, \quad \lambda \to \infty. \tag{35.3.37}$$

This gives for $P_1(a,b)$ the expansion

$$P_1(a,b) \sim \frac{1}{c}\sqrt{\frac{\pi}{2\lambda}}\left(\cos(2\lambda)\sum_{k=0}^{\infty}\frac{\left(\frac{1}{2}\right)_k\left(\frac{3}{2}\right)_k}{k!\,(-2\lambda)^k} - \sin(2\lambda)\right), \quad \lambda\to\infty. \tag{35.3.38}$$

35.3.2 The contribution from $\mathcal{P}_2$

$P_2(a,b)$ can be written in the form

$$P_2(a,b) = \Re\frac{e^{i\lambda(c+1/c)}}{c}\int_{\mathcal{P}_2} e^{\Re\lambda\phi(x)}\left(1+i\frac{dy}{dx}\right)dx, \quad c<x<c+1/c, \tag{35.3.39}$$

where the integration starts at $x=c+1/c$; y is given in (35.3.30) and

$$\Re\phi(x) = -\frac{y(2xc-c^2-1)}{xc}. \tag{35.3.40}$$

We substitute $-\Re\phi(x)=t$, which gives

$$P_2(a,b) = \Re\frac{e^{i\lambda(c+1/c)}}{c}\int_0^{\infty} e^{-\lambda t}f_2(t)\,dt, \quad f_2(t)=\frac{dx}{dt}+i\frac{dy}{dt}. \tag{35.3.41}$$

With the form of y given in (35.3.30) we cannot find an explicit representation of x in terms of t. However, we do not need such a form to derive the asymptotic expansion. We need expansions of x and y in powers of t, which give the expansion of $f_2(t)$ in powers of t. We write

$$x = c+x_1t+x_2t^2+x_3t^3+\ldots, \quad y=y_1t+y_2t^2+y_3t^3+\ldots, \tag{35.3.42}$$

and the coefficients x_k follow from substituting the expansion of x into the equation $-\Re\phi(x)=t$. From the coefficients x_k we can use (35.3.30) to obtain the coefficients y_k. In this way,

$$\begin{aligned} x_1 &= 0, & y_1 &= \frac{c^2}{c^2-1}, \\ x_2 &= \frac{c^3}{(c^2-1)^3}, & y_2 &= -\frac{3c^4\left(c^2+1\right)}{\left(c^2-1\right)^5}, \\ x_3 &= 0, & y_2 &= \frac{c^6\left(c^6+6c^4+6c^2+1\right)}{\left(c^2-1\right)^9}. \end{aligned} \tag{35.3.43}$$

This gives the expansion

$$\begin{aligned} P_2(a,b) \sim{}& \frac{\cos\left(\lambda\left(c+1/c\right)\right)}{c}\sum_{k=0}^{\infty}\frac{(k+1)x_{k+1}k!}{\lambda^{k+1}} - \\ & \frac{\sin\left(\lambda\left(c+1/c\right)\right)}{c}\sum_{k=0}^{\infty}\frac{(k+1)y_{k+1}k!}{\lambda^{k+1}}. \end{aligned} \tag{35.3.44}$$

35.4 A Fresnel-type expansion

We write $P(a, b)$ in the form

$$P(a,b) = \frac{1}{c}\int_0^c \cos(\lambda(t + 1/t))\, dt, \quad \lambda = \sqrt{ab}, \quad c = \sqrt{b/a}, \tag{35.4.45}$$

and introduce a function $w(t)$ by writing

$$t + 1/t = w^2 + 2, \quad \operatorname{sign}(t-1) = \operatorname{sign}(w). \tag{35.4.46}$$

Inverting this we have

$$t(w) = \tfrac{1}{2}\left(w^2 + 2 + w\sqrt{w^2+4}\right), \tag{35.4.47}$$

where the square root is positive for all real w.

Using the substitution in (35.4.46) we obtain

$$P(a,b) = \frac{1}{c}\int_{-\infty}^{\mu} \cos\left(\lambda(w^2+2)\right) f(w)\, dw, \quad \mu = w(c) = \sqrt{c + 1/c - 2}, \tag{35.4.48}$$

with $\operatorname{sign}(\mu) = \operatorname{sign}(c-1)$ and

$$f(w) = \frac{dt}{dw} = w + \frac{w^2+2}{\sqrt{w^2+4}}. \tag{35.4.49}$$

Observe that the sign of μ changes when a/b (or c) crosses the value 1. In that case the asymptotic behavior changes as well. When $a > b$ (or $\mu < 0$) we have $P(a,b) = \mathcal{O}(1/\lambda)$, when $a < b$ (or $\mu > 0$) we have $P(a,b) = \mathcal{O}(1/\sqrt{\lambda})$. For the transition of $a < b$ to $a > b$ we cannot use elementary functions to describe the asymptotic behavior. It appears that we can use a Fresnel-type integral.

We write

$$P(a,b) = \Re \frac{e^{2i\lambda}}{c}\int_{-\infty}^{\mu} e^{\lambda i w^2} f(w)\, dw, \tag{35.4.50}$$

and see the form of the functions that we will consider in Chapter 36, now with a complex exponential. The function f and its derivatives vanish at $-\infty$:

$$f^{(k)}(w) = \mathcal{O}\left(w^{-k-3}\right), \quad w \to -\infty. \tag{35.4.51}$$

We use the same method as for the cumulative distribution functions in §36.1, and obtain

$$P(a,b) \sim \Re\left(\Phi(a,b)\sum_{n=0}^{\infty}\frac{A_n}{(2i\lambda)^n} + \frac{e^{2i\lambda + \lambda i\mu^2}}{c}\sum_{n=0}^{\infty}\frac{B_n}{(2i\lambda)^{n+1}}\right), \tag{35.4.52}$$

where $\Phi(a,b)$ denotes a Fresnel-type integral (see also §14.5.1)

$$\Phi(a,b) = \frac{e^{2i\lambda}}{c}\int_{-\infty}^{\mu} e^{\lambda i w^2}\, dw = \frac{e^{2i\lambda}}{c}\sqrt{\frac{\pi}{2\lambda}}\,\mathcal{F}\left(\sqrt{\frac{2\lambda}{\pi}}\,\mu\right), \tag{35.4.53}$$

and

$$A_n = f_n(0), \quad B_n = \frac{f_n(\mu) - f_n(0)}{\mu}. \tag{35.4.54}$$

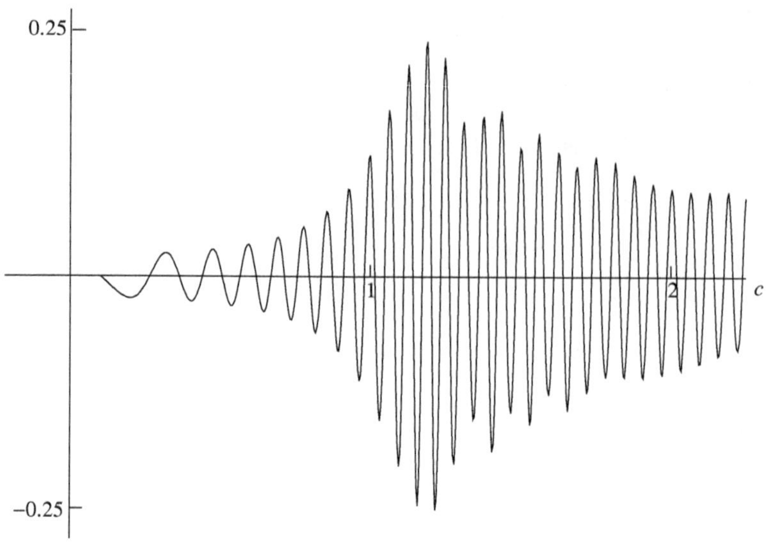

Fig. 35.2 Graph of the approximation (35.4.52) of $P(a,b)$ (taking only the terms with $n = 0$) for $a = 50$ and b-values such that $0.1 \le c \le 2.25$, $c = \sqrt{b/a}$.

The functions f_n are defined by $f_0(w) = f(w)$ and

$$f_{n+1}(w) = -\frac{d}{dw}\frac{f_n(w) - f_n(0)}{w}, \qquad n = 0, 1, 2, \ldots. \tag{35.4.55}$$

The function $\Phi(a,b)$ can be written in terms of the Fresnel sine and cosine integrals

$$C(z) = \int_0^z \cos\left(\tfrac{1}{2}\pi t^2\right) dt, \quad S(z) = \int_0^z \sin\left(\tfrac{1}{2}\pi t^2\right) dt, \tag{35.4.56}$$

and we have

$$\Phi(a,b) = \frac{e^{2i\lambda}}{c}\sqrt{\frac{\pi}{2\lambda}}\left(\tfrac{1}{2}(1+i) + C\left(\mu\sqrt{2\lambda/\pi}\right) + iS\left(\mu\sqrt{2\lambda/\pi}\right)\right). \tag{35.4.57}$$

We can write $\Phi(a,b)$ in terms of the complementary error function:

$$\Phi(a,b) = \frac{e^{(2\lambda+\frac{1}{4}\pi)i}}{2c}\sqrt{\frac{\pi}{\lambda}}\operatorname{erfc}(-\zeta), \quad \zeta = \mu\sqrt{\lambda}e^{-\frac{1}{4}\pi i}. \tag{35.4.58}$$

In Figure 35.2 we show the approximation of $P(a,b)$ by taking only the terms with $n = 0$ in (35.4.52) for $a = 50$ and b-values such that $0.1 \le c \le 2.25$, $c = \sqrt{b/a}$.

PART 7

A Class of Cumulative Distribution Functions

Chapter 36

Expansions of a class of cumulative distribution functions

In this chapter we show how to transform a selection of well-known distribution functions, such as the gamma and beta distributions, into a standard form.[1] We derive asymptotic expansions with respect to one parameter, and the expansion is uniformly valid with respect to a second parameter. The standard form is a convenient starting point in several cases, however, we will see that for some examples it is better to use contour integral representations. This will be explained in Chapter 37 for the incomplete gamma functions, in Chapter 38 for the incomplete beta functions, and in Chapter 39 for the non-central chi-square distribution functions (or Marcum functions). In a final chapter we consider the problem of inverting the cumulative distribution functions by using asymptotic methods.

36.1 Cumulative distribution functions: A standard form

Many cumulative distribution functions can be transformed into the standard form

$$F_a(\eta) = \sqrt{\frac{a}{2\pi}} \int_{-\infty}^{\eta} e^{-\frac{1}{2}a\zeta^2} f(\zeta)\, d\zeta, \quad f(0) = 1, \tag{36.1.1}$$

with f analytic in a neighborhood of $\mathbb{R}$. This is Case 3 of Table 20.1.

The function f can play the role of a probability density function and $F_a(\eta)$ can be viewed as a cumulative distribution function. Then the variable η is related to the random variable of the underlying statistics and the parameter a corresponds to the degrees of freedom.

The problem is to obtain an asymptotic expansion as $a \to \infty$, uniformly valid with respect to $\eta \in \mathbb{R}$. The asymptotic behavior of $F_a(\eta)$ as $a \to \infty$ depends strongly on $\text{sign}(\eta)$. As a rule, we have the non-uniform estimates

[1]This chapter is based on Temme (1982).

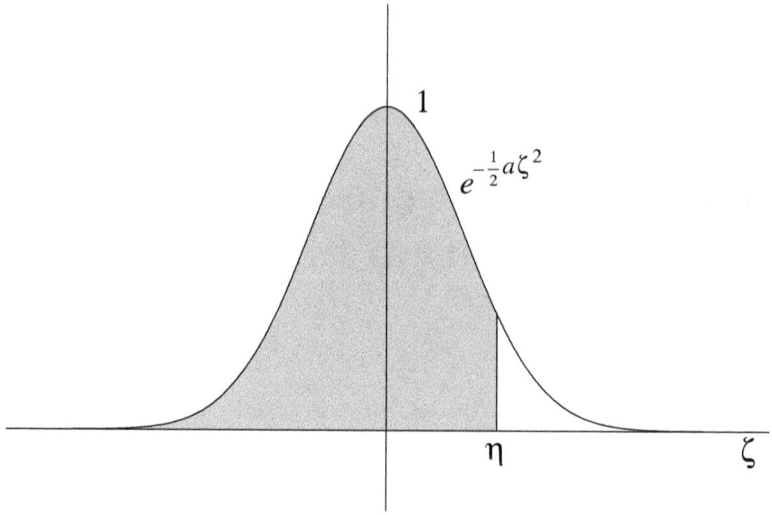

Fig. 36.1 When a is large $F_a(\eta)$ defined in (36.1.1) changes abruptly from 0 to 1 when η crosses the origin in positive direction.

$$F_a(\eta) = \begin{cases} -f(\eta)/(\eta\sqrt{2a\pi})e^{-a\eta^2}(1+\mathcal{O}(1/a)), & \text{if } \eta < 0; \\ \frac{1}{2}(1+\mathcal{O}(1/\sqrt{a})), & \text{if } \eta = 0; \\ 1+\mathcal{O}(1/a), & \text{if } \eta > 0. \end{cases} \tag{36.1.2}$$

The conclusion is: the asymptotic behavior of $F_a(\eta)$ is completely different in the three cases distinguished and, moreover, the asymptotic forms do not pass into one another when η changes from negative values to positive ones. We see that the saddle point at the origin may be inside or outside the interval of integration; see Figure 36.1. The approximations in (36.1.2) are not uniformly valid with respect to small values of $|\eta|$.

We have seen in Chapter 22 another example where the endpoint of integration may coalesce with the saddle point; in that case we assumed an algebraic singularity at the endpoint and we used the parabolic cylinder functions. In the present case we will derive a uniform approximation in terms of the complementary error function; this function is defined in §3.4.

Integrals of the type considered in (36.1.1) occur frequently in the form of distribution functions; in fact the gamma distribution, the beta distribution and many other cumulative distribution functions can be transformed into this standard form.

The basic approximant for (36.1.1) is the normal distribution function, which we can write in terms of the complementary error function:

$$P\left(\eta\sqrt{a}\right) = \sqrt{\frac{a}{2\pi}}\int_{-\infty}^{\eta} e^{-\frac{1}{2}a\zeta^2}\,d\zeta = \tfrac{1}{2}\mathrm{erfc}\left(-\eta\sqrt{a/2}\right). \tag{36.1.3}$$

The procedure to obtain a uniform expansion is based on integration by parts. We write $f(\zeta) = 1 + (f(\zeta) - 1)$ (recall that we assume $f(0) = 1$). Then

$$\begin{aligned} F_a(\eta) &= P(\eta\sqrt{a}) - \frac{1}{\sqrt{2\pi a}} \int_{-\infty}^{\eta} \frac{f(\zeta) - 1}{\zeta} \, d\, e^{-\frac{1}{2}a\zeta^2} \\ &= P(\eta\sqrt{a}) - \frac{1}{\sqrt{2\pi a}} \frac{f(\eta) - 1}{\eta} e^{-\frac{1}{2}a\eta^2} + \frac{1}{\sqrt{2\pi a}} \int_{-\infty}^{\eta} e^{-\frac{1}{2}a\zeta^2} f_1(\zeta)\, d\zeta, \end{aligned} \tag{36.1.4}$$

where

$$f_1(\zeta) = \frac{d}{d\zeta} \frac{f(\zeta) - 1}{\zeta}. \tag{36.1.5}$$

Repeating this process, we obtain

$$F_a(\eta) = P(\eta\sqrt{a}) A(a) - \frac{e^{-\frac{1}{2}a\eta^2}}{\sqrt{2\pi a}} B_a(\eta), \tag{36.1.6}$$

with

$$A(a) \sim \sum_{n=0}^{\infty} \frac{A_n}{a^n}, \quad B_a(\eta) \sim \sum_{n=0}^{\infty} \frac{C_n(\eta)}{a^n}. \tag{36.1.7}$$

The coefficients are given by

$$A_n = f_n(0), \quad C_n(\eta) = \frac{f_n(\eta) - f_n(0)}{\eta}, \tag{36.1.8}$$

with $f_0 = f$ and

$$f_n(\zeta) = \frac{d}{d\zeta} \frac{f_{n-1}(\zeta) - f_{n-1}(0)}{\zeta}, \quad n \geq 1. \tag{36.1.9}$$

The functions f_n defined in (36.1.9) can be represented in the form of Cauchy-type integrals. We have the following theorem.

Theorem 36.1. *Let the rational functions $R_n(\eta, \zeta)$ be defined by*

$$R_0(\eta, \zeta) = \frac{1}{\zeta - \eta}, \quad R_{n+1}(\eta, \zeta) = \frac{-1}{\zeta} \frac{d}{d\zeta} R_n(\eta, \zeta), \quad n = 0, 1, 2, \ldots, \tag{36.1.10}$$

where $\eta, \zeta \in \mathbb{C}, \zeta \neq \eta, \zeta \neq 0$. Let f_n be defined by the recursive scheme (36.1.9), where f_0 is a given analytic function in a domain $\mathcal{D}$ that contains the origin. Then,

$$f_n(\eta) = \frac{1}{2\pi i} \int_C R_n(\eta, \zeta) f_0(\zeta)\, d\zeta, \tag{36.1.11}$$

where C is a simple closed contour in $\mathcal{D}$ that encircles the points η and the origin.

Proof. The proof is similar to the proof of Theorem 25.1, and is left as an exercise. □

In the following sections we show how to apply the method of this chapter to well-known cumulative distribution functions. For more details we refer to Temme (1982), where also representations for the remainder of the expansion of $B_a(\eta)$ in (36.1.7) are given.

Remark 36.1. By differentiating (36.1.6) with respect to η we obtain

$$f(\eta) = A(a) + \eta B_a(\eta) - \frac{1}{a} B_a'(\eta), \tag{36.1.12}$$

and substituting the expansion of $B_a(\eta)$ given in (36.1.7), the following recurrence relation for the coefficients can be obtained:

$$C_0(\eta) = \frac{f(\eta) - 1}{\eta}, \quad \eta C_n(\eta) = C_{n-1}'(\eta) - A_n, \quad n = 1, 2, 3, \ldots. \tag{36.1.13}$$

△

Remark 36.2. By taking $\eta \to \infty$ in (36.1.6), it follows that $F_a(\infty) = A(a)$, and the coefficients A_n of the expansion of $A(a)$ follow from the asymptotic expansion of the complete integral in (36.1.1). It follows that

$$A_n = \left(\tfrac{1}{2}\right)_n 2^n a_{2n}, \quad n = 0, 1, 2, \ldots, \tag{36.1.14}$$

where a_n are the coefficients in the expansion $f(\eta) = \sum_{k=0}^{\infty} a_k \eta^k$. This relation also follows from expanding

$$f_n(\zeta) = \sum_{k=0}^{\infty} a_k^{(n)} \zeta^k, \tag{36.1.15}$$

and by substituting this in (36.1.9). We obtain the relation

$$a_k^{(n)} = (k+1) a_{k+2}^{(n-1)}, \tag{36.1.16}$$

and by iterating:

$$a_k^{(n)} = 2^n \left(\tfrac{1}{2}k + \tfrac{1}{2}\right)_n a_{k+2n}^{(0)}, \quad k, n = 0, 1, 2, \ldots. \tag{36.1.17}$$

Because $A_n = f_n(0) = a_0^{(n)}$ we again find the relation in (36.1.14).

By using (36.1.8) it follows that the coefficients of $c_k^{(n)}$ in the expansion

$$C_n(\eta) = \sum_{k=0}^{\infty} c_k^{(n)} \eta^k, \tag{36.1.18}$$

are given by

$$c_k^{(n)} = a_{k+1}^{(n)} = 2^n \left(\tfrac{1}{2}k + 1\right)_n a_{k+2n+1}^{(0)}, \tag{36.1.19}$$

Expansions of the form (36.1.18) are useful when we have to compute the coefficients $C_n(\eta)$ near the transition point $\eta = 0$. They can also be used to find explicit representations of these coefficients, as we will see in §36.2. △

Problem 36.1. *Consider the expansion*

$$\begin{aligned} G_n(b) &= \int_0^b (\sin t)^{n-1}\,dt \\ &\sim (\sin b)^n \left(\frac{1}{n\cos b} - \frac{1}{n^2\cos^3 b} + \frac{3-2\cos^2 b}{n^3\cos^5 b} + \ldots\right), \end{aligned} \tag{36.1.20}$$

as $n \to \infty, b \in (0, \frac{1}{2}\pi)$. *Clearly, the expansion becomes useless when* $b \to \frac{1}{2}\pi$. *How can we obtain an expansion that holds uniformly with respect to* $b \in [0, \pi]$*?* ♡

36.2 An incomplete normal distribution function

We consider

$$F_a(\eta) = \sqrt{\frac{a}{2\pi}} \int_{-\infty}^{\eta} e^{-\frac{1}{2}a\zeta^2}\,\frac{d\zeta}{\zeta^2+1}. \tag{36.2.21}$$

This function finds applications in probability theory, mathematical statistics and in problems involving the heat conduction equation. Jones (1970/71) used it for describing the asymptotic expansion of a double integral. It can be viewed as a generalization of the complementary error function. To see this, note that

$$F_a(\infty) = \sqrt{2\pi a}\, e^{\frac{1}{2}a} Q\left(\sqrt{a}\right), \tag{36.2.22}$$

where

$$Q(x) = 1 - P(x) = \tfrac{1}{2}\operatorname{erfc}\left(x/\sqrt{2}\right). \tag{36.2.23}$$

We have the representation in (36.1.6) with expansions in (36.1.7). In the present case, with $f(\zeta) = 1/(\zeta^2+1)$, the coefficients $C_n(\eta)$ can be constructed rather easily. From the expansion of the complementary error function given in (3.4.30) we obtain

$$A(a) = F_a(\infty) \sim \sum_{n=0}^{\infty} \frac{A_n}{a^n}, \quad A_n = (-2)^n \left(\tfrac{1}{2}\right)_n, \quad n \ge 0. \tag{36.2.24}$$

This also follows from (see (36.1.14))

$$f_0(\zeta) = \frac{1}{\zeta^2+1} = \sum_{k=0}^{\infty} f_k^{(0)} \zeta^n, \quad f_{2k}^{(0)} = (-1)^k, \quad f_{2k+1}^{(0)} = 0. \tag{36.2.25}$$

Using (36.1.19) we obtain for the series in (36.1.18):

$$C_n(\eta) = \sum_{k=0}^{\infty} c_{2k+1}^{(n)} \eta^{2k+1}, \quad c_{2k+1}^{(n)} = (-1)^{k+n+1} 2^n \left(k+\tfrac{3}{2}\right)_n. \tag{36.2.26}$$

It follows that $C_n(\eta)$ can be expressed in terms of the Gauss hypergeometric function:

$$C_n(\eta) = -(-2)^n \left(\tfrac{3}{2}\right)_n \eta\, {}_2F_1\left(\begin{matrix} 1, n+\frac{3}{2} \\ \frac{3}{2} \end{matrix}; -\eta^2\right), \tag{36.2.27}$$

which can be written in the form (see (12.0.6))

$$C_n(\eta) = -(-2)^n \left(\tfrac{3}{2}\right)_n \frac{\eta}{\eta^2+1}\, {}_2F_1\!\left({1, -n \atop \tfrac{3}{2}} ; \frac{\eta^2}{\eta^2+1}\right). \tag{36.2.28}$$

In this form the hypergeometric function is a polynomial, and we can write

$$C_n(\eta) = -\tfrac{1}{2}(-2)^n \left(\tfrac{3}{2}\right)_n \frac{\eta}{\eta^2+1} \int_0^1 \left(1 - t\frac{\eta^2}{\eta^2+1}\right)^n \frac{dt}{\sqrt{1-t}}, \tag{36.2.29}$$

from which we obtain

$$|C_n(\eta)| \le 2^n \left(\tfrac{3}{2}\right)_n \frac{|\eta|}{\eta^2+1}. \tag{36.2.30}$$

Hence, the $C_n(\eta)$ are bounded functions of η on $\mathbb{R}$.

The first coefficients are

$$C_0(\eta) = -\frac{\eta}{\eta^2+1}, \quad C_1(\eta) = \frac{\eta(\eta^2+3)}{(\eta^2+1)^2}. \tag{36.2.31}$$

In this example other $C_n(\eta)$ can be computed by using recurrence relations of ${}_2F_1$-functions. It follows that

$$\left(\eta^2+1\right) C_{n+1}(\eta) = -2n(2n+1)C_{n-1}(\eta) - \left((2n+1)\eta^2 + 4n + 3\right) C_n(\eta). \tag{36.2.32}$$

For the incomplete normal distribution function in (36.2.21) we have obtained the representation

$$F_a(\eta) = P(\eta\sqrt{a})A(a) - \frac{e^{-\frac{1}{2}a\eta^2}}{\sqrt{2\pi a}} B_a(\eta), \tag{36.2.33}$$

where $A(a) = F_a(\infty)$ is given in (36.2.22), $B_a(\eta)$ has the asymptotic expansion in (36.1.7), and $C_n(\eta)$ are given in (36.2.27) and (36.2.28).

36.3 The Sievert integral

The Sievert integral, named after the Swedish medical physicist Rolf Sievert (Sievert, 1930), is a special function commonly encountered in physics, in particular in radiation problems. The definition is (Abramowitz and Stegun, 1964, §27.4)

$$I(\theta, a) = \int_{-\frac{1}{2}\pi}^{\theta} e^{a(1-1/\cos\phi)}\, d\phi, \quad a > 0, \quad -\tfrac{1}{2}\pi \le \theta \le \tfrac{1}{2}\pi. \tag{36.3.34}$$

The complete integral $I(\frac{1}{2}\pi, a)$ is an integral of the modified Bessel function $K_0(x)$, that is,

$$I\left(\tfrac{1}{2}\pi, a\right) = 2e^a \int_a^\infty K_0(x)\, dx. \tag{36.3.35}$$

We need a transformation in order to bring (36.3.34) into the standard form (36.1.1). The appropriate change of variables is defined by

$$\begin{aligned} -\tfrac{1}{2}\zeta^2 &= 1 - 1/\cos\phi, \quad \zeta = \frac{2\sin(\phi/2)}{\sqrt{\cos\phi}}, \\ -\tfrac{1}{2}\eta^2 &= 1 - 1/\cos\theta, \quad \eta = \frac{2\sin(\theta/2)}{\sqrt{\cos\theta}}, \end{aligned} \tag{36.3.36}$$

and the Sievert integral becomes

$$I(\theta, a) = \int_{-\infty}^{\eta} e^{-\frac{1}{2}a\zeta^2} f(\zeta)\, d\zeta, \quad f(\zeta) = \frac{d\phi}{d\zeta}. \tag{36.3.37}$$

By using (36.3.36), $f(\zeta)$ can be written in the form

$$f(\zeta) = \frac{1}{\left(1 + \frac{1}{2}\zeta^2\right)\sqrt{1 + \frac{1}{4}\zeta^2}}. \tag{36.3.38}$$

In the notation of the standard form we have

$$F_a(\eta) = \sqrt{\frac{a}{2\pi}} I(\theta, a) = \sqrt{\frac{a}{2\pi}} \int_{-\infty}^{\eta} e^{-\frac{1}{2}a\zeta^2} f(\zeta)\, d\zeta, \tag{36.3.39}$$

and $F_a(\infty)$ has the known expansion (Abramowitz and Stegun, 1964, Eq. 11.1.15)

$$F_a(\infty) = 2\sqrt{\frac{a}{2\pi}} e^a \int_a^{\infty} K_0(x)\, dx \sim \sum_{n=0}^{\infty} \frac{A_n}{a^n}, \tag{36.3.40}$$

where

$$A_n = (-1)^n \left(\tfrac{1}{2}\right)_n \sum_{m=0}^{n} \frac{1}{m!\, 2^m} \left(\tfrac{1}{2}\right)_m. \tag{36.3.41}$$

The expansion reads

$$\sqrt{\frac{a}{2\pi}} I(\theta, a) = F_a(\infty) P\left(2 \sin\left(\tfrac{1}{2}\theta \sqrt{a/\cos\theta}\right)\right) - \frac{e^{-\frac{1}{2}a\eta^2}}{\sqrt{2\pi a}} \sum_{n=0}^{\infty} \frac{C_n(\eta)}{a^n}, \tag{36.3.42}$$

as $a \to \infty$, uniformly with respect to $\theta \in [-\frac{1}{2}\pi, \frac{1}{2}\pi]$.

We have

$$f(\eta) = \frac{\cos\theta \sqrt{\cos\theta}}{\cos\frac{1}{2}\theta}, \quad C_0(\eta) = \frac{\cos^2\theta - \cos\left(\frac{1}{2}\theta\right)\sqrt{\cos\theta}}{\sin\theta}. \tag{36.3.43}$$

36.4 The Pearson type IV distribution

We consider the Pearson type IV distribution function in the form (see Fettis (1976))

$$I(\theta, \alpha, \beta) = \int_{-\frac{1}{2}\pi}^{\theta} \cos^{2\alpha}\phi\, e^{2\beta\phi}\, d\phi, \quad -\tfrac{1}{2}\pi \le \theta \le \tfrac{1}{2}\pi, \tag{36.4.44}$$

where we assume that α is a large positive parameter; β is a real parameter (which may be large as well). It is known that

$$I\left(\tfrac{1}{2}\pi, \alpha, \beta\right) = \frac{\pi 2^{-\alpha}\Gamma(2\alpha + 1)}{\Gamma(1 + \alpha + i\beta)\Gamma(1 + \alpha - i\beta)}. \tag{36.4.45}$$

It is possible to write $I(\theta, \alpha, \beta)$ in terms of a Gauss hypergeometric function (in fact as an incomplete beta function) with complex parameters. Another form of the cumulative distribution is

$$I(\theta, \alpha, \beta) = \int_{-\infty}^{\tan^{-1}(\theta)} e^{2\beta \tan^{-1}(t)} \frac{dt}{(1 + t^2)^{\alpha + 1}}. \tag{36.4.46}$$

When α and β are both large, the maximum of the integrand occurs at $\phi = \lambda$, where

$$\lambda = \arctan\gamma, \quad \gamma = \beta/\alpha. \tag{36.4.47}$$

To write (36.4.44) in the standard form, we use the transformations

$$\begin{aligned} -\tfrac{1}{2}\zeta^2 &= \gamma(\phi - \lambda) + \ln\frac{\cos\phi}{\cos\lambda}, \\ -\tfrac{1}{2}\eta^2 &= \gamma(\theta - \lambda) + \ln\frac{\cos\theta}{\cos\lambda}, \end{aligned} \tag{36.4.48}$$

where $\operatorname{sign}\zeta = \operatorname{sign}(\phi - \lambda)$, $\operatorname{sign}\eta = \operatorname{sign}(\theta - \lambda)$. Hence,

$$I(\theta, \alpha, \beta) = \cos^{2\alpha+1}\lambda\, e^{2\beta\lambda} \int_{-\infty}^{\eta} e^{-\frac{1}{2}a\zeta^2} f(\zeta)\, d\zeta, \tag{36.4.49}$$

where

$$f(\zeta) = \frac{1}{\cos\lambda}\frac{d\phi}{d\zeta} = \frac{1}{\cos\lambda}\frac{\zeta}{\tan\phi - \gamma}, \quad f(0) = 1, \quad a = 2\alpha. \tag{36.4.50}$$

After further normalization we can put (36.4.49) into the standard form.

For this we verify the asymptotic expansion of the complete integral in (36.4.45). We can write

$$I\left(\tfrac{1}{2}\pi, \alpha, \beta\right) = \sqrt{\frac{2\pi}{a}}\, \cos^{2\alpha+1}\lambda\, e^{2\beta\lambda}\, G(a), \tag{36.4.51}$$

where

$$G(a) = \frac{\Gamma^*(2\alpha)}{\Gamma^*(\alpha + i\beta)\Gamma^*(\alpha - i\beta)}. \tag{36.4.52}$$

The function $\Gamma^*(z)$ is defined in (25.6.55). The function $G(a)$ has an asymptotic expansion, but it is more convenient to use the expansion of $1/G(a)$. We have

$$\frac{1}{G(a)} = \sum_{n=0}^{\infty} \frac{g_n(\gamma)}{a^n}, \tag{36.4.53}$$

and the first coefficients are

$$\begin{aligned} g_0(\gamma) &= 1, \\ g_1(\gamma) &= -\frac{\gamma^2 - 3}{12(\gamma^2+1)}, \\ g_2(\gamma) &= \frac{\gamma^4 - 6\gamma^2 + 9}{288(\gamma^2+1)^2}, \\ g_3(\gamma) &= \frac{139\gamma^6 + 477\gamma^4 + 7209\gamma^2 - 2025}{51840(\gamma^2+1)^3}, \\ g_4(\gamma) &= -\frac{571\gamma^8 + 60\gamma^6 + 23922\gamma^4 - 96228\gamma^2 + 25515}{2488320(\gamma^2+1)^4}. \end{aligned} \tag{36.4.54}$$

The coefficients can be obtained from the complete integral representation of $I\left(\frac{1}{2}\pi,\alpha,\beta\right)$ that follows from (36.4.44), or from the known expansion of the gamma function, in the same way as explained in §25.6.1. See also §6.7.

This gives the standard form

$$F_a(\eta) = G(a)\frac{I(\theta,\alpha,\beta)}{I\left(\frac{1}{2}\pi,\alpha,\beta\right)} = \sqrt{\frac{a}{2\pi}}\int_{-\infty}^{\eta} e^{-\frac{1}{2}a\zeta^2} f(\zeta)\,d\zeta, \tag{36.4.55}$$

with $F_a(\infty) = G(a)$. This can be written in the form (36.1.6) with $A(a) = G(a)$.

However, it may be convenient to obtain a representation for the ratio. We have

$$\begin{aligned} \frac{I(\theta,\alpha,\beta)}{I\left(\frac{1}{2}\pi,\alpha,\beta\right)} &= \tfrac{1}{2}\operatorname{erfc}\left(-\eta\sqrt{a/2}\right) - R_a(\eta), \\ R_a(\eta) &\sim \frac{e^{-\frac{1}{2}a\eta^2}}{\sqrt{2\pi a}} \sum_{n=0}^{\infty} \frac{c_n(\eta)}{a^n}, \quad a\to\infty. \end{aligned} \tag{36.4.56}$$

Using the method described in Remark 36.1, we conclude that the coefficients $c_n(\eta)$ satisfy the recurrence relation

$$c_0(\eta) = \frac{f(\eta)-1}{\eta}, \quad \eta c_n(\eta) = \frac{d}{d\eta}c_{n-1}(\eta) + g_n(\gamma)f(\eta), \quad n\ge 1, \tag{36.4.57}$$

where $f(\eta)$ is defined in (36.4.50) (with ζ replaced by η and ϕ by θ); $g_n(\gamma)$ are the coefficients in the expansion of $1/G(a)$, see (36.4.53).

36.5 The Von Mises distribution

From the Fourier expansion

$$e^{z\cos t} = \sum_{n=-\infty}^{\infty} \cos nt\; I_n(z), \tag{36.5.58}$$

where $I_n(z)$ is the modified Bessel function, it follows that $I_0(z)$ can be written as

$$I_0(z) = \frac{1}{2\pi}\int_{-\pi}^{\pi} e^{z\cos t}\,dt. \tag{36.5.59}$$

An incomplete version of this integral plays a role as a cumulative distribution function. We define

$$I_0(\theta,\kappa) = \frac{1}{2\pi\, I_0(\kappa)}\int_{-\pi}^{\theta} e^{\kappa\cos t}\,dt, \tag{36.5.60}$$

which is formally equivalent to the cumulative distribution applied by Von Mises (1918) to study deviations of atomic weights from integer values, representable as points on the circumference of a circle or as circular directions. The parameter θ is the angular deviation and κ is the concentration parameter. This distribution of points on a circle is analogous to the normal or Gaussian distribution of points on a line and has application in the study of quantal or periodic data, directions of sedimentary bedding, surface fault lines, wildlife movements, etc. For numerical aspects of the Von Mises distribution we refer to Hill (1977).

In the present case we need a different approach as used earlier in this chapter, because the standard form $F_a(\eta)$ in (36.1.1) would have a finite interval (the exponential function in (36.5.60) does not vanish at $\pm\pi$). First we write

$$I_0(\theta,\kappa) = \tfrac{1}{2} + \frac{1}{2\pi\, I_0(\kappa)} \int_0^\theta e^{\kappa\cos t}\, dt. \tag{36.5.61}$$

The transformation $\frac{1}{2}\zeta^2 = 1-\cos t$, or $\zeta = 2\sin\left(\frac{1}{2}t\right)$, gives

$$I_0(\theta,\kappa) = \tfrac{1}{2} + \frac{e^\kappa}{2\pi\, I_0(\kappa)} \int_0^\eta e^{-\frac{1}{2}\kappa\zeta^2} f(\zeta)\, d\zeta, \tag{36.5.62}$$

where

$$f(\zeta) = \frac{1}{\sqrt{1-\frac{1}{4}\zeta^2}}, \qquad \eta = 2\sin\left(\tfrac{1}{2}\theta\right). \tag{36.5.63}$$

When we assume that $|\theta| \le \theta_0 < \pi$ we can integrate by parts as in the standard form. First we write

$$I_0(\theta,\kappa) = \tfrac{1}{2} + \frac{1}{I_0^*(\kappa)} G_\kappa(\eta), \tag{36.5.64}$$

where

$$G_\kappa(\eta) = \sqrt{\frac{\kappa}{2\pi}} \int_0^\eta e^{-\frac{1}{2}\kappa\zeta^2} f(\zeta)\, d\zeta, \tag{36.5.65}$$

and (see (9.1.4) and (9.3.14))

$$I_0^*(z) = \sqrt{2\pi z}\, e^{-z} I_0(z) \sim 1 + \sum_{n=1}^\infty \frac{\left(\frac{1}{2}\right)_n \left(\frac{1}{2}\right)_n}{2^n\, n!\, z^n}, \qquad z\to\infty. \tag{36.5.66}$$

The integration by parts procedure of §36.1 gives

$$G_\kappa(\eta) = \tfrac{1}{2}\operatorname{erf}\left(\eta\sqrt{\kappa/2}\right) A(\kappa) - \frac{e^{-\frac{1}{2}\kappa\eta^2}}{\sqrt{2\pi\kappa}} B_\kappa(\eta) + \frac{1}{\sqrt{2\pi\kappa}} C(\kappa), \tag{36.5.67}$$

where $A(\kappa)$ and $B_\kappa(\eta)$ have similar expansions as in (36.1.7), with the functions f_n as defined in (36.1.9). The function $C(\kappa)$ has the expansion

$$C(\kappa) \sim \sum_{n=0}^\infty \frac{f_n'(0)}{\kappa^n}. \tag{36.5.68}$$

However, from Remark 36.2 it follows that (see also (36.1.14))

$$A_n = \left(\tfrac{1}{2}\right)_n 2^n a_{2n}, \quad f_n'(0) = 2^n\, n!\, a_{2n+1}, \tag{36.5.69}$$

where a_n are the coefficients of the even function f defined in (36.5.63). Hence, all coefficients of the expansion of $C(\kappa)$ vanish, and for A_n we obtain

$$A_n = \frac{\left(\frac{1}{2}\right)_n \left(\frac{1}{2}\right)_n}{2^n\, n!}, \tag{36.5.70}$$

which are the same as the coefficients in the expansion of $I_0^*(z)$ given in (36.5.66).

Hence, we can write

$$I_0(\theta,\kappa) = \tfrac{1}{2}\mathrm{erfc}\left(-\eta\sqrt{\kappa/2}\right) - \frac{e^{-\frac{1}{2}\kappa\eta^2}}{\sqrt{2\pi\kappa}}\frac{B_\kappa(\eta)}{I_0^*(\kappa)}, \tag{36.5.71}$$

where $B_\kappa(\eta)$ has the expansion given in (36.1.7), and where we have used the relation $\mathrm{erf}\, z + \mathrm{erfc}\, z = 1$; η is defined in (36.5.63). This expansion is valid for large values of κ, uniformly with respect to θ in every closed interval of $(-\pi,\pi)$.

The expansion is not valid when $\theta \to \pm\pi$ because the function f is singular at $\eta = \pm 2$, and all coefficients $C_n(\eta)$ are singular (and unbounded) at these points.

36.5.1 *An expansion near the lower endpoint of integration*

It may be of interest to have information for the cases $\theta \to \pm\pi$, and we give a few steps for $\theta < 0$. For positive θ we can use the same method by using $I_0(\theta,\kappa) = 1 - I_0(-\theta,\kappa)$.

We have, using (36.5.60) and writing $t = \tau - \pi$,

$$I_0(-\theta,\kappa) = \frac{1}{2\pi I_0(\kappa)}\int_{-\pi}^{-\theta} e^{\kappa\cos t}\,dt = \frac{1}{2\pi I_0(\kappa)}\int_0^{\pi-\theta} e^{-\kappa\cos\tau}\,d\tau, \tag{36.5.72}$$

where we assume that $0 < \theta_0 \le \theta \le \pi$. Next we substitute $\zeta = 2\sin\left(\frac{1}{2}\tau\right)$, and obtain

$$I_0(-\theta,\kappa) = \frac{e^{-2\kappa}}{I_0^*(\kappa)}H_\kappa(\zeta), \quad H_\kappa(\sigma) = \sqrt{\frac{\kappa}{2\pi}}\int_0^\sigma e^{\frac{1}{2}\kappa\zeta^2} f(\zeta)\,d\zeta, \tag{36.5.73}$$

where

$$\sigma = 2\cos\tfrac{1}{2}\theta, \tag{36.5.74}$$

$I_0^*(z)$ is defined in (36.5.66) and f in (36.5.63). The main contributions come from the upper limit of integration σ, and if σ is small, there is a nearby saddle point at $\zeta = 0$. This has been considered in Chapter 22, although in a different form.

Integrating by parts as in §36.1 we obtain

$$H_\kappa(\sigma) = e^{\frac{1}{2}\kappa\sigma^2}\left(\frac{\widetilde{A}(\kappa)}{\sqrt{\pi}}F\left(\sigma\sqrt{\kappa/2}\right) + \frac{\widetilde{B}_\kappa(\sigma)}{\sqrt{2\pi\kappa}}\right), \tag{36.5.75}$$

where $F(z)$ is Dawson's integral defined by (see Temme (2010c, Eq. 7.2.5))

$$F(z) = e^{-z^2}\int_0^z e^{t^2}\,dt = -\tfrac{1}{2}i\sqrt{\pi}\,e^{-z^2}\,\mathrm{erf}(iz), \tag{36.5.76}$$

and $\widetilde{A}(\kappa)$, $\widetilde{B}_\kappa(\sigma)$ have the expansions

$$\widetilde{A}(\kappa) \sim \sum_{n=0}^\infty (-1)^n\frac{A_n}{\kappa^n}, \quad \widetilde{B}_\kappa(\sigma) \sim \sum_{n=0}^\infty (-1)^n\frac{C_n(\sigma)}{\kappa^n}, \tag{36.5.77}$$

with coefficients as in (36.1.8) and the f_n as in (36.1.9).

For $I_0(-\theta,\kappa)$ we obtain the representation

$$I_0(-\theta,\kappa) = \frac{e^{-2\kappa\sin^2\frac{1}{2}\theta}}{I_0^*(\kappa)}\left(\frac{\widetilde{A}(\kappa)}{\sqrt{\pi}}F\left(\sigma\sqrt{\kappa/2}\right) + \frac{\widetilde{B}_\kappa(\sigma)}{\sqrt{2\pi\kappa}}\right). \tag{36.5.78}$$

The expansions in (36.5.77) can be used for $0 < \theta_0 \le \theta \le \pi$, where θ_0 is a fixed number.

Chapter 37

Incomplete gamma functions: Uniform expansions

We recall the definitions of the incomplete gamma functions given in Chapter 7:[1]

$$\gamma(a,z)=\int_0^z t^{a-1}e^{-t}\,dt,\quad \Gamma(a,z)=\int_z^\infty t^{a-1}e^{-t}\,dt, \tag{37.0.1}$$

where in the first integral we assume that $\Re a>0$. When z is complex we assume that $|\mathrm{ph}\, z|<\pi$. The ratios are given by

$$P(a,z)=\frac{\gamma(a,z)}{\Gamma(a)},\quad Q(a,z)=\frac{\Gamma(a,z)}{\Gamma(a)}. \tag{37.0.2}$$

Then, $P(a,z)+Q(a,z)=1$.

In Chapter 7 and in many examples at other places, we have given several asymptotic expansions of these functions, and the main feature was that the expansions are not valid near the transition point $z=a$. In this chapter we will derive uniform expansions that are valid at this point, and in fact in a large parameter domain.

In §37.1 we will obtain the uniform expansions with starting point the integrals in (37.0.1) by using the method described in §36.1. In later sections we will use loop integrals.

37.1 Using the standard integral representations

To obtain a representation as in (36.1.1) we consider $P(a,z)$ and first transform $t\to at$. Then we obtain

$$P(a,z)=\frac{1}{\Gamma^*(a)}\sqrt{\frac{a}{2\pi}}\int_0^\lambda e^{-a\phi(t)}t^{-1}\,dt, \tag{37.1.3}$$

where (see also (6.5.63))

$$\lambda=\frac{z}{a},\quad \phi(t)=t-\ln t-1,\quad \Gamma^*(a)=\sqrt{a/(2\pi)}\,e^a a^{-a}\Gamma(a). \tag{37.1.4}$$

Next we transform $\frac12\zeta^2=\phi(t)$, with $\mathrm{sign}\,\zeta=\mathrm{sign}(t-1)$, and obtain the standard form of §36.1:

$$F_a(\eta)=\Gamma^*(a)\,P(a,z)=\sqrt{\frac{a}{2\pi}}\int_{-\infty}^\eta e^{-\frac12 a\zeta^2}f(\zeta)\,d\zeta, \tag{37.1.5}$$

[1]This chapter is based on Temme (1975, 1979, 1996b).

where

$$\eta = \zeta(\lambda), \quad f(\zeta) = t^{-1}\frac{dt}{d\zeta} = \frac{\zeta}{t-1}. \tag{37.1.6}$$

When we use the method described in §36.1 we obtain a representation as in (36.1.6) with $A(a) = \Gamma^*(a)$, because $P(a,\infty) = 1$. Then we can write

$$P(a,z) = P(\eta\sqrt{a}) - \frac{e^{-\frac{1}{2}a\eta^2}}{\sqrt{2\pi a}\Gamma^*(a)}B_a(\eta), \tag{37.1.7}$$

and we can combine the asymptotic expansion of $\Gamma^*(a)$ (see (6.2.31)) with that of $B_a(\eta)$. That is, we write

$$\frac{B_a(\eta)}{\Gamma^*(a)} \sim \sum_{n=0}^{\infty} \frac{C_n(\eta)}{a^n}. \tag{37.1.8}$$

In this way, taking into account the properties $P(a,z) + Q(a,z) = 1$ and $\mathrm{erfc}(z) + \mathrm{erfc}(-z) = 2$, we can write

$$\begin{aligned} Q(a,z) &= \tfrac{1}{2}\,\mathrm{erfc}\left(\eta\sqrt{a/2}\right) + R_a(\eta), \\ P(a,z) &= \tfrac{1}{2}\mathrm{erfc}\left(-\eta\sqrt{a/2}\right) - R_a(\eta), \\ R_a(\eta) &\sim \frac{e^{-\frac{1}{2}a\eta^2}}{\sqrt{2\pi a}}\sum_{n=0}^{\infty}\frac{C_n(\eta)}{a^n}, \quad a\to\infty, \end{aligned} \tag{37.1.9}$$

uniformly with respect to $\eta \in \mathbb{R}$, or $z \ge 0$ (and in large domains of the complex plane).

Details on the coefficients $C_n(\eta)$ are given in §37.2.2.

37.2 Representations by contour integrals

First we verify the contour integral

$$P(a,z) = \frac{1}{2\pi i}\int_{c-i\infty}^{c+i\infty} e^{zs}\frac{ds}{s(s+1)^a}, \quad c > 0. \tag{37.2.10}$$

This representation holds for a wide range of the parameters. We assume $\Re a > 0$ and $|\mathrm{ph}\, z| < \frac{1}{2}\pi,\ z \ne 0$. The contour of integration can be deformed into the Hankel contour (see Figure 2.1), which is used for the reciprocal gamma function in (2.2.17). We assume that the branch cut of $(s+1)^{-a}$ runs from -1 to $-\infty$. The phase of $s+1$ is zero when $s > -1$. By rotating the branch cut and the loop integral around it, we can extend the z-domain to $|\mathrm{ph}\, z| < \pi,\ z \ne 0$.

The representation in (37.2.10) can be proved by using some elementary properties of Laplace transforms. We know the Laplace transform

$$\frac{1}{(s+1)^a} = \int_0^{\infty} e^{-st}\,\frac{d\,P(a,t)}{dt}\,dt. \tag{37.2.11}$$

Because $P(a,0) = 0$, we obtain by integrating by parts

$$\int_0^\infty e^{-st} P(a,t)\,dt = \frac{1}{s(s+1)^a}. \qquad (37.2.12)$$

Upon inverting this Laplace transform we obtain (37.2.10).

The contour in (37.2.10) can be shifted to the left of the origin, but then we have to take into account the residue at this point. It follows that

$$P(a,z) = 1 + \frac{1}{2\pi i}\int_{d-i\infty}^{d+i\infty} e^{zs}\frac{ds}{s(s+1)^a}, \quad -1 < d < 0, \qquad (37.2.13)$$

from which we conclude that

$$Q(a,z) = \frac{-1}{2\pi i}\int_{d-i\infty}^{d+i\infty} e^{zs}\frac{ds}{s(s+1)^a}, \quad -1 < d < 0. \qquad (37.2.14)$$

A simple transformation gives

$$Q(a,z) = \frac{e^{-a\phi(\lambda)}}{2\pi i}\int_{\mathcal{L}} e^{a\phi(t)}\frac{dt}{\lambda - t}, \qquad (37.2.15)$$

where

$$\phi(t) = t - 1 - \ln t, \quad \lambda = \frac{z}{a}, \qquad (37.2.16)$$

and $\mathcal{L}$ is a vertical line that cuts the real axis at $t = c$ with $0 < c < \lambda$.

This representation has a striking relation to the Hankel contour integral representation of the reciprocal gamma function in (6.2.26), where the pole at $t = \lambda$ is not present. Again we can integrate along the path of steepest descent, defined in (6.2.27), but we need to avoid the pole.

Remark 37.1. The loop integrals also follow from representation (7.1.6) by writing the integral as a loop integral, see Remark 2.3. Using this method, and the reflection formula for the gamma function, we obtain

$$e^z\Gamma(1-a)\Gamma(a,z) = \frac{1}{e^{-2\pi i a} - 1}\int_{+\infty}^{(0+)} \frac{e^{-zt}\,t^{-a}}{1+t}\,dt, \qquad (37.2.17)$$

where we assume that $\Re a < 1$ and $\Re z > 0$. The contour starts at $+\infty$ with $\operatorname{ph} t = 0$, encircles the origin anti-clockwise, cuts the negative axis in the interval $(-1,0)$, and returns to $+\infty$, with $\operatorname{ph} t = 2\pi$. A transformation $t = se^{\pi i}$, and the reflection formula for the gamma function (6.0.4), gives a similar integral as in (37.2.14), and the condition $\Re a < 1$ can be removed. △

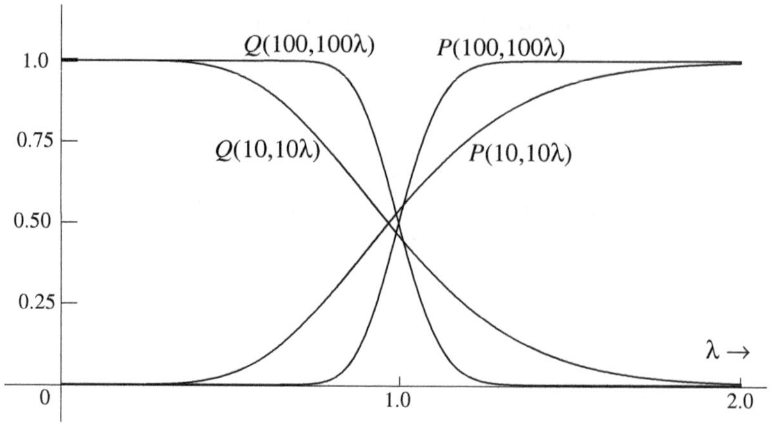

Fig. 37.1 The function $P(a, \lambda a)$ and $Q(a, \lambda a)$ for $\lambda \in [0, 2]$ and $a = 10$ and $a = 100$. As a increases the graphs become steeper when λ passes the transition point $\lambda = 1$.

37.2.1 *Constructing the expansions*

When a is large, the functions $P(a, \lambda a), Q(a, \lambda a)$ change rapidly when λ crosses the value 1; see Figure 37.1. This steep change in behavior can be described by using the complementary error function.

In this section we use the loop integrals to derive asymptotic expansions of $P(a, z), Q(a, z)$; again, a is a large positive parameter. The expansions hold uniformly with respect to $z \in [0, \infty)$, in particular in the neighborhood of $z = a$.

When we take temporarily $z > a$, the pole at $t = \lambda$ in the integral (37.2.15) is located to the right of the saddle point at $t = 1$. In that case we can deform $\mathcal{L}$ into the saddle point contour through $t = 1$ defined by (6.2.27).

On this path $\mathcal{L}$ the function $\phi(t)$ is nonpositive, and we transform

$$\phi(t) = -\tfrac{1}{2}w^2, \tag{37.2.18}$$

with the condition that $t \in \mathcal{L}$ corresponds to $w \in \mathbb{R}$, and $\operatorname{sign}(w) = \operatorname{sign}(\Im t)$. We have

$$w = -i(t-1) + \mathcal{O}\left((t-1)^2\right), \quad t \to 1. \tag{37.2.19}$$

Remark 37.2. The transformation in (37.2.18) is, up to the sign on the right-hand side, the same as the one used to obtain (6.2.28), and for which the details are shown in Figure 6.2. For the present mapping in (37.2.18) the same details follow by rotating the lower figure through $\frac{1}{2}\pi$ clockwise. △

The result of transformation (37.2.18) is

$$Q(a, z) = \frac{e^{-\frac{1}{2}a\eta^2}}{2\pi i}\int_{-\infty}^{\infty} e^{-\frac{1}{2}aw^2}\frac{dt}{dw}\frac{dw}{\lambda - t}, \quad \frac{dt}{dw} = \frac{wt}{1-t}, \tag{37.2.20}$$

where

$$\tfrac{1}{2}\eta^2 = \phi(\lambda) = \lambda - 1 - \ln\lambda. \tag{37.2.21}$$

When taking the square root, we take the branch such that $\text{sign}(\eta) = \text{sign}(\lambda - 1)$ if $\lambda > 0$. Hence,

$$\eta = (\lambda - 1)\sqrt{2\frac{\lambda - 1 - \ln\lambda}{(\lambda - 1)^2}}, \quad \lambda > 0, \tag{37.2.22}$$

where the square root is positive for positive values of the argument.

The mapping in (37.2.18) maps the pole $t = \lambda$ of the integrand in (37.2.15) to a point w_1 defined by $\frac{1}{2}w_1^2 = -\phi(\lambda)$. This is a point on the negative imaginary axis. This follows by observing that when integrating in the t-plane, the pole is on the right of the steepest descent path (recall that we assume $\lambda > 1$). The conformal mapping $t \mapsto w(t)$ preserves this orientation, and hence, when integrating from $-\infty$ to ∞ in the w-plane, the pole should also be on the right of the path of integration. This gives $w_1 = -i\eta$ and we write (37.2.20) in the form

$$\begin{aligned} Q(a,z) &= \frac{e^{-\frac{1}{2}a\eta^2}}{2\pi i}\int_{-\infty}^{\infty} e^{-\frac{1}{2}aw^2} f(w)\frac{dw}{w + i\eta}, \\ f(w) &= \frac{dt}{dw}\frac{w + i\eta}{\lambda - t}. \end{aligned} \tag{37.2.23}$$

When $\lambda \sim 1$ the pole at $w = -i\eta$ is near the saddle point at the origin. As in §21.1 we split off the pole by writing $f(t) = (f(t) - f(-i\eta)) + f(-i\eta)$, and we need to know $f(-i\eta)$. We have by using l'Hôpital's rule

$$\lim_{w \to -i\eta} \frac{dt}{dw}\frac{w + i\eta}{\lambda - t} = -1. \tag{37.2.24}$$

This gives

$$Q(a,z) = \frac{e^{-\frac{1}{2}a\eta^2}}{2\pi i}\int_{-\infty}^{\infty} e^{-\frac{1}{2}aw^2} g(w)\,dw - \frac{e^{-\frac{1}{2}a\eta^2}}{2\pi i}\int_{-\infty}^{\infty} e^{-\frac{1}{2}aw^2}\frac{dw}{w + i\eta}. \tag{37.2.25}$$

For the second integral we use (21.1.3) and obtain the representations (cf. (37.1.9))

$$\begin{aligned} Q(a,z) &= \tfrac{1}{2}\,\text{erfc}\left(\eta\sqrt{a/2}\right) + R_a(\eta), \\ P(a,z) &= \tfrac{1}{2}\,\text{erfc}\left(-\eta\sqrt{a/2}\right) - R_a(\eta), \\ R_a(\eta) &= \frac{e^{-\frac{1}{2}a\eta^2}}{2\pi i}\int_{-\infty}^{\infty} e^{-\frac{1}{2}aw^2} g(w)\,dw, \\ g(w) &= \frac{f(w) - f(-i\eta)}{w + i\eta}. \end{aligned} \tag{37.2.26}$$

The relation for $P(a,z)$ follows by repeating the analysis we used for $Q(a,z)$. Note that the symmetry relation $P(a,z) + Q(a,z) = 1$ is preserved in the above

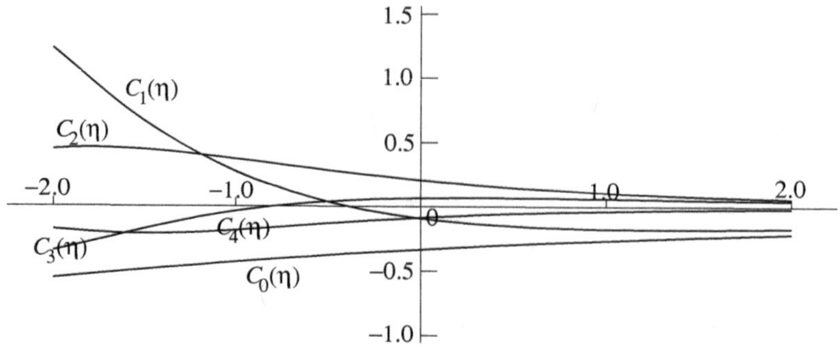

Fig. 37.2 Graphs of the first 5 coefficients $C_n(\eta)$. Because of scaling we have drawn graphs of $\rho C_n(\eta)$, where $\rho = 1, 50, 50, 100, 100$ for $n = 0, 1, 2, 3, 4$, respectively.

representations. The condition $\lambda > 1$ can now be dropped, since the error function and $R_a(\eta)$ are analytic with respect to $\lambda \in (0, \infty)$, in particular at $\lambda = 1$.

By expanding $g(w) = \sum_{n=0}^{\infty} g_n(\eta) w^n$, we obtain the asymptotic expansion

$$R_a(\eta) = \frac{e^{-\frac{1}{2}a\eta^2}}{\sqrt{2\pi a}} S_a(\eta), \quad S_a(\eta) \sim \sum_{n=0}^{\infty} \frac{C_n(\eta)}{a^n}, \quad a \to \infty, \tag{37.2.27}$$

where

$$C_n(\eta) = -i2^n \frac{\Gamma\left(n + \frac{1}{2}\right)}{\Gamma\left(\frac{1}{2}\right)} g_{2n}(\eta), \quad n = 0, 1, 2, \ldots . \tag{37.2.28}$$

Details on the coefficients will be given in §37.2.2.

By considering the location of the singularities of $g(w)$ (for a similar transformation see §6.1.2) it can be concluded that this function is analytic inside the sectors $-\frac{1}{4}\pi < \operatorname{ph} w < \frac{1}{4}\pi$, and the same for $g(-w)$. Hence, from Lemma 3.1 for Laplace's method, it can be concluded that the expansion in (37.2.27) is valid in the sector $-\pi + \delta \le \operatorname{ph} a \le \pi - \delta$.

When $a \sim z$, the parameter η in (37.2.26) is small. When η is small enough to make $\eta\sqrt{a}$ small as well, we have $R_a(\eta) = \mathcal{O}(1/\sqrt{a})$, $a \to \infty$, and both P and Q approach $\frac{1}{2}$.

37.2.2 *Details on the coefficients*

In Figure 37.2 we show the graphs of $C_n(\eta)$ for $n = 0, 1, 2, 3, 4$, properly scaled in order to get them visible in one figure.

We give some details on computing the coefficients. Straightforward evaluation by using the transformation in (37.2.18) and the functions $f(w)$ and $g(w)$ is possible. However, because the functions $P(a, z)$ and $Q(a, z)$ satisfy simple differential equations we can follow a simpler method. In this way we obtain a recurrence relation and a clear structure of the coefficients. This has been explained earlier in Remark 36.1.

The expansion of $S_a(\eta)$ in (37.2.27) can be obtained by differentiating one of the equations in (37.2.26) with respect to η. This gives

$$\frac{d}{d\eta}S_a(\eta) - a\eta S_a(\eta) = a\left(1 - f(\eta)/\Gamma^*(a)\right), \tag{37.2.29}$$

where $\Gamma^*(a)$ is defined in (37.1.4), and

$$f(\eta) = \frac{1}{\lambda}\frac{d\lambda}{d\eta} = \frac{\eta}{\lambda - 1}. \tag{37.2.30}$$

Substituting the asymptotic expansion of $S_a(\eta)$ one finds for the coefficients the relations

$$C_0(\eta) = \frac{1}{\lambda - 1} - \frac{1}{\eta} \tag{37.2.31}$$

and

$$\eta C_n(\eta) = \frac{d}{d\eta}C_{n-1}(\eta) + \gamma_n f(\eta), \quad n \ge 1, \tag{37.2.32}$$

where γ_n are the coefficients in the reciprocal gamma function expansion; see (6.2.31) and (6.2.32).

For $C_1(\eta)$ we have

$$C_1(\eta) = \frac{1}{\eta^3} - \frac{1}{(\lambda-1)^3} - \frac{1}{(\lambda-1)^2} - \frac{1}{12(\lambda-1)}. \tag{37.2.33}$$

It seems that in the paper Tricomi (1950)[2] for the first time an approximation of $\gamma(\pm a, z)$ for large a in terms of the error function was derived. For error bounds of remainders of the expansion (37.2.27), see Paris (2002a). In Paris (2002b) a different type of uniform expansion is given with coefficients that do not possess a removable singularity at $z = a$. See also the papers Dunster (1997), Dunster *et al.* (1998) and Olde Daalhuis (1998).

Remark 37.3. The computation of the coefficients $C_n(\eta)$ needs some special attention, and is not as simple as that of the coefficients we have seen in Chapter 7. In particular near the transition point, that is when $z \sim a$, the removable singularities in the representations of $C_0(\eta)$ and $C_1(\eta)$ as shown in (37.2.33) are inconvenient in numerical computations. All higher coefficients show this type of cancellation. Each $C_n(\eta)$ contains a term with $1/\eta^{2n+1}$, and this term is a removable pole in the representations of $C_n(\eta)$ obtained in this way.

For aspects of the numerical evaluation of the coefficients $C_n(\eta)$, and on the use of the asymptotic representation in (37.2.26), see Gil *et al.* (2007, §8.3) and Gil *et al.* (2012), where, moreover, a method is described based on expanding $S_a(\eta)$ in powers of η. This method is based on the relation in (37.2.29) and can easily be implemented in a numerical algorithm. △

---

[2]In that paper Tricomi called $\gamma(a, z)$ the Cinderella of special functions; see the first page of Chapter 7.

37.2.3 Relations to the coefficients of earlier expansions

The expansions of $P(a,z)$ and $Q(a,z)$ including the complementary error function can be reduced to simpler expansions when η and a are such that the argument of the complementary error function in (37.2.26) is large. If $\eta\sqrt{a/2}\to+\infty$ we can use the expansion in (3.4.30) for $Q(a,z)$, if $\eta\sqrt{a/2}\to-\infty$ we can use that of $P(a,z)$. We assume the latter case, and obtain

$$P(a,z)\sim-\frac{e^{-\frac{1}{2}a\eta^2}}{\eta\sqrt{2\pi a}}\sum_{n=0}^{\infty}(-1)^n\frac{2^n\left(\frac{1}{2}\right)_n}{(a\eta^2)^n}-\frac{e^{-\frac{1}{2}a\eta^2}}{\sqrt{2\pi a}}\sum_{n=0}^{\infty}\frac{C_n(\eta)}{a^n}. \tag{37.2.34}$$

This can be written as

$$\begin{aligned}P(a,z)&\sim-\frac{e^{-\frac{1}{2}a\eta^2}}{\sqrt{2\pi a}}\sum_{n=0}^{\infty}\frac{D_n(\lambda)}{a^n},\\ D_n(\lambda)&=C_n(\eta)+(-1)^n\frac{2^n\left(\frac{1}{2}\right)_n}{\eta^{2n+1}}.\end{aligned} \tag{37.2.35}$$

From the first coefficients given in (37.2.31) and (37.2.33) it follows that

$$D_0(\lambda)=\frac{1}{\lambda-1},\quad D_1(\lambda)=-\frac{1}{(\lambda-1)^3}-\frac{1}{(\lambda-1)^2}-\frac{1}{12(\lambda-1)}. \tag{37.2.36}$$

As observed in Remark 37.3, each $C_n(\eta)$ contains a term with $1/\eta^{2n+1}$, and it is easily verified that these terms are exactly cancelled by the terms in $D_n(\lambda)$ that follow from the expansion of the complementary error function.

In fact, we can write for $D_n(\lambda)$ the recursion (see also (37.2.30))

$$D_n(\lambda)=\frac{\lambda}{\lambda-1}\frac{d}{d\lambda}D_{n-1}(\lambda)+\frac{\gamma_n}{\lambda-1},\quad n\ge1, \tag{37.2.37}$$

with $D_0(\lambda)$ given in (37.2.36).

These coefficients may be compared with those for $\gamma(a,z)$ in §7.3. The inhomogeneous term $\gamma_n/(\lambda-1)$ is not present in the recursion in (7.3.19). These are needed for the coefficients $D_n(\lambda)$ of the expansion in (37.2.35), because here the expansion is for the ratio $P(a,z)$.

A similar relation exists for the coefficients of the expansion of $\Gamma(a,x)$ in §7.4. These also follow from $C_n(\eta)$ when the complementary error function in (37.2.26) is expanded for large positive values of its argument. Observe that the expansion of $\Gamma(a,x)$ in (7.4.41) is derived for $x\gg a$.

37.3 Incomplete gamma functions, negative parameters

The incomplete gamma functions given in (37.0.1) follow from splitting up the complete interval of integration $[0,\infty)$ of the integral for the gamma function.[3]

[3]This section and the next one are based on Temme (1996b).

What happens when we split up the Hankel contour for the reciprocal gamma function defined in (2.2.17)? That is, can we express the integrals

$$G_1(a,z)=\frac{1}{2\pi i}\int_{-\infty}^{z} s^{-a}e^{s}\,ds,\quad G_2(a,z)=\frac{1}{2\pi i}\int_{z}^{-\infty} s^{-a}e^{s}\,ds \qquad (37.3.38)$$

in terms of known special functions? Here, $a\in\mathbb{C}$ and z is a point on the contour given in Figure 2.1, $z\neq 0$, $|\operatorname{ph} z|<\pi$. In the first integral $\operatorname{ph} s=-\pi$ at $-\infty$, in the second one $\operatorname{ph} s=+\pi$ at $-\infty$.

To evaluate the integrals, we start with values of a such that $\Re a<1$. In that case we can deform the path for $G_1(a,z)$ into $(-\infty,0]$ and then from 0 to z. In this way,

$$G_1(a,z)=\frac{e^{\pi ia}}{2\pi i}\Gamma(1-a)+\frac{1}{2\pi i}\int_0^{z} s^{-a}e^{s}\,ds. \qquad (37.3.39)$$

The integral can be expressed in terms of $\gamma(a,z)$, but we like to use the analytic function $\gamma^*(a,z)=\dfrac{z^{-a}}{\Gamma(a)}\gamma(a,z)$ defined in (7.1.4). This gives

$$G_1(a,z)=\frac{\Gamma(1-a)}{2\pi i}\left(e^{\pi ia}+z^{1-a}\gamma^*(1-a,-z)\right). \qquad (37.3.40)$$

In the same way

$$G_2(a,z)=-\frac{\Gamma(1-a)}{2\pi i}\left(e^{-\pi ia}+z^{1-a}\gamma^*(1-a,-z)\right), \qquad (37.3.41)$$

and by using (2.2.29) it follows that indeed $G_1(a,z)+G_2(a,z)=\dfrac{1}{\Gamma(a)}$.

The functions $G_1(a,z)$ and $G_2(a,z)$ are analytic functions of a, as follows from the representations in (37.3.38). In the representations in (37.3.40) and (37.3.41) cancellations of singularities happen when $a=1,2,3,\ldots$.

It is possible to express $G_j(a,z)$ in terms of $\gamma(a,z)$ and $\Gamma(a,z)$, but some care is needed when using these functions with argument $-z$. For example, let us assume for $G_1(a,z)$ that $\Im z<0$ with $\operatorname{ph} z\in(-\pi,0)$. Then the interpretation of $-z$ is $ze^{+\pi i}$. Using this in (7.1.4) and (37.3.40), we have

$$\begin{aligned}G_1(a,z)&=\frac{e^{\pi ia}}{2\pi i}\left(\Gamma(1-a)-\gamma\left(1-a,ze^{+\pi i}\right)\right)\\&=\frac{e^{\pi ia}}{2\pi i}\Gamma\left(1-a,ze^{+\pi i}\right).\end{aligned} \qquad (37.3.42)$$

When using analytic continuation of $\Gamma\left(1-a,ze^{+\pi i}\right)$ we can also use this relation when $\Im z\geq 0$. Similarly for $G_2(a,z)$ where we take $-z=ze^{-\pi i}$. This gives

$$G_2(a,z)=-\frac{e^{-\pi ia}}{2\pi i}\Gamma\left(1-a,ze^{-\pi i}\right). \qquad (37.3.43)$$

Adding up we obtain the known relation (see Paris (2010, Eq. 8.2.10))

$$\frac{e^{\pi ia}}{2\pi i}\Gamma\left(1-a,ze^{+\pi i}\right)-\frac{e^{-\pi ia}}{2\pi i}\Gamma\left(1-a,ze^{-\pi i}\right)=\frac{1}{\Gamma(a)}. \qquad (37.3.44)$$

However, this also follows at once from

$$\Gamma(a,z) = \Gamma(a)\left(1 - z^a\gamma^*(a,z)\right) \tag{37.3.45}$$

and the fact that $\gamma^*(a,z)$ is an analytic function of z.

We observe that, starting from the Hankel contour for the reciprocal gamma function, we obtain the usual incomplete gamma functions with variables with negative signs.

For the incomplete gamma functions in (37.3.44) we can also use different integral representations. We start with (7.1.6) and replace a by $-a$:

$$\Gamma(-a,z) = \frac{e^{-z}}{\Gamma(1+a)}\int_0^\infty \frac{e^{-zt}\,t^a}{1+t}\,dt, \quad \Re a > -1, \quad \Re z > 0. \tag{37.3.46}$$

By rotating the path of integration and invoking the principle of analytic continuation we can enlarge the domain of z. For example, when we consider the path in (37.3.46) along the positive imaginary axis, the integral is defined for $-\pi < \operatorname{ph} z < 0$, and in the overlapping domain $-\frac{1}{2}\pi < \operatorname{ph} z < 0$ its value is the same as in (37.3.46). We rotate the path from the positive imaginary axis $\operatorname{ph} t = \frac{1}{2}\pi$ to the negative axis $\operatorname{ph} t = \pi$, avoiding the pole at $t = -1$ by using a small semi-circle. The integral is then defined for $-\frac{3}{2}\pi < \operatorname{ph} z < -\frac{1}{2}\pi$. We change the variable of integration and the result is:

$$\Gamma(-a, ze^{-i\pi}) = \frac{e^z e^{i\pi a}}{\Gamma(1+a)} \int_0^\infty \frac{e^{-zt}\,t^a}{t-1}\,dt, \tag{37.3.47}$$

valid for $\Re a > -1$ and $-\frac{1}{2}\pi < \operatorname{ph} z < \frac{1}{2}\pi$, and we avoid the pole by integrating below the point $t = 1$.

Rotating the path in (37.3.46) clockwise we obtain

$$\Gamma(-a, ze^{+i\pi}) = \frac{e^z e^{-i\pi a}}{\Gamma(1+a)} \int_0^\infty \frac{e^{-zt}\,t^a}{t-1}\,dt, \tag{37.3.48}$$

again valid for $\Re a > -1$ and $-\frac{1}{2}\pi < \operatorname{ph} z < \frac{1}{2}\pi$, and we avoid the pole by integrating above the point $t = 1$.

An easy consequence is the connection formula in (37.3.44), which follows from computing the residue of the integral over the full circle around the pole at $t = 1$.

In the next section we will use these integral representations for obtaining expansions of $\Gamma(-a,-z)$ and related functions.

37.3.1 *Expansions near the transition point*

We consider the asymptotic behavior of the incomplete gamma functions $\gamma(-a,-z)$ and $\Gamma(-a,-z)$ as $a \to \infty$. Again we need the complementary error function to describe the transition area $z \sim a$.

We assume, for the time being, that a and z are positive. A simple transformation in (37.3.48) gives

$$\begin{aligned}\Gamma\left(-a, ze^{\pi i}\right) &= \frac{e^z \lambda^{-a} e^{-ia\pi}}{\Gamma(a+1)} \int_0^\infty e^{-at} t^a \frac{dt}{t-\lambda} \\ &= \frac{e^{\frac{1}{2}a\eta^2 - ia\pi}}{\Gamma(a+1)} \int_{-\infty}^\infty e^{-\frac{1}{2}a\zeta^2} g(\zeta) \frac{d\zeta}{\zeta-\eta},\end{aligned} \tag{37.3.49}$$

where $\eta=\eta(\lambda)$ is defined in (37.2.22), $\lambda=z/a$, $\zeta=\eta(t)$, that is,

$$\tfrac{1}{2}\zeta^2=t-\ln t-1,\quad \operatorname{sign}\zeta=\operatorname{sign}(t-1), \tag{37.3.50}$$

and

$$g(\zeta)=\frac{dt}{d\zeta}\,\frac{\zeta-\eta}{t-\lambda}=\frac{\zeta t}{t-1}\,\frac{\zeta-\eta}{t-\lambda}. \tag{37.3.51}$$

In the ζ-integral the path passes above the pole at $\zeta=\eta$.

When $z\sim a$, that is, $\lambda\sim 1$, the pole is near the saddle point $\zeta=0$, and for large values of a we need an error function to describe the asymptotic behavior. As in §37.2.1 we can split off the pole by writing $g(\zeta)=(g(\zeta)-g(\eta))+g(\eta)$, where, as is easily verified, $g(\eta)=1$.

Using the representation (see (21.1.2))

$$\tfrac{1}{2}\operatorname{erfc} iz=-\frac{e^{z^2}}{2\pi i}\int_{-\infty}^{\infty}\frac{e^{-t^2}}{t-z}\,dt\quad z\in\mathbb{C}, \tag{37.3.52}$$

where the path passes above the pole at $t=z$, we obtain (introducing a suitable normalization)

$$-\Gamma(a+1)\,\frac{e^{\pi i a}}{2\pi i}\,\Gamma\left(-a,ze^{\pi i}\right)=\tfrac{1}{2}\operatorname{erfc}\left(i\eta\sqrt{a/2}\right)-i\frac{e^{\frac{1}{2}a\eta^2}}{\sqrt{2\pi a}}T_a(\eta), \tag{37.3.53}$$

where

$$T_a(\eta)=-\sqrt{\frac{a}{2\pi}}\int_{-\infty}^{\infty}e^{-\frac{1}{2}a\zeta^2}h(\zeta)\,d\zeta,\quad h(\zeta)=\frac{g(\zeta)-g(\eta)}{\zeta-\eta}. \tag{37.3.54}$$

In a similar way, starting from (37.3.47),

$$\Gamma(a+1)\,\frac{e^{-\pi i a}}{2\pi i}\,\Gamma\left(-a,ze^{-\pi i}\right)=\tfrac{1}{2}\operatorname{erfc}\left(-i\eta\sqrt{a/2}\right)+i\frac{e^{\frac{1}{2}a\eta^2}}{\sqrt{2\pi a}}T_a(\eta). \tag{37.3.55}$$

This also follows from (37.3.44) and (37.3.54).

The asymptotic expansion of $T_a(\eta)$ can be obtained by expanding $h(\zeta)$ in powers of ζ. However, it is easier to use the method of §37.2.2 and to derive a differential equation satisfied by $T_a(\eta)$. In this way we can identify the coefficients of the expansion with those in (37.2.27).

This time the equation is

$$\frac{d}{d\eta}T_a(\eta)+a\eta T_a(\eta)=a\left(f(\eta)\Gamma^*(a)-1\right), \tag{37.3.56}$$

where $f(\eta)$ and $\Gamma^*(a)$ are as in (37.2.29). We use the same coefficients γ_n as in (6.2.31) and (6.2.32). By substituting the expansion of $\Gamma^*(a)$ and an expansion of $T_a(\eta)$ into (37.3.56), it is readily verified that

$$T_a(\eta)\sim\sum_{n=0}^{\infty}(-1)^n\frac{C_n(\eta)}{a^n}, \tag{37.3.57}$$

where the coefficients $C_n(\eta)$ are the same as those in the expansion of $S_a(\eta)$ in (37.2.27).

37.3.2 *A real expansion of* $\gamma^*(-a,-z)$

The function $\gamma^*(-a,-z)$ introduced in (7.1.4) is real for real values of a and z, and it is of interest to have a result for this function with real terms. We have the following asymptotic representation

$$z^{-a}\gamma^*(-a,-z) = \cos\pi a - \frac{2e^{\frac{1}{2}a\eta^2}\sin\pi a}{\sqrt{\pi}}\left(F\left(\eta\sqrt{a/2}\right)+\frac{1}{\sqrt{2a}}T_a(\eta)\right), \quad (37.3.58)$$

where $F(z)$ is Dawson's integral, which we have introduced in (36.5.76), together with the relation to the error function. The representation in (37.3.58) can be verified by using (37.3.45) and (37.3.53), together with relations for the gamma function (for example (6.0.4)).

In (37.3.58) we see that $\gamma^*(-a,-z)$ has an oscillatory behavior if a and z are positive. This is described by two terms, one with $\cos\pi a$ and one with $\exp(\frac{1}{2}a\eta^2)\sin\pi a$ (the factor containing Dawson's integral is slowly varying when the parameters a and z are positive). This complicated oscillatory behavior is one of the problems in writing reliable software for the functions $\gamma(a,z)$ and $\Gamma(a,z)$ when the parameters have large negative real parts. Dawson's integral becomes dominant when we consider complex values of the parameters. The function $F(z)$ is an analytic odd function and has for $\Re z\ge 0$ the asymptotic behavior

$$F(z)\sim\begin{cases}\dfrac{1}{2z}, & \text{if } |\mathrm{ph}\, z|<\frac{1}{4}\pi,\\ \mathrm{sign}(\Im z)\frac{1}{2}i\sqrt{\pi}\,e^{-z^2}, & \text{elsewhere,}\end{cases} \qquad |z|\to\infty. \quad (37.3.59)$$

Remark 37.4. We verify what happens when the parameters a,z in (37.2.26) are taken with negative signs. First let $\Im a,\Im z$ be positive, and replace in $Q(a,z)$ the parameters a,z by $ae^{-\pi i}, ze^{-\pi i}$, respectively. The quantity η defined in (37.2.22) does not change by this operation, whereas the expansion of $S_a(\eta)$ becomes the expansion of $T_a(\eta)$ given in (37.3.57). When we formally write $S_{-a}(\eta)=T_a(\eta)$, which certainly is not true, it follows that the right-hand side of the first line in (37.2.26) formally becomes

$$\tfrac{1}{2}\mathrm{erfc}\left(-i\eta\sqrt{a/2}\right)+i\frac{e^{\frac{1}{2}a\eta^2}}{\sqrt{2\pi a}}T_a(\eta), \quad (37.3.60)$$

which is the right-hand side of (37.3.55). But

$$Q(-a,z^{-\pi i}) = \frac{\Gamma\left(-a,ze^{-\pi i}\right)}{\Gamma(-a)} = -\frac{1}{\pi}\Gamma(a+1)\sin\pi a\,\Gamma(-a,ze^{-\pi i}), \quad (37.3.61)$$

which means that the left-hand side of (37.3.55) equals

$$\Gamma(a+1)\,\frac{e^{-\pi i a}}{2\pi i}\,\Gamma\left(-a,ze^{-\pi i}\right) = \frac{1}{1-e^{2\pi i a}}\,Q\left(-a,ze^{-\pi i}\right). \quad (37.3.62)$$

Hence, by proceeding formally from the relation for $Q(a,z)$ in formula (37.2.26), we miss the factor $1/\left(1-\exp(2\pi i a)\right)$, this factor being negligible when $\Im a$ is positive and large, as we assumed here. A similar conclusion holds when $\Im a,\Im z$ are negative and we replace a,z by $ae^{+\pi i}, ze^{+\pi i}$. △

Chapter 38

Incomplete beta function

The incomplete beta function is considered in §26.3, where a uniform expansion is given by using a certain real incomplete Laplace integral and a loop integral. In this chapter we give uniform expansions in terms of the complementary error function, but we start with simpler expansions.[1]

We recall the definition of the incomplete beta function

$$I_x(p,q) = \frac{1}{B(p,q)} \int_0^x t^{p-1}(1-t)^{q-1}\,dt, \quad x \in [0,1], \quad p > 0, \quad q > 0, \tag{38.0.1}$$

where $B(p,q)$ is Euler's beta integral

$$B(p,q) = \int_0^1 t^{p-1}(1-t)^{q-1}\,dt = \frac{\Gamma(p)\Gamma(q)}{\Gamma(p+q)}. \tag{38.0.2}$$

For certain combinations of x, p, q or to switch parameters, it is convenient to use

$$J_x(p,q) = \frac{1}{B(p,q)} \int_x^1 t^{p-1}(1-t)^{q-1}\,dt = 1 - I_x(p,q) = I_{1-x}(q,p). \tag{38.0.3}$$

We consider asymptotic representations of the incomplete beta function for the following cases:

- a power series expansion: large p; q fixed, $x \in [0,1)$;
- a uniform expansion: large p; q fixed, $x \in [0,1]$;
- the nearly symmetric case: large p; $p = q + \beta$, where β stays bounded;
- the symmetric case: large $p+q$; p/q and q/p both bounded away from zero;
- the incomplete gamma function case: large $p+q$; $x \in [0,1]$.

In the first case elementary functions are used, in the other cases the complementary error function or the incomplete gamma function are the main approximants. Because of the relations in (38.0.3), it is possible to restrict ourselves to $p \ge q$. For the case with both p and q large, we refer to §26.3 and §26.4.1, where expansions are given in terms of the incomplete gamma function for large values of p that hold uniformly with respect to $x \in [0,1]$ and $q \ge 0$.

[1]The expansions in terms of the error function are obtained in Temme (1992b).

38.1 A power series expansion for large p

We use the representation

$$\begin{aligned} I_x(p,q) &= \frac{x^p(1-x)^{q-1}}{pB(p,q)}\,{}_2F_1\!\left(\begin{matrix}1, 1-q\\ p+1\end{matrix};\frac{x}{x-1}\right)\\ &= \frac{x^p(1-x)^{q-1}}{pB(p,q)}\sum_{n=0}^{\infty}\frac{(1-q)_n}{(p+1)_n}\left(\frac{x}{x-1}\right)^n. \end{aligned} \tag{38.1.4}$$

For large values of p we can use this power series, it converges when $0 \le x \le \frac{1}{2}$, and for all fixed $q > 0$ and $x \in [0,1)$ the series has an asymptotic character. We can obtain a Poincaré-type asymptotic expansion in negative powers of p by using the method explained in §12.1.1. See also López and Sesma (1999) for this type of expansion with details on truncation errors.

38.2 A uniform expansion for large p

The expansion in (38.1.4) breaks down when $x \to 1$, and in this section we allow $x \in [0,1]$; q is fixed. We use $t = e^{-u}$ and (38.0.1) becomes

$$I_x(p,q) = \frac{1}{B(p,q)}\int_{\xi}^{\infty} u^{q-1} f(u) e^{-pu}\,du, \tag{38.2.5}$$

where

$$\xi = -\ln x, \quad f(u) = \left(\frac{1-e^{-u}}{u}\right)^{q-1}. \tag{38.2.6}$$

Expanding $f(u) = \sum_{n=0}^{\infty} c_n (u-\xi)^n$, we obtain

$$I_x(p,q) \sim \frac{\Gamma(p+q)}{\Gamma(p)}\sum_{n=0}^{\infty} c_n \Psi_n, \tag{38.2.7}$$

where

$$\begin{aligned} \Psi_n &= \frac{1}{\Gamma(q)}\int_{\xi}^{\infty} u^{q-1}(u-\xi)^n e^{-pu}\,du\\ &= \frac{\xi^{n+q}e^{-p\xi}n!}{\Gamma(q)}U(n+1, n+q+1, p\xi). \end{aligned} \tag{38.2.8}$$

Here we use the integral representation of the Kummer U-function given in (10.1.5).

The Φ_n can be written in terms of the incomplete gamma function ratio, and the first functions are

$$\Psi_0 = p^{-q}Q(q,p\xi), \quad \Psi_1 = \frac{q-p\xi}{p}\Psi_0 + \frac{\xi^q e^{-p\xi}}{p\Gamma(q)}. \tag{38.2.9}$$

The others follow from the recurrence relation

$$p\Psi_{n+1} = (n+q-p\xi)\Psi_n + n\Psi_{n-1}, \quad n = 1,2,3,\ldots. \tag{38.2.10}$$

The first coefficients of the expansion are

$$c_0 = \left(\frac{x-1}{\ln x}\right)^{q-1}, \quad c_1 = (q-1)\frac{x\ln x + 1 - x}{(1-x)\ln x}c_0,$$
$$c_2 = \frac{(q-1)c_0}{2L^2(1-x)^2}\left(xL^2(qx-1-x) + 2x(1-x)(q-1)L + q(1-x)^2\right), \tag{38.2.11}$$

where $L = \ln x$.

In Doman (1996) a related expansion has been given in terms of the incomplete gamma function, with a discussion of numerical applications. In that paper a function related to $f(u)$ in (38.2.6) is expanded in powers of u; this gives a large-p expansion of $I_x(p,q)$ valid near $x = 1$. In §26.3 we have considered the integral in (38.2.5) for the case that q is large as well.

38.3 The nearly symmetric case

We write $q = p + \beta$, where β is fixed. We obtain from (38.0.3)

$$I_x(p, p+\beta) = \frac{4^{-p}}{B(p,p+\beta)}\int_0^x \left(4t(1-t)\right)^p \frac{(1-t)^\beta\, dt}{t(1-t)}. \tag{38.3.12}$$

We transform this into a standard form by writing

$$-\tfrac{1}{2}\zeta^2 = \ln\left(4t(1-t)\right), \quad 0 < t < 1, \quad \operatorname{sign}(\zeta) = \operatorname{sign}(t - \tfrac{1}{2}),$$
$$-\tfrac{1}{2}\eta^2 = \ln\bigl(4x(1-x)\bigr), \quad 0 < x < 1, \quad \operatorname{sign}(\eta) = \operatorname{sign}(x - \tfrac{1}{2}). \tag{38.3.13}$$

Therefore,

$$I_x(p, p+\beta) = \frac{4^{-p}}{B(p,p+\beta)}\int_{-\infty}^{\eta} e^{-\frac{1}{2}p\zeta^2}\frac{(1-t)^\beta}{t(1-t)}\frac{dt}{d\zeta}\, d\zeta. \tag{38.3.14}$$

We can write t as a function of ζ:

$$t = \tfrac{1}{2}\left(1 \pm \sqrt{1 - \exp\left(-\tfrac{1}{2}\zeta^2\right)}\right) = \tfrac{1}{2}\left(1 + \zeta\sqrt{\left(1 - \exp\left(-\tfrac{1}{2}\zeta^2\right)\right)/\zeta^2}\right), \tag{38.3.15}$$

where the second square root is nonnegative for real values of the argument. The same relation holds for x as a function of η. It easily follows that

$$\frac{1}{t(1-t)}\frac{dt}{d\zeta} = \frac{-\zeta}{1-2t}, \tag{38.3.16}$$

and that the following standard form (in the sense of §36.1) can be obtained

$$I_x(p, p+\beta) = \frac{F_p(\eta)}{F_p(\infty)}, \quad F_p(\eta) = \sqrt{\frac{p}{2\pi}}\int_{-\infty}^{\eta} e^{-\frac{1}{2}p\zeta^2} f(\zeta)\, d\zeta, \tag{38.3.17}$$

with

$$F_p(\infty) = \frac{\sqrt{p}\,\Gamma(p)}{\Gamma\left(p + \frac{1}{2}\beta\right)}\frac{\Gamma(p+\beta)}{\Gamma\left(p + \frac{1}{2}\beta + \frac{1}{2}\right)}, \tag{38.3.18}$$

and

$$f(\zeta) = (2(1-t))^\beta \sqrt{\frac{\frac{1}{2}\zeta^2}{1-\exp(-\frac{1}{2}\zeta^2)}}. \tag{38.3.19}$$

The form of $F_p(\infty)$ is obtained by using the duplication formula of the gamma function given in (11.2.16). From the asymptotic expansion of the ratio of gamma functions (see §6.5), we obtain

$$F_p(\infty) \sim A_0 + A_1 p^{-1} + A_2 p^{-2} + \ldots, \quad p \to \infty, \tag{38.3.20}$$

where

$$A_0 = 1, \quad A_1 = \tfrac{1}{8}(2\beta^2 - 2\beta + 1), \quad A_2 = \tfrac{1}{128}(4\beta^4 - 24\beta^3 + 32\beta^2 - 12\beta + 1). \tag{38.3.21}$$

The function $f(\zeta)$ is analytic in a strip containing $\mathbb{R}$; the singularities nearest to the origin occur at $\pm 2\sqrt{\pi}\exp(\pm\frac{1}{4}\pi i)$. The first coefficients of the Taylor expansion

$$f(\zeta) = a_0 + a_1\zeta + a_2\zeta^2 + a_3\zeta^3 + \ldots \tag{38.3.22}$$

are

$$\begin{aligned} a_0 &= 1, \quad a_1 = -\tfrac{1}{2}\beta\sqrt{2}, \quad a_2 = \tfrac{1}{8}\left(2\beta^2 - 2\beta + 1\right), \\ a_3 &= -\tfrac{1}{24}\sqrt{2}\,\beta(\beta-1)(\beta-2), \\ a_4 &= \tfrac{1}{384}\left(4\beta^4 - 24\beta^3 + 32\beta^2 - 12\beta + 1\right), \\ a_5 &= -\tfrac{1}{960}\sqrt{2}\,\beta(\beta-1)(\beta-2)\left(\beta^2 - 7\beta + 2\right). \end{aligned} \tag{38.3.23}$$

From the results in §36.1 it follows that the standard form (38.3.17) can be written in the form

$$I_x(p, p+\beta) = \tfrac{1}{2}\mathrm{erfc}\left(-\eta\sqrt{p/2}\right) - R_p(\eta), \tag{38.3.24}$$

where η is defined in (38.3.13), and $R_p(\eta)$ can be expanded as explained in (36.1.7)–(36.1.9). In the present case we have the expansion

$$R_p(\eta) \sim \frac{1}{F_p(\infty)} \frac{e^{-\frac{1}{2}p\eta^2}}{\sqrt{2\pi p}} \sum_{n=0}^{\infty} \frac{C_n(\eta)}{p^n}, \quad p \to \infty, \tag{38.3.25}$$

where the coefficients follow from the scheme

$$C_n(\eta) = \frac{f_n(\eta) - f_n(0)}{\eta}, \quad f_n(\zeta) = \frac{d}{d\zeta}\frac{f_{n-1}(\zeta) - f_{n-1}(0)}{\zeta}, \tag{38.3.26}$$

$n = 0, 1, 2, \ldots$, with $f_0 = f$ defined in (38.3.19).

For details on evaluating the coefficients $C_n(\eta)$ we refer to Remark 36.1.

38.4 The general error function case

Let us write

$$p = r\sin^2\theta, \quad q = r\cos^2\theta, \quad 0 < \theta < \tfrac{1}{2}\pi. \tag{38.4.27}$$

Then (38.0.1) can be written as

$$I_x(p,q) = \frac{1}{B(p,q)} \int_0^x e^{r\left(\sin^2\theta \ln t + \cos^2\theta \ln(1-t)\right)} \frac{dt}{t(1-t)}. \tag{38.4.28}$$

We consider r as a large parameter, and θ bounded away from 0 and $\frac{1}{2}\pi$. The maximum of the exponential function occurs at $t = \sin^2\theta$. We use the transformation

$$-\tfrac{1}{2}\zeta^2 = \sin^2\theta \ln \frac{t}{\sin^2\theta} + \cos^2\theta \ln \frac{1-t}{\cos^2\theta}, \tag{38.4.29}$$

where the sign of ζ equals the sign of $t - \sin^2\theta$. The same transformation holds for $x \mapsto \eta$ if t and ζ are replaced by x and η, respectively. Using (38.4.29) we obtain

$$-\zeta\frac{d\zeta}{dt} = \frac{\sin^2\theta - t}{t(1-t)}, \tag{38.4.30}$$

and we can write (38.4.28) in the standard form (cf. (38.3.17) and (38.3.18))

$$I_x(p,q) = \frac{F_r(\eta)}{F_r(\infty)}, \quad F_r(\eta) = \sqrt{\frac{r}{2\pi}} \int_{-\infty}^{\eta} e^{-\frac{1}{2}r\zeta^2} f(\zeta)\, d\zeta, \tag{38.4.31}$$

where

$$F_r(\infty) = \frac{\Gamma^*(p)\Gamma^*(q)}{\Gamma^*(r)}, \quad f(\zeta) = \frac{\zeta\sin\theta\cos\theta}{t - \sin^2\theta}. \tag{38.4.32}$$

The function $\Gamma^*(z)$, the slowly varying part of the Euler gamma function, is defined in (6.5.63).

The analogue of the expansion (38.3.20) is now in terms of the large parameter r:

$$F_r(\infty) \sim A_0 + A_1 r^{-1} + A_2 r^{-2} + \ldots, \quad r \to \infty, \tag{38.4.33}$$

where

$$\begin{aligned} A_0 &= 1, \quad A_1 = -\frac{\sin^2\theta\cos^2\theta - 1}{3\sin^2 2\theta}, \quad A_2 = \frac{(\sin^2\theta\cos^2\theta - 1)^2}{18\sin^4 2\theta}, \\ A_3 &= \frac{139(\sin^6\theta\cos^6\theta - \cos^6\theta - \sin^6\theta) + 15\sin^4\theta\cos^4\theta}{810\sin^6 2\theta}. \end{aligned} \tag{38.4.34}$$

The first coefficients of the Taylor expansion

$$f(\zeta) = a_0 + a_1\zeta + a_2\zeta^2 + a_3\zeta^3 + \ldots \tag{38.4.35}$$

are

$$a_0 = 1, \quad a_1 = -\tfrac{2}{3}\cot 2\theta, \quad a_2 = \frac{\sin^4\theta + \cos^4\theta + 1}{6\sin^2 2\theta}. \tag{38.4.36}$$

As in (38.3.24) we write

$$I_x(p,q) = \tfrac{1}{2}\mathrm{erfc}\left(-\eta\sqrt{r/2}\right) - R_r(\eta), \tag{38.4.37}$$

where the relation between x and η follows from (38.4.29) (with ζ replaced by η and t by x), and $R_r(\eta)$ has the expansion

$$R_r(\eta) \sim \frac{1}{F_r(\infty)} \frac{e^{-\frac{1}{2}r\eta^2}}{\sqrt{2\pi r}} \sum_{n=0}^{\infty} \frac{C_n(\eta)}{r^n}, \quad r \to \infty, \tag{38.4.38}$$

as in (38.3.25), and $F_r(\infty)$ is defined in (38.4.32). The coefficients $C_n(\eta)$ can be obtained from the scheme given in (38.3.26) with f defined in (38.4.32).

The expansion in (38.4.38) is uniformly valid for $\theta = \arctan\left(\sqrt{p/q}\right) \in [\delta, \frac{1}{2}\pi - \delta]$, where δ is a small fixed positive number.

Chapter 39

Non-central chi-square, Marcum functions

We consider for positive x, y, μ the functions[1]

$$
\begin{aligned}
P_\mu(x,y) &= e^{-x}\sum_{n=0}^{\infty}\frac{x^n}{n!}P(\mu+n,y),\\
Q_\mu(x,y) &= e^{-x}\sum_{n=0}^{\infty}\frac{x^n}{n!}Q(\mu+n,y),
\end{aligned}
\tag{39.0.1}
$$

in terms of the incomplete gamma functions, see (37.0.1) and (37.0.2). The functions defined in (39.0.1) occur in statistics and probability theory, where they are called non-central chi-square or non-central gamma cumulative distributions. The parameter μ is related to the degrees of freedom and x to the non-centrality.

Because of the complementary relation for the incomplete gamma functions $P(a,x) + Q(a,x) = 1$, we also have a similar relation for the Marcum functions:

$$
P_\mu(x,y) + Q_\mu(x,y) = 1. \tag{39.0.2}
$$

The functions defined in (39.0.1) are also known as generalized Marcum functions. They are named after J. I. Marcum, who introduced the function with $\mu = 1$ in Marcum (1960) (in a different notation). These functions are used in radar detection and communications, where μ is the number of independent samples of the output of a square-law detector. In our analysis μ is not necessarily a positive integer number. For more references to the literature and numerical algorithms we refer to Gil *et al.* (2013).

We derive asymptotic expansions of these functions for large xy and bounded values of μ. These expansions are in particular valid near the transition line $y = x$ in the first quadrant of the (x, y)-plane. When μ is large as well the transition line becomes $y = x + \mu$, and again we can give uniform expansions that are valid near this line, and for other values of x and y. One expansion is in terms of the complementary error function and the other one in terms of the incomplete gamma function. A comparison between these expansions shows that (for the chosen values of x, y, μ) the latter expansion gives a better approximation.

[1]This chapter is based on Temme (1993a) and Gil *et al.* (2013).

39.1 Properties of the Marcum functions

Integral representations in terms of modified Bessel functions follow from replacing the incomplete gamma functions in (39.0.1) by their integral representations, and by using

$$I_\mu(z) = \left(\tfrac{1}{2}z\right)^\mu \sum_{n=0}^{\infty} \frac{(\frac{1}{2}z)^{2n}}{\Gamma(\mu+n+1)\,n!}, \tag{39.1.3}$$

which gives

$$\begin{aligned} P_\mu(x,y) &= e^{-x}\int_0^y \left(\frac{t}{x}\right)^{\frac{1}{2}(\mu-1)} e^{-t} I_{\mu-1}\left(2\sqrt{xt}\right)\,dt, \\ Q_\mu(x,y) &= e^{-x}\int_y^\infty \left(\frac{t}{x}\right)^{\frac{1}{2}(\mu-1)} e^{-t} I_{\mu-1}\left(2\sqrt{xt}\right)\,dt. \end{aligned} \tag{39.1.4}$$

For $x = 0$ the functions $P_\mu(x,y)$ and $Q_\mu(x,y)$ become the standard incomplete gamma functions, which are the standard chi-square distributions.

When we apply the recurrence relations of the incomplete gamma functions

$$\begin{aligned} P(a+1,x) &= P(a,x) - \frac{x^a e^{-x}}{\Gamma(a+1)}, \\ Q(a+1,x) &= Q(a,x) + \frac{x^a e^{-x}}{\Gamma(a+1)} \end{aligned} \tag{39.1.5}$$

to (39.0.1), we obtain the recursions

$$\begin{aligned} P_{\mu+1}(x,y) &= P_\mu(x,y) - \left(\frac{y}{x}\right)^{\frac{1}{2}\mu} e^{-x-y} I_\mu(2\sqrt{xy}), \\ Q_{\mu+1}(x,y) &= Q_\mu(x,y) + \left(\frac{y}{x}\right)^{\frac{1}{2}\mu} e^{-x-y} I_\mu(2\sqrt{xy}). \end{aligned} \tag{39.1.6}$$

Using

$$I_{\mu-1}(z) = I_{\mu+1}(z) + (2\mu/z)\,I_\mu(z), \tag{39.1.7}$$

we can eliminate the Bessel function in (39.1.6). This gives the homogeneous third-order recurrence relation:

$$xQ_{\mu+2}(x,y) = (x-\mu)Q_{\mu+1}(x,y) + (y+\mu)Q_\mu(x,y) - yQ_{\mu-1}(x,y). \tag{39.1.8}$$

Because a constant satisfies this equation, it also holds for $P_\mu(x,y)$.

When x and y are large, and $|x-y|$ is small compared to x and y, the second integral in (39.1.4) has a peculiar behavior. To see this, consider the well-known estimate

$$I_\nu(z) \sim \frac{e^z}{\sqrt{2\pi z}}, \quad z\to\infty, \quad \nu \text{ fixed}. \tag{39.1.9}$$

We see that the quantity $e^{-t-x}I_{\mu-1}(2\sqrt{xt})$ of (39.1.4) is exponentially small, except when $x \ge y$ and $t \sim x$. It follows that, when x and y are large (and μ small with respect to x and y), the behavior of $Q_\mu(x,y)$ significantly changes when y crosses the value x. It will appear that when μ is large as well, this change in behavior occurs when y crosses the value $x+\mu$.

In fact we have

$$Q_\mu(x,y) \sim \begin{cases} 1, & \text{if } x+\mu > y, \\ \frac{1}{2}, & \text{if } x+\mu = y, \\ 0, & \text{if } x+\mu < y, \end{cases} \tag{39.1.10}$$

and complementary behavior for $P_\mu(x,y) = 1 - Q_\mu(x,y)$.

In both cases (that is, for fixed μ and for large μ), the asymptotic behavior can be described by using the error function, or normal distribution function. In both approximations the complementary error function is used to describe the transition from 0 to 1, as shown for $Q_\mu(x,y)$ in (39.1.10).

39.2 More integral representations

We use in the defining series in (39.0.1) the integral representations given in (37.2.10) (37.2.14), and obtain

$$\begin{aligned} P_\mu(x,y) &= \frac{e^{-x-y}}{2\pi i}\int_{c-i\infty}^{c+i\infty} e^{x/s+ys}\,\frac{ds}{(s-1)s^\mu}, \quad c>1, \\ Q_\mu(x,y) &= \frac{e^{-x-y}}{2\pi i}\int_{d-i\infty}^{d+i\infty} e^{x/s+ys}\,\frac{ds}{(1-s)s^\mu}, \quad 0<d<1. \end{aligned} \tag{39.2.11}$$

By moving the second integral over the pole at $s=1$, and taking into account the residue, we again obtain the relation in (39.0.2). The interchange of summation and integration can be verified by observing uniform convergence for fixed values of x of the series $\sum_{n=0}^{\infty}(x/(s+1))^n/n!$ on the path of integration.

First we show that the integrals in (39.1.4) essentially reduce to a sum of two functions defined by

$$F_\mu(\xi,\sigma) = \int_\xi^\infty e^{-(\sigma+1)t} I_\mu(t)\,dt, \quad \sigma>0. \tag{39.2.12}$$

In the second integral of (39.2.11) we substitute $s=t/\rho$ with $\rho=\sqrt{y/x}$ and obtain

$$Q_\mu(x,y) = e^{-x-y+2z\lambda}\rho^\mu R(z), \tag{39.2.13}$$

where λ is free to be chosen, and

$$R(z) = \frac{1}{2\pi i}\int_{c-i\infty}^{c+i\infty} \frac{e^{z(t+1/t-2\lambda)}}{\rho-t}\,\frac{dt}{t^\mu}, \quad z=\sqrt{xy}, \quad 0<c<\rho. \tag{39.2.14}$$

We now assume, for the time being, that $\rho > 1$ and that ρ does not depend on x, y, z. Taking $\lambda = \frac{1}{2}(\rho + 1/\rho)$, and using

$$t + 1/t - \rho - 1/\rho = (t - \rho)(t - 1/\rho)/t, \tag{39.2.15}$$

we obtain

$$\frac{dR(z)}{dz} = -\frac{e^{-2\lambda z}}{2\pi i}\int_{c-i\infty}^{c+i\infty} e^{z(t+1/t)}\left(t - \frac{1}{\rho}\right)\frac{dt}{t^{\mu+1}}. \tag{39.2.16}$$

Invoking the integral representation of the modified Bessel function given in (10.3.25) we derive

$$\frac{dR(z)}{dz} = -e^{-2\lambda z}\left(I_{\mu-1}(2z) - \frac{1}{\rho}I_\mu(2z)\right). \tag{39.2.17}$$

To integrate this we use $R(\infty) = 0$. This follows from a saddle point analysis applied to (39.2.14); observe that the exponential function of the integrand in (39.2.14) has a saddle point at $s = 1$. We obtain

$$Q_\mu(x, y) = \rho^\mu R(z) = \tfrac{1}{2}\rho^\mu\left(F_{\mu-1}(\xi, \sigma) - \frac{1}{\rho}F_\mu(\xi, \sigma)\right), \quad y > x; \tag{39.2.18}$$

ρ has regained its original meaning and the F-function is defined in (39.2.12). Furthermore

$$\xi = 2\sqrt{xy}, \quad \sigma = \frac{(\sqrt{y} - \sqrt{x})^2}{\xi}, \quad \rho = \sqrt{\frac{y}{x}}. \tag{39.2.19}$$

Now let $\rho < 1$. Repeating the analysis that leads to (39.2.18), but now with starting point the first integral of (39.2.11), we obtain for this case

$$Q_\mu(x, y) = 1 - \tfrac{1}{2}\rho^\mu\left(\frac{1}{\rho}F_\mu(\xi, \sigma) - F_{\mu-1}(\xi, \sigma)\right), \quad y < x, \tag{39.2.20}$$

where the parameters are as in (39.2.19). This means

$$P_\mu(x, y) = \tfrac{1}{2}\rho^\mu\left(\frac{1}{\rho}F_\mu(\xi, \sigma) - F_{\mu-1}(\xi, \sigma)\right), \quad y < x. \tag{39.2.21}$$

In §39.3 the large-ξ behavior of $Q_\mu(x, y)$ will be discussed. We have, when $x, y \to \infty$ with bounded μ the transition as in (39.1.10) with $\mu = 0$. It will be shown that a smooth transition can be described in terms of the error function (the normal distribution function). In §39.4 we allow μ to be a large parameter.

Remark 39.1. Note that the integral that defines $F_\mu(\xi, \sigma)$ becomes undetermined when $\sigma = 0$. However, since we use a combination of two F-functions in (39.2.18) and (39.2.21), and ρ tends to unity as $x \to y$, the right-hand sides of (39.2.18) and (39.2.21) are well defined when $x = y$. To verify this, we use the Laplace integral

$$F_\mu(0, \sigma) = \int_0^\infty e^{-(\sigma+1)t} I_\mu(t)\, dt = \frac{\rho^{-\mu}}{\sqrt{(\sigma+1)^2 - 1}}, \quad \sigma > 0, \tag{39.2.22}$$

which follows from the corresponding transform in Watson (1944, p. 386), and where we have used (see the variables given in (39.2.19))

$$\rho = \sigma + 1 + \sqrt{(\sigma+1)^2 - 1}. \tag{39.2.23}$$

Using this result we see that the right-hand sides of (39.2.18) and (39.2.21) remain bounded when $\rho \downarrow 1$, or when $\sigma \downarrow 0$. △

39.3 Asymptotic expansion; μ fixed, ξ large

We concentrate on the function $F_\mu(\xi,\sigma)$ given in (39.2.12). We point out that this function with ξ and σ as in (39.2.19) is symmetric in x and y, and occurs in both (39.2.18) and (39.2.21). Hence, it is sufficient to assume $x < y$. The case $x = y$ follows from the asymptotic results when we let $x \to y$.

The asymptotic feature of the integral in (39.2.12) is that ξ is large, whereas σ tends to zero when $x \to y$. In that case the integrand is not exponentially small as $t \to \infty$. We give an asymptotic expansion that holds uniformly with respect to $\sigma \in [0,\infty)$.

We substitute the well-known expansion (see (9.3.14))

$$e^{-t} I_\mu(t) \sim \frac{1}{\sqrt{2\pi t}} \sum_{n=0}^{\infty} (-1)^n \frac{a_n(\mu)}{t^n}, \quad t \to \infty, \tag{39.3.24}$$

in (39.2.12), where

$$a_n(\mu) = (-1)^n \frac{\left(\frac{1}{2}-\mu\right)_n \left(\frac{1}{2}+\mu\right)_n}{2^n\, n!}, \quad n = 0,1,2,\ldots, \tag{39.3.25}$$

with recursion

$$a_{n+1}(\mu) = -\frac{(2n+1)^2 - 4\mu^2}{8(n+1)} a_n(\mu), \quad n \ge 0, \quad a_0(\mu) = 1. \tag{39.3.26}$$

This gives the expansion

$$F_\mu(\xi,\sigma) \sim \frac{1}{\sqrt{2\pi}} \sum_{n=0}^{\infty} (-1)^n a_n(\mu) \Phi_n, \tag{39.3.27}$$

where Φ_n is an incomplete gamma function (see (37.0.1))

$$\Phi_n = \int_\xi^\infty e^{-\sigma t} t^{-n-\frac{1}{2}}\, dt = \sigma^{n-\frac{1}{2}} \Gamma\left(\tfrac{1}{2} - n, \sigma\xi\right). \tag{39.3.28}$$

The function Φ_0 can be written in terms of the complementary error function (see (3.4.27))

$$\Phi_0 = \sqrt{\pi/\sigma}\, \mathrm{erfc}\, \sqrt{\sigma\xi} = \sqrt{\pi/\sigma}\, \mathrm{erfc}\left(\sqrt{y} - \sqrt{x}\right). \tag{39.3.29}$$

Further terms can be obtained from the recursion

$$\left(n - \tfrac{1}{2}\right) \Phi_n = -\sigma \Phi_{n-1} + e^{-\sigma\xi} \xi^{-n+\frac{1}{2}}, \quad n = 1,2,3,\ldots. \tag{39.3.30}$$

Using (39.2.18) and (39.3.27) we obtain

$$Q_\mu(x,y) \sim \sum_{n=0}^{\infty} \Psi_n, \tag{39.3.31}$$

where

$$\Psi_n = (-1)^n \frac{\rho^\mu}{2\sqrt{2\pi}} \left(a_n(\mu-1) - \frac{1}{\rho} a_n(\mu)\right) \Phi_n. \tag{39.3.32}$$

Information on the asymptotic nature and error bounds of expansion (39.3.27) can be found in Temme (1986a), in which paper the numerical aspects of the recursion (39.3.30) are discussed as well. Expansion (39.3.31) holds for large values of ξ, uniformly with respect to $\sigma \in [0, \infty)$.

The first-term approximation of the series in (39.3.31) reads, because $(\rho - 1)/\sqrt{2\sigma} = \sqrt{\rho}$,

$$Q_\mu(x, y) \sim \Psi_0 = \tfrac{1}{2}\rho^{\mu-\frac{1}{2}} \operatorname{erfc}\left(\sqrt{y} - \sqrt{x}\right). \tag{39.3.33}$$

We remark that the right-hand side reduces to $\frac{1}{2}$ when $x \uparrow y$.

When $y \geq x$ we have for $P_\mu(x, y)$ the expansion

$$P_\mu(x, y) \sim 1 - \sum_{n=0}^{\infty} \Psi_n, \tag{39.3.34}$$

with the same functions Ψ_n as in (39.3.31).

When $x \geq y$ it is better to use the representation in (39.2.21). This gives for $P_\mu(x, y)$ the expansion

$$P_\mu(x, y) \sim \sum_{n=0}^{\infty} \widetilde{\Psi}_n, \tag{39.3.35}$$

where $\widetilde{\Psi}_n = -\Psi_n$, $n \geq 1$, and

$$\widetilde{\Psi}_0 = \tfrac{1}{2}\rho^{\mu-\frac{1}{2}} \operatorname{erfc}\left(\sqrt{x} - \sqrt{y}\right). \tag{39.3.36}$$

Remark 39.2. When deriving the result for $\widetilde{\Psi}_0$, we have needed $\sqrt{\sigma}$, and when $x > y$ we have to interpret this quantity as (see (39.2.19))

$$\sqrt{\sigma} = \frac{\sqrt{x} - \sqrt{y}}{\sqrt{\xi}}. \tag{39.3.37}$$

In this way, see (39.3.32),

$$\widetilde{\Psi}_0 = \frac{\rho^{\mu-1}(1-\rho)}{2\sqrt{2\pi}} \Phi_0, \tag{39.3.38}$$

where $\rho = \sqrt{y/x}$ and Φ_0 is in this case (see (39.3.29))

$$\Phi_0 = \frac{\sqrt{\pi\xi}}{\sqrt{x} - \sqrt{y}} \operatorname{erfc}\left(\sqrt{x} - \sqrt{y}\right), \tag{39.3.39}$$

which gives the correct $\widetilde{\Psi}_0$ of (39.3.36). For other $\widetilde{\Psi}_n$, $n \geq 1$, we have to do the same, but we can use (39.3.32) and the recursion for Φ_n given in (39.3.30), now with starting value Φ_0 of (39.3.39). In the recursion fractional powers of σ do not occur. △

Remark 39.3. We could have used the expansion of the modified Bessel function given in (39.3.24) in the integral representation of $Q_\mu(x, y)$ given in (39.1.4). However, this would require large values of y, whereas in (39.3.27) we require large values of $\xi = 2\sqrt{xy}$. Furthermore, and more importantly, the resulting expansion contains integrals of the form $\int_y^\infty t^\alpha e^{-t+2\sqrt{2xt}}\,dt$, for some α depending on n and μ. These functions cannot be written in terms of the standard special functions. △

39.4 Asymptotic expansion; $\xi + \mu$ large

When one of the parameters ξ and μ is large (or both are large) we consider the representation of $Q_\mu(x,y)$ in (39.2.11). We have, after minor modifications,

$$Q_\mu(\mu x, \mu y) = \frac{\rho^\mu e^{-\mu(x+y)}}{2\pi i}\int_{c-i\infty}^{c+i\infty} e^{\mu\psi(t)}\frac{dt}{\rho - t}, \quad 0 < c < \rho, \tag{39.4.40}$$

again with $\rho = \sqrt{y/x}$. The function $\psi(t)$ is defined by

$$\psi(t) = \tfrac{1}{2}\xi\,(t + 1/t) - \ln t, \quad \xi = 2\sqrt{xy}. \tag{39.4.41}$$

The positive saddle point of $\psi(t)$ is given by

$$t_0 = \frac{1+\sqrt{1+\xi^2}}{\xi}, \tag{39.4.42}$$

and we can take in (39.4.40) $c = t_0$ when $0 < t_0 < \rho$. This condition corresponds to $y > x+1$. The saddle point t_0 coalesces with the pole at ρ when $y = x+1$. For the time being, we assume that $0 < t_0 < \rho$.

The path of steepest descent $\mathcal{L}$ through t_0 follows from the equation $\Im\psi(t) = 0$. Let $t = re^{i\theta}$, then we can describe $\mathcal{L}$ by

$$r = \frac{\theta}{\xi\sin\theta} + \sqrt{1 + \frac{\theta^2}{\xi^2\sin^2\theta}}, \quad -\pi < \theta < \pi. \tag{39.4.43}$$

The transformation

$$\tfrac{1}{2}s^2 = \psi(t) - \psi(t_0) \tag{39.4.44}$$

maps $\mathcal{L}$ to the imaginary axis in the s-plane. We have to define the sign when taking the square root in this relation and we assume that $\operatorname{sign}(s) = \operatorname{sign}(t - t_0)$ when $t > 0$. In this way, the lower part of $\mathcal{L}$ ($\Im t < 0$) is mapped to the negative imaginary axis. We have

$$t = t_0 + c_0\, s\bigl(1 + \mathcal{O}(s)\bigr), \quad s \to 0, \quad c_0 = t_0\sqrt{\frac{t_0^2 - 1}{t_0^2 + 1}}. \tag{39.4.45}$$

We define ζ by writing

$$\tfrac{1}{2}\zeta^2 = \psi(\rho) - \psi(t_0), \tag{39.4.46}$$

that is,

$$\tfrac{1}{2}\zeta^2 = x + y - \sqrt{1+4xy} + \ln\frac{1+\sqrt{1+4xy}}{2y}, \tag{39.4.47}$$

and the sign for ζ has to be determined. We assume that the pole in the t-plane at $t = \rho$ corresponds to $s = -\zeta$ and that $\operatorname{sign}(\zeta) = \operatorname{sign}(t_0 - \rho) = \operatorname{sign}(x + 1 - y)$. In other words, when we travel along $\mathcal{L}$ we see the pole at $t = \rho$ on the right if $t_0 < \rho$; the same should occur in the s-plane. An expansion of ζ in terms of small $\rho - t_0$ follows from (39.4.45) when we substitute $s = -\zeta$ and $t = \rho$.

The transformation (39.4.44) gives

$$Q_\mu(\mu x, \mu y) = \frac{e^{-\frac{1}{2}\mu\zeta^2}}{2\pi i} \int_{-i\infty}^{i\infty} e^{\frac{1}{2}\mu s^2} \frac{dt}{ds} \frac{ds}{\rho - t}, \tag{39.4.48}$$

where we have used $\psi(\rho) = x + y - \ln \rho$.

Observe that, as remarked earlier, when $\rho \to t_0$ (which happens when $y \to x+1$), the pole in the t-plane approaches the saddle point t_0. The same happens in the s-plane: when $\rho \to t_0$, ζ becomes small, and the pole at $-\zeta$ is near the saddle point at the origin. This case has been discussed in Chapter 21.

We split off the pole by writing

$$\frac{dt}{ds} \frac{1}{\rho - t} = f(s) - \frac{1}{s + \zeta}, \quad f(s) = \frac{dt}{ds} \frac{1}{\rho - t} + \frac{1}{s + \zeta}, \tag{39.4.49}$$

and use (see (21.1.2))

$$-\frac{1}{2\pi i} \int_{-i\infty}^{i\infty} e^{\frac{1}{2}\mu s^2} \frac{ds}{\zeta + s} = \tfrac{1}{2} e^{\frac{1}{2}\mu\zeta^2} \operatorname{erfc}\left(-\zeta\sqrt{\mu/2}\right). \tag{39.4.50}$$

This gives the form

$$Q_\mu(\mu x, \mu y) = \tfrac{1}{2} \operatorname{erfc}\left(-\zeta\sqrt{\mu/2}\right) + R_\mu(\zeta), \tag{39.4.51}$$

where

$$R_\mu(\zeta) = \frac{e^{-\frac{1}{2}\mu\zeta^2}}{\sqrt{2\pi\mu}} S_\mu(\zeta), \quad S_\mu(\zeta) = \frac{1}{i}\sqrt{\frac{\mu}{2\pi}} \int_{-i\infty}^{i\infty} e^{\frac{1}{2}\mu s^2} f(s)\, ds. \tag{39.4.52}$$

By expanding

$$f(s) = \sum_{k=0}^{\infty} g_n s^n, \tag{39.4.53}$$

it follows that

$$S_\mu(\zeta) \sim \sum_{n=0}^{\infty} \frac{D_n(\zeta)}{\mu^n}, \quad D_n(\zeta) = (-1)^n 2^n \left(\tfrac{1}{2}\right)_n g_{2n}. \tag{39.4.54}$$

To compute the coefficients we need more terms in the expansion given in (39.4.45). That is, we write

$$t = t_0 + c_0 s \left(1 + c_1 s + c_2 s^2 + c_3 s^3 + \ldots\right), \tag{39.4.55}$$

where c_0 is defined in (39.4.45). A few other coefficients are

$$\begin{aligned} c_1 &= \frac{c_0\left(t_0^2 + 2\right)}{3t_0\left(t_0^2 + 1\right)}, \\ c_2 &= \frac{c_0^2\left(t_0^4 + 4t_0^2 + 13\right)}{36t_0^2\left(t_0^2 + 1\right)^2}, \\ c_3 &= -\frac{c_0^3\left(t_0^2 - 1\right)\left(t_0^4 + 7t_0^2 + 46\right)}{270t_0^3\left(t_0^2 + 1\right)^3}. \end{aligned} \tag{39.4.56}$$

The first $D_n(\zeta)$ are

$$\begin{aligned}
D_0(\zeta) &= \frac{c_0}{r} + \frac{1}{\zeta},\\
D_1(\zeta) &= -\frac{c_0\left(3c_2r^2 + 3c_0c_1r + c_0^2\right)}{r^3} - \frac{1}{\zeta^3},\\
D_2(\zeta) &= \frac{3c_0}{r^5}\Big(5c_4r^4 + 5c_0\left(c_3 + c_1c_2\right)r^3 +\\
&\qquad 5c_0^2\left(c_1^2 + c_2\right)r^2 + 5c_0^3c_1r + c_0^4\Big) + \frac{3}{\zeta^5},
\end{aligned} \tag{39.4.57}$$

where $r = \rho - t_0$.

For small values of ζ we can use expansions. We need the expansion

$$\rho = t_0 - c_0\zeta\left(1 - c_1\zeta + c_2\zeta^2 - c_3\zeta^3 + \ldots\right), \tag{39.4.58}$$

and obtain

$$\begin{aligned}
D_0(\zeta) &= -c_1 + \left(c_2 - c_1^2\right)\zeta + \left(-c_3 + 2c_1c_2 - c_1^3\right)\zeta^2 + \mathcal{O}\left(\zeta^3\right),\\
D_1(\zeta) &= -3c_1c_2 + 3c_3 + c_1^3 + \left(3c_2^2 - 9c_2c_1^2 + 3c_1^4 + 6c_1c_3 - 3c_4\right)\zeta + \mathcal{O}\left(\zeta^2\right),\\
D_2(\zeta) &= 15c_1c_4 + 15c_3c_2 - 15c_3c_1^2 - 15c_1c_2^2 + 15c_2c_1^3 - 15c_5 - 3c_1^5 + \mathcal{O}\left(\zeta\right).
\end{aligned} \tag{39.4.59}$$

We can derive an expansion in terms of the original variables. When ζ is small, $y - x - 1$ is small, and we have

$$D_0(\zeta) = -\frac{3x+1}{3\left(2x+1\right)^{3/2}} + \frac{18x^2+6x+1}{12\left(2x+1\right)^{7/2}}(y-x-1) + \mathcal{O}\left((y-x-1)^2\right). \tag{39.4.60}$$

The domain of validity of these expansions follows from the singularities in the s-plane of the function f defined in (39.4.49). The pole due to the second term on the right-hand side is compensated by the pole of the first term at $t = \rho$. The function f has a singularity for $t = (1 - \sqrt{1+\xi^2})/\xi$, the negative saddle point. The corresponding value in the s-plane is never close to the origin for the considered values of $\xi \geq 0$.

We conclude that the expansion in (39.4.54) is valid for large μ, uniformly with respect to $\xi \in [0, \infty)$. Using a slightly different approach, interchanging the role of ξ and μ, we obtain an expansion of $Q_\mu(x, y)$ (with unscaled x and y) by rearranging the expansion in (39.4.54) term by term. In that way an expansion follows that is valid for large ξ, uniformly with respect to $\mu \in [0, \infty)$. Hence, the method of this section can also be used when μ is fixed (including 0). However, the method of §39.3 gives very simple coefficients in expansion (39.3.31).

Remark 39.4. To derive the expansion in (39.4.54) we have temporarily assumed $t_0 < \rho$, that is $y > x + 1$, but this condition can be dropped. The expansion in (39.4.54) also holds for $y \leq x + 1$. Note that a single error function describes the transition from $y > x + 1$ to $y < x + 1$. Hence, we do not need different representations for $Q_\mu(x, y)$ as in the previous section; cf. (39.2.18) and (39.2.20).

However, for $y \le x+1$, it is better to concentrate on the complementary function $P_\mu(x,y)$. We obtain from (39.4.51)

$$P_\mu(\mu x, \mu y) = \tfrac{1}{2}\text{erfc}\left(\zeta\sqrt{\mu/2}\right) - R_\mu(\zeta), \tag{39.4.61}$$

where all quantities and coefficients in the asymptotic expansion are the same as for the Q-function. △

Remark 39.5. In Gil *et al.* (2013) we have derived a different expansion valid for large μ, starting with the integral representations given in (39.1.4) and using the Debye-type expansion of the Bessel function given in (9.5.57). The obtained expansions of $Q_\mu(x,y)$ and $P_\mu(x,y)$ are more suitable for numerical computations, although the domain of validity is more restricted than that of the expansions that follow from representations (39.4.51) and (39.4.61). △

39.5 An expansion in terms of the incomplete gamma function

The Marcum functions depend on the three variables μ, x, y and in the representation in terms of the complementary error function, the main approximant $\text{erfc}\, x$ is a function of one variable. In this section we will derive a representation in terms of the incomplete gamma function, a function of two variables.

We start with (39.4.40) and we replace the transformation in (39.4.44) by

$$\psi(t) - \psi(t_0) = \chi(s), \quad \chi(s) = s - \ln s - 1, \tag{39.5.62}$$

where t_0 is defined in (39.4.42). This point should correspond to $s = 1$, the point where $\chi'(s) = (s-1)/s$ vanishes. It follows that

$$Q_\mu(\mu x, \mu y) = \frac{e^{-\mu(\psi(\rho)-\psi(t_0))}}{2\pi i}\int_{c-i\infty}^{c+i\infty} e^{\mu\chi(s)} f(s)\,\frac{ds}{\alpha - s}, \quad 0 < c < \alpha, \tag{39.5.63}$$

where the pole at $s = \alpha$ corresponds to $t = \rho$. That is, α is defined by the equation

$$\psi(\rho) - \psi(t_0) = \chi(\alpha). \tag{39.5.64}$$

Also,

$$f(s) = \frac{\alpha - s}{\rho - t}\,\frac{dt}{ds} = \frac{\alpha - s}{\rho - t}\,\frac{\chi'(s)}{\psi'(t)}. \tag{39.5.65}$$

By applying l'Hôpital's rule it follows that $f(\alpha) = 1$ and by using (39.5.64) that

$$Q_\mu(\mu x, \mu y) = \frac{e^{-\mu\chi(\alpha)}}{2\pi i}\int_{c-i\infty}^{c+i\infty} e^{\mu\chi(s)} f(s)\,\frac{ds}{\alpha - s}, \quad 0 < c < \alpha. \tag{39.5.66}$$

For the transformation in (39.5.62) we assume that the integral in (39.4.40) is taken along the steepest descent path $\mathcal{L}$; see (39.4.43). Then $\mathcal{L}$ is mapped to the steepest descent path $\mathcal{P}$ in the s-plane described by

$$r = \frac{\theta}{\sin\theta}, \quad -\pi < \theta < \pi, \quad s = re^{i\theta}, \tag{39.5.67}$$

and we assume that the lower (upper) part of $\mathcal{L}$ is mapped to the lower (upper) part of $\mathcal{P}$. For positive values of t and s this condition becomes $\operatorname{sign}(t-t_0) = \operatorname{sign}(s-1)$, and to determine α from (39.5.64) we need the condition $\operatorname{sign}(\rho-t_0) = \operatorname{sign}(\alpha-1)$. We have $\alpha = 1$ when $\rho = t_0$, that is, when $y = x+1$, and the condition for α can also be written in the form $\operatorname{sign}(\alpha-1) = \operatorname{sign}(y-x-1)$.

We write

$$\frac{f(s)}{\alpha - s} = \frac{1}{\alpha - s} + h(s) \quad \Longrightarrow \quad h(s) = \frac{1}{\rho - t}\frac{dt}{ds} - \frac{1}{\alpha - s}, \tag{39.5.68}$$

and obtain, using the contour integral for the incomplete gamma function derived in §37.2,

$$Q_\mu(\mu x, \mu y) = Q(\mu, \mu\alpha) + \frac{e^{-\mu\chi(\alpha)}}{2\pi i}\int_{c-i\infty}^{c+i\infty} e^{\mu\chi(s)} h(s)\, ds, \quad 0 < c < \alpha. \tag{39.5.69}$$

We expand at the saddle point $h(s) = \sum_{n=0}^{\infty} h_n(\alpha)(s-1)^n$ and obtain

$$Q_\mu(\mu x, \mu y) \sim Q(\mu, \mu\alpha) + e^{-\mu\chi(\alpha)} \sum_{n=0}^{\infty} h_n(\alpha)\Psi_n, \tag{39.5.70}$$

where

$$\Psi_n = \frac{1}{2\pi i}\int_{c-i\infty}^{c+i\infty} e^{\mu\chi(s)}(s-1)^n\, ds, \quad c > 0. \tag{39.5.71}$$

We have the recursion

$$\Psi_{n+1} = -\frac{(n+1)\Psi_n + n\Psi_{n-1}}{\mu}, \quad n \ge 0, \tag{39.5.72}$$

and the first Ψ_n are (see (6.5.63) for the notation of $\Gamma^*(z)$)

$$\begin{aligned}
\Psi_0 &= \frac{e^{-\mu}\mu^{\mu-1}}{\Gamma(\mu)} = \frac{1}{\sqrt{2\pi\mu}\,\Gamma^*(\mu)}, & \Psi_1 &= -\frac{1}{\mu}\Psi_0,\\
\Psi_2 &= -\frac{1}{\mu^2}(\mu-2)\Psi_0, & \Psi_3 &= \frac{1}{\mu^3}(5\mu-6)\Psi_0,\\
\Psi_4 &= \frac{1}{\mu^4}\left(3\mu^2 - 26\mu + 24\right)\Psi_0, & \Psi_5 &= -\frac{1}{\mu^5}\left(35\mu^2 - 154\mu + 120\right)\Psi_0,
\end{aligned} \tag{39.5.73}$$

from which we can conclude about the asymptotic scale of the terms in the expansion.

From (39.4.41) and (39.5.62) we obtain the expansion

$$t = t_0 + c_0(s-1)\left(1 + d_1(s-1) + d_2(s-1)^2 + d_3(s-1)^3 + \ldots\right), \tag{39.5.74}$$

where c_0 is given in (39.4.45) and

$$\begin{aligned}
d_1 &= \frac{c_0\left(t_0^2+2\right) - t_0\left(t_0^2+1\right)}{3\left(t_0^2+1\right)},\\
d_2 &= -\frac{4c_0\left(t_0^2+2\right)\left(t_0^2+1\right)^2 - t_0\left(4t_0^6 + 12t_0^4 + 15t_0^2 - 3\right)}{18t_0\left(t_0^2+1\right)^3},\\
d_3 &= \frac{1}{270t_0\left(t_0^2+1\right)^4}\Bigl(4c_0\left(11t_0^8 + 55t_0^6 + 93t_0^4 + 100t_0^2 + 11\right) -\\
&\quad t_0\left(t_0^2+1\right)\left(44t_0^6 + 132t_0^4 + 177t_0^2 - 61\right)\Bigr).
\end{aligned} \tag{39.5.75}$$

The first coefficients h_n follow from these values and from (39.5.68), and are given by (we write $r = \rho - t_0$)

$$\begin{aligned}
h_0(\alpha) &= \frac{c_0}{r} - \frac{1}{\alpha - 1}, \\
h_1(\alpha) &= \frac{c_0\left(2d_1 r + c_0\right)}{r^2} - \frac{1}{(\alpha-1)^2}, \\
h_2(\alpha) &= \frac{c_0\left(3d_2 r^2 + 3c_0 d_1 r + c_0^2\right)}{r^3} - \frac{1}{(\alpha-1)^3}, \\
h_3(\alpha) &= \frac{c_0\left(4d_3 r^3 + 2c_0\left(2d_2 + d_1^2\right) r^2 + 4c_0^2 d_1 r + c_0^3\right)}{r^4} - \frac{1}{(\alpha-1)^4}, \\
h_4(\alpha) &= \frac{c_0}{r^5}\Big(5d_4 r^4 + 5c_0\left(d_3 + d_1 d_2\right) r^3 + \\
&\quad 5c_0^2\left(d_1^2 + d_2\right) r^2 + 5c_0^3 d_1 r + c_0^4\Big) - \frac{1}{(\alpha-1)^5}.
\end{aligned} \tag{39.5.76}$$

When $\alpha \to 1$ (that is, when $y \to x+1$), we need the expansion

$$\rho = t_0 + c_0\,(\alpha - 1)\left(1 + d_1(\alpha-1) + d_2(\alpha-1)^2 + d_3(\alpha-1)^3 + \ldots\right), \tag{39.5.77}$$

from which we can obtain expansions of the first coefficients $h_n(\alpha)$.

By rearranging the expansion in (39.5.70), we obtain

$$Q_\mu(\mu x, \mu y) \sim Q(\mu, \mu\alpha) + e^{-\mu\chi(\alpha)}\Psi_0 \sum_{n=0}^{\infty} \frac{g_n(\alpha)}{\mu^n}, \tag{39.5.78}$$

where the first coefficients are given by

$$\begin{aligned}
g_0(\alpha) &= h_0(\alpha), \\
g_1(\alpha) &= -h_1(\alpha) - h_2(\alpha), \\
g_2(\alpha) &= 2h_2(\alpha) + 5h_3(\alpha) + 3h_4(\alpha), \\
g_3(\alpha) &= -6h_3(\alpha) - 26h_4(\alpha) - 35h_5(\alpha) - 15h_6(\alpha), \\
g_4(\alpha) &= 24h_4(\alpha) + 154h_5(\alpha) + 340h_6(\alpha) + 315h_7(\alpha) + 105h_8(\alpha).
\end{aligned} \tag{39.5.79}$$

Remark 39.6. For the complementary function $P_\mu(x, y)$ we obtain from (39.5.70)

$$P_\mu(\mu x, \mu y) \sim P(\mu, \mu\alpha) - e^{-\mu\chi(\alpha)}\Psi_0 \sum_{n=0}^{\infty} g_n(\alpha), \tag{39.5.80}$$

where $g_n(\alpha)$ and Ψ_0 are the same as for the Q-function. △

Remark 39.7. We can write the relation for α given in (39.5.64) in the form

$$\chi(\alpha) = \alpha - \ln\alpha - 1 = y - \ln y - 1 + x - \frac{4xy}{1 + \sqrt{1 + 4xy}}, \tag{39.5.81}$$

Table 39.1 Comparing the expansions in terms of the complementary error function and of the incomplete gamma function.

μ	x	y	δ_{erfc}	δ_{incgam}	$Q_\mu(x,y)$
50	25	80	1.467e-09	1.656e-11	0.29762928401887
60	30	95	7.667e-10	6.744e-12	0.31331157669610
80	40	125	2.748e-10	9.228e-12	0.33614831350596
110	55	170	8.813e-11	4.723e-12	0.35870243651449
150	75	230	2.911e-11	1.970e-12	0.37802443729079
200	100	305	1.043e-11	8.060e-13	0.39377535648372
260	130	395	4.091e-12	3.434e-13	0.40647192912157
330	165	500	1.750e-12	1.551e-13	0.41675247541836
410	205	620	8.086e-13	7.437e-14	0.42516429535083
500	250	755	3.994e-13	3.774e-14	0.43213196668205
600	300	905	2.091e-13	2.015e-14	0.43797472157391

from which we conclude that $\alpha \to y$ as $x \to 0$. Observe that in the limit $x \to 0$ the function $Q_\mu(\mu x, \mu y)$ tends to the incomplete gamma function $Q(\mu, \mu y)$ (see the definition in (39.0.1)). This is consistent with the asymptotic relation in (39.5.70) if the function $h(s)$ defined in (39.5.68) vanishes as $x \to 0$. For fixed t and s, the transformation in (39.5.62) becomes $t \sim t_0 s$ as $x \to 0$, from which indeed it follows that $h(s) \to 0$. $\triangle$

39.6 Comparison of the expansions numerically

We compare the expansion including the complementary error function (see (39.4.51) and (39.4.54)) with the expansion in terms of the incomplete gamma function (see (39.5.78)) for values of μ, x and y near the transition line $y = x + \mu$ in the first quadrant of the (x, y) plane. In Table 39.1 we give the computed relative errors δ_{erfc} and δ_{incgam}. We have used the expansions in (39.4.54) and (39.5.78) with three terms. From the table we conclude that the approximations with the incomplete gamma function are better than those obtained from the expansion including the error function. For x and y we use $x = \mu/2$ and $y = x + \mu + 5$.

Chapter 40

A weighted sum of exponentials

We consider the sum

$$F_n(x) = \frac{(-1)^n}{n!} \sum_{j=1}^{n} \binom{n}{j} (-1)^j j^n e^{-x/j}, \tag{40.0.1}$$

for $x \geq 0$ and $n = 1, 2, \ldots$, and we derive an asymptotic expansion of $F_n(x)$ for large n.[1] This sum is the distribution of the sum of independent exponential random variables with different means, that is, it provides the distribution of

$$Y_1 + 2Y_2 + 3Y_3 + \cdots + nY_n, \tag{40.0.2}$$

where the Y's are independent standard exponential. Evaluation of the sum in (40.0.1) straightforwardly may cause severe cancellation of leading digits. For example, the sum for $n = 100$ and $x = 5000$ is

$$F_n(x) = 0.517213188550097 23\ldots, \tag{40.0.3}$$

but the terms in the sum range in magnitude from 10^{-2327} to 10^{27}, and the alternating signs should provide the proper answer.

The Central Limit Theorem applies, and gives

$$F_n(x) \sim 1 - \Phi\left(\frac{x-\mu}{\sigma}\right), \tag{40.0.4}$$

where $\Phi(x)$ is the normal distribution function

$$\Phi(x) = \frac{1}{\sqrt{2\pi}} \int_{-\infty}^{x} e^{-t^2/2}\, dt = \tfrac{1}{2}\mathrm{erfc}\left(-x/\sqrt{2}\right), \tag{40.0.5}$$

and

$$\mu = \tfrac{1}{2}n(n+1), \quad \sigma = \sqrt{\tfrac{1}{6}n(n+1)(2n+1)}. \tag{40.0.6}$$

This approximation is in particular useful for large values of n, with x close to μ. We derive an expansion that also contains the normal distribution function, with a

[1] This chapter is based on Van Leeuwaarden and Temme (2011), where a more general problem is considered.

different argument, and which holds uniformly for $x \ge 0$. We give the first terms in the expansion.

The limiting values at $x = 0$ and $x = \infty$ are given by

$$F_n(0) = 1, \quad F_n(\infty) = 0. \tag{40.0.7}$$

The value $F_n(\infty)$ follows easily from (40.0.1) and $F_n(0)$ follows from a relation to the Stirling numbers of the second kind. We have (see (34.1.7) and (34.1.8))

$$F_n(0) = \frac{(-1)^n}{n!} \sum_{j=1}^{n} \binom{n}{j} (-1)^j j^n = \mathcal{S}_n^n = 1. \tag{40.0.8}$$

40.1 An integral representation

We start by writing

$$F_n(x) = 1 - G_n(x), \tag{40.1.9}$$

where $G_n(x)$ has the integral representation

$$G_n(x) = \frac{1}{2\pi i} \int_{\mathcal{L}} \frac{e^{xs}}{s(s+1)(2s+1)\cdots(ns+1)}\, ds, \tag{40.1.10}$$

and $\mathcal{L}$ is a vertical line with $\Re s > 0$.

We may also write the integrand in the form

$$\frac{e^{xs}}{s(s+1)(2s+1)\cdots(ns+1)} = \frac{e^{sx}}{s^{n+1}} \frac{\Gamma(1+1/s)}{\Gamma(n+1+1/s)} = \frac{e^{sx}}{s^{n+1}} \frac{1}{(1+1/s)_n}. \tag{40.1.11}$$

The integrand has $n+1$ poles. By shifting the contour $\mathcal{L}$ in (40.1.10) to the left, across the $n+1$ poles, we obtain a residue series. The residue at $s = 0$ equals 1, and at $s = -1/j$ it equals

$$e^{-x/j}\, j^n \frac{(-1)^{n+j-1}}{j!\,(n-j)!}, \tag{40.1.12}$$

and using (40.1.9) we obtain (40.0.1). The remaining path integral vanishes when we take $\Re s \to -\infty$.

To obtain an asymptotic representation we write

$$G_n(x) = \frac{1}{2\pi i} \int_{\mathcal{L}} e^{\phi(s)} \frac{ds}{s}, \tag{40.1.13}$$

where

$$\phi(s) = xs - \ln\big((s+1)(2s+1)\cdots(ns+1)\big). \tag{40.1.14}$$

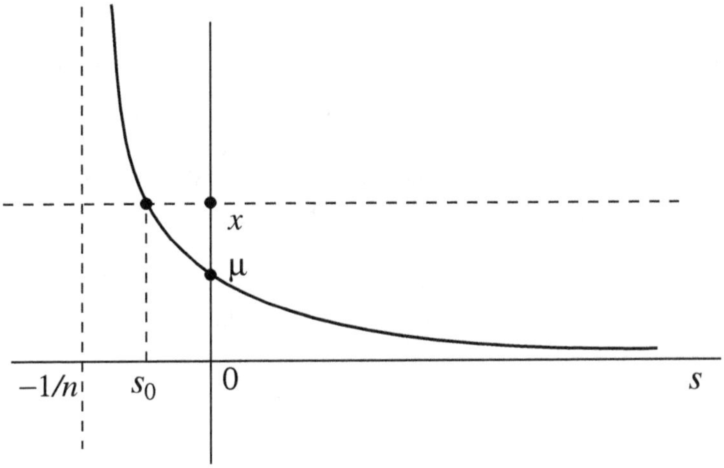

Fig. 40.1 The saddle point s_0 is the solution of the equation in (40.2.15). The curve represents the right-hand side of (40.2.15).

40.2 Saddle point analysis

The saddle point, defined by the s-value satisfying $\phi'(s) = 0$, is governed by the equation

$$x = \frac{1}{s+1} + \frac{2}{2s+1} + \ldots + \frac{n}{ns+1}. \tag{40.2.15}$$

It is not difficult to verify that, if $x > 0$, this equation has a unique solution s_0 in the interval $(-1/n, \infty)$; see Figure 40.1. The curve represents the right-hand side of (40.2.15), and it cuts the vertical axis at a point indicated by $\mu = 1 + 2 + \ldots + n$ defined earlier in (40.0.6).

If $x > \mu$ the saddle point is negative, as shown in the figure. If $x = \mu$ the pole at the origin in (40.1.13) coincides with the saddle point. As explained in Chapter 21 we can use an error function to handle this case. Observe that the argument of $\Phi(x)$ in (40.0.4) changes sign when x crosses the value μ.

We continue the analysis by assuming that $0 < x < \mu$, which means that we have a positive saddle point s_0. We substitute

$$\phi(s) - \phi(s_0) = \tfrac{1}{2}w^2. \tag{40.2.16}$$

We prescribe that $\operatorname{sign}(w) = \operatorname{sign}(s - s_0)$ for real values of s, $(s > -1/n)$.

We obtain from (40.1.13)

$$G_n(x) = \frac{e^{\phi(s_0)}}{2\pi i} \int_{-i\infty}^{i\infty} e^{\frac{1}{2}w^2} f(w) \frac{dw}{w - \xi}, \tag{40.2.17}$$

where

$$f(w) = \frac{w - \xi}{s} \frac{ds}{dw}, \tag{40.2.18}$$

and ξ is the w-value that corresponds to $s = 0$. That is,

$$\tfrac{1}{2}\xi^2 = \phi(0) - \phi(s_0) = -\phi(s_0), \quad \xi = \sqrt{-2\phi(s_0)}, \quad \operatorname{sign}(\xi) = \operatorname{sign}(x - \mu). \quad (40.2.19)$$

It is not difficult to verify that $\phi(s_0)$ is nonpositive. The sign of ξ is prescribed by the transformation in (40.2.16): when $x < \mu$ we have $s_0 > 0$, and in (40.1.13) the saddle point is on the right of the pole at the origin. We obtain the same situation in (40.2.17): the saddle point at $w = 0$ is on the right of the pole at $w = \xi$, if $s_0 > 0$, that is, $\xi < 0$.

Using Van der Waerden's method described in §21.1, we split off the pole at $w = \xi$ by writing $f(w) = (f(w) - f(\xi)) + f(\xi)$ where $f(\xi) = 1$ (as easily follows from (40.2.18)). We define

$$g(w) = \frac{f(w) - 1}{w - \xi}, \quad (40.2.20)$$

and expand

$$g(w) = \sum_{k=0}^{\infty} c_k w^k. \quad (40.2.21)$$

This gives the asymptotic expansion

$$G_n(x) \sim \tfrac{1}{2}\operatorname{erfc}\left(-\xi/\sqrt{2}\right) + \frac{e^{\phi(s_0)}}{\sqrt{2\pi}} \sum_{k=0}^{\infty} (-1)^k \, 2^k \left(\frac{1}{2}\right)_k c_{2k}. \quad (40.2.22)$$

For the function $F_n(x)$ we obtain, using (40.1.9) and $1 - \tfrac{1}{2}\operatorname{erfc} z = \tfrac{1}{2}\operatorname{erfc}(-z)$,

$$F_n(x) \sim \tfrac{1}{2}\operatorname{erfc}\left(\xi/\sqrt{2}\right) - \frac{e^{\phi(s_0)}}{\sqrt{2\pi}} \sum_{k=0}^{\infty} (-1)^k \, 2^k \left(\tfrac{1}{2}\right)_k c_{2k}. \quad (40.2.23)$$

There is no large parameter showing in the asymptotic series, however, the coefficients c_{2k} contain the asymptotic information. The expansion is valid for large values of n, uniformly with respect to $x \ge \delta$, where δ is a fixed positive number.

40.3 Details on the coefficients

We give a few steps in the calculation of the first coefficients. We define

$$A_k^n = \sum_{j=1}^{n} j^k, \quad n \ge 1, \quad k \ge 0. \quad (40.3.24)$$

We can write these quantities in terms of Bernoulli polynomials (see Temme (1996a, p. 23)) and the first few are

$$A_0^n = n, \quad A_1^n = \tfrac{1}{2}n(n+1), \quad A_2^n = \tfrac{1}{6}n(n+1)(2n+1), \quad A_3^n = \tfrac{1}{4}n^2(n+1)^2. \quad (40.3.25)$$

Writing $s = s_0$ in (40.2.15) we can expand for $|s_0| < 1/n$

$$x = A_1^n - A_2^n s_0 + A_3^n s_0^2 + \ldots. \quad (40.3.26)$$

Furthermore, we have

$$\phi'(s) = x - \sum_{j=1}^{n} \frac{j}{js+1}, \quad \phi^{(k)}(s) = (-1)^k (k-1)! \sum_{j=1}^{n} \frac{j^k}{(js+1)^k}, \quad k \geq 2, \tag{40.3.27}$$

and

$$\phi(s) = (x - A_1^n)\, s + \tfrac{1}{2} A_2^n s^2 - \tfrac{1}{3} A_3^n s^3 + \dots\,, \quad |s| < 1/n. \tag{40.3.28}$$

Using (40.3.26) we obtain

$$\phi(s_0) = -\tfrac{1}{2} A_2^n s_0^2 + \tfrac{2}{3} A_3^n s_0^3 - \tfrac{3}{4} A_4^n s_0^4 + \dots\,, \quad |s_0| < 1/n. \tag{40.3.29}$$

Next we need coefficients b_k in the expansion

$$s = \sum_{k=0}^{\infty} b_k w^k, \quad b_0 = s_0, \tag{40.3.30}$$

which follow from the inversion of (40.2.16). Let us write

$$\phi(s) = \sum_{k=0}^{\infty} \frac{a_k}{k!} (s - s_0)^k, \quad a_k = \phi^{(k)}(s_0), \tag{40.3.31}$$

with $a_1 = \phi'(s_0) = 0$. Then the first coefficients b_k of the expansion in (40.3.30) are given by

$$\begin{aligned} &b_0 = s_0, \quad b_1 = \frac{1}{\sqrt{a_2}}, \quad b_2 = -\frac{a_3}{6a_2^2}, \quad b_3 = \frac{5a_3^2 - 3a_2a_4}{72a_2^{7/2}}, \\ &b_4 = \frac{45a_2a_3a_4 - 40a_3^3 - 9a_2^2a_5}{1080a_2^5}, \end{aligned} \tag{40.3.32}$$

where the a_k follow from (40.3.31) and (40.3.27) (with $s = s_0$). By using (40.2.18) and (40.2.19) we find the coefficients c_{2k} of $g(w)$ in (40.2.21). We have

$$\begin{aligned} c_0 &= \frac{s_0\sqrt{a_2} + \xi}{s_0\sqrt{a_2}\,\xi}, \\ c_2 &= \frac{12a_3a_2^3 s_0\xi^3 + 24a_2^4\xi^3 + 5s_0^2\xi^3a_2^2a_3^2 - 3\xi^3 s_0^2a_2^3a_4 + 24s_0^3a_2^{11/2}}{24s_0^3\xi^3a_2^{11/2}}. \end{aligned} \tag{40.3.33}$$

When $x = \mu$ we have $s_0 = 0$, $\xi = 0$, $a_k = (-1)^k(k-1)!\, A_k^n$, and

$$c_0 = -\frac{a_3}{6a_2^{3/2}}, \quad c_2 = \frac{225a_3a_4a_2 - 175a_3^3 - 54a_5a_2^2}{2160a_2^{9/2}}. \tag{40.3.34}$$

Table 40.1 Relative errors in the computation with the expansion in (40.2.23) with the coefficients in (40.3.33) and (40.3.34); $x = \mu(0.5+j/10)$, $j = 0, 1, 2, \ldots, 10$.

j	$n = 10$	$n = 25$	$n = 50$	$n = 100$
0	0.44e-05	0.52e-07	0.21e-09	0.19e-13
1	0.12e-04	0.36e-06	0.76e-08	0.18e-10
2	0.24e-04	0.14e-05	0.82e-07	0.16e-08
3	0.43e-04	0.36e-05	0.40e-06	0.28e-07
4	0.69e-04	0.75e-05	0.12e-05	0.16e-06
5	0.10e-03	0.13e-04	0.25e-05	0.46e-06
6	0.14e-03	0.21e-04	0.45e-05	0.95e-06
7	0.19e-03	0.31e-04	0.72e-05	0.16e-05
8	0.25e-03	0.44e-04	0.11e-04	0.25e-05
9	0.32e-03	0.60e-04	0.15e-04	0.37e-05
10	0.39e-03	0.79e-04	0.20e-04	0.51e-05

40.4 Auxiliary expansions

For small values of s_0 (or ξ) we need expansions in terms of one of these small parameters. We obtain, using (40.3.28),

$$a_k = \phi^{(k)}(s_0) = (-1)^k \sum_{m=0}^{\infty} \frac{(m+k-1)!}{m!}(-s_0)^m A^n_{k+m}, \tag{40.4.35}$$

and, using (40.2.19) and (40.3.29),

$$\xi^2 = 2\sum_{k=2}^{\infty} \frac{k-1}{k}(-s_0)^k A^n_k. \tag{40.4.36}$$

This gives

$$\begin{aligned}\xi = -\sqrt{A^n_2}\, s_0 \Bigg(1 - \frac{n(n+1)}{2n+1} s_0 + \\ \frac{26n^4+52n^3+23n^2-3n-3}{120(2n+1)^2} s_0^2 + \ldots\Bigg).\end{aligned} \tag{40.4.37}$$

It appears to be convenient to expand in terms of powers of ξ. Inverting the relation in (40.4.37) we obtain

$$s_0 = -\widetilde{\xi}\left(1 - \frac{n(n+1)}{2n+1}\widetilde{\xi} + \frac{14n^4+28n^3+17n^2+3n+3}{120(2n+1)^2}\widetilde{\xi}^2 + \ldots\right), \tag{40.4.38}$$

where

$$\widetilde{\xi} = \frac{\xi}{\sqrt{A^n_2}}. \tag{40.4.39}$$

By using (40.4.35) and (40.4.38) we can obtain expansions for the quantities a_k, and finally for c_{2k} of (40.2.23), the first-order approximation being given in (40.3.34).

40.5 Numerical verification

We have used the expansion in (40.2.23) with the coefficients shown in (40.3.33) and (40.3.34), and we give the relative errors in Table 40.1. We have used $n = 10, 25, 50, 100$ and x around the value $\mu = \frac{1}{2}n(n+1)$. We have used $x = \mu(0.5+j/10)$, $j = 0, 1, 2, \ldots, 10$. The relative errors are smaller when $x < \mu$. In that case the complementary error function has a negative argument, and the factor $e^{\phi(s_0)}$ in front of the series in (40.2.23) makes this term exponentially small compared to the term with the complementary error function. This happens always in this kind of approximation of cumulative distribution functions. When $x < \mu$ it is better to perform numerical tests on the expansion of $G_n(x)$ given in (40.2.22).

Chapter 41

A generalized incomplete gamma function

We consider the asymptotic behavior of the function[1]

$$\Gamma(\alpha, x; b) = \int_x^\infty t^{\alpha-1} e^{-t-b/t}\, dt, \quad x > 0,\ \alpha > 0,\ b \geq 0, \tag{41.0.1}$$

as x tends to infinity. We give several expansions for the case that α and b are fixed and we give a uniform expansion in which α and b may range through unbounded intervals.

This function may play a role as main approximant in the asymptotic analysis of the integral

$$F_\lambda(\alpha, \beta, z) = \int_\alpha^\infty t^{\lambda-1} e^{-z(t+\beta^2/t)} f(t)\, dt, \tag{41.0.2}$$

which can be considered as a generalization of the integrals considered in Chapter 26 (with $\beta = 0$) and Chapter 27 (with $\alpha = 0$). We will not investigate this here.

We concentrate on the parameter domains indicated in (41.0.1), but all results to be derived in this chapter can be extended to large complex domains of the parameters. When $x \neq 0$, the function $\Gamma(\alpha, x; b)$ is an analytic function of both variables α and b. The function $\Gamma(\alpha, x; b)$ can be viewed as a generalization of the well-known incomplete gamma function: when we take $b = 0$, it reduces to $\Gamma(\alpha, x)$. When we take $x = 0$ it becomes a modified Bessel function (see (27.1.2)):

$$\Gamma(\alpha, 0; b) = \int_0^\infty t^{\alpha-1} e^{-t-b/t}\, dt = 2b^{\alpha/2} K_\alpha\left(2\sqrt{b}\right), \quad b > 0. \tag{41.0.3}$$

The function $\Gamma(\alpha, x; b)$ has been studied in detail in the paper Chaudhry and Zubair (1994), where it is used to obtain closed-form solutions to several problems in heat conduction with time-dependent boundary conditions. Moreover, the function plays a role as a cumulative distribution function.

We give several expansions of $\Gamma(\alpha, x; b)$ that are valid when the parameter x is large. When α and b are fixed, we give expansions in terms of incomplete gamma functions, Laguerre polynomials, and confluent hypergeometric functions. Although

[1]This chapter is based on Chaudhry *et al.* (1996).

several of these expansions turn out to be convergent, we concentrate on the asymptotic character of the expansions.

We also consider a result in which all three parameters α, x, b may range through unbounded domains. In that case, which is of special interest from the viewpoint of cumulative distribution functions, an error function is used as basic approximant in a uniform asymptotic expansion.

41.1 An expansion in terms of incomplete gamma functions

Using the series expansion of $\exp(-b/t)$ in (41.0.1), we obtain the expansion

$$\Gamma(\alpha, x; b) = \sum_{n=0}^{\infty} \frac{(-b)^n}{n!} \Gamma(\alpha - n, x). \tag{41.1.4}$$

This expansion is the Maclaurin expansion (with respect to b) of $\Gamma(\alpha, x; b)$; recall that this function is an analytic function of the variable b.

For fixed values of n and α we have (see (7.2.9))

$$\Gamma(\alpha - n, x) \sim x^{\alpha-n-1} e^{-x}, \quad x \to \infty. \tag{41.1.5}$$

It follows that (41.1.4) can be viewed as an asymptotic expansion as $x \to \infty$, when b and α are restricted to bounded intervals of the domains indicated in (41.0.1). The incomplete gamma functions in (41.1.4) can be obtained by recursion:

$$(\alpha - n)\Gamma(\alpha - n, x) = \Gamma(\alpha - n + 1, x) - x^{\alpha-n} e^{-x}. \tag{41.1.6}$$

The expansion in (41.1.4) loses its asymptotic character when α and/or b are $\mathcal{O}(x)$ or larger.

41.2 An expansion in terms of Laguerre polynomials

To obtain (41.1.4) the expansion of $\exp(-b/t)$ was used, which can be viewed as an expansion at $t = \infty$. But the main contributions to the integral in (41.0.1) come from an immediate neighborhood on the right of $t = x$. Hence, let us expand

$$t^{\alpha-1} e^{-b/t} = \sum_{n=0}^{\infty} c_n(x)(t - x)^n, \tag{41.2.7}$$

where the coefficients will be given below. Substituting this into (41.0.1), we obtain

$$\Gamma(\alpha, x; b) \sim e^{-x} \sum_{n=0}^{\infty} n!\, c_n(x). \tag{41.2.8}$$

This expansion has an asymptotic character as $x \to \infty$, with α, b fixed, and reduces to the asymptotic expansion of $\Gamma(\alpha, x)$ as $b \to 0$.

The coefficients $c_n(x)$ can be written in terms of Laguerre polynomials. To show this we recall the generating function for the Laguerre polynomials (see (15.0.2)):

$$(1-t)^{-\mu-1}e^{-tx/(1-t)} = \sum_{n=0}^{\infty} L_n^{(\mu)}(x)\, t^n, \quad \mu, x \in \mathbb{C}, \quad |t| < 1. \tag{41.2.9}$$

From this expansion it easily follows that

$$c_n(x) = (-1)^n x^{\alpha-1-n}\, e^{-b/x} L_n^{(-\alpha)}(b/x). \tag{41.2.10}$$

The coefficients can be computed by using the recurrence relation of the Laguerre polynomials:

$$(n+1)L_{n+1}^{(\mu)}(x) = (2n+\mu+1-x)\, L_n^{(\mu)}(x) - (n+\mu)L_{n-1}^{(\mu)}(x), \quad n = 1, 2, 3, \ldots, \tag{41.2.11}$$

with $L_0^{(\mu)}(x) = 1,\ L_1^{(\mu)}(x) = 1 + \mu - x$.

41.3 An expansion in terms of Kummer functions

Substituting in (41.0.1) the expansion

$$e^{-b/t} = e^{-b/x}\, e^{b(t-x)/(xt)} = e^{-b/x} \sum_{n=0}^{\infty} \frac{b^n}{n!} \frac{(t-x)^n}{(xt)^n}, \tag{41.3.12}$$

we obtain

$$\Gamma(\alpha, x; b) = x^\alpha\, e^{-x}\, e^{-b/x} \sum_{n=0}^{\infty} \frac{b^n}{x^n}\, d_n, \tag{41.3.13}$$

where

$$d_n = \frac{1}{n!} \int_0^\infty u^n\, (u+1)^{\alpha-n-1}\, e^{-xu}\, du = U(n+1, \alpha+1, x), \tag{41.3.14}$$

where U denotes the Kummer function introduced in Chapter 10. Since $U(a, c, x) \sim x^{-a}$ as $x \to \infty$, we see that the asymptotic behavior of the terms in the series is $\mathcal{O}(b^n/x^{2n+1})$, which is much better than in the previous cases. Moreover, the series is convergent and has positive terms. The convergence follows from the fact that $d_n = \mathcal{O}(1/n!)$ as $n \to \infty$ (in the integral in (41.3.14) we can use the bound $|u/(1+u)| \le 1$ when $u \ge 0$).

The coefficients d_n can be computed by using a recurrence relation for the U-functions; see (17.2.10).

41.4 An expansion in terms of the error function

In this section we consider large values of α and x and b as free parameters, which may range through the interval $(0, \infty)$. This will not be a restriction on the parameter x, which was the large parameter in the previous sections. For instance, in the expansion of this section we can take α fixed and let x tend to infinity. The expansion has a double (or perhaps triple) asymptotic property. First we scale the parameters in (41.0.1) with respect to α:

$$t = \alpha u, \quad b = \tfrac{1}{4}\alpha^2\beta^2, \quad x = \alpha\xi, \tag{41.4.15}$$

with $\beta > 0$, and obtain

$$\Gamma(\alpha, x; b) = \alpha^\alpha \int_\xi^\infty e^{-\alpha\phi(u)}\frac{du}{u}, \tag{41.4.16}$$

where

$$\phi(u) = u + \tfrac{1}{4}\beta^2/u - \ln u. \tag{41.4.17}$$

The function ϕ assumes its minimal value on $(0,\infty)$ at the point

$$u_0 = \tfrac{1}{2}\left(1 + \sqrt{1+\beta^2}\right), \tag{41.4.18}$$

which will be outside the interval of integration $[\xi, \infty)$ when x and/or b are small with respect to α. When x and/or b grow, the point u_0 will pass the point $u = \xi$, and the asymptotic behavior of the function $\Gamma(\alpha, x; b)$ will change considerably. As we have seen in Chapter 36, for the transition an error function is needed to describe the asymptotic behavior.

We substitute

$$\tfrac{1}{2}v^2 = \phi(u) - \phi(u_0), \quad \operatorname{sign}(v) = \operatorname{sign}(u - u_0), \tag{41.4.19}$$

which gives

$$\Gamma(\alpha, x; b) = \sqrt{2\pi}\,\alpha^{\alpha-\frac{1}{2}}\,e^{-\alpha\phi(u_0)}G_\alpha(\eta), \tag{41.4.20}$$

where

$$G_\alpha(\eta) = \sqrt{\frac{\alpha}{2\pi}}\int_\eta^\infty e^{-\frac{1}{2}\alpha v^2} f(v)\,dv, \quad f(v) = \frac{1}{u}\frac{du}{dv}, \tag{41.4.21}$$

with

$$\frac{du}{dv} = \frac{vu^2}{u^2 - u - \frac{1}{4}\beta^2}, \tag{41.4.22}$$

and η is the value of v that corresponds to $u = \xi$. That is,

$$\tfrac{1}{2}\eta^2 = \phi(\xi) - \phi(u_0), \quad \operatorname{sign}(\eta) = \operatorname{sign}(\xi - u_0). \tag{41.4.23}$$

Note that η is positive when α is large with respect to x and/or b (that is, $u_0 < \xi$), and that η will change sign when x and/or b become large (that is, $u_0 > \xi$).

The function $G_\alpha(\eta)$ has a standard form of integrals considered in §36.1. We can use the complementary relation $F_a(\eta) + G_a(\eta) = A(a)$, where $F_a(\eta)$ has the standard form given in (36.1.1) and the asymptotic representation given in (36.1.6). Using that representation we obtain

$$G_\alpha(\eta) = \tfrac{1}{2}\operatorname{erfc}\left(\eta\sqrt{\alpha/2}\right)\,A(\alpha,\beta) + \frac{e^{-\frac{1}{2}\alpha\eta^2}}{\sqrt{2\pi\alpha}}B(\alpha,\beta,\eta). \tag{41.4.24}$$

The functions $A(\alpha,\beta)$ and $B(\alpha,\beta,\eta)$ have the asymptotic expansions

$$A(\alpha,\beta)\sim\sum_{n=0}^{\infty}\frac{A_n(\beta)}{\alpha^n},\quad B(\alpha,\beta,\eta)\sim\sum_{n=0}^{\infty}\frac{B_n(\beta,\eta)}{\alpha^n}, \tag{41.4.25}$$

where

$$A_n(\beta)=f_n(0),\quad B_n(\beta,\eta)=\frac{f_n(\eta)-f_n(0)}{\eta}. \tag{41.4.26}$$

The sequence of functions $\{f_n\}$ is defined by $f_0=f$ and

$$f_n(v)=\frac{d}{dv}\frac{f_{n-1}(v)-f_{n-1}(0)}{v},\quad n=1,2,3,\ldots. \tag{41.4.27}$$

In §36.1 we have assumed that $f(0)=1$. Here we have

$$f(0)=\left(1+\beta^2\right)^{-\frac{1}{4}}. \tag{41.4.28}$$

The coefficients $A_n(\beta)$ in the first series in (41.4.25) does not depend on x. In fact we can relate the first series with a known expansion. When we let $x\to 0$, we observe that from the definition of η it follows that $\eta\to-\infty$, and that the error function in (41.4.24) tends to the limit 2. Using (41.0.3), we infer that

$$A(\alpha,\beta)=\sqrt{\frac{2\alpha}{\pi}}e^{\alpha\phi(u_0)}(\beta/2)^{\alpha}\,K_\alpha(\alpha\beta), \tag{41.4.29}$$

and that the first series in (41.4.25) corresponds to the Debye-type uniform expansion of the modified Bessel function $K_\alpha(\alpha\beta)$ considered in §9.5. That is, see (9.5.53),

$$\left(1+\beta^2\right)^{\frac{1}{4}}A_n(\beta)=(-1)^n u_n(t),\quad t=\frac{1}{\sqrt{1+\beta^2}}, \tag{41.4.30}$$

and the first $u_n(t)$ are given in (9.5.58).

The coefficients of the expansion of $B(\alpha,\beta,\eta)$ can be obtained from a recurrence relation. By differentiating (41.4.24) with respect to η, we obtain a differential equation for B:

$$\frac{1}{\alpha}\frac{d}{d\eta}B(\alpha,\beta,\eta)-\eta\,B(\alpha,\beta,\eta)=A(\alpha,\beta)-f(\eta). \tag{41.4.31}$$

Substituting the expansions given in (41.4.25), in which the coefficients $A_n(\beta)$ are known from (41.4.30), we obtain the recurrence relation

$$\eta B_{n+1}(\beta,\eta)=\frac{d}{d\eta}B_n(\beta,\eta)-A_{n+1}(\beta),\quad n=0,1,2,\ldots, \tag{41.4.32}$$

with

$$B_0(\beta,\eta)=\frac{f(\eta)-A_0(\beta)}{\eta}=\frac{f(\eta)-f(0)}{\eta}. \tag{41.4.33}$$

Chapter 42

Asymptotic inversion of cumulative distribution functions

The inversion of cumulative distribution functions is an important topic in statistics, probability theory, and econometrics, in particular for computing percentage points of chi-square, F, and Student's t-distributions. In the tails of these distributions the numerical inversion is not very easy, and for the standard distributions asymptotic formulas are available.

In this chapter we use the uniform asymptotic expansions of the incomplete gamma functions and incomplete beta functions, which are the basic functions for several distribution functions, to invert these functions for large values of one or two parameters. The methods have been developed in Temme (1992a,b), and we summarize the main parts of these papers. See also Gil *et al.* (2007, Chapter 10) and Gil *et al.* (2010, 2014). Algorithms for computing the incomplete gamma function ratios and their inverse are given in Gil *et al.* (2012). For the inversion of the complementary error function, we refer to Gil *et al.* (2007, Chapter 10, §10.2).

In the next sections we describe the asymptotic inversion process for the standard form given in (36.1.1), and we give some details for the incomplete gamma, incomplete beta, the hyperbolic cumulative distribution function, and the Marcum Q-function. Asymptotic aspects of the inversion of the Erlang B formula are given in Van Leeuwaarden and Temme (2009).

42.1 The asymptotic inversion method

We consider functions of the form

$$F_a(\eta) = \sqrt{\frac{a}{2\pi}} \int_{-\infty}^{\eta} e^{-\frac{1}{2}a\zeta^2} f(\zeta)\, d\zeta, \tag{42.1.1}$$

where $a > 0$, $\eta \in \mathbb{R}$, and f is analytic and real on $\mathbb{R}$ with $f(0) = 1$. In §37.1 we have shown how we can write the incomplete gamma function in this form, and in other chapters on cumulative distribution functions many examples are given.

The special case $f = 1$ gives the normal distribution

$$P(\eta\sqrt{a}) = \sqrt{\frac{a}{2\pi}} \int_{-\infty}^{\eta} e^{-\frac{1}{2}a\zeta^2}\, d\zeta = \tfrac{1}{2}\mathrm{erfc}\left(-\eta\sqrt{a/2}\right). \tag{42.1.2}$$

To describe the inversion problem, we assume that $p \in (0,1)$ and that a is a large positive parameter. Then we are interested in the value η that solves the equation

$$F_a(\eta) = F_a(\infty)\, p, \tag{42.1.3}$$

where $F_a(\eta)$ has the form (42.1.1). We use the representation (see §36.1)

$$F_a(\eta) = P(\eta\sqrt{a})A(a) \quad - \frac{e^{-\frac12 a\eta^2}}{\sqrt{2\pi a}} B_a(\eta), \quad A(a) = F_a(\infty), \tag{42.1.4}$$

with expansions of $A(a)$ and $B_a(\eta)$ given in (36.1.7).

First we define a number η_0 that solves the reduced equation

$$\tfrac12 \mathrm{erfc}\left(-\eta_0\sqrt{a/2}\right) = p. \tag{42.1.5}$$

Then for the desired value η we assume the expansion

$$\eta \sim \eta_0 + \frac{\eta_1}{a} + \frac{\eta_2}{a^2} + \frac{\eta_3}{a^3} + \ldots, \qquad a \to \infty, \tag{42.1.6}$$

and try to find the coefficients $\eta_1, \eta_2, \eta_3, \ldots$. To obtain the η_j we can substitute the expansion for η into the asymptotic expansion of $B_a(\eta)$ and use formal power series manipulations.

However, here we use the method as explained in our earlier mentioned publications. This method runs as follows. From the earlier equations in this section we obtain

$$\frac{dp}{d\eta_0} = \sqrt{\frac{a}{2\pi}}\, e^{-\frac12 a\eta_0^2}, \qquad \frac{dp}{d\eta} = \sqrt{\frac{a}{2\pi}}\, \frac{f(\eta)}{F_a(\infty)}\, e^{-\frac12 a\eta^2}, \tag{42.1.7}$$

from which we obtain, upon dividing,

$$f(\eta)\frac{d\eta}{d\eta_0} = F_a(\infty) e^{\frac12 a(\eta^2-\eta_0^2)}. \tag{42.1.8}$$

Substituting (42.1.6) and using $F_a(\infty) = 1 + \mathcal{O}(1/a)$, we obtain, after first-order perturbation analysis for large a,

$$f(\eta_0) = e^{\eta_0\eta_1} \quad \Longrightarrow \quad \eta_1 = \frac{1}{\eta_0}\ln f(\eta_0). \tag{42.1.9}$$

Because f is analytic at the origin with $f(0) = 1$, η_1 is well defined as $\eta_0 \to 0$.

For higher-order terms $\eta_j, j \ge 2$, we need in (42.1.8) more coefficients in the asymptotic expansion of $A(a) = F_a(\infty)$ (see (36.1.7)) and we have to expand

$$f(\eta) = f(\eta_0) + (\eta-\eta_0)f'(\eta_0) + \tfrac12(\eta-\eta_0)^2 f''(\eta_0) + \ldots. \tag{42.1.10}$$

Then, the next coefficients are given by

$$\begin{aligned}
\eta_2 &= -\Big(\left(2A_1 + \eta_1^2 - 2\eta_1'\right) f - 2\eta_1 f'\Big) / \left(2\eta_0 f\right), \\
\eta_3 &= -\Big(\left(8\eta_1\eta_2 + 4A_1\eta_1^2 + \eta_1^4 + 4\eta_1^2\eta_0\eta_2 + 4\eta_0^2\eta_2^2 - 8\eta_2'\right) f + \\
&\quad 8\left(A_1\eta_0\eta_2 + A_2\right) f - 8\left(\eta_1\eta_1' + \eta_2\right) f' - 4\eta_1^2 f''\Big) / \left(8\eta_0 f\right),
\end{aligned} \tag{42.1.11}$$

where f, f' and f'' are evaluated at η_0.

For small values of η_0 (that is, when $p \sim \frac{1}{2}$), we need expansions. We have

$$\begin{aligned}
\eta_1 &= a_1 + \tfrac{1}{2}\left(2a_2 - a_1^2\right)\eta_0 + \tfrac{1}{3}\left(3a_3 - 3a_1a_2 + 2a_1^3\right)\eta_0^2 + \mathcal{O}\left(\eta_0^3\right),\\
\eta_2 &= -\tfrac{1}{3}a_1^3 + 2a_3 + \tfrac{1}{8}\left(-12a_2a_1^2 + 5a_1^4 + 24a_4\right)\eta_0 + \mathcal{O}\left(\eta_0^2\right),\\
\eta_3 &= \tfrac{1}{15}\left(4a_1^5 - 5a_2a_1^3 - 15a_3a_1^2 + 120a_5\right) + \mathcal{O}\left(\eta_0\right),
\end{aligned} \tag{42.1.12}$$

where a_k are the coefficients in the expansion $f(\zeta) = \sum_{k=0}^{\infty} a_k\zeta^k$. The a_k are related to the A_k used in (42.1.11); see (36.1.14).

For details on obtaining the coefficients in the expansion in (42.1.6) by using the representation in (42.1.4) and the asymptotic expansion of $B_a(\eta)$, we refer to Van Leeuwaarden and Temme (2009, Appendix D), where this method is explained for the inversion of the incomplete gamma function $Q(a,x)$.

42.2 Asymptotic inversion of the gamma distribution

For the incomplete gamma ratios we want to solve the following two (equivalent) equations

$$P(a,x) = p, \quad Q(a,x) = q, \tag{42.2.13}$$

where $p + q = 1$ and a is a large positive number. We can use representation (see §37.1)

$$F_a(\eta) = F_a(\infty)\,P(a,x) = \sqrt{\frac{a}{2\pi}}\int_{-\infty}^{\eta} e^{-\frac{1}{2}a\zeta^2} f(\zeta)\,d\zeta, \tag{42.2.14}$$

where

$$\begin{aligned}
&\tfrac{1}{2}\zeta^2 = t - \ln t - 1, \quad \operatorname{sign}(\zeta) = \operatorname{sign}(t-1),\\
&\tfrac{1}{2}\eta^2 = \lambda - \ln\lambda - 1, \quad \operatorname{sign}(\eta) = \operatorname{sign}(\lambda - 1), \quad \lambda = \frac{x}{a},\\
&f(\zeta) = t^{-1}\frac{dt}{d\zeta} = \frac{\zeta}{t-1}, \quad F_a(\infty) = A(a) = \Gamma^*(a),
\end{aligned} \tag{42.2.15}$$

with expansion (see (6.1.15)) $\Gamma^*(a)) \sim \sum_{k=0}^{\infty} \frac{A_k}{a^k}$, where

$$A_0 = 1, \quad A_1 = \tfrac{1}{12}, \quad A_2 = \tfrac{1}{288}, \quad A_3 = -\tfrac{139}{51840}. \tag{42.2.16}$$

The coefficients a_k in the expansion of $f(\zeta) = \sum_{k=0}^{\infty} a_k\zeta^k$ are

$$a_0 = 1, \quad a_1 = -\tfrac{1}{3}, \quad a_2 = \tfrac{1}{12}, \quad a_3 = -\tfrac{2}{135}, \quad a_4 = \tfrac{1}{864}. \tag{42.2.17}$$

We obtain η_0 from (42.1.5) and for the expansion in (42.1.6) we can obtain η_1, η_2 and η_3 as in (42.1.9) and (42.1.11). The derivatives of f can be eliminated by using

$$\begin{aligned} f' &= f(1-f^2-f\eta)/\eta, \\ f'' &= f^2(-3\eta-3f+3f^3+5f^2\eta+2\eta^2 f)/\eta^2, \end{aligned} \tag{42.2.18}$$

and so on. These relations follow from the definition of f and the relation between η and λ given in (42.2.15).

Using these relations in η_j, and eliminating the derivatives of previous η_j, it follows that we can write $\eta^{2j-1}\eta_j$ as a polynomial in η, f, η_1. We have

$$\begin{aligned} 12\eta^3\eta_2 &= 12-12f^2-12f\eta-12f^2\eta\eta_1-12f\eta^2\eta_1-\eta^2-6\eta^2\eta_1^2, \\ 12\eta^5\eta_3 &= 12f^2\eta\eta_1+12f\eta^2\eta_1+24f^2\eta^3\eta_1+6\eta_1^3\eta^3-30+ \\ &\quad 60f^3\eta^2\eta_1-12f^2+31f^2\eta^2+72f^3\eta+12\eta_1^2\eta^2f^2+ \\ &\quad 18f^3\eta^3\eta_1^2+6f^2\eta^4\eta_1^2+36f^4\eta\eta_1+12\eta_1^2\eta^3f+42f^4+ \\ &\quad \eta^3\eta_1+-12\eta\eta_1+f\eta^3-12f\eta+12\eta_1^2\eta^2f^4, \end{aligned} \tag{42.2.19}$$

where $\eta=\eta_0$ and η_j $(j\ge 1)$ and f are evaluated at η_0.

For small values of η_0 we can use (cf. (42.1.12))

$$\begin{aligned} \eta_1 &= -\tfrac{1}{3}+\tfrac{1}{36}\eta_0+\tfrac{1}{1620}\eta_0^2-\tfrac{7}{6480}\eta_0^3+\tfrac{5}{18144}\eta_0^4-\tfrac{11}{382725}\eta_0^5+\cdots, \\ \eta_2 &= -\frac{7}{405}-\frac{7}{2592}\eta_0+\frac{533}{204120}\eta_0^2-\frac{1579}{2099520}\eta_0^3+\frac{109}{1749600}\eta_0^4+\cdots, \\ \eta_3 &= \frac{449}{102060}-\frac{63149}{20995200}\eta_0+\frac{29233}{36741600}\eta_0^2+\frac{346793}{5290790400}\eta_0^3+\cdots, \\ \eta_4 &= \frac{319}{183708}-\frac{269383}{4232632320}\eta_0-\frac{449882243}{982102968000}\eta_0^2+\cdots. \end{aligned} \tag{42.2.20}$$

We have taken some extra coefficients from Gil *et al.* (2007, §10.3), where more information about the coefficients can be found. See also Temme (1992a).

Remark 42.1. We have considered the inversion of $P(a,x)=p$, and it will be clear that we obtain the same result when inverting $Q(a,x)=q$ with $p+q=1$. In some applications one may start with the latter problem, with q very small. Then, from a numerical point of view, inverting $P(a,x)=p$, with $p=1-q$ may be a bad start, because in the computation of $p=1-q$ information will be lost. It is better to compute η_0 from the equation (see also (37.1.9))

$$\tfrac{1}{2}\operatorname{erfc}(\eta_0\sqrt{a/2})=q. \tag{42.2.21}$$

When we have η_0, we can use (42.1.6) and the analytical results for $P(a,x)=p$ given in this section. △

Table 42.1 Relative errors $|x_a - x|/x$ and $|P(a, x_a) - p|/p$ for several values of p and a; x_a is obtained by the asymptotic expansion (42.1.6), x is a more accurate value.

p	$a = 1$	$a = 1$	$a = 5$	$a = 5$	$a = 10$	$a = 10$
0.0001	2.3e-04	2.1e-03	1.1e-06	1.6e-05	9.4e-08	1.7e-06
0.1	6.6e-04	1.5e-03	2.0e-06	9.3e-06	1.4e-07	8.8e-07
0.3	8.7e-04	1.0e-03	2.3e-06	6.4e-06	1.6e-07	6.0e-07
0.5	7.0e-04	4.8e-04	6.7e-07	1.2e-06	5.4e-08	1.4e-07
0.7	4.9e-04	1.7e-04	2.7e-06	2.6e-06	1.7e-07	2.6e-07
0.9	1.9e-03	2.0e-04	2.5e-06	8.8e-07	1.8e-07	9.3e-08
0.9999	5.1e-03	5.1e-07	3.9e-06	1.8e-09	6.0e-08	4.8e-11

42.2.1 *Numerical verification*

When $p = q = \frac{1}{2}$, the asymptotics is quite simple. Then η_0 of (42.1.5) equals zero, and from the Taylor expansions for η_j given in (42.2.20), we obtain (42.1.6) in the form

$$\eta \sim -\tfrac{1}{3}a^{-1} - \tfrac{7}{405}a^{-2} + \tfrac{449}{102060}a^{-3} + \dots. \tag{42.2.22}$$

In this case we give an expansion of the desired value x. Inverting

$$\tfrac{1}{2}\eta^2 = \tfrac{1}{2}(\lambda - 1)^2 - \tfrac{1}{3}(\lambda - 1)^3 + \tfrac{1}{4}(\lambda - 1)^4 + \dots \tag{42.2.23}$$

we obtain

$$\lambda = 1 + \eta + \tfrac{1}{3}\eta^2 + \tfrac{1}{36}\eta^3 - \tfrac{1}{270}\eta^4 + \tfrac{1}{4320}\eta^5 + \dots . \tag{42.2.24}$$

Substituting (42.2.22) and using $x = a\lambda$, we have

$$x \sim a\left(1 - \tfrac{1}{3}a^{-1} + \tfrac{8}{405}a^{-2} + \tfrac{184}{25515}a^{-3} + \dots\right). \tag{42.2.25}$$

When $a = 1$, $q = \frac{1}{2}$, the equations in (42.2.13) reduce to $e^{-x} = \frac{1}{2}$, with solution $x = \ln 2 = 0.693147\dots$, while expansion (42.2.25) gives $x \sim 0.6936\dots$, an accuracy of about 3 digits. When $a = 2$, $q = \frac{1}{2}$, the equations in (42.2.13) become $(1 + x)e^{-x} = \frac{1}{2}$, with solution $x = 1.6783469\dots$; in this case our expansion (42.2.25) gives $x \sim 1.6783460\dots$, an accuracy of even 6 significant digits. This shows that (42.2.25) is quite accurate for small values of the (large) parameter a.

In Table 42.1 we give more results of numerical experiments. We have used (42.1.6) up to and including the shown terms. The first column under each a-value gives the relative accuracy $|x_a - x|/x$, where x_a is the result of the asymptotic method, and x is a more accurate value obtained by a Newton method. The second column under each a-value gives the relative errors $|P(a, x_a) - p|/p$.

For more details see Temme (1992a) and Gil *et al.* (2012).

42.2.2 *Other asymptotic inversion methods*

From Gil *et al.* (2012) we summarize a few simpler methods and expansions in which the inversion of the complementary error function and the uniform asymptotic expansions of $P(a,x)$ and $Q(a,x)$ are not needed.

42.2.2.1 *Small values of p*

When p is small we use the representation

$$P(a,x)=\frac{x^a}{\Gamma(a)}\sum_{n=0}^{\infty}\frac{(-1)^n x^n}{(a+n)n!}, \tag{42.2.26}$$

and write the inversion problem $P(a,x)=p$ as

$$x=r\left(1+\sum_{n=1}^{\infty}\frac{a(-1)^n x^n}{(a+n)n!}\right)^{-1/a}, \quad r=\bigl(p\Gamma(1+a)\bigr)^{1/a}, \tag{42.2.27}$$

where we assume that r is small. Inverting this relation, we obtain

$$x=r+\sum_{n=2}^{\infty}c_k r^k, \tag{42.2.28}$$

and the first few coefficients are

$$\begin{aligned}
c_2&=\frac{1}{a+1},\\
c_3&=\frac{3a+5}{2(a+1)^2(a+2)},\\
c_4&=\frac{8a^2+33a+31}{3(a+1)^3(a+2)(a+3)},\\
c_5&=\frac{125a^4+1179a^3+3971a^2+5661a+2888}{24(1+a)^4(a+2)^2(a+3)(a+4)}.
\end{aligned} \tag{42.2.29}$$

It appears that $c_k=\mathcal{O}((a+1)^{-k+1}$ for large values of a, and from numerical experiments we conclude that if $r<0.2(1+a)$, that is, $p<(0.2(1+a))^a/\Gamma(1+a)$ we can obtain 4-digits accuracy in x with the coefficients shown in (42.2.29). This is enough for starting a Newton method for obtaining higher accuracy.

This method also works when a is small, because in that case $r\sim\exp(\frac{1}{a}\ln p)$ becomes small for all fixed $p\in(0,1)$.

42.2.2.2 *Small values of q*

When q is small we use the asymptotic expansion in (7.2.9). A first approximation x_0 of the equation $Q(a,x)=q$ is obtained from the equation

$$e^{-x_0}x_0^a=q\Gamma(a). \tag{42.2.30}$$

Higher approximations of x are obtained in the form

$$x\sim x_0-L+b\sum_{k=1}^{\infty}d_k/x_0^k, \tag{42.2.31}$$

where $b = 1 - a$, $L = \ln(x_0)$, with first coefficients

$$
\begin{aligned}
d_1 &= L - 1, \\
d_2 &= \tfrac{1}{2}(3b - 2bL + L^2 - 2L + 2), \\
d_3 &= \tfrac{1}{6}(24bL - 11b^2 - 24b - 6L^2 + 12L - 12 - 9bL^2 + 6b^2L + 2L^3), \qquad (42.2.32) \\
d_4 &= \tfrac{1}{12}(72 + 36L^2 + 3L^4 - 72L + 162b - 168bL - 12L^3 + \\
&\quad 25b^3 - 22bL^3 + 36b^2L^2 - 12b^3L + 84bL^2 + 120b^2 - 114b^2L).
\end{aligned}
$$

This method works with rather small values of q (large values of x_0). When we assume that $0 < a < 10$ and $q < \min\left(0.2, e^{-\frac{1}{2}a}/\Gamma(a+1)\right)$, we obtain an accuracy of about 4 digits in x, which is enough for starting a Newton method for obtaining higher accuracy. For larger values of a, the method of §42.2 can be used; see also Remark 42.1.

42.2.3 *Asymptotics of the zeros of* $\Gamma(a, z)$

The inversion method of this section can also be used for finding the zeros for large positive values of a of the incomplete gamma function $\Gamma(a, z)$ in the complex z-plane. We find the location of the zeros by using the uniform asymptotic representation given in (37.2.26).[1]

The asymptotic distribution of zeros of the partial sums of the exponential function

$$
s_n(z) = \sum_{j=0}^{n} \frac{z^j}{j!}, \quad n = 0, 1, 2, \ldots, \qquad (42.2.33)
$$

which is a special case of the incomplete gamma function, that is,

$$
Q(n+1, z) = e^{-z} s_n(z), \quad n = 0, 1, 2, \ldots, \qquad (42.2.34)
$$

has received much attention in the literature. For an overview we refer to Varga (1990, Chapter 4).

We express the zeros of $Q(a, z)$ in terms of the zeros of the complementary error function erfc z, because this function is the main approximant in the representation in (37.2.26) and plays a significant role in the inversion process considered earlier.

From Fettis *et al.* (1973) we know that two infinite strings of zeros of erfc z occur in the neighborhood of the diagonals $y = \pm x$ in the left half-plane $x < 0$, $z = x + iy$.

The first few zeros are given in Table 42.2. Numerical values of the first 100 zeros of erfc$z$ and asymptotic approximations of the zeros are also given by Fettis *et al.* A first order approximation reads

$$
z_k^{\pm} \sim (-1 \pm i)\sqrt{(k - 1/8)\pi}, \quad k \to \infty. \qquad (42.2.35)
$$

[1] This section is based on Temme (1995).

Table 42.2 First five pairs $z_k^{\pm} = x_k \pm i\, y_k$ of zeros of erfc z.

k	x_k		$i\, y_k$
1	$-1.35481\ 01281$	$\pm$	$i\ 1.99146\ 68428$
2	$-2.17704\ 49061$	$\pm$	$i\ 2.69114\ 90243$
3	$-2.78438\ 76132$	$\pm$	$i\ 3.23533\ 08684$
4	$-3.28741\ 07894$	$\pm$	$i\ 3.69730\ 97025$
5	$-3.72594\ 87194$	$\pm$	$i\ 4.10610\ 72847$

When we have found a first approximation of the zeros of $Q(a, a\lambda) = q$ with $q = 0$ in terms of η, we need to find the corresponding points in the λ-plane. We find these λ-values by inverting the relation in (37.2.21) for complex values of η near the diagonals $\Re\eta = \pm\Im\eta$, with $\Re\eta < 0$. It is therefore of interest to know the original λ-contours of these diagonals under the mapping (37.2.21). See also the relations in (42.2.15).

To study the mapping and the pre-images of the diagonals, let us write $\eta = \alpha + i\beta$ and $\lambda = \rho e^{i\phi}$. Then the relation between η and λ given in (42.2.15) can be written in terms of the real equations

$$\begin{aligned} \tfrac{1}{2}(\alpha^2 - \beta^2) &= \rho\cos\phi - 1 - \ln\rho, \\ \alpha\beta &= \rho\sin\phi - \phi. \end{aligned} \tag{42.2.36}$$

On the diagonals $\alpha = \pm\beta$ the first equation becomes $\rho\cos\phi = 1 + \ln\rho$. In terms of cartesian coordinates ($\lambda = x + iy$) we have

$$x^2 + y^2 = e^{2(x-1)}. \tag{42.2.37}$$

This equation defines an almond-shaped closed curve between $x = 1$ and $x = -0.278\ldots$, the latter point being the real solution of the equation $-x = \exp(x-1)$; see Figure 42.1. In Varga (1990, Chapter 4) this curve is called the *Szegő curve* and is defined as

$$D_\infty = \left\{\lambda \in \mathbb{C} :\ \left|\lambda e^{1-\lambda}\right| = 1 \quad \text{and} \quad |\lambda| \le 1\right\}. \tag{42.2.38}$$

The function $\Gamma(n, \lambda n)$ (n a positive integer) has $n-1$ zeros in λ, see (42.2.34), and as discussed in detail in Varga (1990, Chapter 4), these zeros approach the Szegő curve D_∞ as $n \to \infty$.

The equation in (42.2.37) also defines solutions for $x > 1$; there are two branches starting in $x = 1, y = 0$ and extending to infinity along the asymptotes $y = \pm\exp(x-1)$. These branches have no meaning in connection with the zeros of $\Gamma(a, \lambda a)$ when a is an integer. However, when a is not an integer $\Gamma(a, \lambda a)$ has an infinity of λ-zeros of which about $[a]$ zeros are located near D_∞ with phases in the interval $(-\pi, \pi)$. An infinite number of zeros is located near the branches

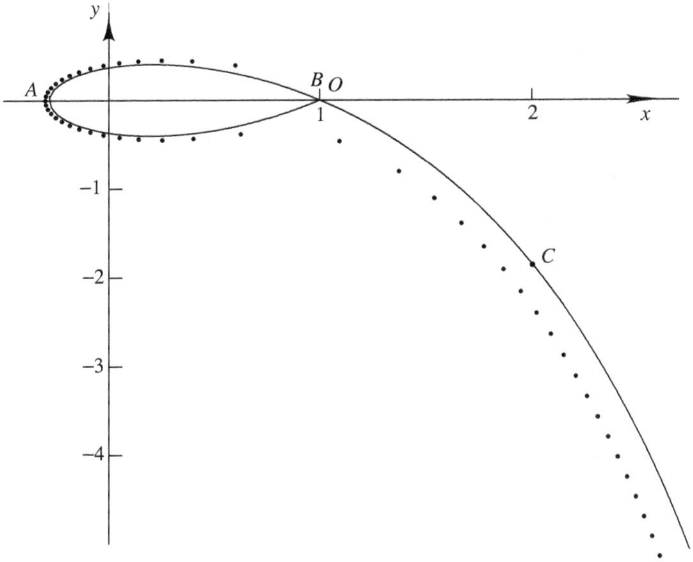

Fig. 42.1 λ-zeros of $\Gamma(a, \lambda a)$ with $a = 30.1$, along curves defined by $x^2+y^2 = \exp[2(x-1)]$; shown are the zeros with phases in the interval $[0, 2\pi]$. There is a conjugate set of zeros with phases in the interval $[-2\pi, 0]$.

of equation (42.2.37) with $x > 1$, but the phases of the zeros are in the intervals $(-2\pi, -\pi)$ and $(\pi, 2\pi)$.

In Figure 42.1 we show 50 zeros of $\Gamma(a, \lambda a)$ for $a = 30.1$ with phases in the interval $(0, 2\pi)$. The phases of the zeros in the upper half-plane belong to the interval $(0, \pi)$, the phases of those in the lower half-plane are in the interval $(\pi, 2\pi)$. A similar picture can be given for zeros with phase in $(-2\pi, 0)$. When $a = n$ is a positive integer the function $\Gamma(a, \lambda a)$ is single-valued, and in that case the zeros along the branches extending to infinity disappear.

The complex λ-zeros of $P(a, \lambda a)$ correspond to the zeros of the complementary error function along the diagonals $\Re\eta = \pm\Im\eta$, with $\Re\eta > 0$. Whether or not a is a positive integer, there are two infinite strings of zeros along the curves defined by (42.2.37) with $x \geq 1$, and the phases of the zeros belong to $(-\frac{1}{2}\pi, \frac{1}{2}\pi)$.

42.3 Asymptotic inversion of the incomplete beta function

We give a few details of the inversion of $I_x(p, q) = \rho \in (0, 1)$ by using the representations given in §38.4 and §26.3 in terms of the complementary error function and of the incomplete gamma function, respectively.[2]

[2]This section is based on Temme (1992b).

42.3.1 *Inverting by using the error function*

We use the representations given in (38.4.31) and (38.4.37). We write $p = r\sin^2\theta$ and $q = r\cos^2\theta$, $0 < \theta < \frac{1}{2}\pi$, and want to solve the equation

$$I_x(p,q) = \rho, \quad 0 < \rho < 1, \tag{42.3.39}$$

for large values of $r = p + q$.

Considering the method of §42.1, we need to find the coefficients η_j in expansion (see (42.1.6))

$$\eta \sim \eta_0 + \frac{\eta_1}{r} + \frac{\eta_2}{r^2} + \frac{\eta_3}{r^3} + \ldots. \tag{42.3.40}$$

The approximation of η obtained in this way is substituted in

$$-\tfrac{1}{2}\eta^2 = \sin^2\theta \ln\frac{x}{\sin^2\theta} + \cos^2\theta\ln\frac{1-x}{\cos^2\theta}, \tag{42.3.41}$$

and we invert this to obtain x.

The coefficient η_0 follows from equation (see (42.1.5))

$$\tfrac{1}{2}\mathrm{erfc}\left(-\eta_0\sqrt{r/2}\right) = \rho, \tag{42.3.42}$$

and the other coefficients can be obtained from (see (42.1.8))

$$f(\eta)\frac{d\eta}{d\eta_0} = F_r(\infty)e^{\frac{1}{2}r(\eta^2-\eta_0^2)}, \tag{42.3.43}$$

where $F_r(\infty)$ is defined in (38.4.32) and $f(\eta) = \eta\sin\theta\cos\theta/(x-\sin^2\theta)$.

The first coefficients η_j are as in (42.1.9) and (42.1.11), with the first A_k given in (38.4.34).

For small values of η_0 we can expand

$$\begin{aligned}\eta_1 = {}& \frac{2s^2-1}{3sc} - \frac{5s^4-5s^2-1}{36s^2c^2}\eta_0 + \frac{46s^6-69s^4+21s^2+1}{1620s^3c^3}\eta_0^2 + \\ & \frac{2s^2+62s^6-31s^8-33s^4-7}{6480s^4c^4}\eta_0^3 + \ldots,\end{aligned} \tag{42.3.44}$$

$$\begin{aligned}\eta_2 = {}& \frac{-52s^6+78s^4-12s^2-7}{405s^3c^3} + \\ & \frac{2s^2-370s^6+185s^8+183s^4-7}{2592s^4c^4}\eta_0 - \\ & \frac{776s^2+10240s^6-13525s^8-533+5410s^{10}-1835s^4}{204120s^5c^5}\eta_0^2 + \ldots,\end{aligned} \tag{42.3.45}$$

$$\begin{aligned}\eta_3 = {}& \frac{3704s^{10}-9260s^8+6686s^6-769s^4-1259s^2+449}{102060s^5c^5} - \\ & \frac{750479s^{12}-2251437s^{10}+2239932s^8-727469s^6}{20995200s^6c^6}\eta_0 - \\ & \frac{140052s^4-151557s^2+63149}{20995200s^6c^6}\eta_0 + \ldots\ ,\end{aligned} \tag{42.3.46}$$

where $s = \sin\theta$, $c = \cos\theta$.

42.3.2 *Inverting by using the incomplete gamma function*

We consider the asymptotic condition that the sum $p+q$ should be large, and we allow that $\mu = q/p$ may tend to zero or to infinity. We concentrate on the case $p \ge q$ ($\mu \le 1$). In the other case we can use (38.0.3); see also Remark 26.2.

In §26.3 we have given asymptotic representations of the incomplete beta function in terms of the incomplete gamma function, and for solving equation (42.3.39) we use these representations. We have (see (26.3.25), (26.3.26) and (26.3.31))

$$I_x(p,q) = \frac{e^{-pA(\mu)}}{B(p,q)} \int_\eta^\infty t^{q-1} e^{-pt} f(t,\mu)\, dt, \tag{42.3.47}$$

and

$$I_x(p,q) = Q(q,\eta p) + R_{p,q}(\eta), \tag{42.3.48}$$

and for $R_{p,q}(\eta)$ we have given an expansion, see (26.3.32), which holds for $p \to \infty$, uniformly with respect to $x \in [0,1]$ and $q \in [0,\infty)$.

The parameter η is given by the equation (in §26.3 η is called α)

$$\eta - \mu \ln \eta + A(\mu) = -\ln x - \mu \ln(1-x), \tag{42.3.49}$$

with

$$\mu = \frac{q}{p}, \quad A(\mu) = (1+\mu)\ln(1+\mu) - \mu. \tag{42.3.50}$$

Corresponding points in the relation between x and η are

$$x = 0 \Longleftrightarrow \eta = +\infty, \quad x = \frac{1}{1+\mu} \Longleftrightarrow \eta = \mu, \quad x = 1 \Longleftrightarrow \eta = 0. \tag{42.3.51}$$

From (42.3.49) it follows that

$$\frac{dx}{d\eta} = \frac{\eta-\mu}{\eta} \, \frac{x(1-x)}{(1+\mu)x - 1}. \tag{42.3.52}$$

We obtain the solution of equation (42.3.39) for large values of p, by first determining η_0, the solution of the reduced equation

$$Q(q,\eta_0 p) = \rho. \tag{42.3.53}$$

This involves an inversion of the incomplete gamma function, which is considered in §42.2, especially for large values of q.

As in the previous sections, the exact solution of (42.3.39) is expanded in the form

$$\eta \sim \eta_0 + \frac{\eta_1}{p} + \frac{\eta_2}{p^2} + \frac{\eta_3}{p^3} + \ldots, \quad p \to \infty. \tag{42.3.54}$$

We have from (42.3.47) and (42.3.53)

$$\frac{d\rho}{d\eta_0} = -\frac{p^q}{\eta_0 \Gamma(q)}\, e^{p(-\eta_0 + \mu \ln \eta_0)}, \quad \frac{d\rho}{d\eta} = -\frac{f(\eta,\mu)}{\eta B(p,q)} e^{p(-\eta+\mu \ln \eta - A(\mu))}, \tag{42.3.55}$$

where

$$f(\eta,\mu) = \frac{\eta-\mu}{1-x(1+\mu)}. \tag{42.3.56}$$

Upon dividing these equations, we obtain

$$f(\eta,\mu)\frac{d\eta}{d\eta_0} = \frac{\eta}{\eta_0}\frac{p^q\Gamma(p)}{\Gamma(p+q)}e^{p(\eta-\eta_0-\mu\ln(\eta/\eta_0)+A(\mu))}. \tag{42.3.57}$$

We write this in the form (cf. the form in (42.1.8) for inverting with the complementary error function)

$$\phi(\eta)\frac{d\eta}{d\eta_0} = \Phi(p)\frac{\eta}{\eta_0}e^{p(\eta-\eta_0-\mu\ln(\eta/\eta_0))}, \tag{42.3.58}$$

where

$$\phi(\eta) = \frac{\eta-\mu}{1-x(1+\mu)}\,\frac{1}{\sqrt{1+\mu}}, \qquad \Phi(p) = \frac{\Gamma^*(p)}{\Gamma^*(p+q)} \sim \sum_{k=0}^{\infty}\frac{d_k(\mu)}{p^k}. \tag{42.3.59}$$

This expansion is valid for large values of p, uniformly with respect to $\mu \geq 0$. See (25.6.56), with the first coefficients given in (25.6.57). The function $\phi(\eta)$ has the expansion

$$\phi(\eta) = c_0 + c_1(\eta-\mu) + c_2(\eta-\mu)^2 + \ldots, \tag{42.3.60}$$

with coefficients

$$\begin{aligned}
c_0 &= 1, \quad c_1 = \frac{w+2}{3(w+1)w}, \quad c_2 = \frac{1}{12w^2}, \quad c_3 = \frac{8w^3+9w^2-9w-8}{540w(w+1)^3},\\
c_4 &= \frac{15w^4-68w^3-182w^2-68w+15}{12960w^4(w+1)^4},\\
c_5 &= -\frac{32w^5+265w^4+253w^3-253w^2-265w-32}{90720w^5(w+1)^5},
\end{aligned} \tag{42.3.61}$$

where

$$w = \sqrt{1+\mu}. \tag{42.3.62}$$

The first coefficients of the expansion of x in powers of $\eta-\mu$ follow from (26.3.41), with $\rho = w$.

Substituting the expansion (42.3.54) into (42.3.58), we find the first coefficient:

$$\eta_1 = \frac{\ln\phi(\eta_0)}{1-\mu/\eta_0}, \tag{42.3.63}$$

a regular function at $\eta_0 = \mu$, as follows from the expansion (see (42.3.60)) of $\phi(\eta_0)$ at $\eta_0 = \mu$.

The next term is

$$\eta_2 = \frac{2\eta_0^2\left(\eta_1'\phi(\eta_0)+\eta_1\phi'(\eta_0)-d_1(\mu)\phi(\eta_0)\right)-2\eta_0\eta_1\phi(\eta_0)-\mu\eta_1^2\phi(\eta_0)}{2\eta_0(\eta_0-\mu)\phi(\eta_0)}. \tag{42.3.64}$$

Table 42.3 Relative errors $|I_x(a,b)-\rho|/\rho$ for $r=p+q=100$, and several values of ρ and $\sin\theta=\sqrt{p/r}$; the asymptotic inversion is based on the method of §42.3.1.

ρ	$\sin\theta$	0.5	0.4	0.3	0.2
0.0001		6.3e-04	1.1e-03	6.1e-04	7.5e-02
0.1		3.1e-04	5.9e-04	3.2e-03	1.1e-03
0.3		2.1e-04	4.0e-04	2.4e-03	1.0e-02
0.5		1.4e-04	2.8e-04	1.8e-03	9.8e-03
0.7		8.9e-05	1.8e-04	1.2e-03	7.1e-03
0.9		3.5e-05	7.0e-05	4.7e-04	3.0e-03
0.9999		6.7e-08	1.4e-07	9.1e-07	5.1e-06

For small values of $|\eta_0-\mu|$ we can expand

$$\begin{aligned}\eta_1 = &\frac{(w+2)(w-1)}{3w}+\frac{w^3+9w^2+21w+5}{36w^2(w+1)}(\eta_0-\mu)-\\ &\frac{w^4-13w^3+69w^2+167w+46}{1620(w+1)^2w^3}(\eta_0-\mu)^2-\\ &\frac{7w^5+21w^4+70w^3+26w^2-93w-31}{6480(w+1)^3w^4}(\eta_0-\mu)^3+\ldots,\end{aligned} \tag{42.3.65}$$

$$\begin{aligned}\eta_2 = &\frac{\left(28w^4+131w^3+402w^2+581w+208\right)(w-1)}{1620(w+1)w^3}-\\ &\frac{35w^6-154w^5-623w^4-1636w^3-3983w^2-3514w-925}{12960(w+1)^2w^4}(\eta_0-\mu)-\\ &\frac{2132w^7+7915w^6+16821w^5+35066w^4}{816480w^5(w+1)^3}(\eta_0-\mu)^2-\\ &\frac{87490w^3+141183w^2+95993w+21640}{816480w^5(w+1)^3}(\eta_0-\mu)^2+\ldots,\end{aligned} \tag{42.3.66}$$

where w is given by (42.3.62). More details on the coefficients η_k can be found in Temme (1992b).

When we have obtained η from (42.3.54), and solved (42.3.49) for x, we obtain the desired approximation of the solution of equation (42.3.39).

In this section, the functions $\Phi(p),\phi(\eta),\eta$ have expansions with coefficients $d_k(\mu),c_k,\eta_k$, in which the parameter $\mu=q/p$ may assume every value in $[0,\infty)$. This aspect demonstrates the uniform character (with respect to μ) of the present approach.

Remark 42.2. In applications the inversion of $J_x(p,q)=\sigma$ may be needed with $\sigma\in(0,1)$. The complementary function $J_x(p,q)$ is introduced in (26.3.16) with asymptotic representation in (26.3.38). We can consider the inversion of $I_x(p,q)=\rho$

Table 42.4 Relative errors $|I_x(a,b) - \rho|/\rho$ for $p = 100$ and several values of ρ and $\mu = q/p$; the asymptotic inversion is based on the method of §42.3.2.

ρ | μ	0.1	0.5	2.0	10
0.0001	3.1e-04	3.8e-04	3.0e-04	1.6e-04
0.1	1.1e-04	1.6e-04	1.3e-04	7.4e-05
0.3	7.0e-05	1.0e-04	8.8e-05	4.9e-05
0.5	4.6e-05	7.0e-05	5.4e-05	9.6e-06
0.7	2.7e-05	4.3e-05	3.7e-05	2.1e-05
0.9	9.8e-06	1.6e-05	1.5e-05	8.3e-06
0.9999	1.5e-08	3.0e-08	2.9e-08	1.7e-08

with $\rho = 1 - \sigma$, but when σ is very small, it is better to use the representation (cf. (42.3.48))

$$J_x(p,q) = P(q, \eta p) - R_{p,q}(\eta), \tag{42.3.67}$$

and start the inversion with determining η_0 from $P(q, \eta_0 p) = \sigma$. When we have η_0 we can proceed as before, because the analysis remains the same. △

42.3.3 *Numerical verification*

We explain a few steps in the numerical inversion process for the error function case of §42.3.1. We compute η_0 from (42.3.42). Then we invert equation (42.3.41) to obtain x_0. With x_0 and η_0 we compute (see (42.1.9))

$$\eta_1 = \frac{1}{\eta_0} \ln \frac{\eta_0 - \sin\theta\cos\theta}{x_0 - \sin^2\theta}, \tag{42.3.68}$$

and other coefficients η_j. Then we compute $\eta = \eta_0 + \eta_1/r + \ldots$, where $r = p + q$, and with this η we invert (42.3.41) for the second time to obtain our approximation x.

In Table 42.3 and Table 42.4 we show the relative errors $|(I_x(p,q) - \rho)/\rho|$, where x is obtained by the asymptotic inversion methods of §42.3.1 and §42.3.2. We have used the expansions in (42.3.40) and (42.3.54) with two terms. In almost all cases the smaller values of θ give less accuracy in the results in Table 42.3, which shows the non-uniform character of the method as $\theta \to 0$.

From Table 42.4 it follows that the results are not influenced by large or small values of μ. This shows the uniform character of the method of §42.3.2 with respect to $\mu \geq 0$. In fact, this method can be used in extreme situations: the ratio p/q may be very small and very large, and ρ may assume values quite close to zero or to unity.

In both tables we see better performances as ρ approaches unity. An explanation is that in representations (38.4.37) and (42.3.48) the R-terms become exponentially small compared to the terms with the error function and the incomplete gamma function. This has an important effect in the nature of the approximation of the incomplete beta function, and, consequentially, in the accuracy of the inversion process. For these values of ρ it is more realistic to perform and check the inversion by using the representation of the complementary function $J_x(p,q)$; see Remark 42.2.

42.4 The hyperbolic cumulative distribution

The hyperbolic distribution was introduced in Barndorff-Nielsen (1977) and is given by

$$F(y) = C\int_{-\infty}^{y} e^{-\alpha\sqrt{\delta^2+(x-\mu)^2}+\beta(x-\mu)}\,dx, \quad y\in\mathbb{R}, \tag{42.4.69}$$

where $\alpha>0$, $|\beta|<\alpha$, δ and μ are arbitrarily real constants, and C is the normalizing constant which gives $F(\infty)=1$. The value of C is given by

$$C = \frac{\omega}{2\alpha\delta^2 K_1(\omega)}, \quad \omega = \delta\sqrt{\alpha^2-\beta^2}, \tag{42.4.70}$$

where $K_1(\omega)$ denotes the modified Bessel function.[3]

We transform the function $F(y)$ into the standard form. Because $|\beta|<\alpha$, we can write $\beta=\alpha\tanh\theta$. We substitute in (42.4.69) $x=\mu+\delta\sinh(\theta+t)$, and obtain

$$F(y) = \frac{1}{2K_1(\omega)}\int_{-\infty}^{\tau} e^{-\omega\cosh t}\frac{\cosh(t+\theta)}{\cosh\theta}\,dt, \tag{42.4.71}$$

where ω is given in (42.4.70) and

$$\tau = \operatorname{arcsinh}\frac{y-\mu}{\delta}-\theta, \quad \cosh\theta = \frac{\alpha}{\sqrt{\alpha^2-\beta^2}}. \tag{42.4.72}$$

Next we use the transformation

$$\cosh t = 1+2\zeta^2 \quad \Longrightarrow \quad t = 2\operatorname{arcsinh}\zeta, \tag{42.4.73}$$

which gives

$$F(y) = \frac{e^{-\omega}}{K_1(\omega)}\int_{-\infty}^{\eta} e^{-\frac12 a\zeta^2} f(\zeta)\,d\zeta, \tag{42.4.74}$$

where

$$a = 4\omega, \quad \eta = \sinh\tfrac12\tau, \tag{42.4.75}$$

and

$$f(\zeta) = \frac{1+2\zeta^2+2\zeta\sqrt{\zeta^2+1}\,\tanh\theta}{\sqrt{\zeta^2+1}}. \tag{42.4.76}$$

We see that $f(0)=1$ and it follows that we can write $F(y)$ in the form

$$F(y) = \frac{F_a(\eta)}{F_a(\infty)}, \quad F_a(\infty) = \sqrt{\frac{2\omega}{\pi}}e^{\omega}K_1(\omega), \tag{42.4.77}$$

[3]This section is based on Gil *et al.* (2010).

Table 42.5 Values y and relative errors $\Delta = |F(y) - p|/p$ of the inversion $F(y) = p$, where $F(y)$ is given in (42.4.69) for $\alpha = 5$, $\beta = 3$, $\mu = 0$, and several values of δ and p.

δ	1		10		100	
p	y	Δ	y	Δ	y	Δ
0.0001	−1.1087	0.43e-01	1.2413	0.82e-03	53.110	0.14e-04
0.1	0.1646	0.10e-01	5.2635	0.17e-03	67.317	0.12e-04
0.3	0.5963	0.25e-02	6.6627	0.24e-03	71.931	0.95e-05
0.5	0.9465	0.70e-02	7.6884	0.23e-03	75.188	0.72e-05
0.7	1.3565	0.74e-02	8.7664	0.18e-03	78.496	0.49e-05
0.9	2.0826	0.40e-02	10.426	0.90e-04	83.367	0.21e-05
0.9999	5.8365	0.79e-05	16.767	0.28e-06	99.863	0.57e-08

where $F_a(\eta)$ has the standard form (42.1.1). We have (see (9.1.3))

$$F_a(\infty) = 1 + \frac{3}{8\omega} + \mathcal{O}(1/\omega^2), \quad \omega \to \infty. \tag{42.4.78}$$

It also follows that the inversion problem $F(y) = p$ when a is large can be written in the form (42.1.3). When we have found η from the expansion (42.1.6), we compute $\tau = 2\,\mathrm{arcsinh}\,\eta$ and finally (see (42.4.72))

$$y = \mu + \delta \sinh(\theta + \tau), \quad \theta = \mathrm{arctanh}\,\frac{\beta}{\alpha}. \tag{42.4.79}$$

42.4.1 *Numerical verification*

In Leobacher and Pillichshammer (2002) analytic approximations for small values of p and $1-p$ are constructed for the inverse function F^{-1} of $F(y)$ given in (42.4.69). With these approximations a numerical algorithm from Mathematica is used to compute the inverse F^{-1} from the differential equation satisfied by this function.

We demonstrate our approach by taking $\alpha = 5$, $\beta = 3$, $\mu = 0$ and $\delta = 1, 10, 100$. These values give $\omega = 4, 40, 400$ and $a = 16, 160, 1600$, respectively.

First we compute η_0 from (42.1.5) and next η_1 from (42.1.9), with $f(\eta)$ given in (42.4.76). The computed value η then follows from (42.1.6) (with two terms). Next we compute $\tau = 2\,\mathrm{arcsinh}\,\eta$ (see (42.4.75)), and with τ we can compute y by inverting the first equation in (42.4.72) with $\theta = \mathrm{arctanh}(\beta/\alpha)$.

In Table 42.5 we give for several values of p and δ the computed value y, and the relative error $|F(y) - p|/p$. We observe that the approximations of y become indeed better when the large parameter $a = 4\delta\sqrt{\alpha^2 - \beta^2}$ increases. Moreover, the approximations are better when $p \sim 1$; we can give the same explanation as for the incomplete beta function in Table 42.4.

42.5 The Marcum functions

In §39.1 we have given properties of these functions. We include here relations for the derivatives, which are important for the inversion problem.[4]

Taking the derivative with respect to y in (39.1.4) and using (39.1.6), we have

$$\frac{\partial Q_\mu(x,y)}{\partial y} = Q_{\mu-1}(x,y) - Q_\mu(x,y), \tag{42.5.80}$$

and similarly

$$\frac{\partial Q_\mu(x,y)}{\partial x} = Q_{\mu+1}(x,y) - Q_\mu(x,y). \tag{42.5.81}$$

By using the relations in (39.1.6) it follows that

$$\frac{\partial Q_\mu(x,y)}{\partial x} = -\frac{\partial Q_{\mu+1}(x,y)}{\partial y} = \left(\frac{y}{x}\right)^{\mu/2} e^{-x-y} I_\mu\left(2\sqrt{xy}\right), \tag{42.5.82}$$

and we see that $Q_\mu(x,y)$ ($P_\mu(x,y)$) is an increasing (decreasing) function of x and a decreasing (increasing) function of y. With respect to μ, $Q_\mu(x,y)$ is increasing and $P_\mu(x,y)$ is decreasing.

To discuss the inversion of the Marcum Q-function, we follow the process described by Helstrom (1998), where the inversions are linked to a specific problem in radiometry. In this reference, the inversion of the Q-function is performed in two steps; for the interpretation of these steps with respect to applications in radiometry we refer to Helstrom's paper.

In the two steps described by Helstrom we need two given numbers q_0, q_1, satisfying $0 < q_0 \le q_1 < 1$. For the asymptotic inversion we assume that μ is a large parameter. The two steps are:

Step 1: Find y from the equation

$$Q_\mu(0,y) = q_0, \tag{42.5.83}$$

and denote this value by y_0. Recall, see (39.0.1), that $Q_\mu(0,y) = Q_\mu(y)$ (the normalized incomplete gamma function).

Step 2: Find x from the equation

$$Q_\mu(x,y_0) = q_1, \tag{42.5.84}$$

and denote this value with x_1. The value y_0 is obtained in Step 1.

For Step 1 we refer to §42.2. For Step 2 we use an asymptotic representation of $Q_\mu(x,y)$, and for these details we refer to §42.5.2. This inversion process is with respect to x with $y = y_0$ a fixed value obtained in Step 1. In §42.5.3 we consider the inversion with respect to y with fixed x.

In statistics, the inversion of $Q_\mu(x,y)$ with respect to x corresponds to the problem of inverting the distribution function with respect to the non-centrality parameter given the upper tail probability. On the other hand, the inversion of $P_\mu(x,y)$ with respect to y with fixed x corresponds to the problem of computing the p-quantiles of the distribution function.

---

[4]This section is based on Gil *et al.* (2014).

42.5.1 *Asymptotic inversion*

For the asymptotic inversion methods we consider the representations in (39.4.51) and (39.4.61), with ζ defined in (39.4.47). We do not have a standard form as in (42.1.1), but we use elements of the representation in terms of the complementary error function given in §39.4. However, we change the sign of ζ. That is, we write

$$Q_\mu(\mu x, \mu y) = \tfrac{1}{2}\mathrm{erfc}\left(\zeta\sqrt{\mu/2}\right) + R_\mu(\zeta), \tag{42.5.85}$$

and

$$\tfrac{1}{2}\zeta^2 = x + y - \sqrt{1+4xy} + \ln\frac{1+\sqrt{1+4xy}}{2y}, \tag{42.5.86}$$

with $\mathrm{sign}(\zeta) = \mathrm{sign}(y - x - 1)$. In our inversion analysis the form of $R_\mu(\zeta)$ is not relevant.

42.5.2 *Asymptotic inversion with respect to x*

This corresponds to Step 2 described in §42.5. Throughout this case we take $y = y_0$, the value obtained in Step 1, that satisfies $q_0 = Q_\mu(0, y_0) = Q_\mu(y_0)$, and the present inversion problem is

$$Q_\mu(x, y) = q_1, \quad q_0 < q_1 < 1, \quad y = y_0. \tag{42.5.87}$$

We use the method described for the incomplete gamma functions; see §42.2. We use representation (42.5.85) and start with solving the equation

$$\tfrac{1}{2}\mathrm{erfc}\left(\zeta\sqrt{\mu/2}\right) = q_1. \tag{42.5.88}$$

We call the solution ζ_0, and considering q_1 as a function of ζ_0, we have

$$\frac{dq_1}{d\zeta_0} = -\sqrt{\frac{\mu}{2\pi}}\, e^{-\frac{1}{2}\mu\zeta_0^2}. \tag{42.5.89}$$

Considering q_1 in (42.5.87) as a function of ζ, we have (see (42.5.82))

$$\frac{dq_1}{d\zeta} = \frac{\partial Q_\mu(x,y)}{\partial x}\frac{\partial x}{\partial \zeta} = \left(\frac{y}{x}\right)^{\mu/2} e^{-x-y} I_\mu\left(2\sqrt{xy}\right)\frac{\partial x}{\partial \zeta}. \tag{42.5.90}$$

Upon dividing, and replacing x, y with $\mu x, \mu y$, we obtain

$$\frac{d\zeta}{d\zeta_0} = -\frac{1}{\sqrt{2\mu\pi}}\,\frac{e^{-\mu\left(\frac{1}{2}\zeta_0^2 - x - y + \ln\rho\right)}}{I_\mu(\mu\xi)}\,\frac{\partial\zeta}{\partial x}, \tag{42.5.91}$$

where

$$\rho = \sqrt{\frac{y}{x}}, \quad \xi = 2\sqrt{xy}. \tag{42.5.92}$$

Next we use the Debye asymptotic representation of the modified Bessel function that is valid for large values of μ, uniformly with respect to $\xi \ge 0$. We have, see (9.5.57),

$$I_\mu(\mu\xi) = \frac{e^{\mu\eta}}{\sqrt{2\pi\mu}\,(1+\xi^2)^{\frac{1}{4}}}U_\mu(\xi), \quad U_\mu(\xi) \sim \sum_{k=0}^{\infty}\frac{U_k(p)}{\mu^k}, \tag{42.5.93}$$

where

$$\eta = \sqrt{1+\xi^2} + \ln \frac{\xi}{1+\sqrt{1+\xi^2}}, \quad p = \frac{1}{\sqrt{1+\xi^2}}, \tag{42.5.94}$$

and the coefficients $U_k(p)$ are polynomials in p. The first two are

$$U_0(p) = 1, \quad U_1(p) = \tfrac{1}{24}\left(3 - 5p^2\right) p. \tag{42.5.95}$$

Using the first part of (42.5.93) in (42.5.91) we obtain

$$f(\zeta)\frac{d\zeta}{d\zeta_0} = e^{-\frac{1}{2}\mu(\zeta_0^2-\zeta^2)}, \tag{42.5.96}$$

where

$$\tfrac{1}{2}\zeta^2 = x + y - \ln\rho - \eta, \tag{42.5.97}$$

which is the same as the relation in (42.5.86), and

$$f(\zeta) = -\frac{U_\mu(\xi)}{\frac{\partial\zeta}{\partial x}(1+\xi^2)^{\frac{1}{4}}}, \tag{42.5.98}$$

in which

$$\frac{\partial\zeta}{\partial x} = \frac{1+\sqrt{1+\xi^2}-2y}{\zeta\left(1+\sqrt{1+\xi^2}\right)}. \tag{42.5.99}$$

As in the asymptotic inversion of the incomplete gamma functions (see §42.2), where we obtained a similar equation as in (42.5.96), we solve this differential equation by substituting an expansion of the form

$$\zeta \sim \zeta_0 + \sum_{n=1}^{\infty} \frac{\zeta_n}{\mu^n}, \tag{42.5.100}$$

where ζ_0 is the solution of equation (42.5.88). After substituting this expansion in (42.5.96) and considering the first order of approximation for large μ, we obtain for ζ_1 the relation

$$f_0(\zeta_0) = e^{\zeta_0\zeta_1} \quad \Longrightarrow \quad \zeta_1 = \frac{1}{\zeta_0}\ln\bigl(f_0(\zeta_0)\bigr), \tag{42.5.101}$$

where $f_0(\zeta)$ is the part of $f(\zeta)$ defined in (42.5.98) by putting $U_\mu(\xi) = 1$. That is,

$$f_0(\zeta) = -\frac{\zeta\left(1+\sqrt{1+\xi^2}\right)}{\left(1+\sqrt{1+\xi^2}-2y\right)(1+\xi^2)^{\frac{1}{4}}}, \tag{42.5.102}$$

which can be written in the form

$$f_0(\zeta) = \frac{\zeta}{y-x-1}\,\frac{1+2x+\sqrt{1+4xy}}{2(1+4xy)^{\frac{1}{4}}}. \tag{42.5.103}$$

In equation (42.5.103) we may write x also with subscript 0, because it is related to ζ_0 via equation (42.5.86) with ζ replaced by ζ_0. We have to invert this equation to find x when ζ_0 is available. This can be done by using standard equation solvers, such as Newton's method.

When we have the value x corresponding to ζ_0, we can compute $f_0(\zeta_0)$ by using (42.5.103), and then ζ_1 from (42.5.101). This gives the second-order approximation $\zeta \sim \zeta_0 + \zeta_1/\mu$. When we have higher-order coefficients ζ_n in the expansion of ζ, we use this ζ to find the final requested value x from (42.5.86).

Note that, after (42.5.90), we have worked with scaled variables x and y, and that we obtain the real world variables by changing $x, y \to \mu x, \mu y$.

The higher coefficients ζ_n can be obtained by expanding $f(\zeta)$ in negative powers of μ (after substitution of (42.5.100), and also by expanding the exponential function in (42.5.96)). The comparison of the coefficients of equal powers of μ gives the relations for the ζ_j.

To explain a few steps, we write

$$f(\zeta) = \sum_{k=0}^{\infty} \frac{f_k(\zeta)}{\mu^k}, \quad f_k(\zeta) = U_k(p) f_0(\zeta), \tag{42.5.104}$$

with $f_0(\zeta)$ defined in (42.5.102). Next we write

$$f_k(\zeta) = \sum_{j=0}^{\infty} f_{j,k}(\zeta - \zeta_0)^j, \quad k = 0, 1, 2, \ldots. \tag{42.5.105}$$

Then we find

$$\zeta_2 = -\frac{f_{0,0}\zeta_1^2\zeta_0 - 2f_{0,1}\zeta_0 + 2\zeta_1 f_{0,0} - 2f_{1,0} - 2f_{1,0}\zeta_1\zeta_0}{2\zeta_0^2 f_{0,0}}, \tag{42.5.106}$$

where we have used

$$\zeta_1' = \frac{f_{1,0} - \zeta_1 f_{0,0}}{\zeta_0 f_{0,0}}. \tag{42.5.107}$$

For small values of ζ_0 (that is, when $q_1 \sim \frac{1}{2}$ in (42.5.88)), we need expansions. First we write

$$x = y - 1 + \sum_{k=1}^{\infty} a_k(y)\zeta^k, \tag{42.5.108}$$

where the relation between x, y, ζ is given in (42.5.86). Taking into account the relation $\operatorname{sign}(\zeta) = \operatorname{sign}(y - x - 1)$ and assuming $2y > 1$, we find

$$a_1(y) = -\sqrt{2y-1}, \quad a_2(y) = \frac{3y-1}{3(2y-1)}, \quad a_3(y) = \frac{6y-1}{36(2y-1)^{5/2}}. \tag{42.5.109}$$

The expansion of x is used to obtain the following expansions:

$$f_0(\zeta) = \sum_{k=0}^{\infty} c_k(y)\zeta^k, \quad \zeta_1 = \sum_{k=0}^{\infty} z_k^{(1)}(y)\zeta_0^k, \quad \zeta_2 = \sum_{k=0}^{\infty} z_k^{(2)}(y)\zeta_0^k, \tag{42.5.110}$$

where $f_0(\zeta_0)$, ζ_1 and ζ_2 are defined in (42.5.102), (42.5.101), and(42.5.106), respectively.

The first coefficients are

$$
\begin{aligned}
&c_0(y) = 1, \quad c_1(y) = \frac{3a_1^2 - 1}{6a_1^3}, \quad c_2(y) = \frac{5 - 3a_1^4}{24a_1^6}, \\
&c_3(y) = \frac{135a_1^6 + 117a_1^4 - 67a_1^2 - 625}{2160a_1^9},
\end{aligned}
\tag{42.5.111}
$$

$$
\begin{aligned}
&z_0^{(1)}(y) = \frac{3a_1^2 - 1}{6a_1^3}, \quad z_1^{(1)}(y) = -\frac{9a_1^4 - 3a_1^2 - 7}{36a_1^6}, \\
&z_2^{(1)}(y) = \frac{540a1^6 - 27a1^4 - 1305a1^2 - 830}{3240a_1^9},
\end{aligned}
\tag{42.5.112}
$$

$$
\begin{aligned}
&z_0^{(2)}(y) = -\frac{135a_1^6 - 81a_1^4 + 45a_1^2 - 155}{3240a_1^9}, \\
&z_1^{(2)}(y) = \frac{243a_1^8 - 108a_1^6 - 81a_1^4 - 420a_1^2 - 505}{2592a_1^{12}}.
\end{aligned}
\tag{42.5.113}
$$

42.5.3 *Asymptotic inversion with respect to y*

In this case the inversion problem is defined by

$$
Q_\mu(x, y) = q, \quad 0 < q < 1, \tag{42.5.114}
$$

with x a given fixed value. Observe that this time q may be any value in the interval $(0,1)$, because for any positive x, we have have $Q_\mu(x,0) = 1$ and $Q_\mu(x,y)$ is monotonically decreasing to 0 as $y \to \infty$; see (42.5.82).

We proceed as in the previous case, computing ζ_0 from equation (42.5.88) with $q_1 = q$, and by using (39.1.4) and (42.5.82). We obtain the relation (cf. (42.5.91))

$$
\frac{d\zeta}{d\zeta_0} = \rho \frac{1}{\sqrt{2\mu\pi}} \frac{e^{-\mu(\frac{1}{2}\zeta_0^2 - x - y + \ln\rho)}}{I_{\mu-1}(\mu\xi)} \frac{\partial\zeta}{\partial y}, \tag{42.5.115}
$$

again with scaled variables.

We replace the Bessel function by using $I_{\mu-1}(\mu\xi) = I'_\mu(\mu\xi) + (1/\xi)I_\mu(\mu\xi)$ and we use the expansion for the derivative

$$
I'_\mu(\mu\xi) = \frac{e^{\mu\eta}}{\xi\sqrt{2\pi\mu}}(1+\xi^2)^{\frac{1}{4}} V_\mu(\xi), \quad V_\mu(\xi) \sim \sum_{k=0}^{\infty} \frac{V_k(p)}{\mu^k}, \tag{42.5.116}
$$

where η and p are the same as in (42.5.94) and the coefficients $V_k(p)$ are polynomials in p. The first two are

$$
V_0(p) = 1, \quad V_1(p) = \tfrac{1}{24}\left(7p^2 - 9\right)p. \tag{42.5.117}
$$

This gives the analogue of (42.5.96) in the form

$$
g(\zeta)\frac{d\zeta}{d\zeta_0} = e^{-\frac{1}{2}\mu(\zeta_0^2 - \zeta^2)}, \tag{42.5.118}
$$

where

$$g(\zeta) = g_0(\zeta) W_\mu(\xi), \quad g_0(\zeta) = \frac{1+\sqrt{1+\xi^2}}{2y\frac{\partial\zeta}{\partial y}(1+\xi^2)^{\frac14}}, \tag{42.5.119}$$

in which

$$\frac{\partial\zeta}{\partial y} = \frac{y-2xy-1+(y-1)\sqrt{1+\xi^2}}{y\zeta\left(1+\sqrt{1+\xi^2}\right)}. \tag{42.5.120}$$

and

$$W_\mu(\xi) = \frac{U_\mu(\xi)+\sqrt{1+\xi^2}V_\mu(\xi)}{1+\sqrt{1+\xi^2}} \sim \sum_{k=0}^{\infty} \frac{W_k(p)}{\mu^k}. \tag{42.5.121}$$

We have

$$W_0(p) = 1, \quad W_1(p) = -\tfrac{1}{24}\left(5p^2-12p+9\right)p. \tag{42.5.122}$$

In (42.5.118) we replace $g(\zeta)$ by $g_0(\zeta)$. It turns out that $g_0(\zeta) = f_0(\zeta)$, see (42.5.103). As a consequence, ζ_1 in the approximation $\zeta \sim \zeta_0 + \zeta_1/\mu + \zeta_2/\mu^2 + \ldots$ is given in (42.5.101). To obtain coefficients ζ_k, $k \ge 2$, we need expansions for $g(\zeta)$ as in (42.5.104), (42.5.105), with $U_k(p)$ replaced by $W_k(p)$. This gives ζ_2 as in (42.5.106) with f_{jk} replaced by g_{jk}.

For the expansions for small ζ_0, we first expand

$$y = x + 1 + \sum_{k=1}^{\infty} b_k(x)\zeta^k, \tag{42.5.123}$$

with first coefficients

$$b_1(x) = \sqrt{2x+1}, \quad b_2(x) = \frac{3x+1}{3(2x+1)}, \quad b_3(x) = \frac{6x+1}{36(2x+1)^{5/2}}. \tag{42.5.124}$$

The expansion of y is used to obtain the following expansions:

$$g_0(\zeta) = \sum_{k=0}^{\infty} c_k(x)\zeta^k, \quad \zeta_1 = \sum_{k=0}^{\infty} z_k^{(1)}(x)\zeta_0^k, \quad \zeta_2 = \sum_{k=0}^{\infty} z_k^{(2)}(x)\zeta_0^k. \tag{42.5.125}$$

The first coefficients are

$$\begin{aligned} c_0(x) &= 1, \quad c_1(x) = -\frac{3b_1^2-1}{6b_1^3}, \quad c_2(x) = \frac{9b_1^4-12b_1^2+5}{24b_1^6}, \\ c_3(x) &= -\frac{675b_1^6-1863b_1^4+1845b_1^2-625}{2160b_1^9}, \end{aligned} \tag{42.5.126}$$

$$\begin{aligned} z_0^{(1)}(x) &= -\frac{3b_1^2-1}{6b_1^3}, \quad z_1^{(1)}(x) = \frac{9b_1^4-15b_1^2+7}{36b_1^6}, \\ z_2^{(2)}(x) &= -\frac{540b_1^6-1917b_1^4+2205b_1^2-830}{3240b_1^9}, \end{aligned} \tag{42.5.127}$$

$$\begin{aligned}
z_0^{(2)}(x) &= -\frac{675b_1^6 - 999b_1^4 + 225b_1^2 + 155}{3240b_1^9},\\
z_1^{(2)}(y) &= \frac{891b_1^8 - 2808b_1^6 + 2403b_1^4 + 12b_1^2 - 505}{2592b_1^{12}}.
\end{aligned} \tag{42.5.128}$$

For further details and for numerical test we refer to Gil *et al.* (2014).

Remark 42.3. For applications in mathematical statistics it is of interest to consider the inversion for the P-function in the form

$$P_\mu(x, y) = p, \quad 0 < p < 1. \tag{42.5.129}$$

By using $q = 1 - p$ we can use the inversion of the Q-function, but when p is very small the evaluation $q = 1 - p$ does not make sense. We can repeat the analysis for the P-function, and the only change we have to make is to change the sign of ζ_0, and to assume as earlier $\mathrm{sign}(\zeta_0) = \mathrm{sign}(y - x - 1)$. This follows from the representation in Remark 39.4, see (39.4.61). △

Example 42.1. As an application we use $q = \frac{1}{2}$. Then the equation

$$\tfrac{1}{2}\mathrm{erfc}\left(\zeta_0\sqrt{\mu/2}\right) = \tfrac{1}{2} \tag{42.5.130}$$

gives $\zeta_0 = 0$ and we have

$$\zeta \sim \frac{\zeta_1}{\mu} = \frac{z_0^{(1)}(x)}{\mu}, \tag{42.5.131}$$

where $z_0^{(1)}(x)$ is shown in the expansion (42.5.127), and using this in expansion (42.5.123) we find

$$y \sim x + 1 + \frac{b_1(x) z_0^{(1)}(x)}{\mu} = x + 1 - \frac{3x + 1}{3\mu(2x + 1)}, \quad \mu \to \infty. \tag{42.5.132}$$

This is for the scaled variables in $Q_\mu(\mu x, \mu y)$. For the real life variables x, y in $Q_\mu(x, y)$ we have: $Q_\mu(x, y) = \frac{1}{2}$ when

$$y \sim x + \mu - \frac{3x + 1}{3(2x + 1)}, \quad \mu \to \infty. \tag{42.5.133}$$

This gives a description of the transition in the quadrant $(x \geq 0, y \geq 0)$ from small values to values near unity of the Marcum functions. ♢

Bibliography

Abad, J. and Sesma, J. (1997). A new expansion of the confluent hypergeometric function in terms of modified Bessel functions, *J. Comput. Appl. Math.* **78**, 1, pp. 97–101.

Abad, J. and Sesma, J. (1999). Buchholz polynomials: a family of polynomials relating solutions of confluent hypergeometric and Bessel equations, *J. Comput. Appl. Math.* **101**, 1-2, pp. 237–241.

Abramowitz, M. and Stegun, I. A. (1964). *Handbook of mathematical functions with formulas, graphs, and mathematical tables, National Bureau of Standards Applied Mathematics Series*, Vol. 55 (U.S. Government Printing Office, Washington).

Aldaya, V., Bisquert, J., and Navarro-Salas, J. (1991). The quantum relativistic harmonic oscillator: generalized Hermite polynomials, *Phys. Lett. A* **156**, 7-8, pp. 381–385.

Alkauskas, G. (2012). The Minkowski $?(x)$ function and Salem's problem, *C. R. Math. Acad. Sci. Paris* **350**, 3-4, pp. 137–140.

Askey, R. A. and Olde Daalhuis, A. B. (2010). Chapter 16, Generalized hypergeometric functions and Meijer G–function, in *NIST Handbook of Mathematical Functions* (Cambridge University Press, Cambridge, UK), pp. 403–418, http://dlmf.nist.gov/16.

Barndorff-Nielsen, O. E. (1977). Exponentially decreasing distributions for the logarithm of particle size, *Proc. Roy. Soc. London Ser. A* **353**, pp. 401–419.

Berry, M. V. (1989). Uniform asymptotic smoothing of Stokes's discontinuities, *Proc. Roy. Soc. London Ser. A* **422**, 1862, pp. 7–21.

Berry, M. V. (1991). Infinitely many Stokes smoothings in the gamma function, *Proc. Roy. Soc. London Ser. A* **434**, 1891, pp. 465–472.

Berry, M. V. and Howls, C. J. (1991). Hyperasymptotics for integrals with saddles, *Proc. Roy. Soc. London Ser. A* **434**, 1892, pp. 657–675.

Berry, M. V. and Howls, C. J. (1993). Unfolding the high orders of asymptotic expansions with coalescing saddles: singularity theory, crossover and duality, *Proc. Roy. Soc. London Ser. A* **443**, 1917, pp. 107–126.

Berry, M. V. and Howls, C. J. (1994). Overlapping Stokes smoothings: survival of the error function and canonical catastrophe integrals, *Proc. Roy. Soc. London Ser. A* **444**, 1920, pp. 201–216.

Bleick, W. E. and Wang, P. C. C. (1974). Asymptotics of Stirling numbers of the second kind, *Proc. Amer. Math. Soc.* **42**, pp. 575–580, Erratum in the same journal, Vol. 48, 518 (1975).

Bleistein, N. (1966). Uniform asymptotic expansions of integrals with stationary point near algebraic singularity, *Comm. Pure Appl. Math.* **19**, pp. 353–370.

Bleistein, N. and Handelsman, R. A. (1975). *Asymptotic expansions of integrals* (Holt,

Rinehart, and Winston, New York), reprinted with corrections by Dover Publications Inc., New York, 1986.

Boersma, J. (1991). Uniform asymptotics of a Bessel-function series occurring in a transmission-line problem, *J. Comput. Appl. Math.* **37**, 1-3, pp. 143–159.

Boersma, J., Indenkleef, J. J. E., and Kuiken, H. K. (1984). A diffusion problem in semiconductor technology, *J. Engrg. Math.* **18**, 4, pp. 315–333.

Bornemann, F., Laurie, D., Wagon, S., and Waldvogel, J. (2004). *The SIAM 100-digit challenge* (Society for Industrial and Applied Mathematics (SIAM), Philadelphia, PA).

Borwein, D., Borwein, J. M., and Crandall, R. E. (2008). Effective Laguerre asymptotics, *SIAM J. Numer. Anal.* **46**, 6, pp. 3285–3312.

Boyd, W. G. C. (1990). Stieltjes transforms and the Stokes phenomenon, *Proc. Roy. Soc. London Ser. A* **429**, 1876, pp. 227–246.

Boyd, W. G. C. (1994). Gamma function asymptotics by an extension of the method of steepest descents, *Proc. Roy. Soc. London Ser. A* **447**, 1931, pp. 609–630.

Bressoud, D. M. (2010). Chapter 26, Combinatorial analysis, in *NIST Handbook of Mathematical Functions* (Cambridge University Press, Cambridge), pp. 618–636, http://dlmf.nist.gov/10.

Broder, A. Z. (1984). The r-Stirling numbers, *Discrete Math.* **49**, 3, pp. 241–259.

Buchholz, H. (1969). *The confluent hypergeometric function with special emphasis on its applications*, Translated from the German by H. Lichtblau and K. Wetzel. Springer Tracts in Natural Philosophy, Vol. 15 (Springer-Verlag New York Inc., New York).

Butzer, P. L. and Hauss, M. (1991). On Stirling functions of the second kind, *Stud. Appl. Math.* **84**, 1, pp. 71–91.

Campos, R. G. and Calderón, M. L. (2012). Approximate closed-form formulas for the zeros of the Bessel polynomials, *Int. J. Math. Math. Sci.* , pp. Art. ID 873078, 10.

Carlitz, L. (1958). On some polynomials of Tricomi, *Boll. Un. Mat. Ital. (3)* **13**, pp. 58–64.

Catalan, E. (1887). Nombres de Segner, *Rend. Circ. Mat. Pal.* **1**, pp. 190–201.

Chaudhry, M. A., Temme, N. M., and Veling, E. J. M. (1996). Asymptotics and closed form of a generalized incomplete gamma function, *J. Comput. Appl. Math.* **67**, 2, pp. 371–379.

Chaudhry, M. A. and Zubair, S. M. (1994). Generalized incomplete gamma functions with applications, *J. Comput. Appl. Math.* **55**, 1, pp. 99–124.

Chelluri, R., Richmond, L. B., and Temme, N. M. (2000). Asymptotic estimates for generalized Stirling numbers, *Analysis (Munich)* **20**, 1, pp. 1–13.

Chester, C., Friedman, B., and Ursell, F. (1957). An extension of the method of steepest descents, *Proc. Cambridge Philos. Soc.* **53**, pp. 599–611.

Chiccoli, C., Lorenzutta, S., and Maino, G. (1989). A note on a Tricomi expansion for the generalized exponential integral and related functions, *Nuovo Cimento B (11)* **103**, 5, pp. 563–568.

Clark, L. (2004). An asymptotic expansion for the Catalan-Larcombe-French sequence, *J. Integer Seq.* **7**, 2, pp. Article 04.2.1, 5 pp. (electronic).

Clemmow, P. C. (1950). Some extensions to the method of integration by steepest descents, *Quart. J. Mech. Appl. Math.* **3**, pp. 241–256.

Comtet, L. (1974). *Advanced combinatorics* (D. Reidel Publishing Co., Dordrecht).

Corcino, C. B., Hsu, L. C., and Tan, E. L. (1999). Asymptotic approximations of r-Stirling numbers, *Approx. Theory Appl. (N.S.)* **15**, 3, pp. 13–25.

De Bruijn, N. G. (1958). *Asymptotic methods in analysis*, Bibliotheca Mathematica. Vol. 4 (North-Holland Publishing Co., Amsterdam), Third edition by Dover (1981).

Deaño, A. and Temme, N. M. (2009). On modified asymptotic series involving confluent hypergeometric functions, *Electron. Trans. Numer. Anal.* **35**, pp. 88–103.

Debye, P. (1909). Näherungsformeln für die Zylinderfunktionen für große Werte des Arguments und unbeschränkt veränderliche Werte des Index, *Math. Ann.* **67**, 4, pp. 535–558.

Dilcher, K. (2010). Chapter 24, Bernoulli and Euler polynomials, in *NIST Handbook of Mathematical Functions* (Cambridge University Press, Cambridge, UK), pp. 587–599, http://dlmf.nist.gov/24.

Dilcher, K., Skula, L., and Slavutskiĭ, I. S. (1991). *Bernoulli numbers, Queen's Papers in Pure and Applied Mathematics*, Vol. 87 (Queen's University, Kingston, ON), Bibliography (1713–1990).

Dingle, R. B. (1973). *Asymptotic expansions: their derivation and interpretation* (Academic Press [A subsidiary of Harcourt Brace Jovanovich, Publishers], London-New York,).

Dočev, K. (1962). On the generalized Bessel polynomials, *Bŭlgar. Akad. Nauk. Izv. Mat. Inst.* **6**, pp. 89–94.

Doman, B. G. S. (1996). An asymptotic expansion for the incomplete beta function, *Math. Comp.* **65**, 215, pp. 1283–1288.

Dunster, T. M. (1989). Uniform asymptotic expansions for Whittaker's confluent hypergeometric functions, *SIAM J. Math. Anal.* **20**, 3, pp. 744–760.

Dunster, T. M. (1990). Bessel functions of purely imaginary order, with an application to second-order linear differential equations having a large parameter, *SIAM J. Math. Anal.* **21**, 4, pp. 995–1018.

Dunster, T. M. (1991). Conical functions with one or both parameters large, *Proc. Roy. Soc. Edinburgh Sect. A* **119**, 3-4, pp. 311–327.

Dunster, T. M. (1997). Error analysis in a uniform asymptotic expansion for the generalised exponential integral, *J. Comput. Appl. Math.* **80**, 1, pp. 127–161.

Dunster, T. M. (2001). Uniform asymptotic expansions for the reverse generalized Bessel polynomials, and related functions, *SIAM J. Math. Anal.* **32**, 5, pp. 987–1013 (electronic).

Dunster, T. M. (2010). Chapter 14, Legendre and related functions, in *NIST Handbook of Mathematical Functions* (Cambridge University Press, Cambridge), pp. 351–381, http://dlmf.nist.gov/14.

Dunster, T. M., Gil, A., Segura, J., and Temme, N. M. (2014). Computation of a numerically satisfactory pair of solutions of the differential equation for conical functions of non-negative integer orders, *Numer. Algorithms* .

Dunster, T. M. and Lutz, D. A. (1991). Convergent factorial series expansions for Bessel functions, *SIAM J. Math. Anal.* **22**, 4, pp. 1156–1172.

Dunster, T. M., Paris, R. B., and Cang, S. (1998). On the high-order coefficients in the uniform asymptotic expansion for the incomplete gamma function, *Methods Appl. Anal.* **5**, 3, pp. 223–247.

Eckhaus, W. and de Jager, E. M. (1966). Asymptotic solutions of singular perturbation problems for linear differential equations of elliptic type, *Arch. Rational Mech. Anal.* **23**, pp. 26–86.

Elbert, Á. and Laforgia, A. (1992). Asymptotic formulas for ultraspherical polynomials $P_n^\lambda(x)$ and their zeros for large values of λ, *Proc. Amer. Math. Soc.* **114**, 2, pp. 371–377.

Erdélyi, A. (1955). Asymptotic representations of Fourier integrals and the method of stationary phase, *J. Soc. Indust. Appl. Math.* **3**, pp. 17–27.

Erdélyi, A. (1960). Asymptotic forms for Laguerre polynomials, *J. Indian Math. Soc. (N.S.)* **24**, pp. 235–250 (1961).

Erdélyi, A. (1974). Asymptotic evaluation of integrals involving a fractional derivative, *SIAM J. Math. Anal.* **5**, pp. 159–171.

Erdélyi, A., Magnus, W., Oberhettinger, F., and Tricomi, F. (1953–1955). *Higher transcendental functions* (McGraw-Hill Book Company, Inc., New York-Toronto-London), 3 Volumes. Based, in part, on notes left by Harry Bateman.

Farid Khwaja, S. and Olde Daalhuis, A. (2013). Exponentially accurate uniform asymptotic approximations for integrals and Bleistein's method revisited, *Proc. Roy. Soc. London Ser. A* **469**, 2153, pp. 1–12.

Farid Khwaja, S. and Olde Daalhuis, A. (2014). Uniform asymptotic expansions for hypergeometric functions with large parameters IV, *Anal. Appl. (Singap.)* **12**, 6.

Ferreira, C., Lopez, J. L., and Mainar, E. (2003). Asymptotic relations in the Askey scheme for hypergeometric orthogonal polynomials, *Adv. in Appl. Math.* **31**, 1, pp. 61–85.

Ferreira, C., López, J. L., and Pagola, P. J. (2008a). Asymptotic approximations between the Hahn-type polynomials and Hermite, Laguerre and Charlier polynomials, *Acta Appl. Math.* **103**, 3, pp. 235–252.

Ferreira, C., López, J. L., and Sinusía, E. P. (2008b). Asymptotic relations between the Hahn-type polynomials and Meixner-Pollaczek, Jacobi, Meixner and Krawtchouk polynomials, *J. Comput. Appl. Math.* **217**, 1, pp. 88–109.

Fettis, H. E. (1976). Fourier series expansions for Pearson type IV distributions and probabilities, *SIAM J. Appl. Math.* **31**, 3, pp. 511–518.

Fettis, H. E., Caslin, J. C., and Cramer, K. R. (1973). Complex zeros of the error function and of the complementary error function, *Math. Comp.* **27**, pp. 401–407.

Fields, J. L. (1966). A note on the asymptotic expansion of a ratio of gamma functions, *Proc. Edinburgh Math. Soc. (2)* **15**, pp. 43–45.

Flajolet, P. and Prodinger, H. (1999). On Stirling numbers for complex arguments and Hankel contours, *SIAM J. Discrete Math.* **12**, 2, pp. 155–159 (electronic).

Franklin, J. and Friedman, B. (1957). A convergent asymptotic representation for integrals, *Proc. Cambridge Philos. Soc.* **53**, pp. 612–619.

Frenzen, C. L. (1987). Error bounds for asymptotic expansions of the ratio of two gamma functions, *SIAM J. Math. Anal.* **18**, 3, pp. 890–896.

Frenzen, C. L. and Wong, R. (1988). Uniform asymptotic expansions of Laguerre polynomials, *SIAM J. Math. Anal.* **19**, 5, pp. 1232–1248.

Garcia, E. and López, J. L. (2010). The Appell's function F_2 for large values of its variables, *Quart. Appl. Math.* **68**, 4, pp. 701–712.

Garoni, T. M., Frankel, N. E., and Glasser, M. L. (2001). Complete asymptotic expansions of the Fermi-Dirac integrals $\mathcal{F}_p(\eta) = 1/\Gamma(p+1)\int_0^\infty [\epsilon^p/(1+e^{\epsilon-\eta})]d\epsilon$, *J. Math. Phys.* **42**, 4, pp. 1860–1868.

Gawronski, W. and Van Assche, W. (2003). Strong asymptotics for relativistic Hermite polynomials, *Rocky Mountain J. Math.* **33**, 2, pp. 489–524.

Gil, A., Segura, J., and Temme, N. M. (2000). Computing toroidal functions for wide ranges of the parameters, *J. Comput. Phys.* **161**, 1, pp. 204–217.

Gil, A., Segura, J., and Temme, N. M. (2001). On nonoscillating integrals for computing inhomogeneous Airy functions, *Math. Comp.* **70**, 235, pp. 1183–1194.

Gil, A., Segura, J., and Temme, N. M. (2002). Evaluation of the modified Bessel function of the third kind of imaginary orders, *J. Comput. Phys.* **175**, 2, pp. 398–411.

Gil, A., Segura, J., and Temme, N. M. (2004). Computing solutions of the modified Bessel differential equation for imaginary orders and positive arguments, *ACM Trans. Math. Software* **30**, 2, pp. 145–158.

Gil, A., Segura, J., and Temme, N. M. (2006). Computing the real parabolic cylinder functions $U(a,x)$, $V(a,x)$, *ACM Transactions on Mathematical Software* **32**, 1, pp.

70–101.

Gil, A., Segura, J., and Temme, N. M. (2007). *Numerical methods for special functions* (Society for Industrial and Applied Mathematics (SIAM), Philadelphia, PA).

Gil, A., Segura, J., and Temme, N. M. (2009). Computing the conical function $P^{\mu}_{-1/2+i\tau}(x)$, *SIAM J. Sci. Comput.* **31**, 3, pp. 1716–1741.

Gil, A., Segura, J., and Temme, N. M. (2010). The asymptotic inversion of certain cumulative distribution functions. in *Progress in industrial mathematics at ECMI 2008. Proceedings of the 15th European conference on mathematics for industry, London, UK, June 30 - July 4, 2008.* (Berlin: Springer), pp. 117–122.

Gil, A., Segura, J., and Temme, N. M. (2012). Efficient and accurate algorithms for the computation and inversion of the incomplete gamma function ratios, *SIAM J. Sci. Comput.* **34**, 6, pp. A2965–A2981.

Gil, A., Segura, J., and Temme, N. M. (2013). Algorithm 939: Computation of the Marcum Q-function, *ACM Trans. Math. Software* **40**, 3.

Gil, A., Segura, J., and Temme, N. M. (2014). The asymptotic and numerical inversion of the Marcum $Q-$function, *Stud. Appl. Math.* Accepted.

Godoy, E., Ronveaux, A., Zarzo, A., and Area, I. (1998). On the limit relations between classical continuous and discrete orthogonal polynomials, *J. Comput. Appl. Math.* **91**, 1, pp. 97–105.

Goh, W. M. Y. and Wimp, J. (1994). On the asymptotics of the Tricomi-Carlitz polynomials and their zero distribution. I, *SIAM J. Math. Anal.* **25**, 2, pp. 420–428.

Goh, W. M. Y. and Wimp, J. (1997). The zero distribution of the Tricomi-Carlitz polynomials, *Comput. Math. Appl.* **33**, 1-2, pp. 119–127.

Graham, R. L., Knuth, D. E., and Patashnik, O. (1994). *Concrete mathematics*, 2nd edn. (Addison-Wesley Publishing Company, Reading, MA).

Grosswald, E. (1978). *Bessel polynomials, Lecture Notes in Mathematics*, Vol. 698 (Springer, Berlin).

Hahn, E. (1980). Asymptotik bei Jacobi-Polynomen und Jacobi-Funktionen, *Math. Z.* **171**, 3, pp. 201–226.

Handelsman, R. A. and Bleistein, N. (1969). Uniform asymptotic expansions of integrals that arise in the analysis of precursors, *Arch. Rational Mech. Anal.* **35**, pp. 267–283.

Hankel, H. (1863). *Dissertation* (Leipzig), see also Die Euler'schen Integrale bei unbeschränkter Variabilität des Argumentes, *Zeitschrift für Math. und Phys.* 9:1–21, 1864.

Helstrom, C. W. (1998). Approximate inversion of Marcum's $Q-$function, *IEEE Trans. Aerospace and Electron. Systems* **34**, 1, pp. 317–319.

Hill, G. (1977). Incomplete Bessel function I_0: The Von Mises distribution, *ACM Trans. Math. Softw.* **3**, pp. 279–284.

Howls, C. J. (1992). Hyperasymptotics for integrals with finite endpoints, *Proc. Roy. Soc. London Ser. A* **439**, 1906, pp. 373–396.

Hsu, L. C. (1948). Note on an asymptotic expansion of the nth difference of zero, *Ann. Math. Statistics* **19**, pp. 273–277.

Ismail, M. E. H. (2005). *Classical and quantum orthogonal polynomials in one variable, Encyclopedia of Mathematics and its Applications*, Vol. 98 (Cambridge University Press, Cambridge), with two chapters by Walter Van Assche; with a foreword by Richard A. Askey.

Jarvis, A. F., Larcombe, P. J., and French, D. R. (2004). Linear recurrences between two recent integer sequences, *Congr. Numer.* **169**, pp. 79–99.

Jin, X.-S. and Wong, R. (1998). Uniform asymptotic expansions for Meixner polynomials, *Constr. Approx.* **14**, 1, pp. 113–150.

Jones, D. S. (1970/71). A uniform asymptotic expansion for a certain double integral, *Proc. Roy. Soc. Edinburgh Sect. A* **69**, pp. 205–226.

Jones, D. S. (1990). Uniform asymptotic remainders, in *Asymptotic and computational analysis (Winnipeg, MB, 1989), Lecture Notes in Pure and Appl. Math.*, Vol. 124 (Dekker, New York), pp. 241–264.

Jones, D. S. (2006). Parabolic cylinder functions of large order, *J. Comput. Appl. Math.* **190**, 1-2, pp. 453–469.

Jordan, C. (1947). *The calculus of finite differences*, 2nd edn. (Chelsea Publishing Cmpany, New York).

Karlin, S. and McGregor, J. (1959). Random walks, *Illinois J. Math.* **3**, pp. 66–81.

Knessl, C. and Keller, J. B. (1991). Stirling number asymptotics from recursion equations using the ray method, *Stud. Appl. Math.* **84**, 1, pp. 43–56.

Knopp, K. (1964). *Theorie and Anwendung der unendlichen Reihen*, Fünfte berichtigte Auflage. Die Grundlehren der Mathematischen Wissen schen Wissenschaften, Band 2 (Springer-Verlag, Berlin), English translation: *Theory and application of infinite series*, Blackie & Son, London, 2nd ed. 1951.

Koekoek, R., Lesky, P. A., and Swarttouw, R. F. (2010). *Hypergeometric orthogonal polynomials and their q-analogues*, Springer Monographs in Mathematics (Springer-Verlag, Berlin), with a foreword by Tom H. Koornwinder.

Koekoek, R. and Swarttouw, R. F. (1998). The Askey-scheme of hypergeometric orthogonal polynomials and its -analogue,, Research reports 98–17, Faculty of Information Technology and Systems, Department of Technical Mathematics and Informatics, Delft University of Technology.

Koornwinder, T. H., Wong, R., Koekoek, R., and Swarttouw, R. F. (2010). Chapter 18, Orthogonal polynomials, in *NIST Handbook of Mathematical Functions* (U.S. Dept. Commerce, Washington, DC), pp. 435–484, http://dlmf.nist.gov/18.

Krall, H. L. and Frink, O. (1949). A new class of orthogonal polynomials: The Bessel polynomials, *Trans. Amer. Math. Soc.* **65**, pp. 100–115.

Larcombe, P. J. (2006). Formal proofs of the limiting behaviour of two finite series using dominated convergence, in *Proceedings of the Thirty-Seventh Southeastern International Conference on Combinatorics, Graph Theory and Computing*, Vol. 178, pp. 125–146.

Larcombe, P. J. and French, D. R. (2004). A new generating function for the Catalan-Larcombe-French sequence: proof of a result by Jovovic, *Congr. Numer.* **166**, pp. 161–172.

Lauwerier, H. A. (1974). *Asymptotic analysis* (Mathematisch Centrum, Amsterdam), Mathematical Centre Tracts, No. 54.

Lebedev, N. N. (1972). *Special functions and their applications* (Dover Publications Inc., New York), Revised edition, translated from the Russian and edited by Richard A. Silverman.

Lee, K. F. and Wong, R. (2010). Uniform asymptotic expansions of the Tricomi-Carlitz polynomials, *Proc. Amer. Math. Soc.* **138**, 7, pp. 2513–2519.

Lee, K. F. and Wong, R. (2014). Asymptotic expansion of the Tricomi–Carlitz polynomials and their zeros, *J. Comput. Appl. Math.* **265**, pp. 220–242.

Lekner, J. (1987). *Theory of reflection of electromagnetic and particle waves, Developments in Electromagnetic Theory and Applications*, Vol. 3 (Martinus Nijhoff Publishers, Dordrecht).

Leobacher, G. and Pillichshammer, F. (2002). A method for approximate inversion of the hyperbolic CDF, *Computing* **69**, 4, pp. 291–303.

Li, X. and Wong, R. (2001). On the asymptotics of the Meixner-Pollaczek polynomials and their zeros, *Constr. Approx.* **17**, 1, pp. 59–90.

Li, X.-C. and Wong, R. (2000). A uniform asymptotic expansion for Krawtchouk polynomials, *J. Approx. Theory* **106**, 1, pp. 155–184.

Lin, Y. and Wong, R. (2013). Global asymptotics of the discrete Chebyshev polynomials, *Asymptot. Anal.* **82**, 1-2, pp. 39–64.

Liu, S.-Y., Wong, R., and Zhao, Y.-Q. (2013). Uniform treatment of Darboux's method and the Heisenberg polynomials, *Proc. Amer. Math. Soc.* **141**, 8, pp. 2683–2691.

López, J. L. (2008). Asymptotic expansions of Mellin convolution integrals, *SIAM Rev.* **50**, 2, pp. 275–293.

López, J. L. and Pagola, P. J. (2011a). An explicit formula for the coefficients of the saddle point method. *Constr. Approx.* **33**, 2, pp. 145–162.

López, J. L. and Pagola, P. J. (2011b). A systematic "saddle point near a pole" asymptotic method with application to the Gauss hypergeometric function, *Stud. Appl. Math.* **127**, 1, pp. 24–37.

López, J. L., Pérez Sinusía, E., and Temme, N. M. (2006). First order approximation of an elliptic 3D singular perturbation problem, *Stud. Appl. Math.* **116**, 3, pp. 303–319.

López, J. L., Pérez Sinusía, E., and Temme, N. M. (2007). Asymptotic behaviour of three-dimensional singularly perturbed convection-diffusion problems with discontinuous data, *J. Math. Anal. Appl.* **328**, 2, pp. 931–945.

López, J. L. and Sesma, J. (1999). Asymptotic expansion of the incomplete beta function for large values of the first parameter, *Integral Transform. Spec. Funct.* **8**, 3-4, pp. 233–236.

López, J. L. and Temme, N. M. (1999a). Approximation of orthogonal polynomials in terms of Hermite polynomials, *Methods Appl. Anal.* **6**, 2, pp. 131–146.

López, J. L. and Temme, N. M. (1999b). Hermite polynomials in asymptotic representations of generalized Bernoulli, Euler, Bessel, and Buchholz polynomials, *J. Math. Anal. Appl.* **239**, 2, pp. 457–477.

López, J. L. and Temme, N. M. (2002). Two-point Taylor expansions of analytic functions, *Stud. Appl. Math.* **109**, 4, pp. 297–311.

López, J. L. and Temme, N. M. (2004). Multi-point Taylor expansions of analytic functions, *Trans. Amer. Math. Soc.* **356**, 11, pp. 4323–4342 (electronic).

López, J. L. and Temme, N. M. (2010a). Asymptotics and numerics of polynomials used in Tricomi and Buchholz expansions of Kummer functions, *Numer. Math.* **116**, 2, pp. 269–289.

López, J. L. and Temme, N. M. (2010b). Large degree asymptotics of generalized Bernoulli and Euler polynomials, *J. Math. Anal. Appl.* **363**, 1, pp. 197–208.

López, J. L. and Temme, N. M. (2011). Large degree asymptotics of generalized Bessel polynomials, *J. Math. Anal. Appl.* **377**, 1, pp. 30–42.

López, J. L. and Temme, N. M. (2012). New series expansions of the Gauss hypergeometric function, *Adv. Comput. Math.* .

Luke, Y. L. (1969). *The special functions and their approximations, Vol. I*, Mathematics in Science and Engineering, Vol. 53 (Academic Press, New York).

Luke, Y. L. (1970). Evaluation of the gamma function by means of Padé approximations, *SIAM J. Math. Anal.* **1**, pp. 266–281.

Magnus, W., Oberhettinger, F., and Soni, R. P. (1966). *Formulas and theorems for the special functions of mathematical physics*, Third enlarged edition. Die Grundlehren der mathematischen Wissenschaften, Band 52 (Springer-Verlag New York, Inc., New York).

Mahler, K. (1986). On a special integral, I, II, Research reports 37, 38, Department of Mathematics, The Australian National University.

Maino, G., Menapace, E., and Ventura, A. (1981). Computation of parabolic cylinder functions by means of a Tricomi expansion, *J. Comput. Phys.* **40**, 2, pp. 294–304.

Marcum, J. I. (1960). A statistical theory of target detection by pulsed radar, *Trans. IRE* **IT-6**, pp. 59–267.

Martínez-Finkelshtein, A., Martínez-González, P., and Orive, R. (2001). On asymptotic zero distribution of Laguerre and generalized Bessel polynomials with varying parameters, *J. Comput. Appl. Math.* **133**, 1-2, pp. 477–487.

Meijer, H. G. (1985). Asymptotic expansion of Jacobi polynomials, in *Orthogonal polynomials and applications (Bar-le-Duc, 1984), Lecture Notes in Math.*, Vol. 1171 (Springer, Berlin), pp. 380–389.

Meyer, R. E. (1989). A simple explanation of the Stokes phenomenon, *SIAM Rev.* **31**, 3, pp. 435–445.

Miller, J. C. P. (1955). *Tables of Weber parabolic cylinder functions, giving solutions of the differential equation* $d^2y/dx^2 + (\frac{1}{4}x^2 - a)y = 0$ (Her Majesty's Stationery Office, London (British Information Services, 30 Rockefeller Plaza, New York 20, N. Y.)).

Milne-Thomson, L. M. (1951). *The Calculus of Finite Differences* (Macmillan and Co., Ltd., London).

Moser, L. and Wyman, M. (1957). Stirling numbers of the second kind, *Duke Math. J.* **25**, pp. 29–43.

Moser, L. and Wyman, M. (1958). Asymptotic development of the Stirling numbers of the first kind, *J. London Math. Soc.* **33**, pp. 133–146.

Nagel, B. (1994). The relativistic Hermite polynomial is a Gegenbauer polynomial, *J. Math. Phys.* **35**, 4, pp. 1549–1554.

Nagel, B. (2004). Confluence expansions of the generalized hypergeometric function, *J. Math. Phys.* **45**, 1, pp. 495–508.

Nemes, G. (2013). An explicit formula for the coefficients in Laplace's method. *Constr. Approx.* **38**, 3, pp. 471–487.

Nemes, G. (2014). The resurgence properties of the large order asymptotics of the Hankel and Bessel functions, *Anal. Appl. (Singap.)* **12**, 403–462.

Nielsen, N. (1906). *Handbuch der Theorie der Gammafunktion* (B. G. Teubner, Leipzig, Germany).

Nörlund, N. (1924). *Vorlesungen über Differenzenrechnung* (Springer, Berlin).

Nörlund, N. E. (1961). Sur les valeurs asymptotiques des nombres et des polynômes de Bernoulli, *Rend. Circ. Mat. Palermo (2)* **10**, pp. 27–44.

Oberhettinger, F. (1957). *Tabellen zur Fourier Transformation* (Springer-Verlag, Berlin).

Oberhettinger, F. (1959). On a modification of Watson's lemma, *J. Res. Nat. Bur. Standards Sect. B* **63B**, pp. 15–17.

Olde Daalhuis, A. B. (1992). Hyperasymptotic expansions of confluent hypergeometric functions, *IMA J. Appl. Math.* **49**, 3, pp. 203–216.

Olde Daalhuis, A. B. (1993). Hyperasymptotics and the Stokes' phenomenon, *Proc. Roy. Soc. Edinburgh Sect. A* **123**, 4, pp. 731–743.

Olde Daalhuis, A. B. (1995). Hyperasymptotic solutions of second-order linear differential equations. II, *Methods Appl. Anal.* **2**, 2, pp. 198–211.

Olde Daalhuis, A. B. (1998). On the resurgence properties of the uniform asymptotic expansion of the incomplete gamma function, *Methods Appl. Anal.* **5**, 4, pp. 425–438.

Olde Daalhuis, A. B. (2003). Uniform asymptotic expansions for hypergeometric functions with large parameters. I, *Anal. Appl. (Singap.)* **1**, 1, pp. 111–120.

Olde Daalhuis, A. B. (2010a). Chapter 13, Confluent hypergeometric functions, in *NIST Handbook of Mathematical Functions* (Cambridge University Press, Cambridge), pp.

321–349, http://dlmf.nist.gov/13.

Olde Daalhuis, A. B. (2010b). Chapter 15, Hypergeometric function, in *NIST Handbook of Mathematical Functions* (Cambridge University Press, Cambridge), pp. 383–401, http://dlmf.nist.gov/15.

Olde Daalhuis, A. B. and Olver, F. W. J. (1995). Hyperasymptotic solutions of second-order linear differential equations. I, *Methods Appl. Anal.* **2**, 2, pp. 173–197.

Olde Daalhuis, A. B. and Temme, N. M. (1994). Uniform Airy-type expansions of integrals, *SIAM J. Math. Anal.* **25**, 2, pp. 304–321.

Olver, F. W. J. (1959). Uniform asymptotic expansions for Weber parabolic cylinder functions of large orders, *J. Res. Nat. Bur. Standards Sect. B* **63B**, pp. 131–169.

Olver, F. W. J. (1968). Error bounds for the Laplace approximation for definite integrals, *J. Approximation Theory* **1**, pp. 293–313.

Olver, F. W. J. (1970a). A paradox in asymptotics, *SIAM J. Math. Anal.* **1**, pp. 533–534.

Olver, F. W. J. (1970b). Why steepest descents? *SIAM Rev.* **12**, pp. 228–247.

Olver, F. W. J. (1974). Error bounds for stationary phase approximations, *SIAM J. Math. Anal.* **5**, pp. 19–29.

Olver, F. W. J. (1975). Legendre functions with both parameters large, *Philos. Trans. Roy. Soc. London Ser. A* **278**, pp. 175–185.

Olver, F. W. J. (1980). Whittaker functions with both parameters large: uniform approximations in terms of parabolic cylinder functions, *Proc. Roy. Soc. Edinburgh Sect. A* **86**, 3-4, pp. 213–234.

Olver, F. W. J. (1991a). Uniform, exponentially improved, asymptotic expansions for the confluent hypergeometric function and other integral transforms, *SIAM J. Math. Anal.* **22**, 5, pp. 1475–1489.

Olver, F. W. J. (1991b). Uniform, exponentially improved, asymptotic expansions for the generalized exponential integral, *SIAM J. Math. Anal.* **22**, 5, pp. 1460–1474.

Olver, F. W. J. (1994). The generalized exponential integral, in *Approximation and computation (West Lafayette, IN, 1993), Internat. Ser. Numer. Math.*, Vol. 119 (Birkhäuser Boston, Boston, MA), pp. 497–510.

Olver, F. W. J. (1995). On an asymptotic expansion of a ratio of gamma functions, *Proc. Roy. Irish Acad. Sect. A* **95**, 1, pp. 5–9.

Olver, F. W. J. (1997). *Asymptotics and special functions*, AKP Classics (A K Peters Ltd., Wellesley, MA), reprint of the 1974 original [Academic Press, New York].

Olver, F. W. J. (2010). Chapter 9, Airy and related functions, in *NIST Handbook of Mathematical Functions* (U.S. Dept. Commerce, Washington, DC), pp. 193–213, http://dlmf.nist.gov/9.

Olver, F. W. J. and Maximon, L. C. (2010). Chapter 10, Bessel functions, in *NIST Handbook of Mathematical Functions* (Cambridge University Press, Cambridge), pp. 215–286, http://dlmf.nist.gov/10.

Olver, F. W. J. and Wong, R. (2010). Chapter 2, Asymptotic approximations, in *NIST Handbook of Mathematical Functions* (U.S. Dept. Commerce, Washington, DC), pp. 41–70, http://dlmf.nist.gov/2.

Ott, H. (1943). Die Sattelpunktsmethode in der Umgebung eines Pols mit Anwendungen auf die Wellenoptik und Akustik, *Ann. Physik (5)* **43**, pp. 393–403.

Paris, R. B. (2002a). Error bounds for the uniform asymptotic expansion of the incomplete gamma function, *J. Comput. Appl. Math.* **147**, 1, pp. 215–231.

Paris, R. B. (2002b). A uniform asymptotic expansion for the incomplete gamma function, *J. Comput. Appl. Math.* **148**, 2, pp. 323–339.

Paris, R. B. (2004a). On the use of Hadamard expansions in hyperasymptotic evaluation of Laplace-type integrals. I. Real variable, *J. Comput. Appl. Math.* **167**, 2, pp. 293–319.

Paris, R. B. (2004b). On the use of Hadamard expansions in hyperasymptotic evaluation of Laplace-type integrals. II. Complex variable, *J. Comput. Appl. Math.* **167**, 2, pp. 321–343.

Paris, R. B. (2010). Chapter 8, Incomplete gamma and related functions, in *NIST Handbook of Mathematical Functions* (U.S. Dept. Commerce, Washington, DC), pp. 173–192, http://dlmf.nist.gov/8.

Paris, R. B. (2011). *Hadamard expansions and hyperasymptotic evaluation, Encyclopedia of Mathematics and its Applications*, Vol. 141 (Cambridge University Press, Cambridge).

Paris, R. B. (2013a). Asymptotics of the Gauss hypergeometric function with large parameters, I, *J. Classical Analysis* **2**, 2, pp. 183–203.

Paris, R. B. (2013b). Asymptotics of the Gauss hypergeometric function with large parameters, II, *J. Classical Analysis* **3**, 1, pp. 1–15.

Paris, R. B. and Kaminski, D. (2001). *Asymptotics and Mellin-Barnes integrals, Encyclopedia of Mathematics and its Applications*, Vol. 85 (Cambridge University Press, Cambridge).

Paris, R. B. and Wood, A. D. (1992). Exponentially-improved asymptotics for the gamma function, *J. Comput. Appl. Math.* **41**, 1-2, pp. 135–143.

Paris, R. B. and Wood, A. D. (1995). Stokes phenomenon demystified, *Bull. Inst. Math. Appl.* **31**, 1-2, pp. 21–28.

Pasquini, L. (2000). Accurate computation of the zeros of the generalized Bessel polynomials, *Numer. Math.* **86**, 3, pp. 507–538.

Perron, O. (1921). Über das Verhalten einer ausgearteten hypergeometrischen Reihe bei unbegrenztem Wachstum eines Parameters, *J. Reine Angew. Math.* **151**, pp. 63–78.

Petrova, S. S. and Solov′ev, A. D. (1997). The origin of the method of steepest descent, *Historia Math.* **24**, 4, pp. 361–375.

Poincaré, H. (1886). Sur les intégrales irrégulières des équations linéaires, *Acta Math.* **8**, pp. 295–344.

Prudnikov, A. P., Brychkov, Y. A., and Marichev, O. I. (1988). *Integrals and series. Vol. 2*, 2nd edn. (Gordon & Breach Science Publishers, New York), Special functions, Translated from the Russian by N. M. Queen.

Prudnikov, A. P., Brychkov, Y. A., and Marichev, O. I. (1990). *Integrals and series. Vol. 3* (Gordon and Breach Science Publishers, New York), More special functions, Translated from the Russian by G. G. Gould.

Qu, C. K. and Wong, R. (1990). Transformation to canonical form for uniform asymptotic expansions, *J. Math. Anal. Appl.* **149**, 1, pp. 210–219.

Ronveaux, A., Zarzo, A., Area, I., and Godoy, E. (1998). Transverse limits in the Askey tableau, *J. Comput. Appl. Math.* **99**, 1-2, pp. 327–335.

Roy, R. and Olver, F. W. J. (2010). Chapter 4, Elementary functions, in *NIST Handbook of Mathematical Functions* (U.S. Dept. Commerce, Washington, DC), pp. 103–134, http://dlmf.nist.gov/1.

Roy, R., Olver, F. W. J., Askey, R. A., and Wong, R. (2010). Chapter 1, Algebraic and analytic methods, in *NIST Handbook of Mathematical Functions* (U.S. Dept. Commerce, Washington, DC), pp. 1–39, http://dlmf.nist.gov/1.

Schell, H.-J. (1987). Über das asymptotische Verhalten des Fermi-Dirac-Integrals, *Z. Anal. Anwendungen* **6**, 5, pp. 421–438.

Schmidt, P. W. (1978). An asymptotic approximation for a type of Fourier integral, *Math. Comp.* **32**, 144, pp. 1171–1182.

Shi, W. (2008). A globally uniform asymptotic expansion of the Hermite polynomials, *Acta Math. Sci. Ser. B Engl. Ed.* **28**, 4, pp. 834–842.

Shi, W. and Wong, R. (2010). Error bounds for uniform asymptotic expansions-modified Bessel function of purely imaginary order. *Chin. Ann. Math., Ser. B* **31**, 5, pp. 759–780.

Shivakumar, P. N. and Wong, R. (1988). Error bounds for a uniform asymptotic expansion of the Legendre function $P_n^{-m}(\cosh z)$, *Quart. Appl. Math.* **46**, 3, pp. 473–488.

Sievert, R. (1930). Die v-Strahlungsintensität an der Oberfläche und der nächsten Umgebung von Radiumnadeln, *Acta Radiologica* **11**, pp. 239–301.

Skovgaard, H. (1959). Asymptotic forms of Hermite polynomials, Report 18, Deptartment of Mathematics, Cal. Inst. Technology, Pasadena.

Slater, L. J. (1960). *Confluent hypergeometric functions* (Cambridge University Press, New York).

Soni, K. (1983). A note on uniform asymptotic expansion of incomplete Laplace integrals, *SIAM J. Math. Anal.* **14**, 5, pp. 1015–1018.

Soni, K. and Sleeman, B. D. (1987). On uniform asymptotic expansions and associated polynomials, *J. Math. Anal. Appl.* **124**, 2, pp. 561–583.

Spira, R. (1971). Calculation of the gamma function by Stirling's formula, *Math. Comp.* **25**, pp. 317–322.

Stieltjes, T. J. (1890). Sur la valeur asymptotique des polynômes de legendre, *C. R. Acad. Sci. Paris* **110**, pp. 1026–1027, Oeuvre Complètes, Vol. 2, pp. 234–235.

Stirling, J. (1730). *Methodus differentialis: Sive tractatus de summation et interpolation serierum infinitarium* (Typis Gul. Bowyer; impensis G. Strahan, London), English translation by J. Holliday, The Differential Method: A Treatise of the Summation and Interpolation of Infinite Series. 1749.

Szegő, G. (1975). *Orthogonal polynomials* (American Mathematical Society, Providence, RI).

Temme, N. M. (1975). Uniform asymptotic expansions of the incomplete gamma functions and the incomplete beta function, *Math. Comp.* **29**, 132, pp. 1109–1114.

Temme, N. M. (1976). Remarks on a paper of A. Erdélyi: "Asymptotic evaluation of integrals involving a fractional derivative" (SIAM J. Math. Anal. **5** (1974), 159–171), *SIAM J. Math. Anal.* **7**, 5, pp. 767–770.

Temme, N. M. (1978). Uniform asymptotic expansions of confluent hypergeometric functions, *J. Inst. Math. Appl.* **22**, 2, pp. 215–223.

Temme, N. M. (1979). The asymptotic expansion of the incomplete gamma functions, *SIAM J. Math. Anal.* **10**, 4, pp. 757–766.

Temme, N. M. (1981). On the expansion of confluent hypergeometric functions in terms of Bessel functions, *J. Comput. Appl. Math.* **7**, 1, pp. 27–32.

Temme, N. M. (1982). The uniform asymptotic expansion of a class of integrals related to cumulative distribution functions, *SIAM J. Math. Anal.* **13**, 2, pp. 239–253.

Temme, N. M. (1983). Uniform asymptotic expansions of Laplace integrals, *Analysis* **3**, 1-4, pp. 221–249.

Temme, N. M. (1985). Laplace type integrals: transformation to standard form and uniform asymptotic expansions, *Quart. Appl. Math.* **43**, 1, pp. 103–123.

Temme, N. M. (1986a). A double integral containing the modified Bessel function: asymptotics and computation, *Math. Comp.* **47**, 176, pp. 683–691.

Temme, N. M. (1986b). Laguerre polynomials: Asymptotics for large degree, Research report R 8610, Department of Applied Mathematics, CWI, Amsterdam.

Temme, N. M. (1986c). Uniform asymptotic expansion for a class of polynomials biorthogonal on the unit circle, *Constr. Approx.* **2**, 4, pp. 369–376.

Temme, N. M. (1987). Incomplete Laplace integrals: uniform asymptotic expansion with application to the incomplete beta function, *SIAM J. Math. Anal.* **18**, 6, pp. 1638–

1663.

Temme, N. M. (1989). Asymptotic expansion of a special integral, *CWI Quarterly* **2**, 1, pp. 67–72.

Temme, N. M. (1990a). Asymptotic estimates for Laguerre polynomials, *Z. Angew. Math. Phys.* **41**, 1, pp. 114–126.

Temme, N. M. (1990b). Polynomial asymptotic estimates of Gegenbauer, Laguerre, and Jacobi polynomials, in *Asymptotic and computational analysis (Winnipeg, MB, 1989)*, *Lecture Notes in Pure and Appl. Math.*, Vol. 124 (Dekker, New York), pp. 455–476.

Temme, N. M. (1990c). Uniform asymptotic expansions of a class of integrals in terms of modified Bessel functions, with application to confluent hypergeometric functions, *SIAM J. Math. Anal.* **21**, 1, pp. 241–261.

Temme, N. M. (1992a). Asymptotic inversion of incomplete gamma functions, *Math. Comp.* **58**, 198, pp. 755–764.

Temme, N. M. (1992b). Asymptotic inversion of the incomplete beta function, *J. Comput. Appl. Math.* **41**, 1-2, pp. 145–157.

Temme, N. M. (1993a). Asymptotic and numerical aspects of the noncentral chi-square distribution, *Comput. Math. Appl.* **25**, 5, pp. 55–63.

Temme, N. M. (1993b). Asymptotic estimates of Stirling numbers, *Stud. Appl. Math.* **89**, 3, pp. 233–243.

Temme, N. M. (1994a). Computational aspects of incomplete gamma functions with large complex parameters, in R. V. M. Zahar (ed.), *Approximation and Computation. A Festschrift in Honor of Walter Gautschi.*, *International Series of Numerical Mathematics*, Vol. 119 (Birkhäuser Boston, Boston, MA), pp. 551–562.

Temme, N. M. (1994b). Steepest descent paths for integrals defining the modified Bessel functions of imaginary order, *Methods Appl. Anal.* **1**, 1, pp. 14–24.

Temme, N. M. (1995). Asymptotics of zeros of incomplete gamma functions, *Ann. Numer. Math.* **2**, 1-4, pp. 415–423.

Temme, N. M. (1996a). *Special functions: An introduction to the classical functions of mathematical physics*, A Wiley-Interscience Publication (John Wiley & Sons Inc., New York).

Temme, N. M. (1996b). Uniform asymptotics for the incomplete gamma functions starting from negative values of the parameters, *Methods Appl. Anal.* **3**, 3, pp. 335–344.

Temme, N. M. (1997). Numerical algorithms for uniform Airy-type asymptotic expansions, *Numer. Algorithms* **15**, 2, pp. 207–225.

Temme, N. M. (2000). Numerical and asymptotic aspects of parabolic cylinder functions, *J. Comput. Appl. Math.* **121**, 1-2, pp. 221–246.

Temme, N. M. (2003). Large parameter cases of the Gauss hypergeometric function, *J. Comput. Appl. Math.* **153**, 1-2, pp. 441–462.

Temme, N. M. (2006). Asymptotics of a ${}_3F_2$ polynomial associated with the Catalan-Larcombe-French sequence, *Anal. Appl. (Singap.)* **4**, 4, pp. 335–344.

Temme, N. M. (2007). Analytical methods for an elliptic singular perturbation problem in a circle, *J. Comput. Appl. Math.* **207**, 2, pp. 301–322.

Temme, N. M. (2010a). Chapter 12, Parabolic cylinder functions, in *NIST Handbook of Mathematical Functions* (U.S. Dept. Commerce, Washington, DC), pp. 303–319, http://dlmf.nist.gov/12.

Temme, N. M. (2010b). Chapter 6, Exponential, logarithmic, sine, and cosine integrals, in *NIST Handbook of Mathematical Functions* (U.S. Dept. Commerce, Washington, DC), pp. 149–157, http://dlmf.nist.gov/6.

Temme, N. M. (2010c). Chapter 7, Error functions, Dawson's and Fresnel integrals, in *NIST Handbook of Mathematical Functions* (U.S. Dept. Commerce, Washington,

DC), pp. 159–171, http://dlmf.nist.gov/7.

Temme, N. M. (2012). The role of special functions as leading term approximations in asymptotic expansions, in A. K. Agarwal (ed.), *Proceedings of Tenth and Eleventh Annual Conferences, ICSFA 2012, Surat, India, 27–29 June 2012*, Vol. 10–11 (Society for Special Functions & Their Applications, Aligarh, India), pp. 67–92.

Temme, N. M. (2013a). Remarks on Slater's asymptotic expansions of Kummer functions for large values of the $a-$parameter, *Adv. Dyn. Syst. Appl.* **8**, 2, pp. 365–377.

Temme, N. M. (2013b). Uniform asymptotic methods for integrals, *Indag. Math. (N.S.)* **24**, 4, pp. 739–765.

Temme, N. M. and Olde Daalhuis, A. B. (1990). Uniform asymptotic approximation of Fermi-Dirac integrals, *J. Comput. Appl. Math.* **31**, 3, pp. 383–387.

Thompson, I. J. (2010). Chapter 33, Coulomb functions, in *NIST Handbook of Mathematical Functions* (U.S. Dept. Commerce, Washington, DC), pp. 741–756, http://dlmf.nist.gov.

Tricomi, F. G. (1947). Sulle funzioni ipergeometriche confluenti, *Ann. Mat. Pura Appl. (4)* **26**, pp. 141–175.

Tricomi, F. G. (1948). *Equazioni Differenziali* (Giulio Einaudi, Torino).

Tricomi, F. G. (1950). Asymptotische Eigenschaften der unvollständigen Gammafunktion, *Math. Z.* **53**, pp. 136–148.

Tricomi, F. G. (1954). *Funzioni ipergeometriche confluenti* (Edizioni Cremonese, Roma).

Tricomi, F. G. and Erdélyi, A. (1951). The asymptotic expansion of a ratio of gamma functions, *Pacific J. Math.* **1**, pp. 133–142.

Tweddle, I. (2003). *James Stirling's Methodus differentialis*, Sources and Studies in the History of Mathematics and Physical Sciences (Springer-Verlag London Ltd., London), an annotated translation of Stirling's text.

Ursell, F. (1984). Integrals with a large parameter: Legendre functions of large degree and fixed order, *Math. Proc. Cambridge Philos. Soc.* **95**, 2, pp. 367–380.

Ursell, F. (2007). Integrals with nearly coincident branch points: Gegenbauer polynomials of large degree, *Proc. R. Soc. Lond. Ser. A Math. Phys. Eng. Sci.* **463**, 2079, pp. 697–710.

Van der Corput, J. G. (1948). On the method of critical points. I, *Nederl. Akad. Wetensch., Proc.* **51**, pp. 650–658; *Indagationes Math.* **10**, 201–209 (1948).

Van der Waerden, B. L. (1951). On the method of saddle points, *Appl. Sci. Research B.* **2**, pp. 33–45.

Van Leeuwaarden, J. S. H. and Temme, N. M. (2009). Asymptotic inversion of the Erlang B formula, *SIAM J. Appl. Math.* **70**, 1, pp. 1–23.

Van Leeuwaarden, J. S. H. and Temme, N. M. (2011). A uniform asymptotic expansion for weighted sums of exponentials, *Statist. Probab. Lett.* **81**, 11, pp. 1571–1579.

Varga, R. S. (1990). *Scientific computation on mathematical problems and conjectures, CBMS-NSF Regional Conference Series in Applied Mathematics*, Vol. 60 (Society for Industrial and Applied Mathematics (SIAM), Philadelphia, PA).

Vidunas, R. and Temme, N. M. (2002). Symbolic evaluation of coefficients in Airy-type asymptotic expansions, *J. Math. Anal. Appl.* **269**, 1, pp. 317–331.

Vignat, C. (2011). Old and new results about relativistic Hermite polynomials, *J. Math. Phys.* **52**, 9, pp. 093503, 16.

Von Mises, R. (1918). Über die 'Ganzzahligkeit' der Atomgewicht und verwandte Fragen, *Physikalische Z.* **19**, pp. 490–500.

Wagner, E. (1984). Asymptotische Entwicklungen der hypergeometrischen Funktionen $F(a, b, c; z)$ für $|a| \to \infty$ und konstante b, c, z, *Z. Anal. Anwendungen* **3**, 3, pp. 213–226.

Wagner, E. (1988). Asymptotische Entwicklungen der hypergeometrischen Funktion $F(a,b,c,z)$ für $|c| \to \infty$ und konstante Werte a, b und z, *Demonstratio Math.* **21**, 2, pp. 441–458.

Watson, G. N. (1918a). Asymptotic expansions of hypergeometric functions, *Trans. Cambridge Philos. Soc.* **22**, pp. 277–308.

Watson, G. N. (1918b). The harmonic functions associated with the parabolic cylinder, *Proc. London Math. Soc. (3)* **17**, pp. 116–148.

Watson, G. N. (1944). *A treatise on the theory of Bessel functions*, Cambridge Mathematical Library (Cambridge University Press, Cambridge), second edition.

Weber, H. F. (1869). Über die Integration der partiellen Differentialgleichung $\partial^2 u/\partial x^2 + \partial^2 u/\partial y^2 + k^2 u = 0$, *Math. Annal.* **1**, pp. 1–36.

Weniger, E. J. (2010). Summation of divergent power series by means of factorial series. *Appl. Numer. Math.* **60**, 12, pp. 1429–1441.

Whittaker, E. T. and Watson, G. N. (1996). *A course of modern analysis*, Cambridge Mathematical Library (Cambridge University Press, Cambridge), reprint of the fourth (1927) edition.

Wojdylo, J. (2006a). Computing the coefficients in Laplace's method, *SIAM Rev.* **48**, 1, pp. 76–96 (electronic).

Wojdylo, J. (2006b). On the coefficients that arise from Laplace's method, *J. Comput. Appl. Math.* **196**, 1, pp. 241–266.

Wong, R. (2001). *Asymptotic approximations of integrals*, *Classics in Applied Mathematics*, Vol. 34 (Society for Industrial and Applied Mathematics (SIAM), Philadelphia, PA), Corrected reprint of the 1989 original.

Wong, R. and Wyman, M. (1972). A generalization of Watson's lemma, *Canad. J. Math.* **24**, pp. 185–208.

Wong, R. and Zhang, J.-M. (1997). Asymptotic expansions of the generalized Bessel polynomials, *J. Comput. Appl. Math.* **85**, 1, pp. 87–112.

Wong, R. and Zhao, Y.-Q. (2003). Estimates for the error term in a uniform asymptotic expansion of the Jacobi polynomials, *Anal. Appl. (Singap.)* **1**, 2, pp. 213–241.

Wong, R. and Zhao, Y.-Q. (2005). On a uniform treatment of Darboux's method, *Constr. Approx.* **21**, 2, pp. 225–255.

Yakubovich, S. B. (1996). *Index transforms* (World Scientific Publishing Co. Inc., River Edge, NJ).

Zil′bergleĭt, A. S. (1976). Uniform asymptotic expansions of certain definite integrals, *Ž. Vyčisl. Mat. i Mat. Fiz.* **16**, 1, pp. 40–47, 275, English translation: U.S.S.R. Computational Math. and Math. Phys. 16 (1976), no. 1, 36–44 (1977).

Zil′bergleĭt, A. S. (1977). Uniform asymptotic expansion of the Dirichlet integral, *Ž. Vyčisl. Mat. i Mat. Fiz.* **17**, 6, pp. 1588–1592, 1630, English translation: U.S.S.R. Computational Math. and Math. Phys. 17 (1977), no. 6, 237–242 (1978).

Index

www.ingramcontent.com/pod-product-compliance
Lightning Source LLC
LaVergne TN
LVHW081355011026
849161LV00013BA/2814